The *ABCD* Parameters of Some Useful Two-Port Circuits

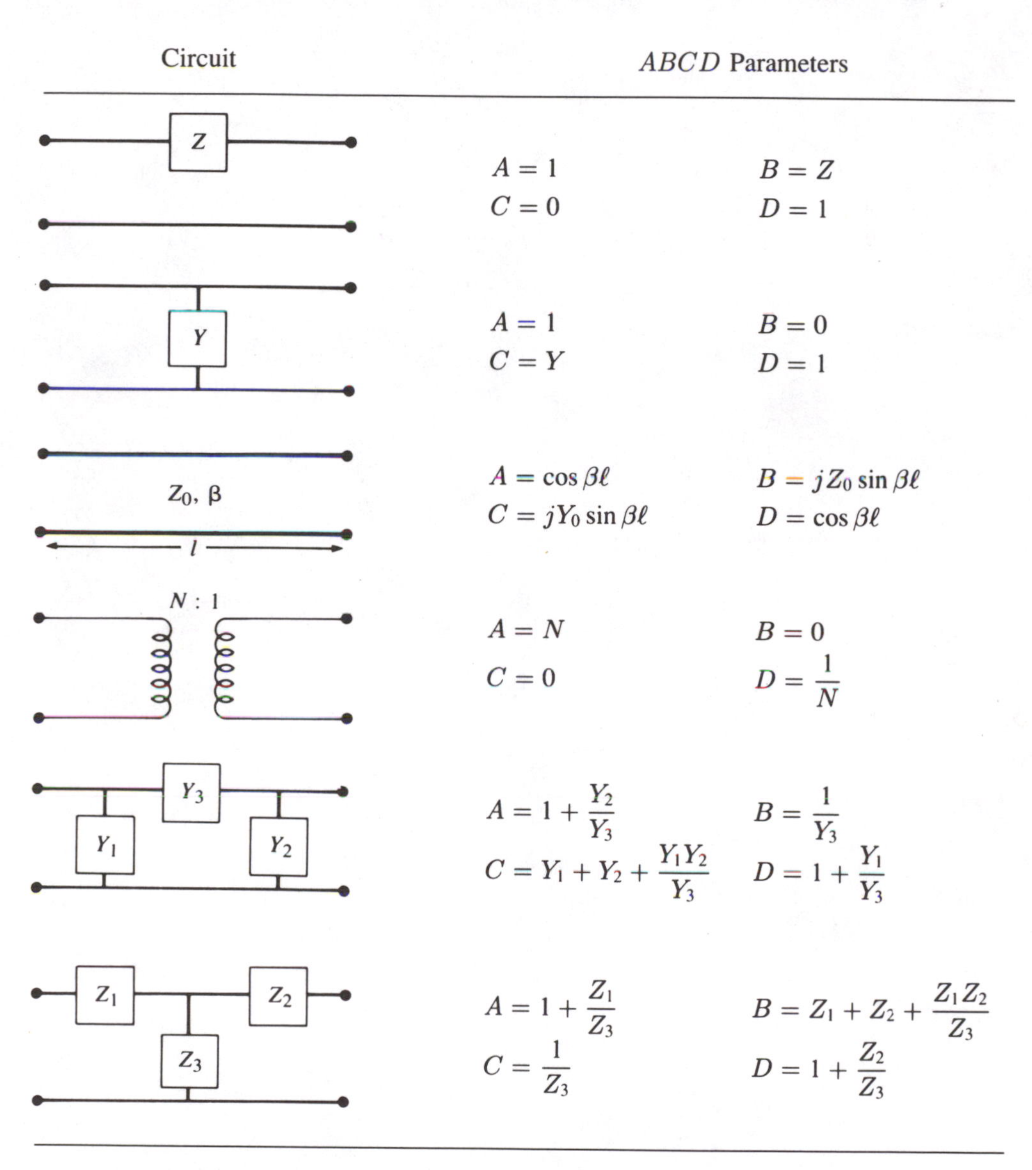

Circuit	*ABCD* Parameters	
Series impedance Z	$A = 1$ $C = 0$	$B = Z$ $D = 1$
Shunt admittance Y	$A = 1$ $C = Y$	$B = 0$ $D = 1$
Transmission line Z_0, β, length l	$A = \cos \beta\ell$ $C = jY_0 \sin \beta\ell$	$B = jZ_0 \sin \beta\ell$ $D = \cos \beta\ell$
Transformer $N : 1$	$A = N$ $C = 0$	$B = 0$ $D = \frac{1}{N}$
π-network Y_1, Y_2, Y_3	$A = 1 + \frac{Y_2}{Y_3}$ $C = Y_1 + Y_2 + \frac{Y_1 Y_2}{Y_3}$	$B = \frac{1}{Y_3}$ $D = 1 + \frac{Y_1}{Y_3}$
T-network Z_1, Z_2, Z_3	$A = 1 + \frac{Z_1}{Z_3}$ $C = \frac{1}{Z_3}$	$B = Z_1 + Z_2 + \frac{Z_1 Z_2}{Z_3}$ $D = 1 + \frac{Z_2}{Z_3}$

Microwave Engineering

Microwave Engineering

David M. Pozar
University of Massachusetts at Amherst

Addison-Wesley Publishing Company

Reading, Massachusetts · Menlo Park, California · New York
Don Mills, Ontario · Wokingham, England · Amsterdam · Bonn
Sydney · Singapore · Tokyo · Madrid · San Juan

This book is in the Addison-Wesley Series in Electrical and Computer Engineering

Many of the designations used by manufacturers and sellers to distinguish their products are claimed as trademarks. Where those designations appear in this book, and Addison-Wesley was aware of a trademark claim, the designations have been printed in initial caps or all caps.

The programs and applications presented in this book have been included for their instructional value. They have been tested with care, but are not guaranteed for any particular purpose. The publisher does not offer any warranties or representations, nor does it accept any liabilities with respect to the programs or applications.

Library of Congress Cataloging-in-Publication Data

Pozar, David M.
Microwave engineering / by David M. Pozar.
p. cm.
ISBN 0-201-50418-9
1. Microwaves 2. Microwave devices. 3. Microwave circuits.
I. Title.
TK7876.P69 1990
621.381'3--dc20 89-38233
CIP

Reprinted with corrections September, 1993

8 9 10–DOC–97969594

Preface

Education should be the accumulation of understanding, not just an accumulation of facts. Thus, I have tried to write a textbook that emphasizes the basic concepts and phenomenon of Maxwell's equations, wave propagation, network theory, and related topics, as applied to modern microwave engineering. I have avoided the handbook approach, where a large amount of detailed material (possibly quite useful material) is presented with little or no analysis or explanation. Yet, a considerable amount of material in this book is related to the design of specific microwave components, due to their motivational and practical value. I have tried to present the analysis and logic behind these designs so that the reader can see and understand the process of applying fundamental principles to arrive at useful results. The engineer who has a firm grasp of the basic concepts and principles of microwave engineering and has seen how these concepts and principles can be applied toward a specific design objective is the engineer who is most likely to be rewarded with a creative and productive career. As the proverb says, "Give a man a fish and he will eat for a day; teach a man to fish and he will eat for a lifetime."

This book is an outgrowth of class notes that have been used for several years for a two-semester course in microwave engineering at the University of Massachusetts. It is intended for senior or first-year graduate students in electrical engineering, and it is assumed that students have had a junior-year course in electromagnetics and transmission line theory. Students with less background may need to spend more time reviewing Chapters 2 and 3. Overall, the book contains more than enough material for a full-year course, so the instructor can choose how much material to cover on topics such as couplers, filters, ferrite components, active circuit design, and systems. (A sample curriculum for a one-year course is given below.)

Chapter 1 provides a brief introduction to the field of microwave engineering and a short history of some of the key developments in the field. Hopefully some students will find the historical perspective not only interesting in itself, but also of value in relation to later material.

Chapter 2 gives an overview of electromagnetic theory, including Maxwell's equations, boundary conditions, energy and power relations, and the effect of dielectric and magnetic materials. Then, some basic results for plane wave propagation and reflection from dielectric and conducting media are presented, followed by a discussion of reciprocity and image theory.

One of the primary goals of this book is to show that many problems in microwave engineering can be simplified by reducing a field analysis solution to an equivalent circuit. Transmission line theory is probably the most important link in this process, and is the subject of Chapter 3.

It presents the basic properties and phenomenon of wave propagation on transmission lines from both field analysis and circuit theory points of view. Chapters 2 and 3 provide a reference for the material in electromagnetic and transmission line theory that will be needed throughout the rest of the book. Chapters 2 and 3 also include several topics that the student may not have seen before, such as anisotropic media, the multiple reflection interpretation of the quarter-wave transformer, conjugate matching, perturbation theory for attenuation, and the Wheeler incremental inductance rule.

Chapter 4 discusses TEM, TE, and TM wave propagation and the propagation of such waves on several of the most common types of waveguides and transmission lines, including microstrip lines, stripline, and surface waveguides. An important point to note is that planar transmission lines are not as amenable to simple closed form analysis as are traditional components, such as rectangular waveguides. In addition to the usual design formulas for planar lines, we also discuss some approximate numerical solutions to these problems and emphasize the similarity between propagation on a planar transmission line and propagation on other common waveguides.

Chapter 5 further develops the use of circuit theory in microwave engineering by discussing various network matrices for multiport circuits and their properties. Signal flowgraphs are also presented, as is some material on the excitation of waveguides.

Chapters 2 through 5 can be considered as laying the necessary theoretical foundations for the specific microwave components, circuits, and systems that are discussed in Chapters 6 through 12. Thus, we begin in Chapter 6 with impedance matching and tuning, including lumped-element sections, multisection transmission line transformers, tapered lines, and the Bode-Fano criteria. If desired, Chapter 6 can be covered before Chapter 5, without difficulty.

Chapter 7 discusses a variety of microwave resonators and relates the operation of a distributed microwave resonator to a lumped-element equivalent circuit. Chapter 8 is devoted to a number of practical power divider and directional coupler designs, including the theory of coupled lines and the design of couplers using such elements. Chapter 9 similarly presents the theory and design of some typical microwave filters from both the image parameter and the insertion loss viewpoints. It could be argued that the latter method is of most interest today, but the image parameter method has an interesting interpretation in terms of periodic structures and k-β diagrams.

Chapter 10 discusses wave propagation in ferrite materials and the design of components using ferrites. The effect of the ferrite material requires that we revert to a field theory type of analysis. After a discussion of the (microscopic) cause of the magnetic properties of ferrite material, we characterize the material from a macroscopic point of view using tensor permeability, and then we treat several practical ferrite components, including isolators, phase shifters, and circulators.

Chapter 11 treats active microwave circuit design, a topic that could easily fill a textbook in its own right. We begin with a discussion of noise and its effect on microwave systems. Then, we treat diode detectors, mixers, and intermodulation distortion. The section on transistor amplifier design covers amplifier stability, gain, and noise figure with several practical design examples. This important topic also serves as an excellent application of S-parameters. Then, we give an introductory discussion of one- and two-port oscillator circuits, followed by a brief treatment of PIN diode control circuits. Finally, we give a qualitative summary of microwave integrated circuit technology and microwave sources. The solid state physics of active devices is not discussed since for our purposes such components are adequately characterized by their terminal parameters. Also space limitations prevent us from giving a more thorough treatment of these topics. Most of the material in Chapter 11 could be covered after Chapter 6, if desired.

Applications of microwaves at a systems level are introduced in Chapter 12, including antennas and communication, radar, and radiometry systems. I feel a brief treatment of these subjects is relevant to the rest of the material in this book and shows how components can be put together to perform a useful function.

The following outline shows the material covered in a typical two-semester course in microwave engineering, with the approximate number of hour lectures. The material is self-contained enough so that much of it could be covered earlier, if desired.

First Semester:

Topic	**Text Sections**	**Lecture Hours**
Review of EM Theory and Transmission Lines	2.1–2.7, 3.1–3.9	5
Analysis of Transmission Lines and Waveguides	4.1–4.8, 4.10	9
Microwave Network Analysis	5.1–5.7	8
Impedance Matching	6.1–6.9	7
Power Dividers, Couplers, and Hybrids	8.1–8.8	11

Second Semester:

Topic	**Text Sections**	**Lecture Hours**
Microwave Resonators	7.1–7.7	5
Active Microwave Circuits	11.1, 11.3–11.4	9
Filter Theory and Design	9.1–9.7	10
Ferrite Components	10.1–10.6	9
Detectors and Mixers	11.2	4
Introduction to Microwave Systems	12.1–12.3	4

I suggest that students have access to computer-aided design software for microwave circuit analysis. There are several mainframe- and PC-based software packages that are available that can be used to evaluate the performance of matching networks, couplers, filters, amplifiers, and other circuits. Students can then design a circuit or component as a homework exercise and check its response with the CAD software. This immediate feedback builds confidence and makes the effort more rewarding. Since the drudgery of repetitive calculation is eliminated, students can easily try alternative approaches and explore problems in a more detailed way.

I also recommend that this course be taught with a laboratory. This is the best way for students to develop an intuition and physical feeling for microwave phenomenon, and to get hands-on experience with microwave components and equipment. A lab with the first-semester course should cover the microwave measurement of power, frequency,

SWR, impedance, and S-parameters, as well as the characterization of basic microwave components such as Gunn diode sources, couplers, circulators, and loads. The type of experiments that can be offered is heavily dependent on the type of microwave equipment that is available. New microwave test equipment is very expensive, but worthwhile experiments can be performed with surplus waveguide components.

Many people deserve acknowledgement for their help in completing this book, but my foremost appreciation must go to my Microwave Engineering classes of 1988–1990 for using a draft copy of the book that was far from complete. These students found many of the mistakes and inconsistencies that were in the draft and also made a number of valuable suggestions for improving the text. Next, I would like to give my sincere thanks to Ms. Linda Parsons for typing almost the entire manuscript in TEX. It was a pleasure working with Linda, and she made things a lot easier for me. I would also like to thank my friend and colleague, Professor Robert Jackson, who deserves credit for developing several of the analyses that I have used here. He also read and critiqued several of the chapters. In addition, I received useful comments from Professor Karl Stephan, Professor Nick Buris, and from my graduate students Nirod Das and Jim Aberle. I would also like to thank Professors Linda Katehi (University of Michigan), Glenn Smith (Georgia Tech), and Jayanti Venkataraman (Rochester Institute of Technology) for thoughtful reviews of the manuscript. Several people were very helpful in supplying photographs for this book, in particular, Dr. Naresh Deo of Millitech Corp., Dr. Allan Love of Rockwell International, Dr. John Bryant of the University of Michigan, Dr. Eli Brookner of Raytheon Co., Mr. John Mather of Raytheon Co., Mr. Hugo Vifian of Hewlett-Packard Co., and Mr. Harlan Howe of Adams-Russell Corp. Finally, it is a pleasure to acknowledge the professional and enthusiastic engineering group at Addison-Wesley. I am especially appreciative of the interest and energy that my editor, Ms. Eileen Bernadette Moran, has had from the beginning of this project. And I would like to thank Ms. Karen Myer for her special attention to the production of this book.

D. Pozar
Amherst

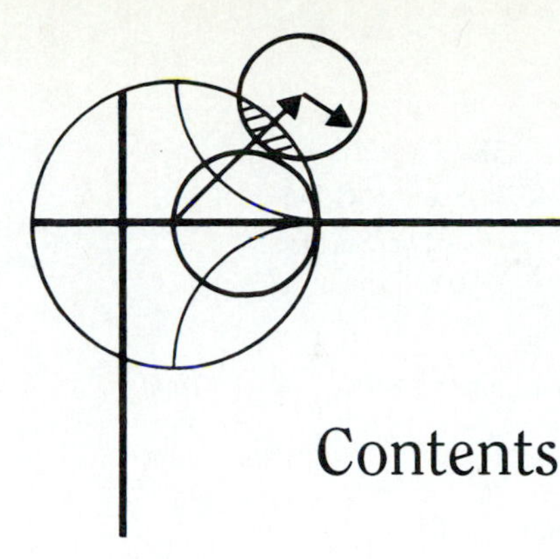

Contents

5 MICROWAVE NETWORK ANALYSIS 205

6 IMPEDANCE MATCHING AND TUNING 281

CHAPTER 1

Introduction

This chapter is intended to provide a brief overview of microwave engineering, including some of the major applications of microwaves. Also included is a short history of the microwave field. The student should thus get a glimpse of the larger context of the material that follows.

1.1 WHAT IS MICROWAVE ENGINEERING?

The term *microwave* refers to alternating current signals with frequencies between 300 MHz (3×10^8 Hz) and 300 GHz (3×10^{11} Hz). See Figure 1.1 for the location of the microwave frequency band in the electromagnetic spectrum. The period, $T = 1/f$, of a microwave signal then ranges from 3 ns (3×10^{-9} sec) to 3 ps (3×10^{-12} sec), respectively, and the corresponding electrical wavelength ranges from $\lambda = c/f = 1$ m to $\lambda = 1$ mm, respectively, where $c = 3 \times 10^8$ m/sec, the speed of light in a vacuum. Signals with wavelengths on the order of millimeters are called *millimeter* waves. It is really the values of the above quantities that make microwave engineering different from other areas of electrical engineering. Because of the high frequencies (and short wavelengths), standard circuit theory cannot be used directly to solve microwave network problems. In a sense, standard circuit theory is an approximation or special case of the broader theory of electromagnetics as described by Maxwell's equations. This is due to the fact that, in general, the lumped circuit element approximations of circuit theory are not valid at microwave frequencies. Microwave components usually are *distributed* elements, where the phase of a voltage or current changes signficantly over the physical length of the device because the device dimensions are on the order of the microwave wavelength. At much lower frequencies, the wavelength is so large that there is little variation in phase across the dimensions of a component.

At the other extreme is optical engineering, where the wavelength is much shorter than the dimensions of the components. In this case Maxwell's equations can be simplified to the geometrical optics regime, and optical systems can be designed with the theory of geometrical optics. Such techniques are sometimes even applicable to millimeter wave systems, where they are referred to as "quasi-optical."

In microwave engineering, then, one must begin with Maxwell's equations and their solutions. It is in the nature of these equations that mathematical complexity arises, as Maxwell's equations involve vector differential or integral operations on vector field

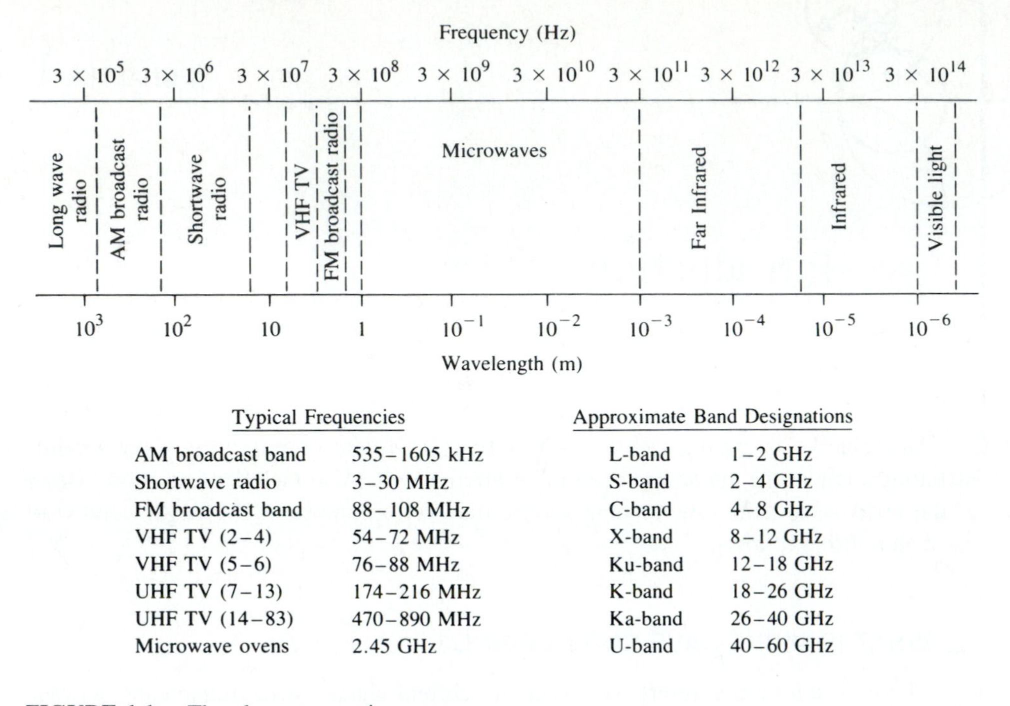

Typical Frequencies	
AM broadcast band	535–1605 kHz
Shortwave radio	3–30 MHz
FM broadcast band	88–108 MHz
VHF TV (2–4)	54–72 MHz
VHF TV (5–6)	76–88 MHz
UHF TV (7–13)	174–216 MHz
UHF TV (14–83)	470–890 MHz
Microwave ovens	2.45 GHz

Approximate Band Designations	
L-band	1–2 GHz
S-band	2–4 GHz
C-band	4–8 GHz
X-band	8–12 GHz
Ku-band	12–18 GHz
K-band	18–26 GHz
Ka-band	26–40 GHz
U-band	40–60 GHz

FIGURE 1.1 The electromagnetic spectrum.

quantities, and these fields are functions of spatial coordinates. One of the goals of this book, however, is to try to reduce the complexity of a field theory solution to a result which can be expressed in terms of circuit theory. The field theory solution generally provides a complete description of the electromagnetic field at every point in space—usually much more information than we really need. We usually are interested in terminal quantities, such as power, impedance, voltage, current, etc., which can often be expressed in terms of circuit theory concepts. This complexity adds to the difficulty, as well as to the challenge and reward, of microwave engineering.

1.2 APPLICATIONS OF MICROWAVES

Just as the high frequencies and small wavelengths of microwave energy make for difficult analysis and design of microwave components, these same factors provide unique applications for microwave systems. This is because of the following considerations:

- Antenna gain is proportional to the electrical size of the antenna. At higher frequencies, more antenna gain is possible for a given physical antenna size.
- More bandwidth (and hence information-carrying capacity) can be realized at higher frequencies. A 1% bandwidth at 600 MHz is 6 MHz (the bandwidth of one television channel), while at 60 GHz a 1% bandwidth is 600 MHz (about 100 television channels).

- Microwave signals travel by line-of-sight, and are not bent by the ionosphere as are lower frequency signals. Communication links (both terrestrial and with orbiting satellites) with high capacities are thus possible. (Millimeter wave frequencies can, however, be highly attenuated by the atmosphere or rain.)
- The effective reflection area (radar cross-section) of a radar target is usually proportional to the target's electrical size. This fact, coupled with the characteristics of antenna gain, often make microwave frequencies the preferred band for radar applications.
- Various molecular, atomic, and nuclear resonances occur at microwave frequencies, creating a variety of applications in the areas of basic science, remote sensing, medical diagnostics and treatment, and cooking methods.

Today, the majority of applications of microwaves are related to radar and communication systems. Radar systems are used for detecting and locating air, ground, or sea-going targets, by airport traffic-control radars, missile tracking radars, fire-control radars, and other weapons systems. Radar is also used for weather prediction and remote sensing applications. Microwave communications systems handle a large fraction of international and other long-haul telephone traffic, in addition to television programs and military communications.

1.3 A SHORT HISTORY OF MICROWAVE ENGINEERING

The field of microwave engineering, as well as other areas of applied electromagnetics, is often referred to as a "mature discipline" because of the fact that the fundamental concepts of electromagnetics were laid down by James Clerk Maxwell in 1873 [1], and probably because radar, being the first major application of microwaves, was intensively developed during World War II. Thus, although exciting new developments are still being made today, microwave engineering had its beginnings in the last century.

A brief history of some of the more significant developments in microwave engineering is presented in the following sections. This outline, of course, does not come close to a complete description of the field, or give credit to all the workers who have contributed. The reader who is interested in further details is encouraged to begin with the special Centennial Issue of the *IEEE Transactions on Microwave Theory and Techniques*, September 1984, which is devoted to the historical perspectives of microwave technology.

Early History

The foundations of modern electromagnetic theory were laid down in Maxwell's treatise [1] in the late nineteeth century. In this theory Maxwell hypothesized, from mathematical considerations, electromagnetic wave propagation and the notion that light was a form of electromagnetic energy. Heinrich Hertz, a German professor of physics, was a gifted experimenter who also understood the theory published by Maxwell. In the period 1887–1891, Hertz carried out a series of experiments that completely validated Maxwell's theory of electromagnetic waves [2]. It is interesting to note that this is an instance of a discovery occuring after a prediction had been made on theoretical grounds; this is characteristic of many of the major discoveries throughout the history of science. In a sense, all of the practical applications of electromagnetic theory, including radio, television, communications satellites, radar, and others, owe their existence to the electromagnetic theory of Maxwell.

In addition to Hertz's work, Oliver Heaviside contributed significantly with a series of papers between 1885 and 1887 that made the theory of Maxwell more accessible to practicing scientists [3]. Heaviside was a reclusive genius whose work removed a lot of the mathematical complexities of Maxwell's theory, introduced vector notation, and provided a foundation for practical applications in guided waves and transmission line theory.

Hertz began his experimental work by using a high-voltage spark discharge (a source rich in high-frequency harmonics) to excite a half-wave dipole antenna at a frequency of about 60 MHz. A receive antenna consisted of an adjustable loop of wire with another spark gap. When both transmit and receive antennas were adjusted for the same resonant frequency, Hertz was able to demonstrate propagation of electromagnetic waves. He also studied standing waves set up by reflection from a metal surface.

Continuing his experiments, Hertz used a coaxial line to show that electromagnetic waves propagated with a finite velocity, and he discovered basic transmission line effects such as the existence of nodes in a standing wave pattern a quarter wavelength from an open circuit and a half wavelength from a short circuit. He then went on to develop cylindrical parabolic reflectors for directional antennas, as well as a number of other radio frequency (RF) and microwave devices and techniques. A photograph of the original equipment used by Hertz in shown in Figure 1.2. Hertz died at the age of 36 in 1894, but he can justifiably be called the first microwave engineer.

At this point, major effort was directed toward the commercial realization, by Marconi and others, of radio. This was at a much lower frequency than the work of Hertz, for long-distance communication. In addition, high-frequency alternators, vacuum tubes, and other devices were developed for radio use.

The Development of Waveguide

Probably the next major milestone in the history of microwave engineering was the development of waveguide. Hertz never reported on the possibility of propagation of electromagnetic waves inside a hollow metal tube, and although Heaviside in 1893 considered the possibility, he rejected the concept as he felt that two conductors were necessary for the transfer of electromagnetic energy [4].

In 1897, Lord Rayleigh (John William Strutt) [5] mathematically proved that wave propagation in waveguides was possible, for both circular and rectangular cross-sections. Rayleigh also noted the infinite set of modes of the TE or TM type which were possible, and the existence of a cutoff frequency. Lord Rayleigh had an eminent career as a mathematical physicist, and made contributions in the areas of acoustics, fluids, and optics.

Although no experimental verification of Rayleigh's work on waveguide was made, Sir Oliver Lodge in 1894 noted that directional radiation was obtained when he surrounded a spark oscillator with a metal tube. Waveguide was essentially forgotten,

FIGURE 1.2 Original apparatus built by Hertz for his electromagnetics experiments along with laboratory items. (1) First oscillator/radiator transmitter, signal source, 6 m wavelength. (2) Wooden frame and parallel wires for polarization demonstration, both transmission and reflection. (3) Possibly a demountable vacuum apparatus for cathode-ray experiments. (4) Hot-wire galvanometer. (5) A pair of Reiss or Knochenhauer spirals. (6) Rolled-paper electroscope. (7) Metal sphere with insulated handle. (8) Reiss's spark micrometer. (9) Receiver/detector used with coaxial line, 6 m wavelength. (10, 11, and 13) Apparatus to demonstrate dielectric polarization effects in insulators, 3 m wavelength. (12) Mercury interrupter. (14) Meidinger cell (primary battery), same chemistry as the Daniell cell. (15) Bell jar, photoelectric effect experiments. (16) Induction coil. (17) Bunsen cells (primary batteries). (18) Large-area conductor of a high-voltage electrostatic machine, or as a "capacitor." (19) Circular-loop receiver/detector, 6 m wavelength. (20) Receiver/detector, not otherwise identified. (21) Rotating mirror and mercury interrupter assembly. (22) Square-loop receiver/detector, 6 m wavelength. (23) Stack of three wedge-shaped wooden boxes to hold dielectric material for refraction demonstration and dielectric constant measurement. (24) An assembly of two square-loop receiver/detectors, 6 m wavelength. (25) Square-loop receiver/detector, 6 m wavelength. (26) Transmitter dipole, 70 cm wavelength. (27) Induction coil. (28) Coaxial line. (29) Discharging table similar to a Henley's discharger. (30) Cylindrical parabolic reflector, receiver/detector, 3 m wavelength. (31) Cylindrical parabolic reflector, transmitter, 70 cm wavelength. (32) Circular-loop receiver/detector, 3 m wavelength. (33) Plane metal reflector. Photographed on 1 October 1913 in the auditory of Bavarian Academy of Science, Munich, Germany, with Julius Amman, who assisted Hertz. (Courtesy of Museum of Science and Industry, Chicago, and University of Karlsruhe, Federal Republic of Germany. Identification of objects by J. H. Bryant, University of Michigan, *IEEE Transactions on Microwave Theory and Techniques*, Vol. 36, No. 5, May 1988, Fig. 2.)

however, until it was rediscovered independently in 1936 by two men [6]. George C. Southworth was working at the AT&T Company in New York and demonstrated propagation through a water-filled copper pipe in 1932. After this preliminary experiment, interest was generated at AT&T for using waveguide as a high-bandwidth transmission line, and patents for such an application were filed. Southworth soon received theoretical support from S. A. Schelkunoff and others at AT&T, and analysis showed that a dielectric (e.g., water) was not needed for waveguide operation. This theoretical work also treated attenuation, and the subsequent discovery that the TE_{01} mode of the circular guide would have an attenuation factor that decreased with frequency appeared to offer great potential for low-loss, wideband communications. (This application never proved to be practical, however.)

Because of business concerns, Southworth was not able to publish his work until a technical meeting in May 1936. At this same meeting, W. L. Barrow presented a paper on the same topic. Barrow had been working at MIT on antennas and electromagnetic propagation at short wavelengths, and had studied reflectors of various shapes. This led him to the idea of a hollow tube to guide electromagnetic waves. His first experiment in 1935 on waveguide propagation was not successful, however. As he had no theoretical model at that time to support his experiments, Barrow was not familiar with the concept of a cutoff frequency. In his experiment, the free-space wavelength was 50 cm and the circular waveguide diameter was 4.5 cm, so the waveguide was well below cutoff, and no propagation was observed. Soon Barrow developed a theory, and redid the experiment with a tube 18 inches in diameter. In March 1936 his tests were successful, and he reported on his work at the URSI/IRE meeting in Washington on May 1, 1936. Barrow's paper was recently reprinted in the *IEEE Proceedings* [7].

Radar

Prior to World War II, the magnetron was developed in Great Britain as a reliable source of high-power microwave energy. This was critical to the development of radar (RAdio Detection And Ranging), which was under development in the United States, Great Britain, and other countries in the 1930s [8]. Early radar work was at very high frequencies (VHF), but by the 1940s microwaves were introduced. Higher frequencies allowed the use of reasonable sized antennas with high gain, and thus good angular resolution could be obtained. Radar was quickly developed in Great Britain and the United States, and played an important role in the war.

In the United States, the Radiation Laboratory was established at MIT to develop radar theory and practice. A number of top scientists, including N. Marcuvitz, I.I. Rabi, J.S. Schwinger, H.A. Bethe, E.M. Purcell, C.G. Montgomery, and R.H. Dicke, among others, were gathered for what turned out to be a very intensive period of development in the microwave field. Their work included the theoretical and experimental treatment of waveguide discontinuities, small aperture coupling theory, and the beginnings of microwave network theory. Many of these researchers were physicists who went back to physics research after the war (many received Nobel Prizes), but their microwave work is summarized in the classic 28-volume Radiation Laboratory Series of books that still finds application today [9].

The work on microwave network theory begun at the Radiation Laboratory was continued and extended at the Microwave Research Institute, which was organized at the Polytechnic Institute of Brooklyn in 1942. The researchers there included E. Weber, N. Marcuvitz, A. A. Oliner, L. B. Felsen, A. Hessel, and others. An interesting history of microwave field theory, with a number of personal anecdotes concerning the early work in the area, is given by Oliner [10].

Planar Transmission Lines

Through World War II, practically all microwave systems employed rectangular waveguide and/or coaxial lines as transmission line media. Waveguide had the advantage of being able to handle the relatively high power needed for radar systems, but had limited bandwidth, was bulky, and was expensive. Coax was broadband, and convenient for many microwave applications [11], but was a difficult medium in which to fabricate complex microwave networks.

Planar transmission lines refer to media such as stripline, microstrip, slotline, coplanar waveguide, and related structures. Such transmission lines are compact, low-cost, and capable of being integrated with active devices such as diodes and transistors. According to R. M. Barrett [12], the first planar transmission line may have been a flat-strip coaxial line, similar to a stripline, used in a production power divider network for an antenna system in World War II. It is also reported by H. Howe, Jr., [13] that H. Wheeler fabricated a line with two flat coplanar strips in 1936.

Planar lines did not receive intensive development, however, until the 1950s. R. Barrett is generally credited with the invention of stripline, while important theoretical work on the properties of stripline was carried out by S. Cohn [14], and others. Soon a variety of couplers, hybrids, filters, antennas, and other components were being fabricated in stripline. Airborne Instruments Laboratory (AIL) coined the term stripline, while other companies (such as Sanders Associates) came up with different tradenames, such as triplate. Sanders Associates published a book called *A Handbook of Tri-plate Microwave Components* [15], intended as a stripline version of the well-known *Waveguide Handbook* of the *Radiation Laboratory Series* [9].

Microstrip line was developed at ITT laboratories [16], and was a competitor of stripline. The first microstrip lines used a relatively thick dielectric substrate, which accentuated the non-TEM behavior and frequency dispersion of microstrip line. This characteristic made it less desirable than stripline until the 1960s, when the substrate was made very thin. This reduced the frequency dependence of the line, and now microstrip is often the preferred medium for microwave integrated circuits. Stripline is still in use, however, particularly for components such as couplers, hybrids, and filters.

Current work is underway to develop other planar lines such as coplanar waveguide, finline, image guide, and others, with emphasis on monolithic integration and/or millimeter wave frequencies.

Passive Microwave Components

Microwave filter theory and practice began in the years preceeding World War II, by pioneers such as Mason, Sykes, Darlington, Fano, Lawson, and Richards. The image parameter method of filter design was developed in the late 1930s, and was useful for filters of a noncritical specification. Today, however, most filter design is done according to the insertion loss method, which is based on network synthesis techniques. Because of the continuing advancement of such techniques, particularly in relation to synthesis with distributed elements, filter design remains an active research area.

In 1948 P. I. Richards contributed an important concept to microwave filter design by relating the well-known lumped element filter theory to filters using distributed transmission lines [17]. Richards' transformations coupled with K. Kuroda's four identities then allowed lumped-element filter prototypes to be physically realized with open or short-circuited transmission line stubs [18].

In the early 1950s, R. Levy and S. Cohn [19] developed design formulas for maximally flat and Chebyshev responses for multicavity filters. Around this time a group at Stanford Research Institute, consisting of G. Matthaei, L. Young, E. Jones, S. Cohn, and others became very active in filter and coupler work. A voluminous handbook on filters and couplers resulted from this work and remains a valuable reference [20]. This group developed a variety of devices using coupled lines and interdigitated lines.

Directional couplers form another class of passive devices similar in development to microwave filters [21], and many of the early researchers in filter theory also contributed to coupler work [19]. The Radiation Laboratory Series [9] contains descriptions of numerous waveguide couplers, including the Bethe-hole coupler, the Schwinger coupler, multihole couplers, and probe couplers. After the war, Riblet and others published designs for general multihole (or multiprobe) coupler design, as well as related designs for hybrid junctions. Chebyshev couplers were quickly developed after the publication by C. Dolph in 1946 of the Dolph-Chebyshev antenna array synthesis technique—array synthesis being closely related to coupler and matching transformer synthesis. In the mid-1950s through the 1960s, coupled planar transmission lines received attention as directional coupler components, and considerable theory and design techniques were developed at the Stanford Research Institute and at other laboratories. The even-odd mode analysis of J. Reed and G. Wheeler [22] proved to be of fundamental importance in this work.

Ferrite devices constitute another class of passive microwave devices, and although the first microwave ferrite device was not demonstrated until 1949 [23], they soon became indispensable in isolator and phase shifter (critical for phased array antennas) applications. Ferrites are also used in various other nonreciprocal devices such as circulators, and in magnetically tunable cavities and filters.

Active Microwave Devices and Circuits

Active microwave devices include tubes as well as solid state devices, and are used for sources, detectors, amplifiers, and other functions. The magnetron tube was the first reliable source of microwave power at centimeter wavelengths, and led to the advancement of radar. The Klystron, invented by the Varian brothers in the United States in 1937 [24], was even more versatile, as it could be used as an amplifier as well as an oscillator.

The earliest solid state microwave device was probably the point-contact detector diode. This took the form of the "cat-whisker" detector used in early radio work, but the advent of tubes as detectors and amplifiers eliminated this component in most radio systems. The crystal diode was later used by Southworth in his 1930s experiments with waveguides, as the tube detectors could not operate at such high frequencies. Co-workers at Bell Labs, including A. P. King and R. S. Ohl, greatly improved the crystal detector with better materials and a rugged cartridge package.

Frequency conversion and heterodyning were developed in the 1920s for radio applications, and these techniques were applied to radar receivers during World War II using crystal diodes as mixers. These mixers and detectors are discussed in Volume 15 of the Radiation Laboratory Series [9]. Around this time, theoretical work by Friis [25] on noise greatly increased the understanding of receiver sensitivity and the proper design of low-noise circuits. Another important contribution was the theory developed by Peterson and Llewellyn [26] on small-signal frequency conversion in nonlinear devices.

In the 1950s, traveling wave tubes (TWTs) were developed as low-noise amplifiers, and the maser was also being developed as a low-noise amplifier [27]. Such components were usually quite large and bulky, however, and often required supporting equipment like high-voltage power supplies and refrigeration units. There was clearly a need for compact devices for microwave amplifier and oscillator circuits.

In the early 1950s, J. M. Manley and H. E. Rowe at Bell Labs derived quite general relations for parametric amplification using nonlinear circuit elements [28]. This theory has shown to be very useful in a variety of applications, including maser and laser devices. It is interesting to note that the work of Manley and Rowe, like that of Southworth, was delayed in publication because of some skepticism at Bell Labs [27]. The Manley-Rowe theory, however, soon led to the development of parametric amplifiers using diodes and varactors as nonlinear reactive elements. Such techniques were also demonstrated as being capable of frequency multiplication, and by 1960 200 mW of power at X-band was generated using varactors [27].

The 1960s witnessed a large growth in solid-state microwave devices and circuitry. PIN diodes were developed and used in switches and phase shifters [29]. The basic theory for the field effect transistor was developed by W. Shockley in 1952 [30], and the first field-effect transistors (FETs) used silicon. The first gallium arsenide Schottky barrier FET was made by C. A. Mead at Cal Tech in 1965 [31], and microwave gallium arsenide FETs were developed in the late 1960s. Because of the desire for faster computers, gallium arsenide integrated circuits began to be constructed in the late 1970s [32], and this application promises to be a large-volume market for gallium arsenide technology.

The logical trend for microwave circuits was to integrate transmission lines, active devices, and other components on a single semiconductor substrate, to form a monolithic microwave integrated circuit (MMIC). The first MMICs were developed in the late 1960s [31]–[33], and constituted relatively simple functions. Soon, however, sophisticated functions such as multistage FET amplifiers, 3- or 4-bit phase shifters, complete transmit/receive modules, and other circuits were being fabricated as MMICs, and the present trend is toward more complexity, higher power, and higher frequencies. Monolithic technology promises to lower the cost of microwave circuitry, while enhancing performance and reliability, and reducing size.

It is interesting to observe that this type of design requires more complete and accurate modeling and simulation of the circuit than did older, discrete designs. This is because the discrete designs could usually be tuned and adjusted on a component or module level, after fabrication, to achieve optimum performance, while monolithic designs are less amenable to such trimming. The development of improved computer-aided design (CAD) techniques is thus of considerable current importance.

REFERENCES

[1] J. C. Maxwell, *A Treatise on Electricity and Magnetism*, Dover, N.Y., 1954.

[2] H. Hertz, *Electric Waves, being researchs on the propagation of electric action with finite velocity through space*, D. E. Jones translation. Macmillan, N.Y., 1893 and Dover 1962.

[3] O. Heaviside, "Electromagnetic Induction and Its Propagation," *The Electrician*, vols. 15–20. Reprinted in Oliver Heaviside, *Electrical Papers* (2 vols.), Chelsea, N.Y., 1970.

[4] O. Heaviside, *Electromagnetic Theory*, vol. 1, 1893. Reprinted by Dover, N.Y., 1950.

[5] Lord Rayleigh, "On the Passage of Electric Waves Through Tubes," *Phil. Mag.*, vol. 43, pp. 125–132, 1897. Reprinted in *Collected Papers*, Cambridge Univ. Press, 1903.

[6] K. S. Packard, "The Origin of Waveguides: A Case of Multiple Rediscovery," *IEEE Trans. Microwave Theory and Techniques*, vol. MTT-32, pp. 961–969, September 1984.

[7] W. L. Barrow, "Transmission of Electromagnetic Waves in Hollow Tubes of Metal," *IEEE Proc.*, vol. 72, pp. 1064–1076, August 1984.

[8] M. I. Skolnik, "Fifty Years of Radar," *IEEE Proc.*, vol. 73, pp. 182–197, February 1985.

[9] *Radiation Laboratory Series*, Massachusetts Institute of Technology, McGraw-Hill, N.Y., 1948–1950.

[10] A. A. Oliner, "Historical Perspective on Microwave Field Theory," *IEEE Trans. Microwave Theory and Techniques*, vol. MTT-32, pp. 1022–1045, September 1984.

[11] J. H. Bryant, "Coaxial Transmission Lines, Related Two-Conductor Transmission Lines, Connectors, and Components: A U.S. Historical Perspective," *IEEE Trans. Microwave Theory and Techniques*, vol. MTT-32, pp. 970–983, September 1984.

[12] R. M. Barrett, "Microwave Printed Circuits—The Early Years," *IEEE Trans. Microwave Theory and Technqiues*, vol. MTT-32, pp. 983–990, September 1984.

[13] H. Howe, Jr., "Microwave Integrated Circuits—An Historical Perspective," *IEEE Trans. Microwave Theory and Techniques*, vol. MTT-32, pp. 991–996, September 1984.

[14] S. Cohn, "Problems in Strip Transmission Lines," *IEEE Trans. Microwave Theory and Techniques*, vol. MTT-3, pp. 119–126, March 1955.

[15] R. W. Peters et al., *Handbook of Tri-plate Microwave Components*, Sanders Associates, N.H. 1956.

[16] D. D. Grieg and H. F. Engelmann, "Microstrip—A New Transmission Technique for the Kilomegacycle Range," *Proc. IRE*, vol. 40, pp. 1644–1650, December 1952.

[17] P. I. Richards, "Resistor-Transmission-Line Circuits," *Proc. IRE*, vol. 36, pp. 217–220, February 1948.

[18] H. Ozaki and J. Ishii, "Synthesis of a Class of Stripline Filters," *IRE Trans. Circuit Theory*, vol. CT-5, pp. 104–109, June 1958.

[19] R. Levy and S. B. Cohn, "A History of Microwave Filter Research, Design, and Development," *IEEE Trans. Microwave Theory and Techniques*, vol. MTT-32, pp. 1055–1067, September 1984.

[20] G. L. Matthaei, L. Young, and E.M.T. Jones, *Microwave Filters, Impedance-Matching Networks and Coupling Structures*, McGraw-Hill, N.Y., 1964.

[21] S. B. Cohn and R. Levy, "History of Microwave Passive Components with Particular Attention to Directional Couplers," *IEEE Trans. Microwave Theory and Techniques*, vol. MTT-32, pp. 1046–1054, September 1984.

[22] J. Reed and G. J. Wheeler, "A Method of Analysis of Symmetrical Four-Port Networks," *IRE Trans. Microwave Theory Tech.*, vol. MTT-4, pp. 246–252, October 1956.

[23] K. J. Button, "Microwave Ferrite Devices: The First Ten Years," *IEEE Trans. Microwave Theory and Techniques*, vol. MTT-32, pp. 1088–1096, September 1984.

[24] E. L. Ginzton, "The $100 Idea," *IEEE Spectrum*, vol. 12, pp. 30–39, February 1975.

[25] H. T. Friis, "Noise Figures of Radio Receivers," *Proc. IRE*, vol. 32, pp. 419–422, July 1944.

[26] L. C. Peterson and F. B. Llewellyn, "The Performance and Measurement of Mixers in Terms of Linear-Network Theory," *Proc. IRE*, vol. 33, pp. 458–476, July 1954.

[27] M. E. Hines, "The Virtues of Nonlinearity-Detection, Frequency Conversion, Parametric Amplification and Harmonic Generation," *IEEE Trans. Microwave Theory and Technqiues*, vol. MTT-32, pp. 1097-1104, September 1984.

[28] J. M. Manley and H. E. Rowe, "Some General Properties of Non-Linear Elements—Part I, General Energy Relations," *Proc. IRE*, vol. 44, pp. 904–913, July 1956.

[29] J. F. White, "Origins of High-Power Diode Switching," *IEEE Trans. Microwave Theory and Techniques*, vol. MTT-32, pp. 1105–1117, September 1984.

[30] W. Shockley, "A Unipolar 'Field-Effect' Transistor," *Proc. IRE*, vol. 40, pp. 1365–1376, November 1952.

[31] C. A. Mead, "Schottky-Barrier Gate Field Effect Transistors," *Proc. IEEE*, vol. 54, pp. 307–308, February 1966.

[32] P. Greiling, "The Historical Development of GaAs FET Digital IC Technology," *IEEE Trans. Microwave Theory and Techniques*, vol. MTT-32, pp. 1144–1156, September 1984.

[33] D. N. McQuiddy, Jr., J. W. Wassel, J. B. Lagrange, and W. R. Wisseman, "Monolithic Microwave Integrated Circuits: An Historical Perspective," *IEEE Trans. Microwave Theory and Techniques*, vol. MTT-32, pp. 997–1008, September 1984.

CHAPTER 2

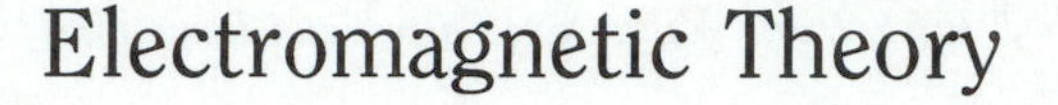

Electromagnetic Theory

2.1 INTRODUCTION—MAXWELL'S EQUATIONS

Electric and magnetic phenomenon at the macroscopic level is described by Maxwell's equations, as published by James C. Maxwell in 1873 [1]. This work summarized the state of electromagnetic science at that time, and hypothesized from theoretical considerations the existence of the electrical displacement current, which led to the discovery by Hertz and Marconi of electromagnetic wave propagation. Maxwell's work was based on a large body of empirical and theoretical knowledge developed by Gauss, Ampere, Faraday, and others. A first course in electromagnetics usually follows this historical (or deductive) approach, and it is assumed that the reader has had such a course as a prerequisite to the present material. Several books are available, [2]–[9], that provide a good treatment of electromagnetic theory at the undergraduate or graduate level.

This chapter will outline the fundamental concepts of electromagnetic theory which we will require for the rest of the book. Maxwell's equations will be presented, and boundary conditions and the effect of dielectric and magnetic materials will be discussed. Wave phenomenon is of essential importance in microwave engineering, and so, much of the chapter is spent on plane wave topics. Plane waves are the simplest form of electromagnetic waves, and so serve to illustrate a number of basic properties associated with wave propagation. Although it is assumed that the reader has studied plane waves before, the present material should help to reinforce many of the basic principles in the reader's mind, and perhaps to introduce some concepts which the reader has not seen previously. This material will also serve as a useful reference for later chapters.

With an awareness of the historical perspective, it is usually advantageous from a pedagogical point of view to present electromagnetic theory from the "inductive," or axiomatic, approach by beginning with Maxwell's equations. The general form of time-varying Maxwell equations, then, can be written in "point," or differential, form as

$$\nabla \times \bar{\mathcal{E}} = \frac{-\partial \bar{\mathcal{B}}}{\partial t} - \bar{\mathcal{M}}, \tag{2.1a}$$

$$\nabla \times \bar{\mathcal{H}} = \frac{\partial \bar{\mathcal{D}}}{\partial t} + \bar{\mathcal{J}}, \tag{2.1b}$$

$$\nabla \cdot \bar{\mathcal{D}} = \rho, \tag{2.1c}$$

$$\nabla \cdot \bar{\mathcal{B}} = 0. \tag{2.1d}$$

The MKS system of units is used throughout this book. The script quantities represent time-varying vector fields and are real functions of spatial coordinates x,y,z, and the time coordinate t. These quantities are defined as follows:

$\bar{\mathcal{E}}$ is the electric field intensity, in V/m.

$\bar{\mathcal{H}}$ is the magnetic field intensity, in A/m.

$\bar{\mathcal{D}}$ is the electric flux density, in Coul/m^2.

$\bar{\mathcal{B}}$ is the magnetic flux density, in Wb/m^2.

$\bar{\mathcal{M}}$ is the (fictitious) magnetic current density, in V/m^2.

$\bar{\mathcal{J}}$ is the electric current density, in A/m^2.

ρ is the electric charge density, in Coul/m^3.

The sources of the electromagnetic field are the currents $\bar{\mathcal{M}}$ and $\bar{\mathcal{J}}$, and the electric charge ρ. The magnetic current $\bar{\mathcal{M}}$ is a fictitious source in the sense that it is only a mathematical convenience: the real source of a magnetic current is always a loop of electric current or some similar type of magnetic dipole, as opposed to the flow of an actual magnetic charge (magnetic monopole charges are not known to exist). The magnetic current is included here for completeness, as we will have occasion to use it in Chapter 5 when dealing with apertures. Since electric current is really the flow of charge, it can be said that the electric charge density ρ is the ultimate source of the electromagnetic field.

In free-space, the following simple relations hold between the electric and magnetic field intensities and flux densities:

$$\bar{\mathcal{B}} = \mu_0 \bar{\mathcal{H}}, \tag{2.2a}$$

$$\bar{\mathcal{D}} = \epsilon_0 \bar{\mathcal{E}}, \tag{2.2b}$$

where $\mu_0 = 4\pi \times 10^{-7}$ henry/m is the permeability of free-space, and $\epsilon_0 = 8.854 \times 10^{-12}$ farad/m is the permittivity of free-space. We will see in the next section how media other than free-space affect these constitutive relations.

Equations (2.1a)–(2.1d) are linear, but are not independent of each other. For instance, consider the divergence of (2.1a). Since the divergence of the curl of any vector is zero [vector identity (B.12), from Appendix B], we have

$$\nabla \cdot \nabla \times \bar{\mathcal{E}} = 0 = -\frac{\partial}{\partial t}(\nabla \cdot \bar{\mathcal{B}}) - \nabla \cdot \bar{\mathcal{M}}.$$

Since there is no free magnetic charge, $\nabla \cdot \bar{\mathcal{M}} = 0$, which leads to $\nabla \cdot \bar{\mathcal{B}} = 0$, or (2.1d). The continuity equation can be similarly derived by taking the divergence of (2.1b), giving

$$\nabla \cdot \bar{\mathcal{J}} + \frac{\partial \rho}{\partial t} = 0, \tag{2.3}$$

where (2.1c) was used. This equation states that charge is conserved, or that current is continuous, since $\nabla \cdot \bar{\mathcal{J}}$ represents the outflow of current at a point, and $\partial\rho/\partial t$ represents the charge buildup with time at the same point. It is this result that led Maxwell to the conclusion that the displacement current density $\partial\bar{\mathcal{D}}/\partial t$ was necessary in (2.1b), which can be seen by taking the divergence of this equation.

The above differential equations can be converted to integral form through the use of various vector integral theorems. Thus, applying the divergence theorem (B.15) to (2.1c) and (2.1d) yields

$$\oint_S \bar{\mathcal{D}} \cdot d\bar{s} = \int_V \rho dv = Q, \tag{2.4}$$

$$\oint_S \bar{\mathcal{B}} \cdot d\bar{s} = 0, \tag{2.5}$$

where Q in (2.4) represents the total charge contained in the closed volume V (enclosed by a closed surface S). Applying Stoke's theorem (B.16) to (2.1a) gives

$$\oint_C \bar{\mathcal{E}} \cdot d\bar{l} = -\frac{\partial}{\partial t}\int_S \bar{\mathcal{B}} \cdot d\bar{s} - \int_S \bar{\mathcal{M}} \cdot d\bar{s}, \tag{2.6}$$

which, without the $\bar{\mathcal{M}}$ term, is the usual form of Faraday's law, and forms the basis for Kirchhoff's voltage law. In (2.6), C represents a closed contour around the surface S, as shown in Figure 2.1. Ampere's law can be derived by applying Stoke's theorem to (2.1b):

$$\oint_C \bar{\mathcal{H}} \cdot d\bar{l} = \frac{\partial}{\partial t}\int_S \bar{\mathcal{D}} \cdot d\bar{s} + \int_S \bar{\mathcal{J}} \cdot d\bar{s} = \frac{\partial}{\partial t}\int_S \bar{\mathcal{D}} \cdot d\bar{s} + \mathcal{I}, \tag{2.7}$$

where $\mathcal{I} = \int_S \bar{\mathcal{J}} \cdot d\bar{s}$ is the total electric current flow through the surface S. Equations (2.4)–(2.7) constitute the integral forms of Maxwell's equations.

The above equations are valid for arbitrary time dependence, but most of our work will be involved with fields having a sinusoidal, or harmonic, time dependence, with steady-state conditions assumed. In this case phasor notation is very convenient, and so all field quantities will be assumed to be complex vectors with an implied $e^{j\omega t}$ time dependence, and written with roman (rather than script) letters. Thus, a sinusoidal electric field in the $\hat{x}$ direction of the form

$$\bar{\mathcal{E}}(x, y, z, t) = \hat{x}A(x, y, z)\cos(\omega t + \phi), \tag{2.8}$$

where A is the (real) amplitude, ω is the radian frequency, and ϕ is the phase reference of the wave at $t = 0$, has the phasor form

$$\bar{E}(x, y, z) = \hat{x}A(x, y, z)e^{j\phi}. \tag{2.9}$$

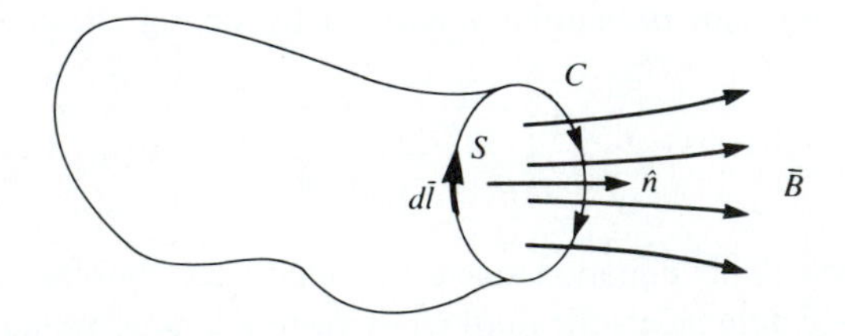

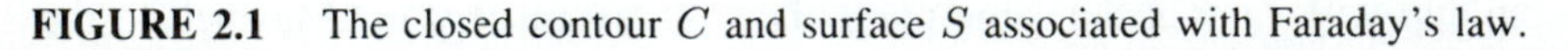

FIGURE 2.1 The closed contour C and surface S associated with Faraday's law.

We will assume cosine-based phasors in this book, so the conversion from phasor quantities to real time-varying quantities is accomplished by multiplying the phasor by $e^{j\omega t}$ and taking the real part:

$$\bar{\mathcal{E}}(x, y, z, t) = \text{Re}[\bar{E}(x, y, z)e^{j\omega t}], \tag{2.10}$$

as substituting (2.9) into (2.10) to obtain (2.8) demonstrates. When working in phasor notation, it is customary to suppress the common $e^{j\omega t}$ factor on all terms.

When dealing with power and energy, we will often be interested in the time-average of a quadratic quantity. This can be found very easily for time harmonic fields. For example, the average of the square of the magnitude of an electric field given by

$$\bar{\mathcal{E}} = \hat{x}E_1 \cos(\omega t + \phi_1) + \hat{y}E_2 \cos(\omega t + \phi_2) + \hat{z}E_2 \cos(\omega t + \phi_3), \tag{2.11}$$

which has the phasor form

$$\bar{E} = \hat{x}E_1 e^{j\phi_1} + \hat{y}E_2 e^{j\phi_2} + \hat{z}E_3 e^{j\phi_3}, \tag{2.12}$$

can be calculated as

$$\begin{aligned} |\bar{E}|^2_{\text{av}} &= \frac{1}{T}\int_0^T \bar{\mathcal{E}} \cdot \bar{\mathcal{E}} dt \\ &= \frac{1}{T}\int_0^T [E_1^2 \cos^2(\omega t + \phi_1) + E_2^2 \cos^2(\omega t + \phi_2) + E_3^2 \cos^2(\omega t + \phi_3)]dt \\ &= \frac{1}{2}(E_1^2 + E_2^2 + E_3^2) = \frac{1}{2}|\bar{E}|^2 = \frac{1}{2}\bar{E} \cdot \bar{E}^*. \end{aligned} \tag{2.13}$$

Then the rms value is $|\bar{E}|_{\text{rms}} = |\bar{E}|/\sqrt{2}$.

Assuming an $e^{j\omega t}$ time dependence, the time derivatives in (2.1a)–(2.1d) can be replaced by $j\omega$. Maxwell's equations in phasor form then become

$$\nabla \times \bar{E} = -j\omega\bar{B} - \bar{M}, \tag{2.14a}$$

$$\nabla \times \bar{H} = j\omega\bar{D} + \bar{J}, \tag{2.14b}$$

$$\nabla \cdot \bar{D} = \rho, \tag{2.14c}$$

$$\nabla \cdot \bar{B} = 0. \tag{2.14d}$$

The Fourier transform can be used to convert a solution to Maxwell's equations for an arbitrary frequency ω to a solution for arbitrary time dependence.

The electric and magnetic current sources, $\bar{J}$ and $\bar{M}$, in (2.14) are volume current densities with units A/m^2 or V/m^2, respectively. In many cases, however, the actual currents will be in the form of a current sheet, a line current, or an infinitesimal dipole current. These special types of current distributions can always be written as volume current densities through the use of delta functions. Figure 2.2 shows examples of this procedure for electric and magnetic currents.

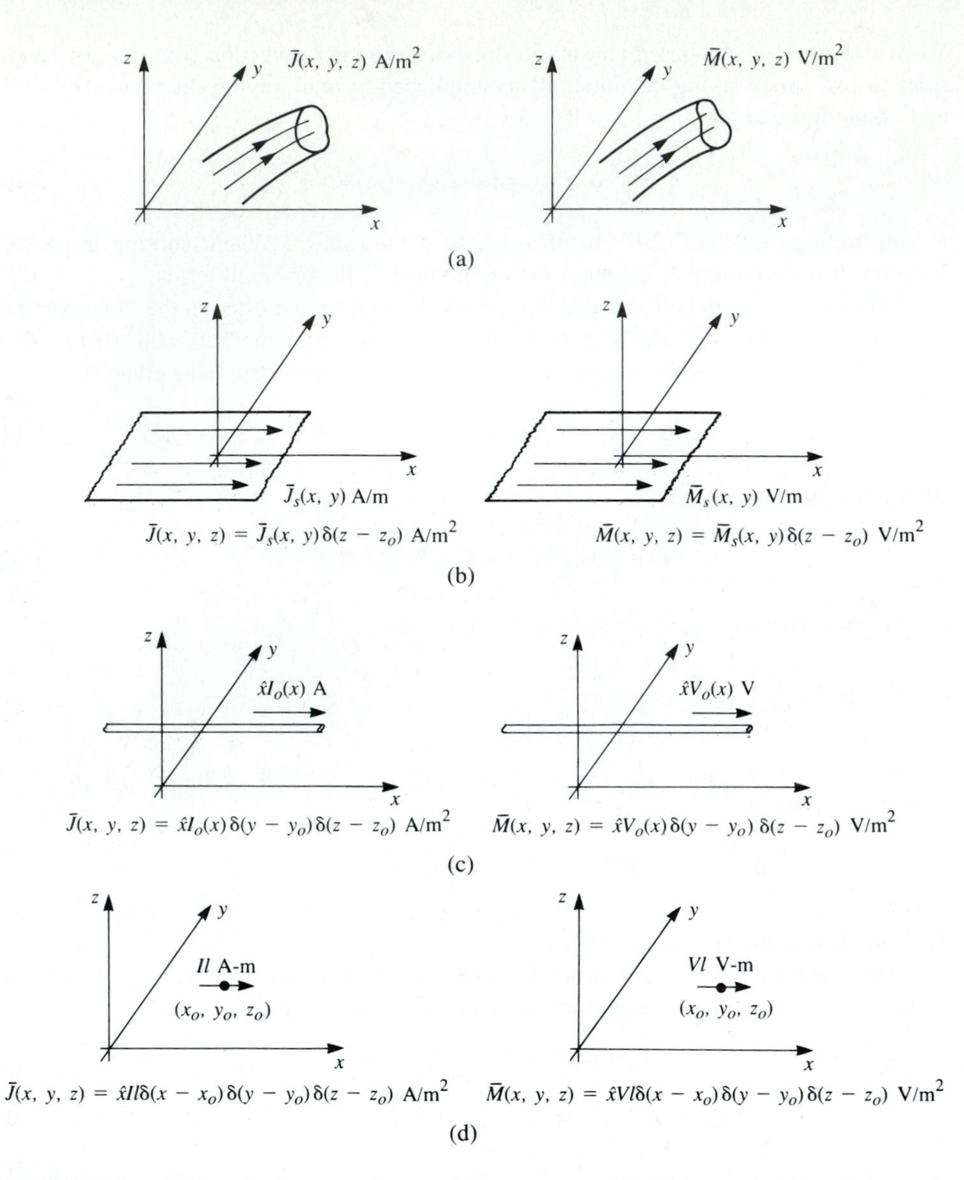

FIGURE 2.2 Arbitrary volume, surface, and line currents. (a) Arbitrary electric and magnetic volume current densities. (b) Arbitrary electric and magnetic surface current densities in the $z = z_0$ plane. (c) Arbitrary electric and magnetic line currents. (d) Infinitesimal electric and magnetic dipoles parallel to the x-axis.

2.2 FIELDS IN MEDIA

In the preceding section it was assumed that the electric and magnetic fields were in free-space, with no material bodies present. In practice, material bodies are often present; this complicates the analysis, but also allows the useful application of material properties to microwave components. When electromagnetic fields exist in material media, the field vectors are related to each other by the constitutive relations.

For a dielectric material, an applied electric field $\bar{E}$ causes the polarization of the atoms or molecules of the material to create electric dipole moments that augment the total displacement flux, $\bar{D}$. This additional polarization vector is called $\bar{P}_e$, the electric polarization, where

$$\bar{D} = \epsilon_0 \bar{E} + \bar{P}_e. \tag{2.15}$$

In a linear medium, the electric polarization is linearly related to the applied electric field as

$$\bar{P}_e = \epsilon_0 \chi_e \bar{E}, \tag{2.16}$$

where χ_e, which may be complex, is called the electric susceptibility. Then,

$$\bar{D} = \epsilon_0 \bar{E} + \bar{P}_e = \epsilon_0 (1 + \chi_e) \bar{E} = \epsilon \bar{E}, \tag{2.17}$$

where

$$\epsilon = \epsilon' - j\epsilon'' = \epsilon_0 (1 + \chi_e) \tag{2.18}$$

is the complex permittivity of the medium. The imaginary part of ϵ accounts for loss in the medium (heat) due to damping of the vibrating dipole moments. (Free-space, having a real ϵ, is lossless.) Due to energy conservation, as we will see in Section 2.6, the imaginary part of ϵ must be negative (ϵ'' positive). The loss of a dielectric material may also be considered as an equivalent conductor loss. In a material with conductivity σ, a conduction current density will exist:

$$\bar{J} = \sigma \bar{E}, \tag{2.19}$$

which is Ohm's law from an electromagnetic field's point of view. Maxwell's curl equation for $\bar{H}$ (2.14b) then becomes

$$\begin{aligned} \nabla \times \bar{H} &= j\omega \bar{D} + \bar{J} \\ &= j\omega\epsilon \bar{E} + \sigma \bar{E} \\ &= j\omega\epsilon' \bar{E} + (\omega\epsilon'' + \sigma) \bar{E} \\ &= j\omega \left(\epsilon' - j\epsilon'' - j\frac{\sigma}{\omega} \right) \bar{E}, \end{aligned} \tag{2.20}$$

where it is seen that loss due to dielectric damping ($\omega\epsilon''$) is indistinguishable from conductivity loss (σ). The term $\omega\epsilon'' + \sigma$ can then be considered as the total effective conductivity. A related quantity of interest is the loss tangent, defined as

$$\tan \delta = \frac{\omega\epsilon'' + \sigma}{\omega\epsilon'}, \tag{2.21}$$

which is seen to be the ratio of the real to the imaginary parts of the total displacement current. Microwave materials are usually characterized by specifying the real permittivity, $\epsilon' = \epsilon_r \epsilon_0$, and the loss tangent at a certain frequency. These constants are listed in Appendix G for several types of materials. It is useful to note that, after a problem has been solved assuming a lossless dielectric, loss can be easily introduced by replacing the real ϵ with a complex $\epsilon = \epsilon' - j\epsilon'' = \epsilon'(1 - j \tan \delta)$.

In the above discussion it was assumed that $\bar{P}_e$ was a vector in the same direction as $\bar{E}$. Such materials are called isotropic materials, but not all materials have this property. Some materials are anisotropic, and are characterized by a more complicated relation between $\bar{P}_e$ and $\bar{E}$, or $\bar{D}$ and $\bar{E}$. The most general linear relation between these vectors takes the form of a tensor of rank two (a dyad), which can be written in matrix form as

$$\begin{bmatrix} D_x \\ D_y \\ D_z \end{bmatrix} = \begin{bmatrix} \epsilon_{xx} & \epsilon_{xy} & \epsilon_{xz} \\ \epsilon_{yx} & \epsilon_{yy} & \epsilon_{yz} \\ \epsilon_{zx} & \epsilon_{zy} & \epsilon_{zz} \end{bmatrix} \begin{bmatrix} E_x \\ E_y \\ E_z \end{bmatrix} = [\epsilon] \begin{bmatrix} E_x \\ E_y \\ E_z \end{bmatrix}. \qquad 2.22$$

It is thus seen that a given vector component of $\bar{E}$ gives rise, in general, to three components of $\bar{D}$. Crystal structures and ionized gases are examples of anisotropic dielectrics. For a linear isotropic material, the matrix of (2.22) would reduce to a diagonal matrix with elements, ϵ.

An analogous situation occurs for magnetic materials. An applied magnetic field may align magnetic dipole moments in a magnetic material to produce a magnetic polarization (or magnetization) $\bar{P}_m$. Then,

$$\bar{B} = \mu_0(\bar{H} + \bar{P}_m). \qquad 2.23$$

For a linear magnetic material, $\bar{P}_m$ is linearly related to $\bar{H}$ as

$$\bar{P}_m = \chi_m \bar{H}, \qquad 2.24$$

where χ_m is a complex magnetic susceptibility. From (2.23) and (2.24),

$$\bar{B} = \mu_0(1 + \chi_m)\bar{H} = \mu\bar{H}, \qquad 2.25$$

where $\mu = \mu_0(1 + \chi_m) = \mu' - j\mu''$ is the permeability of the medium. Again, the imaginary part of χ_m or μ accounts for loss due to damping forces; there is no magnetic conductivity, since there is no real magnetic current. As in the electric case, magnetic materials may be anisotropic, in which case a tensor permeability can be written as

$$\begin{bmatrix} B_x \\ B_y \\ B_z \end{bmatrix} = \begin{bmatrix} \mu_{xx} & \mu_{xy} & \mu_{xz} \\ \mu_{yx} & \mu_{yy} & \mu_{yz} \\ \mu_{zx} & \mu_{zy} & \mu_{zz} \end{bmatrix} \begin{bmatrix} H_x \\ H_y \\ H_z \end{bmatrix} = [\mu] \begin{bmatrix} H_x \\ H_y \\ H_z \end{bmatrix}. \qquad 2.26$$

An important example of anisotropic magnetic materials in microwave engineering is the class of ferrimagnetic materials known as ferrites; these materials and their applications will be discussed further in Chapter 10.

If linear media are assumed (ϵ, μ not depending on $\bar{E}$ or $\bar{H}$), then Maxwell's equations can be written in phasor form as

$$\nabla \times \bar{E} = -j\omega\mu\bar{H} - \bar{M}, \qquad 2.27a$$

$$\nabla \times \bar{H} = j\omega\epsilon\bar{E} + \bar{J}, \qquad 2.27b$$

$$\nabla \cdot \bar{D} = \rho, \qquad 2.27c$$

$$\nabla \cdot \bar{B} = 0. \qquad 2.27d$$

The constitutive relations are

$$\bar{D} = \epsilon \bar{E}, \tag{2.28a}$$

$$\bar{B} = \mu \bar{H}, \tag{2.28b}$$

where ϵ and μ may be complex, and may be tensors. Note that relations like (2.28a) and (2.28b) generally cannot be written in time domain form, even for linear media, because of the possible phase shift between $\bar{D}$ and $\bar{E}$, or $\bar{B}$ and $\bar{H}$. The phasor representation accounts for this phase shift by the complex form of ϵ and μ.

2.3 BOUNDARY CONDITIONS

Maxwell's equations (2.27a)–(2.27d) in differential form require known boundary values for a complete and unique solution. A general method used throughout this book is to solve the source-free Maxwell's equations in a certain region to obtain solutions with unknown coefficients, and then apply boundary conditions to solve for these coefficients. A number of specific cases of boundary conditions arise, as discussed below.

Fields at a General Material Interface

Consider a plane interface between two media, as shown in Figure 2.3. Maxwell's equations in integral form can be used to deduce conditions involving the normal and tangential fields at this interface. The time-harmonic version of (2.4), where S is the closed "pillbox" shaped surface shown in Figure 2.4, can be written as

$$\oint_S \bar{D} \cdot d\bar{s} = \int_V \rho dv. \tag{2.29}$$

In the limit as $h \to 0$, the contribution of D_{tan} through the sidewalls goes to zero, so that (2.29) reduces to

$$\Delta S D_{2n} - \Delta S D_{1n} = \Delta S \rho_s,$$

or

$$D_{2n} - D_{1n} = \rho_s, \tag{2.30}$$

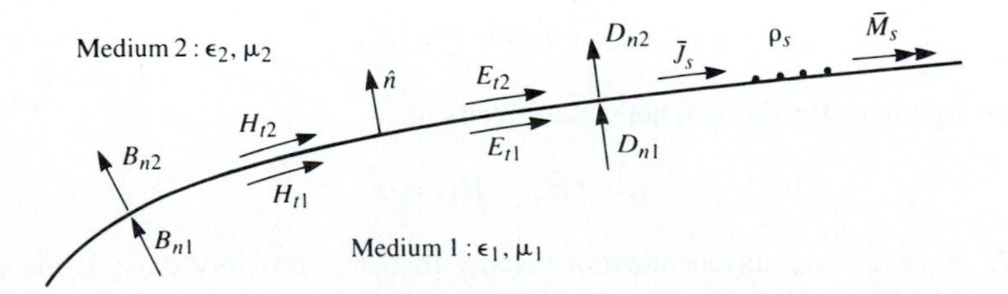

FIGURE 2.3 Fields, currents, and surface charge at a general interface between two media.

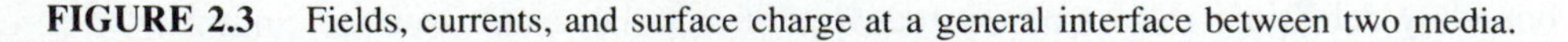

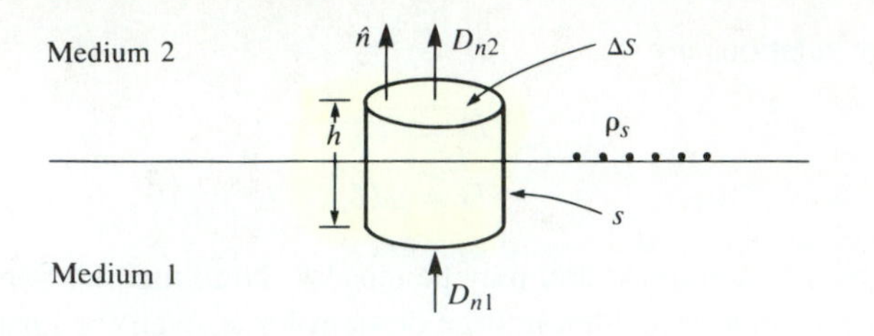

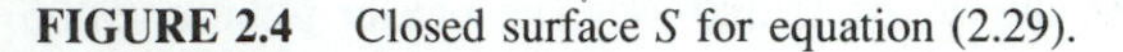

FIGURE 2.4 Closed surface S for equation (2.29).

where ρ_s is the surface charge density on the interface. In vector form, we can write

$$\hat{n} \cdot (\bar{D}_2 - \bar{D}_1) = \rho_s. \tag{2.31}$$

A similar argument for $\bar{B}$ leads to the result that

$$\hat{n} \cdot \bar{B}_2 = \hat{n} \cdot \bar{B}_1, \tag{2.32}$$

since there is no free magnetic charge.

For the tangential components of the electric field we use the phasor form of (2.6),

$$\oint_C \bar{E} \cdot d\bar{l} = -j\omega \int_S \bar{B} \cdot d\bar{s} - \int_S \bar{M} \cdot d\bar{s}, \tag{2.33}$$

in connection with the closed contour C shown in Figure 2.5. In the limit as $h \to 0$, the surface integral of $\bar{B}$ vanishes (since $S = h\Delta\ell$ vanishes). The contribution from the surface integral of $\bar{M}$, however, may be nonzero if a magnetic surface current density $\bar{M}_s$ exists on the surface. The Dirac delta function can then be used to write

$$\bar{M} = \bar{M}_s \delta(h), \tag{2.34}$$

where h is a coordinate measured normal from the interface. Equation (2.33) then gives

$$\Delta\ell E_{t1} - \Delta\ell E_{t2} = -\Delta\ell M_s,$$

or

$$E_{t1} - E_{t2} = - M_s, \tag{2.35}$$

which can be generalized in vector form as

$$(\bar{E}_2 - \bar{E}_1) \times \hat{n} = \bar{M}_s. \tag{2.36}$$

A similar argument for the magnetic field leads to

$$\hat{n} \times (\bar{H}_2 - \bar{H}_1) = \bar{J}_s, \tag{2.37}$$

where $\bar{J}_s$ is an electric surface current density that may possibly exist at the interface. Equations (2.31), (2.32), (2.36), and (2.37) are the most general expressions for the boundary conditions at an arbitrary interface of materials and/or surface currents.

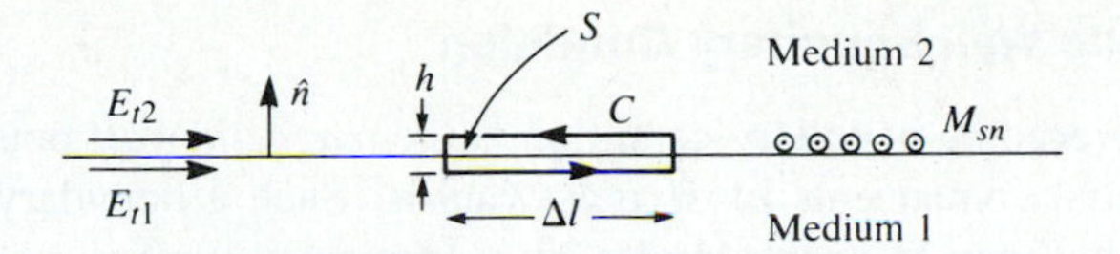

FIGURE 2.5 Closed contour C for equation (2.33).

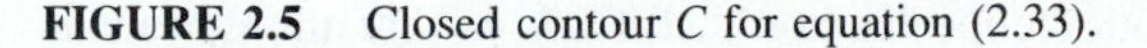

Fields at a Dielectric Interface

At an interface between two lossless dielectric materials, no charge or surface current densities will ordinarily exist. Equations (2.31), (2.32), (2.36), and (2.37) then reduce to

$$\hat{n} \cdot \bar{D}_1 = \hat{n} \cdot \bar{D}_2, \tag{2.38a}$$

$$\hat{n} \cdot \bar{B}_1 = \hat{n} \cdot \bar{B}_2, \tag{2.38b}$$

$$\hat{n} \times \bar{E}_1 = \hat{n} \times \bar{E}_2, \tag{2.38c}$$

$$\hat{n} \times \bar{H}_1 = \hat{n} \times \bar{H}_2. \tag{2.38d}$$

In words, these equations state that the normal components of $\bar{D}$ and $\bar{B}$ are continuous across the interface, and the tangential components of $\bar{E}$ and $\bar{H}$ are equal across the interface. Because Maxwell's equations are not all linearly independent, the six boundary conditions contained in the above equations are not all linearly independent. Thus, the enforcement of (2.38c) and (2.38d) for the four tangential field components, for example, will automatically force the satisfaction of the equations for the continuity of the normal components.

Fields at the Interface with a Perfect Conductor (Electric Wall)

Many problems in microwave engineering involve boundaries with good conductors (e.g., metals), which can often be assumed as lossless ($\sigma \to \infty$). In this case of a perfect conductor, all field components must be zero inside the conducting region. This result can be seen by considering a conductor with finite conductivity ($\sigma < \infty$) and noting that the skin depth (the depth to which most of the microwave power penetrates) goes to zero as $\sigma \to \infty$. (Such an analysis will be performed in Section 2.7.) If we also assume here that $\bar{M}_s = 0$, which would be the case if the perfect conductor filled all the space on one side of the boundary, then (2.31), (2.32), (2.36), and (2.37) reduce to the following:

$$\hat{n} \cdot \bar{D} = \rho_s, \tag{2.39a}$$

$$\hat{n} \cdot \bar{B} = 0, \tag{2.39b}$$

$$\hat{n} \times \bar{E} = 0, \tag{2.39c}$$

$$\hat{n} \times \bar{H} = \bar{J}_s, \tag{2.39d}$$

where ρ_s and $\bar{J}_s$ are the electric surface charge density and current density, respectively, on the interface, and $\hat{n}$ is the normal unit vector pointing out of the perfect conductor. Such a boundary is also known as an *electric wall*, since the tangential components of $\bar{E}$ are "shorted out," as seen from (2.39c), and must vanish at the surface of the conductor.

The Magnetic Wall Boundary Condition

Dual to the preceding boundary condition is the *magnetic wall* boundary condition, where the tangential components of $\bar{H}$ must vanish. Such a boundary does not really exist in practice, but may be approximated by a corrugated surface, or in certain planar transmission line problems. In addition, the idealization that $\hat{n} \times \bar{H} = 0$ at an interface is often a convenient simplification, as we will see in later chapters. We will also see that the magnetic wall boundary condition is analogous to the relations between the voltage and current at the end of an open-circuited transmission line, while the electric wall boundary condition is analogous to the voltage and current at the end of a short-circuited transmission line. The magnetic wall condition, then, provides a degree of completeness in our formulation of boundary conditions, and is a useful approximation in several cases of practical interest.

The fields at a magnetic wall satisfy the following conditions:

$$\hat{n} \cdot \bar{D} = 0, \qquad 2.40a$$

$$\hat{n} \cdot \bar{B} = 0, \qquad 2.40b$$

$$\hat{n} \times \bar{E} = -\bar{M}_s, \qquad 2.40c$$

$$\hat{n} \times \bar{H} = 0, \qquad 2.40d$$

where $\hat{n}$ is the normal unit vector pointing out of the magnetic wall region.

The Radiation Condition

When dealing with problems that have one or more infinite boundaries, such as plane waves in an infinite medium, or infinitely long transmission lines, a condition on the fields at infinity must be enforced. This boundary condition is known as the *radiation condition*, and is essentially a statement of energy conservation. It states that, at an infinite distance from a source, the fields must either be vanishingly small (i.e., zero), or propagating in an outward direction. This result can easily be seen by allowing the infinite medium to contain a small loss factor (as any physical medium would have). Incoming waves (from infinity) of finite amplitude would then require an infinite source at infinity, and so are disallowed.

2.4 THE WAVE EQUATION AND BASIC PLANE WAVE SOLUTIONS

The Helmholtz Equations

In a source-free, linear, isotropic, homogeneous region, Maxwell's curl equations in phasor form are

$$\nabla \times \bar{E} = -j\omega\mu\bar{H}, \qquad 2.41a$$

$$\nabla \times \bar{H} = j\omega\epsilon\bar{E}, \qquad 2.41b$$

and constitute two equations for the two unknowns, $\bar{E}$ and $\bar{H}$. As such, they can be solved for either $\bar{E}$ or $\bar{H}$. Thus, taking the curl of (2.41a) and using (2.41b) gives

$$\nabla \times \nabla \times \bar{E} = -j\omega\mu\nabla \times \bar{H} = \omega^2\mu\epsilon\bar{E},$$

which is an equation for $\bar{E}$. This result can be simplified through the use of vector identity (B.14), $\nabla \times \nabla \times \bar{A} = \nabla(\nabla \cdot \bar{A}) - \nabla^2\bar{A}$, which is valid for the rectangular components of an arbitrary vector $\bar{A}$. Then,

$$\nabla^2\bar{E} + \omega^2\mu\epsilon\bar{E} = 0, \tag{2.42}$$

since $\nabla \cdot \bar{E} = 0$ in a source-free region. Equation (2.42) is the wave equation, or Helmholtz equation, for $\bar{E}$. An identical equation for $\bar{H}$ can be derived in the same manner:

$$\nabla^2\bar{H} + \omega^2\mu\epsilon\bar{H} = 0. \tag{2.43}$$

A constant $k = \omega\sqrt{\mu\epsilon}$ is defined and called the wavenumber, or propagation constant, of the medium; its units are 1/m.

As a way of introducing wave behavior, we will next study the solutions to the above wave equations in their simplest forms, first for a lossless medium and then for a lossy (conducting) medium.

Plane Waves in a Lossless Medium

In a lossless medium, ϵ and μ are real numbers, so k is real. A basic plane wave solution to the above wave equations can be found by considering an electric field with only an $\hat{x}$ component, and uniform (no variation) in the x and y directions. Then, $\partial/\partial x = \partial/\partial y = 0$, and the Helmholtz equation of (2.42) reduces to

$$\frac{\partial^2 E_x}{\partial z^2} + k^2 E_x = 0. \tag{2.44}$$

The solutions to this equation are easily seen, by substitution, to be of the form

$$E_x(z) = E^+ e^{-jkz} + E^- e^{jkz}, \tag{2.45}$$

where E^+ and E^- are arbitrary amplitude constants.

The above solution is for the time harmonic case at frequency ω. In the time domain, this result is written as

$$\mathcal{E}_x(z,t) = E^+ \cos(\omega t - kz) + E^- \cos(\omega t + kz), \tag{2.46}$$

where we have assumed that E^+ and E^- are real constants. Consider the first term in (2.46). This term represents a wave traveling in the $+z$ direction, since, to maintain a fixed point on the wave ($\omega t - kz$ = constant), one must move in the $+z$ direction as time increases. Similarly, the second term in (2.46) represents a wave traveling in the negative z direction; hence the notation E^+ and E^- for these wave amplitudes. The velocity of the wave in this sense is called the phase velocity, since it is the velocity at which a fixed phase point on the wave travels, and it is given by

$$v_p = \frac{dz}{dt} = \frac{d}{dt}\left(\frac{\omega t - \text{constant}}{k}\right) = \frac{\omega}{k} = \frac{1}{\sqrt{\mu\epsilon}} \tag{2.47}$$

In free-space, we have $v_p = 1/\sqrt{\mu_0\epsilon_0} = c = 2.998 \times 10^8$ m/sec, which is the speed of light.

The wavelength, λ, is defined as the distance between two successive maxima (or minima, or any other reference points) on the wave, at a fixed instant of time. Thus,

$$[\omega t - kz] - [\omega t - k(z + \lambda)] = 2\pi,$$

so,

$$\lambda = \frac{2\pi}{k} = \frac{2\pi v_p}{\omega} = \frac{v_p}{f}. \tag{2.48}$$

A complete specification of the plane wave electromagnetic field must include the magnetic field. In general, whenever $\bar{E}$ or $\bar{H}$ is known, the other field vector can be readily found by using one of Maxwell's curl equations. Thus, applying (2.41a) to the electric field of (2.45) gives $H_x = H_z = 0$, and

$$H_y = \frac{1}{\eta}[E^+ e^{-jkz} - E^- e^{jkz}], \tag{2.49}$$

where $\eta = \omega\mu/k = \sqrt{\mu/\epsilon}$ is the wave impedance for the plane wave, defined as the ratio of the $\bar{E}$ and $\bar{H}$ fields. For plane waves, this impedance is also the intrinsic impedance of the medium. In free-space we have $\eta_0 = \sqrt{\mu_0/\epsilon_0} = 377\ \Omega$. Note that the $\bar{E}$ and $\bar{H}$ vectors are orthogonal to each other, and orthogonal to the direction of propagation ($\pm\hat{z}$); this is a characteristic of transverse electromagnetic (TEM) waves.

EXAMPLE 2.1

A plane wave with a frequency of 3 GHz is propagating in an unbounded material with $\epsilon_r = 7$ and $\mu_r = 3$. Compute the wavelength, phase velocity, and wave impedance for this wave.

Solution
From (2.47) the phase velocity is

$$v_p = \frac{1}{\sqrt{\mu\epsilon}} = \frac{c}{\sqrt{\mu_r\epsilon_r}} = \frac{3 \times 10^8}{\sqrt{(7)(3)}} = 6.55 \times 10^7 \text{ m/sec}.$$

This is slower than the speed of light in free-space by a factor of $\sqrt{21} = 4.58$. From (2.48) the wavelength is

$$\lambda = \frac{v_p}{f} = \frac{6.55 \times 10^7}{3 \times 10^9} = 0.0218 \text{ m}.$$

The wave impedance is

$$\eta = \sqrt{\frac{\mu}{\epsilon}} = \eta_0\sqrt{\frac{\mu_r}{\epsilon_r}} = 377\sqrt{\frac{3}{7}} = 246.8\ \Omega.$$

○

Plane Waves in a General Lossy Medium

Now consider the effect of a lossy medium. If the medium is conductive, with a conductivity, σ, Maxwell's curl equations can be written, from (2.41a) and (2.20) as

$$\nabla \times \bar{E} = -j\omega\mu\bar{H}, \tag{2.50a}$$

$$\nabla \times \bar{H} = j\omega\epsilon\bar{E} + \sigma\bar{E}. \tag{2.50b}$$

The resulting wave equation for $\bar{E}$ then becomes

$$\nabla^2\bar{E} + \omega^2\mu\epsilon\left(1 - j\frac{\sigma}{\omega\epsilon}\right)\bar{E} = 0, \tag{2.51}$$

where we see a similarity with (2.42), the wave equation for $\bar{E}$ in the lossless case. The difference is that the wavenumber $k^2 = \omega^2\mu\epsilon$ of (2.42) is replaced by $\omega^2\mu\epsilon[1 - j(\sigma/\omega\epsilon)]$ in (2.51). We then define a complex propagation constant for the medium as

$$\gamma = \alpha + j\beta = j\omega\sqrt{\mu\epsilon}\sqrt{1 - j\frac{\sigma}{\omega\epsilon}} \tag{2.52}$$

If we again assume an electric field with only an $\hat{x}$ component, and uniform in x and y, the wave equation of (2.51) reduces to

$$\frac{\partial^2 E_x}{\partial z^2} - \gamma^2 E_x = 0, \tag{2.53}$$

which has solutions

$$E_x(z) = E^+e^{-\gamma z} + E^-e^{\gamma z}. \tag{2.54}$$

The positive traveling wave then has a propagation factor of the form

$$e^{-\gamma z} = e^{-\alpha z}e^{-j\beta z},$$

which in the time domain is of the form

$$e^{-\alpha z}\cos(\omega t - \beta z).$$

We see then that this represents a wave traveling in the $+z$ direction with a phase velocity $v_p = \omega/\beta$, a wavelength $\lambda = 2\pi/\beta$, and an exponential damping factor. The rate of decay with distance is given by the attenuation constant, α. The negative traveling wave term of (2.54) is similarly damped along the $-z$ axis. If the loss is removed, $\sigma = 0$, and we have $\gamma = jk$ and $\alpha = 0$, $\beta = k$.

As discussed in Section 2.2, loss can also be treated through the use of a complex permittivity. From (2.20) with $\sigma = 0$ but $\epsilon = \epsilon' - j\epsilon''$ complex, we have that

$$\gamma = j\omega\sqrt{\mu\epsilon} = jk = j\omega\sqrt{\mu\epsilon'(1 - j\tan\delta)}, \tag{2.55}$$

where $\tan\delta = \epsilon''/\epsilon'$ is the loss tangent of the material.

Next, the associated magnetic field can be calculated as

$$H_y = \frac{j}{\omega\mu}\frac{\partial E_x}{\partial z} = \frac{-j\gamma}{\omega\mu}(E^+ e^{-\gamma z} - E^- e^{\gamma z}). \tag{2.56}$$

As with the lossless case, a wave impedance can be defined to relate the electric and magnetic fields:

$$\eta = \frac{j\omega\mu}{\gamma}. \tag{2.57}$$

Then (2.56) can be rewritten as

$$H_y = \frac{1}{\eta}(E^+ e^{-\gamma z} - E^- e^{\gamma z}). \tag{2.58}$$

Note that η is, in general, complex and reduces to the lossless case of $\eta = \sqrt{\mu/\epsilon}$ when $\gamma = jk = j\omega\sqrt{\mu\epsilon}$.

Plane Waves in a Good Conductor

Many problems of practical interest involve loss or attenuation due to good (but not perfect) conductors. A good conductor is a special case of the preceding analysis, where the conductive current is much greater than the displacement current, which means $\sigma \gg \omega\epsilon$. Most metals can be categorized as good conductors. In terms of a complex ϵ, rather than conductivity, this condition is equivalent to $\epsilon'' \gg \epsilon'$. The propagation constant of (2.52) can then be adequately approximated by ignoring the displacement current term, to give

$$\gamma = \alpha + j\beta \simeq j\omega\sqrt{\mu\epsilon}\sqrt{\frac{\sigma}{j\omega\epsilon}} = (1+j)\sqrt{\frac{\omega\mu\sigma}{2}}. \tag{2.59}$$

The skin depth, or characteristic depth of penetration, is defined as

$$\delta_s = \frac{1}{\alpha} = \sqrt{\frac{2}{\omega\mu\sigma}}. \tag{2.60}$$

Then the amplitude of the fields in the conductor decay by an amount $1/e$ or 36.8%, after traveling a distance of one skin depth, since $e^{-\alpha z} = e^{-\alpha\delta_s} = e^{-1}$. At microwave frequencies, for a good conductor, this distance is very small. The practical importance of this result is that only a thin plating of a good conductor (e.g., silver or gold) is necessary for low-loss microwave components.

EXAMPLE 2.2

Compute the skin depth of aluminum, copper, gold, and silver at a frequency of 10 GHz.

Solution

The conductivities for these metals are listed in Appendix F. Equation (2.60) gives the skin depths as

$$\delta_s = \sqrt{\frac{2}{\omega\mu\sigma}} = \sqrt{\frac{1}{\pi f \mu_0 \sigma}} = \sqrt{\frac{1}{\pi(10^{10})(4\pi \times 10^{-7})}}\sqrt{\frac{1}{\sigma}}$$

$$= 5.03 \times 10^{-3}\sqrt{\frac{1}{\sigma}}.$$

For aluminum: $\delta_s = 5.03 \times 10^{-3}\sqrt{\dfrac{1}{3.816 \times 10^7}} = 8.14 \times 10^{-7}$ m.

For copper: $\delta_s = 5.03 \times 10^{-3}\sqrt{\dfrac{1}{5.813 \times 10^7}} = 6.60 \times 10^{-7}$ m.

For gold: $\delta_s = 5.03 \times 10^{-3}\sqrt{\dfrac{1}{4.098 \times 10^7}} = 7.86 \times 10^{-7}$ m.

For silver: $\delta_s = 5.03 \times 10^{-3}\sqrt{\dfrac{1}{6.173 \times 10^7}} = 6.40 \times 10^{-7}$ m.

These results show that most of the current flow in a good conductor occurs in an extremely thin region near the surface of the conductor. ○

The wave impedance inside a good conductor can be obtained from (2.57) and (2.59). The result is

$$\eta = \frac{j\omega\mu}{\gamma} \simeq (1+j)\sqrt{\frac{\omega\mu}{2\sigma}} = (1+j)\frac{1}{\sigma\delta_s}. \tag{2.61}$$

Notice that the phase angle of this impedance is 45°, a characteristic of good conductors. The phase angle of the impedance for a lossless material is 0°, and the phase angle of the impedance of an arbitrary lossy medium is somewhere between 0° and 45°.

Table 2.1 summarizes the results for plane wave propagation in lossless and lossy homogeneous media.

2.5 GENERAL PLANE WAVE SOLUTIONS

Some specific features of plane waves were discussed in Section 2.4. Here we will look at plane waves again, from a more general point of view, and solve the wave equation by the method of separation of variables. This technique will find application in succeeding chapters. We will also discuss circularly polarized plane waves, which will be important for the discussion of ferrites in Chapter 10.

In free-space, the Helmholtz equation for $\bar{E}$ can be written as

$$\nabla^2\bar{E} + k_0^2\bar{E} = \frac{\partial^2\bar{E}}{\partial x^2} + \frac{\partial^2\bar{E}}{\partial y^2} + \frac{\partial^2\bar{E}}{\partial z^2} + k_0^2\bar{E} = 0, \tag{2.62}$$

TABLE 2.1 Summary of Results for Plane Wave Propagation in Various Media

Quantity	Type of Medium: Lossless ($\epsilon'' = \sigma = 0$)	General Lossy	Good Conductor $\epsilon'' >> \epsilon'$ or $\sigma >> \omega\epsilon'$
Complex propagation constant	$\gamma = j\omega\sqrt{\mu\epsilon}$	$\gamma = j\omega\sqrt{\mu\epsilon}$ $= j\omega\sqrt{\mu\epsilon'}\sqrt{(1 - j\sigma)/\omega\epsilon'}$	$\gamma = (1 + j)\sqrt{\omega\mu\sigma/2}$
Phase constant (wavenumber)	$\beta = k = \omega\sqrt{\mu\epsilon}$	$\beta = Im(\gamma)$	$\beta = Im(\gamma) = \sqrt{\omega\mu\sigma/2}$
Attenuation constant	$\alpha = 0$	$\alpha = \mathrm{Re}(\gamma)$	$\alpha = \mathrm{Re}(\gamma) = \sqrt{\omega\mu\sigma/2}$
Impedance	$\eta = \sqrt{\mu/\epsilon} = \omega\mu/k$	$\eta = j\omega\mu/\gamma$	$\eta = (1 + j)\sqrt{\omega\mu/2\sigma}$
Skin depth	$\delta_s = \infty$	$\delta_s = 1/\alpha$	$\delta_s = \sqrt{2/\omega\mu\sigma}$
Wavelength	$\lambda = 2\pi/\beta$	$\lambda = 2\pi/\beta$	$\lambda = 2\pi/\beta$
Phase velocity	$v_p = \omega/\beta$	$v_p = \omega/\beta$	$v_p = \omega/\beta$

and this vector wave equation holds for each rectangular component of $\bar{E}$:

$$\frac{\partial^2 E_i}{\partial x^2} + \frac{\partial^2 E_i}{\partial y^2} + \frac{\partial^2 E_i}{\partial z^2} + k_0^2 E_i = 0, \tag{2.63}$$

where the index $i = x, y,$ or z. This equation will now be solved by the method of separation of variables, a standard technique for treating such partial differential equations. The method begins by assuming that the solution to (2.63) for, say E_x, can be written as a product of three functions for each of the three coordinates:

$$E_x(x, y, z) = f(x)\ g(y)\ h(z). \tag{2.64}$$

Substituting this form into (2.63) and dividing by fgh gives

$$\frac{f''}{f} + \frac{g''}{g} + \frac{h''}{h} + k_0^2 = 0, \tag{2.65}$$

where the double primes denote the second derivative. Now the key step in the argument is to recognize that each of the terms in (2.65) must be equal to a constant, since they are independent of each other. That is, f''/f is only a function of x, and the remaining terms in (2.65) do not depend on x, so f''/f must be a constant, and similarly for the other terms in (2.65). Thus, we define three separation constants, k_x, k_y and k_z, such that

$$f''/f = -k_x^2; \quad g''/g = -k_y^2; \quad h''/h = -k_z^2;$$

or $$\frac{d^2 f}{dx^2} + k_x^2 f = 0; \quad \frac{d^2 g}{dy^2} + k_y^2 g = 0; \quad \frac{d^2 h}{dz^2} + k_z^2 h = 0. \tag{2.66}$$

Combining (2.65) and (2.66) shows that

$$k_x^2 + k_y^2 + k_z^2 = k_0^2. \tag{2.67}$$

The partial differential equation of (2.63) has now been reduced to three separate ordinary differential equations in (2.66). Solutions to these equations are of the form $e^{\pm jk_x x}, e^{\pm jk_y y}$, and $e^{\pm jk_z z}$, respectively. As we have seen in the previous section, the terms with $+$ signs result in waves traveling in the negative x, y, or z direction, while the terms with $-$ signs result in waves traveling in the positive direction. Both solutions are possible and are valid; the amount to which these various terms are excited is dependent on the source of the fields. For our present discussion, we will select a plane wave traveling in the positive direction for each coordinate, and write the complete solution for E_x as

$$E_x(x, y, z) = Ae^{-j(k_x x + k_y y + k_z z)}, \tag{2.68}$$

where A is an arbitrary amplitude constant. Now define a wave number vector $\bar{k}$ as

$$\bar{k} = k_x \hat{x} + k_y \hat{y} + k_z \hat{z} = k_0 \hat{n}. \tag{2.69}$$

Then from (2.67) $|\bar{k}| = k_0$, and so $\hat{n}$ is a unit vector in the direction of propagation. Also define a position vector as

$$\bar{r} = x\hat{x} + y\hat{y} + z\hat{z}, \tag{2.70}$$

then (2.68) can be written as

$$E_x(x, y, z) = Ae^{-j\bar{k}\cdot\bar{r}}. \tag{2.71}$$

Solutions to (2.63) for E_y and E_z are, of course, similar in form to E_x of (2.71), but with different amplitude constants:

$$E_y(x, y, z) = Be^{-j\bar{k}\cdot\bar{r}}, \tag{2.72}$$

$$E_z(x, y, z) = Ce^{-j\bar{k}\cdot\bar{r}}. \tag{2.73}$$

The x, y, and z dependencies of the three components of $\bar{E}$ in (2.71)–(2.73) must be the same (same k_x, k_y, k_z), because the divergence condition that

$$\nabla \cdot \bar{E} = \frac{\partial E_x}{\partial x} + \frac{\partial E_y}{\partial y} + \frac{\partial E_z}{\partial z} = 0$$

must also be applied in order to satisfy Maxwell's equations, which implies that E_x, E_y, and E_z must each have the same variation in x, y, and z. (Note that the solutions in the preceding section automatically satisfied the divergence condition since E_x was the only component of $\bar{E}$, and E_x did not vary with x). This condition also imposes a constraint on the amplitudes A, B, and C, since if

$$\bar{E}_0 = A\hat{x} + B\hat{y} + C\hat{z},$$

we have

$$\bar{E} = \bar{E}_0 e^{-j\bar{k}\cdot\bar{r}},$$

and
$$\nabla \cdot \bar{E} = \nabla \cdot (\bar{E}_0 e^{-j\bar{k}\cdot\bar{r}}) = \bar{E}_0 \nabla e^{-j\bar{k}\cdot\bar{r}} = -j\bar{k} \cdot \bar{E}_0 e^{-j\bar{k}\cdot\bar{r}} = 0,$$

where vector identity (B.7) was used. Thus, we must have

$$\bar{k} \cdot \bar{E}_0 = 0, \tag{2.74}$$

which means that the electric field amplitude vector $\bar{E}_0$ must be perpendicular to the direction of propagation, $\bar{k}$. This condition is a general result for plane waves, and implies that only two of the three amplitude constants, A, B, and C, can be chosen independently.

The magnetic field can be found from Maxwell's equation,

$$\nabla \times \bar{E} = -j\omega\mu_0 \bar{H}, \tag{2.75}$$

to give

$$\begin{aligned}
\bar{H} &= \frac{j}{\omega\mu_0} \nabla \times \bar{E} = \frac{j}{\omega\mu_0} \nabla \times (\bar{E}_0 e^{-j\bar{k}\cdot\bar{r}}) \\
&= \frac{-j}{\omega\mu_0} \bar{E}_0 \times \nabla e^{-j\bar{k}\cdot\bar{r}} \\
&= \frac{-j}{\omega\mu_0} \bar{E}_0 \times (-j\bar{k}) e^{-j\bar{k}\cdot\bar{r}} \\
&= \frac{k_0}{\omega\mu_0} \hat{n} \times \bar{E}_0 e^{-j\bar{k}\cdot\bar{r}} \\
&= \frac{1}{\eta_0} \hat{n} \times \bar{E}_0 e^{-j\bar{k}\cdot\bar{r}} \\
&= \frac{1}{\eta_0} \hat{n} \times \bar{E},
\end{aligned} \tag{2.76}$$

where vector identity (B.9) was used in obtaining the second line. This result shows that the magnetic field intensity vector $\bar{H}$ lies in a plane normal to $\bar{k}$, the direction of propagation, and that $\bar{H}$ is perpendicular to $\bar{E}$. See Figure 2.6 for an illustration of these vector relations. The quantity $\eta_0 = \sqrt{\mu_0/\epsilon_0} = 377\ \Omega$ in (2.76) is the intrinsic impedance of free-space.

The time-domain expression for the electric field can be found as

$$\begin{aligned}
\bar{\mathcal{E}}(x, y, z, t) &= \text{Re}\{\bar{E}(x, y, z) e^{j\omega t}\} \\
&= \text{Re}\{\bar{E}_0 e^{-j\bar{k}\cdot\bar{r}} e^{j\omega t}\} \\
&= \bar{E}_0 \cos(\bar{k} \cdot \bar{r} - \omega t),
\end{aligned} \tag{2.77}$$

assuming that the amplitude constants A, B, and C contained in $\bar{E}_0$ are real. If these constants are not real, their phases should be included inside the cosine term of (2.77).

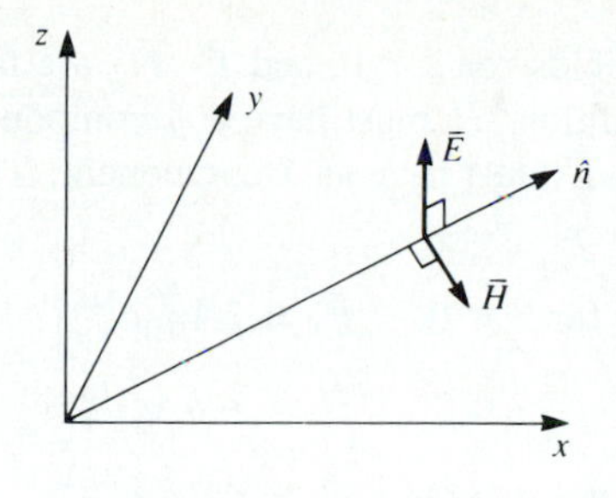

FIGURE 2.6 Orientation of the $\bar{E}$, $\bar{H}$, and $\bar{k} = k_0\hat{n}$ vectors for a plane wave.

From (2.77) we see that a wavelength, defined as the distance the wave must travel to undergo a phase shift of 2π, can be found as

$$|\bar{k}|\lambda_0 = k_0\lambda_0 = 2\pi,$$

or $\lambda_0 = 2\pi/k_0$, which is identical to the result obtained in Section 2.4. The phase velocity, or the speed at which one would have to travel to maintain a constant phase point on the wave, can be found from the condition that

$$\bar{k} \cdot \bar{r} - \omega t = \text{constant}.$$

Taking the derivative with respect to time gives

$$k_0 \frac{dr}{dt} - \omega = 0,$$

or

$$k_0 v_p = \omega,$$

so that

$$v_p = \omega/k_0 = \frac{1}{\sqrt{\mu_0 \epsilon_0}} = c = 2.998 \times 10^8 \text{ m/s},$$

as in Section 2.4.

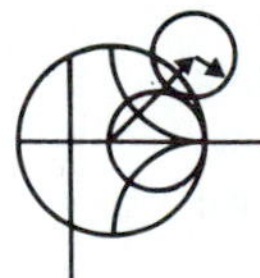

EXAMPLE 2.3

An infinite sheet of surface current can be considered as a source for plane waves. If an electric surface current density $\bar{J}_s = J_o\hat{x}$ exists on the $z = 0$ plane in free-space, find the resulting fields by assuming plane waves on either side of the current sheet and enforcing the boundary conditions.

Solution

Since the source does not vary with x or y, the fields will not vary with x or y, but will propagate away from the source in the $\pm z$ direction. The boundary conditions to be satisfied at $z = 0$ are

$$\hat{n} \times (\bar{E}_2 - \bar{E}_1) = \hat{z} \times (\bar{E}_2 - \bar{E}_1) = 0,$$

$$\hat{n} \times (\bar{H}_2 - \bar{H}_1) = \hat{z} \times (\bar{H}_2 - \bar{H}_1) = J_o\hat{x},$$

where $\bar{E}_1, \bar{H}_1$ are the fields for $z < 0$, and $\bar{E}_2, \bar{H}_2$ are the fields for $z > 0$. To satisfy the second condition, $\bar{H}$ must have a $\hat{y}$ component. Then for $\bar{E}$ to be orthogonal to $\bar{H}$ and $\hat{z}$, $\bar{E}$ must have an $\hat{x}$ component. Thus the fields will have the following form:

$$\text{for } z < 0, \quad \bar{E}_1 = \hat{x} A \eta_0 e^{j k_0 z},$$
$$\bar{H}_1 = -\hat{y} A e^{j k_0 z},$$
$$\text{for } z > 0, \quad \bar{E}_2 = \hat{x} B \eta_0 e^{-j k_0 z},$$
$$\bar{H}_2 = \hat{y} B e^{-j k_0 z},$$

where A and B are arbitrary amplitude constants. The first boundary condition, that E_x is continuous at $z = 0$, yields $A = B$, while the boundary condition for $\bar{H}$ yields the equation

$$-B - A = J_o.$$

Solving for A, B gives

$$A = B = -J_o/2,$$

which completes the solution. ○

Circularly Polarized Plane Waves

The plane waves discussed above all had their electric field vector pointing in a fixed direction, and so are called linearly polarized waves. In general, the polarization of a plane wave refers to the orientation of the electric field vector, which may be in a fixed direction, or may change with time.

Consider the superposition of an $\hat{x}$ linearly polarized wave with amplitude E_1, and a $\hat{y}$ linearly polarized wave with amplitude E_2, both traveling in the positive $\hat{z}$ direction. The total electric field can be written as

$$\bar{E} = (E_1 \hat{x} + E_2 \hat{y}) e^{-j k_0 z}. \tag{2.78}$$

A number of possibilities now arise. If $E_1 \neq 0$, and $E_2 = 0$, we have a plane wave linearly polarized in the $\hat{x}$ direction. Similarly, if $E_1 = 0$, and $E_2 \neq 0$, we have a plane wave linearly polarized in the $\hat{y}$ direction. If E_1 and E_2 are both real and nonzero, we have a plane wave linearly polarized at the angle

$$\phi = \tan^{-1} \frac{E_2}{E_1}.$$

For example, if $E_1 = E_2 = E_0$, we have

$$\bar{E} = E_0 (\hat{x} + \hat{y}) e^{-j k_0 z},$$

which represents an electric field vector at a 45° angle from the x-axis.

Now consider the case where $E_1 = j E_2 = E_0$, where E_0 is real, so that

$$\bar{E} = E_0 (\hat{x} - j \hat{y}) e^{-j k_0 z}. \tag{2.79}$$

The time domain form of this field is

$$\bar{\mathcal{E}}(z,t) = E_0\{\hat{x}\cos(\omega t - k_0 z) + \hat{y}\cos(\omega t - k_0 z - \pi/2)\}. \qquad 2.80$$

This expression shows that the electric field vector changes with time or, equivalently, with distance along the z-axis. To see this, pick a fixed position, say $z = 0$. Equation (2.80) then reduces to

$$\bar{\mathcal{E}}(0,t) = E_0\{\hat{x}\cos\omega t + \hat{y}\sin\omega t\}, \qquad 2.81$$

so as ωt increases from zero, the electric field vector rotates counterclockwise from the x-axis. The resulting angle from the x-axis of the electric field vector at time t, at $z = 0$, is then

$$\phi = \tan^{-1}\left(\frac{\sin\omega t}{\cos\omega t}\right) = \omega t,$$

which shows that the polarization rotates at the uniform angular velocity ω. Since the fingers of the right hand point in the direction of rotation when the thumb points in the direction of propagation, this type of wave is referred to as a right hand circularly polarized (RHCP) wave. Similarly, a field of the form

$$\bar{E} = E_0(\hat{x} + j\hat{y})e^{-jk_0 z} \qquad 2.82$$

constitutes a left hand circularly polarized (LHCP) wave, where the electric field vector rotates in the opposite direction. See Figure 2.7 for a sketch of the polarization vectors for RHCP and LHCP plane waves.

The magnetic field associated with a circularly polarized wave may be found from Maxwell's equations, or by using the wave impedance applied to each component of the electric field. For example, applying (2.76) to the electric field of a RHCP wave as

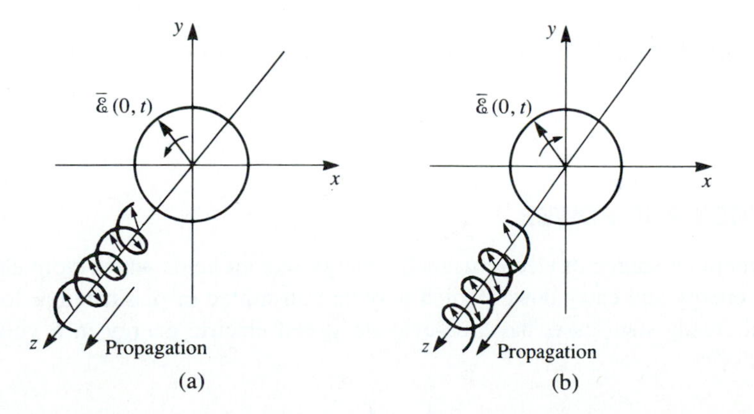

FIGURE 2.7 Electric field polarization for (a) RHCP and (b) LHCP plane waves.

given in (2.79) yields

$$\bar{H} = \frac{E_0}{\eta_0}\hat{z} \times (\hat{x} - j\hat{y})e^{-jk_0z} = \frac{E_0}{\eta_0}(\hat{y} + j\hat{x})e^{-jk_0z} = \frac{jE_0}{\eta_0}(\hat{x} - j\hat{y})e^{-jk_0z},$$

which is also seen to be a circularly polarized wave with a polarization vector of the RHCP sense.

EXAMPLE 2.4

Show that a linearly polarized plane wave of the form,

$$\bar{E} = E_0(\hat{x} + 2\hat{y})e^{-jk_0z},$$

can be decomposed into the sum of a RHCP wave and a LHCP wave.

Solution
Let

$$\bar{E} = A(\hat{x} - j\hat{y})e^{-jk_0z} + B(\hat{x} + j\hat{y})e^{-jk_0z},$$

where A is the amplitude of the RHCP component and B is the amplitude of the LHCP component. Equating this to the given linearly polarized wave expression gives,

$$A + B = E_0,$$

$$-jA + jB = 2E_0.$$

Solving for A, B gives

$$A = \left(\frac{1}{2} + j\right)E_0,$$

$$B = \left(\frac{1}{2} - j\right)E_0.$$

Any linearly polarized wave can be decomposed into two circularly polarized waves. ○

2.6 ENERGY AND POWER

In general, a source of electromagnetic energy sets up fields which store electric and magnetic energy and carry power which may be transmitted or dissipated as loss. In the sinusoidal steady-state case, the time-average stored electric energy in a volume V is given by,

$$W_e = \frac{1}{4}\text{Re}\int_V \bar{E} \cdot \bar{D}^* dv, \tag{2.83}$$

which in the case of simple lossless isotropic, homogeneous, linear media, where ϵ is a real scalar constant, reduces to

$$W_e = \frac{\epsilon}{4}\int_V \bar{E}\cdot\bar{E}^* dv. \tag{2.84}$$

Similarly, the time-average magnetic energy stored in the volume V is

$$W_m = \frac{1}{4}\text{Re}\int_V \bar{H}\cdot\bar{B}^* dv, \tag{2.85}$$

which becomes

$$W_m = \frac{\mu}{4}\int_V \bar{H}\cdot\bar{H}^* dv, \tag{2.86}$$

for a real, constant, scalar μ.

We can now derive Poynting's theorem, which leads to energy conservation for electromagnetic fields and sources. If we have an electric source current, $\bar{J}_s$, and a conduction current $\sigma\bar{E}$, as defined in (2.19), then the total electric current density is $\bar{J} = \bar{J}_s + \sigma\bar{E}$. Then multiplying (2.27a) by $\bar{H}^*$, and multiplying the conjugate of (2.27b) by $\bar{E}$, yields

$$\bar{H}^*\cdot(\nabla\times\bar{E}) = -j\omega\mu|\bar{H}|^2 - \bar{H}^*\cdot\bar{M}_s,$$
$$\bar{E}\cdot(\nabla\times\bar{H}^*) = \bar{E}\cdot\bar{J}^* - j\omega\epsilon^*|\bar{E}|^2 = \bar{E}\cdot\bar{J}_s^* + \sigma|\bar{E}|^2 - j\omega\epsilon^*|\bar{E}|^2,$$

where $\bar{M}_s$ is the magnetic source current. Using these two results in vector identity (B.8) gives

$$\begin{aligned}\nabla\cdot(\bar{E}\times\bar{H}^*) &= \bar{H}^*\cdot(\nabla\times\bar{E}) - \bar{E}\cdot(\nabla\times\bar{H}^*)\\ &= \sigma|\bar{E}|^2 + j\omega(\epsilon^*|\bar{E}|^2 - \mu|\bar{H}|^2) - (\bar{E}\cdot\bar{J}_s^* + \bar{H}^*\cdot\bar{M}_s).\end{aligned}$$

Now integrate over a volume V and use the divergence theorem:

$$\begin{aligned}\int_V \nabla\cdot(\bar{E}\times\bar{H}^*)dv &= \oint_S \bar{E}\times\bar{H}^*\cdot d\bar{s}\\ &= \sigma\int_V |\bar{E}|^2 dv + j\omega\int_V (\epsilon^*|\bar{E}|^2 - \mu|\bar{H}|^2 dv - \int_V (\bar{E}\cdot\bar{J}_s^* + \bar{H}^*\cdot\bar{M}_s)dv,\end{aligned} \tag{2.87}$$

where S is a closed surface enclosing the volume V, as shown in Figure 2.8. Allowing $\epsilon = \epsilon' - j\epsilon''$ and $\mu = \mu' - j\mu''$ to be complex to allow for loss, and rewriting (2.87) gives

$$\begin{aligned}-\frac{1}{2}\int_V(\bar{E}\cdot\bar{J}_s^* + \bar{H}^*\cdot\bar{M}_s)dv &= \frac{1}{2}\oint_S \bar{E}\times\bar{H}^*\cdot d\bar{s}\\ &+ \frac{\sigma}{2}\int_V |\bar{E}|^2 dv + \frac{\omega}{2}\int_V(\epsilon''|\bar{E}|^2 + \mu''|\bar{H}|^2)dv + j\frac{\omega}{2}\int_V(\mu'|\bar{H}|^2 - \epsilon'|\bar{E}|^2)dv.\end{aligned} \tag{2.88}$$

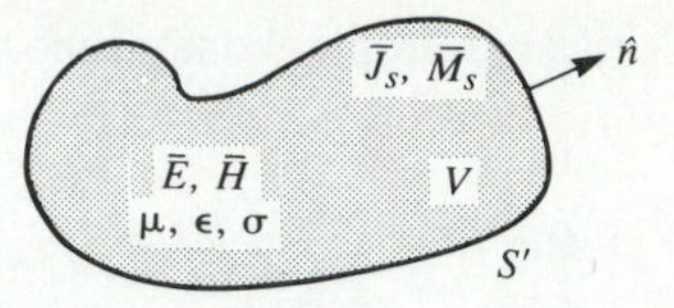

FIGURE 2.8 A volume V, enclosed by the closed surface S, containing fields $\bar{E}$, $\bar{H}$, and current sources $\bar{J}_s$, $\bar{M}_s$.

This result is known as Poynting's theorem, after the physicist J.H. Poynting (1852–1914), and is basically a power balance equation. Thus, the integral on the left-hand side represents the complex power, P_s, delivered by the sources $\bar{J}_s$ and $\bar{M}_s$, inside S:

$$P_s = -\frac{1}{2}\int_V (\bar{E} \cdot \bar{J}_s^* + \bar{H}^* \cdot \bar{M}_s)dv. \tag{2.89}$$

The first integral on the right-hand side of (2.88) represents complex power flow out of the closed surface S. If we define a quantity called the Poynting vector, $\bar{S}$, as

$$\bar{S} = \bar{E} \times \bar{H}^*, \tag{2.90}$$

then this power can be expressed as

$$P_o = \frac{1}{2}\oint_S \bar{E} \times \bar{H}^* \cdot d\bar{s} = \frac{1}{2}\oint_S \bar{S} \cdot d\bar{s}. \tag{2.91}$$

The surface S in (2.91) must be a closed surface in order for this interpretation to be valid. The real parts of P_s and P_o in (2.89) and (2.91) represent time-average powers.

The second and third integrals in (2.88) are real quantities representing the time-average power dissipated in the volume V due to conductivity, dielectric, and magnetic losses. If we define this power as P_ℓ we have that

$$P_\ell = \frac{\sigma}{2}\int_V |\bar{E}|^2 dv + \frac{\omega}{2}\int_V (\epsilon''|\bar{E}|^2 + \mu''|\bar{H}|^2)dv, \tag{2.92}$$

which is sometimes referred to as Joule's law. The last integral in (2.88) can be seen to be related to the stored electric and magnetic energies, as defined in (2.84) and (2.86).

With the above definitions, Poynting's theorem can be rewritten as

$$P_s = P_o + P_\ell + 2j\omega(W_m - W_e). \tag{2.93}$$

In words, this complex power balance equation states that the power delivered by the sources (P_s) is equal to the sum of the power transmitted through the surface (P_o), the power lost to heat in the volume (P_ℓ), and 2ω times the net reactive energy stored in the volume.

EXAMPLE 2.5

Compute the Poynting vector for the plane wave field of (2.76).

Solution
From (2.90), the Poynting vector is

$$\bar{S} = \bar{E} \times \bar{H}^* = \bar{E}_0 e^{-j\bar{k}\cdot\bar{r}} \times \frac{1}{\eta_0}(\hat{n} \times \bar{E}_0 e^{-j\bar{k}\cdot\bar{r}})^* = |E_0|^2 \frac{\hat{n}}{\eta_0},$$

which shows that a power density of $|E_0|^2/\eta_0\ W/m^2$ is flowing in the direction of propagation. ○

Power Absorbed by a Good Conductor

To calculate attenuation and loss due to an imperfect conductor, one must find the power dissipated in the conductor. We will show that this can be done using only the fields at the surface of the conductor, which is a very helpful simplification when calculating attenuation.

Consider the geometry of Figure 2.9, which shows the interface between a lossless medium and a good conductor. We assume that a field is incident from $z < 0$, and that the field penetrates into the conducting region $z > 0$. The real average power entering the conductor volume defined by the cross-sectional surface S_0 at the interface and the surface S is given from (2.91) as

$$P_{\text{av}} = \frac{1}{2}\text{Re}\int_{S_0+S} \bar{E} \times \bar{H}^* \cdot \hat{n}\, ds, \tag{2.94}$$

where $\hat{n}$ is a unit normal vector pointing into the closed surface $S_0 + S$, and $\bar{E}, \bar{H}$ are the fields at this surface. The contribution to the integral in (2.94) from the surface S can be made zero by proper selection of this surface. For example, if the field is a normally incident plane wave, the Poynting vector $\bar{S} = \bar{E} \times \bar{H}^*$ will be in the $\hat{z}$ direction, and so tangential to the top, bottom, front, and back of S, if these walls are made parallel to the z-axis. If the wave is obliquely incident, these walls can be slanted to obtain the same

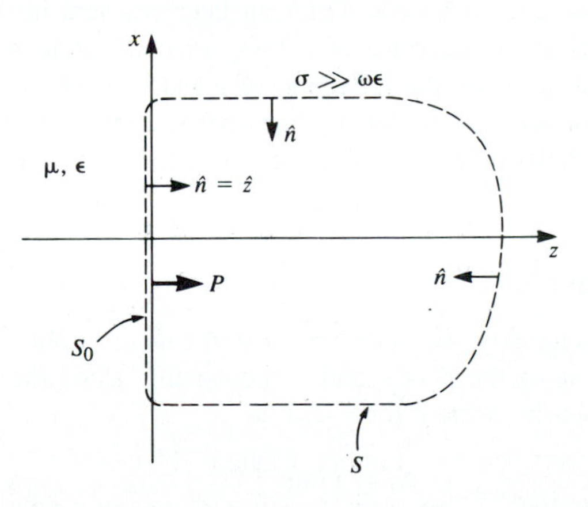

FIGURE 2.9 An interface between a lossless medium and a good conductor with a closed surface $S_0 + S$ for computing the power dissipated in the conductor.

result. And, if the conductor is good, the decay of the fields from the interface at $z = 0$ will be very rapid, so that the right-hand end of S can be made far enough away from $z = 0$ so that there is negligible contribution to the integral from this part of the surface S. The power entering the conductor through S_0 can then be written as

$$P_{\text{av}} = \frac{1}{2}\text{Re}\int_{S_0} \bar{E} \times \bar{H}^* \cdot \hat{z}\, ds. \tag{2.95}$$

From vector identity (B.3) we have

$$\hat{z} \cdot (\bar{E} \times \bar{H}^*) = (\hat{z} \times \bar{E}) \cdot \bar{H}^* = \eta \bar{H} \cdot \bar{H}^*, \tag{2.96}$$

since $\bar{H} = \hat{n} \times \bar{E}/\eta$, as generalized from (2.76) for conductive media, where η is the intrinsic wave impedance of the conductor. Equation (2.95) can then be written as

$$P_{\text{av}} = \frac{R_s}{2}\int_{S_0} |\bar{H}|^2 ds, \tag{2.97}$$

where

$$R_s = \text{Re}(\eta) = \text{Re}\left[(1+j)\sqrt{\frac{\omega\mu}{2\sigma}}\right] = \sqrt{\frac{\omega\mu}{2\sigma}} = \frac{1}{\sigma\delta_s} \tag{2.98}$$

is called the surface resistivity of the conductor. The magnetic field $\bar{H}$ in (2.97) is tangential to the conductor surface, and needs only to be evaluated at the surface of the conductor; since H_t is continuous at $z = 0$, it doesn't matter whether this field is evaluated just outside the conductor, or just inside the conductor. In the next section we will show how (2.97) can be evaluated in terms of a surface current density flowing on the surface of the conductor, where the conductor can be assumed to be perfect.

2.7 PLANE WAVE REFLECTION FROM A CONDUCTING MEDIUM

A number of problems to be considered in later chapters involve the behavior of electromagnetic fields at the interface of a lossy or conducting medium, and so it is beneficial at this time to study the reflection of a plane wave normally incident from free-space onto the surface of a conducting half-space. The geometry is shown in Figure 2.10 where the lossy half-space $z > 0$ is characterized by the parameters ϵ, μ, and σ.

General Medium

With no loss of generality, we can assume that the incident plane wave has an electric field vector oriented along the x-axis, and is propagating along the positive z-axis. The incident fields can then be written, for $z < 0$, as

$$\bar{E}_i = \hat{x} E_0 e^{-jk_0 z}, \tag{2.99a}$$

$$\bar{H}_i = \hat{y}\frac{1}{\eta_0} E_0 e^{-jk_0 z}, \tag{2.99b}$$

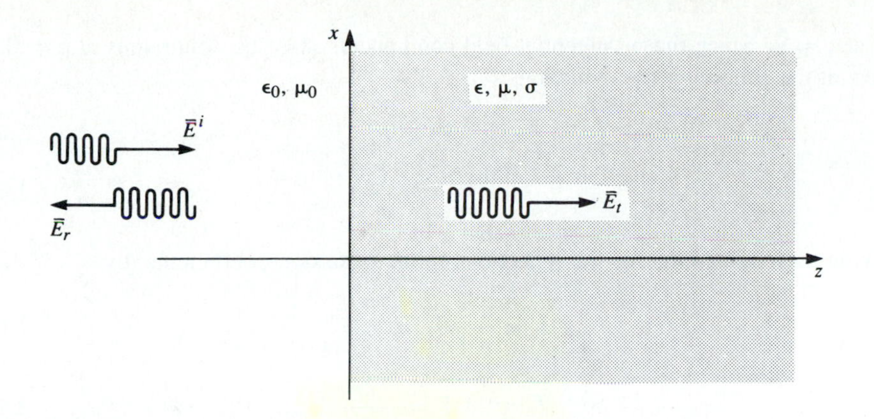

FIGURE 2.10 Plane wave reflection from a lossy medium; normal incidence.

where η_0 is the wave impedance of free-space, and E_0 is an arbitrary amplitude. Also in the region $z < 0$, a reflected wave may exist with the form

$$\bar{E}_r = \hat{x}\Gamma E_0 e^{+jk_0 z}, \tag{2.100a}$$

$$\bar{H}_r = -\hat{y}\frac{\Gamma}{\eta_0} E_0 e^{+jk_0 z}, \tag{2.100b}$$

where Γ is the unknown reflection coefficient of the electric field. Note that in (2.100), the sign in the exponential terms has been chosen as positive, to represent waves traveling in the $-\hat{z}$ direction of propagation, as derived in (2.49). This is also consistent with the Poynting vector $\bar{S}_r = \bar{E}_r \times \bar{H}_r^* = -|\Gamma|^2 |E_0|^2 \hat{z}/\eta_0$ which shows power to be traveling in the $-\hat{z}$ direction.

As shown in Section 2.4, from equations (2.54) and (2.58), the transmitted fields for $z > 0$ in the lossy medium can be written as

$$\bar{E}_t = \hat{x} T E_0 e^{-\gamma z}, \tag{2.101a}$$

$$\bar{H}_t = \frac{\hat{y} T E_0}{\eta} e^{-\gamma z}, \tag{2.101b}$$

where T is the transmission coefficient of the electric field and η is the intrinsic impedance of the lossy medium in the region $z > 0$. From (2.57) and (2.52) the intrinsic impedance is

$$\eta = \frac{j\omega\mu}{\gamma}, \tag{2.102}$$

and the propagation constant is

$$\gamma = \alpha + j\beta = j\omega\sqrt{\mu\epsilon}\sqrt{1 - j\sigma/\omega\epsilon}. \tag{2.103}$$

We now have a boundary value problem where the general form of the fields are known via (2.99)–(2.101) on either side of the material discontinuity at $z = 0$. The two unknown constants, Γ and T, are found by applying two boundary conditions on E_x and

H_y at $z = 0$. Since these tangential field components must be continuous at $z = 0$, we arrive at the following two equations:

$$(1 + \Gamma) = T, \tag{2.104a}$$

$$\frac{1 - \Gamma}{\eta_0} = \frac{T}{\eta}. \tag{2.104b}$$

Solving these equations for the reflection and transmission coefficients gives

$$\Gamma = \frac{\eta - \eta_0}{\eta + \eta_0}, \tag{2.105a}$$

$$T = 1 + \Gamma = \frac{2\eta}{\eta + \eta_0}. \tag{2.105b}$$

This is a general solution for reflection and transmission of a normally incident wave at the interface of a lossy material, where η is the impedance of the material. We now consider three special cases of the above result.

Lossless Medium

If the region for $z > 0$ is a lossless dielectric, then $\sigma = 0$, and μ and ϵ are real quantities. The propagation constant in this case is purely imaginary and can be written as

$$\gamma = j\beta = j\omega\sqrt{\mu\epsilon} = jk_0\sqrt{\mu_r\epsilon_r}, \tag{2.106}$$

where $k_0 = \omega\sqrt{\mu_0\epsilon_0}$ is the wavenumber of a plane wave in free-space. The wavelength in the dielectric is

$$\lambda = \frac{2\pi}{\beta} = \frac{2\pi}{\omega\sqrt{\mu\epsilon}} = \frac{\lambda_0}{\sqrt{\mu_r\epsilon_r}}, \tag{2.107}$$

which is seen to be shorter than the wavelength in free-space (λ_0). The corresponding phase velocity is

$$v_p = \frac{\omega}{\beta} = \frac{1}{\sqrt{\mu\epsilon}} = \frac{c}{\sqrt{\mu_r\epsilon_r}}, \tag{2.108}$$

which is slower than the speed of light in free-space (c). The wave impedance of the dielectric is

$$\eta = \frac{j\omega\mu}{\gamma} = \sqrt{\frac{\mu}{\epsilon}} = \eta_0\sqrt{\frac{\mu_r}{\epsilon_r}}, \tag{2.109}$$

which may be greater or lesser than the impedance of free-space (η_0), depending on whether μ_r is greater or lesser than ϵ_r. In the lossless case, η is real, so both Γ and T from (2.105) are real, and $\bar{E}$ and $\bar{H}$ are in phase with each other in both mediums.

Power conservation for the incident, reflected, and transmitted waves can be demonstrated by computing the Poynting vectors in the two regions. Thus, for $z < 0$, the complex Poynting vector is

$$\begin{aligned}\bar{S}^- &= \bar{E} \times \bar{H}^* = (\bar{E}_i + \bar{E}_r) \times (\bar{H}_i + \bar{H}_r)\\ &= \hat{z}|E_0|^2 \frac{1}{\eta_0}(e^{-jk_0z} + \Gamma e^{jk_0z})(e^{-jk_0z} - \Gamma e^{jk_0z})^*\\ &= \hat{z}|E_0|^2 \frac{1}{\eta_0}(1 - |\Gamma|^2 + \Gamma e^{2jk_0z} - \Gamma^* e^{-2jk_0z})\\ &= \hat{z}|E_0|^2 \frac{1}{\eta_0}(1 - |\Gamma|^2 + 2j\Gamma \sin 2k_0 z),\end{aligned} \quad 2.110a$$

since Γ is real. For $z > 0$, the complex Poynting vector is

$$\bar{S}^+ = \bar{E}_t \times \bar{H}_t^* = \hat{z}\frac{|E_0|^2|T|^2}{\eta},$$

which can be rewritten, using (2.105), as

$$\bar{S}^+ = \hat{z}|E_0|^2 \frac{4\eta}{(\eta + \eta_0)^2} = \hat{z}|E_0|^2 \frac{1}{\eta_0}(1 - |\Gamma|^2). \quad 2.110b$$

Now observe that at $z = 0$, $\bar{S}^- = \bar{S}^+$ so that complex power flow is conserved across the interface. Now consider the time-average power flow in the two regions. For $z < 0$, the time-average power flow through a 1m^2 cross-section is

$$P^- = \frac{1}{2}\text{Re}(\bar{S}^- \cdot \hat{z}) = \frac{1}{2}|E_0|^2 \frac{1}{\eta_0}(1 - |\Gamma|^2). \quad 2.111a$$

and for $z > 0$, the time-average power flow through a $1\ \text{m}^2$ cross-section is

$$P^+ = \frac{1}{2}\text{Re}(\bar{S}^+ \cdot \hat{z}) = \frac{1}{2}|E_0|^2 \frac{1}{\eta_0}(1 - |\Gamma|^2) = P^-, \quad 2.111b$$

so real power flow is conserved.

We now note a subtle point. When computing the complex Poynting vector for $z < 0$ in (2.110a), we used the total $\bar{E}$ and $\bar{H}$ fields. If we compute separately the Poynting vectors for the incident and reflected waves, we obtain

$$\bar{S}_i = \bar{E}_i \times \bar{H}_i^* = \hat{z}\frac{|E_0|^2}{\eta_0}, \quad 2.112a$$

$$\bar{S}_r = \bar{E}_r \times \bar{H}_r^* = -\hat{z}\frac{|E_0|^2|\Gamma|^2}{\eta_0}, \quad 2.112b$$

and we see that $\bar{S}_i + \bar{S}_r \neq \bar{S}^-$ of (2.110a). The missing cross-product terms account for stored reactive energy in the standing wave in the $z < 0$ region. Thus, the decomposition of a Poynting vector into incident and reflected components is not, in general, meaningful. Some books define a time-average Poynting vector as $(1/2)\text{Re}(\bar{E} \times \bar{H}^*)$, and in this case such a definition applied to the individual incident and reflected components will give the correct result, since $P_i = (1/2)\text{Re}|\bar{E}_0|^2/\eta_0$, and $P_r = -(1/2)|E_0|^2|\Gamma|^2/\eta_0$, so $P_i + P_r =$

P^-. But even this definition will fail to provide meaningful results when the medium for $z < 0$ is lossy.

Good Conductor

If the region for $z > 0$ is a good (but not perfect) conductor, the propagation constant can be written as discussed in Section 2.4:

$$\gamma = \alpha + j\beta = (1+j)\sqrt{\frac{\omega\mu\sigma}{2}} = (1+j)\frac{1}{\delta_s}. \tag{2.113}$$

Similarly, the intrinsic impedance of the conductor simplifies to

$$\eta = (1+j)\sqrt{\frac{\omega\mu}{2\sigma}} = (1+j)\frac{1}{\sigma\delta_s}. \tag{2.114}$$

Now the impedance is complex, with a phase angle of 45°, so $\bar{E}$ and $\bar{H}$ will be 45°out of phase, and Γ and T will be complex. In (2.113) and (2.114), $\delta_s = 1/\alpha$ is the skin depth, as defined in (2.60).

Let us check the complex power balance in this case. For $z < 0$, the complex Poynting vector is

$$\begin{aligned} \bar{S}^- &= \bar{E} \times \bar{H}^* = (\bar{E}_i + \bar{E}_r) \times (\bar{H}_i + \bar{H}_r)^* \\ &= \hat{z}|E_0|^2 \frac{1}{\eta_0}[1 - |\Gamma|^2 + 2jIm(\Gamma e^{2jk_0 z})], \end{aligned} \tag{2.115a}$$

which can be evaluated at $z = 0$ to give

$$\bar{S}^-(z=0) = \hat{z}|E_0|^2 \frac{1}{\eta_0}(1 - |\Gamma|^2 + \Gamma - \Gamma^*).$$

For $z > 0$, the complex Poynting vector is

$$\bar{S}^+ = \bar{E}_t \times \bar{H}_t^* = \hat{z}|E_0|^2 |T|^2 \frac{1}{\eta^*} e^{-2\alpha z},$$

and using (2.105) for T and Γ gives

$$\bar{S}^+ = \hat{z}|E_0|^2 \frac{4\eta}{|\eta + \eta_0|^2} e^{-2\alpha z} = \hat{z}|E_0|^2 \frac{1}{\eta_0}(1 - |\Gamma|^2 + \Gamma - \Gamma^*) e^{-2\alpha z}. \tag{2.115b}$$

So at the interface at $z = 0$, $\bar{S}^- = \bar{S}^+$, and complex power is conserved.

Observe that if we were to compute the separate incident and reflected Poynting vectors for $z < 0$ as

$$\bar{S}_i = \bar{E}_i \times \bar{H}_i^* = \hat{z}\frac{|E_0|^2}{\eta_0}, \tag{2.116a}$$

$$\bar{S}_r = \bar{E}_r \times \bar{H}_r^* = -\hat{z}\frac{|E_0|^2|\Gamma|^2}{\eta_0}, \tag{2.116b}$$

we do not obtain $\bar{S}_i + \bar{S}_r = \bar{S}^-$ of (2.115a), even for $z = 0$. It is possible, however, to consider real power flow in terms of the individual traveling wave components. Thus, the time-average power flows through a 1 m^2 cross-section are

$$P^- = \frac{1}{2}\text{Re}(\bar{S}^- \cdot \hat{z}) = \frac{1}{2}|E_0|^2 \frac{1}{\eta_0}(1 - |\Gamma|^2), \tag{2.117a}$$

$$P^+ = \frac{1}{2}\text{Re}(\bar{S}^+ \cdot \hat{z}) = \frac{1}{2}|E_0|^2 \frac{1}{\eta_0}(1 - |\Gamma|^2)e^{-2\alpha z}, \tag{2.117b}$$

which shows power balance at $z = 0$. In addition, $P_i = |E_0|^2/2\eta_0$, and $P_r = -|E_0|^2 |\Gamma|^2/2\eta_0$ so that $P_i + P_r = P^-$, showing that the real power flow for $z < 0$ can be decomposed into incident and reflected wave components.

Now notice that $\bar{S}^+$, the power density in the lossy conductor, decays exponentially according to the $e^{-2\alpha z}$ attenuation factor. This means that power is being dissipated in the lossy material as the wave propagates into the medium in the $+z$ direction. The power, and also the fields, decay to a negligibly small value within a few skin depths of the material, which for a reasonably good conductor is an extremely small distance at microwave frequencies.

The electric volume current density flowing in the conducting region is given as

$$\bar{J} = \sigma \bar{E}_t = \hat{x}\sigma E_0 T e^{-\gamma z} \text{A/m}^2, \tag{2.118}$$

and so the average power dissipated in (or transmitted into) a 1 m^2 cross-sectional volume of the conductor can be calculated from the conductor loss term of (2.92) (Joule's Law) as

$$\begin{aligned} P^t &= \frac{1}{2}\int_V \bar{E} \cdot \bar{J}^* dv = \frac{1}{2}\int_{x=0}^{1}\int_{y=0}^{1}\int_{z=0}^{\infty} (\hat{x}E_0 T e^{-\gamma z}) \cdot (\hat{x}\sigma E_0 T e^{-\gamma z})^* dz\, dy\, dx \\ &= \frac{1}{2}\sigma |E_0|^2\, |T|^2 \int_{z=0}^{\infty} e^{-2\alpha z} dz = \frac{\sigma |E_0|^2 |T|^2}{4\alpha}. \end{aligned} \tag{2.119}$$

Since $1/\eta = \sigma\delta_s/(1+j) = (\sigma/2\alpha)(1-j)$, the real power entering the conductor through a 1 m^2 cross-section (as given by $1/2\text{Re}(\bar{S}^+ \cdot \hat{z})$ at $z = 0$) can be expressed using (2.115b) as, $P^t = |E_0|^2|T|^2(\sigma/4\alpha)$, which is in agreement with (2.119).

Perfect Conductor

Now assume that the region $z > 0$ contains a perfect conductor. The above results can be specialized to this case by allowing $\sigma \to \infty$. Then, from (2.113) $\alpha \to \infty$; from (2.114) $\eta \to 0$; from (2.60) $\delta_s \to 0$; and from (2.105a,b) $T \to 0$, and $\Gamma \to -1$. The fields for $z > 0$ thus decay infinitely fast, and are identically zero in the perfect conductor. The perfect conductor can be thought of as "shorting out" the incident electric field. For $z < 0$, from (2.99) and (2.100), the total $\bar{E}$ and $\bar{H}$ fields are, since $\Gamma = -1$,

$$\bar{E} = \bar{E}_i + \bar{E}_r = \hat{x}E_0(e^{-jk_0 z} - e^{jk_0 z}) = -\hat{x}2jE_0 \sin k_0 z, \tag{2.120a}$$

$$\bar{H} = \bar{H}_i + \bar{H}_r = \hat{y}\frac{1}{\eta_0}E_0(e^{-jk_0 z} + e^{jk_0 z}) = \hat{y}\frac{2}{\eta_0}E_0 \cos k_0 z. \tag{2.120b}$$

Observe that at $z = 0$, $\bar{E} = 0$ and $\bar{H} = \hat{y}(2/\eta_0)E_0$. The Poynting vector for $z < 0$ is

$$\bar{S}^- = \bar{E} \times \bar{H}^* = \hat{z}j\frac{4}{\eta_0}|E_0|^2 \sin k_0 z \cos k_0 z, \tag{2.121}$$

which has a zero real part and thus indicates that no real power is delivered to the perfect conductor.

The volume current density of (2.118) for the lossy conductor reduces to an infinitely thin sheet of surface current in the limit of infinite conductivity:

$$\bar{J}_s = \hat{n} \times \bar{H} = -\hat{z} \times \left(\hat{y}\frac{2}{\eta_0}E_0 \cos k_0 z\right)\Big|_{z=0} = \hat{x}\frac{2}{\eta_0}E_0 \text{ A/m}. \tag{2.122}$$

The Surface Impedance Concept

In many problems, particularly those where the effect of attenuation or conductor loss is needed, the presence of an imperfect conductor must be taken into account. The surface impedance concept allows us to do this in a very convenient way. We will develop this method from the theory presented in the previous sections.

Consider a good conductor in the region $z > 0$. As we have seen, a plane wave normally incident on this conductor is mostly reflected, and the power that is transmitted into the conductor is dissipated as heat within a very short distance from the surface. There are three ways to compute this power.

First, we can use Joule's law, as in (2.119):

$$P^t = \frac{1}{2}\int_V \bar{E} \cdot \bar{J}^* dv = \frac{1}{2\sigma}\int_V |\bar{J}|^2 dv. \tag{2.123}$$

For a 1 m^2 area of conductor surface, the power transmitted through this surface and dissipated as heat is, from (2.119),

$$P^t = \frac{\sigma |E_0|^2 |T|^2}{4\alpha}.$$

Using (2.150b) for T, (2.114) for η, and the fact that $\alpha = 1/\delta_s$, gives the following result:

$$\frac{\sigma |T|^2}{\alpha} = \frac{\sigma \delta_s 4|\eta|^2}{|\eta + \eta_0|^2} \simeq \frac{8}{\sigma \delta_s \eta_0^2} = \frac{8R_s}{\eta_0^2},$$

where we have assumed $\eta << \eta_0$, which is true for a good conductor. Then the above power can be written as

$$P^t = \frac{2|E_0|^2 R_s}{\eta_0^2}, \tag{2.124}$$

where

$$R_s = \text{Re}(\eta) = \text{Re}\left(\frac{1+j}{\sigma\delta_s}\right) = \frac{1}{\sigma\delta_s} = \sqrt{\frac{\omega\mu}{2\sigma}} \tag{2.125}$$

is the surface resistance of the metal.

Another way to find the power loss is to compute the power flow into the conductor using the Poynting vector, since all power entering the conductor at $z = 0$ is dissipated.

As in (2.115b), we have

$$P^t = \frac{1}{2}\text{Re}(\bar{S}^+ \cdot \hat{z})_{z=0} = \frac{2|E_0|^2\text{Re}(\eta)}{|\eta + \eta_0|^2},$$

which for large conductivity becomes, since $\eta \ll \eta_0$,

$$P^t = \frac{2|E_0|^2 R_s}{\eta_0^2}, \tag{2.126}$$

which agrees with (2.124).

A third method uses an effective surface current density and the surface impedance, without the need for the fields inside the conductor. From (2.118), the volume current density in the conductor is

$$\bar{J} = \hat{x}\sigma T E_0 e^{-\gamma z} \text{ A/m}^2, \tag{2.127}$$

so the total current flow per unit width in the x direction is

$$\bar{J}_s = \int_0^\infty \bar{J} dz = \hat{x}\sigma T E_0 \int_0^\infty e^{-\gamma z} dz = \frac{\hat{x}\sigma T E_0}{\gamma} \text{ A/m},$$

and taking the limit of $\sigma T/\gamma$ for large σ gives

$$\frac{\sigma T}{\gamma} = \frac{\sigma\delta_s}{(1+j)}\frac{2\eta}{(\eta+\eta_0)} \simeq \frac{\sigma\delta_s}{(1+j)}\frac{2(1+j)}{\sigma\delta_s\eta_0} = \frac{2}{\eta_0},$$

so

$$\bar{J}_s = \hat{x}\frac{2E_0}{\eta_0} \text{ A/m}. \tag{2.128}$$

If the conductivity were infinite, a true surface current density of

$$\bar{J}_s = \hat{n} \times \bar{H}|_{z=0} = -\hat{z} \times (\bar{H}_i + \bar{H}_r)|_{z=0} = \hat{x}E_0\frac{1}{\eta_0}(1-\Gamma) = \hat{x}\frac{2E_0}{\eta_0} \text{ A/m}$$

would flow, which is identical to the total current in (2.128).

Now replace the exponentially decaying volume current of (2.127) in Joule's law of (2.123) with a uniform volume current extending a distance of one skin depth. Thus, let

$$\bar{J} = \begin{cases} \bar{J}_s/\delta_s & \text{for } 0 < z < \delta_s \\ 0 & \text{for } z > \delta_s, \end{cases} \tag{2.129}$$

so that the total current flow is the same. Then use (2.123) to find the power lost:

$$P^t = \frac{1}{2\sigma}\int_S \int_{z=0}^{\delta_s} \frac{|\bar{J}_s|^2}{\delta_s^2} dz\, ds = \frac{R_s}{2}\int_S |\bar{J}_s|^2 ds = \frac{2|E_0|^2 R_s}{\eta_0^2}, \tag{2.130}$$

where $\int_S$ denotes a surface integral over the conductor surface, in this case chosen as 1 m^2. The result of (2.130) agrees with our previous results for P^t in (2.126) and (2.124), and shows that the power loss can be calculated as

$$P^t = \frac{R_s}{2} \int_S |\bar{J}_s|^2 ds = \frac{R_s}{2} \int_S |\bar{H}_t|^2 ds, \tag{2.131}$$

in terms of the surface resistance R_s and the surface current $\bar{J}_s$, or tangential magnetic field $\bar{H}_t$. It is important to realize that the surface current can be found from $\bar{J}_s = \hat{n} \times \bar{H}$, as if the metal were a perfect conductor. This method is very general, applying to fields other than plane waves, and to conductors of arbitrary shape, as long as bends or corners have radii on the order of a skin depth or larger. The method is also quite accurate, as the only approximation in the above was that $\eta \ll \eta_0$, which is a good approximation. As an example, copper at 1 GHz has $|\eta| = 0.012\ \Omega$, which is indeed much less than $\eta_0 = 377\ \Omega$.

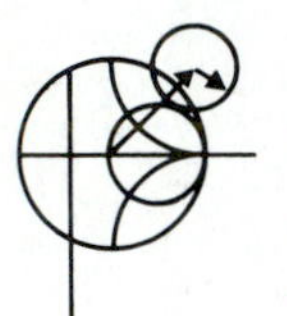

EXAMPLE 2.6

Consider a plane wave normally incident on a half-space of copper. If $f = 1$ GHz, compute the propagation constant, impedance, and skin depth for the conductor. Also compute the reflection and transmission coefficients.

Solution

For copper, $\sigma = 5.813 \times 10^7 \text{S/m}$, so from (2.60) the skin depth is

$$\delta_s = \sqrt{\frac{2}{\omega\mu\sigma}} = 2.088 \times 10^{-6} \text{ m},$$

and the propagation constant is, from (2.113),

$$\gamma = \frac{1+j}{\delta_s} = (4.789 + j4.789) \times 10^5 \text{ m}^{-1}.$$

The intrinsic impedance is, from (2.114),

$$\eta = \frac{1+j}{\sigma\delta_s} = (8.239 + j8.239) \times 10^{-3}\ \Omega,$$

which is quite small relative to the impedance of free-space ($\eta_0 = 377\ \Omega$). The reflection coefficient is then

$$\Gamma = \frac{\eta - \eta_0}{\eta + \eta_0} = 1\angle 179.99^\circ$$

(practically that of an ideal short circuit), and the transmission coefficient is

$$T = \frac{2\eta}{\eta + \eta_0} = 6.181 \times 10^{-5} \angle 45^\circ.$$

○

2.8 OBLIQUE INCIDENCE AT A DIELECTRIC INTERFACE

We continue our discussion of plane waves by considering the problem of a plane wave obliquely incident on a plane interface between two lossless dielectric regions, as shown in Figure 2.11. There are two canonical cases of this problem: the electric field is either in the xz plane (parallel polarization), or normal to the xz plane (perpendicular polarization). An arbitrary incident plane wave, of course, may have a polarization which is neither of these, but it can be expressed as a linear combination of these two individual cases.

The general method of solution is similar to the problem of normal incidence: we will write expressions for the incident, reflected, and transmitted fields in each region, and match boundary conditions to find the unknown amplitude coefficients and angles.

Parallel Polarization

In this case, the electric field vector lies in the xz plane, and the incident fields can be written as

$$\bar{E}_i = E_0(\hat{x}\cos\theta_i - \hat{z}\sin\theta_i)e^{-jk_1(x\sin\theta_i + z\cos\theta_i)}, \quad 2.132a$$

$$\bar{H}_i = \frac{E_0}{\eta_1}\hat{y}e^{-jk_1(x\sin\theta_i + z\cos\theta_i)}, \quad 2.132b$$

where $k_1 = \omega\sqrt{\mu_0\epsilon_1}$, and $\eta_1 = \sqrt{\mu_0/\epsilon_1}$ are the wave number and wave impedance of

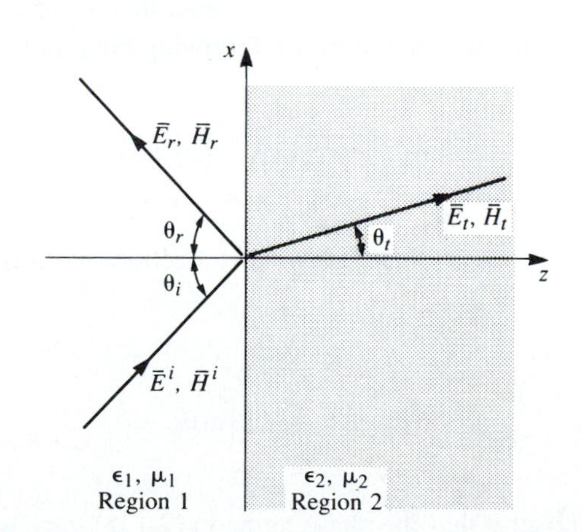

FIGURE 2.11 Geometery for a plane wave obliquely incident at the interface between two dielectric regions.

region 1. The reflected and transmitted fields can be written as

$$\bar{E}_r = E_0\Gamma(\hat{x}\cos\theta_r + \hat{z}\sin\theta_r)e^{-jk_1(x\sin\theta_r - z\cos\theta_r)}, \tag{2.133a}$$

$$\bar{H}_r = \frac{-E_0\Gamma}{\eta_1}\hat{y}e^{-jk_1(x\sin\theta_r - z\cos\theta_r)}, \tag{2.133b}$$

$$\bar{E}_t = E_0T(\hat{x}\cos\theta_t - \hat{z}\sin\theta_t)e^{-jk_2(x\sin\theta_t + z\cos\theta_t)}, \tag{2.134a}$$

$$\bar{H}_t = \frac{E_0T}{\eta_2}\hat{y}e^{-jk_2(x\sin\theta_t + z\cos\theta_t)}. \tag{2.134b}$$

In the above, Γ and T are the reflection and transmission coefficients, and k_2, η_2 are the wavenumber and wave impedance of region 2, defined as

$$k_2 = \omega\sqrt{\mu_0\epsilon_2}, \qquad \eta_2 = \sqrt{\mu_0/\epsilon_2}.$$

At this point, we have Γ, T, θ_r, and θ_t as unknowns.

We can obtain two complex equations for these unknowns by enforcing the continuity of E_x and H_y, the tangential field components, at the interface at $z = 0$. We then obtain

$$\cos\theta_i e^{-jk_1x\sin\theta_i} + \Gamma\cos\theta_r e^{-jk_1x\sin\theta_r} = T\cos\theta_t e^{-jk_2x\sin\theta_t}, \tag{2.135a}$$

$$\frac{1}{\eta_1}e^{-jk_1x\sin\theta_i} - \frac{\Gamma}{\eta_1}e^{-jk_1x\sin\theta_r} = \frac{T}{\eta_2}e^{-jk_2x\sin\theta_t}. \tag{2.135b}$$

Both sides of (2.135a) and (2.135b) are functions of the coordinate x. If E_x and H_y are to be continuous at the interface $z = 0$ for all x, then this x variation must be the same on both sides of the equations, leading to the following condition:

$$k_1\sin\theta_i = k_1\sin\theta_r = k_2\sin\theta_t,$$

which results in the well-known Snell's laws of reflection and refraction:

$$\theta_i = \theta_r, \tag{2.136a}$$

$$k_1\sin\theta_i = k_2\sin\theta_t. \tag{2.136b}$$

The above argument ensures that the phase terms in (2.135) vary with x at the same rate on both sides of the interface, and so is often called the *phase matching condition.*

Using (2.136) in (2.135) allows us to solve for the reflection and transmission coefficients as

$$\Gamma = \frac{\eta_2 \cos\theta_t - \eta_1 \cos\theta_i}{\eta_2 \cos\theta_t + \eta_1 \cos\theta_i}, \tag{2.137a}$$

$$T = \frac{2\eta_2 \cos\theta_i}{\eta_2 \cos\theta_t + \eta_1 \cos\theta_i}. \tag{2.137b}$$

Observe that for normal incidence, we have $\theta_i = \theta_r = \theta_t = 0$, so then

$$\Gamma = \frac{\eta_2 - \eta_1}{\eta_2 + \eta_1}, \text{ and } T = \frac{2\eta_2}{\eta_2 + \eta_1},$$

in agreement with the results of Section 2.7.

For this polarization, a special angle of incidence, θ_b, called the Brewster angle, exists where $\Gamma = 0$. This occurs when the numerator of (2.137a) goes to zero ($\theta_i = \theta_b$):

$$\eta_2 \cos\theta_t = \eta_1 \cos\theta_b,$$

which can be reduced using

$$\cos\theta_t = \sqrt{1 - \sin^2\theta_t} = \sqrt{1 - \frac{k_1^2}{k_2^2}\sin^2\theta_b},$$

to give

$$\sin\theta_b = \frac{1}{\sqrt{1 + \epsilon_1/\epsilon_2}}. \tag{2.138}$$

Perpendicular Polarization

In this case, the electric field vector is perpendicular to the xz plane. The incident field can be written as

$$\bar{E}_i = E_0 \hat{y} e^{-jk_1(x \sin\theta_i + z \cos\theta_i)}, \tag{2.139a}$$

$$\bar{H}_i = \frac{E_0}{\eta_1}(-\hat{x}\cos\theta_i + \hat{z}\sin\theta_i) e^{-jk_1(x \sin\theta_i + z \cos\theta_i)}, \tag{2.139b}$$

where $k_1 = \omega\sqrt{\mu_0 \epsilon_1}$ and $\eta_1 = \sqrt{\mu_0/\epsilon_1}$ are the wavenumber and wave impedance for region 1, as before. The reflected and transmitted fields can be expressed as

$$\bar{E}_r = E_0 \Gamma \hat{y} e^{-jk_1(x \sin\theta_r - z \cos\theta_r)}, \tag{2.140a}$$

$$\bar{H}_r = \frac{E_0 \Gamma}{\eta_1}(\hat{x}\cos\theta_r + \hat{z}\sin\theta_r) e^{-jk_1(x \sin\theta_r - z \cos\theta_r)}, \tag{2.140b}$$

$$\bar{E}_t = E_0 T \hat{y} e^{-jk_2(x \sin\theta_t + z \cos\theta_t)}, \tag{2.141a}$$

$$\bar{H}_t = \frac{E_0 T}{\eta_2}(-\hat{x}\cos\theta_t + \hat{z}\sin\theta_t) e^{-jk_2(x \sin\theta_t + z \cos\theta_t)}, \tag{2.141b}$$

with $k_2 = \omega\sqrt{\mu_0\epsilon_2}$ and $\eta_2 = \sqrt{\mu_0/\epsilon_2}$ being the wavenumber and wave impedance in region 2.

Equating the tangential field components E_y and H_x at $z = 0$ gives

$$e^{-jk_1x\sin\theta_i} + \Gamma e^{-jk_1x\sin\theta_r} = Te^{-jk_2x\sin\theta_t}, \tag{2.142a}$$

$$\frac{-1}{\eta_1}\cos\theta_i e^{-jk_1x\sin\theta_i} + \frac{\Gamma}{\eta_1}\cos\theta_r e^{-jk_2x\sin\theta_r} = \frac{-T}{\eta_2}\cos\theta_t e^{-jk_2x\sin\theta_t}. \tag{2.142b}$$

By the same phase matching argument that was used in the parallel case, we obtain Snell's laws

$$k_1\sin\theta_i = k_1\sin\theta_r = k_2\sin\theta_t$$

identical to (2.136).

Using (2.136) in (2.142) allows us to solve for the reflection and transmission coefficients as

$$\Gamma = \frac{\eta_2\cos\theta_i - \eta_1\cos\theta_t}{\eta_2\cos\theta_i + \eta_1\cos\theta_t}, \tag{2.143a}$$

$$T = \frac{2\eta_2\cos\theta_i}{\eta_2\cos\theta_i + \eta_1\cos\theta_t}. \tag{2.143b}$$

Again, for the normally incident case, these results reduce to those of Section 2.7.

For this polarization no Brewster angle exists where $\Gamma = 0$, as we can see by examining the numerator of (2.143a),

$$\eta_2\cos\theta_i = \eta_1\cos\theta_t,$$

and using Snell's law to give

$$k_2^2(\eta_2^2 - \eta_1^2) = (k_2^2\eta_2^2 - k_1^2\eta_1^2)\sin^2\theta_i.$$

But this leads to a contradiction, since the term in parentheses on the right-hand side is identically zero for dielectric media. Thus, no Brewster angle exists for perpendicular polarization, for dielectric media.

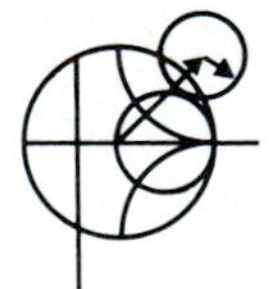

EXAMPLE 2.7

Plot the reflection coefficients for parallel and perpendicular polarized plane waves incident from free-space onto a dielectric region with $\epsilon_r = 2.55$, versus incidence angle.

Solution

The wave impedances are

$$\eta_1 = 377\,\Omega,$$

$$\eta_2 = \frac{\eta_0}{\sqrt{\epsilon_r}} = \frac{377}{\sqrt{2.55}} = 236\,\Omega.$$

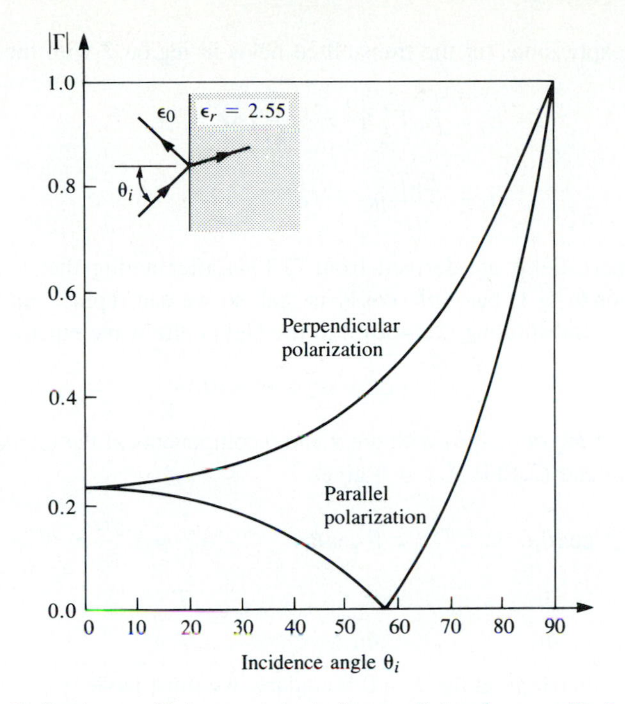

FIGURE 2.12 Reflection coefficient magnitude for parallel and perpendicular polarizations of a plane wave obliquely incident on a dielectric half-space.

We then evaluate (2.137a) and (2.143a) versus incidence angle; the results are shown in Figure 2.12. ○

Total Reflection and Surface Waves

Snell's law of (2.136b) can be rewritten as

$$\sin\theta_t = \sqrt{\frac{\epsilon_1}{\epsilon_2}}\sin\theta_i. \tag{2.144}$$

Now consider the case (for either parallel or perpendicular polarization), where $\epsilon_1 > \epsilon_2$. As θ_i increases from 0° to 90°, the refraction angle θ_t will increase from 0° to 90°, but at a faster rate than θ_i increases. The incidence angle θ_i for which $\theta_t = 90°$ is called the critical angle, θ_c, thus

$$\sin\theta_c = \sqrt{\frac{\epsilon_2}{\epsilon_1}}. \tag{2.145}$$

At this angle and beyond, the incident wave will be totally reflected, as the transmitted wave is not propagating into region 2. Let us look at this situation more closely, for the case of $\theta_i > \theta_c$ with parallel polarization.

When $\theta_i > \theta_c$ (2.144) shows that $\sin\theta_t > 1$, so that $\cos\theta_t = \sqrt{1-\sin^2\theta_t}$ must be imaginary, and so the angle θ_t loses its physical significance. At this point, it is better to replace the expressions for the transmitted fields in region 2 with the following:

$$\bar{E}_t = E_0 T\left(\frac{j\alpha}{k_2}\hat{x} - \frac{\beta}{k_2}\hat{z}\right)e^{-j\beta x}e^{-\alpha z}, \tag{2.146a}$$

$$\bar{H}_t = \frac{E_0 T}{\eta_2}\hat{y}e^{-j\beta x}e^{-\alpha z}. \tag{2.146b}$$

The form of these fields are derived from (2.134) after noting that $-jk_2 \sin\theta_t$ is still imaginary for $\sin\theta_t > 1$, but $-jk_2 \cos\theta_t$ is real, so we can replace $\sin\theta_t$ by β/k_2, and $\cos\theta_t$ by $j\alpha/k_2$. Substituting (2.146b) into the Helmholtz wave equation for $\bar{H}$ gives

$$-\beta^2 + \alpha^2 + k_2^2 = 0. \tag{2.147}$$

Matching E_x and H_y of (2.146) with the $\hat{x}$ and $\hat{y}$ components of the incident and reflected fields of (2.132) and (2.133) at $z = 0$ gives

$$\cos\theta_i e^{-jk_1 x \sin\theta_i} + \Gamma \cos\theta_r e^{-jk_1 x \sin\theta_r} = T\frac{j\alpha}{k_2}e^{-j\beta x}, \tag{2.148a}$$

$$\frac{1}{\eta_1}e^{-jk_1 x \sin\theta_i} - \frac{\Gamma}{\eta_1}e^{-jk_1 x \sin\theta_r} = \frac{T}{\eta_2}e^{-j\beta x}. \tag{2.148b}$$

To obtain phase matching at the $z = 0$ boundary, we must have

$$k_1 \sin\theta_i = k_1 \sin\theta_r = \beta,$$

which leads again to Snell's law for reflection, $\theta_i = \theta_r$, and to $\beta = k_1 \sin\theta_i$. Then α is determined from (2.147) as

$$\alpha = \sqrt{k_2^2 + \beta^2} = \sqrt{k_2^2 - k_1^2 \sin^2\theta_i}, \tag{2.149}$$

which is seen to be a positive real number. The reflection and transmission coefficients can be obtained from (2.148) as

$$\Gamma = \frac{(j\alpha/k_2)\eta_2 - \eta_1 \cos\theta_i}{(j\alpha/k_2)\eta_2 + \eta_1 \cos\theta_i}, \tag{2.150a}$$

$$T = \frac{2\eta_2 \cos\theta_i}{(j\alpha/k_2)\eta_2 + \eta_1 \cos\theta_i}. \tag{2.150b}$$

Since Γ is of the form $(a - jb)/(a + jb)$, its magnitude is unity, indicating that all incident power is reflected.

The transmitted fields of (2.146) show propagation in the x direction, along the interface, but exponential decay in the z direction. Such a field is known as a *surface wave,** since it is tightly bound to the interface. A surface wave is an example of a nonuniform plane wave, so called because it has an amplitude variation in the z direction, apart from the propagation factor in the x direction.

* Some authors argue that the term "surface wave" should not be used for a field of this type, since it only exists when plane wave fields exist in the $z < 0$ region, and so prefer to call it a "surface wave-like" field, or a "forced surface wave."

Finally, it is of interest to calculate the complex Poynting vector for the surface wave fields of (2.146):

$$\bar{S}_t = \bar{E}_t \times \bar{H}_t^* = \frac{|E_0|^2|T|^2}{\eta_2}\left(\hat{z}\frac{j\alpha}{k_2} + \hat{x}\frac{\beta}{k_2}\right)e^{-2\alpha z}. \qquad 2.151$$

This shows that no real power flow occurs in the z direction. The real power flow in the x direction is that of the surface wave field, and decays exponentially with distance into region 2. So even though no real power is transmitted into region 2, a nonzero field does exist there, in order to satisfy the boundary conditions at the interface.

2.9 PLANE WAVE PROPAGATION IN ANISOTROPIC MEDIA

As discussed in Section 2.2, anisotropic media is characterized by nonparallel $\bar{E}$ and $\bar{D}$ vectors, and/or nonparallel $\bar{H}$ and $\bar{B}$ vectors. This is equivalent to saying that the media has different electrical properties in different directions, and thus the permittivity and/or permeability has a matrix form. Such materials include crystals, ferrites, and plasmas (ionized gases), as well as a variety of semiconductor and other composite materials, which are very useful in a number of microwave devices. Ferrites, for example, are used in many nonreciprocal components which we will study in greater detail in Chapter 10. It also appears that anisotropic media will play an important role in future microwave component development.

In this section we will study plane wave propagation in two special types of media with an electric (permittivity) anisotropy: a uniaxial medium and a gyrotropic medium. Besides introducing the unique properties of wave propagation in an anisotropic medium, this section will help to familiarize us with the methods of solution for problems involving such media.

A Uniaxial Medium

A uniaxial medium is an anisotropic material that is characterized by a diagonal permittivity tensor $[\epsilon]$ of the form

$$[\epsilon] = \epsilon_0 \begin{bmatrix} \epsilon_r & 0 & 0 \\ 0 & 1 & 0 \\ 0 & 0 & 1 \end{bmatrix}, \qquad 2.152$$

where two of the diagonal elements are identical (here chosen as ϵ_0). Similarly, a biaxial medium is one where none of the diagonal elements of the permittivity tensor are the same. Uniaxial or biaxial magnetic media can also be typified by permeability tensors of the same form.

Examples of uniaxial materials include certain birefringent crystals and plasmas. The direction in which the one different permittivity element occurs is called the optical axis. Note that if the material has its optical axis nonaligned with any single coordinate axis, the permittivity matrix will not be diagonal. A rotation of coordinate axes, however, can be performed to diagonalize the permittivity matrix to the form of (2.152).

Now consider plane wave propagation in an infinite uniaxial medium. We will assume propagation in the $+z$ direction, and no variation in the x or y directions. The fields may have x or y (transverse) components, but no z (longitudinal) components. Maxwell's equations then reduce to

$$\frac{-\partial E_y}{\partial z} = -j\omega\mu_0 H_x, \tag{2.153a}$$

$$\frac{\partial E_x}{\partial z} = -j\omega\mu_0 H_y, \tag{2.153b}$$

$$\frac{-\partial H_y}{\partial z} = j\omega\epsilon_0\epsilon_r E_x, \tag{2.154a}$$

$$\frac{\partial H_x}{\partial z} = j\omega\epsilon_0 E_y. \tag{2.154b}$$

Now, H_y can be eliminated using (2.153b) and (2.154a) to give a wave equation for E_x:

$$\frac{\partial^2 E_x}{\partial z^2} + \epsilon_r k_0^2 E_x = 0. \tag{2.155a}$$

Similarly, a wave equation for E_y can be found as

$$\frac{\partial^2 E_y}{\partial z^2} + k_0^2 E_y = 0. \tag{2.155b}$$

Positive traveling wave solutions can then be written as

$$\bar{E} = \hat{x}E_{0x}e^{-jk_e z} + \hat{y}E_{0y}e^{-jk_0 z}, \tag{2.156}$$

where $k_e = \sqrt{\epsilon_r}k_0$. We see that the x and y components of the electric field travel with different propagation constants. The y component, with wavenumber k_0, is called the ordinary wave, while the x component, with wavenumber k_e, is called the extraordinary wave.

One application of such an effect is to generate a circularly polarized wave from a linearly polarized wave, with a plate of uniaxial material. Thus, let $E_{0x} = E_{0y} = E_0$, and consider a slab of a uniaxial material parallel to the xy plane, extending from $z = 0$ to $z = d$. Ignoring reflections, the wave entering the slab at $z = 0$ is linearly polarized at 45° from the x-axis:

$$\bar{E}(z = 0) = E_0(\hat{x} + \hat{y}). \tag{2.157}$$

The different components of the wave travel with different phase velocities in the slab, so that at $z = d$ the field is

$$\bar{E}(z = d) = E_0(\hat{x}e^{-jk_e d} + \hat{y}e^{-jk_0 d}). \tag{2.158}$$

Now if we choose the slab thickness d such that

$$(k_e - k_0)d = \frac{\pi}{2},$$

or

$$d = \frac{\lambda_0}{4(\sqrt{\epsilon_r} - 1)}, \tag{2.159}$$

then a 90° phase shift will be introduced between the x and y field components, so that a LHCP wave will be generated with the field

$$\bar{E}(z = d) = E_0(\hat{x} + j\hat{y})e^{-jk_e d}. \tag{2.160}$$

RHCP can be generated using a slab of thickness

$$d = \frac{3\lambda_0}{4(\sqrt{\epsilon_r} - 1)}$$

to give a $-90°$ (actually 270°) phase shift. Such polarizers are called quarter-wave plates, and are frequently used in optics.

A Gyrotropic Medium

A gyrotropic medium is an anisotropic material that is characterized by a permittivity tensor $[\epsilon]$ of the form

$$[\epsilon] = \epsilon_0 \begin{bmatrix} \epsilon_r & j\kappa & 0 \\ -j\kappa & \epsilon_r & 0 \\ 0 & 0 & 1 \end{bmatrix}, \tag{2.161}$$

or a permeability tensor $[\mu]$ of similar form. Examples of gyrotropic media include ferrites (permeability), and magnetized plasmas (permittivity). We will consider plane wave propagation along the z-axis in an infinite gyrotropic medium with permittivity as in (2.161). As before, we assume no variation in x or y, and only transverse field components. Maxwell's equations can then be written as

$$\frac{-\partial E_y}{\partial z} = -j\omega\mu_0 H_x, \tag{2.162a}$$

$$\frac{\partial E_x}{\partial z} = -j\omega\mu_0 H_y, \tag{2.162b}$$

$$\frac{-\partial H_y}{\partial z} = j\omega\epsilon_0[\epsilon_r E_x + j\kappa E_y], \tag{2.163a}$$

$$\frac{\partial H_x}{\partial z} = j\omega\epsilon_0[-j\kappa E_x + \epsilon_r E_y]. \tag{2.163b}$$

In this case, because of cross-coupling between E_x and E_y, it is not possible to obtain a single second-degree wave equation for E_x or E_y. Instead, we will assume an $e^{-j\beta z}$ propagation factor and use Maxwell's equations directly. Equations (2.162)–(2.163) then reduce to

$$\beta E_y = -\omega\mu_0 H_x, \tag{2.164a}$$

$$\beta E_x = \omega\mu_0 H_y, \tag{2.164b}$$

$$\beta H_y = \omega\epsilon_0[\epsilon_r E_x + j\kappa E_y], \tag{2.165a}$$

$$\beta H_x = -\omega\epsilon_0[-j\kappa E_x + \epsilon_r E_y]. \tag{2.165b}$$

Using (2.164) in (2.165) to eliminate H_x and H_y gives a set of homogeneous equations for E_x and E_y:

$$\begin{bmatrix} \beta^2 - \epsilon_r k_0^2 & -j\kappa k_0^2 \\ j\kappa k_0^2 & \beta^2 - \epsilon_r k_0^2 \end{bmatrix} \begin{bmatrix} E_x \\ E_y \end{bmatrix} = 0. \tag{2.166}$$

Obtaining a nontrivial solution to these equations requires that the determinant of the coefficient matrix must vanish; thus,

$$(\beta^2 - \epsilon_r k_0^2)^2 - (\kappa k_0^2)^2 = 0,$$

or

$$\beta^2 = \epsilon_r k_0^2 \pm \kappa k_0^2 = k_0^2(\epsilon_r \pm \kappa) = \beta_\pm^2. \tag{2.167}$$

Thus we see that two distinct propagation constants, β_+ and β_-, are possible in the gyrotropic medium. The next step is to determine the form of the fields for each of these propagation constants. For $\beta_+^2 = k_0^2(\epsilon_r + \kappa)$, substitution into the first row of (2.166) yields $E_x - jE_y = 0$, so the electric field associated with this propagation constant is of the form

$$\bar{E}^+ = E_0(\hat{x} - j\hat{y})e^{-j\beta_+ z}, \tag{2.168a}$$

which clearly represents a RHCP plane wave propagating in the $+z$ direction. Similarly, the electric field associated with the propagation constant β_- is of the form

$$\bar{E}^- = E_0(\hat{x} + j\hat{y})e^{-j\beta_- z}, \tag{2.168b}$$

which represents a LHCP wave propagating in the $+z$ direction. The propagation constants, β_+ and β_-, are known as the eigenvalues of the system of equations in (2.166), and $\bar{E}^+$ and $\bar{E}^-$ are called the eigenvectors. Through the use of (2.164), the associated magnetic fields can be found as

$$\bar{H}^+ = \frac{jE_0}{\eta^+}(\hat{x} - j\hat{y})e^{-j\beta_+ z}, \tag{2.169a}$$

$$\bar{H}^- = \frac{jE_0}{\eta^-}(\hat{x} + j\hat{y})e^{-j\beta_- z}, \tag{2.169b}$$

where $\eta^\pm = \omega\mu_0/\beta^\pm$ is the wave impedance of the + or − mode.

Observe then that the RHCP (+) and the LHCP (−) modes propagate with different phase velocities. This property leads to a number of useful microwave devices, which we will study further in Chapter 10.

2.10 SOME USEFUL THEOREMS

Finally, we discuss several theorems in electromagnetics which we will find useful for later discussions.

The Reciprocity Theorem

Reciprocity is a general concept that occurs in many areas of physics and engineering, and the reader may already be familiar with the reciprocity theorem of circuit theory.

Here we will derive the Lorentz reciprocity theorem for electromagnetic fields in two different forms. This theorem will be used later in the book to obtain general properties of network matrices representing microwave circuits, and to evaluate the coupling of waveguides from current probes and loops, and the coupling of waveguides through apertures. There are a number of other important uses of this powerful concept.

Consider the two separate sets of sources, $\bar{J}_1, \bar{M}_1$ and $\bar{J}_2, \bar{M}_2$, which generate the fields $\bar{E}_1, \bar{H}_1$, and $\bar{E}_2, \bar{H}_2$, respectively, in the volume V enclosed by the closed surface S, as shown in Figure 2.13. Maxwell's equations are satisfied individually for these two sets of sources and fields, so we can write

$$\nabla \times \bar{E}_1 = -j\omega\mu\bar{H}_1 - \bar{M}_1, \qquad 2.170a$$

$$\nabla \times \bar{H}_1 = j\omega\epsilon\bar{E}_1 + \bar{J}_1, \qquad 2.170b$$

$$\nabla \times \bar{E}_2 = -j\omega\mu\bar{H}_2 - \bar{M}_2, \qquad 2.171a$$

$$\nabla \times \bar{H}_2 = j\omega\epsilon\bar{E}_2 + \bar{J}_2. \qquad 2.171b$$

Now consider the the quantity $\nabla \cdot (\bar{E}_1 \times \bar{H}_2 - \bar{E}_2 \times \bar{H}_1)$, which can be expanded using vector identity (B.8) to give

$$\nabla \cdot (\bar{E}_1 \times \bar{H}_2 - \bar{E}_2 \times \bar{H}_1) = \bar{J}_1 \cdot \bar{E}_2 - \bar{J}_2 \cdot \bar{E}_1 + \bar{M}_2 \cdot \bar{\Pi}_1 - \bar{M}_1 \cdot \bar{H}_2. \qquad 2.172$$

Integrating over the volume V, and applying the divergence theorem (B.15), gives

$$\int_V \nabla \cdot (\bar{E}_1 \times \bar{H}_2 - \bar{E}_2 \times \bar{H}_1) dv = \oint_S (\bar{E}_1 \times \bar{H}_2 - \bar{E}_2 \times \bar{H}_1) \cdot ds \qquad 2.173$$

$$= \int_V (\bar{E}_2 \cdot \bar{J}_1 - \bar{E}_1 \cdot \bar{J}_2 + \bar{H}_1 \cdot \bar{M}_2 - \bar{H}_2 \cdot \bar{M}_1) dv$$

Equation (2.173) represents a general form of the reciprocity theorem, but in practice a number of special situations often occur leading to some simplification. We will consider three cases.

S encloses no sources. Then $\bar{J}_1 = \bar{J}_2 = \bar{M}_1 = \bar{M}_2 = 0$, and the fields $\bar{E}_1, \bar{H}_1$, and $\bar{E}_2, \bar{H}_2$ are source-free fields. In this case, the right-hand side of (2.173) vanishes with

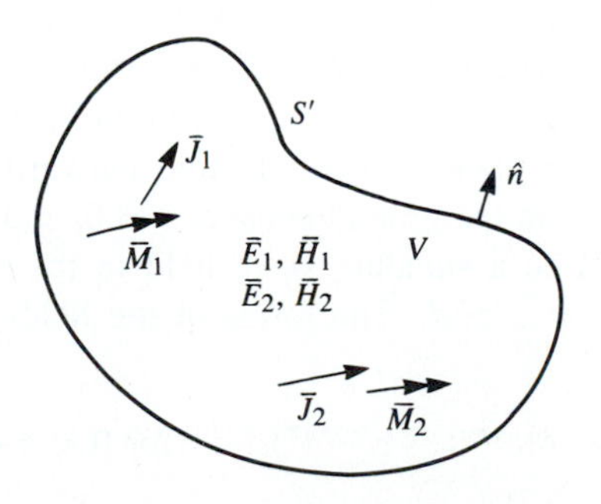

FIGURE 2.13 Geometery for the Lorentz reciprocity theorem.

the result that

$$\oint_S \bar{E}_1 \times \bar{H}_2 \cdot d\bar{s} = \oint_S \bar{E}_2 \times \bar{H}_1 \cdot d\bar{s}. \tag{2.174}$$

This result will be used in Chapter 5, when demonstrating the symmetry of the impedance matrix for a reciprocal microwave network.

S bounds a perfect conductor. For example, S may be the inner surface of a closed, perfectly conducting cavity. Then the surface integral of (2.173) vanishes, since $\bar{E}_1 \times \bar{H}_1 \cdot \hat{n} = (\hat{n} \times \bar{E}_1) \cdot \bar{H}_2$ (by vector identity B.3), and $\hat{n} \times \bar{E}_1$ is zero on the surface of a perfect conductor (similarly for $\bar{E}_2$). The result is

$$\int_V (\bar{E}_1 \cdot \bar{J}_2 - \bar{H}_1 \cdot \bar{M}_2) dv = \int_V (\bar{E}_2 \cdot \bar{J}_1 - \bar{H}_2 \cdot \bar{M}_1) dv. \tag{2.175}$$

This result is analogous to the reciprocity theorem of circuit theory. In words, this result states that the system response $\bar{E}_1$ or $\bar{E}_2$ is not changed when the source and observation points are interchanged. That is, $\bar{E}_2$ (caused by $\bar{J}_2$) at $\bar{J}_1$ is the same as $\bar{E}_1$ (caused by $\bar{J}_1$) at $\bar{J}_2$.

S is a sphere at infinity. In this case, the fields evaluated on S are very far from the sources, and so can be considered locally as plane waves. Then the impedance relation $\bar{H} = \hat{n} \times \bar{E}/\eta$ applies to (2.173) to give

$$\begin{aligned}(\bar{E}_1 \times \bar{H}_2 - \bar{E}_2 \times \bar{H}_1) \cdot \hat{n} &= (\hat{n} \times \bar{E}_1) \cdot \bar{H}_2 - (\hat{n} \times \bar{E}_2) \cdot \bar{H}_1 \\ &= \frac{1}{\eta} \bar{H}_1 \cdot \bar{H}_2 - \frac{1}{\eta} \bar{H}_2 \cdot \bar{H}_1 = 0,\end{aligned}$$

so that the result of (2.175) is again obtained. This result can also be obtained for the case of a closed surface S where the surface impedance boundary condition applies.

Image Theory

In many problems a current source is located in the vicinity of a conducting ground plane. Image theory permits the removal of the ground plane by placing a virtual image source of the other side of the ground plane. The reader should be familiar with this concept from electrostatics [2], so we will prove the result for an infinite current sheet next to an infinite ground plane, and then summarize the other possible cases.

Consider the surface current density $\bar{J}_s = J_{s0}\hat{x}$ parallel to a ground plane, as shown in Figure 2.14a. Because the current source is of infinite extent, and is uniform in the x, y directions, it will excite plane waves traveling outward from it. The negatively traveling wave will reflect from the ground plane at $z = 0$, and then travel in the positive direction. Thus, there will be a standing wave field in the region $0 < z < d$, and a positively traveling wave for $z > d$. The forms of the fields in these two regions can thus be written as

$$E_x^s = A(e^{jk_0 z} - e^{-jk_0 z}), \quad \text{for } 0 < z < d, \tag{2.176a}$$

$$H_y^s = \frac{-A}{\eta_0}(e^{jk_0 z} + e^{-jk_0 z}), \quad \text{for } 0 < z < d, \tag{2.176b}$$

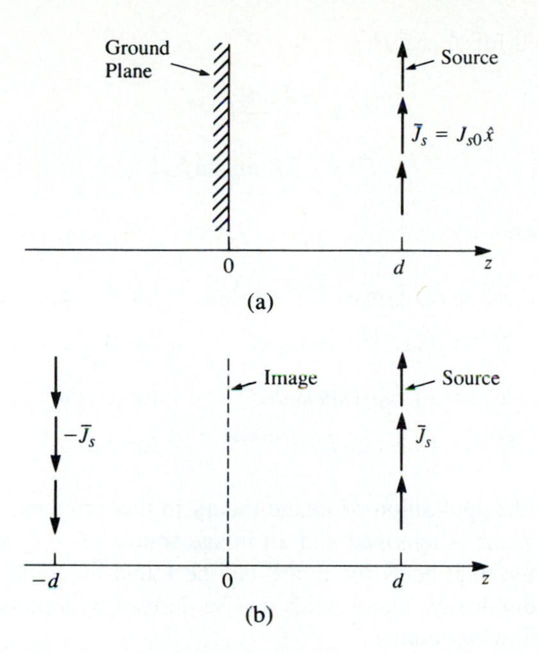

FIGURE 2.14 Illustration of image theory as applied to an electric current source next to a ground plane. (a) An electric surface current density parallel to a ground plane. (b) The ground plane of (a) replaced with image current at $z = -d$.

$$E_x^+ = Be^{-jk_0 z}, \quad \text{for } z > d, \qquad 2.177a$$

$$H_y^+ = \frac{B}{\eta_0}e^{-jk_0 z}, \quad \text{for } z > d, \qquad 2.177b$$

where η_0 is the wave impedance of free-space. Note that the standing wave fields of (2.176) have been constructed to satisfy the boundary condition that $E_x = 0$ at $z = 0$. The remaining boundary conditions to satisfy are continuity of $\bar{E}$ at $z = d$, and the discontinuity in the $\bar{H}$ field at $z = d$ due to the current sheet. From (2.36), since $\bar{M}_s = 0$,

$$E_x^s = E_x^+|_{z=0}, \qquad 2.178a$$

while from (2.37) we have

$$\bar{J}_s = \hat{z} \times \hat{y}(H_y^+ - H_y^s)|_{z=0}. \qquad 2.178b$$

Using (2.176) and (2.177) then gives

$$2jA \sin k_0 d = Be^{-jk_0 d}$$

and

$$J_{s0} = -\frac{B}{\eta_0}e^{-jk_0 d} - \frac{2A}{\eta_0}\cos k_0 d,$$

which can be solved for A and B:

$$A = \frac{-J_{s0}\eta_0}{2} e^{-jk_0 d},$$
$$B = -jJ_{s0}\eta_0 \sin k_0 d.$$

So the total fields are

$$E_x^s = -jJ_{s0}\eta_0 e^{-jk_0 d} \sin k_0 z, \quad \text{for } 0 < z < d, \tag{2.179a}$$

$$H_y^s = J_{s0} e^{-jk_0 d} \cos k_0 z, \quad \text{for } 0 < z < d, \tag{2.179b}$$

$$E_x^+ = -jJ_{s0}\eta_0 \sin k_0 d e^{-jk_0 z}, \quad \text{for } z > d, \tag{2.180a}$$

$$H_y^+ = -jJ_{s0} \sin k_0 d e^{-jk_0 z}, \quad \text{for } z > d. \tag{2.180b}$$

Now consider the application of image theory to this problem. As shown in Figure 2.14b, the ground plane is removed and an image source of $-\bar{J}_s$ is placed at $z = -d$. By superposition, the total fields for $z > 0$ can be found by combining the fields from the two sources individually. These fields can be derived by a procedure similar to that above, with the following results:

Fields due to source at $z = d$:

$$E_x = \begin{cases} \dfrac{-J_{s0}\eta_0}{2} e^{-jk_0(z-d)} & \text{for } z > d \\ \dfrac{-J_{s0}\eta_0}{2} e^{jk_0(z-d)} & \text{for } z < d, \end{cases} \tag{2.181a}$$

$$H_y = \begin{cases} \dfrac{-J_{s0}}{2} e^{-jk_0(z-d)} & \text{for } z > d \\ \dfrac{J_{s0}}{2} e^{jk_0(z-d)} & \text{for } z < d. \end{cases} \tag{2.181b}$$

Fields due to source at $z = -d$:

$$E_x = \begin{cases} \dfrac{J_{s0}\eta_0}{2} e^{-jk_0(z+d)} & \text{for } z > -d \\ \dfrac{J_{s0}\eta_0}{2} e^{jk_0(z+d)} & \text{for } z < -d, \end{cases} \tag{2.182a}$$

$$H_y = \begin{cases} \dfrac{J_{s0}}{2} e^{-jk_0(z+d)} & \text{for } z > -d \\ \dfrac{-J_{s0}}{2} e^{jk_0(z+d)} & \text{for } z < -d, \end{cases} \tag{2.182b}$$

The reader can verify that the solution is identical to that of (2.179) for $0 < z < d$, and to (2.180) for $z > d$, thus verifying the validity of the image theory solution. Note that image theory only gives the correct fields to the right of the conducting plane. Figure 2.15 shows more general image theory results for electric and magnetic dipoles.

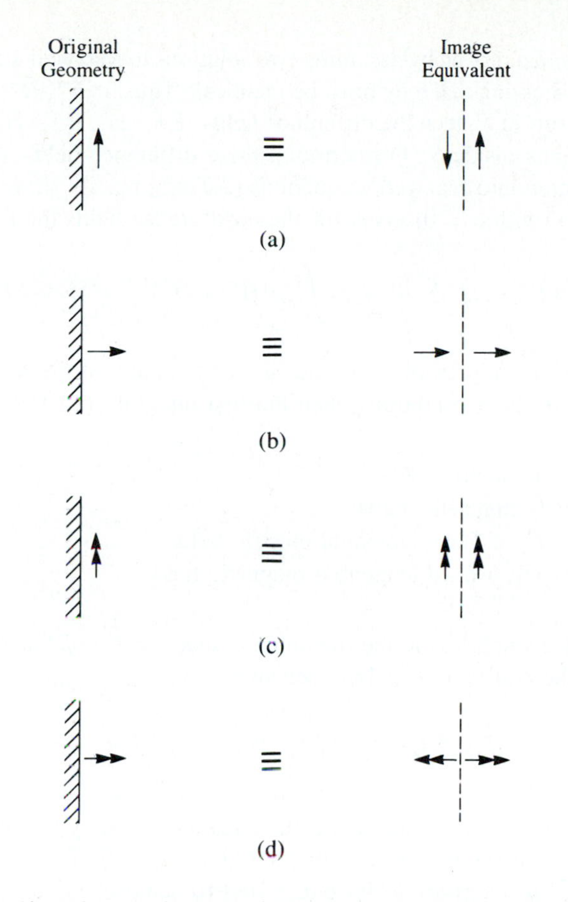

FIGURE 2.15 Electric and magnetic current images. (a) An electric current parallel to a ground plane. (b) An electric current normal to a ground plane. (c) A magnetic current parallel to a ground plane. (d) A magnetic current normal to a ground plane.

The Uniqueness Theorem

Once we have found a solution to Maxwell's equations and the appropriate boundary conditions, the uniqueness theorem assures us that, under the proper conditions, this solution is the only possible solution. This is a particularly useful result when, as in Chapter 5, we can find the fields due to a source by postulating the form of the fields, and then enforce boundary conditions by adjusting some amplitude constants. The uniqueness theorem then guarantees that this procedure gives the correct and unique solution.

Although it can be expressed in various ways, we will state the uniqueness theorem in the following form: In a region bounded by a closed surface S and completely filled with dissipative media, the field $\bar{E}, \bar{H}$ is uniquely determined by the source currents in the region and the tangential components of $\bar{E}$ or $\bar{H}$ on S.

This result can be proved by assuming two solutions to Maxwell's equations, $\bar{E}^a, \bar{H}^a$, and $\bar{E}^b, \bar{H}^b$, and showing that they must be identical. Thus, if $\bar{E}^a, \bar{H}^a$ and $\bar{E}^b, \bar{H}^b$ satisfy Maxwell's equations in S, then the difference fields, $\bar{E}^a - \bar{E}^b, \bar{H}^a - \bar{H}^b$, must also satisfy Maxwell's equations inside S. Furthermore, these difference fields must be source-free fields, as substitution into Maxwell's equations (2.27a,b) readily shows. Then Poynting's theorem of (2.88) (with $\sigma = 0$) gives for these difference fields the following result:

$$\oint_S (\bar{E}^a - \bar{E}^b) \times (\bar{H}^a - \bar{H}^b)^* \cdot d\bar{s} + j\omega \int_V (\mu|\bar{H}^a - \bar{H}^b|^2 - \epsilon^*|\bar{E}^a - \bar{E}^b|^2)dv = 0. \quad 2.183$$

Now, if $\bar{E}$ and/or $\bar{H}$ are prescribed on the surface S in any of the following ways, or in any combination of these conditions, then the first integral in (2.183) will vanish:

1. $\hat{n} \times \bar{E} = 0$, electric walls.
2. $\hat{n} \times \bar{H} = 0$, magnetic walls.
3. $\hat{n} \times \bar{E} = \bar{E}_t$, a fixed tangential electric field.
4. $\hat{n} \times \bar{H} = \bar{H}_t$, a fixed tangential magnetic field.

Now assume a small loss in the medium, so that $\epsilon = \epsilon' - j\epsilon''$ and $\mu = \mu' - j\mu''$ are complex. Then the real part of (2.183) becomes

$$\int_V (\mu''|\bar{H}^a - \bar{H}^b|^2 + \epsilon''|\bar{E}^a - \bar{E}^b|^2)dv = 0. \quad 2.184$$

Since all of these terms are nonnegative, the equation can only be satisfied if $\bar{E}^a = \bar{E}^b$, and $\bar{H}^a = \bar{H}^b$, which shows that only one solution is possible.

Note that it was necessary to introduce loss to achieve this result; loss can always be introduced into a problem by making ϵ and/or μ complex. This result also suggests that problems involving lossless materials may not have unique solutions. This is indeed the case as the following examples point out.

Resonant modes in a lossless cavity. Such source-free fields are defined as having equal time-average electric and magnetic stored energies, as given by (2.84) and (2.86). Then, if there is no loss, μ and ϵ are real, and the second integral in (2.183) is identically zero for a difference field equal to the field of such a resonant mode. Such source-free fields would rapidly dissipate in the presence of loss, however.

Plane wave incident on a lossless dielectric slab. In Chapter 3 we will solve the problem of plane wave transmission and reflection from a dielectric slab of finite thickness. This solution will satisfy Maxwell's equations and the boundary conditions, but it may not be the only solution. This is because a surface wave field can also exist on the dielectric slab. This field is source-free, and also satisfies Maxwell's equations and the boundary conditions. For a lossy slab, such a field would quickly be dissipated.

Plane waves in free-space. In infinite lossless free-space, source-free plane wave solutions are possible for any polarization, and any direction of propagation. If loss is introduced, such fields would quickly decay to zero.

REFERENCES

[1] J. C. Maxwell, *A Treatise on Electricity and Magnetism*, Dover, N.Y., 1954.

[2] G. G. Skitek and S. V. Marshall, *Electromagnetic Concepts and Applications*, Prentice-Hall, N.J., 1982.

[3] C. A. Balanis, *Advanced Engineering Electromagnetics*, John Wiley & Sons, N.Y., 1989.

[4] R. E. Collin, *Foundations for Microwave Engineering*, McGraw-Hill, N.Y., 1966.

[5] L. C. Shen and J. A. Kong, *Applied Electromagnetism*, Brooks-Cole, Calif., 1983.

[6] D. K. Cheng, *Field and Wave Electromagnetics*, Addison-Wesley, Reading, Mass., 1983.

[7] S. Ramo, T. R. Whinnery, and T. van Duzer, *Fields and Waves in Communication Electronics*, John Wiley & Sons, N.Y., 1965.

[8] C. G. Montgomery, R. H. Dicke, and E. M. Purcell, *Principles of Microwave Circuits*, vol. 8 of MIT Rad. Lab. Series, McGraw-Hill, N.Y., 1948.

[9] R.F. Harrington, *Time-Harmonic Electromagnetic Fields*, McGraw-Hill, N.Y., 1961.

PROBLEMS

2.1 Assume that an infinite sheet of electric surface current density $\bar{J}_s = J_o\hat{x}$ A/m is placed on the $z = 0$ plane between free-space for $z < 0$, and a dielectric with $\epsilon = \epsilon_r\epsilon_0$ for $z > 0$, as shown below. Find the resulting $\bar{E}$ and $\bar{H}$ fields in the two regions. HINT: Assume plane wave solutions propagating away from the current sheet, and match boundary conditions to find the amplitudes, as in Example 2.3.

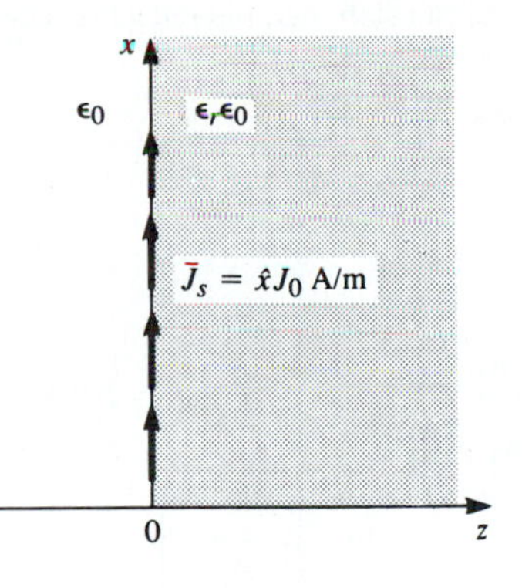

2.2 Let $\bar{E} = E_\rho\hat{\rho} + E_\phi\hat{\phi} + E_z\hat{z}$ be an electric field vector in cylindrical coordinates. Demonstrate that the $\nabla^2\bar{E}$ term in the vector identity $\nabla \times \nabla \times \bar{E} = \nabla(\nabla \cdot \bar{E}) - \nabla^2\bar{E}$ does not apply to the cylindrical components of $\bar{E}$, by expanding both sides of this relation for the given electric field.

2.3 An anisotropic material has a tensor permittivity $[\epsilon]$ as given below, and a permeability of $4\mu_0$. At a certain point in the material, the electric field is known to be $\bar{E} = 3\hat{x} + 4\hat{y} + 6\hat{z}$. What is $\bar{D}$ at this point?

$$[\epsilon] = \epsilon_0 \begin{bmatrix} 1 & -2j & 0 \\ 2j & 3 & 0 \\ 0 & 0 & 4 \end{bmatrix}$$

2.4 Consider a permanent magnet with a steady magnetic field $\bar{H} = H_0\hat{y}$, and a parallel plate capacitor with an electric field $\bar{E} = E_0\hat{x}$, arranged as shown below. Calculate the Poynting vector at a point between both the magnet poles and the capacitor plates. This nonzero result seems to imply real power flow in the z direction, but clearly there is no wave propagation or power delivered from the sources. How do you explain this apparent paradox?

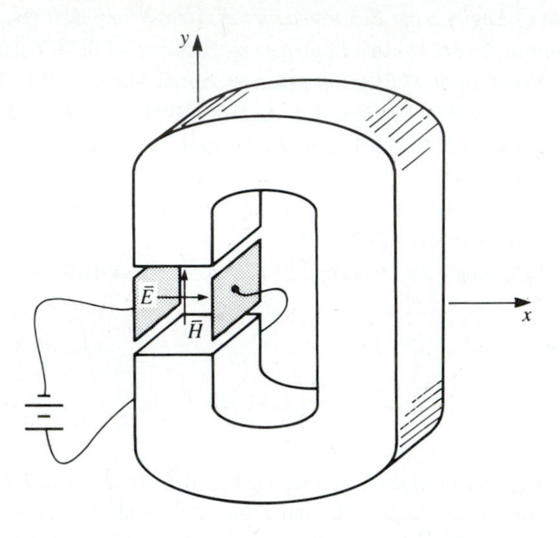

2.5 A plane wave is normally incident on a dielectric slab of permittivity ϵ_r and thickness d, where $d = \lambda_0/(4\sqrt{\epsilon_r})$, and λ_0 is the free-space wavelength of the incident wave, as shown below. If free-space exists on both sides of the slab, find the reflection coefficient of the wave reflected from the front of the slab.

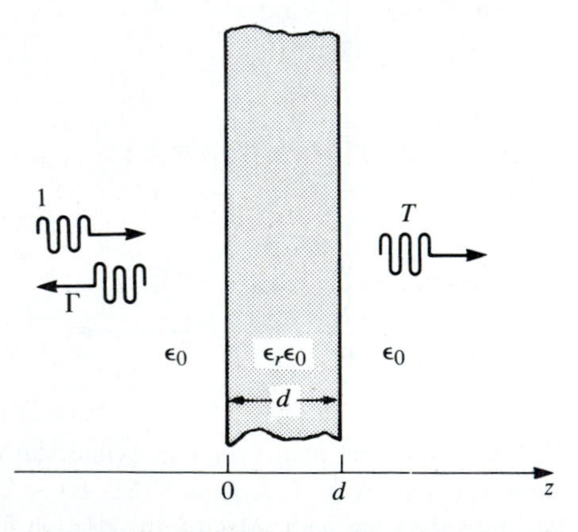

2.6 Consider a RHCP plane wave normally incident from free-space ($z < 0$) onto a half-space ($z > 0$) consisting of a good conductor. Let the incident electric field be of the form

$$\bar{E}_i = E_0(\hat{x} - j\hat{y})e^{-jk_0z},$$

and find the electric and magnetic fields in the region $z > 0$. Compute the Poynting vectors for $z < 0$ and $z > 0$, and show that complex power is conserved. What is the polarization of the reflected wave?

2.7 Consider a plane wave propagating in a lossy dielectric medium for $z < 0$, with a perfectly conducting plate at $z = 0$. Assume that the lossy medium is characterized by $\epsilon = (5 - j2)\epsilon_0$, $\mu = \mu_0$, and that the frequency of the plane wave is 1.0 GHz, and let the amplitude of the incident electric field be 4 V/m at $z = 0$. Find the reflected electric field for $z < 0$, and plot the magnitude of the total electric field for $-0.5 \leq z \leq 0$.

2.8 A plane wave in free-space is normally incident on a thin copper sheet of thickness t. What is the approximate required thickness if the copper sheet is to be used as a shield to reduce the level of the transmitted electric field by 90 dB? Do this calculation for $f = 100$ MHz, 1 GHz, and 10 GHz. HINT: Simplify the problem by ignoring reflections at the interfaces.

2.9 A uniform lossy medium with $\epsilon_r = 3.0, \tan\delta = 0.1$, and $\mu = \mu_0$ fills the region between $z = 0$ and $z = 20$ cm, with a ground plane at $z = 20$ cm, as shown below. An incident plane wave with an electric field,

$$\bar{E}_i = \hat{x}100e^{-\gamma z} \text{ V/m},$$

is present at $z = 0$ and propagates in the $+z$ direction. The frequency is $f = 3.0$ GHz.

(a) Compute P_i, the power density of the incident wave, and P_r, the power density of the reflected wave, at $z = 0$.

(b) Compute the input power density, P_{in}, at $z = 0$, from the total fields at $z = 0$. Does $P_{in} = P_i - P_r$?

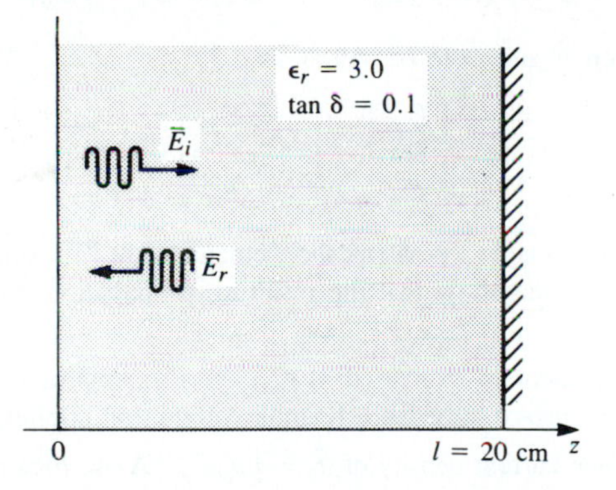

2.10 Redo Problem 2.1, but with an electric surface current density of $\bar{J}_s = J_o\hat{x}e^{-j\beta x}$ A/m, where $\beta < k_0$.

2.11 A parallel polarized plane wave is obliquely incident from free-space onto a magnetic material with permittivity ϵ_0 and permeability $\mu_0\mu_r$. Find the reflection and transmission coefficients. Does a Brewster angle exist for this case, where the reflection coefficient vanishes for a particular angle of incidence?

2.12 Repeat Problem 2.11 for the perpendicularly polarized case.

2.13 Consider a uniaxial medium for $z > 0$, of the same form as in Section 2.9, and free-space for $z < 0$. If a LHCP wave of the form

$$\bar{E}_i = E_0(\hat{x} + j\hat{y})e^{-jk_0 z}$$

is incident on this medium from free-space ($z < 0$), compute the reflected electric field $\bar{E}_r$ in the region $z < 0$. What is the polarization of this reflected wave?

2.14 Consider an anisotropic dielectric material with a permittivity tensor of the form

$$[\epsilon] = \begin{bmatrix} \epsilon_x & a & 0 \\ b & \epsilon_0 & 0 \\ 0 & 0 & \epsilon_0 \end{bmatrix}.$$

What is the relation between a and b for the medium to be lossless? Assume ϵ_x is real, and a, b complex.

2.15 Consider the gyrotropic permittivity tensor of Section 2.9:

$$[\epsilon] = \epsilon_0 \begin{bmatrix} \epsilon_r & j\kappa & 0 \\ -j\kappa & \epsilon_r & 0, \\ 0 & 0 & 1 \end{bmatrix}.$$

The $\bar{D}$ and $\bar{E}$ fields are then related as

$$\begin{bmatrix} D_x \\ D_y \\ D_z \end{bmatrix} = [\epsilon] \begin{bmatrix} E_x \\ E_y, \\ E_z \end{bmatrix}.$$

Show that the transformations,

$$E_+ = E_x - jE_y, \qquad D_+ = D_x - jD_y,$$

$$E_- = E_x + jE_y, \qquad D_- = D_x + jD_y,$$

allow the relation between $\bar{E}$ and $\bar{D}$ to be written as,

$$\begin{bmatrix} D_+ \\ D_- \\ D_z \end{bmatrix} = [\epsilon'] \begin{bmatrix} E_+ \\ E_-, \\ E_z \end{bmatrix},$$

where $[\epsilon']$ is now a diagonal matrix. What are the elements of $[\epsilon']$? Using this result, derive wave equations for E_+ and E_-, and show that the resulting propagation constants agree with those obtained in Section 2.9.

2.16 Show that the reciprocity theorem expressed in (2.175) also applies to a region enclosed by a closed surface S, where a surface impedance boundary condition applies.

2.17 Consider an electric surface current density of $\bar{J}_s = \hat{y} J_o e^{-\beta x}$ A/m, located on the $z = d$ plane. If a perfectly conducting ground plane is placed at $z = 0$, use image theory to find the total fields for $z > 0$.

CHAPTER 3

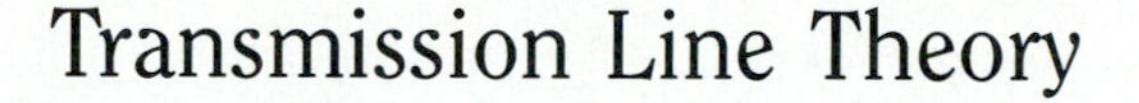

Transmission Line Theory

In many ways transmission line theory bridges the gap between field analysis and basic circuit theory, and so is of significant importance in microwave network analysis. As we will see, the phenomenon of wave propagation on transmission lines can be approached from an extension of circuit theory, or from a specialization of Maxwell's equations; we shall present both viewpoints, and show how this wave propagation is described by equations very similar to those used in Chapter 2 for plane wave propagation.

3.1 THE LUMPED-ELEMENT CIRCUIT MODEL FOR A TRANSMISSION LINE

The key difference between circuit theory and transmission line theory is electrical size. Circuit analysis assumes that the physical dimensions of a network are much smaller than the electrical wavelength, while transmission lines may be a considerable fraction of a wavelength, or many wavelengths, in size. Thus a transmission line is a distributed-parameter network, where voltages and currents can vary in magnitude and phase over its length.

As shown in Figure 3.1a, a transmission line is often schematically represented as a two-wire line, since transmission lines (for TEM wave propagation) always have at least two conductors. The short piece of line of length Δz of Figure 3.1a can be modeled as a lumped-element circuit, as shown in Figure 3.1b, where R, L, G, C are per unit length quantities defined as follows:

R = series resistance per unit length, for both conductors, in Ω/m.

L = series inductance per unit length, for both conductors, in H/m.

G = shunt conductance per unit length, in S/m.

C = shunt capacitance per unit length, in F/m.

The series inductance L represents the total self-inductance of the two conductors, while the shunt capacitance C is due to the close proximity of the two conductors. The series resistance R represents the resistance due to the finite conductivity of the conductors, while the shunt conductance G is due to dielectric loss in the material between the

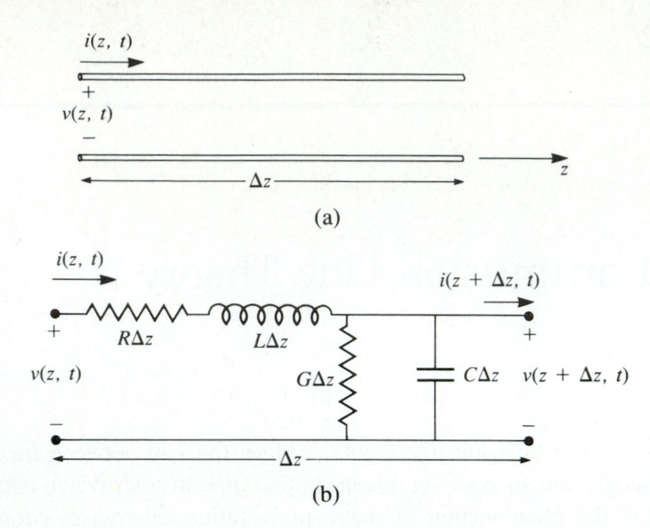

FIGURE 3.1 Voltage and current definitions and equivalent circuit for an incremental length of transmission line. (a) Voltage and current definitions. (b) Lumped-element equivalent circuit.

conductors. R and G, therefore, represent loss. A finite length of transmission line can be viewed as a cascade of sections of the form of Figure 3.1b.

From the circuit of Figure 3.1b, Kirchhoff's voltage law can be applied to give

$$v(z,t) - R\Delta z i(z,t) - L\Delta z \frac{\partial i(z,t)}{\partial t} - v(z+\Delta z, t) = 0, \tag{3.1a}$$

while Kirchhoff's current law leads to

$$i(z,t) - G\Delta z v(z+\Delta z,t) - C\Delta z \frac{\partial v(z+\Delta z,t)}{\partial t} - i(z+\Delta z,t) = 0. \tag{3.1b}$$

Dividing (3.1a) and (3.1b) by Δz and taking the limit as $\Delta z \to 0$ gives the following differential equations:

$$\frac{\partial v(z,t)}{\partial z} = -Ri(z,t) - L\frac{\partial i(z,t)}{\partial t}, \tag{3.2a}$$

$$\frac{\partial i(z,t)}{\partial z} = -Gv(z,t) - C\frac{\partial v(z,t)}{\partial t}. \tag{3.2b}$$

These equations are the time-domain form of the transmission line, or telegrapher, equations.

For the sinusoidal steady-state condition, with cosine-based phasors, (3.2) simplify to

$$\frac{dV(z)}{dz} = -(R + j\omega L)I(z), \tag{3.3a}$$

$$\frac{dI(z)}{dz} = -(G + j\omega C)V(z). \tag{3.3b}$$

Note the similarity in the form of (3.3) and Maxwell's curl equations of (2.41a) and (2.41b).

Wave Propagation on a Transmission Line

The two equations of (3.3) can be solved simultaneously to give wave equations for $V(z)$ and $I(z)$:

$$\frac{d^2V(z)}{dz^2} - \gamma^2 V(z) = 0, \tag{3.4a}$$

$$\frac{d^2I(z)}{dz^2} - \gamma^2 I(z) = 0, \tag{3.4b}$$

where

$$\gamma = \alpha + j\beta = \sqrt{(R + j\omega L)(G + j\omega C)} \tag{3.5}$$

is the complex propagation constant, which is a function of frequency. Traveling wave solutions to (3.4) can be found as

$$V(z) = V_o^+ e^{-\gamma z} + V_o^- e^{\gamma z}, \tag{3.6a}$$

$$I(z) = I_o^+ e^{-\gamma z} + I_o^- e^{\gamma z}, \tag{3.6b}$$

where the $e^{-\gamma z}$ term represents wave propagation in the $+z$ direction, and the $e^{\gamma z}$ term represents wave propagation in the $-z$ direction. Applying (3.3a) to the voltage of (3.6a) gives the current on the line:

$$I(z) = \frac{\gamma}{R + j\omega L}\left[V_o^+ e^{-\gamma z} - V_o^- e^{\gamma z}\right].$$

Comparison with (3.6b) shows that a characteristic impedance, Z_0, can be defined as

$$Z_0 = \frac{R + j\omega L}{\gamma} = \sqrt{\frac{R + j\omega L}{G + j\omega C}}, \tag{3.7}$$

to relate the voltage and current on the line as

$$\frac{V_o^+}{I_o^+} = Z_0 = \frac{-V_o^-}{I_o^-}.$$

Then (3.6b) can be rewritten in the following form:

$$I(z) = \frac{V_o^+}{Z_0} e^{-\gamma z} - \frac{V_o^-}{Z_0} e^{\gamma z}. \tag{3.8}$$

Converting back to the time domain, the voltage waveform can be expressed as

$$\begin{aligned} v(z,t) = &|V_o^+| \cos(\omega t - \beta z + \phi^+) e^{-\alpha z} \\ &+ |V_o^-| \cos(\omega t + \beta z + \phi^-) e^{\alpha z}, \end{aligned} \tag{3.9}$$

where $\phi^{\pm}$ is the phase angle of the complex voltage $V_o^{\pm}$. Using arguments similar to those in Section 2.4, we find that the wavelength on the line is

$$\lambda = \frac{2\pi}{\beta}, \tag{3.10}$$

and the phase velocity is

$$v_p = \frac{\omega}{\beta} = \lambda f. \tag{3.11}$$

The Lossless Line

The above solution was for a general transmission line, including loss effects, and it was seen that the propagation constant and characteristic impedance were complex. In many practical cases, however, the loss of the line is very small and so can be neglected, resulting in a simplification of the above results. Setting $R = G = 0$ in (3.5) gives the propagation constant as

$$\gamma = \alpha + j\beta = j\omega\sqrt{LC},$$

or

$$\beta = \omega\sqrt{LC}, \tag{3.12a}$$

$$\alpha = 0. \tag{3.12b}$$

As expected for the lossless case, the attenuation constant α is zero. The characteristic impedance of (3.7) reduces to

$$Z_0 = \sqrt{\frac{L}{C}}, \tag{3.13}$$

which is now a real number. The general solutions for voltage and current on a lossless transmission line can then be written as

$$V(z) = V_o^+ e^{-j\beta z} + V_o^- e^{j\beta z}, \tag{3.14a}$$

$$I(z) = \frac{V_o^+}{Z_0} e^{-j\beta z} - \frac{V_o^-}{Z_0} e^{j\beta z}. \tag{3.14b}$$

The wavelength is

$$\lambda = \frac{2\pi}{\beta} = \frac{2\pi}{\omega\sqrt{LC}}, \tag{3.15}$$

and the phase velocity is

$$v_p = \frac{\omega}{\beta} = \frac{1}{\sqrt{LC}}. \tag{3.16}$$

3.2 FIELD ANALYSIS OF TRANSMISSION LINES

In this section we will rederive the time-harmonic form of the telegrapher's equations, starting with Maxwell's equations. We will begin by deriving the transmission line parameters (R, L, G, C) in terms of the electric and magnetic fields of the transmission line, and then derive the telegrapher equations using these parameters for the specific case of a coaxial line.

Transmission Line Parameters

Consider a 1 m section of a uniform transmission line with fields $\bar{E}$ and $\bar{H}$, as shown in Figure 3.2, where S is the cross-sectional surface area of the line. Let the voltage between the conductors be $V_o e^{\pm j\beta z}$ and the current be $I_o e^{\pm j\beta z}$. The time-average stored magnetic energy for this 1 m section of line can be written, from (2.86), as

$$W_m = \frac{\mu}{4} \int_S \bar{H} \cdot \bar{H}^* ds,$$

while circuit theory gives $W_m = L|I_o|^2/4$, in terms of the current on the line. We can thus identify the self-inductance per unit length as

$$L = \frac{\mu}{|I_o|^2} \int_S \bar{H} \cdot \bar{H}^* ds \text{ H/m.} \tag{3.17}$$

Similarly, the time-average stored electric energy per unit length can be found from (2.84) as

$$W_e = \frac{\epsilon}{4} \int_S \bar{E} \cdot \bar{E}^* ds,$$

while circuit theory gives $W_e = C|V_o|^2/4$, resulting in the following expression for the capacitance per unit length:

$$C = \frac{\epsilon}{|V_o|^2} \int_S \bar{E} \cdot \bar{E}^* ds \text{ F/m.} \tag{3.18}$$

From (2.130), the power loss per unit length due to the finite conductivity of the metallic conductors is

$$P_c = \frac{R_s}{2} \int_{C_1+C_2} \bar{H} \cdot \bar{H}^* d\ell$$

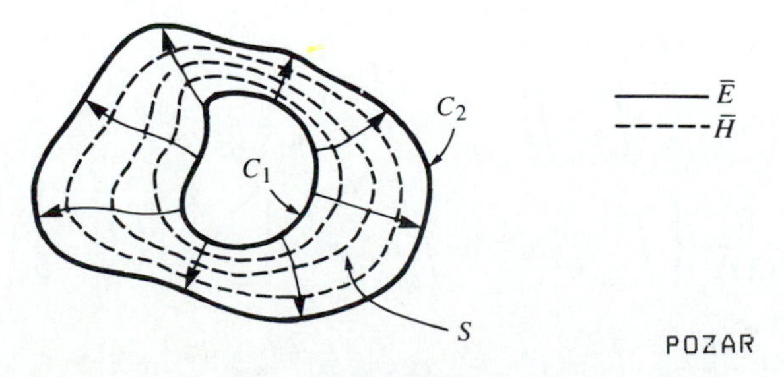

POZAR

FIGURE 3.2 Field lines on an arbitrary TEM transmission line.

(assuming $\bar{H}$ is tangential to S), while circuit theory gives $P_c = R|I_o|^2/2$, so the series resistance R per unit length of line is

$$R = \frac{R_s}{|I_o|^2} \int_{C_1+C_2} \bar{H} \cdot \bar{H}^* dl \ \Omega/\text{m}. \qquad 3.19$$

In (3.19), $R_s = 1/\sigma\delta_s$ is the surface resistance of the conductors, and $C_1 + C_2$ represent integration paths over the conductor boundaries. From (2.92), the time-average power dissipated per unit length in a lossy dielectric is

$$P_d = \frac{\omega\epsilon''}{2} \int_S \bar{E} \cdot \bar{E}^* ds,$$

where ϵ'' is the imaginary part of the complex dielectric constant $\epsilon = \epsilon' - j\epsilon'' = \epsilon'(1 - j\tan\delta)$. Circuit theory gives $P_d = G|V_o|^2/2$, so the shunt conductance per unit length can be written as

$$G = \frac{\omega\epsilon''}{|V_o|^2} \int_S \bar{E} \cdot \bar{E}^* ds \ \text{S/m}. \qquad 3.20$$

EXAMPLE 3.1

The fields of a traveling TEM wave inside the coaxial line shown in Figure 3.3 can be expressed as

$$\bar{E} = \frac{V_o \hat{\rho}}{\rho \ln b/a} e^{-\gamma z},$$

$$\bar{H} = \frac{I_o \hat{\phi}}{2\pi\rho} e^{-\gamma z},$$

where γ is the propagation constant of the line. The conductors are assumed to have a surface resistivity R_s, and the material filling the space between the conductors is assumed to have a complex permittivity $\epsilon = \epsilon' - j\epsilon''$ and a permeability $\mu = \mu_0\mu_r$. Determine the transmission line parameters.

Solution

From (3.17)–(3.20) and the above fields the parameters of the coaxial line can be calculated as

$$L = \frac{\mu}{(2\pi)^2} \int_{\phi=\phi}^{2\pi} \int_{\rho=a}^{b} \frac{1}{\rho^2} \rho d\rho d\phi = \frac{\mu}{2\pi} \ln b/a \quad \text{H/m},$$

$$C = \frac{\epsilon'}{(\ln b/a)^2} \int_{\phi=0}^{2\pi} \int_{\rho=a}^{b} \frac{1}{\rho^2} \rho d\rho d\phi = \frac{2\pi\epsilon'}{\ln b/a} \quad \text{F/m},$$

$$R = \frac{R_s}{(2\pi)^2} \left\{ \int_{\phi=0}^{2\pi} \frac{1}{a^2} a d\phi + \int_{\phi=0}^{2\pi} \frac{1}{b^2} b d\phi \right\} = \frac{R_s}{2\pi} \left(\frac{1}{a} + \frac{1}{b} \right) \ \Omega/\text{m},$$

$$G = \frac{\omega\epsilon''}{(\ln b/a)^2} \int_{\phi=0}^{2\pi} \int_{\rho=a}^{b} \frac{1}{\rho^2} \rho d\rho d\phi = \frac{2\pi\omega\epsilon''}{\ln b/a} \quad \text{S/m}.$$

○

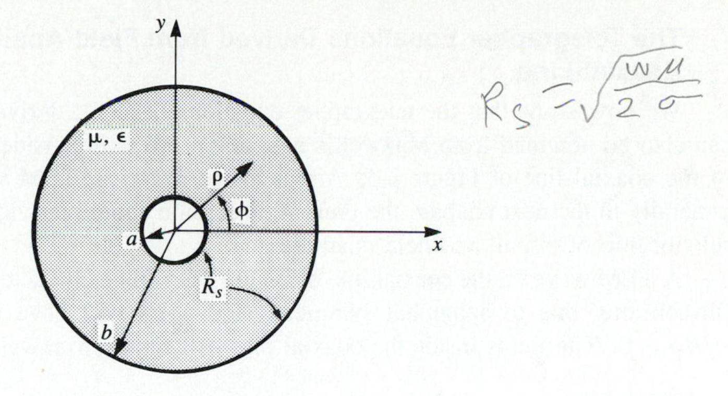

FIGURE 3.3 Geometery of a coaxial line with surface resistance R_s on the inner and outer conductors.

Table 3.1 summarizes the parameters for coaxial, two-wire, and parallel plate lines. As we will see in the next chapter, the propagation constant, characteristic impedance, and attenuation of most transmission lines are derived directly from a field theory solution; the approach here of first finding the equivalent circuit parameters (L, C, R, G) is only useful for relatively simple lines. Nevertheless, it provides a helpful intuitive concept, and relates a line to its equivalent circuit model.

TABLE 3.1 Transmission Line Parameters for Some Common Lines

	COAX	TWO-WIRE	PARALLEL PLATE
	(a, b)	(a, D)	(w, d)
L	$\frac{\mu}{2\pi}\ln\frac{b}{a}$	$\frac{\mu}{\pi}\cosh^{-1}\left(\frac{D}{2a}\right)$	$\frac{\mu d}{w}$
C	$\frac{2\pi\epsilon'}{\ln b/a}$	$\frac{\pi\epsilon'}{\cosh^{-1}(D/2a)}$	$\frac{\epsilon' w}{d}$
R	$\frac{R_s}{2\pi}\left(\frac{1}{a}+\frac{1}{b}\right)$	$\frac{R_s}{\pi a}$	$\frac{2R_s}{w}$
G	$\frac{2\pi\omega\epsilon''}{\ln b/a}$	$\frac{\pi\omega\epsilon''}{\cosh^{-1}(D/2a)}$	$\frac{\omega\epsilon'' w}{d}$

The Telegrapher Equations Derived from Field Analysis of a Coaxial Line

We now show that the telegrapher equations of (3.3), derived using circuit theory, can also be obtained from Maxwell's equations. We will consider the specific geometry of the coaxial line of Figure 3.3. Although we will treat TEM wave propagation more generally in the next chapter, the present discussion should provide some insight into the relationship of circuit and field quantities.

A TEM wave on the coaxial line of Figure 3.3 will be characterized by $E_z = H_z = 0$; furthermore, due to azimuthal symmetry, the fields will have no ϕ-variation, and so $\partial/\partial\phi = 0$. The fields inside the coaxial line will satisfy Maxwell's curl equations,

$$\nabla \times \bar{E} = -j\omega\mu\bar{H}, \tag{3.21a}$$

$$\nabla \times \bar{H} = j\omega\epsilon\bar{E}, \tag{3.21b}$$

where $\epsilon = \epsilon' - j\epsilon''$ may be complex to allow for a lossy dielectric filling. Conductor loss will be ignored here. A rigorous field analysis of conductor loss can be carried out, but at this point would tend to obscure our purpose; the interested reader is referred to Ramo, Winnery, and Van Duzer [1] or Stratton [2].

Expanding (3.21a) and (3.21b) then gives the following vector equations:

$$-\hat{\rho}\frac{\partial E_\phi}{\partial z} + \hat{\phi}\frac{\partial E_\rho}{\partial z} + \hat{z}\frac{1}{\rho}\frac{\partial}{\partial \rho}(\rho E_\phi) = -j\omega\mu\left(\hat{\rho}H_\rho + \hat{\phi}H_\phi\right), \tag{3.22a}$$

$$-\hat{\rho}\frac{\partial H_\phi}{\partial z} + \hat{\phi}\frac{\partial H_\rho}{\partial z} + \hat{z}\frac{1}{\rho}\frac{\partial}{\partial \rho}(\rho H_\phi) = j\omega\epsilon\left(\hat{\rho}E_\rho + \hat{\phi}E_\phi\right). \tag{3.22b}$$

Since the $\hat{z}$ components of these two equations must vanish, it is seen that E_ϕ and H_ϕ must have the forms

$$E_\phi = \frac{f(z)}{\rho}, \tag{3.23a}$$

$$H_\phi = \frac{g(z)}{\rho}. \tag{3.23b}$$

To satisfy the boundary condition that $E_\phi = 0$ at $\rho = a, b$, we must have $E_\phi = 0$ everywhere, due to the form of E_ϕ in (3.23a). Then from the $\hat{\rho}$ component of (3.22a), it is seen that $H_\rho = 0$. With these results, (3.22) can be reduced to

$$\frac{\partial E_\rho}{\partial z} = -j\omega\mu H_\phi, \tag{3.24a}$$

$$\frac{\partial H_\phi}{\partial z} = -j\omega\epsilon E_\rho. \tag{3.24b}$$

From the form of H_ϕ in (3.23b) and (3.24a), E_ρ must be of the form

$$E_\rho = \frac{h(z)}{\rho}. \tag{3.25}$$

Using (3.23b) and (3.25) in (3.24) gives

$$\frac{\partial h(z)}{\partial z} = -j\omega\mu g(z), \tag{3.26a}$$

$$\frac{\partial g(z)}{\partial z} = -j\omega\epsilon h(z). \tag{3.26b}$$

Now the voltage between the two conductors can be evaluated as

$$V(z) = \int_{\rho=a}^{b} E_\rho(\rho, z) d\rho = h(z) \int_{\rho=a}^{b} \frac{d\rho}{\rho} = h(z) \ln \frac{b}{a}, \tag{3.27a}$$

and the total current on the inner conductor at $\rho = a$ can be evaluated using (3.23b) as

$$I(z) = \int_{\phi=0}^{2\pi} H_\phi(a, z) a d\phi = 2\pi g(z). \tag{3.27b}$$

Then $h(z)$ and $g(z)$ can be eliminated from (3.26) by using (3.27) to give

$$\frac{\partial V(z)}{\partial z} = -j\frac{\omega\mu \ln b/a}{2\pi} I(z),$$

$$\frac{\partial I(z)}{\partial z} = -j\omega(\epsilon' - j\epsilon'')\frac{2\pi V(z)}{\ln b/a}.$$

Finally, using the results for L, G, and C for a coaxial line as derived above, we obtain the telegrapher equations as

$$\frac{\partial V(z)}{\partial z} = -j\omega L I(z), \tag{3.28a}$$

$$\frac{\partial I(z)}{\partial z} = -(G + j\omega C) V(z) \tag{3.28b}$$

(excluding R, the series resistance, since the conductors were assumed to have perfect conductivity). A similar analysis can be carried out for other simple transmission lines.

Propagation Constant, Impedance, and Power Flow for the Lossless Coaxial Line

Equations (3.24a) and (3.24b) for E_ρ and H_ϕ can be simultaneously solved to yield a wave equation for E_ρ (or H_ϕ):

$$\frac{\partial^2 E_\rho}{\partial z^2} + \omega^2 \mu\epsilon E_\rho = 0, \tag{3.29}$$

from which it is seen that the propagation constant is $\gamma^2 = -\omega^2\mu\epsilon$, which, for lossless media, reduces to

$$\beta = \omega\sqrt{\mu\epsilon} = \omega\sqrt{LC}, \tag{3.30}$$

where the last result is from (3.12). Observe that this propagation constant is of the same form as that for plane waves in a lossless dielectric medium. This is a general result for TEM transmission lines.

The wave impedance is defined as $Z_w = E_\rho / H_\phi$, which can be calculated from (3.24a) assuming an $e^{-j\beta z}$ dependence to give

$$Z_w = \frac{E_\rho}{H_\phi} = \frac{\omega\mu}{\beta} = \sqrt{\mu/\epsilon} = \eta. \tag{3.31}$$

This wave impedance is then seen to be identical to the intrinsic impedance of the medium, η, and again is a general result for TEM transmission lines.

The characteristic impedance of the coaxial line is defined as

$$Z_0 = \frac{V_o}{I_o} = \frac{E_\rho \ln b/a}{2\pi H_\phi} = \frac{\eta \ln b/a}{2\pi} = \sqrt{\frac{\mu}{\epsilon}} \frac{\ln b/a}{2\pi}, \tag{3.32}$$

where the forms for E_ρ and H_ϕ from Example 3.1 have been used. The characteristic impedance is geometry dependent, and will be different for other transmission line configurations.

Finally, the power flow (in the z direction) on the coaxial line may be computed from the Poynting vector as

$$P = \frac{1}{2}\int_s \bar{E} \times \bar{H}^* \cdot d\bar{s} = \frac{1}{2}\int_{\phi=0}^{2\pi}\int_{\rho=a}^{b} \frac{V_o I_o^*}{2\pi\rho^2 \ln b/a} \rho d\rho d\phi = \frac{1}{2} V_o I_o^*, \tag{3.33}$$

a result which is in clear agreement with circuit theory. This shows that the flow of power in a transmission line takes place entirely via the electric and magnetic fields between the two conductors; power is not transmitted through the conductors themselves. As we will see later, for the case of finite conductivity, power may enter the conductors, but this power is then lost as heat and is not delivered to the load.

3.3 THE TERMINATED LOSSLESS TRANSMISSION LINE

Figure 3.4 shows a lossless transmission line terminated in an arbitrary load impedance Z_L. This problem will illustrate wave reflection on transmission lines, a fundamental property of distributed systems.

Assume that an incident wave of the form $V_o^+ e^{-j\beta z}$ is generated from a source at $z < 0$. We have seen that the ratio of voltage to current for such a traveling wave is

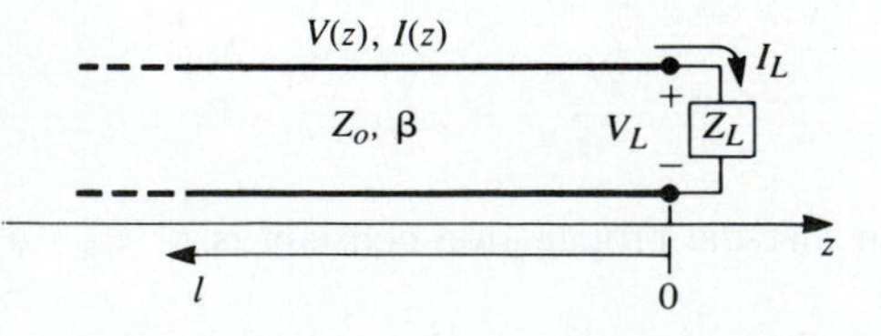

FIGURE 3.4 A transmission line terminated in a load impedance Z_L.

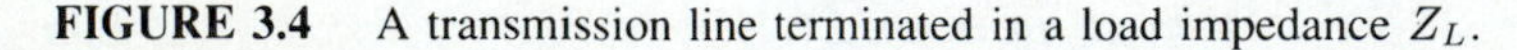

Z_0, the characteristic impedance. But when the line is terminated in an arbitrary load $Z_L \neq Z_0$, the ratio of voltage to current at the load must be Z_L. Thus, a reflected wave must be excited with the appropriate amplitude to satisfy this condition. The total voltage on the line can then be written as in (3.14a), as a sum of incident and reflected waves:

$$V(z) = V_o^+ e^{-j\beta z} + V_o^- e^{j\beta z}. \tag{3.34a}$$

Similarly, the total current on the line is described by (3.14b):

$$I(z) = \frac{V_o^+}{Z_0} e^{-j\beta z} - \frac{V_o^-}{Z_0} e^{j\beta z}. \tag{3.34b}$$

The total voltage and current at the load are related by the load impedance, so at $z = 0$ we must have

$$Z_L = \frac{V(0)}{I(0)} = \frac{V_o^+ + V_o^-}{V_o^+ - V_o^-} Z_0.$$

Solving for V_o^- gives

$$V_o^- = \frac{Z_L - Z_0}{Z_L + Z_0} V_o^+.$$

The amplitude of the reflected voltage wave normalized to the amplitude of the incident voltage wave is known as the voltage reflection coefficient, Γ:

$$\Gamma = \frac{V_o^-}{V_o^+} = \frac{Z_L - Z_0}{Z_L + Z_0}. \tag{3.35}$$

A current reflection coefficient, giving the normalized amplitude of the reflected current wave, can also be defined. But because such a current reflection coefficient is just the negative of the voltage reflection coefficient (as seen from (3.34)), we will avoid confusion by using only the voltage reflection coefficient in this book.

The total voltage and current waves on the line can then be written as

$$V(z) = V_o^+ \left[e^{-j\beta z} + \Gamma e^{j\beta z}\right], \tag{3.36a}$$

$$I(z) = \frac{V_o^+}{Z_0} \left[e^{-j\beta z} - \Gamma e^{j\beta z}\right]. \tag{3.36b}$$

From these equations it is seen that the voltage and current on the line consists of a superposition of an incident and reflected wave; such waves are called standing waves. Only when $\Gamma = 0$ is there no reflected wave. To obtain $\Gamma = 0$, the load impedance Z_L must be equal to the characteristic impedance Z_0 of the transmission line, as seen from (3.35). Such a load is then said to be *matched* to the line, since there is no reflection of the incident wave.

Now consider the time-average power flow along the line at the point z:

$$P_{\text{av}} = \frac{1}{2}\text{Re}\left[V(z)I(z)^*\right] = \frac{1}{2}\frac{|V_o^+|^2}{Z_0}\text{Re}\left\{1 - \Gamma^* e^{-2j\beta z} + \Gamma e^{2j\beta z} - |\Gamma|^2\right\},$$

where (3.36) has been used. The middle two terms in the brackets are of the form $A - A^* = 2jIm(A)$, and so are purely imaginary. This simplifies the result to

$$P_{\text{av}} = \frac{1}{2} \frac{|V_o^+|^2}{Z_0} \left(1 - |\Gamma|^2\right), \tag{3.37}$$

which shows that the average power flow is constant at any point on the line, and that the total power delivered to the load (P_{av}) is equal to the incident power ($|V_o^+|^2/2Z_0$), minus the reflected power ($|V_o|^2|\Gamma|^2/2Z_0$). If $\Gamma = 0$, maximum power is delivered to the load, while no power is delivered for $|\Gamma| = 1$. The above discussion assumes that the generator is matched, so that there is no re-reflection of the reflected wave from $z < 0$.

When the load is mismatched, then, not all of the available power from the generator is delivered to the load. This "loss" is called *return loss* (RL), and is defined (in dB) as

$$\text{RL} = -20 \log |\Gamma| \text{ dB}, \tag{3.38}$$

so that a matched load ($\Gamma = 0$) has a return loss of ∞ dB (no reflected power), while a total reflection ($|\Gamma| = 1$) has a return loss of 0 dB (all incident power is reflected).

If the load is matched to the line, Γ=0 and the magnitude of the voltage on the line is $|V(z)| = |V_o^+|$, which is a constant. Such a line is sometimes said to be "flat." When the load is mismatched, however, the presence of a reflected wave leads to standing waves where the magnitude of the voltage on the line is not constant. Thus, from (3.36a),

$$\begin{aligned} |V(z)| &= |V_o^+||1 + \Gamma e^{2j\beta z}| = |V_o^+||1 + \Gamma e^{-2j\beta \ell}| \\ &= |V_o^+||1 + |\Gamma| e^{j(\theta - 2\beta\ell)}|, \end{aligned} \tag{3.39}$$

where $\ell = -z$ is the positive distance measured from the load at $z = 0$, and θ is the phase of the reflection coefficient ($\Gamma = |\Gamma|e^{\theta}$). This result shows that the voltage magnitude oscillates with position z along the line. The maximum value occurs when the phase term $e^{j(\theta - 2\beta\ell)} = 1$, and is given by

$$V_{\text{max}} = |V_o^+|(1 + |\Gamma|). \tag{3.40a}$$

The minimum value occurs when the phase term $e^{j(\theta - 2\beta\ell)} = -1$, and is given by

$$V_{\text{min}} = |V_o^+|(1 - |\Gamma|). \tag{3.40b}$$

As $|\Gamma|$ increases, the ratio of V_{max} to V_{min} increases, so a measure of the mismatch of a line, called the *standing wave ratio* (SWR), can be defined as

$$\text{SWR} = \frac{V_{\text{max}}}{V_{\text{min}}} = \frac{1 + |\Gamma|}{1 - |\Gamma|}. \tag{3.41}$$

This quantity is also known as the *voltage standing wave ratio*, and is sometimes identified as VSWR. From (3.41) it is seen that SWR is a real number such that $1 \leq \text{SWR} \leq \infty$, where SWR = 1 implies a matched load.

From (3.39), it is seen that the distance between two successive voltage maxima (or minima) is $\ell = 2\pi/2\beta = \pi\lambda/2\pi = \lambda/2$, while the distance between a maxima and a minima is $\ell = \pi/2\beta = \lambda/4$, where λ is the wavelength on the transmission line.

The reflection coefficient of (3.35) was defined as the ratio of the reflected to the incident voltage wave amplitudes at the load ($\ell = 0$), but this quantity can be generalized to any point ℓ on the line as follows. From (3.34a), with $z = -\ell$, the ratio of the reflected component to the incident component is

$$\Gamma(\ell) = \frac{V_o^- e^{-j\beta\ell}}{V_o^+ e^{j\beta\ell}} = \Gamma(0)e^{-2j\beta\ell}, \tag{3.42}$$

where $\Gamma(0)$ is the reflection coefficient at $z = 0$, as given by (3.35). This form is useful when transforming the effect of a load mismatch down the line.

We have seen that the real power flow on the line is a constant, but that the voltage amplitude, at least for a mismatched line, is oscillatory with position on the line. The perceptive reader may therefore have concluded that the impedance seen looking into the line must vary with position, and this is indeed the case. At a distance $\ell = -z$ from the load, the input impedance seen looking toward the load is

$$Z_{\text{in}} = \frac{V(-\ell)}{I(-\ell)} = \frac{V_o^+\left[e^{j\beta\ell} + \Gamma e^{-j\beta\ell}\right]}{V_o^+\left[e^{j\beta\ell} - \Gamma e^{-j\beta\ell}\right]} Z_0 = \frac{1 + \Gamma e^{-2j\beta\ell}}{1 - \Gamma e^{-2j\beta\ell}} Z_0, \tag{3.43}$$

where (3.36a,b) have been used for $V(z)$ and $I(z)$. A more usable form may be obtained by using (3.35) for Γ in (3.43):

$$\begin{aligned} Z_{\text{in}} &= Z_0 \frac{(Z_L + Z_0)e^{j\beta\ell} + (Z_L - Z_0)e^{-j\beta\ell}}{(Z_L + Z_0)e^{j\beta\ell} - (Z_L - Z_0)e^{-j\beta\ell}} \\ &= Z_0 \frac{Z_L \cos\beta\ell + jZ_0 \sin\beta\ell}{Z_0 \cos\beta\ell + jZ_L \sin\beta\ell} \\ &= Z_0 \frac{Z_L + jZ_0 \tan\beta\ell}{Z_0 + jZ_L \tan\beta\ell}. \end{aligned} \tag{3.44}$$

This is an important result giving the input impedance of a length of transmission line with an arbitrary load impedance. We will refer to this result as the transmission line impedance equation; some special cases will be considered next.

Special Cases of Lossless Terminated Lines

A number of special cases of lossless terminated transmission lines will frequently appear in our work, so it is appropriate to consider the properties of such cases here.

Consider first the transmission line circuit shown in Figure 3.5, where a line is terminated in a short circuit, $Z_L = 0$. From (3.35) it is seen that the reflection coefficient for a short circuit load is $\Gamma = -1$; it then follows from (3.41) that the standing wave ratio is infinite. From (3.36) the voltage and current on the line are

$$V(z) = V_o^+\left[e^{-j\beta z} - e^{j\beta z}\right] = -2jV_o^+ \sin\beta z, \tag{3.45a}$$

$$I(z) = \frac{V_o^+}{Z_0}\left[e^{-j\beta z} + e^{j\beta z}\right] = \frac{2V_o^+}{Z_0} \cos\beta z, \tag{3.45b}$$

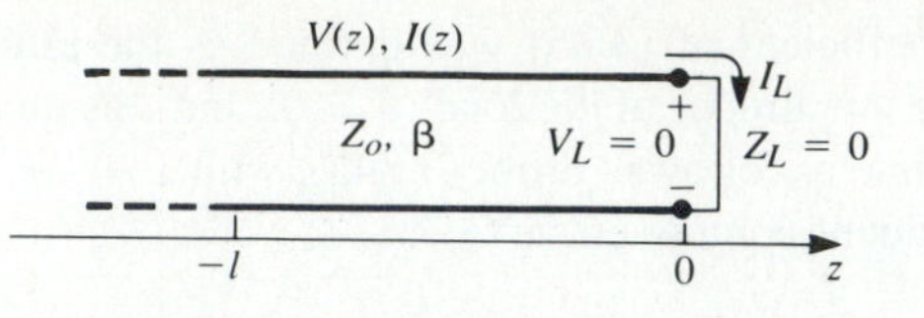

FIGURE 3.5 A transmission line terminated in a short circuit.

which shows that $V = 0$ at the load (as it should, for a short circuit), while the current is a maximum there. From (3.44), or the ratio $V(-\ell)/I(-\ell)$, the input impedance is

$$Z_{\text{in}} = jZ_0 \tan \beta\ell, \tag{3.45c}$$

which is seen to be purely imaginary for any length, ℓ, and to take on all values between $+j\infty$ and $-j\infty$. For example, when $\ell = 0$ we have $Z_{\text{in}} = 0$, but for $\ell = \lambda/4$ we have $Z_{\text{in}} = \infty$ (open-circuit). Equation (3.45c) also shows that the impedance is periodic in ℓ, repeating for multiples of $\lambda/2$. The voltage, current, and input reactance for the short-circuited line are plotted in Figure 3.6.

Next consider the open-circuited line shown in Figure 3.7, where $Z_L = \infty$. Dividing the numerator and denominator of (3.35) by Z_L and allowing $Z_L \rightarrow \infty$ shows that the reflection coefficient for this case is $\Gamma = 1$, and the standing wave ratio is again infinite. From (3.36) the voltage and current on the line are

$$V(z) = V_o^+ \left[e^{-j\beta z} + e^{j\beta z}\right] = 2V_o^+ \cos \beta z, \tag{3.46a}$$

$$I(z) = \frac{V_o^+}{Z_0} \left[e^{-j\beta z} - e^{j\beta z}\right] = \frac{-2jV_o^+}{Z_0} \sin \beta z, \tag{3.46b}$$

which shows that now $I = 0$ at the load, as expected for an open circuit, while the voltage is a maximum. The input impedance is

$$Z_{\text{in}} = -jZ_0 \cot \beta\ell, \tag{3.46c}$$

which is also purely imaginary for any length, ℓ. The voltage, current, and input reactance of the open-circuited line are plotted in Figure 3.8.

Now consider terminated transmission lines with some special lengths. If $\ell = \lambda/2$, (3.44) shows that

$$Z_{\text{in}} = Z_L, \tag{3.47}$$

meaning that a half-wavelength line (or any multiple of $\lambda/2$) does not alter or transform the load impedance, regardless of the characteristic impedance.

If the line is a quarter-wavelength long or, more generally, $\ell = \lambda/4 + n\lambda/2$, for $n = 1, 2, 3 \cdots$, (3.44) shows that the input impedance is given by

$$Z_{\text{in}} = \frac{Z_0^2}{Z_L}. \tag{3.48}$$

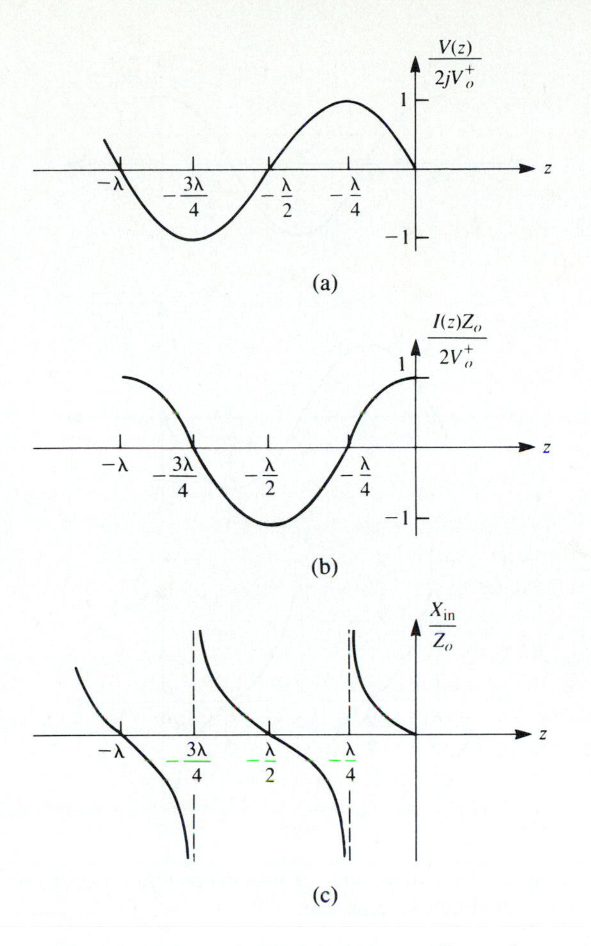

FIGURE 3.6 (a) Voltage, (b) current, and (c) impedance ($R_{\text{in}} = 0$ or ∞) variation along a short-circuited transmission line.

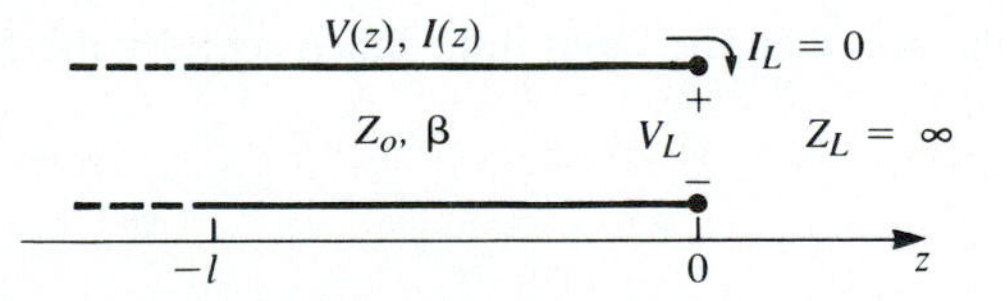

FIGURE 3.7 A transmission line terminated in an open circuit.

Such a line is known as a *quarter-wave transformer* since it has the effect of transforming the load impedance, in an inverse manner, depending on the characteristic impedance of the line. We will study this case more thoroughly in Section 3.5.

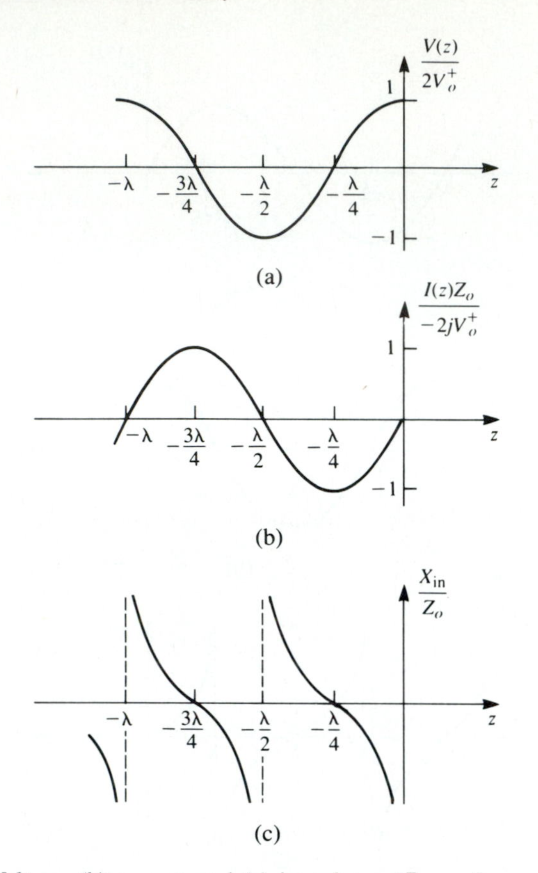

FIGURE 3.8 (a) Voltage, (b) current, and (c) impedance ($R_{in} = 0$ or ∞) variation along an open-circuited transmission line.

Now consider a transmission line of characteristic impedance Z_0 feeding a line of different characteristic impedance, Z_1, as shown in Figure 3.9. If the load line is infinitely long, or if it is terminated in its own characteristic impedance, so that there are no reflections from its end, then the input impedance seen by the feed line is Z_1, so that

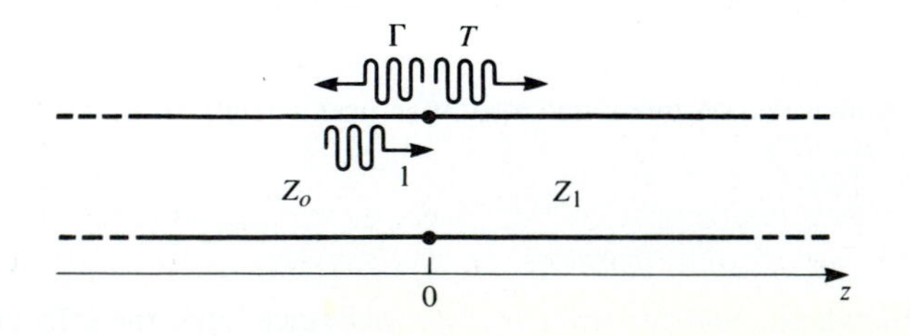

FIGURE 3.9 Reflection and transmission at the junction of two transmission lines with different characteristic impedances.

the reflection coefficient Γ is

$$\Gamma = \frac{Z_1 - Z_0}{Z_1 + Z_0}. \tag{3.49}$$

Not all of the incident wave is reflected; some of it is transmitted onto the second line with a voltage amplitude given by a transmission coefficient, T.

From (3.36a), the voltage for $z < 0$ is

$$V(z) = V_o^+(e^{-j\beta z} + \Gamma e^{j\beta z}), \qquad z < 0, \tag{3.50a}$$

where V_o^+ is the amplitude of the incident voltage wave on the feed line. The voltage wave for $z > 0$, in the absence of reflections, is outgoing only, and can be written as

$$V(z) = V_o^+ T e^{-j\beta z}, \qquad \text{for } z > 0. \tag{3.50b}$$

Equating these voltages at $z = 0$ gives the transmission coefficient, T, as

$$T = 1 + \Gamma = 1 + \frac{Z_1 - Z_0}{Z_1 + Z_0} = \frac{2Z_1}{Z_1 + Z_0}. \tag{3.51}$$

The transmission coefficient between two points in a circuit is often expressed in dB as the *insertion loss*, IL,

$$IL = -20 \log |T| \text{ dB}. \tag{3.52}$$

POINT OF INTEREST: Decibels and Nepers

Often the ratio of two power levels, P_1 and P_2, in a microwave system is expressed in decibels (dB) as

$$10 \log \frac{P_1}{P_2} \text{ dB} .$$

Thus, a power ratio of 2 is equivalent to 3 dB, while a power ratio of 0.1 is equivalent to -10 dB. Using power ratios in dB makes it easy to calculate power loss or gain through a series of components, since multiplicative loss or gain factors can be accounted for by adding the loss or gain in dB for each stage. For example, a signal passing through a 6 dB attenuator followed by a 23 dB amplifier will have an overall gain of $23 - 6 = 17$ dB.

Decibels are only used to represent power ratios, but if $P_1 = V_1^2/R_1$ and $P_2 = V_2^2/R_2$, then the result in terms of voltage ratios is

$$10 \log \frac{V_1^2 R_2}{V_2^2 R_1} = 20 \log \frac{V_1}{V_2} \sqrt{\frac{R_2}{R_1}} \text{ dB} ,$$

where R_1, R_2 are the load resistances and V_1, V_2 are the voltages appearing across these loads. If the load resistances are equal, then this formula simplifies to

$$20 \log \frac{V_1}{V_2} \text{ dB} .$$

The ratio of voltages across equal load resistances can also be expressed in terms of nepers (Np) as

$$\ln \frac{V_1}{V_2} \text{ Np.}$$

The corresponding expression in terms of powers is

$$\frac{1}{2} \ln \frac{P_1}{P_2} \text{ Np,}$$

since voltage is proportional to the square root of power. Transmission line attenuation is often expressed in nepers. Since 1 Np corresponds to a power ratio of e^2, the conversion between nepers and decibels is

$$1\text{Np} = 10 \log e^2 = 8.686 \text{ dB.}$$

Absolute powers can also be expressed in decibel notation if a reference power level is assumed. If we let $P_2 = 1$ mW, then the power P_1 can be expressed in dBm as

$$10 \log \frac{P_1}{1 \text{ mW}} \text{ dBm.}$$

Thus a power of 1 mW is 0 dBm, while a power of 1W is 30 dBm, etc.

3.4 THE SMITH CHART

The Smith chart, shown in Figure 3.10, is a graphical aid that is very useful when solving transmission line problems. Although there are a number of other impedance and reflection coefficient charts that can be used for such problems [3], the Smith chart is probably the best known and most widely used. It was developed in 1939 by P. Smith at the Bell Telephone Laboratories. The reader may feel that, in this day of scientific calculators and powerful computers, graphical solutions have no place in modern engineering. The Smith chart, however, is more than just a graphical technique. Besides being an integral part of much of the current computer-aided design (CAD) software and test equipment for microwave design, the Smith chart provides an extremely useful way of visualizing transmission line phenomenon, and so is also important for pedagogical reasons. A microwave engineer can develop his intuition about transmission line and impedance-matching problems by learning to think in terms of the Smith chart.

At first glance the Smith chart of Figure 3.10 may seem intimidating, but the key to its understanding is to realize that it is essentially a polar plot of the voltage reflection coefficient, Γ. Let the reflection coefficient be expressed in magnitude and phase (polar) form as $\Gamma = |\Gamma|e^{j\theta}$. Then the magnitude $|\Gamma|$ is plotted as a radius ($|\Gamma| \leq 1$) from the center of the chart, and the angle $\theta(-180° \leq \theta \leq 180°)$ is measured from the right-hand side of the horizontal diameter. Any passively realizable ($|\Gamma| \leq 1$) reflection coefficient can then be plotted as a unique point on the Smith chart.

The real utility of the Smith chart, however, lies in the fact that it can be used to convert from reflection coefficients to normalized impedances (or admittances), and vice versa, using the impedance (or admittance) circles printed on the chart. When

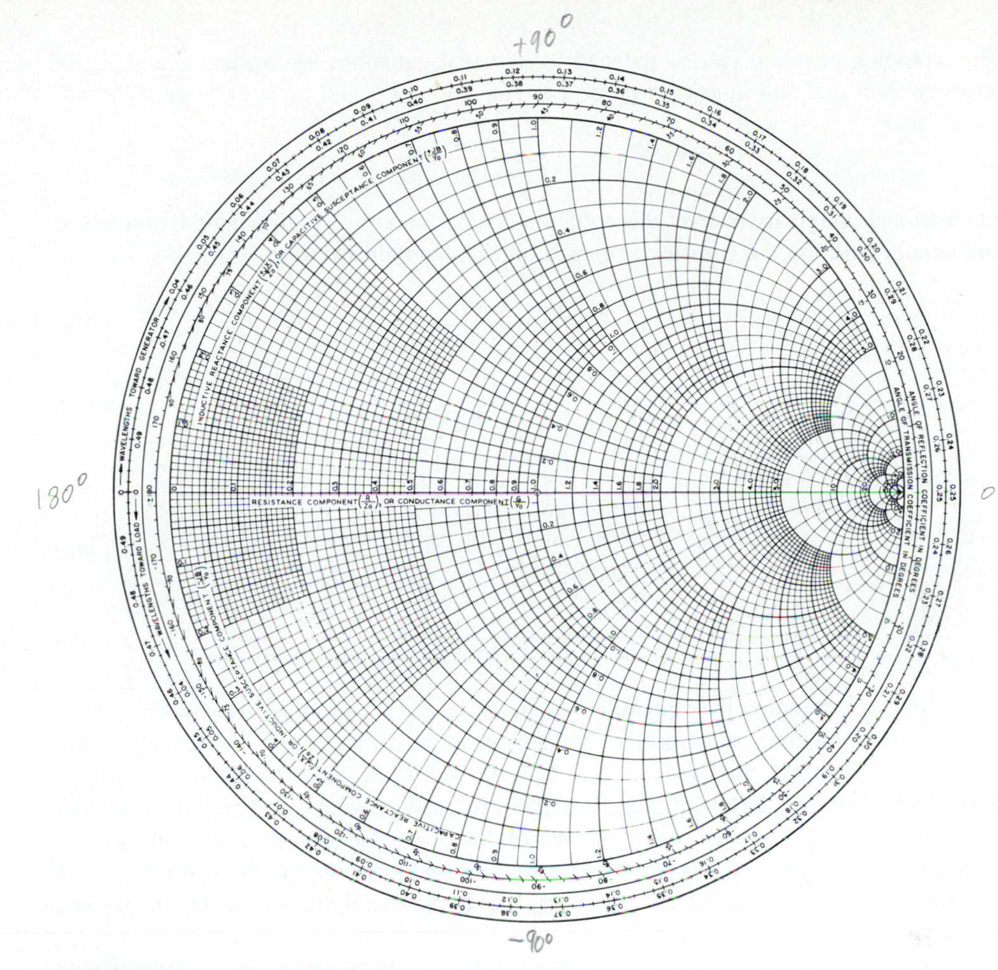

FIGURE 3.10 The Smith chart.

dealing with impedances on a Smith chart, normalized quantitites are generally used, which we will denote by lower case letters. The normalization constant is usually the characteristic impedance of the line. Thus, $z = Z/Z_0$ represents the normalized version of the impedance Z.

If a lossless line of characteristic impedance Z_0 is terminated with a load impedance Z_L, the reflection coefficient at the load can be written from (3.35) as

$$\Gamma = \frac{z_L - 1}{z_L + 1} = |\Gamma| e^{j\theta}, \tag{3.53}$$

where $z_L = Z_L/Z_0$ is the normalized load impedance. This relation can be solved for z_L in terms of Γ to give (or, from (3.43) with $\ell = 0$)

$$z_L = \frac{1 + |\Gamma| e^{j\theta}}{1 - |\Gamma| e^{j\theta}}. \tag{3.54}$$

This complex equation can be reduced to two real equations by writing Γ and z_L in terms of their real and imaginary parts. Let $\Gamma = \Gamma_r + j\Gamma_i$, and $z_L = r_L + jx_L$. Then,

$$r_L + jx_L = \frac{(1+\Gamma_r) + j\Gamma_i}{(1-\Gamma_r) - j\Gamma_i}.$$

The real and imaginary parts of this equation can be found by multiplying the numerator and denominator by the complex conjugate of the denominator to give

$$r_L = \frac{1 - \Gamma_r^2 - \Gamma_i^2}{(1-\Gamma_r)^2 + \Gamma_i^2}, \tag{3.55a}$$

$$x_L = \frac{2\Gamma_i^2}{(1-\Gamma_r)^2 + \Gamma_i^2}. \tag{3.55b}$$

Rearranging (3.55) gives

$$\left(\Gamma_r - \frac{r_L}{1+r_L}\right)^2 + \Gamma_i^2 = \left(\frac{1}{1+r_L}\right)^2, \tag{3.56a}$$

$$(\Gamma_r - 1)^2 + \left(\Gamma_i - \frac{1}{x_L}\right)^2 = \left(\frac{1}{x_L}\right)^2, \tag{3.56b}$$

which are seen to represent two families of circles in the Γ_r, Γ_i plane. Resistance circles are defined by (3.56a), and reactance circles are defined by (3.56b). For example, the $r_L = 1$ circle has its center at $\Gamma_r = 0.5$, $\Gamma_i = 0$, and has a radius of 0.5, and so passes through the center of the Smith chart. All of the resistance circles of (3.56a) have centers on the horizontal $\Gamma_i = 0$ axis, and pass through the $\Gamma = 1$ point on the right-hand side of the chart. The centers of all of the reactance circles of (3.56b) lie on the vertical $\Gamma_r = 1$ line (off the chart), and these circles also pass through the $\Gamma = 1$ point. The resistance and reactance circles are orthogonal.

The Smith chart can also be used to graphically solve the transmission line impedance equation of (3.44), since this can be written in terms of the generalized reflection coefficient as

$$Z_{\text{in}} = Z_0 \frac{1 + \Gamma e^{-2j\beta\ell}}{1 - \Gamma e^{-2j\beta\ell}}, \tag{3.57}$$

where Γ is the reflection coefficient at the load, and ℓ is the (positive) length of transmission line. We then see that (3.57) is of the same form as (3.54), differing only by the phase angles of the Γ terms. Thus, if we have plotted the reflection coefficient $|\Gamma|e^{j\theta}$ at the load, the normalized input impedance seen looking into a length ℓ of transmission line terminated with z_L can be found by rotating the point clockwise an amount $2\beta\ell$ (subtracting $2\beta\ell$ from θ) around the center of the chart. The radius stays the same, since the magnitude of Γ does not change with position along the line.

To facilitate such rotations, the Smith chart has scales around its periphery calibrated in electrical wavelengths, toward ~~and away from~~ the "generator" (which just means the direction away from the load). These scales are relative, so only the difference in wave lengths between two points on the Smith chart is meaningful. The scales cover a range of 0 to 0.5 wavelengths, which reflects the fact that the Smith chart automatically includes the periodicity of transmission line phenomenon. Thus, a line of length $\lambda/2$ (or any multiple) requires a rotation of $2\beta\ell = 2\pi$ around the center of the chart, bringing the point back to its original position, showing that the input impedance of a load seen through a $\lambda/2$ line is unchanged.

We will now illustrate the use of the Smith chart for a variety of typical transmission line problems through examples.

EXAMPLE 3.2

At the load of a terminated transmission line having a characteristic impedance of 100 Ω, the reflection coefficient is $\Gamma = 0.560 + j0.215$. What is the load impedance?

Solution

To solve this problem with the Smith chart, we first convert the reflection coefficient to polar form, $\Gamma = 0.60\angle 21°$, and then plot this point on the chart (Figure 3.11). The magnitude is specified with a compass set to 0.60 from the voltage reflection coefficient scale below the chart, while the angle can be specified by drawing a radial line from the center of the chart out to the 21° phase angle point at the outer edge of the chart. The intersection of this line and a circle of radius 0.60 give the normalized load impedance as

$$z_L = 2.6 + j1.8.$$

The actual load impedance is then,

$$Z_L = Z_0 z_L = 260 + j180\ \Omega.$$

○

EXAMPLE 3.3

A load impedance of $Z_L = 80 - j40\,\Omega$ terminates a $50\,\Omega$ line. Find the return loss in dB, standing wave ratio on the line, and the reflection coefficient at the load.

Solution

The normalized load impedance is

$$z_L = \frac{Z_L}{Z_0} = 1.60 - j0.80,$$

which can be plotted on the Smith chart (Figure 3.12). Using a compass and the voltage reflection coefficient scale below the chart, the reflection coefficient magnitude is found to be $|\Gamma| = 0.36$. This same compass setting can be applied to the standing wave ratio (SWR) scale to give SWR = 2.2, and to the return loss (in dB) scale to give RL = 8.7 dB. The angle of the reflection coefficient is read from the outer scale of the chart as −36°.

If a circle is drawn through the load impedance point, the standing wave ratio can be read from the intersection with the horizontal axis for $r > 1$. Such a circle is called an SWR circle, since the SWR is constant everywhere on this circle.

○

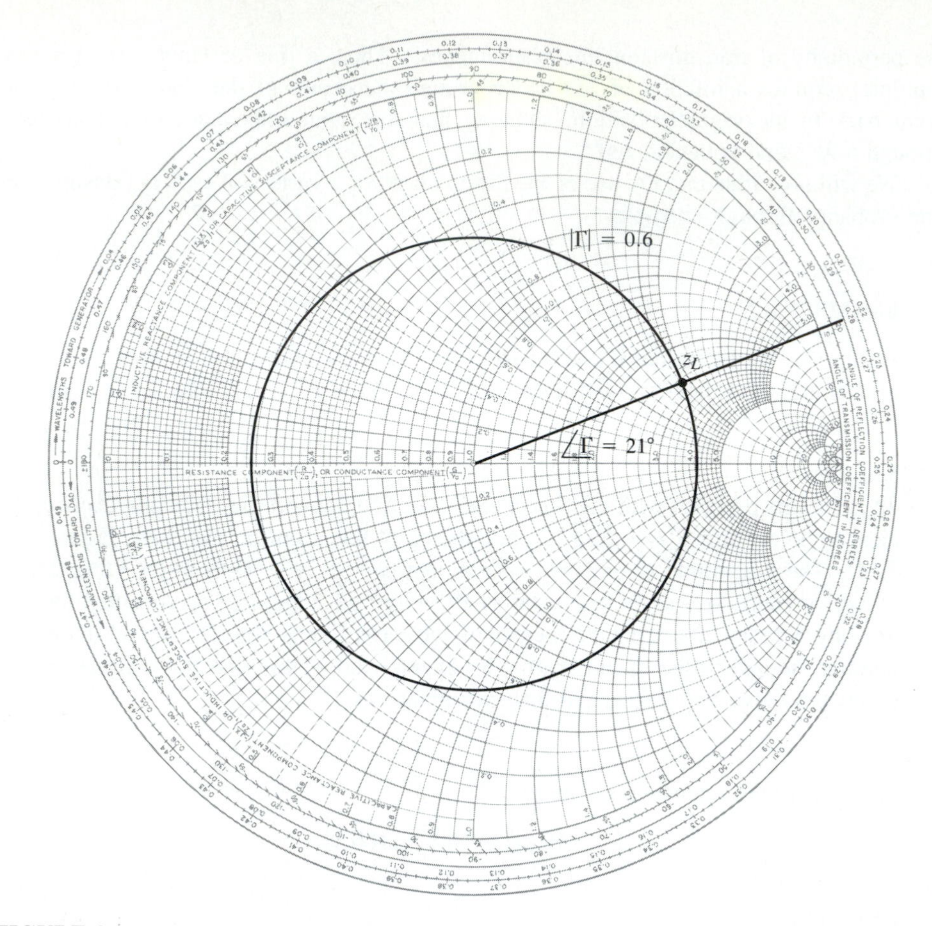

FIGURE 3.11 Smith chart for Example 3.2.

EXAMPLE 3.4

A coaxial transmission line with a characteristic impedance of $Z_0 = 75\ \Omega$ is of length $\ell = 2.0$ cm, and is terminated with a load impedance $Z_L = 37.5 + j75\ \Omega$. If the dielectric constant of the line is 2.56, and the frequency is 3.0 GHz, find the input impedance of this line, and the SWR on the line.

Solution

The normalized load impedance is $z_L = 0.5 + j1.0$, which can be plotted on the Smith chart (Figure 3.13). The SWR circle is then drawn through this point, and the SWR read as 4.3. At this point we know that the desired normalized input impedance lies somewhere along the SWR circle. Drawing a radial line through the load impedance point gives us a reference position for the load, on the wavelengths-toward-generator (WTG) scale, of 0.135λ. We must now move toward the generator (clockwise) an electrical distance equal to the length of the line.

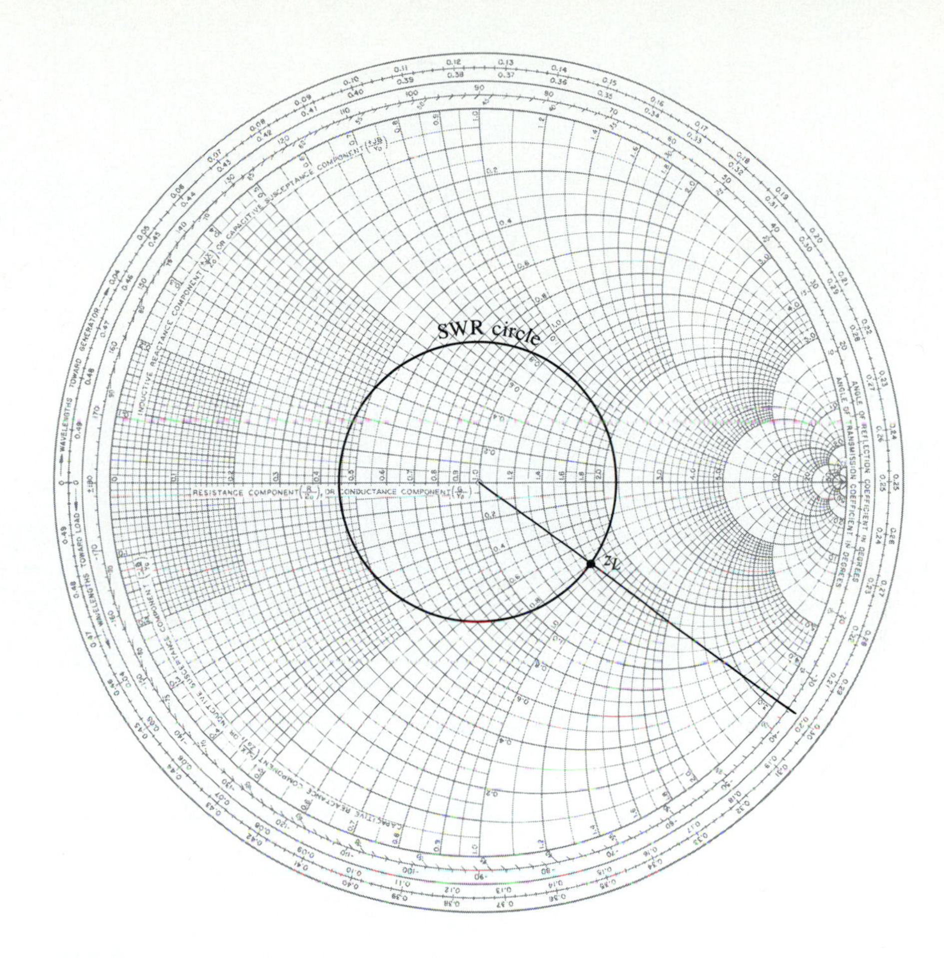

FIGURE 3.12 Smith chart for Example 3.3.

The wavelength on the coaxial line is

$$\lambda = \frac{v_p}{f} = \frac{3 \times 10^8}{3 \times 10^9 \sqrt{2.56}} = 6.25 \text{ cm}.$$

The electrical length of the line is then

$$\ell = \frac{2.0}{6.25} = 0.32\lambda.$$

Adding this amount to the starting position of 0.135λ gives 0.455λ. A radial line through this position on the WTG scale intersects the SWR circle at $z_{\text{in}} = 0.25 - j0.28$. The input impedance is then

$$Z_{\text{in}} = Z_0 z_{\text{in}} = 18.75 - j21.0 \ \Omega.$$

○

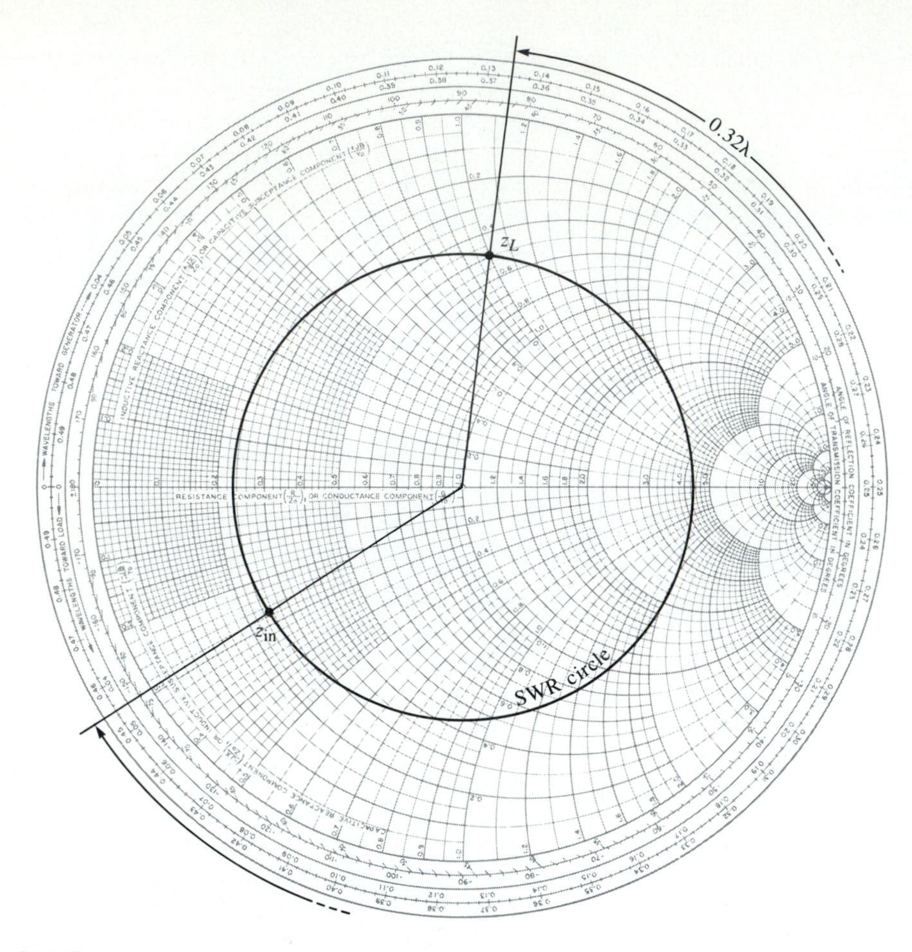

FIGURE 3.13 Smith chart for Example 3.4.

The Combined Impedance-Admittance Smith Chart

The Smith chart can be used for normalized admittance in the same way that it is used for normalized impedances, and it can be used to convert between impedance and admittance. The latter technique is based on the fact that, in normalized form, the input impedance of a load z_L connected to a $\lambda/4$ line is, from (3.44),

$$z_{\text{in}} = 1/z_L,$$

which has the effect of converting a normalized impedance to a normalized admittance.

Since a complete revolution around the Smith chart corresponds to a length of $\lambda/2$, a $\lambda/4$ transformation is equivalent to rotating the chart by 180°; this is also equivalent to imaging a given impedance (or admittance) point across the center of the chart to obtain the corresponding admittance (or impedance) point.

Thus, the same Smith chart can be used for both impedance and admittance calculations during the solution of a given problem. At different stages of the solution, then, the chart may be either an *impedance Smith chart* or an *admittance Smith chart*. This procedure can be made less confusing by using a Smith chart that has a superposition of the scales for a regular Smith chart and the scales of a Smith chart which has been rotated 180°, as shown in Figure 3.14. Such a chart is referred to as an *impedance and admittance Smith chart*, and usually has different colored scales for impedance and admittance.

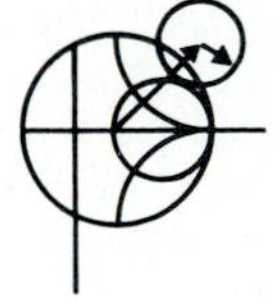

EXAMPLE 3.5

A load of $Z_L = 100 + j50\ \Omega$ terminates a $50\ \Omega$ line. What is the load admittance and the input admittance if the line is 0.15λ long?

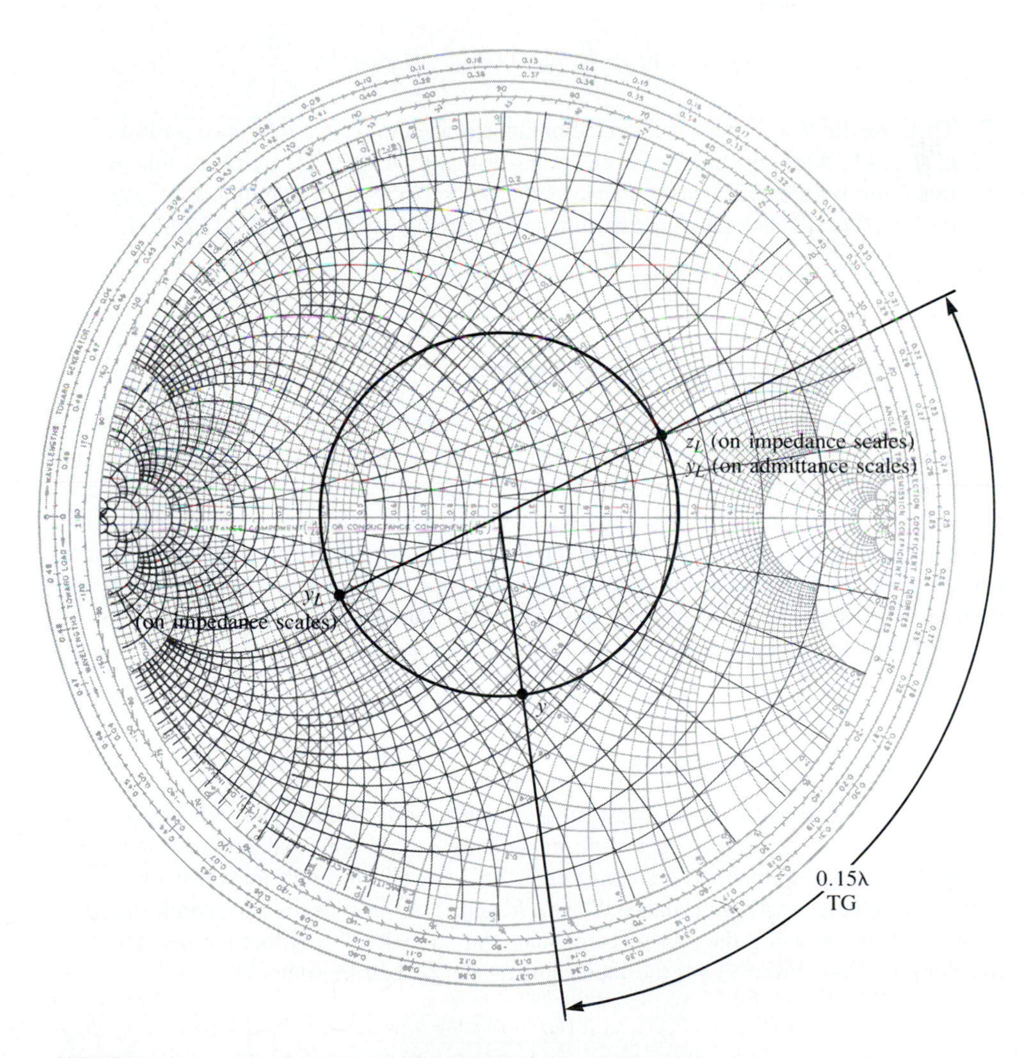

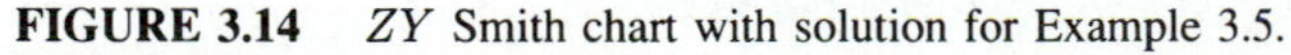

FIGURE 3.14 *ZY* Smith chart with solution for Example 3.5.

Solution

The normalized load impedance is $z_L = 2 + j1$. A standard Smith chart can be used for this problem by initially considering it as an impedance chart and plotting z_L and the SWR circle. Conversion to admittance can be accomplished with a $\lambda/4$ rotation of z_L (easily obtained by drawing a straight line through z_L and the center of the chart to intersect the SWR circle). The chart can now be considered as an admittance chart, and the input admittance can be found by rotating 0.15λ from y_L.

Alternatively, we can use the combined zy chart of Figure 3.14, where conversion between impedance and admittance is accomplished merely by reading the appropriate scales. Plotting z_L on the impedances scales and reading the admittance scales at this same point gives $y_L = 0.40 - j0.20$. The actual load admittance is then

$$Y_L = y_L Y_0 = \frac{y_L}{Z_0} = 0.0080 - j0.0040 \text{ S}.$$

Then, on the WTG scale, the load admittance is seen to have a reference position of 0.214λ. Moving 0.15λ past this point brings us to 0.364λ. A radial line at this point on the WTG scale intersects the SWR circle at an admittance of $y = 0.61 + j0.66$. The actual input admittance is then $Y = 0.0122 + j0.0132$ S. ○

3.5 THE QUARTER-WAVE TRANSFORMER

The quarter-wave transformer is a useful and practical circuit for impedance matching, and also provides a simple transmission line circuit that further illustrates the properties of standing waves on a mismatched line. Although we will study the design and performance of quarter-wave matching transformers more extensively in Chapter 5, the main purpose here is the application of the previously developed transmission line theory to a basic transmission line circuit. We will first approach the problem from the impedance viewpoint, and then show how this result can also be interpreted in terms of an infinite set of multiple reflections on the matching section.

The Impedance Viewpoint

Figure 3.15 shows a circuit employing a quarter-wave transformer. The load resistance R_L, and the feedline characteristic impedance Z_0, are both real and assumed to be given. These two components are connected with a lossless piece of transmission line of (unknown) characteristic impedance Z_1 and length $\lambda/4$. It is desired to match the load to the Z_0 line, by using the $\lambda/4$ piece of line, and so make $\Gamma = 0$ looking into the $\lambda/4$ matching section. From (3.44) the input impedance Z_{in} can be found as

$$Z_{in} = Z_1 \frac{R_L + jZ_1 \tan \beta\ell}{Z_1 + jR_L \tan \beta\ell}. \qquad 3.58$$

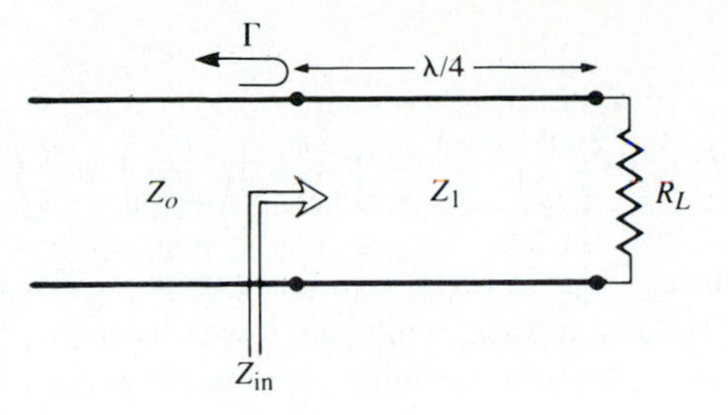

FIGURE 3.15 The quarter-wave matching transformer.

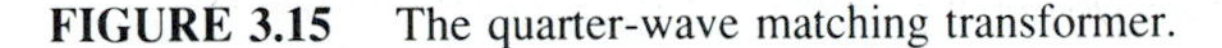

To evaluate this for $\beta\ell = (2\pi/\lambda)(\lambda/4) = \pi/2$, we can divide the numerator and denominator by $\tan\beta\ell$ and take the limit as $\beta\ell \rightarrow \pi/2$ to get

$$Z_{in} = \frac{Z_1^2}{R_L}. \tag{3.59}$$

In order for $\Gamma = 0$, we must have $Z_{in} = Z_0$, which yields the characteristic impedance Z_1 as

$$Z_1 = \sqrt{Z_0 R_L}, \tag{3.60}$$

the geometric mean of the load and source impedances. Then there will be no standing waves on the feedline (SWR = 1), although there will be standing waves on the $\lambda/4$ matching section. Also, the above condition applies only when the length of the matching section is $\lambda/4$, or an odd multiple ($(2n+1)\lambda/4$) of $\lambda/4$ long, so that a perfect match may be achieved at one frequency, but mismatch will occur at other frequencies.

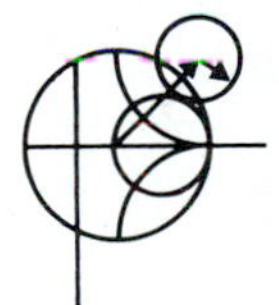

EXAMPLE 3.6

Consider a load resistance $R_L = 100\ \Omega$, to be matched to a 50 Ω line with a quarter-wave transformer. Find the characteristic impedance of the matching section and plot the magnitude of the reflection coefficient versus normalized frequency, f/f_o, where f_o is the frequency at which the line is $\lambda/4$ long.

Solution
From (3.60), the necessary characteristic impedance is

$$Z_1 = \sqrt{(50)(100)} = 70.71\ \Omega.$$

The reflection coefficient magnitude is given as

$$|\Gamma| = \left|\frac{Z_{in} - Z_0}{Z_{in} + Z_0}\right|,$$

where the input impedance Z_{in} is a function of frequency as given by (3.44). The frequency dependence in (3.44) comes from the $\beta\ell$ term, which can be written in terms of f/f_o as

$$\beta\ell = \left(\frac{2\pi}{\lambda}\right)\left(\frac{\lambda_0}{4}\right) = \left(\frac{2\pi f}{v_p}\right)\left(\frac{v_p}{4f_o}\right) = \frac{\pi f}{2f_o},$$

where it is seen that $\beta\ell = \pi/2$ for $f = f_o$, as expected. For higher frequencies the line looks electrically longer, while for lower frequencies it looks shorter. The magnitude of the reflection coefficient is plotted versus f/f_o in Figure 3.16.

○

This method of impedance matching is limited to real load impedances, although a complex load impedance can easily be made real, at a single frequency, by transformation through an appropriate length of line.

The above analysis shows how useful the impedance concept can be when solving transmission line problems, and this method is probably the preferred method in practice. It may aid our understanding of the quarter-wave transformer (and other transmission line circuits), however, if we now look at it from the viewpoint of multiple reflections.

The Multiple Reflection Viewpoint

Figure 3.17 shows the quarter-wave transformer circuit with reflection and transmission coefficients defined as follows:

Γ = overall, or total, reflection coefficient of a wave incident on the $\lambda/4$-transformer (same as Γ in Example 3.6).

Γ_1 = partial reflection coefficient of a wave incident on a load Z_1, from the Z_0 line.

Γ_2 = partial reflection coefficient of a wave incident on a load Z_0, from the Z_1 line.

Γ_3 = partial reflection coefficient of a wave incident on a load R_L, from the Z_1 line.

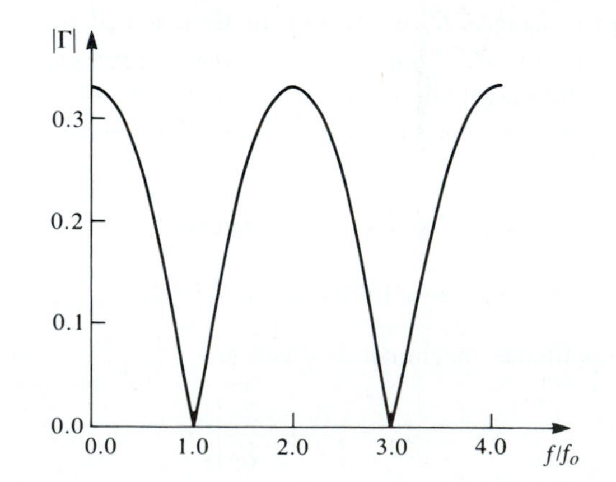

FIGURE 3.16 Reflection coefficient versus normalized frequency for the quarter-wave transformer of Example 3.6.

T_1 = partial transmission coefficient of a wave from the Z_0 line into the Z_1 line.

T_2 = partial transmission coefficient of a wave from the Z_1 line into the Z_0 line.

These coefficients can then be expressed as

$$\Gamma_1 = \frac{Z_1 - Z_0}{Z_1 + Z_0}, \tag{3.61a}$$

$$\Gamma_2 = \frac{Z_0 - Z_1}{Z_0 + Z_1} = -\Gamma_1, \tag{3.61b}$$

$$\Gamma_3 = \frac{R_L - Z_1}{R_L + Z_1}, \tag{3.61c}$$

$$T_1 = \frac{2Z_1}{Z_1 + Z_0}, \tag{3.61d}$$

$$T_2 = \frac{2Z_0}{Z_1 + Z_0}. \tag{3.61e}$$

Now think of the quarter-wave transformer of Figure 3.17 in the time domain, and imagine a wave traveling down the Z_0 feed line toward the transformer. When the wave first hits the junction with the Z_1 line, it sees only an impedance Z_1 since it has not yet traveled to the load R_L and can't see that effect. Part of the wave is reflected with a coefficient Γ_1, and part is transmitted onto the Z_1 line with a coefficient T_1. The

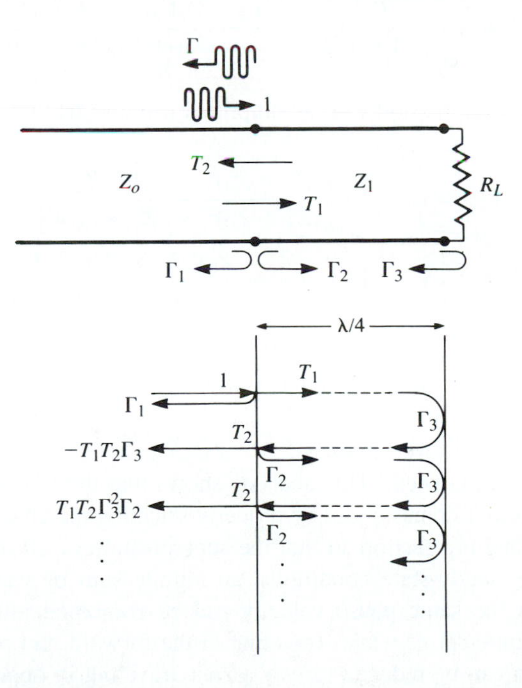

FIGURE 3.17 Multiple reflection analysis of the quarter-wave transformer.

transmitted wave then travels $\lambda/4$ to the load, gets reflected with a coefficient Γ_3, and travels another $\lambda/4$ back to the junction with the Z_0 line. Part of this wave is transmitted through (to the left) to the Z_0 line, with coefficient T_2, and part is reflected back toward the load with coefficient Γ_2. Clearly, this process continues with an infinite number of bouncing waves, and the total reflection coefficient, Γ, is the sum of all of these partial reflections. Since each round trip path up and down the $\lambda/4$ transformer section results in a 180° phase shift (90° up and 90° down), the total reflection coefficient can be expressed as

$$\begin{aligned}\Gamma &= \Gamma_1 - T_1T_2\Gamma_3 + T_1T_2\Gamma_2\Gamma_3^2 - T_1T_2\Gamma_2^2\Gamma_3^3 + \cdots \\ &= \Gamma_1 - T_1T_2\Gamma_3 \sum_{n=0}^{\infty} (-\Gamma_2\Gamma_3)^n .\end{aligned} \tag{3.62}$$

Since $|\Gamma_3| < 1$ and $|\Gamma_2| < 1$, the infinite series in (3.62) can be summed using the geometric series result that

$$\sum_{n=0}^{\infty} x^n = \frac{1}{1-x}, \qquad \text{for } |x| < 1,$$

to give

$$\Gamma = \Gamma_1 - \frac{T_1T_2\Gamma_3}{1+\Gamma_2\Gamma_3} = \frac{\Gamma_1 + \Gamma_1\Gamma_2\Gamma_3 - T_1T_2\Gamma_3}{1+\Gamma_2\Gamma_3}. \tag{3.63}$$

The numerator of this expression can be simplified using (3.61) to give

$$\begin{aligned}\Gamma_1 - \Gamma_3\left(\Gamma_1^2 + T_1T_2\right) &= \Gamma_1 - \Gamma_3\left[\frac{(Z_1-Z_0)^2}{(Z_1+Z_0)^2} + \frac{4Z_1Z_0}{(Z_1+Z_0)^2}\right] \\ &= \Gamma_1 - \Gamma_3 = \frac{(Z_1-Z_0)(R_L+Z_1) - (R_L-Z_1)(Z_1+Z_0)}{(Z_1+Z_0)(R_L+Z_1)} \\ &= \frac{2(Z_1^2 - Z_0R_L)}{(Z_1+Z_0)(R_L+Z_1)},\end{aligned}$$

which is seen to vanish if we choose $Z_1 = \sqrt{Z_0R_L}$, as in (3.60). Then Γ of (3.63) is zero, and the line is matched. This analysis shows that the matching property of the quarter-wave transformer comes about by properly selecting the characteristic impedance and length of the matching section so that the superposition of all the partial reflections add to zero. Under steady-state conditions, an infinite sum of waves traveling in the same direction with the same phase velocity can be combined into a single traveling wave. Thus, the infinite set of waves traveling in the forward and reverse directions on the matching section can be reduced to two waves, traveling in opposite directions. See Problem 3.24.

3.6 GENERATOR AND LOAD MISMATCHES

In Section 3.3 we treated the terminated (mismatched) transmission line assuming that the generator was matched, so that no reflections occurred at the generator. In general, however, both generator and load may present mismatched impedances to the transmission line. We will study this case, and also see that the condition for maximum power transfer from the generator to the load may, in some situations, require a standing wave on the line.

Figure 3.18 shows a transmission line circuit with arbitrary generator and load impedances, Z_g and Z_ℓ, which may be complex. The transmission line is assumed to be lossless, with a length ℓ and characteristic impedance Z_0. This circuit is general enough to model most passive and active networks that occur in practice.

Because both the generator and load are mismatched, multiple reflections can occur on the line, as in the problem of the quarter-wave transformer. The present circuit could thus be analyzed using an infinite series to represent the multiple bounces, as in Section 3.5, but we will use the easier and more useful method of impedance transformation. The input impedance looking into the terminated transmission line from the generator end is, from (3.43) and (3.44),

$$Z_{\text{in}} = Z_0 \frac{1 + \Gamma_\ell e^{-2j\beta\ell}}{1 - \Gamma_\ell e^{-2j\beta\ell}} = Z_0 \frac{Z_\ell + jZ_0 \tan\beta\ell}{Z_0 + jZ_\ell \tan\beta\ell}, \tag{3.64}$$

where Γ_ℓ is the reflection coefficient of the load:

$$\Gamma_\ell = \frac{Z_\ell - Z_0}{Z_\ell + Z_0}. \tag{3.65}$$

The voltage on the line can be written as

$$V(z) = V_o^+ \left(e^{-j\beta z} + \Gamma_\ell e^{j\beta z}\right), \tag{3.66}$$

and we can find V_o^+ from the voltage at the generator end of the line, where $z = -\ell$:

$$V(-\ell) = V_g \frac{Z_{\text{in}}}{Z_{\text{in}} + Z_g} = V_o^+ \left(e^{j\beta\ell} + \Gamma_\ell e^{-j\beta\ell}\right),$$

so that

$$V_o^+ = V_g \frac{Z_{\text{in}}}{Z_{\text{in}} + Z_g} \frac{1}{\left(e^{j\beta\ell} + \Gamma_\ell e^{-j\beta\ell}\right)}. \tag{3.67}$$

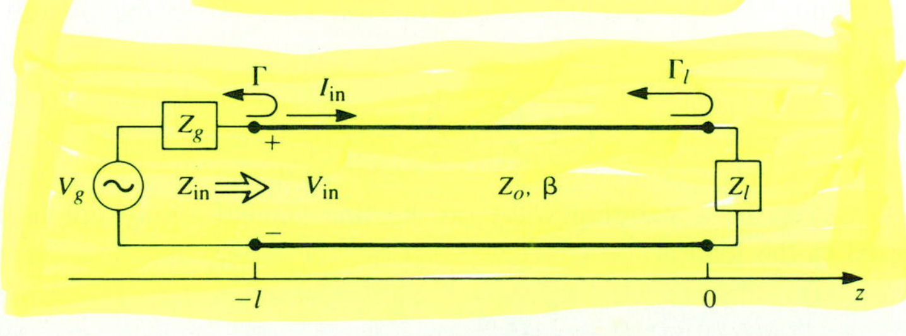

FIGURE 3.18 Transmission line circuit for mismatched load and generator.

This can be rewritten, using (3.64), as

$$V_o^+ = V_g \frac{Z_0}{Z_0 + Z_g} \frac{e^{-j\beta\ell}}{(1 - \Gamma_\ell \Gamma_g e^{-2j\beta\ell})}, \tag{3.68}$$

where Γ_g is the reflection coefficient seen looking into the generator:

$$\Gamma_g = \frac{Z_g - Z_0}{Z_g + Z_0}. \tag{3.69}$$

The standing wave ratio on the line is then

$$\text{SWR} = \frac{1 + |\Gamma_\ell|}{1 - |\Gamma_\ell|}. \tag{3.70}$$

The power delivered to the load is

$$P = \frac{1}{2}\text{Re}\{V_{\text{in}} I_{\text{in}}^*\} = \frac{1}{2}|V_{\text{in}}|^2 \text{Re}\left\{\frac{1}{Z_{\text{in}}}\right\} = \frac{1}{2}|V_g|^2 \left(\frac{Z_{\text{in}}}{Z_{\text{in}} + Z_g}\right)^2 \text{Re}\left\{\frac{1}{Z_{\text{in}}}\right\}. \tag{3.71}$$

Now let $Z_{\text{in}} = R_{\text{in}} + jX_{\text{in}}$ and $Z_g = R_g + jX_g$; then (3.71) can be reduced to

$$P = \frac{1}{2}|V_g|^2 \frac{R_{\text{in}}}{(R_{\text{in}} + R_g)^2 + (X_{\text{in}} + X_g)^2}. \tag{3.72}$$

We now assume that the generator impedance, Z_g, is fixed, and consider three cases of load impedance.

Load Matched to Line ($Z_\ell = Z_0$)

In this case we have $\Gamma_\ell = 0$, and SWR $= 1$, from (3.65) and (3.70), since the line is terminated in a matched load. Then the input impedance is $Z_{\text{in}} = Z_0$, and the power delivered to the load is, from (3.72),

$$P = \frac{1}{2}|V_g|^2 \frac{Z_0}{(Z_0 + R_g)^2 + X_g^2}. \tag{3.73}$$

Generator Matched to Loaded Line ($Z_{\text{in}} = Z_g$)

In this case the load impedance Z_ℓ and/or the transmission line parameters $\beta\ell$, Z_0 are chosen to make the input impedance $Z_{\text{in}} = Z_g$, so that the generator is matched to the load presented by the terminated transmission line. Then the overall reflection coefficient, Γ, is zero:

$$\Gamma = \frac{Z_{\text{in}} - Z_g}{Z_{\text{in}} + Z_g} = 0. \tag{3.74}$$

There may, however, be a standing wave on the line since Γ_ℓ may not be zero. The power delivered to the load is

$$P = \frac{1}{2}|V_g|^2 \frac{R_g}{4(R_g^2 + X_g^2)}. \tag{3.75}$$

Now observe that even though the loaded line is matched to the generator, the power delivered to the load may be less than the power delivered to the load from (3.73), where the loaded line was not necessarily matched to the generator. Thus, we are led to the question of what is the optimum load impedance, or equivalently, what is the optimum input impedance, to achieve maximum power transfer to the load for a given generator impedance.

Input Impedance for Maximum Power Transfer (Conjugate Matching)

Assuming that the generator series impedance, Z_g, is fixed, we may vary the input impedance Z_{in} until we achieve the maximum power delivered to the load. Knowing Z_{in}, it is then easy to find the corresponding load impedance Z_ℓ via an impedance transformation along the line. To maximize P, we differentiate with respect to the real and imaginary parts of Z_{in}. Using (3.72) gives

$$\frac{\partial P}{\partial R_{\text{in}}} = 0 \rightarrow \frac{1}{(R_{\text{in}} + R_g)^2 + (X_{\text{in}} + X_g)^2} + \frac{-2R_{\text{in}}(R_{\text{in}} + R_g)}{\left[(R_{\text{in}} + R_g)^2 + (X_{\text{in}} + X_g)^2\right]^2} = 0,$$

or,

$$R_g^2 - R_{\text{in}}^2 + (X_{\text{in}} + X_g)^2 = 0, \tag{3.76a}$$

$$\frac{\partial P}{\partial X_{\text{in}}} = 0 \rightarrow \frac{-2X_{\text{in}}(X_{\text{in}} + X_g)}{\left[(R_{\text{in}} + R_g)^2 + (X_{\text{in}} + X_g)^2\right]^2} = 0,$$

or,

$$X_{\text{in}}(X_{\text{in}} + X_g) = 0. \tag{3.76b}$$

Solving (3.76a,b) simultaneously for R_{in} and X_{in} gives

$$R_{\text{in}} = R_g, \qquad X_{\text{in}} = -X_g,$$

or

$$Z_{\text{in}} = Z_g^* . \tag{3.77}$$

This condition is known as conjugate matching, and results in maximum power transfer to the load, for a fixed generator impedance. The power delivered is, from (3.72) and (3.77),

$$P = \frac{1}{2}|V_g|^2 \frac{1}{4R_g}, \tag{3.78}$$

which is seen to be greater than or equal to the powers of (3.73) or (3.75). Also note that the reflection coefficients Γ_ℓ, Γ_g, and Γ may be nonzero. Physically, this means that in some cases the power in the multiple reflections on a mismatched line may add in phase to deliver more power to the load than would be delivered if the line were flat (no reflections). If the generator impedance is real ($X_g = 0$), then the last two cases reduce to the same result, which is that maximum power is delivered to the load when the loaded line is matched to the generator ($R_{\text{in}} = R_g$, with $X_{\text{in}} = X_g = 0$).

Finally, note that neither matching for zero reflection ($Z_\ell = Z_0$) or conjugate matching ($Z_{\text{in}} = Z_g^*$) necessarily yields a system with the best efficiency. For example, if $Z_g = Z_\ell = Z_0$ then both load and generator are matched (no reflections), but only half the power produced by the generator is delivered to the load (half is lost in Z_g), for a transmission efficiency of 50%. This efficiency can only be improved by making Z_g as small as possible.

3.7 THE SLOTTED LINE

A slotted line is a transmission line configuration (usually waveguide or coax) that allows the sampling of the electric field amplitude of a standing wave on a terminated line. With this device the SWR and the distance of the first voltage minimum from the load can be measured, and from this data the load impedance can be determined. Note that because the load impedance is in general a complex number (with two degrees of freedom), two distinct quantities must be measured with the slotted line to uniquely determine this impedance. A typical waveguide slotted line is shown in Figure 3.19.

Although the slotted line used to be the principal way of measuring an unknown impedance at microwave frequencies, it has been largely superseded by the modern vector network analyzer in terms of accuracy, versatility, and convenience. The slotted line is still of some use, however, in certain applications such as high-millimeter wave frequencies, or where it is desired to avoid connector mismatches by connecting the unknown load directly to the slotted line, thus avoiding the use of imperfect transitions. Another reason for studying the slotted line is that it provides an unexcelled tool for learning basic concepts of standing waves and mismatched transmission lines. We will derive expressions for finding the unknown load impedance from slotted line measurements, and also show how the Smith chart can be used for the same purpose.

FIGURE 3.19 An X-band waveguide slotted line.
Courtesy of Hewlett-Packard Company, Santa Rosa, Calif.

Assume that, for a certain terminated line, we have measured the SWR on the line and ℓ_{min}, the distance from the load to the first voltage minimum on the line. The load impedance Z_L can then be determined as follows. From (3.41) the magnitude of the reflection coefficient on the line is found from the standing wave ratio as

$$|\Gamma| = \frac{\text{SWR} - 1}{\text{SWR} + 1}. \tag{3.79}$$

From Section 3.3, we know that a voltage minimum occurs when $e^{j(\theta - 2\beta\ell)} = -1$, where θ is the phase angle of the reflection coefficient, $\Gamma = |\Gamma|e^{j\theta}$. The phase of the reflection coefficient is then

$$\theta = \pi + 2\beta\ell_{min}, \tag{3.80}$$

where ℓ_{min} is the distance from the load to the first voltage minimum. Actually, since the voltage minimums repeat every $\lambda/2$, where λ is the wavelength on the line, any multiple of $\lambda/2$ can be added to ℓ_{min} without changing the result in (3.80), because this just amounts to adding $2\beta n\lambda/2 = 2\pi n$ to θ, which will not change Γ. Thus, the two quantities SWR and ℓ_{min} can be used to find the complex reflection coefficient Γ at the load. It is then straightforward to use (3.43) with $\ell = 0$ to find the load impedance from Γ:

$$Z_L = Z_0 \frac{1 + \Gamma}{1 - \Gamma}. \tag{3.81}$$

The use of the Smith chart in solving this problem is best illustrated by an example.

EXAMPLE 3.7

The following two-step procedure has been carried out with a 50 Ω slotted line to determine an unknown load impedance:

1. A short circuit is placed at the load plane, resulting in a standing wave on the line with infinite SWR, and sharply defined voltage minima, as shown in Figure 3.20a. On the arbitrarily positioned scale on the slotted line, voltage minima are recorded at

$$z = 0.2 \text{ cm}, 2.2 \text{ cm}, 4.2 \text{ cm}.$$

2. The short circuit is removed, and replaced with the unknown load. The standing wave ratio is measured as SWR = 1.5, and voltage minima, which are not as sharply defined as those in step 1, are recorded at

$$z = 0.72 \text{ cm}, 2.72 \text{ cm}, 4.72 \text{ cm},$$

as shown in Figure 3.20b. Find the load impedance.

Solution

Knowing that voltage minima repeat every $\lambda/2$, we have from the data of step 1 above that $\lambda = 4.0$ cm. In addition, because the reflection coefficient and

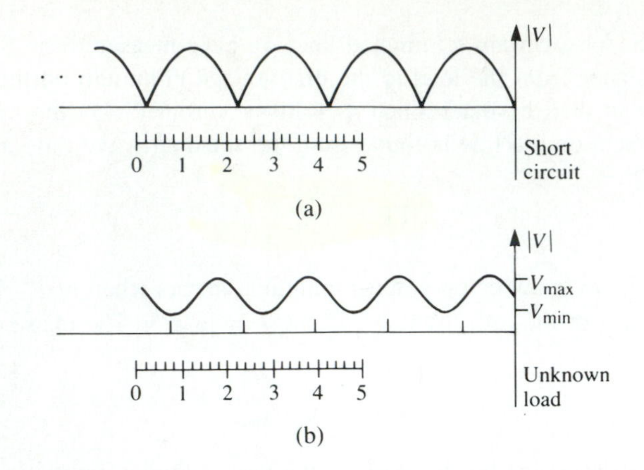

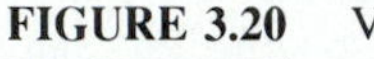

FIGURE 3.20 Voltage standing wave patterns for Example 3.7. (a) Standing wave for short-circuit load. (b) Standing wave for unknown load.

input impedance also repeat every $\lambda/2$, we can consider the load terminals to be effectively located at any of the voltage minima locations listed in step 1. Thus, if we say the load is at 4.2 cm, then the data from step 2 shows that the next voltage minimum away from the load occurs at 2.72 cm, giving $\ell_{\text{min}} = 4.2 - 2.72 = 1.48 \text{ cm} = 0.37\lambda$.

Applying (3.79)–(3.81) to this data gives

$$|\Gamma| = \frac{1.5 - 1}{1.5 + 1} = 0.2,$$

$$\theta = \pi + \frac{4\pi}{4.0}(1.48) = 86.4^\circ,$$

so

$$\Gamma = 0.2e^{j86.4^\circ} = 0.0126 + j0.1996.$$

The load impedance is then

$$Z_L = 50\left(\frac{1+\Gamma}{1-\Gamma}\right) = 47.3 + j19.7 \ \Omega.$$

For the Smith chart version of the solution, we begin by drawing the SWR circle for SWR = 1.5, as shown in Figure 3.21; the unknown normalized load impedance must lie on this circle. The reference that we have is that the load is 0.37λ away from the first voltage minimum. On the Smith chart, the position of a voltage minimum corresponds to the minimum impedance point (minimum voltage, maximum current), which is the horizontal axis (zero reactance) to the left of the origin. Thus, we begin at the voltage minimum point and move 0.37λ toward the load (counterclockwise), to the normalized load impedance point, $z_L = 0.95 + j0.4$, as shown in Figure 3.21. The actual load impedance is then $Z_L = 47.5 + j20 \ \Omega$, in close agreement with the above result using the equations.

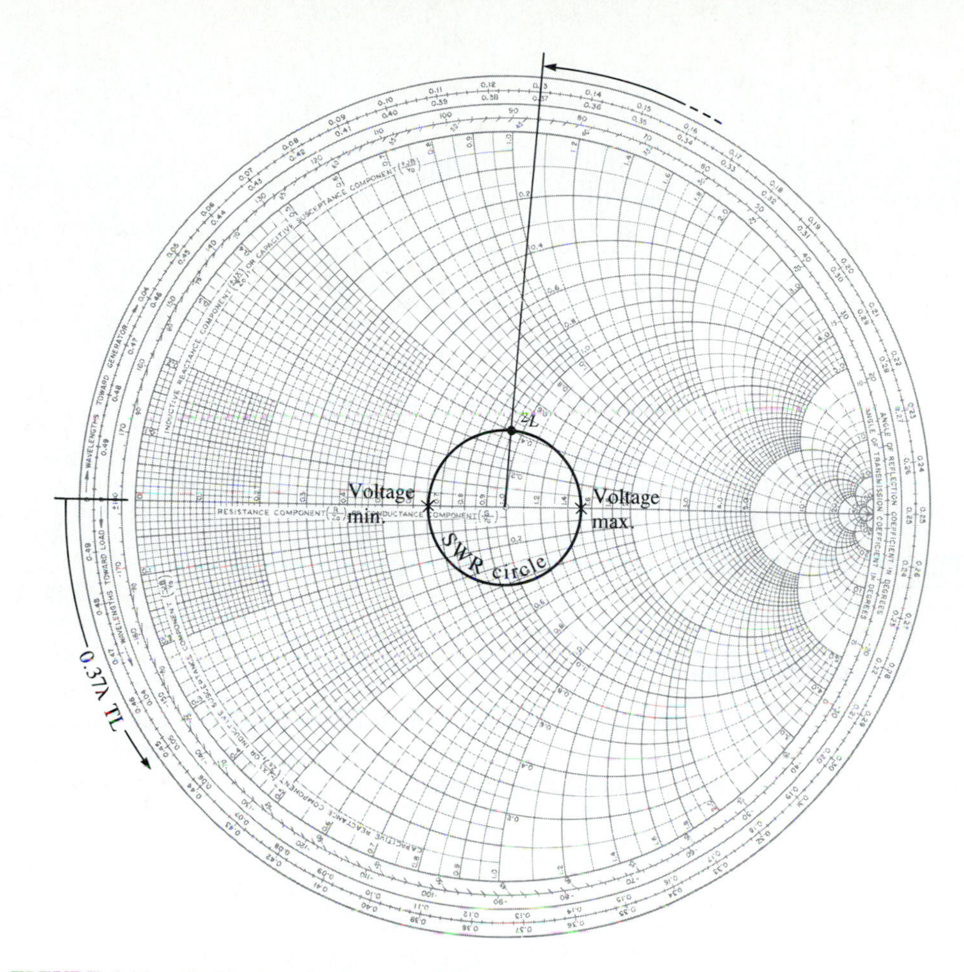

FIGURE 3.21 Smith chart for Example 3.7.

Note that, in principle, voltage maxima locations could be used as well as voltage minima positions, but that voltage minima are more sharply defined than voltage maxima, and so usually result in greater accuracy. ○

3.8 THE TRANSMISSION LINE ANALOGY FOR PLANE WAVE REFLECTION

By now the reader has observed the similarity between the equations and phenomenon of plane waves, as discussed in Chapter 2, and the corresponding features of transmission lines presented in this chapter. This similarity is due to the fact that both are cases of basic wave phenomenon as derived from Maxwell's equations, and so concepts such as reflection, impedance, standing waves, and the Smith chart developed for one system can be equally applied to the other. In this section we will formalize this equivalence by developing the transmission line analogy for plane wave propagation and reflection, for both normal and oblique incidence angles.

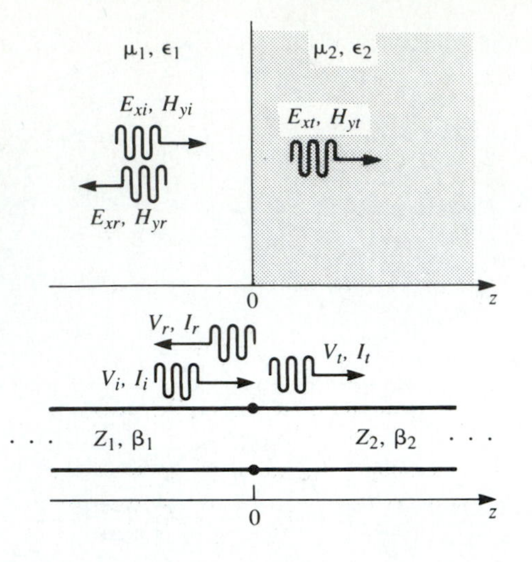

FIGURE 3.22 Transmission line analogy for plane wave reflection.

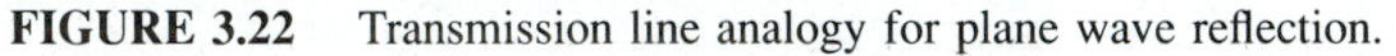

Normal Incidence

We can begin by considering a plane wave in a lossless medium, and a wave on a lossless transmission line, as shown in Figure 3.22. From Chapter 2, the plane wave problem is described by the following equations:

$$\begin{aligned} E_x(z) &= E_{xi}e^{-j\beta_1 z} + E_{xr}e^{j\beta_1 z} && \text{for } z < 0 \\ &= E_{xt}e^{-j\beta_2 z} && \text{for } z > 0, \end{aligned} \tag{3.82a}$$

$$\begin{aligned} H_y(z) &= H_{yi}e^{-j\beta_1 z} + H_{yr}e^{j\beta_1 z} && \text{for } z < 0 \\ &= H_{yt}e^{-j\beta_2 z} && \text{for } z < 0. \end{aligned} \tag{3.82b}$$

The propagation constants are

$$\beta_1 = \omega\sqrt{\mu_1\epsilon_1}, \tag{3.82c}$$

$$\beta_2 = \omega\sqrt{\mu_2\epsilon_2}, \tag{3.82d}$$

and the wave impedances are

$$Z_1 = \sqrt{\mu_1/\epsilon_1} = \eta_1, \tag{3.82e}$$

$$Z_2 = \sqrt{\mu_2/\epsilon_2} = \eta_2. \tag{3.82f}$$

From Chapter 2, reflection and transmission coefficients may be defined as

$$\Gamma = \frac{E_{xr}}{E_{xi}} = \frac{Z_2 - Z_1}{Z_2 + Z_1}, \tag{3.82g}$$

$$T = \frac{E_{xt}}{E_{xi}} = \frac{2Z_2}{Z_2 + Z_1}. \tag{3.82h}$$

The corresponding relations for the voltage and current waves on the transmission line of Figure 3.22 are as follows:

$$V(z) = V_i e^{-j\beta_1 z} + V_r e^{j\beta_1 z} \quad \text{for } z < 0$$
$$= V_t e^{-j\beta_2 z} \quad \text{for } z > 0, \qquad 3.83a$$

$$I(z) = I_i e^{-j\beta_1 z} - I_r e^{j\beta_1 z} \quad \text{for } z < 0$$
$$= I_t e^{-j\beta_2 z} \quad \text{for } z > 0. \qquad 3.83b$$

The reflection and transmission coefficients are, from (3.50) and (3.52),

$$\Gamma = \frac{V_r}{V_i} = \frac{Z_2 - Z_1}{Z_2 + Z_1}, \qquad 3.83c$$

$$T = \frac{V_t}{V_i} = \frac{2Z_2}{Z_2 + Z_1}, \qquad 3.83d$$

where Z_1, Z_2 are the characteristic impedances of the transmission lines. Thus, if the characteristic impedances Z_1 and Z_2 of the transmission line circuit of Figure 3.22 are made equal to the wave impedances η_1 and η_2 of the plane wave problem of Figure 3.22, then the reflection and transmission coefficients as computed from (3.83c,d) and (3.82g,h) will be identical, so that the transmission line circuit can be considered as an equivalent circuit for the plane wave problem. Then all of the concepts and techniques, which we have developed for transmission lines, such as SWR and the Smith chart, can be applied to plane wave problems. A particularly useful concept in this regard is the generalized impedance defined when a standing wave is present. From (3.44), the impedance seen on the transmission line at any point $z = -\ell \leq 0$ is

$$Z_{\text{in}} = \frac{V(-\ell)}{I(-\ell)} = Z_1 \frac{Z_2 + jZ_1 \tan \beta_1 \ell}{Z_1 + jZ_2 \tan \beta_1 \ell}. \qquad 3.84$$

This same formula can now be used for plane wave reflection, and is of great utility when solving matching problems.

The transmission line analogy is thus seen to be a good example of how a problem in field analysis can be reduced to a distributed circuit problem.

Oblique Incidence

The above discussion for plane waves at normal incidence follows transmission line theory quite closely, but the analogy can also be made for plane waves at oblique incidence on an interface between two lossless dielectric mediums. The basic difference is that effective propagation constants and characteristic impedances, that depend on the angles of incidence, reflection, and transmission, must be defined for use in the transmission line analog.

Consider the plane wave reflection problem with oblique incidence, as previously shown in Figure 2.10 and discussed in Section 2.8. For either parallel or perpendicular polarization, the incident fields propagate according to the following phase factor:

$$e^{-j\beta_1(x \sin\theta_i + z \cos\theta_i)}, \tag{3.85}$$

where β_1 is the propagation constant along the direction of propagation, as defined in (3.82c), and θ_i is the angle of incidence measured from the normal of the dielectric interface. To satisfy boundary conditions on the $z = 0$ plane, it is useful to define phase, or propagation, constants tangential and normal to the interface. Thus, let

$$\beta_{x1} = \beta_1 \sin\theta_i, \tag{3.86a}$$

$$\beta_{z1} = \beta_1 \cos\theta_i. \tag{3.86b}$$

Similar definitions can be made for the reflected and transmitted waves. Thus,

$$\beta_{x2} = \beta_2 \sin\theta_t, \tag{3.87a}$$

$$\beta_{z2} = \beta_2 \cos\theta_t, \tag{3.87b}$$

are the corresponding definitions for the transmitted wave. Since, from Snell's laws, $\theta_i = \theta_r$, (3.86) also applies to the reflected wave.

To satisfy continuity of the tangential field components at the interface, the β_x phase constants for each of the wave components must be identical (the phase matching condition), leading to Snell's laws as in Section 2.8. The normal phase constants, β_{z1} and β_{z2} of (3.86b) and (3.87b), then correspond to β_1 and β_2 in the transmission line analogy.

Next, we must define a wave impedance as the ratio of the electric and magnetic field components tangential to the interface, as it is this ratio or impedance that is continuous across the interface. Thus, from (2.132a,b) and (2.134a,b) for parallel polarization, we define a transverse wave impedance as

$$Z_{1t} = \frac{E_{xi}}{H_{yi}} = \eta_1 \cos\theta_i, \tag{3.88a}$$

$$Z_{2t} = \frac{E_{xt}}{H_{yt}} = \eta_2 \cos\theta_t. \tag{3.88b}$$

For perpendicular polarization, we have, from (2.139a,b) and (2.141a,b),

$$Z_{1t} = \frac{-E_{yi}}{H_{xi}} = \eta_1 \sec\theta_i, \tag{3.89a}$$

$$Z_{2t} = \frac{-E_{yt}}{H_{xt}} = \eta_2 \sec\theta_t, \tag{3.89b}$$

where the negative signs are due to power flow in the $+z$ direction. In (3.88) and (3.89), η_1 and η_2 are the intrinsic impedances of the media as defined in (3.82). The transmission line analog can then be formed by using Z_{1t}, Z_{2t}, β_{z1}, and β_{z2} for Z_1, Z_2, β_1, and β_2, respectively, in (3.83) and (3.84).

3.9 LOSSY TRANSMISSION LINES

In practice, all transmission lines have loss due to finite conductivity and/or lossy dielectric, but these losses are usually small. In many practical problems, then, loss may be neglected, but at times the effect of loss may be of interest. Such is the case when dealing with the attenuation of a transmission line, or the Q of a resonant cavity, for example. In this section we will study the effects of loss on transmission line behavior, and show how the attenuation constant can be calculated.

The Low-Loss Line

In most practical microwave transmission lines the loss is small—if this were not the case, the line would be of little practical value. When the loss is small, some approximations can be made that simplify the expressions for the general transmission line parameters of $\gamma = \alpha + j\beta$ and Z_0.

The general expression for the complex propagation constant is, from (3.5),

$$\gamma = \sqrt{(R + j\omega L)(G + j\omega C)}, \tag{3.90}$$

which can be rearranged as

$$\begin{aligned} \gamma &= \sqrt{(j\omega L)(j\omega C)\left(1 + \frac{R}{j\omega L}\right)\left(1 + \frac{G}{j\omega C}\right)} \\ &= j\omega\sqrt{LC}\sqrt{1 - j\left(\frac{R}{\omega L} + \frac{G}{\omega C}\right) - \frac{RG}{\omega^2 LC}}. \end{aligned} \tag{3.91}$$

If the line is low-loss we can assume that $R << \omega L$ and $G << \omega C$, which means that both the conductor loss and dielectric loss are small. Then, $RG << \omega^2 LC$, and (3.91) reduces to

$$\gamma = j\omega\sqrt{LC}\sqrt{1 - j\left(\frac{R}{\omega L} + \frac{G}{\omega C}\right)}. \tag{3.92}$$

If we were to ignore the $(R/\omega L + G/\omega C)$ term, we would obtain the result that γ was purely imaginary (no loss), so we will instead use the first two terms of the Taylor series expansion for $\sqrt{1+x} \simeq 1 + x/2 + \cdots$, to give the first higher-order real term for γ:

$$\gamma \simeq j\omega\sqrt{LC}\left[1 - \frac{j}{2}\left(\frac{R}{\omega L} + \frac{G}{\omega C}\right)\right],$$

so that

$$\alpha \simeq \frac{1}{2}\left(R\sqrt{\frac{C}{L}} + G\sqrt{\frac{L}{C}}\right) = \frac{1}{2}\left(\frac{R}{Z_0} + GZ_0\right), \tag{3.93a}$$

$$\beta \simeq \omega\sqrt{LC}, \tag{3.93b}$$

where $Z_0 = \sqrt{L/C}$ is the characteristic impedance of the line in the absence of loss. Note from (3.93b) that the propagation constant β is the same as the lossless case of (3.12). By the same order of approximation, the characteristic impedance Z_0 can be approximated as a real quantity:

$$Z_0 = \sqrt{\frac{R + j\omega L}{G + j\omega C}} \simeq \sqrt{\frac{L}{C}}. \tag{3.94}$$

Equations (3.93)–(3.94) are known as the high-frequency, low-loss approximations for transmission lines, and are important because they show that the propagation constant and characteristic impedance for a low-loss line can be closely approximated by considering the line as lossless.

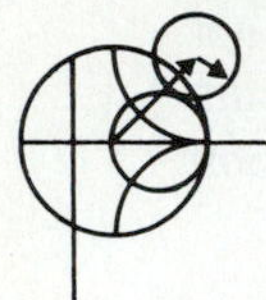

EXAMPLE 3.8

In Example 3.1 the L, C, R, and G parameters were derived for a lossy coaxial line. Assuming the loss is small, calculate the attenuation constant from (3.93a) and the results of Example 3.1.

Solution
From (3.93a),

$$\alpha = \frac{1}{2}\left(R\sqrt{\frac{C}{L}} + G\sqrt{\frac{L}{C}}\right).$$

Using the results derived in Example 3.1 gives

$$\alpha = \frac{1}{2}\left[\frac{R_s}{\eta \ln b/a}\left(\frac{1}{a} + \frac{1}{b}\right) + \omega\epsilon''\eta\right],$$

where $\eta = \sqrt{\mu/\epsilon'}$ is the intrinsic impedance of the dielectric material filling the coaxial line. Also, $\beta = \omega\sqrt{LC} = \omega\sqrt{\mu\epsilon'}$, and $Z_0 = \sqrt{L/C} = (\eta/2\pi)\ln b/a$. ○

The above method for the calculation of attenuation requires that the line parameters L, C, R, and G be known. These can often be derived using the formulas of (3.17)–(3.20), but a more direct and versatile procedure is to use the perturbation method, to be discussed shortly.

The Distortionless Line

As can be seen from the exact equations (3.90)–(3.91) for the propagation constant of a lossy line, the phase term β is generally a complicated function of frequency, ω, when loss is present. In particular, we note that β is generally not exactly a linear function of frequency, as in (3.93b), unless the line is lossless. If β is not a linear function of frequency (of the form $\beta = a\omega$), then the phase velocity $v_p = \omega/\beta$ will be different for different frequencies ω. The implication is that the various frequency components of a wideband signal will travel with different phase velocities, and so arrive at the receiver end of the transmission line at slightly different times. This will lead to *dispersion*, or distortion of the signal, and is generally an undesirable effect. Granted, as we have argued above, the departure of β from a linear function may be quite small, but the effect can be significant if the line is very long. This effect leads to the concept of group velocity, which we will address in detail in Section 4.10.

There is a special case, however, of a lossy line that has a linear phase factor as a function of frequency. Such a line is called a distortionless line, and is characterized by line parameters that satisfy the relation

$$\frac{R}{L} = \frac{G}{C}. \tag{3.95}$$

From (3.91) the exact complex propagation constant, under the condition specified by (3.95), reduces to

$$\begin{aligned}\gamma &= j\omega\sqrt{LC}\sqrt{1 - 2j\frac{R}{\omega L} - \frac{R^2}{\omega^2 L^2}} \\ &= j\omega\sqrt{LC}\left(1 - j\frac{R}{\omega L}\right) \\ &= R\sqrt{\frac{C}{L}} + j\omega\sqrt{LC} = \alpha + j\beta,\end{aligned} \tag{3.96}$$

which shows that $\beta = \omega\sqrt{LC}$ is a linear function of frequency. Equation (3.96) also shows that the attenuation constant, $\alpha = R\sqrt{C/L}$, is not a function of frequency, so that all frequency components will be attenuated by the same amount (actually, R is usually a weak function of frequency). Thus, the distortionless line is not loss-free, but is capable of passing a pulse or modulation envelope without distortion. To obtain a transmission line with parameters that satisfy (3.95) often requires that L be increased by adding series loading coils spaced periodically along the line.

The above theory for the distortionless line was first developed by Oliver Heaviside (1850–1925), a reclusive genius who, with no formal education, solved many problems in transmission line theory and worked Maxwell's original theory of electromagnetism into the modern and more usable version that we are familiar with today [5].

The Terminated Lossy Line

Figure 3.23 shows a length ℓ of a lossy transmission line terminated in a load impedance Z_L. Thus, $\gamma = \alpha + j\beta$ is complex, but we assume the loss is small so that Z_0 is approximately real, as in (3.94).

In (3.36), expressions for the voltage and current wave on a lossless line are given. The analogous expressions for the lossy case are

$$V(z) = V_o^+\left[e^{-\gamma z} + \Gamma e^{\gamma z}\right], \tag{3.97a}$$

$$I(z) = \frac{V_o^+}{Z_0}\left[e^{-\gamma z} - \Gamma e^{\gamma z}\right], \tag{3.97b}$$

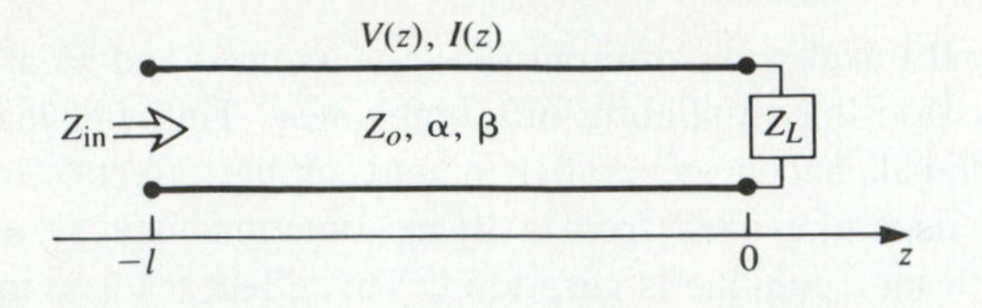

FIGURE 3.23 A lossy transmission line terminated in the impedance Z_L.

where Γ is the reflection coefficient of the load, as given in (3.35), and V_o^+ is the incident voltage amplitude referenced at $z = 0$. From (3.42), the reflection coefficient at a distance ℓ from the load is

$$\Gamma(\ell) = \Gamma e^{-2j\beta\ell} e^{-2\alpha\ell} = \Gamma e^{-2\gamma\ell}. \tag{3.98}$$

The input impedance Z_{in} at a distance ℓ from the load is then

$$Z_{\text{in}} = \frac{V(-\ell)}{I(-\ell)} = Z_0 \frac{Z_L + Z_0 \tanh \gamma\ell}{Z_0 + Z_L \tanh \gamma\ell}. \tag{3.99}$$

We can compute the power delivered to the input of the terminated line at $z = -\ell$ as

$$\begin{aligned} P_{\text{in}} &= \frac{1}{2}\text{Re}\left\{V(-\ell)I^*(-\ell)\right\} = \frac{|V_o^+|^2}{2Z_0}\left[e^{2\alpha\ell} - |\Gamma|^2 e^{-2\alpha\ell}\right] \\ &= \frac{|V_o^+|^2}{2Z_0}\left[1 - |\Gamma(\ell)|^2\right] e^{2\alpha\ell}, \end{aligned} \tag{3.100}$$

where (3.97) have been used for $V(-\ell)$ and $I(-\ell)$. The power actually delivered to the load is

$$P_L = \frac{1}{2}\text{Re}\left\{V(0)I^*(0)\right\} = \frac{|V_o^+|^2}{2Z_0}(1 - |\Gamma|^2). \tag{3.101}$$

The difference in these powers corresponds to the power lost in the line:

$$P_{\text{loss}} = P_{\text{in}} - P_L = \frac{|V_o^+|^2}{2Z_0}\left[\left(e^{2\alpha\ell} - 1\right) + |\Gamma|^2\left(1 - e^{-2\alpha\ell}\right)\right]. \tag{3.102}$$

The first term in (3.102) accounts for the power loss of the incident wave, while the second term accounts for the power loss of the reflected wave; note that both terms increase as α increases.

The Perturbation Method for Calculating Attenuation

Here we derive a useful and standard technique for finding the attenuation constant of a low-loss line. The method avoids the use of the transmission line parameters L, C, R, and G, and instead uses the fields of the lossless line, with the assumption that the fields of the lossy line are not greatly different from the fields of the lossless line—hence the term, *perturbation*.

We have seen that the power flow along a lossy transmission line, in the absence of reflections, is of the form

$$P(z) = P_o e^{-2\alpha z}, \tag{3.103}$$

where P_o is the power at the $z = 0$ plane, and α is the attenuation constant we wish to determine. Now define the power loss per unit length along the line as

$$P_\ell = \frac{-\partial P}{\partial z} = 2\alpha P_o e^{-2\alpha z} = 2\alpha P(z),$$

where the negative sign on the derivative was chosen so that P_ℓ would be a positive quantity. From this, the attenuation constant can be determined as

$$\alpha = \frac{P_\ell(z)}{2P(z)} = \frac{P_\ell(z=0)}{2P_o}. \tag{3.104}$$

This equation states that α can be determined from P_o, the power on the line, and P_ℓ, the power loss per unit length of line. It is important to realize that P_ℓ can be computed from the fields of the lossless line, and can account for both conductor loss (using 2.131) and dielectric loss (using 2.92).

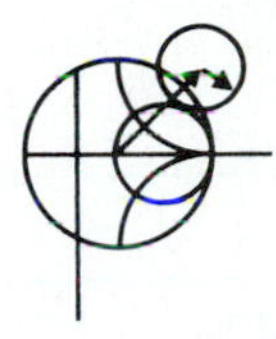

EXAMPLE 3.9

Use the perturbation method to find the attenuation constant of a coaxial line having a lossy dielectric and lossy conductors.

Solution

From Example 3.1 and (3.32), the fields of the lossless coaxial line are, for $a < \rho < b$,

$$\bar{E} = \frac{V_o \hat{\rho}}{\rho \ln b/a} e^{-j\beta z},$$

$$\bar{H} = \frac{V_o \hat{\phi}}{2\pi \rho Z_0} e^{-j\beta z},$$

where $Z_0 = (\eta/2\pi) \ln b/a$ is the characteristic impedance of the coaxial line and V_o is the voltage across the line at $z = 0$. The first step is to find P_o, the power flowing on the lossless line:

$$P_o = \frac{1}{2}\text{Re}\int_S \bar{E} \times \bar{H}^* \cdot d\bar{s} = \frac{|V_o|^2}{2Z_0}\int_{\rho=a}^{b}\int_{\phi=0}^{2\pi} \frac{\rho\, d\rho\, d\phi}{2\pi\rho^2 \ln b/a} = \frac{|V_o|^2}{2Z_0},$$

as expected from basic circuit theory.

The loss per unit length, P_ℓ, comes from conductor loss ($P_{\ell c}$) and dielectric loss ($P_{\ell d}$). From (2.131), the conductor loss in a 1 m length of line can be found as

$$P_{\ell c} = \frac{R_s}{2}\int_S |\bar{H}_t|^2 ds = \frac{R_s}{2}\int_{z=0}^{1}\left\{\int_{\phi=0}^{2\pi} |H_\phi(\rho = a)|^2 a d\phi + \int_{\phi=0}^{2\pi} |H_\phi(\rho = b)|^2 b d\phi\right\} dz$$

$$= \frac{R_s |V_o|^2}{4\pi Z_0^2}\left(\frac{1}{a} + \frac{1}{b}\right).$$

The dielectric loss in a 1 m length of line is, from (2.92),

$$P_{\ell d} = \frac{\omega\epsilon''}{2}\int_V |\bar{E}|^2 ds = \frac{\omega\epsilon''}{2}\int_{\rho=a}^{b}\int_{\phi=0}^{2\pi}\int_{z=0}^{1} |E_\rho|^2 \rho d\rho d\phi\, dz = \frac{\pi\omega\epsilon''}{\ln b/a}|V_o|^2,$$

where ϵ'' is the imaginary part of the complex dielectric constant, $\epsilon = \epsilon' - j\epsilon''$. Finally, applying (3.104) gives

$$\alpha = \frac{P_{\ell c} + P_{\ell d}}{2P_o} = \frac{R_s}{4\pi Z_0}\left(\frac{1}{a} + \frac{1}{b}\right) + \frac{\pi\omega\epsilon'' Z_o}{\ln b/a}$$

$$= \frac{R_s}{2\eta \ln b/a}\left(\frac{1}{a} + \frac{1}{b}\right) + \frac{\omega\epsilon''\eta}{2},$$

where $\eta = \sqrt{\mu/\epsilon'}$. This result is seen to agree with that of Example 3.8. ○

The Wheeler Incremental Inductance Rule

Another useful technique for the practical evaluation of attenuation due to conductor loss for TEM or quasi-TEM lines is the Wheeler incremental inductance rule [6]. This method is based on the similarity of the equations for the inductance per unit length and resistance per unit length of a transmission line, as given by (3.17) and (3.19), respectively. In other words, the conductor loss of a line is due to current flow inside the conductor which, as was shown in Section 2.7, is related to the tangential magnetic field at the surface of the conductor, and thus to the inductance of the line.

From (2.131), the power loss into a cross-section S of a good (but not perfect) conductor is

$$P_\ell = \frac{R_s}{2}\int_S |\bar{J}_s|^2 ds = \frac{R_s}{2}\int_S |\bar{H}_t|^2 ds \text{ W/m}^2, \qquad 3.105$$

so the power loss per unit length of a uniform transmission line is

$$P_\ell = \frac{R_s}{2}\int_C |\bar{H}_t|^2 d\ell \text{ W/m}, \qquad 3.106$$

where the line integral of (3.106) is over the cross-sectional contours of both conductors. Now, from (3.17), the inductance per unit length of the line is

$$L = \frac{\mu}{|I|^2}\int_S |\bar{H}|^2 ds, \qquad 3.107$$

which is computed assuming the conductors are lossless. When the conductors have a small loss, the $\bar{H}$ field in the conductor is no longer zero, and this field contributes a small additional "incremental" inductance, ΔL, to that of (3.107). As discussed in Chapter 2, the fields inside the conductor decay exponentially so that the integration into the conductor dimension can be evaluated as

$$\Delta L = \frac{\mu_0 \delta_s}{2|I|^2} \int_C |\bar{H}_t|^2 d\ell, \tag{3.108}$$

since $\int_0^\infty e^{-2z/\delta_s} dz = \delta_s/2$. (The skin depth is $\delta_s = \sqrt{2/\omega\mu\sigma}$.) Then P_ℓ from (3.106) can be written in terms of ΔL as

$$P_\ell = \frac{R_s |I|^2 \Delta L}{\mu_0 \delta_s} = \frac{|I|^2 \Delta L}{\sigma \mu_0 \delta_s^2} = \frac{|I|^2 \omega \Delta L}{2} \text{ W/m}, \tag{3.109}$$

since $R_s = \sqrt{(\omega\mu_0)/2\sigma} = 1/(\sigma\delta_s)$. Then from (3.104) the attenuation due to conductor loss can be evaluated as

$$\alpha_c = \frac{P_\ell}{2P_o} = \frac{\omega \Delta L}{2Z_0}, \tag{3.110}$$

since P_o, the total power flow down the line, is $P_o = |I|^2 Z_0/2$, where Z_0 is the characteristic impedance of the line. In (3.110), ΔL is evaluated as the change in inductance when all conductor walls are receded by an amount $\delta_s/2$.

Equation (3.110) can also be written in terms of the change in characteristic impedance, since

$$Z_0 = \sqrt{\frac{L}{C}} = \frac{L}{\sqrt{LC}} = L v_p, \tag{3.111}$$

so that

$$\alpha_c = \frac{\beta \Delta Z_0}{2Z_0}, \tag{3.112}$$

where ΔZ_0 is the change in characteristic impedance when all conductor walls are receded by an amount $\delta_s/2$. Yet another form of the incremental inductance rule can be obtained by using the first two terms of a Taylor series expansion for Z_0. Thus,

$$Z_0\left(\frac{\delta_s}{2}\right) \simeq Z_0 + \frac{\delta_s}{2} \frac{dZ_0}{d\ell}, \tag{3.113}$$

so that

$$\Delta Z_0 = Z_0\left(\frac{\delta_s}{2}\right) - Z_0 = \frac{\delta_s}{2} \frac{dZ_0}{d\ell},$$

where $Z_0\left(\delta_s/2\right)$ refers to the characteristic impedance of the line when the walls are receded by $\delta_s/2$, and ℓ refers to a distance into the conductors. Then (3.112) can be written as

$$\alpha_c = \frac{\beta \delta_s}{4Z_0} \frac{dZ_0}{d\ell} = \frac{R_s}{2Z_0 \eta} \frac{dZ_0}{d\ell}, \tag{3.114}$$

where $\eta = \sqrt{\mu_0/\epsilon}$ is the intrinsic impedance of the dielectric, and R_s is the surface resistivity of the conductor. Equation (3.114) is one of the most practical forms of the incremental inductance rule, because the characteristic impedance is known for a wide variety of transmission lines.

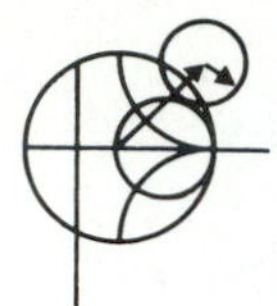

EXAMPLE 3.10

Calculate the attenuation due to conductor loss of a coaxial line using the incremental inductance rule.

Solution

From (3.32) the characteristic impedance of the coaxial line is

$$Z_0 = \frac{\eta}{2\pi} \ln \frac{b}{a}.$$

Then, using the incremental inductance rule of the form in (3.114), the attenuation due to conductor loss is

$$\alpha_c = \frac{R_s}{2Z_0\eta}\frac{dZ_0}{d\ell} = \frac{R_s}{4\pi Z_0}\left\{\frac{d \ln b/a}{db} - \frac{d \ln b/a}{da}\right\} = \frac{R_s}{4\pi Z_0}\left(\frac{1}{b} + \frac{1}{a}\right),$$

which is seen to be in agreement with the result of Example 3.9. The negative sign on the second differentiation in the above equation is because the derivative for the inner conductor is in the $-\rho$ direction (receding wall). ○

Regardless of how attenuation is calculated, measured attenuation constants for practical lines are usually higher. The main reason for this discrepancy is the fact that realistic transmission lines have metallic surfaces which are somewhat rough, which increases the loss, while our theoretical calculations assume perfectly smooth conductors. A quasi-empirical formula that can be used to correct for surface roughness for any transmission line is [7]

$$\alpha_c' = \alpha_c \left[1 + \frac{2}{\pi} \tan^{-1} 1.4 \left(\frac{\Delta}{\delta_s}\right)^2\right], \tag{3.115}$$

where α_c is the attenuation due to perfectly smooth conductors, α_c' is the attenuation corrected for surface roughness, Δ is the rms surface roughness, and δ_s is the skin depth of the conductors.

3.10 TRANSIENTS ON TRANSMISSION LINES

So far we have concentrated on the behavior of transmission lines at a single frequency. In the majority of microwave systems and applications such a viewpoint is entirely satisfactory, but in some situations, where pulse transmission or wideband signal propagation is present, it is useful to consider wave propagation from the transient point of view.

The student is probably aware that an arbitrary signal in the time domain can be expressed as a superposition of single frequency signals, via the Fourier or Laplace transform. Thus, we will demonstrate a transient solution using the transform technique later in this section. We will begin, however, with an intuitive approach to transient propagation on lossless transmission lines.

Bounce Diagrams

Consider the circuit shown in Figure 3.24, where a 12-volt DC source is switched on at $t = 0$. For $t < 0$, the voltage at the input to the transmission line is zero:

$$v(z,t) = v(0,0) = 0, \qquad \text{for } t < 0.$$

Because there is no signal on the line, the input impedance of the line is unaffected by the 200 Ω load, and essentially looks as if its input impedance were 100 Ω, the characteristic impedance of the line. Thus, when the switch is closed at $t = 0$, a voltage divider network consisting of the 100 Ω source impedance and the 100 Ω line gives a line voltage of 6 volts, at $z = 0$. As time progresses, this voltage propagates along the line with velocity v_p, as shown in Figure 3.25a. The pulse reaches the load at time $t = \ell/v_p$, and is partly reflected with a coefficient of

$$\Gamma_\ell = \frac{200 - 100}{200 + 100} = \frac{1}{3}.$$

The superposition of the incident voltage and reflected voltage is then

$$v^+ + v^- = 6 + 6\Gamma_\ell = 8 \text{ V},$$

as shown in Figure 3.25b. At time $t = 2\ell/v_p$, the pulse has made a full round trip on the line, and is absorbed by the 100 Ω generator impedance, which is matched to the line. The system is now in a steady state condition, with 8 volts everywhere on the line, as shown in Figure 3.25c. Of course, this is the DC value that one would expect from the voltage divider consisting of the 100 Ω generator impedance and the 200 Ω load:

$$12 \cdot \frac{200}{200 + 100} = 8 \text{ V}.$$

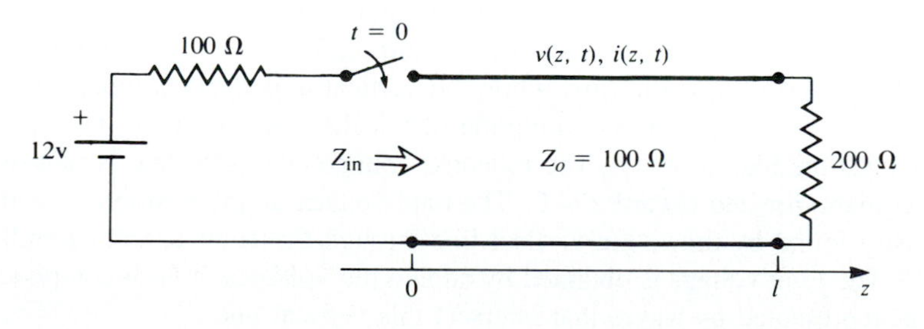

FIGURE 3.24 A terminated transmission line driven by a step function voltage source. Generator is matched to transmission line.

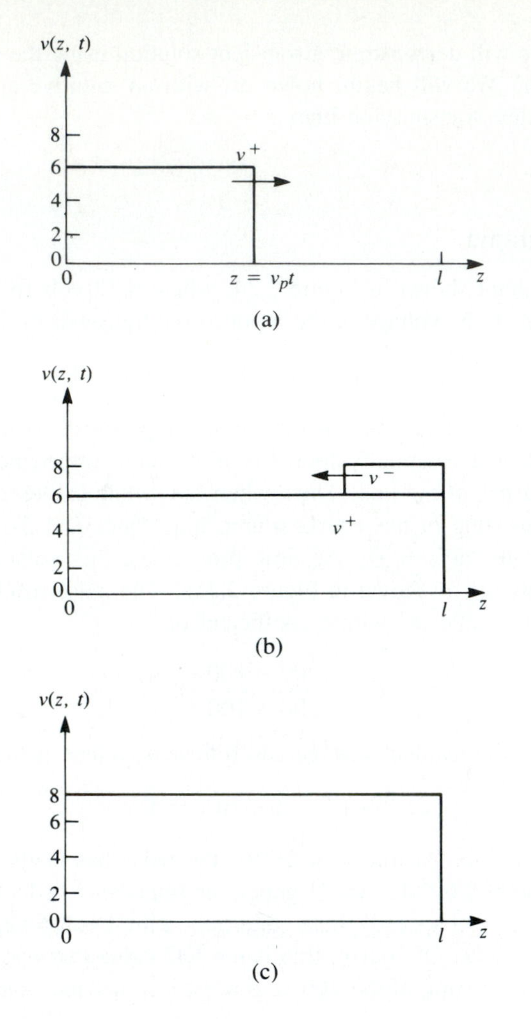

FIGURE 3.25 Transient response of the circuit of Figure 3.24. (a) For $0 < t < \ell/v_p$. (b) For $\ell/v_p < t < 2\ell/v_p$. (c) For $2\ell/v_p < t$.

Another way of viewing the progress of the pulse propagating in time and position is to use the *bounce diagram*, as shown in Figure 3.26. The horizontal axis represents position along the transmission line, while the vertical axis denotes time. The incident wave starts at $z = t = 0$ with an amplitude of 6 volts, and travels in the $+z$ direction until it reaches the load at $z = \ell$. The reflected voltage is 2 volts, and is represented by the line segment directed toward $z = 0$. The total voltage at any position, z, and time, t, can be easily found by drawing a vertical line through the point z, and extending from $t = 0$ to t. The total voltage is obtained by adding the voltages of each component wave present, as represented by waves that intersect this vertical line.

Such analysis can be readily extended to circuits with multiple reflections. Figure 3.27 shows a transient circuit with a mismatched generator. The reflection coefficient at

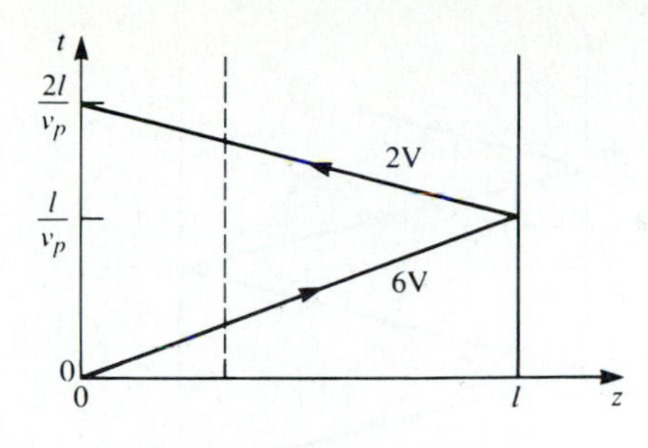

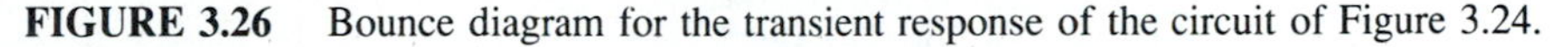

FIGURE 3.26 Bounce diagram for the transient response of the circuit of Figure 3.24.

the generator is

$$\Gamma_g = \frac{50 - 100}{50 + 100} = -\frac{1}{3},$$

while the reflection coefficient at the load is

$$\Gamma_\ell = \frac{200 - 100}{200 + 100} = \frac{1}{3}.$$

The amplitude of the incident wave on the transmission line is

$$v^+ = 12\frac{100}{100 + 50} = 8 \text{ V}.$$

Figure 3.28 shows the first several reflected waves for this circuit.

A Transform Solution

Although the bounce diagrams of the previous section provide an intuitive picture of transient behavior of transmission lines, it is sometimes necessary or desirable to obtain an analytic solution to such problems. In this section we show how the Laplace transform method of circuit theory can be applied to determine the time domain response of a lossless transmission line.

The problem is shown in Figure 3.29. A short-circuited length of lossless transmission line ($R = G = 0$) is connected to a matched generator which delivers a step voltage of amplitude V_o. From (3.2a,b), the equations that describe the space and time variation

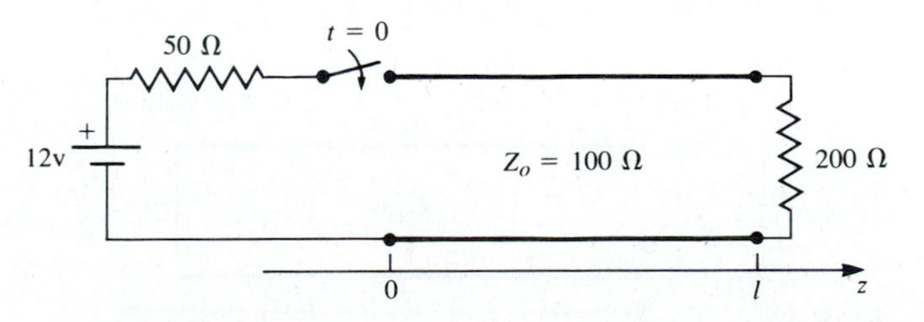

FIGURE 3.27 A terminated transmission line driven by a step function voltage source; mismatched generator.

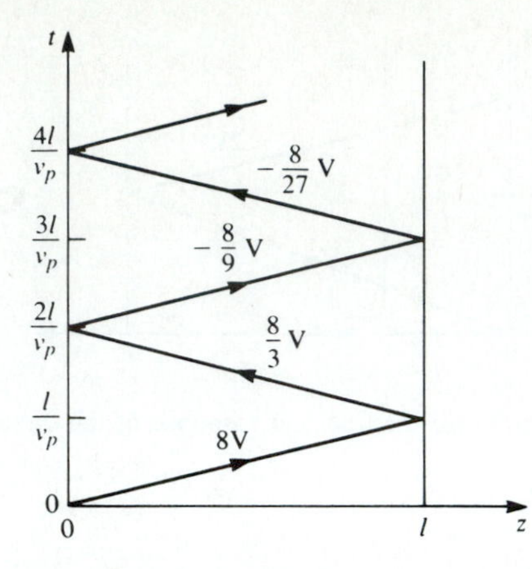

FIGURE 3.28 Bounce diagram for the transient response of the circuit of Figure 3.27.

of the voltage and current on the line are

$$\frac{\partial v(z,t)}{\partial z} + L\frac{\partial i(z,t)}{\partial t} = 0, \qquad 3.116a$$

$$\frac{\partial i(z,t)}{\partial z} + C\frac{\partial v(z,t)}{\partial t} = 0. \qquad 3.116b$$

From circuit theory, the Laplace transform $F(s)$ of a function $f(t)$ is defined as

$$F(s) = \mathcal{L}\{f(t)\} = \int_0^\infty f(t)e^{-st}dt. \qquad 3.117$$

For reference, some of the basic Laplace transform results which we will need are listed below:

$$\mathcal{L}\left\{\frac{\partial f(t)}{\partial t}\right\} = SF(s) - f(0^+), \qquad 3.118$$

$$\mathcal{L}\{U(t)\} = \frac{1}{s}, \qquad 3.119$$

$$\mathcal{L}\{f(t-b)\} = F(s)e^{-sb}, \qquad 3.120$$

In the above, it is assumed that $f(t)$ is zero for $t < 0$.

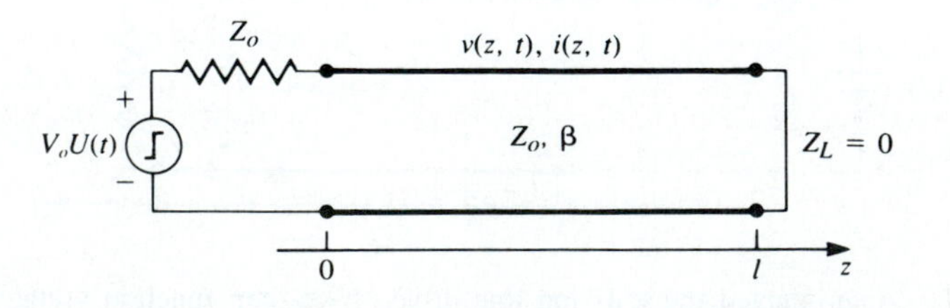

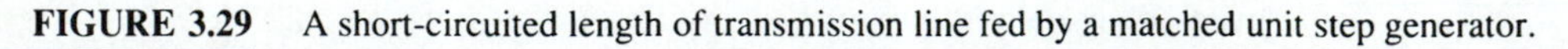

FIGURE 3.29 A short-circuited length of transmission line fed by a matched unit step generator.

Applying (3.118) to (3.116a,b) gives the transformed form of the telegrapher equations:

$$\frac{\partial V(z,s)}{\partial z} + sLI(z,s) - Li(z,0^+) = 0, \qquad 3.121a$$

$$\frac{\partial I(z,s)}{\partial z} + sCV(z,s) - Cv(z,0^+) = 0, \qquad 3.121b$$

where $V(z,s)$ and $I(z,s)$ are the Laplace transforms of $v(z,t)$ and $i(z,t)$, respectively. To solve these equations for V or I, which are functions of the independent variables z and s, we need boundary conditions at $z = 0$ and ℓ:

$$\text{at } z = 0, \quad -V_o U(t) + i(0,t)Z_0 + v(0,t) = 0, \qquad 3.122a$$

$$\text{at } z = \ell, \quad v(\ell,t) = 0. \qquad 3.122b$$

In the transform domain these equations can be written as

$$\frac{-V_o}{s} + I(0,s)Z_0 + V(0,s) = 0, \qquad 3.123a$$

$$V(\ell,s) = 0. \qquad 3.123b$$

We also require initial conditions. Because of the finite velocity of propagation, we can say that the voltage and current are both zero at $t = 0^+$, at all points z on the transmission line for $0 < z \leq \ell$. Thus,

$$v(z,0^+) = i(z,0^+) = 0. \qquad 3.124$$

Equations (3.121) then reduce to

$$\frac{\partial V}{\partial z} + sLI = 0, \qquad 3.125a$$

$$\frac{\partial I}{\partial z} + sCV = 0, \qquad 3.125b$$

which can be solved simultaneously for $V(z,s)$ to give

$$\frac{\partial^2 V}{\partial z^2} - s^2 LCV = 0. \qquad 3.126$$

The general solution to this wave equation for V as a function of z can be written as

$$V(z,s) = Ae^{-sz/v_p} + Be^{sz/v_p}, \qquad 3.127a$$

where $v_p = 1\sqrt{LC}$ is the velocity of propagation on the line. As before, these two terms represent traveling waves in the $\pm z$ directions. Using (3.125a) and (3.127a) to solve for

I gives

$$I(z,s) = \frac{1}{sL}\frac{\partial V}{\partial z} = \frac{A}{Z_0}e^{-sz/v_p} - \frac{B}{Z_0}e^{sz/v_p}. \qquad 3.127b$$

We must now determine the unknown constants A and B by applying the two boundary conditions. From (3.123), we have

$$\text{at } z = 0, \quad \frac{-V_o}{s} + A - B + A + B = 0,$$

or

$$A = \frac{V_o}{2s}. \qquad 3.128a$$

$$\text{At } z = \ell, \quad Ae^{-s\ell/v_p} + Be^{s\ell/v_p} = 0,$$

or

$$B = \frac{-V_o}{2s}e^{-2s\ell/v_p}. \qquad 3.128b$$

The solution for $V(z,s)$ is then

$$V(z,s) = \frac{V_o}{2s}\left[e^{-sz/v_p} - e^{-s(2\ell-z)/v_p}\right]. \qquad 3.129$$

Using (3.119) and (3.120) to find the inverse transform of this result gives the time-domain solution as

$$v(z,t) = \frac{V_o}{2}\left[U\left(\frac{t-z}{v_p}\right) - U\left(\frac{t-2\ell}{v_p} + \frac{z}{v_p}\right)\right], \qquad 3.130$$

which is seen to consist of two traveling pulses. For $0 < t < \ell/v_p$, only the first term is present, and represents the incident voltage wave of $V_o/2$ traveling toward the load, as shown in Figure 3.30a. For $t > \ell/v_p$, the incident wave has reflected off of the short circuit load with a reflection coefficient of -1, and cancels the incident wave as it travels back toward the generator, as shown in Figure 3.30b. The voltage seen at the input to the transmission line looks like a pulse, since from (3.130) we have for $z = 0$,

$$v(0,t) = \frac{V_o}{2}\left[U(t) - U\left(\frac{t-2\ell}{v_p}\right)\right], \qquad 3.131$$

which is plotted in Figure 3.30c. The duration of the pulse is $2\ell/v_p$, which may be quite small. Thus, such circuits are sometimes used for the generation of short pulses.

REFERENCES

[1] S. Ramo, J. R. Winnery, and T. Van Duzer, *Fields and Waves in Communication Electronics*, John Wiley & Sons, N.Y., 1965.

[2] J. A. Stratton, *Electromagnetic Theory*, McGraw-Hill, N.Y., 1941.

[3] H. A. Wheeler, "Reflection Charts Relating to Impedance Matching," *IEEE Trans. Microwave Theory and Techniques*, vol. MTT-32, pp. 1008–1021, September 1984.

[4] P. H. Smith, "Transmission Line Calculator," *Electronics*, vol. 12, No. 1, pp. 29–31, January 1939.

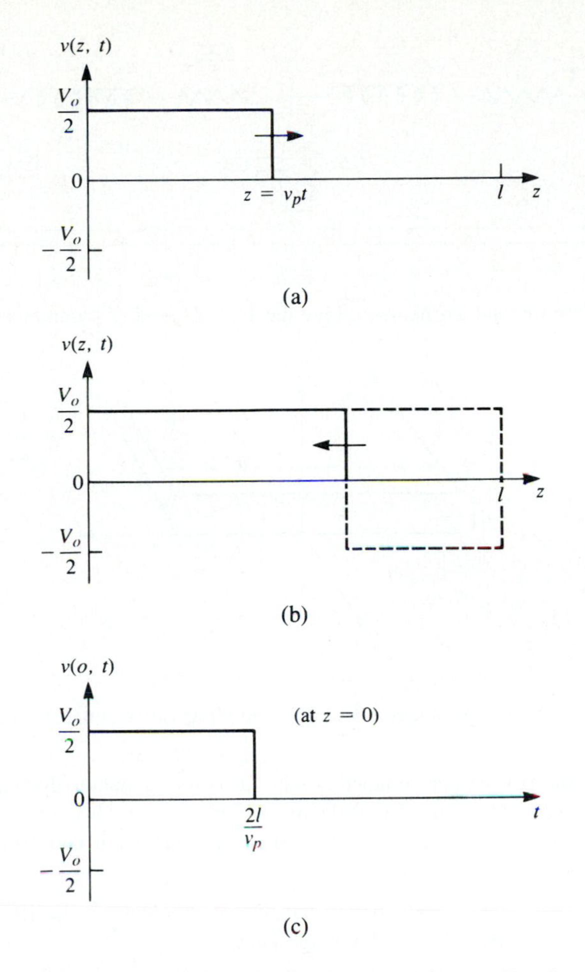

FIGURE 3.30 Transient response of the circuit of Figure 3.29. (a) For $0 < t < \ell/v_p$. (b) For $\ell/v_p < t < 2\ell/v_p$. (c) For $z = 0$.

[5] P. J. Nahin, *Oliver Heaviside: Sage in Solitude*, IEEE Press, N.Y., 1988.

[6] H. A. Wheeler, "Formulas for the Skin Effect," *Proc. IRE*, vol. 30, pp. 412–424, September 1942.

[7] T. C. Edwards, *Foundations for Microstrip Circuit Design*, John Wiley & Sons, N.Y., 1987.

PROBLEMS

3.1 A transmission line has the following per unit length parameters: $L = 0.2\ \mu$H/m, $C = 300$ pF/m, $R = 5\ \Omega$/m, and $G = 0.01$ S/m. Calculate the propagation constant and characteristic impedance of this line at 500 MHz. Recalculate these quantities in the absence of loss ($R = G = 0$).

3.2 Show that the following T-model of a transmission line also yields the telegrapher equations derived in Section 3.1.

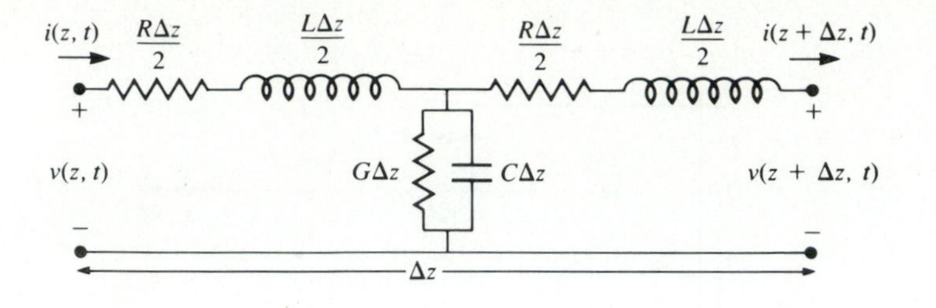

3.3 For the parallel plate line shown below, derive the R, L, G, and C parameters. Assume $w >> d$.

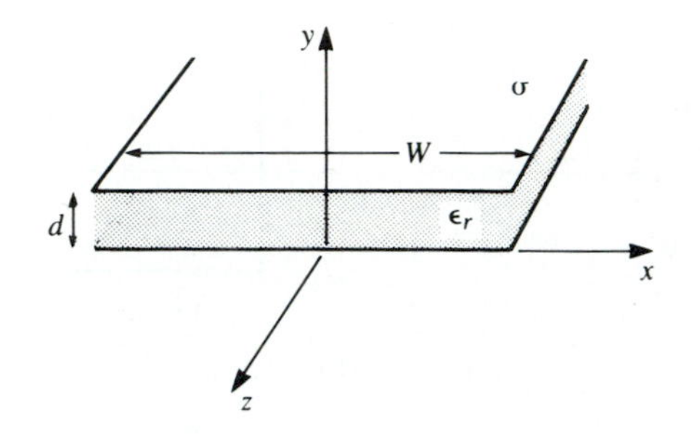

3.4 For the parallel plate line of Problem 3.3, derive the telegrapher equations using the field theory approach.

3.5 A certain coaxial line has copper conductors with an inner conductor diameter of 1 mm and an outer conductor diameter of 3 mm. The dielectric filling has $\epsilon_r = 2.8$ with a loss tangent of 0.005. Compute the R, L, G, and C parameters of this line at 3 GHz, and the characteristic impedance and phase velocity.

3.6 Compute and plot the attenuation of the coaxial line of Problem 3.5, in dB/m, over a frequency range of 10 MHz to 10 GHz. Use log-log graph paper.

3.7 A lossless transmission line of electrical length $\ell = 0.3\lambda$ is terminated with a complex load impedance as shown below. Find the reflection coefficient at the load, the SWR on the line, and the input impedance to the line.

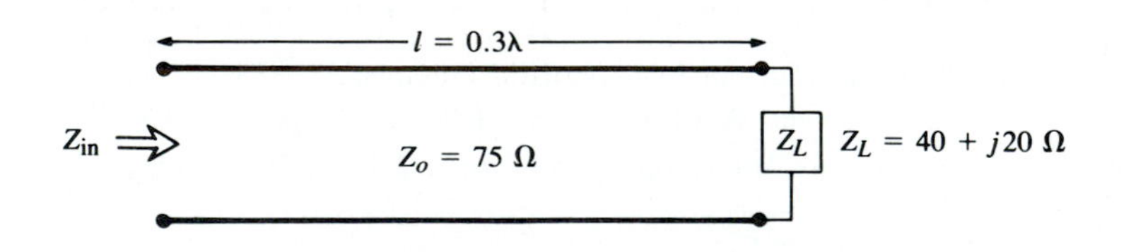

3.8 A lossless transmission line is terminated with a 100 Ω load. If the SWR on the line is 1.5, find the two possible values for the characteristic impedance of the line.

3.9 A radio transmitter is connected to an antenna having an impedance $80 + j40\ \Omega$ with a 50 Ω coaxial cable. If the transmitter can deliver 30 W to a 50 Ω load, how much power is delivered to the antenna?

3.10 A load impedance of $40 - j80\ \Omega$ is connected to a 100 Ω line. Calculate the reflection coefficient at the load and the reflection coefficient at the input to the line, if the line is 0.7λ long.

3.11 Calculate SWR, reflection coefficient magnitude, and return loss values to complete the entries in the following table:

SWR	$\|\Gamma\|$	*RL* (dB)
1.00	0.00	∞
1.01		
	0.01	
1.05		
		30.0
1.10		
1.20		
	0.10	
1.50		
		10.0
2.00		
2.50		

3.12 A 10 V source with an impedance of $Z_g = 50\ \Omega$ is connected to a load impedance $Z_L = 80+j40\ \Omega$ with a $50\ \Omega$ transmission line. Calculate the power delivered to the load if the line is 0.3λ long. What is the power if the line length is increased to 0.6λ?

3.13 For a purely reactive load impedance of the form $Z_L = jX$, show that the reflection coefficient magnitude $|\Gamma|$ is always unity. Assume the characteristic impedance Z_0 is real.

3.14 Consider the transmission line circuit shown below. Compute the incident power, the reflected power, and the power transmitted into the infinite $75\ \Omega$ line. Show that power conservation is satisfied.

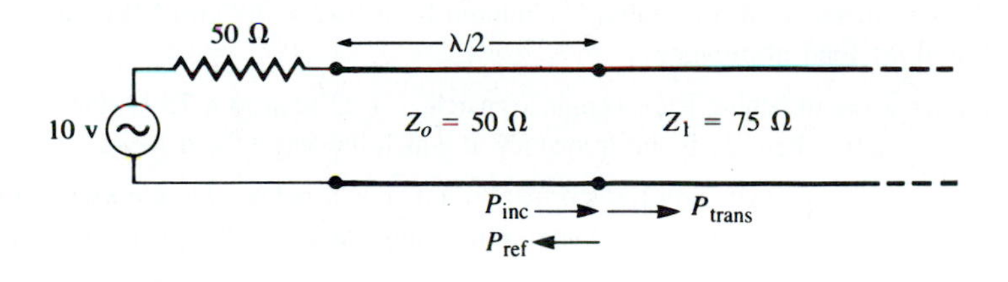

3.15 A generator is connected to a transmission line as shown below. Find the voltage as a function of z along the transmission line. Plot the magnitude of this voltage for $-\ell \leq z \leq 0$.

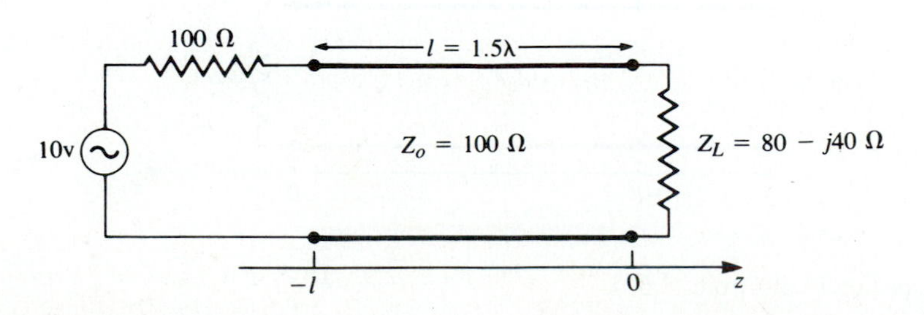

3.16 A load impedance of $Z_L = 80 + j20\ \Omega$ is to be matched to a $Z_0 = 100\ \Omega$ line using a length ℓ of lossless line of characteristic impedance Z_1. Find the required Z_1 (real) and ℓ.

3.17 Use the Smith chart to find the following quantities for the transmission line circuit below:
(a) The SWR on the line.
(b) The reflection coefficient at the load.
(c) The load admittance.
(d) The input impedance of the line.
(e) The distance from the load to the first voltage minimum.
(f) The distance from the load to the first voltage maximum.

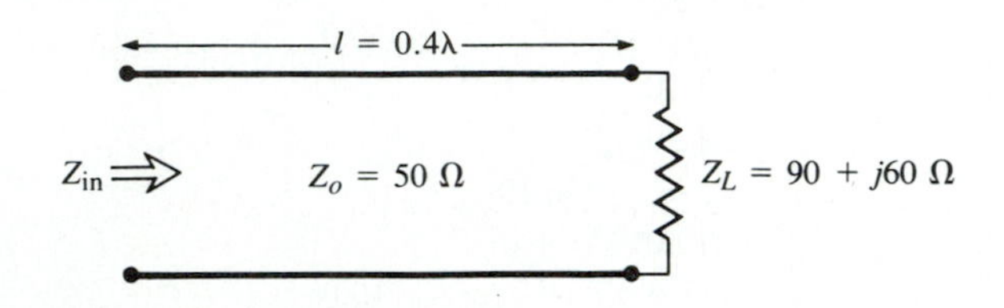

3.18 Repeat problem 3.17 for $Z_L = 30 - j20\ \Omega$.

3.19 Repeat problem 3.17 for $\ell = 1.7\lambda$.

3.20 Use the Smith chart to find the shortest lengths of a short-circuited 75 Ω line to give the following input impedance:
(a) $Z_{in} = 0$.
(b) $Z_{in} = \infty$.
(c) $Z_{in} = j75\ \Omega$.
(d) $Z_{in} = -j50\ \Omega$.
(e) $Z_{in} = j10\ \Omega$.

3.21 Repeat Problem 3.20 for an open-circuited length of 75 Ω line.

3.22 A slotted-line experiment is performed with the following results: distance between succesive minima = 2.1 cm; distance of first voltage minimum from load = 0.9 cm; SWR of load = 2.5. If $Z_0 = 50\ \Omega$, find the load impedance.

3.23 Design a quarter-wave matching transformer to match a 40 Ω load to a 75 Ω line. Plot the SWR for $0.5 \leq f/f_o \leq 2.0$, where f_o is the frequency at which the line is $\lambda/4$ long.

3.24 Consider the quarter-wave matching transformer circuit shown below. Derive expressions for V^+ and V^-, the amplitude of the forward and reverse traveling waves on the quarter-wave line section, in terms of V^i, the incident voltage amplitude.

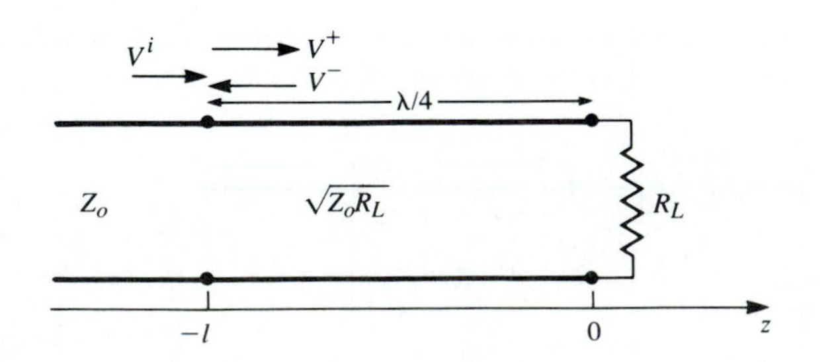

3.25 Derive equation (3.68) from (3.67).

3.26 In Example 3.9, the attenuation of a coaxial line due to finite conductivity is

$$\alpha_c = \frac{R_s}{2\eta \ln b/a}\left(\frac{1}{a}+\frac{1}{b}\right).$$

Show that α_c is minimized for conductor radii such that $x \ln x = 1 + x$, where $x = b/a$.

3.27 Compute and plot the factor by which attenuation is increased due to surface roughness, for rms roughness ranging from zero to 0.01 mm. Assume copper conductors at 10 GHz.

3.28 A 50 Ω transmission line is matched to a 10 W source, and feeds a load $Z_L = 100$ Ω. If the line is 2.3λ long and has an attenuation constant $\alpha = 0.5$ dB/λ, find the powers that are delivered by the source, lost in the line, and delivered to the load.

3.29 Plot the bounce diagram for the transient response of the circuit shown below.

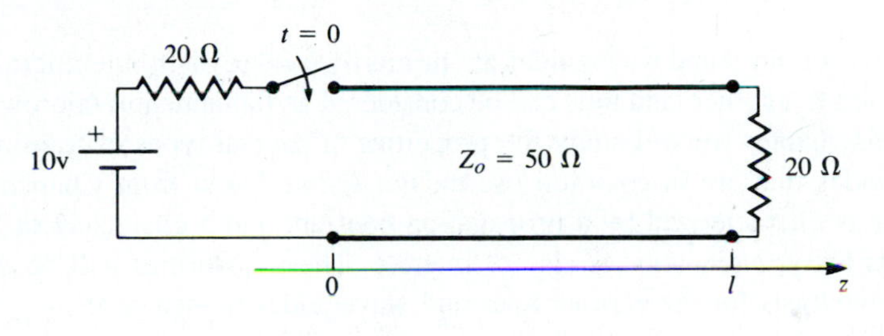

CHAPTER 4

Transmission Lines and Waveguides

Transmission lines and waveguides are primarily used to distribute microwave power from one point to another, and thus can be considered as fundamental microwave components. In this chapter we will study the properties of several types of transmission lines and waveguides that are in common use today. As we know from Chapter 3, a transmission line is characterized by a propagation constant and a characteristic impedance; if the line is lossy, attenuation is also of interest. These quantities will be derived by a field theory analysis for the various lines and waveguides treated here.

We will begin with a general discussion of the different types of wave propagation and modes that can exist on transmission lines and waveguides. Transmission lines that consist of two or more conductors may support transverse electromagnetic (TEM) waves, characterized by the lack of longitudinal field components. TEM waves have a uniquely defined voltage, current, and characteristic impedance. Waveguides, often consisting of a single conductor, support transverse electric (TE) and/or transverse magnetic (TM) waves, characterized by the presence of longitudinal magnetic or electric, respectively, field components. As we will see in Chapter 5, a unique definition of characteristic impedance is not possible for such waves, although definitions can be chosen so that the characteristic impedance concept can be used for waveguides with meaningful results.

4.1 GENERAL SOLUTIONS FOR TEM, TE, AND TM WAVES

In this section we will find general solutions to Maxwell's equations for the specific cases of TEM, TE, and TM wave propagation in cylindrical transmission lines or waveguides. The geometry of an arbitrary transmission line or waveguide is shown in Figure 4.1, and is characterized by conductor boundaries that are parallel to the z-axis. These structures are assumed to be uniform in the z direction, and infinitely long. The conductors will initially be assumed to be perfectly conducting, but attenuation can be found by the perturbation method discussed in Chapter 3.

We assume time-harmonic fields with an $e^{j\omega t}$ dependence, and wave propagation along the z-axis. The electric and magnetic fields can then be written as

$$\bar{E}(x, y, z) = [\bar{e}(x, y) + \hat{z}e_z(x, y)]e^{-j\beta z}, \tag{4.1a}$$

$$\bar{H}(x, y, z) = [\bar{h}(x, y) + \hat{z}h_z(x, y)]e^{-j\beta z}, \tag{4.1b}$$

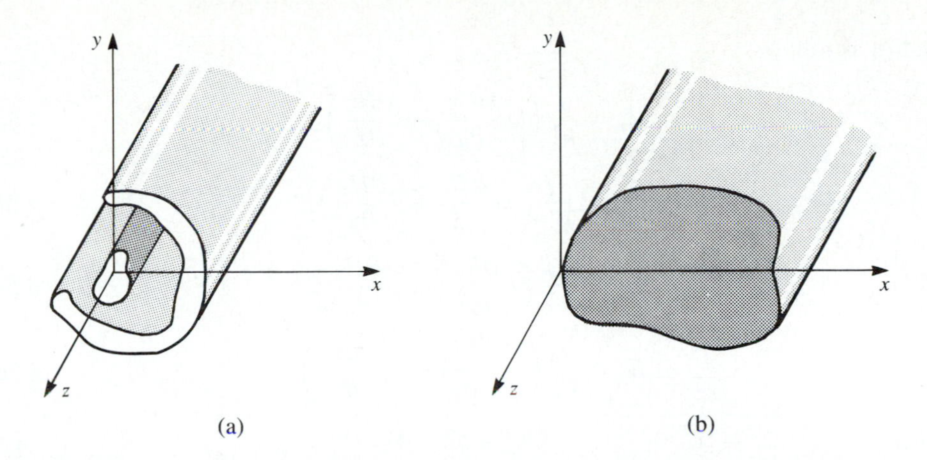

FIGURE 4.1 (a) General two-conductor transmission line and (b) closed waveguide.

where $\bar{e}(x, y)$ and $\bar{h}(x, y)$ represent the transverse $(\hat{x}, \hat{y})$ electric and magnetic field components, while e_z and h_z are the longitudinal electric and magnetic field components. In the above, the wave is propagating in the $+z$ direction; $-z$ propagation can be obtained by replacing β by $-\beta$. Also, if conductor or dielectric loss is present, the propagation constant will be complex; $j\beta$ should then be replaced with $\gamma = \alpha + j\beta$.

Assuming that the transmission line or waveguide region is source-free, Maxwell's equations can be written as

$$\nabla \times \bar{E} = -j\omega\mu\bar{H}, \tag{4.2a}$$

$$\nabla \times \bar{H} = j\omega\epsilon\bar{E}. \tag{4.2b}$$

With an $e^{-j\beta z}$ z dependence, the three components of each of the above vector equations can be reduced to the following:

$$\frac{\partial E_z}{\partial y} + j\beta E_y = -j\omega\mu H_x, \tag{4.3a}$$

$$-j\beta E_x - \frac{\partial E_z}{\partial x} = -j\omega\mu H_y, \tag{4.3b}$$

$$\frac{\partial E_y}{\partial x} - \frac{\partial E_x}{\partial y} = -j\omega\mu H_z, \tag{4.3c}$$

$$\frac{\partial H_z}{\partial y} + j\beta H_y = j\omega\epsilon E_x, \tag{4.4a}$$

$$-j\beta H_x - \frac{\partial H_z}{\partial x} = j\omega\epsilon E_y, \tag{4.4b}$$

$$\frac{\partial H_y}{\partial x} - \frac{\partial H_x}{\partial y} = j\omega\epsilon E_z. \tag{4.4c}$$

The above six equations can be solved for the four transverse field components in terms of E_z and H_z (for example, H_x can be derived by eliminating E_y from (4.3a) and

(4.4b)) as follows:

$$H_x = \frac{j}{k_c^2}\left(\omega\epsilon\frac{\partial E_z}{\partial y} - \beta\frac{\partial H_z}{\partial x}\right), \tag{4.5a}$$

$$H_y = \frac{-j}{k_c^2}\left(\omega\epsilon\frac{\partial E_z}{\partial x} + \beta\frac{\partial H_z}{\partial y}\right), \tag{4.5b}$$

$$E_x = \frac{-j}{k_c^2}\left(\beta\frac{\partial E_z}{\partial x} + \omega\mu\frac{\partial H_z}{\partial y}\right), \tag{4.5c}$$

$$E_y = \frac{j}{k_c^2}\left(-\beta\frac{\partial E_z}{\partial y} + \omega\mu\frac{\partial H_z}{\partial x}\right), \tag{4.5d}$$

where

$$k_c^2 = k^2 - \beta^2, \tag{4.6}$$

has been defined as the cutoff wavenumber; the reason for this terminology will become clear later. As in previous chapters,

$$k = \omega\sqrt{\mu\epsilon} = 2\pi/\lambda \tag{4.7}$$

is the wavenumber of the material filling the transmission line or waveguide region. If dielectric loss is present, ϵ can be made complex by using $\epsilon = \epsilon_o\epsilon_r(1 - j\tan\delta)$, where $\tan\delta$ is the loss tangent of the material.

Equations (4.5a–d) are very useful general results that can be applied to a variety of waveguiding systems. We will now specialize these results to specific wave types.

TEM Waves

Transverse electromagnetic (TEM) waves are characterized by $E_z = H_z = 0$. Observe from (4.5) that if $E_z = H_z = 0$, then the transverse fields are also all zero, unless $k_c^2 = 0$ ($k^2 = \beta^2$), in which case we have an indeterminate result. Thus, we can return to (4.3)–(4.4) and apply the condition that $E_z = H_z = 0$. Then from (4.3a) and (4.4b), we can eliminate H_z to obtain

$$\beta^2 E_y = \omega^2\mu\epsilon E_y,$$

or

$$\beta = \omega\sqrt{\mu\epsilon} = k, \tag{4.8}$$

as noted earlier. (This result can also be obtained from (4.3b) and (4.4a).) The cutoff wavenumber, $k_c = \sqrt{k^2 - \beta^2}$, is thus zero for TEM waves.

Now the Helmholtz wave equation for E_x is, from (2.42),

$$\left(\frac{\partial^2}{\partial x^2} + \frac{\partial^2}{\partial y^2} + \frac{\partial^2}{\partial z^2} + k^2\right)E_x = 0, \tag{4.9}$$

but for $e^{-j\beta z}$ dependence, $(\partial^2/\partial z^2)E_x = -\beta^2 E_x = -k^2 E_x$, so (4.9) reduces to

$$\left(\frac{\partial^2}{\partial x^2} + \frac{\partial^2}{\partial y^2}\right)E_x = 0. \tag{4.10}$$

A similar result also applies to E_y, so using the form of $\bar{E}$ assumed in (4.1a) we can write

$$\nabla_t^2 \bar{e}(x, y) = 0, \quad 4.11$$

where $\nabla_t^2 = \partial^2/\partial x^2 + \partial^2/\partial y^2$ is the Laplacian operator in the two transverse dimensions.

The result of (4.11) shows that the transverse electric fields, $\bar{e}(x, y)$, of a TEM wave satisfy Laplace's equation. It is easy to show in the same way that the transverse magnetic fields also satisfy Laplace's equation:

$$\nabla_t^2 \bar{h}(x, y) = 0. \quad 4.12$$

The transverse fields of a TEM wave are thus the same as the static fields that can exist between the conductors. In the electrostatic case, we know that the electric field can be expressed as the gradient of a scalar potential, $\Phi(x, y)$:

$$\bar{e}(x, y) = -\nabla_t \Phi(x, y), \quad 4.13$$

where $\nabla_t = \hat{x}(\partial/\partial x) + \hat{y}(\partial/\partial y)$ is the transverse gradient operator in two dimensions. In order for the relation in (4.13) to be valid, the curl of $\bar{e}$ must vanish, and this is indeed the case here since

$$\nabla_t \times \bar{e} = -j\omega\mu h_z \hat{z} = 0.$$

Using the fact that $\nabla \cdot \bar{D} = \epsilon \nabla_t \cdot \bar{e} = 0$ with (4.13) shows that $\Phi(x, y)$ also satisfies Laplace's equation,

$$\nabla_t^2 \Phi(x, y) = 0, \quad 4.14$$

as expected from electrostatics. The voltage between two conductors can be found as

$$V_{12} = \Phi_1 - \Phi_2 = \int_1^2 \bar{E} \cdot d\bar{\ell}, \quad 4.15$$

where Φ_1 and Φ_2 represent the potential at conductors 1 and 2, respectively. The current flow on a conductor can be found from Ampere's law as

$$I = \oint_C \bar{H} \cdot d\bar{\ell}, \quad 4.16$$

where C is the cross-sectional contour of the conductor.

TEM waves can exist when two or more conductors are present. Plane waves are also examples of TEM waves, since there are no field components in the direction of propagation; in this case the transmission line conductors may be considered to be two infinitely large plates separated to infinity. The above results show that a closed conductor (such as a rectangular waveguide) cannot support TEM waves, since the corresponding static potential in such a region would be zero (or possibly a constant), leading to $\bar{e} = 0$.

The wave impedance of a TEM mode can be found as the ratio of the transverse electric and magnetic fields:

$$Z_{\text{TEM}} = \frac{E_x}{H_y} = \frac{\omega\mu}{\beta} = \sqrt{\frac{\mu}{\epsilon}} = \eta, \tag{4.17a}$$

where (4.4a) was used. The other pair of transverse field components, from (4.3a), give

$$Z_{\text{TEM}} = \frac{-E_y}{H_x} = \sqrt{\frac{\mu}{\epsilon}} = \eta. \tag{4.17b}$$

Combining the results of (4.17a) and (4.17b) gives a general expression for the transverse fields as

$$\bar{h}(x, y) = \frac{1}{Z_{\text{TEM}}} \hat{z} \times \bar{e}(x, y). \tag{4.18}$$

Note that the wave impedance is the same as that for a plane wave in a lossless medium, as derived in Chapter 2; the reader should not confuse this impedance with the characteristic impedance, Z_0, of a transmission line. The latter relates an incident voltage and current and is a function of the line geometry as well as the material filling the line, while the wave impedance relates transverse field components and is dependent only on the material constants. From (3.32), the characteristic impedance of the TEM line is $Z_0 = V/I$, where V and I are the amplitudes of the incident voltage and current waves.

The procedure for analyzing a TEM line can be summarized as follows:

1. Solve Laplace's equation, (4.14), for $\Phi(x, y)$. The solution will contain several unknown constants.
2. Find these constants by applying the boundary conditions for the known voltages on the conductors.
3. Compute $\bar{e}$ and $\bar{E}$ from (4.13), (4.1a). Compute $\bar{h}$, $\bar{H}$ from (4.18), (4.1b).
4. Compute V from (4.15), I from(4.16).
5. The propagation constant is given by (4.8), and the characteristic impedance is given by $Z_0 = V/I$.

TE Waves

Transverse electric (TE) waves, (also referred to as H-waves) are characterized by $E_z = 0$ and $H_z \neq 0$. Equations (4.5) then reduce to

$$H_x = \frac{-j\beta}{k_c^2} \frac{\partial H_z}{\partial x}, \tag{4.19a}$$

$$H_y = \frac{-j\beta}{k_c^2} \frac{\partial H_z}{\partial y}, \tag{4.19b}$$

$$E_x = \frac{-j\omega\mu}{k_c^2} \frac{\partial H_z}{\partial y}, \tag{4.19c}$$

$$E_y = \frac{j\omega\mu}{k_c^2} \frac{\partial H_z}{\partial x}. \tag{4.19d}$$

In this case, $k_c \neq 0$, and the propagation constant $\beta = \sqrt{k^2 - k_c^2}$ is generally a function of frequency and the geometry of the line or guide. To apply (4.19), one must first find H_z from the Helmholtz wave equation,

$$\left(\frac{\partial^2}{\partial x^2} + \frac{\partial^2}{\partial y^2} + \frac{\partial^2}{\partial z^2} + k^2\right) H_z = 0, \tag{4.20}$$

which, since $H_z(x, y, z) = h_z(x, y)e^{-j\beta z}$, can be reduced to a two-dimensional wave equation for h_z:

$$\left(\frac{\partial^2}{\partial x^2} + \frac{\partial^2}{\partial y^2} + k_c^2\right) h_z = 0, \tag{4.21}$$

since $k_c^2 = k^2 - \beta^2$. This equation must be solved subject to the boundary conditions of the specific guide geometry.

The TE wave impedance can be found as

$$Z_{\text{TE}} = \frac{E_x}{H_y} = \frac{-E_y}{H_x} = \frac{\omega\mu}{\beta} = \frac{k\eta}{\beta}, \tag{4.22}$$

which is seen to be frequency dependent. TE waves can be supported inside closed conductors, as well as between two or more conductors.

TM Waves

Transverse magnetic (TM) waves, (also referred to as E-waves) are characterized by $E_z \neq 0$ and $H_z = 0$. Equations (4.5) then reduce to

$$H_x = \frac{j\omega\epsilon}{k_c^2} \frac{\partial E_z}{\partial y}, \tag{4.23a}$$

$$H_y = \frac{-j\omega\epsilon}{k_c^2} \frac{\partial E_z}{\partial x}, \tag{4.23b}$$

$$E_x = \frac{-j\beta}{k_c^2} \frac{\partial E_z}{\partial x}, \tag{4.23c}$$

$$E_y = \frac{-j\beta}{k_c^2} \frac{\partial E_z}{\partial y}. \tag{4.23d}$$

As in the TE case, $k_c \neq 0$, and the propagation constant $\beta = \sqrt{k^2 - k_c^2}$ is a function of frequency and the geometry of the line or guide. E_z is found from the Helmholtz wave equation,

$$\left(\frac{\partial^2}{\partial x^2} + \frac{\partial^2}{\partial y^2} + \frac{\partial^2}{\partial z^2} + k^2\right) E_z = 0, \tag{4.24}$$

which, since $E_z(x, y, z) = e_z(x, y)e^{-j\beta z}$, can be reduced to a two-dimensional wave equation for e_z:

$$\left(\frac{\partial^2}{\partial x^2} + \frac{\partial^2}{\partial y^2} + k_c^2\right) e_z = 0, \tag{4.25}$$

since $k_c^2 = k^2 - \beta^2$. This equation must be solved subject to the boundary conditions of the specific guide geometry.

The TM wave impedance can be found as

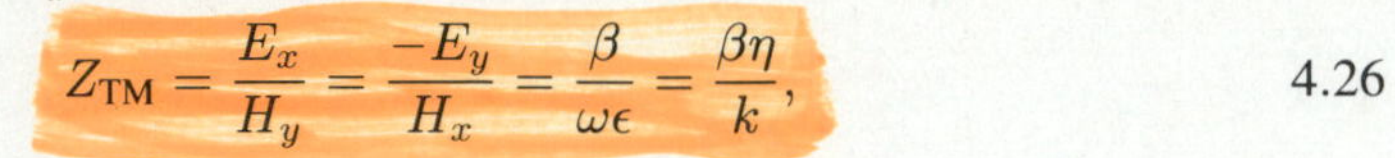

$$Z_{\text{TM}} = \frac{E_x}{H_y} = \frac{-E_y}{H_x} = \frac{\beta}{\omega\epsilon} = \frac{\beta\eta}{k}, \qquad 4.26$$

which is frequency dependent. As for TE waves, TM waves can be supported inside closed conductors, as well as between two or more conductors.

The procedure for analyzing TE and TM waveguides can be summarized as follows:

1. Solve the reduced Helmholtz equation, (4.21) or (4.25), for h_z or e_z. The solution will contain several unknown constants, and the unknown cutoff wavenumber, k_c.
2. Use (4.19) or (4.23) to find the transverse fields from h_z or e_z.
3. Apply the boundary conditions to the appropriate field components to find the unknown constants and k_c.
4. The propagation constant is given by (4.6), and the wave impedance by (4.22) or (4.26).

Attenuation Due to Dielectric Loss

Attenuation in a transmission line or waveguide can be caused by either dielectric loss or conductor loss. If α_d is the attenuation constant due to dielectric loss, and α_c is the attenuation constant due to conductor loss, then the total attenuation constant is $\alpha = \alpha_d + \alpha_c$.

Attenuation caused by conductor loss can be calculated using the perturbation method of Section 3.9; this loss depends on the field distribution in the guide, and so must be evaluated separately for each type of transmission line or waveguide. But if the line or guide is completely filled with a homogeneous dielectric, the attenuation due to lossy dielectric can be calculated from the propagation constant, and this result will apply to any guide or line with a homogeneous dielectric filling.

Thus, using the complex dielectric constant allows the complex propagation constant to be written as

$$\begin{aligned}\gamma = \alpha_d + j\beta &= \sqrt{k_c^2 - k^2} \\ &= \sqrt{k_c^2 - \omega^2\mu_0\epsilon_0\epsilon_r(1 - j\tan\delta)}.\end{aligned} \qquad 4.27$$

In practice, most dielectric materials have a very small loss ($\tan\delta << 1$), so this expression can be simplified by using the first two terms of the Taylor expansion,

$$\sqrt{a^2 + x^2} \simeq a + \frac{1}{2}\left(\frac{x^2}{a}\right), \quad \text{for } x << a.$$

Then (4.27) reduces to

$$\gamma = \sqrt{k_c^2 - k^2 + jk^2 \tan\delta}$$
$$\simeq \sqrt{k_c^2 - k^2} + \frac{jk^2 \tan\delta}{2\sqrt{k_c^2 - k^2}}$$
$$= \frac{k^2 \tan\delta}{2\beta} + j\beta, \qquad 4.28$$

since $\sqrt{k_c^2 - k^2} = j\beta$. In these results, $k^2 = \omega^2 \mu_0 \epsilon_0 \epsilon_r$ is the (real) wavenumber in the absence of loss. Equation (4.28) shows that when the loss is small the phase constant, β, is unchanged, while the attenuation constant due to dielectric loss is given by

$$\alpha_d = \frac{k^2 \tan\delta}{2\beta} \text{Np/m (TE or TM waves).} \qquad 4.29$$

This result applies to any TE or TM wave, as long as the guide is completely filled with the dielectric. It can also be used for TEM lines, where $k_c = 0$, by letting $\beta = k$:

$$\alpha_d = \frac{k \tan\delta}{2} \text{Np/m (TEM waves).} \qquad 4.30$$

4.2 PARALLEL PLATE WAVEGUIDE

The parallel plate waveguide is probably the simplest type of guide that can support TM and TE modes; it can also support a TEM mode, since it is formed from two flat plates, or strips, as shown in Figure 4.2. Although an idealization, this guide is also important for practical reasons, since its operation is quite similar to a variety of other waveguides, and models the propagation of higher order modes in stripline.

In the geometry of the parallel plate waveguide in Figure 4.2, the strip width W is assumed to be much greater than the separation, d, so that fringing fields and any x-variation can be ignored. A material with permittivity ϵ and permeability μ is assumed to fill the region between the two plates. We will discuss solutions for TEM, TM, and TE waves.

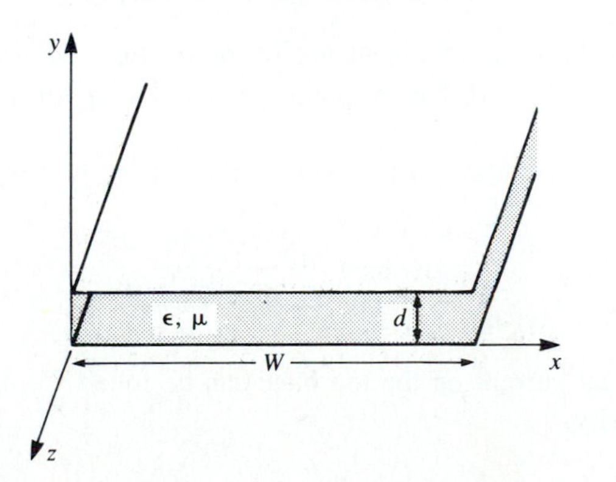

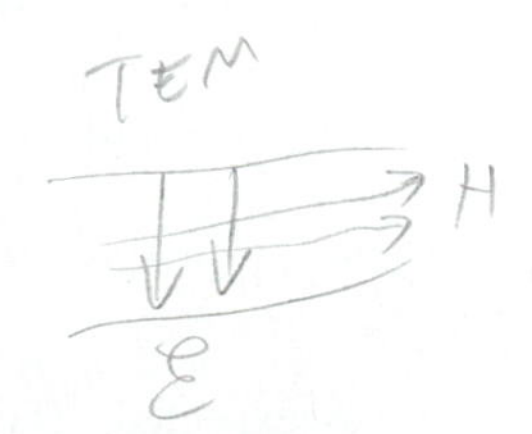

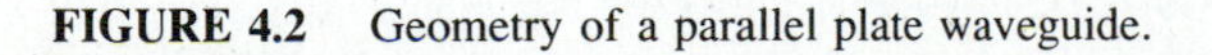

FIGURE 4.2 Geometry of a parallel plate waveguide.

TEM Modes

As discussed in Section 4.1, the TEM mode solution can be obtained by solving Laplace's equation, (4.14), for the electrostatic potential $\Phi(x, y)$ between the two plates. Thus,

$$\nabla_t^2 \Phi(x, y) = 0, \qquad \text{for } 0 \leq x \leq W, \quad 0 \leq y \leq d. \tag{4.31}$$

If we assume that the bottom plate is at ground (zero) potential and the top plate at a potential of V_o, then the boundary conditions for $\Phi(x, y)$ are

$$\Phi(x, 0) = 0, \tag{4.32a}$$

$$\Phi(x, d) = V_o. \tag{4.32b}$$

Since there is no variation in x, the general solution to (4.31) for $\Phi(x, y)$ is

$$\Phi(x, y) = A + By,$$

and the constants A, B can be evaluated from the boundary conditions of (4.32) to give the final solution as

$$\Phi(x, y) = V_o y / d. \tag{4.33}$$

The transverse electric field is, from (4.13),

$$\bar{e}(x, y) = -\nabla_t \Phi(x, y) = -\hat{y}\frac{V_o}{d}, \tag{4.34}$$

so that the total electric field is

$$\bar{E}(x, y, z) = \bar{e}(x, y)e^{-jkz} = -\hat{y}\frac{V_o}{d}e^{-jkz}, \tag{4.35}$$

where $k = \omega\sqrt{\mu\epsilon}$ is the propagation constant of the TEM wave, as in (4.8). The magnetic field, from (4.18), is

$$\bar{H}(x, y, z) = \frac{1}{\eta}\hat{z} \times \bar{E}(x, y, z) = \hat{x}\frac{V_o}{\eta d}e^{-jkz}, \tag{4.36}$$

where $\eta = \sqrt{\mu/\epsilon}$ is the intrinsic impedance of the medium between the parallel plates. Note that $E_z = H_z = 0$, and that the fields are similar in form to a plane wave in a homogeneous region.

The voltage of the top plate with respect to the bottom plate can be calculated from (4.15) and (4.35) as

$$V = -\int_{y=0}^{d} E_y dy = V_o e^{-jkz}, \tag{4.37}$$

as expected. The total current on the top plate can be found from Ampere's law or the surface current density:

$$I = \int_{x=0}^{w} \bar{J}_s \cdot \hat{z} dx = \int_{x=0}^{w} (-\hat{y} \times \bar{H}) \cdot \hat{z} dx = \int_{x=0}^{w} H_x dx = \frac{wV_o}{\eta d}e^{-jkz}. \tag{4.38}$$

Thus the characteristic impedance can be found as

$$Z_0 = \frac{V}{I} = \frac{\eta d}{w}, \tag{4.39}$$

which is seen to be a constant dependent only on the geometry and material parameters of the guide. The phase velocity is also a constant:

$$v_p = \frac{\omega}{\beta} = \frac{1}{\sqrt{\mu\epsilon}}, \tag{4.40}$$

which is the speed of light in the material medium.

Attenuation due to dielectric loss is given by (4.30). The formula for conductor attenuation will be derived in the next subsection, as a special case of TM mode attenuation.

TM Modes

As discussed in Section 4.1, TM waves are characterized by $H_z = 0$ and a nonzero E_z field which satisfies the reduced wave equation of (4.25), with $\partial/\partial x = 0$:

$$\left(\frac{\partial^2}{\partial y^2} + k_c^2\right) e_z(x, y) = 0, \tag{4.41}$$

where $k_c^2 = k^2 - \beta^2$ is the cutoff wavenumber, and $E_z(x, y, z) = e_z(x, y)e^{-j\beta z}$. The general solution to (4.41) is of the form

$$e_z(x, y) = A \sin k_c y + B \cos k_c y, \tag{4.42}$$

subject to the boundary conditions that

$$e_z(x, y) = 0, \qquad \text{at } y = 0, d. \tag{4.43}$$

This implies that $B = 0$ and $k_c d = n\pi$, for $n = 0, 1, 2, 3 \cdots$, or

$$k_c = \frac{n\pi}{d}, \qquad n = 0, 1, 2, 3 \cdots. \tag{4.44}$$

Thus the cutoff wavenumber k_c is constrained to discrete values as given by (4.44); this implies that the propagation constant β is given by

$$\beta = \sqrt{k^2 - k_c^2} = \sqrt{k^2 - (n\pi/d)^2}. \tag{4.45}$$

The solution for $e_z(x, y)$ is then

$$e_z(x, y) = A_n \sin \frac{n\pi y}{d}, \tag{4.46}$$

thus,

$$E_z(x, y, z) = A_n \sin \frac{n\pi y}{d} e^{-j\beta z}. \tag{4.47}$$

The transverse field components can be found, using (4.23), to be

$$H_x = \frac{j\omega\epsilon}{k_c} A_n \cos \frac{n\pi y}{d} e^{-j\beta z}, \tag{4.48a}$$

$$E_y = \frac{-j\beta}{k_c} A_n \cos \frac{n\pi y}{d} e^{-j\beta z}, \tag{4.48b}$$

$$E_x = H_y = 0. \tag{4.48c}$$

Observe that for $n = 0$, $\beta = k = \omega\sqrt{\mu\epsilon}$, and that $E_z = 0$. The E_y and H_x fields are then constant in y, so that the TM_o mode is actually identical to the TEM mode. For $n \geq 1$, however, the situation is different. Each value of n corresponds to a different TM mode, denoted as the TM_n mode, and each mode has its own propagation constant given by (4.45), and field expressions as given by (4.48).

From (4.45) it can be seen that β is real only when $k > k_c$. Since $k = \omega\sqrt{\mu\epsilon}$ is proportional to frequency, the TM_n modes (for $n > 0$) exhibit a cutoff phenomenon, whereby no propagation will occur until the frequency is such that $k > k_c$. The cutoff frequency of the TM_n mode can then be deduced as

$$f_c = \frac{k_c}{2\pi\sqrt{\mu\epsilon}} = \frac{n}{2d\sqrt{\mu\epsilon}}. \tag{4.49}$$

Thus, the TM mode which propagates at the lowest frequency is the TM_1 mode, with a cutoff frequency of $f_c = 1/2d\sqrt{\mu\epsilon}$; the TM_2 mode has a cutoff frequency equal to twice this value, and so on. At frequencies below the cutoff frequency of a given mode, the propagation constant is purely imaginary, corresponding to a rapid exponential decay of the fields. Such modes are referred to as cutoff, or evanescent, modes. TM_n mode propagation is analogous to a high-pass filter response.

The wave impedance of the TM modes, from (4.26), is a function of frequency:

$$Z_{TM} = \frac{-E_y}{H_x} = \frac{\beta}{\omega\epsilon} = \frac{\beta\eta}{k}, \tag{4.50}$$

which we see is pure real for $f > f_c$, but pure imaginary for $f < f_c$. The phase velocity is also a function of frequency:

$$v_p = \frac{\omega}{\beta}, \tag{4.51}$$

and is seen to be greater than $1/\sqrt{\mu\epsilon} = \omega/k$, the speed of light in the medium, since $\beta < k$. The guide wavelength is defined as

$$\lambda_g = \frac{2\pi}{\beta}, \tag{4.52}$$

and is the distance between equi-phase planes along the z-axis. Note that $\lambda_g > \lambda = 2\pi/k$, the wavelength of a plane wave in the material. The phase velocity and guide wavelength are only defined for a propagating mode, for which β is real. One may also define a cutoff wavelength as

$$\lambda_c = 2d. \tag{4.53}$$

It is instructive to compute the Poynting vector to see how power propagates in the TM_n mode. From (2.91), the time-average power passing a transverse cross-section of the parallel plate guide is

$$P_o = \frac{1}{2}\text{Re}\int_{x=0}^{w}\int_{y=0}^{d} \bar{E}\times\bar{H}^*\cdot\hat{z}dy\,dx = -\frac{1}{2}\text{Re}\int_{x=0}^{w}\int_{y=0}^{d} E_y H_x^* dy\,dx$$

$$= \frac{w\text{Re}(\beta)\omega\epsilon}{2k_c^2}|A_n|^2\int_{y=0}^{d}\cos^2\frac{n\pi y}{d}dy = \begin{cases} \dfrac{w\text{Re}(\beta)\omega\epsilon d}{4k_c^2}|A_n|^2 & \text{for } n>0 \\ \dfrac{w\text{Re}(\beta)\omega\epsilon d}{2k_c^2}|A_n|^2 & \text{for } n=0 \end{cases} \tag{4.54}$$

where (4.48,a,b) were used for E_y, H_x. Thus, P_o is positive and nonzero when β is real, which occurs for $f > f_c$. When the mode is below cutoff, β is imaginary and so $P_o = 0$.

The TM (or TE) waveguide mode propagation has an interesting interpretation when viewed as a pair of bouncing plane waves. For example, consider the dominant TM_1 mode which has a propagation constant,

$$\beta_1 = \sqrt{k^2 - (\pi/d)^2}, \tag{4.55}$$

and E_z field,

$$E_z = A_1 \sin\frac{\pi y}{d}e^{-j\beta_1 z},$$

which can be rewritten as

$$E_z = \frac{A_1}{2j}\{e^{j[\pi y/d - \beta_1 z]} - e^{-j[\pi y/d + \beta_1 z]}\}. \tag{4.56}$$

This result is in the form of two plane waves traveling obliquely, in the $-y, +z$ and $+y, +z$ directions, respectively, as shown in Figure 4.3. By comparison with the phase factor of (2.132), the angle θ that each plane wave makes with the z-axis satisfies the relations

$$k\sin\theta = \frac{\pi}{d}, \tag{4.57a}$$

$$k\cos\theta = \beta_1, \tag{4.57b}$$

so that $(\pi/d)^2 + \beta_1^2 = k^2$, as in (4.55). For $f > f_c, \beta$ is real and less than k_1, so θ is some angle between 0° and 90°, and the mode can be thought of as two plane waves alternately bouncing off of the top and bottom plates.

The phase velocity of each plane wave along its direction of propagation (θ direction) is $\omega/k = 1/\sqrt{\mu\epsilon}$, which is the speed of light in the material filling the guide. But the phase velocity of the plane waves in the z direction is $\omega/\beta_1 = 1/\sqrt{\mu\epsilon}\cos\theta$, which is greater than the speed of light in the material. (This situation is analogous to ocean waves hitting a shore line: the intersection point of the shore and an obliquely incident

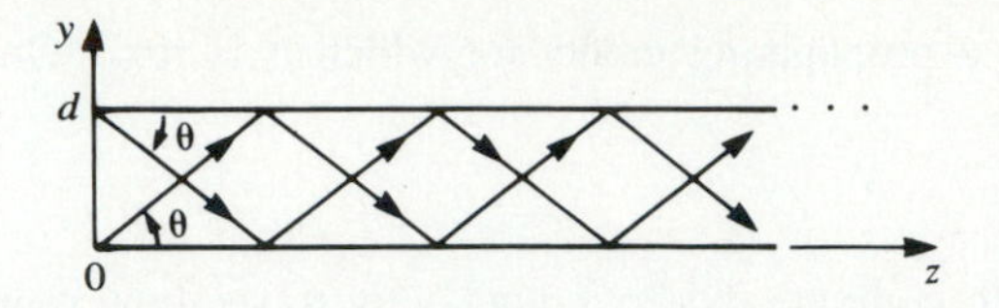

FIGURE 4.3 Bouncing plane wave interpretation of the TM_1 parallel plate waveguide mode.

wave crest moves faster than the wave crest itself.) The superposition of the two plane wave fields is such that complete cancellation occurs at $y = 0$ and $y = d$, to satisfy the boundary condition that $E_z = 0$ at these planes. As f decreases to f_c, β_1 approaches zero so that, by (4.57b), θ approaches 90°. The two plane waves are then bouncing up and down with no motion in the $+z$ direction, and no real power flow occurs in the z direction.

Attenuation due to dielectric loss can be found from (4.29). Conductor loss can be treated using the perturbation method. Thus,

$$\alpha_c = \frac{P_\ell}{2P_o}, \tag{4.58}$$

where P_o is the power flow down the guide in the absence of conductor loss, as given by (4.54). P_ℓ is the power dissipated per unit length in the two lossy conductors, and can be found from (3.105) as

$$P_\ell = 2\left(\frac{R_s}{2}\right)\int_{x=0}^{w} |\bar{J}_s|^2 dx = \frac{\omega^2\epsilon^2 R_s w}{k_c^2}|A_n|^2, \tag{4.59}$$

where R_s is the surface resistivity of the conductors. Using (4.54) and (4.59) in (4.58) gives the attenuation due to conductor loss as

$$\alpha_c = \frac{2\omega\epsilon R_s}{\beta d} = \frac{2kR_s}{\beta\eta d} \text{ Np/m}, \qquad \text{for } n > 0. \tag{4.60}$$

As discussed previously, the TEM mode is identical to the TM_0 mode for the parallel plate waveguide, so the above attenuation results for the TM_n mode can be used to obtain the TEM mode attenuation by letting $n = 0$. For attenuation due to conductor loss for the TEM mode, the $n = 0$ result of (4.54) must be used in (4.58), to obtain

$$\alpha_c = \frac{R_s}{\eta d} \text{ Np/m}. \tag{4.61}$$

TE Modes

TE modes, characterized by $E_z = 0$, can also propagate on the parallel plate waveguide. From (4.21), with $\partial/\partial x = 0, H_z$ must satisfy the reduced wave equation,

$$\left(\frac{\partial^2}{\partial y^2} + k_c^2\right)h_z(x, y) = 0, \tag{4.62}$$

where $k_c^2 = k^2 - \beta^2$ is the cutoff wavenumber, and $H_z(x,y,z) = h_z(x,y)e^{-j\beta z}$. The general solution to (4.62) is

$$h_z(x,y) = A\sin k_c y + B\cos k_c y. \tag{4.63}$$

The boundary conditions are that $E_x = 0$ at $y = 0$, d; E_z is identically zero for TE modes. From (4.19c), we have

$$E_x = \frac{-j\omega\mu}{k_c}\left[A\cos k_c y - B\sin k_c y\right]e^{-j\beta z}, \tag{4.64}$$

and applying the boundary conditions shows that $A = 0$ and,

$$k_c = \frac{n\pi}{d}, \qquad n = 1, 2, 3\cdots, \tag{4.65}$$

as for the TM case. The final solution for H_z is then

$$H_z(x,y) = B_n \cos\frac{n\pi y}{d}e^{-j\beta z}. \tag{4.66}$$

The transverse fields can be computed from (4.19) as

$$E_x = \frac{j\omega\mu}{k_c}B_n \sin\frac{n\pi y}{d}e^{-j\beta z}, \tag{4.67a}$$

$$H_y = \frac{j\beta}{k_c}B_n \sin\frac{n\pi y}{d}e^{-j\beta z}, \tag{4.67b}$$

$$E_y = H_x = 0. \tag{4.67c}$$

The propagation constant of the TE_n mode is thus,

$$\beta = \sqrt{k^2 - \left(\frac{n\pi}{d}\right)^2}, \tag{4.68}$$

which is the same as the propagation constant of the TM_n mode. The cutoff frequency of the TE_n mode is

$$f_c = \frac{n}{2d\sqrt{\mu\epsilon}}, \tag{4.69}$$

which is also identical to that of the TM_n mode. The wave impedance of the TE_n mode is, from (4.22),

$$Z_{\text{TE}} = \frac{E_x}{H_y} = \frac{\omega\mu}{\beta} = \frac{k\eta}{\beta}, \tag{4.70}$$

which is seen to be real for propagating modes, and imaginary for nonpropagating, or cutoff, modes. The phase velocity, guide wavelength, and cutoff wavelength are similar to the results for the TM modes.

The power flow down the guide for a TE_n mode can be calculated as

$$P_o = \frac{1}{2}\text{Re}\int_{x=0}^{w}\int_{y=0}^{d} \bar{E}\times\bar{H}^*\cdot\hat{z}dy\,dx = \frac{1}{2}\text{Re}\int_{x=0}^{w}\int_{y=0}^{d} E_x H_y^* dy\,dx$$
$$= \frac{\omega\mu d w}{4k_c^2}|B_n|^2\text{Re}(\beta), \qquad \text{for } n > 0, \tag{4.71}$$

which is zero if the operating frequency is below the cutoff frequency (β imaginary).

Note that if $n = 0$, then $E_x = H_y = 0$ from (4.67), and thus $P_o = 0$, implying that there is no TE_o mode.

Attenuation can be calculated in the same way as for the TM modes. The attenuation due to dielectric loss is given by (4.29). It is left as a problem to show that the attenuation due to conductor loss for the TE modes is given by

$$\alpha_c = \frac{2k_c^2 R_s}{\omega\mu\beta d} = \frac{2k_c^2 R_s}{k\beta\eta d}\ \text{Np/m}. \tag{4.72}$$

Figure 4.4 shows the attenuation due to conductor loss for the TEM, TM_1, and TE_1 modes. Observe that $\alpha_c \to \infty$ as cutoff is approached for the TM and TE modes.

Table 4.1 summarizes a number of useful results for TEM, TM, and TE mode propagation on parallel plate waveguides. Field lines for the TEM, TM_1, and TE_1 modes are shown in Figure 4.5.

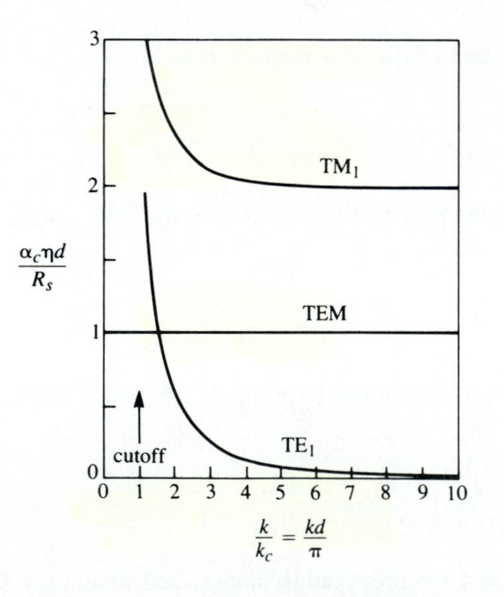

FIGURE 4.4 Attenuation due to conductor loss for the TEM, TM_1, and TE_1 modes of a parallel plate waveguide.

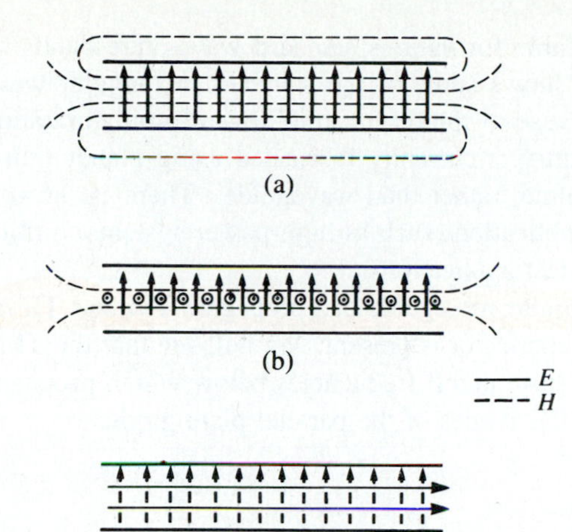

FIGURE 4.5 Field lines for the (a) TEM, (b) TM_1, and (c) TE_1 modes of a parallel plate waveguide. No variation across width.

TABLE 4.1 Summary of Results for Parallel Plate Waveguide

Quantity	TEM Mode	TM_n Mode	TE_n Mode
k	$\omega\sqrt{\mu\epsilon}$	$\omega\sqrt{\mu\epsilon}$	$\omega\sqrt{\mu\epsilon}$
k_c	0	$n\pi/d$	$n\pi/d$
β	$k=\omega\sqrt{\mu\epsilon}$	$\sqrt{k^2-k_c^2}$	$\sqrt{k^2-k_c^2}$
λ_c	∞	$2\pi/k_c = 2d/n$	$2\pi/k_c = 2d/n$
λ_g	$2\pi/k$	$2\pi/\beta$	$2\pi/\beta$
v_p	$\omega/k = 1/\sqrt{\mu\epsilon}$	ω/β	ω/β
α_d	$(k\tan\delta)/2$	$(k^2\tan\delta)/2\beta$	$(k^2\tan\delta)/2\beta$
α_c	$R_s/\eta d$	$2kR_s/\beta\eta d$	$2k_c^2R_s/k\beta\eta d$
E_z	0	$A_n\sin(n\pi y/d)e^{-j\beta z}$	0
H_z	0	0	$B_n\cos(n\pi y/d)e^{-j\beta z}$
E_x	0	0	$(j\omega\mu/k_c)B_n\sin(n\pi y/d)e^{-j\beta z}$
E_y	$(-V_o/d)e^{-j\beta z}$	$(-j\beta/k_c)A_n\cos(n\pi y/d)e^{-j\beta z}$	0
H_x	$(V_o/\eta d)e^{-j\beta z}$	$(j\omega\epsilon/k_c)A_n\cos(n\pi y/d)e^{-j\beta z}$	0
H_y	0	0	$(j\beta/k_c)B_n\sin(n\pi y/d)e^{-j\beta z}$
Z	$Z_{TEM} = \eta d/w$	$Z_{TM} = \beta\eta/k$	$Z_{TE} = k\eta/\beta$

4.3 RECTANGULAR WAVEGUIDE

Rectangular waveguides were one of the earliest types of transmission lines used to transport microwave signals, and are still used today for many applications. A large variety of components such as couplers, detectors, isolators, attenuators, and slotted lines are commercially available for various standard waveguide bands from 1 GHz to over 220 GHz. Figure 4.6 shows some of the standard rectangular waveguide components that are available. Because of the recent trend toward miniaturization and integration, a lot of microwave circuitry is currently fabricated using planar transmission lines, such as microstrip and stripline, rather than waveguide. There is, however, still a need for waveguides in many applications such as high-power systems, millimeter wave systems, and in some precision test applications.

The hollow rectangular waveguide can propagate TM and TE modes, but not TEM waves, since only one conductor is present. We will see that the TM and TE modes of a rectangular waveguide have cutoff frequencies below which propagation is not possible, similar to the TM and TE modes of the parallel plate guide.

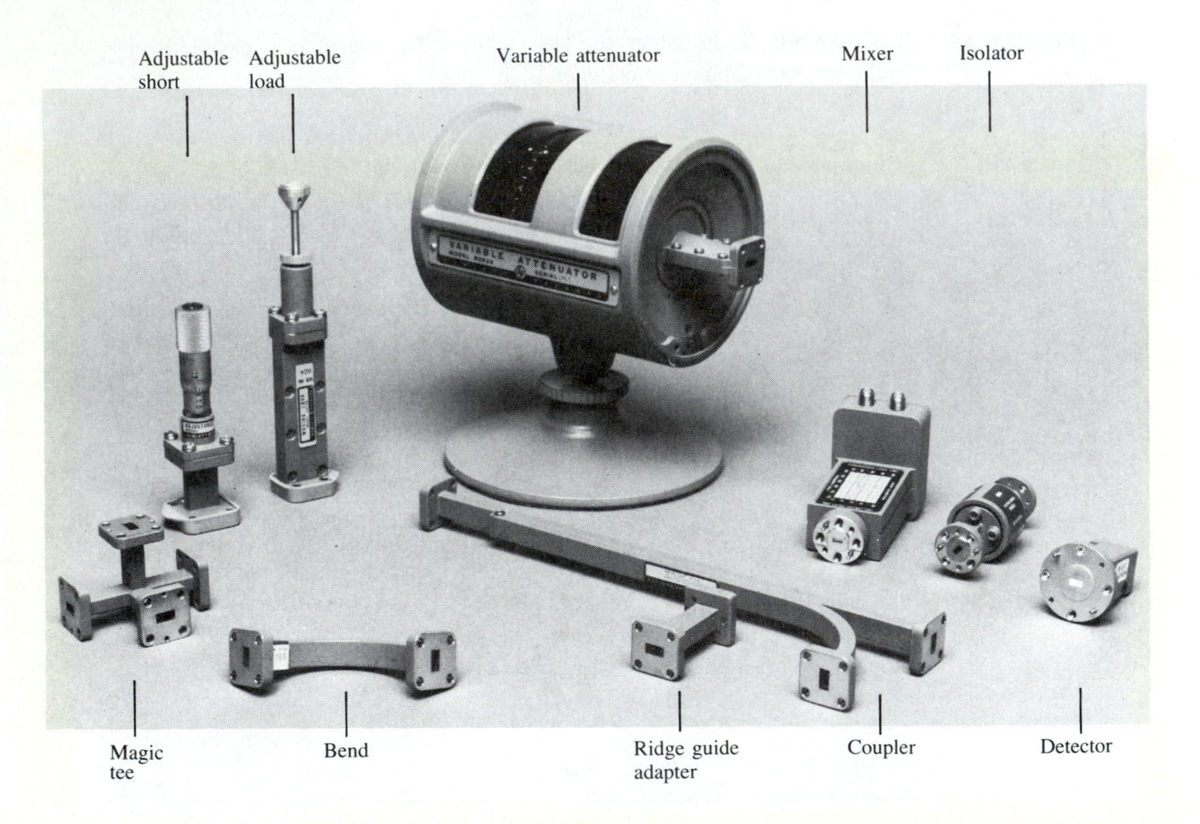

FIGURE 4.6 Photograph of Ka-band (WR-28) rectangular waveguide components. Clockwise from top: a variable attenuator, a waveguide mixer, an isolator, a detector, a directional coupler, an adapter to ridge waveguide, a waveguide bend, an E-H (magic) tee junction, an adjustable short, and a sliding matched load.

Courtesy of Hewlett-Packard Company, Santa Rosa, Calif.

TE Modes

The geometry of a rectangular waveguide is shown in Figure 4.7, where it is assumed that the guide is filled with a material of permittivity ϵ and permeability μ. It is standard convention to have the longest side of the waveguide along the x-axis, so that $a > b$.

The TE modes are characterized by fields with $E_z = 0$, while H_z must satisfy the reduced wave equation of (4.21):

$$\left(\frac{\partial^2}{\partial x^2} + \frac{\partial^2}{\partial y^2} + k_c^2 \right) h_z(x, y) = 0, \qquad 4.73$$

with $H_z(x, y, z) = h_z(x, y)e^{-j\beta z}$, and $k_c^2 = k^2 - \beta^2$ is the cutoff wavenumber. The partial differential equation of (4.73) can be solved by the method of separation of variables by letting

$$h_z(x, y) = X(x)Y(y), \qquad 4.74$$

and substituting into (4.73) to obtain

$$\frac{1}{X}\frac{d^2X}{dx^2} + \frac{1}{Y}\frac{d^2Y}{dy^2} + k_c^2 = 0. \qquad 4.75$$

Then, by the usual separation of variables argument, each of the terms in (4.75) must be equal to a constant, so we define separation constants k_x and k_y, such that

$$\frac{d^2X}{dx^2} + k_x^2 X = 0, \qquad 4.76a$$

$$\frac{d^2Y}{dy^2} + k_y^2 Y = 0, \qquad 4.76b$$

and

$$k_x^2 + k_y^2 = k_c^2. \qquad 4.77$$

The general solution for h_z can then be written as

$$h_z(x, y) = (A \cos k_x x + B \sin k_x x)(C \cos k_y y + D \sin k_y y). \qquad 4.78$$

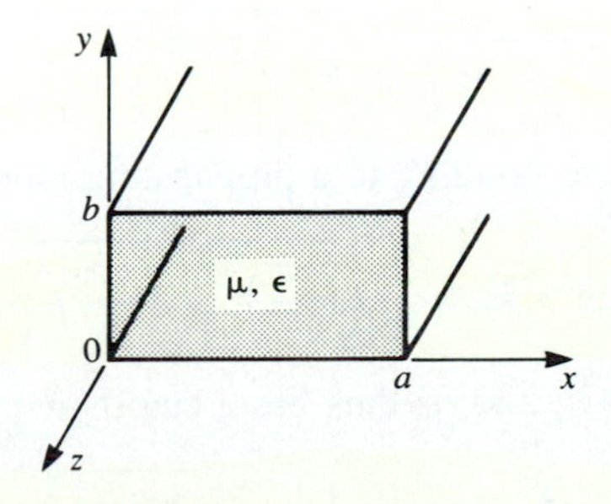

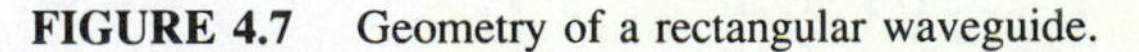

FIGURE 4.7 Geometry of a rectangular waveguide.

To evaluate the constants in (4.78) we must apply the boundary conditions on the electric field components tangential to the waveguide walls. That is,

$$e_x(x, y) = 0, \quad \text{at } y = 0, b, \tag{4.79a}$$

$$e_y(x, y) = 0, \quad \text{at } x = 0, a. \tag{4.79b}$$

We thus cannot use h_z of (4.78) directly, but must first use (4.19c) and (4.19d) to find e_x and e_y from h_z:

$$e_x = \frac{-j\omega\mu}{k_c^2} k_y (A \cos k_x x + B \sin k_x x)(-C \sin k_y y + D \cos k_y y), \tag{4.80a}$$

$$e_y = \frac{-j\omega\mu}{k_c^2} k_x (-A \sin k_x x + B \cos k_x x)(C \cos k_y y + D \sin k_y y). \tag{4.80b}$$

Then from (4.79a) and (4.80a), we see that $D = 0$, and $k_y = n\pi/b$ for $n = 0, 1, 2 \ldots$. From (4.79b) and (4.80b) we have that $B = 0$ and $k_x = m\pi/a$ for $m = 0, 1, 2 \ldots$. The final solution for H_z is then

$$H_z(x, y, z) = A_{mn} \cos \frac{m\pi x}{a} \cos \frac{n\pi y}{b} e^{-j\beta z}, \tag{4.81}$$

where A_{mn} is an arbitrary amplitude constant composed of the remaining constants A and C of (4.78).

The transverse field components of the TE_{mn} mode can then be found using (4.19) and (4.81):

$$E_x = \frac{j\omega\mu n\pi}{k_c^2 b} A_{mn} \cos \frac{m\pi x}{a} \sin \frac{n\pi y}{b} e^{-j\beta z}, \tag{4.82a}$$

$$E_y = \frac{-j\omega\mu m\pi}{k_c^2 a} A_{mn} \sin \frac{m\pi x}{a} \cos \frac{n\pi y}{b} e^{-j\beta z}, \tag{4.82b}$$

$$H_x = \frac{j\beta m\pi}{k_c^2 a} A_{mn} \sin \frac{m\pi x}{a} \cos \frac{n\pi y}{b} e^{-j\beta z}, \tag{4.82c}$$

$$H_y = \frac{j\beta n\pi}{k_c^2 b} A_{mn} \cos \frac{m\pi x}{a} \sin \frac{n\pi y}{b} e^{-j\beta z}. \tag{4.82d}$$

The propagation constant is

$$\beta = \sqrt{k^2 - k_c^2} = \sqrt{k^2 - \left(\frac{m\pi}{a}\right)^2 - \left(\frac{n\pi}{b}\right)^2}, \tag{4.83}$$

which is seen to be real, corresponding to a propagating mode, when

$$k > k_c = \sqrt{\left(\frac{m\pi}{a}\right)^2 + \left(\frac{n\pi}{b}\right)^2}.$$

Each mode (combination of m and n) thus has a cutoff frequency $f_{c_{mn}}$ given by

$$f_{c_{mn}} = \frac{k_c}{2\pi\sqrt{\mu\epsilon}} = \frac{1}{2\pi\sqrt{\mu\epsilon}} \sqrt{\left(\frac{m\pi}{a}\right)^2 + \left(\frac{n\pi}{b}\right)^2}. \tag{4.84}$$

The mode with the lowest cutoff frequency is called the dominant mode; since we have assumed $a > b$, the lowest f_c occurs for the TE_{10} $(m = 1, n = 0)$ mode:

$$f_{c_{10}} = \frac{1}{2a\sqrt{\mu\epsilon}}. \tag{4.85}$$

Thus the TE_{10} mode is the dominant TE mode and, as we will see, the overall dominant mode of the rectangular waveguide. Observe that the field expressions for $\bar{E}$ and $\bar{H}$ in (4.82) are all zero if both $m = n = 0$; thus there is no TE_{00} mode.

At a given operating frequency f, only those modes having $f_c < f$ will propagate; modes with $f_c > f$ will lead to an imaginary β (or real α), meaning that all field components will decay exponentially away from the source of excitation. Such modes are referred to as cutoff, or evanescent, modes. If more than one mode is propagating, the waveguide is said to be *overmoded.*

From (4.22) the wave impedance that relates the transverse electric and magnetic fields is

$$Z_{\mathrm{TE}} = \frac{E_x}{H_y} = \frac{-E_y}{H_x} = \frac{k\eta}{\beta}, \tag{4.86}$$

where $\eta = \sqrt{\mu/\epsilon}$ is the intrinsic impedance of the material filling the waveguide. Note that Z_{TE} is real when β is real (a propagating mode), but is imaginary when β is imaginary (an evanescent mode).

The guide wavelength is defined as the distance between two equal phase planes along the waveguide, and is equal to

$$\lambda_g = \frac{2\pi}{\beta} > \frac{2\pi}{k} = \lambda, \tag{4.87}$$

which is thus greater than λ, the wavelength of a plane wave in the filling medium. The phase velocity is

$$v_p = \frac{\omega}{\beta} > \frac{\omega}{k} = 1/\sqrt{\mu\epsilon}, \tag{4.88}$$

which is greater than $1/\sqrt{\mu\epsilon}$, the speed of light (plane wave) in the filling material.

In the vast majority of applications the operating frequency and guide dimensions are chosen so that only the dominant TE_{10} mode will propagate. Because of the practical importance of the TE_{10} mode, then, we will list the field components and derive the attenuation due to conductor loss for this case.

Specializing (4.81) and (4.82) to the $m = 1, n = 0$ case gives the following results for the TE_{10} mode fields:

$$H_z = A_{10} \cos \frac{\pi x}{a} e^{-j\beta z}, \tag{4.89a}$$

$$E_y = \frac{-j\omega\mu a}{\pi} A_{10} \sin \frac{\pi x}{a} e^{-j\beta z}, \tag{4.89b}$$

$$H_x = \frac{j\beta a}{\pi} A_{10} \sin \frac{\pi x}{a} e^{-j\beta z}, \tag{4.89c}$$

$$E_x = E_z = H_y = 0. \tag{4.89d}$$

In addition, for the TE_{10} mode,

$$k_c = \pi/a, \tag{4.90}$$

and

$$\beta = \sqrt{k^2 - (\pi/a)^2}. \tag{4.91}$$

The power flow down the guide for the TE_{10} mode is calculated as

$$\begin{aligned} P_{10} &= \frac{1}{2}\text{Re}\int_{x=0}^{a}\int_{y=0}^{b} \bar{E}\times\bar{H}^*\cdot\hat{z}\, dy\, dx \\ &= \frac{1}{2}\text{Re}\int_{x=0}^{a}\int_{y=0}^{b} E_y H_x^*\, dy\, dx \\ &= \frac{\omega\mu a^2}{2\pi^2}\text{Re}(\beta)|A_{10}|^2 \int_{x=0}^{a}\int_{y=0}^{b} \sin^2\frac{\pi x}{a}\, dx \\ &= \frac{\omega\mu a^3|A_{10}|^2 b}{4\pi^2}\text{Re}(\beta). \end{aligned} \tag{4.92}$$

Note that this result gives nonzero real power only when β is real, corresponding to a propagating mode.

Attenuation in a rectangular waveguide can occur because of dielectric loss, or conductor loss. Dielectric loss can be treated by making ϵ complex and using a Taylor series approximation, with the general result given in (4.29).

Conductor loss is best treated using the perturbation method. The power lost per unit length due to finite wall conductivity is, from (2.131),

$$P_\ell = \frac{R_s}{2}\int_C |\bar{J}_s|^2 d\ell, \tag{4.93}$$

where R_s is the wall surface resistance, and the integration contour C encloses the perimeter of the guide walls. There are surface currents on all four walls, but from symmetry the currents on the top and bottom walls are identical, as are the currents on the left and right side walls. So we can compute the power lost in the walls at $x = 0$ and $y = 0$, and double their sum to obtain the total power loss. The surface current on the $x = 0$ (left) wall is

$$\bar{J}_s = \hat{n}\times\bar{H}|_{x=0} = \hat{x}\times\hat{z}H_z|_{x=0} = -\hat{y}H_z|_{x=0} = -\hat{y}A_{10}e^{-j\beta z}, \tag{4.94a}$$

while the surface current on the $y = 0$ (bottom) wall is

$$\begin{aligned} \bar{J}_s &= \hat{n}\times\bar{H}|_{y=0} = \hat{y}\times(\hat{x}H_x|_{y=0} + \hat{z}H_z|_{y=0}) \\ &= -\hat{z}\frac{j\beta a}{\pi}A_{10}\sin\frac{\pi x}{a}e^{-j\beta z} + \hat{x}A_{10}\cos\frac{\pi x}{a}e^{-j\beta z}. \end{aligned} \tag{4.94b}$$

Substituting (4.94) into (4.93) gives

$$P_\ell = R_s \int_{y=0}^{b} |J_{sy}|^2 dy + R_s \int_{x=0}^{a} \left[|J_{sx}|^2 + |J_{sz}|^2 \right] dx$$
$$= R_s |A_{10}|^2 \left(b + \frac{a}{2} + \frac{\beta^2 a^3}{2\pi^2} \right). \quad 4.95$$

The attenuation due to conductor loss for the TE_{10} mode is then

$$\alpha_c = \frac{P_\ell}{2P_{10}} = \frac{2\pi^2 R_s [b + a/2 + (\beta^2 a^3)/2\pi^2]}{\omega\mu a^3 b\beta}$$
$$= \frac{R_s}{a^3 b\beta k\eta}(2b\pi^2 + a^3 k^2) \text{ Np/m}. \quad 4.96$$

TM Modes

The TM modes are characterized by fields with $H_z = 0$, while E_z must satisfy the reduced wave equation of (4.25):

$$\left(\frac{\partial^2}{\partial x^2} + \frac{\partial^2}{\partial y^2} + k_c^2 \right) e_z(x, y) = 0, \quad 4.97$$

with $E_z(x, y, z) = e_z(x, y)e^{-j\beta z}$ and $k_c^2 = k^2 - \beta^2$. Equation (4.97) can be solved by the separation of variables procedure that was used for the TE modes. The general solution is then

$$e_z(x, y) = (A \cos k_x x + B \sin k_x x)(C \cos k_y y + D \sin k_y y). \quad 4.98$$

The boundary conditions can be applied directly to e_z:

$$e_z(x, y) = 0, \quad \text{at } x = 0, a, \quad 4.99a$$

$$e_z(x, y) = 0, \quad \text{at } y = 0, b. \quad 4.99b$$

We will see that satisfaction of the above conditions on e_z will lead to satisfaction of the boundary conditions by e_x and e_y.

Applying (4.99a) to (4.98) shows that $A = 0$ and $k_x = m\pi/a$, for $m = 1, 2, 3 \ldots$. Similarly, applying (4.99b) to (4.98) shows that $C = 0$ and $k_y = n\pi/b$, for $n = 1, 2, 3 \ldots$. The solution for E_z then reduces to

$$E_z(x, y) = B_{mn} \sin \frac{m\pi x}{a} \sin \frac{n\pi y}{b} e^{-j\beta z}, \quad 4.100$$

where B_{mn} is an arbitrary amplitude constant.

The transverse field components for the TM_{mn} mode can be computed from (4.23) and (4.100) as

$$E_x = \frac{-j\beta m\pi}{ak_c^2} B_{mn} \cos \frac{m\pi x}{a} \sin \frac{n\pi y}{b} e^{-j\beta z}, \quad 4.101a$$

$$E_y = \frac{-j\beta n\pi}{bk_c^2} B_{mn} \sin \frac{m\pi x}{a} \cos \frac{n\pi y}{b} e^{-j\beta z}, \quad 4.101b$$

$$H_x = \frac{j\omega\epsilon n\pi}{bk_c^2} B_{mn} \sin\frac{m\pi x}{a} \cos\frac{n\pi y}{b} e^{-j\beta z}, \quad 4.101c$$

$$H_y = \frac{-j\omega\epsilon m\pi}{ak_c^2} B_{mn} \cos\frac{m\pi x}{a} \sin\frac{n\pi y}{b} e^{-j\beta z}. \quad 4.101d$$

As for the TE modes, the propagation constant is

$$\beta = \sqrt{k^2 - k_c^2} = \sqrt{k^2 - \left(\frac{m\pi}{a}\right)^2 - \left(\frac{n\pi}{b}\right)^2}, \quad 4.102$$

and is real for propagating modes, and imaginary for evanescent modes. The cutoff frequency for the TM_{mn} modes is also the same as that of the TE_{mn} modes, as given in (4.84). The guide wavelength and phase velocity for TM modes are also the same as those for TE modes.

Observe that the field expressions for $\bar{E}$ and $\bar{H}$ in (4.101) are identically zero if either m or n is zero. Thus there are no TM_{00}, TM_{01}, or TM_{10} modes, and the lowest order TM mode to propagate (lowest f_c) is the TM_{11} mode, having a cutoff frequency of

$$f_{c_{11}} = \frac{1}{2\pi\sqrt{\mu\epsilon}} \sqrt{\left(\frac{\pi}{a}\right)^2 + \left(\frac{\pi}{b}\right)^2}, \quad 4.103$$

which is seen to be larger than $f_{c_{10}}$ for the cutoff frequency of the TE_{10} mode.

The wave impedance relating the transverse electric and magnetic fields is, from (4.26),

$$Z_{TM} = \frac{E_x}{H_y} = \frac{-E_y}{H_x} = \frac{\beta\eta}{k}. \quad 4.104$$

Attenuation due to dielectric loss is computed in the same way as for the TE modes, with the same result. The calculation of attenuation due to conductor loss is left as a problem, while Figure 4.8 shows the attenuation versus frequency for some TE and TM modes in a rectangular waveguide. Table 4.2 summarizes results for TE and TM wave propagation in rectangular waveguides, and Figure 4.9 shows the field lines for several of the lowest-order TE and TM modes.

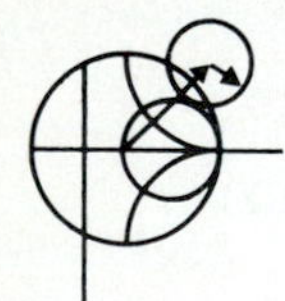

EXAMPLE 4.1

Consider a length of air-filled copper X-band waveguide, with dimensions $a =$ 2.286 cm, $b =$ 1.016 cm. Find the cutoff frequencies of the first four propagating modes. What is the attenuation in dB of a 1 m length of this guide when operating at $f =$ 10 GHz?

Solution

From (4.84), the cutoff frequencies are given by

$$f_{c_{mn}} = \frac{c}{2\pi} \sqrt{\left(\frac{m\pi}{a}\right)^2 + \left(\frac{n\pi}{b}\right)^2}.$$

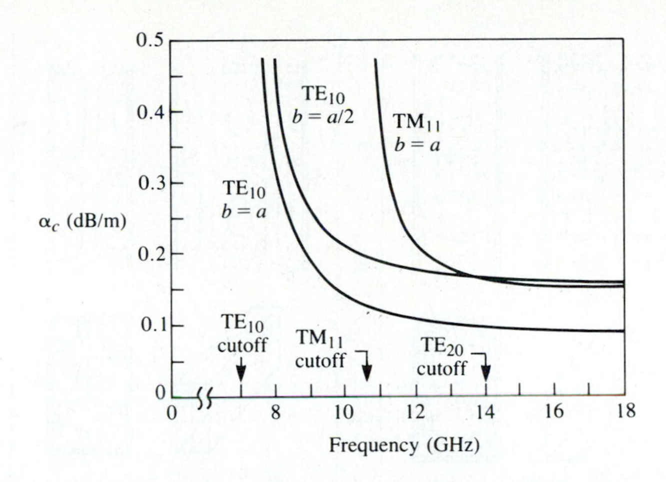

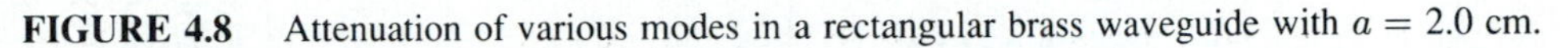

FIGURE 4.8 Attenuation of various modes in a rectangular brass waveguide with $a = 2.0$ cm.

Computing f_c for the first few values of m and n gives:

Mode	m	n	fc_{mn}(GHz)
TE	1	0	6.562
TE	2	0	13.123
TE	0	1	14.764
TE, TM	1	1	16.156
TE, TM	1	2	30.248
TE, TM	2	1	19.753

Thus the TE_{10}, TE_{20}, TE_{01}, and TE_{11} modes will be the first four modes to propagate (the TE_{11} and TM_{11} modes have the same cutoff frequency).

At 10 GHz, $k = 209.44\ \text{m}^{-1}$, and the propagation constant of the TE_{10} mode (the only propagating mode) is

$$\beta = \sqrt{k^2 - \left(\frac{\pi}{a}\right)^2} = \sqrt{\left(\frac{2\pi f}{c}\right)^2 - \left(\frac{\pi}{a}\right)^2} = 158.05\ \text{m}^{-1}.$$

The surface resistivity of the copper walls is ($\sigma = 5.8 \times 10^7 \text{S/m}$)

$$R_s = \sqrt{\frac{\omega\mu}{2\sigma}} = 0.026\ \Omega,$$

so the attenuation constant, from (4.96), is

$$\alpha_c = \frac{R_s}{a^3 b\beta k\eta}(2b\pi^2 + a^3k^2) = 0.0125\ \text{Np/m},$$

$$\alpha_c(\text{dB}) = -20\log e^{-\alpha_c} = 0.11\ \text{dB/m},$$

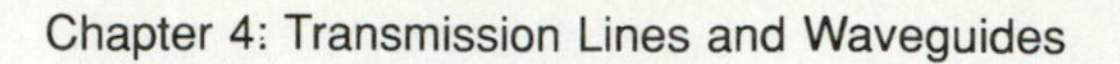

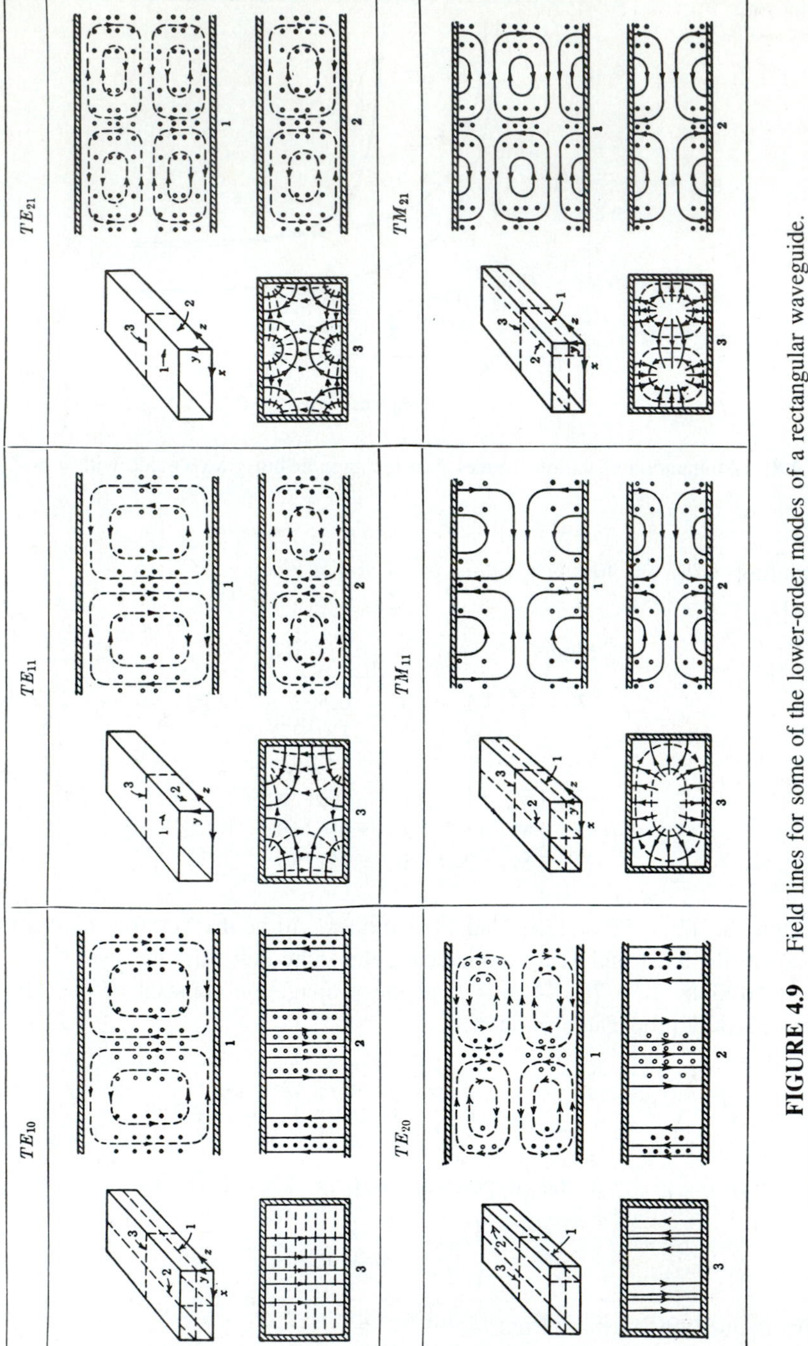

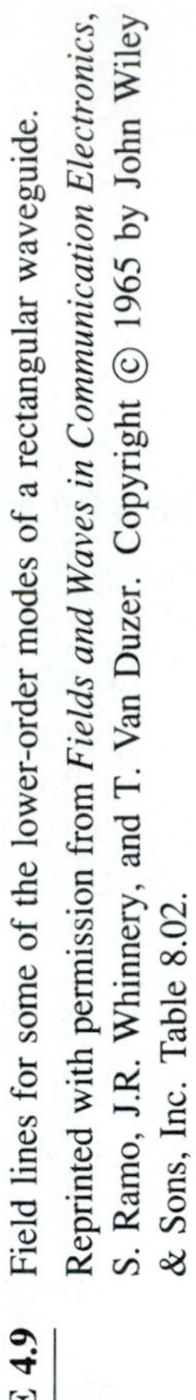

FIGURE 4.9 Field lines for some of the lower-order modes of a rectangular waveguide.
Reprinted with permission from *Fields and Waves in Communication Electronics*, S. Ramo, J.R. Whinnery, and T. Van Duzer. Copyright © 1965 by John Wiley & Sons, Inc. Table 8.02.

TABLE 4.2 Summary of Results for Rectangular Waveguide

Quantity	TE_{mn} Mode	TM_{mn} Mode
k	$\omega\sqrt{\mu\epsilon}$	$\omega\sqrt{\mu\epsilon}$
k_c	$\sqrt{(m\pi/a)^2+(n\pi/b)^2}$	$\sqrt{(m\pi/a)^2+(n\pi/b)^2}$
β	$\sqrt{k^2-k_c^2}$	$\sqrt{k^2-k_c^2}$
λ_c	$\dfrac{2\pi}{k_c}$	$\dfrac{2\pi}{k_c}$
λ_g	$\dfrac{2\pi}{\beta}$	$\dfrac{2\pi}{\beta}$
v_p	$\dfrac{\omega}{\beta}$	$\dfrac{\omega}{\beta}$
α_d	$\dfrac{k^2\tan\delta}{2\beta}$	$\dfrac{k^2\tan\delta}{2\beta}$
E_z	0	$B_{mn}\sin\dfrac{m\pi x}{a}\sin\dfrac{n\pi y}{b}e^{-j\beta z}$
H_z	$A_{mn}\cos\dfrac{m\pi x}{a}\cos\dfrac{n\pi y}{b}e^{-j\beta z}$	0
E_x	$\dfrac{j\omega\mu n\pi}{k_c^2 b}A_{mn}\cos\dfrac{m\pi x}{a}\sin\dfrac{n\pi y}{b}e^{-j\beta z}$	$\dfrac{-j\beta m\pi}{k_c^2 a}B_{mn}\cos\dfrac{m\pi x}{a}\sin\dfrac{n\pi y}{b}e^{-j\beta z}$
E_y	$\dfrac{-j\omega\mu m\pi}{k_c^2 a}A_{mn}\sin\dfrac{m\pi x}{a}\cos\dfrac{n\pi y}{b}e^{-j\beta z}$	$\dfrac{-j\beta n\pi}{k_c^2 b}B_{mn}\sin\dfrac{m\pi x}{a}\cos\dfrac{n\pi y}{b}e^{-j\beta z}$
H_x	$\dfrac{j\beta m\pi}{k_c^2 a}A_{mn}\sin\dfrac{m\pi x}{a}\cos\dfrac{n\pi y}{b}e^{-j\beta z}$	$\dfrac{j\omega\epsilon n\pi}{k_c^2 b}B_{mn}\sin\dfrac{m\pi x}{a}\cos\dfrac{n\pi y}{b}e^{-j\beta z}$
H_y	$\dfrac{j\beta n\pi}{k_c^2 b}A_{mn}\cos\dfrac{m\pi x}{a}\sin\dfrac{n\pi y}{b}e^{-j\beta z}$	$\dfrac{-j\omega\epsilon m\pi}{k_c^2 a}B_{mn}\cos\dfrac{m\pi x}{a}\sin\dfrac{n\pi y}{b}e^{-j\beta z}$
Z	$Z_{\text{TE}}=\dfrac{k\eta}{\beta}$	$Z_{\text{TM}}=\dfrac{\beta\eta}{k}$

$\beta = \frac{w}{v_p}$

TE_{mo} Modes of a Partially Loaded Waveguide

The above results also apply for a rectangular waveguide filled with a homogeneous dielectric or magnetic material, but in many cases of practical interest (such as impedance matching or phase-shifting sections) a waveguide is used with only a partial filling. Then an additional set of boundary conditions are introduced at the material interface, necessitating a new analysis. To illustrate the technique we will consider the TE_{mo} modes of a rectangular waveguide that is partially loaded with a dielectric slab, as shown in Figure 4.10. The analysis still follows the basic procedure outlined at the end of Section 4.1.

Since the geometry is uniform in the y direction and $n=0$, the TE_{mo} modes have no y dependence. Then the wave equation of (4.21) for h_z can be written separately for the dielectric and air regions as

$$\left(\frac{\partial^2}{\partial x^2}+k_d^2\right)h_z=0, \qquad \text{for } 0\le x\le t, \tag{4.105a}$$

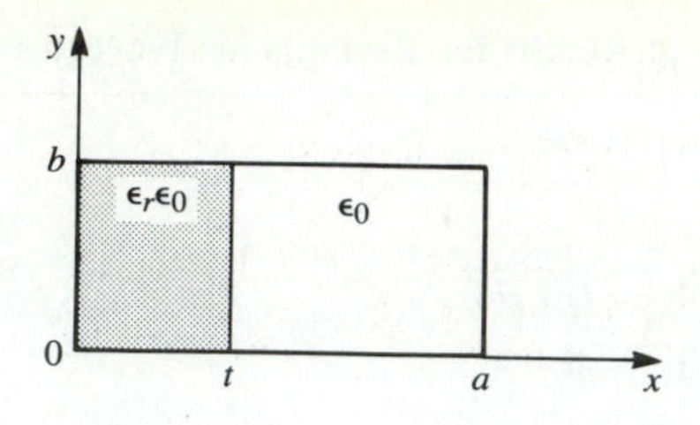

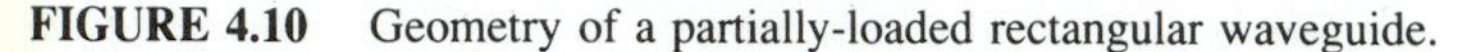
FIGURE 4.10 Geometry of a partially-loaded rectangular waveguide.

$$\left(\frac{\partial^2}{\partial x^2} + k_a^2\right)h_z = 0, \qquad \text{for } t \leq x \leq a, \tag{4.105b}$$

where k_d and k_a are the cutoff wavenumbers for the dielectric and air regions, defined as follows:

$$\beta = \sqrt{\epsilon_r k_0^2 - k_d^2}, \tag{4.106a}$$

$$\beta = \sqrt{k_0^2 - k_a^2}. \tag{4.106b}$$

These relations incorporate the fact that the propagation constant, β, must be the same in both regions to ensure phase matching of the fields along the interface at $x = t$. The solutions to (4.105) can be written as

$$h_z = \begin{cases} A\cos k_d x + B\sin k_d x & \text{for } 0 \leq x \leq t \\ C\cos k_a(a-x) + D\sin k_a(a-x) & \text{for } t \leq x \leq a, \end{cases} \tag{4.107}$$

where the form of the solution for $t < x < a$ was chosen to simplify the evaluation of boundary conditions at $x = a$.

Now we need $\hat{y}$ and $\hat{z}$ field components to apply the boundary conditions at $x = 0, t$, and a. $E_z = 0$ for TE modes, and $H_y = 0$ since $\partial/\partial y = 0$. E_y is found from (4.23d) as

$$e_y = \begin{cases} \dfrac{j\omega\mu_0}{k_d}[-A\sin k_d x + B\cos k_d x] & \text{for } 0 \leq x \leq t \\ \dfrac{j\omega\mu_0}{k_a}[C\sin k_a(a-x) - D\cos k_a(a-x)] & \text{for } t \leq x \leq a. \end{cases} \tag{4.108}$$

To satisfy the boundary conditions that $E_y = 0$ at $x = 0$ and $x = a$ requires that $B = D = 0$. Next, we must enforce continuity of tangential fields (E_y, H_x) at $x = t$. Equations (4.107) and (4.108) then give the following:

$$\frac{-A}{k_d}\sin k_d t = \frac{C}{k_a}\sin k_a(a-t),$$

$$A\cos k_d t = C\cos k_a(a-t).$$

Since this is a homogeneous set of equations, the determinant must vanish in order to have a nontrivial solution. Thus,

$$k_a \tan k_d t + k_d \tan k_a (a - t) = 0. \tag{4.109}$$

Using (4.106) allows k_a and k_d to be expressed in terms of β, so (4.109) can be solved numerically for β. There are an infinite number of solutions to (4.109), corresponding to the propagation constants of the TE_{mo} modes.

This technique can be applied to many other waveguide geometries involving dielectric or magnetic inhomogeneities, such as the surface waveguide of Section 4.6 or the ferrite-loaded waveguide of Section 10.3. In some cases, however, it will be impossible to satisfy all the necessary boundary conditions with only TE- or TM-type modes, and a hybrid combination of both types of modes will be required.

POINT OF INTEREST: Waveguide Flanges

There are two commonly used waveguide flanges: the cover flange, and the choke flange. As shown in the figure, two waveguides with cover-type flanges can be bolted together to form a contacting joint. To avoid reflections and resistive loss at this joint, it is necessary that the contacting surfaces be smooth, clean, and square, because RF currents must flow across this discontinuity. In high-power applications voltage breakdown may occur at this joint. Otherwise, the simplicity of the cover-to-cover connection makes it preferable for general use. The SWR from such a joint is typically less than 1.03.

An alternative waveguide connection uses a cover flange against a choke flange, as shown in the figure. The choke flange is machined to form an effective radial transmission line in the narrow gap between the two flanges; this line is approximately $\lambda_g/4$ in length between the guide and the point of contact for the two flanges. Another $\lambda_g/4$ line is formed by a circular axial groove in the choke flange. So the short circuit at the right-hand end of this groove is transformed to an open circuit at the contact point of the flanges. Any resistance in this contact is in series with an infinite (or very high) impedance, and thus has little effect. Then this high impedance is transformed back to a short circuit (or very low impedance) at the edges of the waveguides, to provide an effective low-resistance path for current flow across the joint. Since there is a negligible voltage drop across the ohmic contact between the flanges, voltage breakdown is avoided. Thus, the cover-to-choke

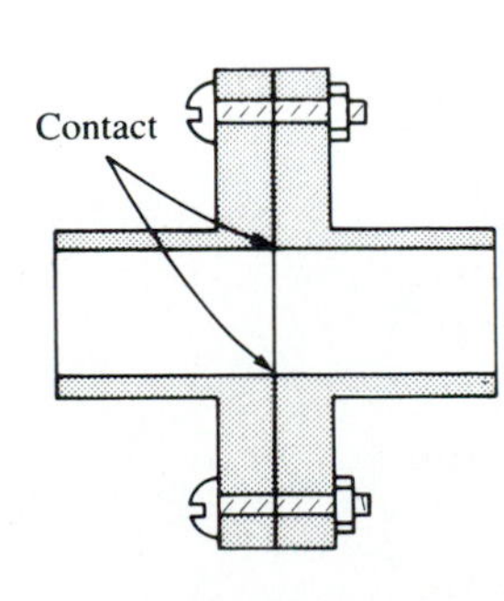

Cover-to-cover connection

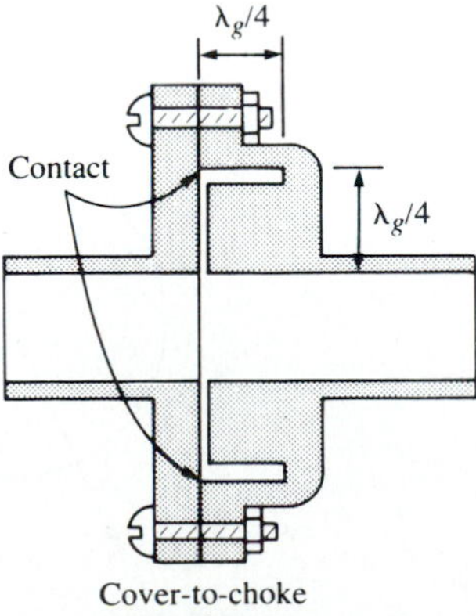

Cover-to-choke connection

connection can be useful for high-power applications. The SWR for this joint is typically less than 1.05, but is more frequency dependent than the cover-to-cover joint.

Reference: C. G. Montgomery, R. H. Dicke, and E. M. Purcell, *Principles of Microwave Circuits*, McGraw-Hill, New York, 1948.

first midterm

4.4 CIRCULAR WAVEGUIDE

A hollow metal tube of circular cross-section also supports TE and TM waveguide modes. Figure 4.11 shows the cross-section geometry of such a circular waveguide of inner radius a. Since a cylindrical geometry is involved, it is appropriate to employ cylindrical coordinates. As in the rectangular coordinate case, the transverse fields in cylindrical coordinates can be derived from E_z or H_z field components, for TM and TE modes, respectively. Paralleling the development of Section 4.1, the cylindrical components of the transverse fields can be derived from the longitudinal components as

$$E_\rho = \frac{-j}{k_c^2}\left(\beta\frac{\partial E_z}{\partial \rho} + \frac{\omega\mu}{\rho}\frac{\partial H_z}{\partial \phi}\right), \tag{4.110a}$$

$$E_\phi = \frac{-j}{k_c^2}\left(\frac{\beta}{\rho}\frac{\partial E_z}{\partial \phi} - \omega\mu\frac{\partial H_z}{\partial \rho}\right), \tag{4.110b}$$

$$H_\rho = \frac{j}{k_c^2}\left(\frac{\omega\epsilon}{\rho}\frac{\partial E_z}{\partial \phi} - \beta\frac{\partial H_z}{\partial \rho}\right), \tag{4.110c}$$

$$H_\phi = \frac{-j}{k_c^2}\left(\omega\epsilon\frac{\partial E_z}{\partial \rho} + \frac{\beta}{\rho}\frac{\partial H_z}{\partial \phi}\right), \tag{4.110d}$$

where $k_c^2 = k^2 - \beta^2$, and $e^{-j\beta z}$ propagation has been assumed. For $e^{+j\beta z}$ propagation, replace β with $-\beta$ in all expressions.

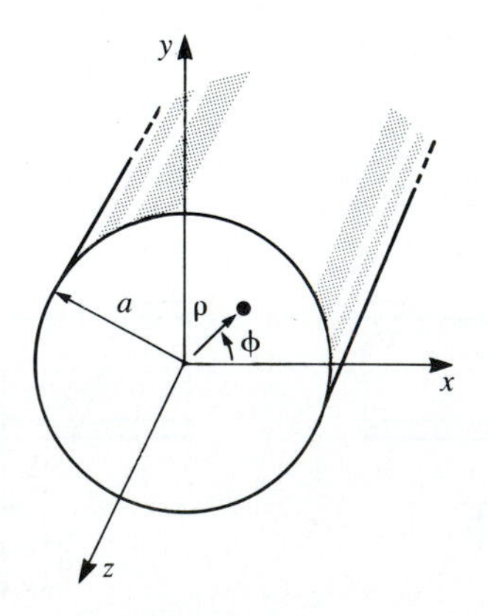

FIGURE 4.11 Geometry of a circular waveguide.

TE Modes

For TE modes, $E_z = 0$, and H_z is a solution to the wave equation,

$$\nabla^2 H_z + k^2 H_z = 0. \tag{4.111}$$

If $H_z(\rho, \phi, z) = h_z(\rho, \phi)e^{-j\beta z}$, (4.111) can be expressed in cylindrical coordinates as,

$$\left(\frac{\partial^2}{\partial \rho^2} + \frac{1}{\rho}\frac{\partial}{\partial \rho} + \frac{1}{\rho^2}\frac{\partial^2}{\partial \phi^2} + k_c^2\right) h_z(\rho, \phi) = 0. \tag{4.112}$$

Again, a solution can be derived using the method of separation of variables. Thus, we let

$$h_z(\rho, \phi) = R(\rho)P(\phi), \tag{4.113}$$

and substitute into (4.112) to obtain

$$\frac{1}{R}\frac{d^2R}{d\rho^2} + \frac{1}{\rho R}\frac{dR}{d\rho} + \frac{1}{\rho^2 P}\frac{d^2P}{d\phi^2} + k_c^2 = 0, \tag{4.114}$$

or

$$\frac{\rho^2}{R}\frac{d^2R}{d\rho^2} + \frac{\rho}{R}\frac{dR}{d\rho} + \rho^2 k_c^2 = \frac{-1}{P}\frac{d^2P}{d\phi^2}.$$

The left side of this equation depends on ρ (not ϕ), while the right side depends only on ϕ. Thus, each side must be equal to a constant, which we will call k_ϕ^2. Then,

$$\frac{-1}{P}\frac{d^2P}{d\phi^2} = k_\phi^2,$$

or

$$\frac{d^2P}{d\phi^2} + k_\phi^2 P = 0. \tag{4.115}$$

Also,

$$\rho^2\frac{d^2R}{d\rho^2} + \rho\frac{dR}{d\rho} + \left(\rho^2 k_c^2 - k_\phi^2\right) R = 0. \tag{4.116}$$

The general solution to (4.115) is

$$P(\phi) = A\sin k_\phi \phi + B\cos k_\phi \phi. \tag{4.117}$$

Since the solution to h_z must be periodic in ϕ (that is, $h_z(\rho, \phi) = h_z(\rho, \phi \pm 2m\pi)$), k_ϕ must be an integer, n. Thus (4.117) becomes

$$P(\phi) = A\sin n\phi + B\cos n\phi, \tag{4.118}$$

while (4.116) becomes

$$\rho^2\frac{d^2R}{d\rho^2} + \rho\frac{dR}{d\rho} + \left(\rho^2 k_c - n^2\right) R = 0, \tag{4.119}$$

which is recognized as Bessel's differential equation. The solution is

$$R(\rho) = CJ_n(k_c\rho) + DY_n(k_c\rho), \tag{4.120}$$

where $J_n(x)$ and $Y_n(x)$ are the Bessel functions of first and second kinds, respectively. Since $Y_n(k_c\rho)$ becomes infinite at $\rho = 0$, this term is physically unacceptable for the circular waveguide problem, so that $D = 0$. The solution for h_z can then be written as

$$h_z(\rho, \phi) = (A \sin n\phi + B \cos n\phi) J_n(k_c \rho), \qquad 4.121$$

where the constant C of (4.120) has been absorbed into the constants A and B of (4.121). We must still determine the cutoff wavenumber k_c, which we can do by enforcing the boundary condition that $E_{\tan} = 0$ on the waveguide wall. Since $E_z = 0$, we must have that

$$E_\phi(\rho, \phi) = 0, \qquad \text{at } \rho = a. \qquad 4.122$$

From (4.110b), we find E_ϕ from H_z as

$$E_\phi(\rho, \phi, z) = \frac{j\omega\mu}{k_c}(A \sin n\phi + B \cos n\phi) J_n'(k_c \rho) e^{-j\beta z}, \qquad 4.123$$

where the notation $J_n'(k_c\rho)$ refers to the derivative of J_n with respect to its argument. For E_ϕ to vanish at $\rho = a$, we must have

$$J_n'(k_c a) = 0. \qquad 4.124$$

If the roots of $J_n'(x)$ are defined as p'_{nm}, so that $J_n'(p'_{nm}) = 0$, where p'_{nm} is the mth root of J_n', then k_c must have the value

$$k_{c_{nm}} = \frac{p'_{nm}}{a}. \qquad 4.125$$

Values of p'_{nm} are given in mathematical tables; the first few values are listed in Table 4.3.

The TE$_{nm}$ modes are thus defined by the cutoff wavenumber, $k_{c_{nm}} = p'_{nm}/a$, where n refers to the number of circumferential (ϕ) variations, and m refers to the number of radial (ρ) variations. The propagation constant of the TE$_{nm}$ mode is

$$\beta_{nm} = \sqrt{k^2 - k_c^2} = \sqrt{k^2 - \left(\frac{p'_{nm}}{a}\right)^2}, \qquad 4.126$$

with a cutoff frequency of

$$f_{c_{nm}} = \frac{k_c}{2\pi\sqrt{\mu\epsilon}} = \frac{p'_{nm}}{2\pi a\sqrt{\mu\epsilon}}. \qquad 4.127$$

TABLE 4.3 Values of p'_{nm} for TE Modes of a Circular Waveguide

n	p'_{n1}	p'_{n2}	p'_{n3}
0	3.832	7.016	10.174
1	1.841	5.331	8.536
2	3.054	6.706	9.970

The first TE mode to propagate is the mode with the smallest p'_{nm}, which from Table 4.3 is seen to be the TE_{11} mode. This mode is then the dominant circular waveguide mode, and the one most frequently used. Because $m \geq 1$, there is no TE_{10} mode, but there is a TE_{01} mode.

The transverse field components are, from (4.110) and (4.121),

$$E_\rho = \frac{-j\omega\mu n}{k_c^2 \rho} (A \cos n\phi - B \sin n\phi)\, J_n(k_c\rho)\, e^{-j\beta z}, \tag{4.128a}$$

$$E_\phi = \frac{j\omega\mu}{k_c} (A \sin n\phi + B \cos n\phi)\, J'_n(k_c\rho) e^{-j\beta z}, \tag{4.128b}$$

$$H_\rho = \frac{-j\beta}{k_c} (A \sin n\phi + B \cos n\phi) J'_n(k_c\rho) e^{-j\beta z}, \tag{4.128c}$$

$$H_\phi = \frac{-j\beta n}{k_c^2 \rho} (A \cos n\phi - B \sin n\phi) J_n(k_c\rho) e^{-j\beta z}. \tag{4.128d}$$

The wave impedance is

$$Z_{\rm TE} = \frac{E_\rho}{H_\phi} = \frac{-E_\phi}{H_\rho} = \frac{\eta k}{\beta}. \tag{4.129}$$

In the above solutions there are two remaining arbitrary amplitude constants, A and B. These constants control the amplitude of the $\sin n\phi$ and $\cos n\phi$ terms, each of which are independent. That is, because of the azimuthal symmetry of the circular waveguide, both the $\sin n\phi$ and $\cos n\phi$ terms are valid solutions, and can be present in a specific problem to any degree. The actual amplitudes of these terms will be dependent on the excitation of the waveguide. From a different viewpoint, the coordinate system can be rotated about the z-axis to obtain an h_z with either $A = 0$ or $B = 0$.

Now consider the dominant TE_{11} mode with an excitation such that $B = 0$. The fields can be written as

$$H_z = A \sin\phi J_1(k_c\rho) e^{-j\beta z}, \tag{4.130a}$$

$$E_\rho = \frac{-j\omega\mu}{k_c^2 \rho} A \cos\phi J_1(k_c\rho) e^{-j\beta z}, \tag{4.130b}$$

$$E_\phi = \frac{j\omega\mu}{k_c} A \sin\phi J'_1(k_c\rho) e^{-j\beta z}, \tag{4.130c}$$

$$H_\rho = \frac{-j\beta}{k_c} A \sin\phi J'_1(k_c\rho) e^{-j\beta z}, \tag{4.130d}$$

$$H_\phi = \frac{-j\beta}{k_c^2 \rho} A \cos\phi J_1(k_c\rho) e^{-j\beta z}, \tag{4.130e}$$

$$E_z = 0. \tag{4.130f}$$

The power flow down the guide can be computed as

$$
\begin{aligned}
P_o &= \frac{1}{2}\text{Re}\int_{\rho=0}^{a}\int_{\phi=0}^{2\pi} \bar{E}\times\bar{H}^*\cdot\hat{z}\rho\,d\phi\,d\rho \\
&= \frac{1}{2}\text{Re}\int_{\rho=0}^{a}\int_{\phi=0}^{2\pi} \left[E_\rho H_\phi^* - E_\phi H_\rho^*\right]\rho\,d\phi\,d\rho \\
&= \frac{\omega\mu|A|^2\text{Re}(\beta)}{2k_c^4}\int_{\rho=0}^{a}\int_{\phi=0}^{2\pi}\left[\frac{1}{\rho^2}\cos^2\phi J_1^2(k_c\rho) + k_c^2\sin^2\phi J_1'^2(k_c\rho)\right]\rho\,d\phi\,d\rho \\
&= \frac{\pi\omega\mu|A|^2\text{Re}(\beta)}{2k_c^4}\int_{\rho=0}^{a}\left[\frac{1}{\rho}J_1^2(k_c\rho) + \rho k_c^2 J_1'^2(k_c\rho)\right]d\rho \\
&= \frac{\pi\omega\mu|A|^2\text{Re}(\beta)}{4k_c^4}\left(p_{11}'^2 - 1\right)J_1^2(k_c a),
\end{aligned}
\qquad 4.131
$$

which is seen to be nonzero only when β is real, corresponding to a propagating mode. (The required integral for this result is given in Appendix C.)

Attenuation due to dielectric loss is given by (4.29). The attenuation due to a lossy waveguide conductor can be found by computing the power loss per unit length of guide:

$$
\begin{aligned}
P_\ell &= \frac{R_s}{2}\int_{\phi=0}^{2\pi}|\bar{J}_s|^2 a\,d\phi \\
&= \frac{R_s}{2}\int_{\phi=0}^{2\pi}\left[|H_\phi|^2 + |H_z|^2\right]a\,d\phi \\
&= \frac{|A|^2 R_s}{2}\int_{\phi=0}^{2\pi}\left[\frac{\beta^2}{k_c^4 a^2}\cos^2\phi + \sin^2\phi\right]J_1^2(k_c a)\,a\,d\phi \\
&= \frac{\pi|A|^2 R_s a}{2}\left(1 + \frac{\beta^2}{k_c^4 a^2}\right)J_1^2(k_c a).
\end{aligned}
\qquad 4.132
$$

The attenuation constant is then

$$
\begin{aligned}
\alpha_c &= \frac{P_\ell}{2P_o} = \frac{R_s\left(k_c^4 a^2 + \beta^2\right)}{\eta k\beta a(p_{11}'^2 - 1)} \\
&= \frac{R_s}{ak\eta\beta}\left(k_c^2 + \frac{k^2}{p_{11}'^2 - 1}\right)\text{Np/m}.
\end{aligned}
\qquad 4.133
$$

TM Modes

For the TM modes of the circular waveguide, we must solve for E_z from the wave equation in cylindrical coordinates:

$$
\left(\frac{\partial^2}{\partial\rho^2} + \frac{1}{\rho}\frac{\partial}{\partial\rho} + \frac{1}{\rho^2}\frac{\partial^2}{\partial\phi^2} + k_c^2\right)e_z = 0, \qquad 4.134
$$

where $E_z(\rho, \phi, z) = e_z(\rho, \phi)e^{-j\beta z}$, and $k_c^2 = k^2 - \beta^2$. Since this equation is identical to (4.107), the general solutions are the same. Thus, from (4.121),

$$e_z(\rho, \phi) = (A \sin n\phi + B \cos n\phi) J_n(k_c \rho). \tag{4.135}$$

The difference between the TE solution and the present solution is that the boundary conditions can now be applied directly to e_z of (4.135), since

$$E_z(\rho, \phi) = 0, \qquad \text{at } \rho = a. \tag{4.136}$$

Thus, we must have

$$J_n(k_c a) = 0, \tag{4.137}$$

or

$$k_c = p_{nm}/a, \tag{4.138}$$

where p_{nm} is the mth root of $J_n(x)$; that is, $J_n(p_{nm}) = 0$. Values of p_{nm} are given in mathematical tables; the first few values are listed in Table 4.4.

The propagation constant of the TM_{nm} mode is

$$\beta_{nm} = \sqrt{k^2 - k_c^2} = \sqrt{k^2 - (p_{nm}/a)^2}. \tag{4.139}$$

The cutoff frequency is

$$f_{c_{nm}} = \frac{k_c}{2\pi\sqrt{\mu\epsilon}} = \frac{p_{nm}}{2\pi a\sqrt{\mu\epsilon}}. \tag{4.140}$$

Thus, the first TM mode to propagate is the TM_{01} mode, with $p_{01} = 2.405$. Since this is greater than $p'_{11} = 1.841$ of the lowest order TE_{11} mode, the TE_{11} mode is the dominant mode of the circular waveguide. As with the TE modes, $m \geq 1$, so there is no TM_{10} mode.

From (4.110), the transverse fields can be derived as

$$E_\rho = \frac{-j\beta}{k_c}(A \sin n\phi + B \cos n\phi) J'_n(k_c \rho) e^{-j\beta z}, \tag{4.141a}$$

$$E_\phi = \frac{-j\beta n}{k_c^2 \rho}(A \cos n\phi - B \sin n\phi) J_n(k_c \rho) e^{-j\beta z}, \tag{4.141b}$$

$$H_\rho = \frac{j\omega\epsilon n}{k_c^2 \rho}(A \cos n\phi - B \sin n\phi) J_n(k_c \rho) e^{-j\beta z}, \tag{4.141c}$$

$$H_\phi = \frac{-j\omega\epsilon}{k_c}(A \sin n\phi + B \cos n\phi) J'_n(k_c \rho) e^{-j\beta z}. \tag{4.141d}$$

TABLE 4.4 Values of p_{nm} for TM Modes of a Circular Waveguide

n	p_{n1}	p_{n2}	p_{n3}
0	2.405	5.520	8.654
1	3.832	7.016	10.174
2	5.135	8.417	11.620

The wave impedance is

$$Z_{\text{TM}} = \frac{E_\rho}{H_\phi} = \frac{-E_\phi}{H_\rho} = \frac{\eta\beta}{k}. \tag{4.142}$$

Calculation of the attenuation for TM modes is left as a problem. Figure 4 12 shows the attenuation due to conductor loss versus frequency for various modes of a circular waveguide. Observe that the attenuation of the TE_{01} mode decreases to a very small value with increasing frequency. This property makes the TE_{01} mode of interest for low-loss transmission over long distances. Unfortunately, this mode is not the dominant mode of the circular waveguide, so in practice power can be lost from the TE_{01} mode to lower-order propagating modes.

Figure 4.13 shows the relative cutoff frequencies of the TE and TM modes, and Table 4.5 summarizes results for wave propagation in circular waveguide. Field lines for some of the lowest-order TE and TM modes are shown in Figure 4.14.

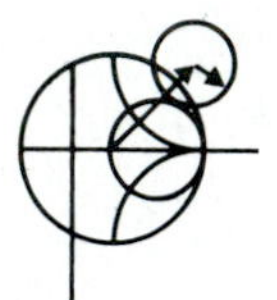

EXAMPLE 4.2

Find the cutoff frequencies of the first two propagating modes of a circular waveguide with $a = 0.5$ cm and $\epsilon_r = 2.25$. If the guide is silver-plated and

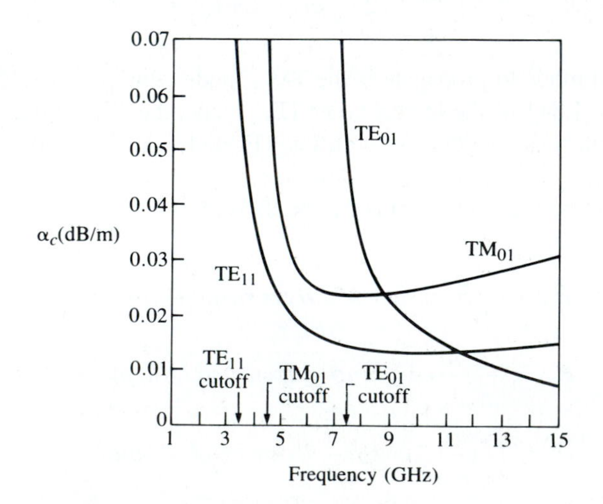

FIGURE 4.12 Attenuation of various modes in a circular copper waveguide with $a = 2.54$ cm.

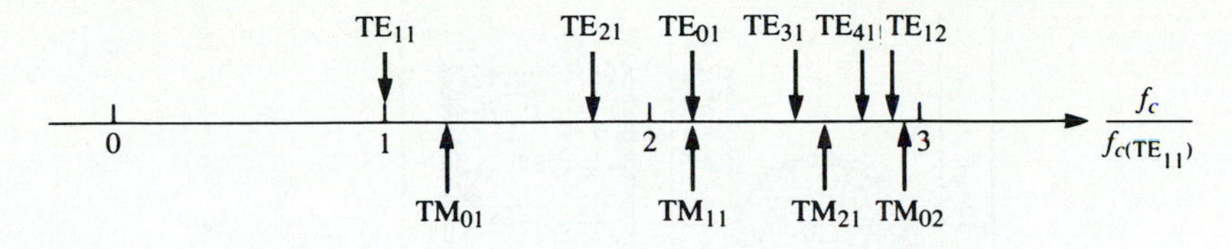

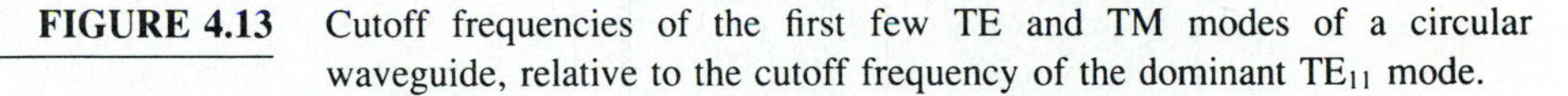

FIGURE 4.13 Cutoff frequencies of the first few TE and TM modes of a circular waveguide, relative to the cutoff frequency of the dominant TE_{11} mode.

TABLE 4.5 Summary of Results for Circular Waveguide

Quantity	TE_{nm} Mode	TM_{nm} Mode
k	$\omega\sqrt{\mu\epsilon}$	$\omega\sqrt{\mu\epsilon}$
k_c	$\frac{p'_{nm}}{a}$	$\frac{p_{nm}}{a}$
β	$\sqrt{k^2 - k_c^2}$	$\sqrt{k^2 - k_c^2}$
λ_c	$\frac{2\pi}{k_c}$	$\frac{2\pi}{k_c}$
λ_g	$\frac{2\pi}{\beta}$	$\frac{2\pi}{\beta}$
v_p	$\frac{\omega}{\beta}$	$\frac{\omega}{\beta}$
α_d	$\frac{k^2 \tan\delta}{2\beta}$	$\frac{k^2 \tan\delta}{2\beta}$
E_z	0	$(A\sin n\phi + B\cos n\phi)J_n(k_c\rho)e^{-j\beta z}$
H_z	$(A\sin n\phi + B\cos n\phi)J_n(k_c\rho)e^{-j\beta z}$	0
E_ρ	$\frac{-j\omega\mu n}{k_c^2\rho}(A\cos n\phi - B\sin n\phi)J_n(k_c\rho)e^{-j\beta z}$	$\frac{-j\beta}{k_c}(A\sin n\phi + B\cos n\phi)J'_n(k_c\rho)e^{-j\beta z}$
E_ϕ	$\frac{j\omega\mu}{k_c}(A\sin n\phi + B\cos n\phi)J'_n(k_c\rho)e^{-j\beta z}$	$\frac{-j\beta n}{k_c^2\rho}(A\cos n\phi - B\sin n\phi)J_n(k_c\rho)e^{-j\beta z}$
H_ρ	$\frac{-j\beta}{k_c}(A\sin n\phi + B\cos n\phi)J'_n(k_c\rho)e^{-j\beta z}$	$\frac{j\omega\epsilon n}{k_c^2\rho}(A\cos n\phi - B\sin n\phi)J_n(k_c\rho)e^{-j\beta z}$
H_ϕ	$\frac{-j\beta n}{k_c^2\rho}(A\cos n\phi - B\sin n\phi)J_n(k_c\rho)e^{-j\beta z}$	$\frac{-j\omega\epsilon}{k_c}(A\sin n\phi + B\cos n\phi)J'_n(k_c\rho)e^{-j\beta z}$
Z	$Z_{TE} = \frac{k\eta}{\beta}$	$Z_{TM} = \frac{\beta\eta}{k}$

the dielectric loss tangent is 0.001, calculate the attenuation in dB for a 50 cm length of guide operating at 13.0 GHz.

Solution

From Figure 4.13, the first two propagating modes of a circular waveguide are the TE_{11} and TM_{01} modes. The cutoff frequencies can be found using (4.127)

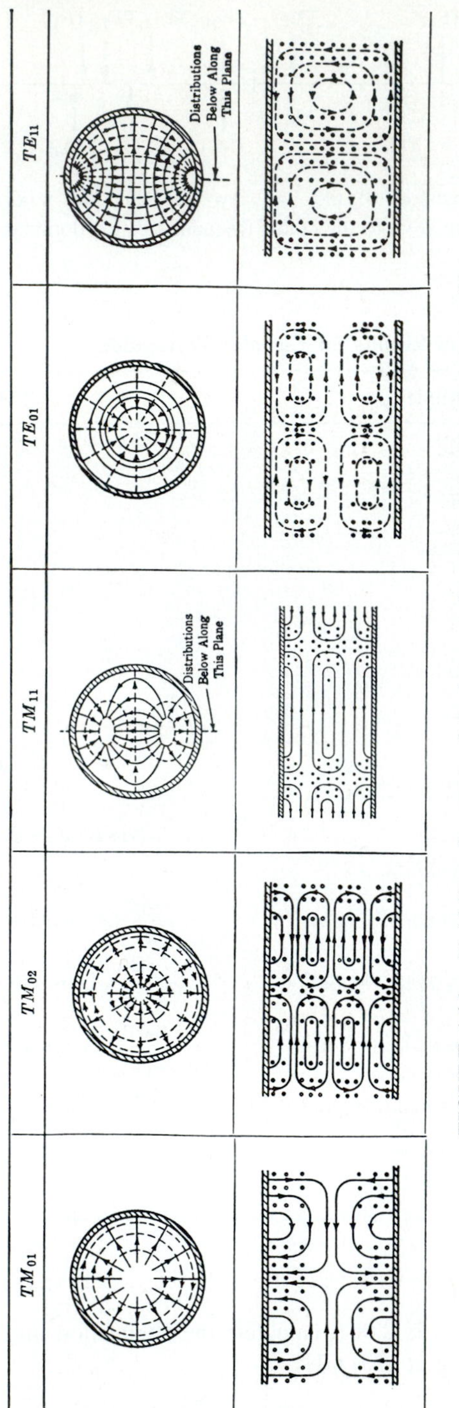

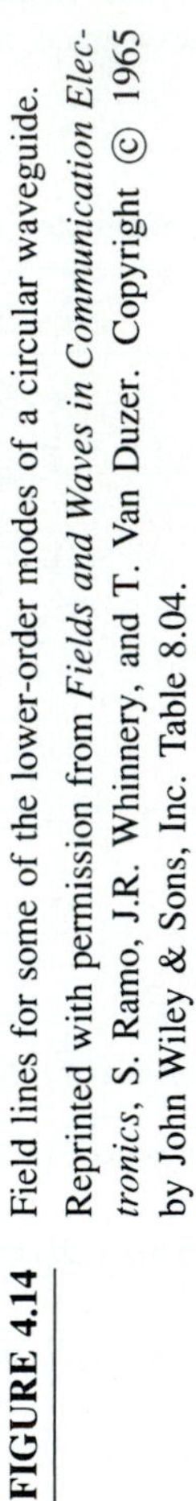

FIGURE 4.14 Field lines for some of the lower-order modes of a circular waveguide. Reprinted with permission from *Fields and Waves in Communication Electronics*, S. Ramo, J.R. Whinnery, and T. Van Duzer. Copyright © 1965 by John Wiley & Sons, Inc. Table 8.04.

and (4.140):

$$\text{TE}_{11}: \quad f_c = \frac{p'_{11}c}{2\pi a\sqrt{\epsilon_r}} = \frac{1.841(3 \times 10^8)}{2\pi(0.005)\sqrt{2.25}} = 11.72 \text{ GHz},$$

$$\text{TM}_{01}: \quad f_c = \frac{p_{01}c}{2\pi a\sqrt{\epsilon_r}} = \frac{2.405(3 \times 10^8)}{2\pi(0.005)\sqrt{2.25}} = 15.31 \text{ GHz}.$$

So only the TE_{11} mode is propagating at 13.0 GHz. The wavenumber is

$$k = \frac{2\pi f\sqrt{\epsilon_r}}{c} = \frac{2\pi(13 \times 10^9)\sqrt{2.25}}{3 \times 10^8} = 408.4 \text{ m}^{-1},$$

and the propagation constant of the TE_{11} mode is

$$\beta = \sqrt{k^2 - \left(\frac{p'_{11}}{a}\right)^2} = \sqrt{(408.4)^2 - \left(\frac{1.841}{0.005}\right)^2} = 176.7 \text{ m}^{-1}.$$

The attenuation due to dielectric loss is calculated from (4.29) as

$$\alpha_c = \frac{k^2 \tan\delta}{2\beta} = \frac{(408.4)^2(0.001)}{2(176.7)} = 0.47 \text{ Np/m}.$$

The conductivity of silver is $\sigma = 6.17 \times 10^7$ S/m, so the surface resistance is

$$R_s = \sqrt{\frac{\omega\mu_0}{2\sigma}} = 0.029\,\Omega.$$

Then from (4.133) the attenuation due to metallic loss is

$$\alpha_c = \frac{R_s}{ak\eta\beta}\left(k_c^2 + \frac{k^2}{p'^2_{11} - 1}\right) = 0.066 \text{ Np/m}.$$

So the total attenuation factor is

$$\alpha = \alpha_c + \alpha_d = 0.54 \text{ Np/m}.$$

Note that the dielectric loss dominates this result. The attenuation in the 50 cm-long guide is

$$\text{attenuation (dB)} = -20\log e^{-\alpha\ell} = -20\log e^{-(0.547)(0.5)} = 2.38 \text{ dB}. \qquad \bigcirc$$

4.5 COAXIAL LINE

TEM Modes

Although we have already discussed TEM mode propagation on a coaxial line in Chapter 3, we will reconsider it here in the context of the general framework which was developed in the previous section.

The coaxial line geometry is shown in Figure 4.15, where the inner conductor is at a potential of V_o volts and the outer conductor is at zero volts. From Section 4.1, we know that the fields can be derived from a scalar potential function, $\Phi(\rho, \phi)$, which is a solution to Laplace's equation (4.14); in cylindrical coordinates Laplace's equation takes

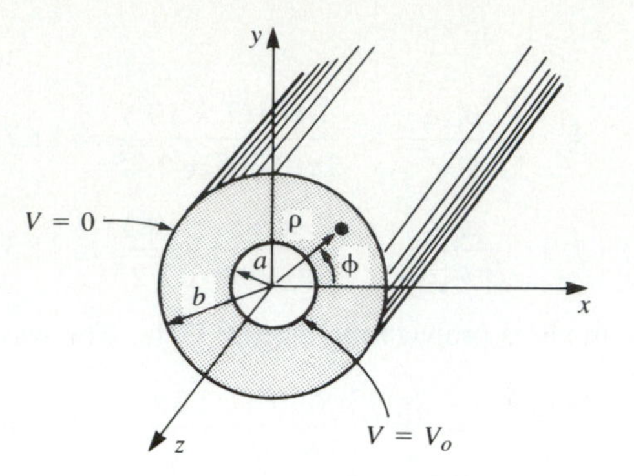

FIGURE 4.15 Coaxial line geometry.

the form

$$\frac{1}{\rho}\frac{\partial}{\partial \rho}\left(\rho \frac{\partial \Phi(\rho, \phi)}{\partial \rho}\right) + \frac{1}{\rho^2}\frac{\partial^2 \Phi(\rho, \phi)}{\partial \phi^2} = 0. \tag{4.143}$$

This equation must be solved for $\Phi(\rho, \phi)$ subject to the boundary conditions that

$$\Phi(a, \phi) = V_o, \tag{4.144a}$$

$$\Phi(b, \phi) = 0. \tag{4.144b}$$

Using the method of separation of variables, we let $\Phi(\rho, \phi)$ be expressed in product form as

$$\Phi(\rho, \phi) = R(\rho)P(\phi). \tag{4.145}$$

Substitution of (4.145) into (4.143) gives

$$\frac{\rho}{R}\frac{\partial}{\partial \rho}\left(\rho \frac{dR}{d\rho}\right) + \frac{1}{\rho}\frac{d^2 P}{d\phi^2} = 0. \tag{4.146}$$

By the usual separation of variables argument, the two terms in (4.146) must be equal to constants, so that

$$\frac{\rho}{R}\frac{\partial}{\partial \rho}\left(\rho \frac{dR}{d\rho}\right) = -k_\rho^2, \tag{4.147}$$

$$\frac{1}{P}\frac{d^2 P}{d\phi^2} = -k_\phi^2, \tag{4.148}$$

and

$$k_\rho^2 + k_\phi^2 = 0. \tag{4.149}$$

The general solution to (4.148) is

$$P(\phi) = A\cos n\phi + B\sin n\phi, \tag{4.150}$$

where $k_\phi = n$ must be an integer, since increasing ϕ by a multiple of 2π should not change the result. Now, because of the fact that the boundary conditions of (4.144) do not vary with ϕ, the potential $\Phi(\rho, \phi)$ should not vary with ϕ. Thus, n must be zero. By (4.149), this implies that k_ρ must also be zero, so that the equation for $R(\rho)$ in (4.147) reduces to

$$\frac{\partial}{\partial \rho}\left(\rho \frac{dR}{d\rho}\right) = 0.$$

The solution for $R(\rho)$ is then

$$R(\rho) = C \ln \rho + D,$$

and so

$$\Phi(\rho, \phi) = C \ln \rho + D. \tag{4.151}$$

Applying the boundary conditions of (4.144) gives two equations for the constants C and D:

$$\Phi(a, \phi) = V_o = C \ln a + D, \tag{4.152a}$$

$$\Phi(b, \phi) = 0 = C \ln b + D. \tag{4.152b}$$

After solving for C and D, the final solution for $\Phi(\rho, \phi)$ can be written as

$$\Phi(\rho, \phi) = \frac{V_o \ln b/\rho}{\ln b/a}. \tag{4.153}$$

From (4.13), the transverse electric field can be found as

$$\bar{e}(\rho, \phi) = -\nabla_t \Phi(\rho, \phi) = -\left(\hat{\rho}\frac{\partial \Phi}{\partial \rho} + \frac{\hat{\phi}}{\rho}\frac{\partial \Phi}{\partial \phi}\right) = \frac{V_o \hat{\rho}}{\rho \ln b/a}. \tag{4.154}$$

The electric field including the propagation factor is then

$$\bar{E}(\rho, \phi, z) = \bar{e}(\rho, \phi) e^{-j\beta z} = \frac{V_o \hat{\rho} e^{-j\beta z}}{\rho \ln b/a}, \tag{4.155}$$

where the propagation constant is $\beta = k = \omega\sqrt{\mu\epsilon}$, from (4.7). From (4.18) the transverse magnetic field is

$$\bar{h}(\rho, \phi) = \frac{1}{\eta}\hat{z} \times \bar{e}(\rho, \phi) = \frac{V_o \hat{\phi}}{\eta \rho \ln b/a}, \tag{4.156}$$

where $\eta = \sqrt{\mu/\epsilon}$ is the intrinsic impedance of the medium.

The magnetic field with its propagation factor is

$$\bar{H}(\rho, \phi, z) = \frac{V_o \hat{\phi} e^{-j\beta z}}{\eta \rho \ln b/a}. \tag{4.157}$$

As a check, we can compute the potential difference between the two conductors by using (4.15) and (4.155):

$$V_{ab} = \int_{\rho=a}^{b} E_\rho(\rho,\phi)d\rho = V_o e^{-j\beta z}, \tag{4.158}$$

which is the expected result for a forward-traveling voltage wave. The total current on the inner conductor can be found from (4.16) and (4.157) as

$$I_a = \int_{\phi=0}^{2\pi} H_\phi(a,\phi)ad\phi = \frac{2\pi V_o e^{-j\beta z}}{\eta \ln b/a}. \tag{4.159}$$

The current on the outer conductor is the negative of this, which can be shown by applying Ampere's law to a contour that encloses both conductors to obtain zero total current, or by finding the surface current density on the outer conductor. Thus, at $\rho = b$,

$$\bar{J}_s = -\hat{\rho} \times \bar{H}(b,\phi) = \frac{-\hat{z} V_o e^{-j\beta z}}{\eta b \ln b/a}. \tag{4.160}$$

The total current on the outer conductor is then

$$I_b = \int_{\phi=0}^{2\pi} J_{sz} b d\phi = \frac{-2\pi V_o e^{-j\beta z}}{\eta \ln b/a} = -I_a. \tag{4.161}$$

The characteristic impedance can be calculated as

$$Z_0 = \frac{V_o}{I_a} = \frac{\eta \ln b/a}{2\pi}. \tag{4.162}$$

Attenuation due to dielectric or conductor loss has already been treated in Chapter 3, and will not be repeated here.

Higher-Order Modes

The coaxial line, like the parallel plate waveguide, can also support TE and TM waveguide modes in addition to a TEM mode. In practice these modes are usually cutoff (evanescent), and so have only a reactive effect near discontinuities or sources, where they are excited. It is important in practice, however, to be aware of the cutoff frequency of the lowest-order waveguide-type modes, to avoid the propagation of these modes. Deleterious effects may otherwise occur, due to the superposition of two or more propagating modes with different propagation constants. Avoiding the propagation of higher-order modes sets an upper limit on the size of a coaxial cable; this ultimately limits the power handling capacity of a coaxial line (see the Point of Interest on power capacity of transmission lines).

We will derive the solution for the TE modes of the coaxial line; the TE_{11} mode is the dominant waveguide mode of the coaxial line, and so is of primary importance.

For TE modes, $E_z = 0$, and H_z satisfies the wave equation of (4.112):

$$\left(\frac{\partial^2}{\partial\rho^2} + \frac{1}{\rho}\frac{\partial}{\partial\rho} + \frac{1}{\rho^2}\frac{\partial^2}{\partial\phi^2} + k_c^2\right) h_z(\rho,\phi) = 0, \tag{4.163}$$

where $H_z(\rho, \phi, z) = h_z(\rho, \phi)e^{-j\beta z}$, and $k_c^2 = k^2 - \beta^2$. The general solution to this equation, as derived in Section 4.4, is given by the product of (4.118) and (4.120):

$$h_z(\rho, \phi) = (A \sin n\phi + B \cos n\phi)(CJ_n(k_c\rho) + DY_n(k_c\rho)). \qquad 4.164$$

In this case, $a \le \rho \le b$, so we have no reason to discard the Y_n term. The boundary conditions are that

$$E_\phi(\rho, \phi) = 0, \qquad \text{for } \rho = a, b. \qquad 4.165$$

Using (4.110b) to find E_ϕ from H_z gives

$$E_\phi = \frac{j\omega\mu}{k_c}(A \sin n\phi + B \cos n\phi)(CJ'_n(k_c\rho) + DY'_n(k_c\rho))e^{-j\beta z}. \qquad 4.166$$

Applying (4.165) to (4.166) gives two equations:

$$CJ'_n(k_c a) + DY'_n(k_c a) = 0, \qquad 4.167a$$

$$CJ'_n(K_c b) + DY'_n(k_c b) = 0. \qquad 4.167b$$

Since this is a homogeneous set of equations, the only nontrivial ($C \neq 0, D \neq 0$) solution occurs when the determinant is zero. Thus we must have

$$J'_n(k_c a)Y'_n(k_c b) = J'_n(k_c b)Y'_n(k_c a). \qquad 4.168$$

This is a characteristic (or eigenvalue) equation for k_c. The values of k_c that satisfy (4.168) then define the TE_{nm} modes of the coaxial line.

Equation (4.168) is a transcendental equation, which must be solved numerically for k_c. Figure 4.16 shows the result of such a solution for $n = 1$, for various b/a ratios. An approximate solution that is often used in practice is

$$k_c = \frac{2}{a + b}.$$

Once k_c is known, the propagation constant or cutoff frequency can be determined. Solutions for the TM modes can be found in a similar manner; the required determinantal equation is the same as (4.168), except for the derivatives. Field lines for the TEM and TE_{11} modes of the coaxial line are shown in Figure 4.17.

EXAMPLE 4.3

Consider a piece of RG-142 coaxial cable, with $a = 0.035$" and $b = 0.116$", and a dielectric with $\epsilon_r = 2.2$. What is the highest usable frequency, before the TE_{11} waveguide mode starts to propagate?

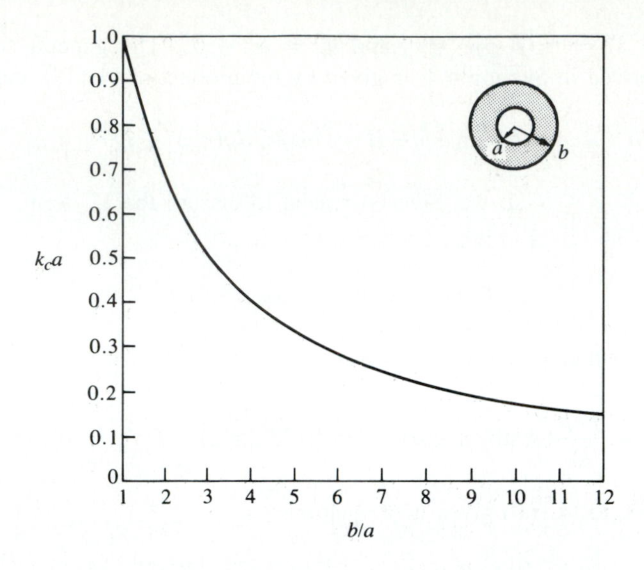

FIGURE 4.16 Normalized cutoff frequency of the dominant TE_{11} waveguide mode for a coaxial line.

Solution
We have

$$\frac{b}{a} = \frac{0.116}{0.035} = 3.3.$$

From Figure 4.16 this value of b/a gives $k_c a = 0.47$ (the approximate result is $k_c a = 2/(1 + b/a) = 0.465$). Thus, the cutoff frequency of the TE_{11} mode is

$$f_c = \frac{ck_c}{2\pi\sqrt{\epsilon_r}} = 17 \text{ GHz}.$$

In practice a 5% safety margin is usually recommended, so $f_{\max} = 0.95(17 \text{ GHz}) = 16 \text{ GHz}$. ○

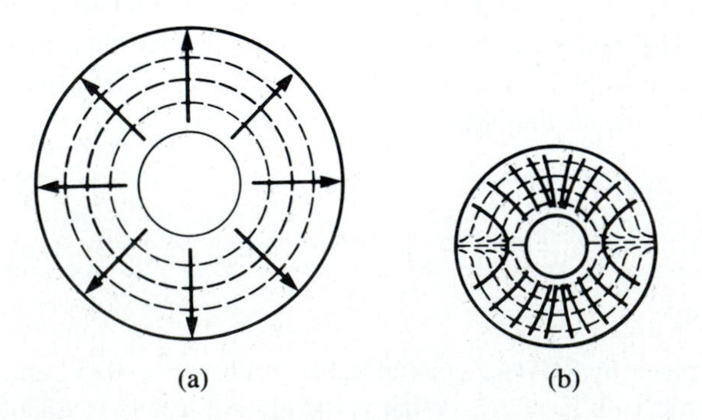

FIGURE 4.17 Field lines for the (a) TEM and (b) TE_{11} modes of a coaxial line.

POINT OF INTEREST: Coaxial Connectors

Most coaxial cables and connectors in common use have a 50 Ω characteristic impedance, with an exception being the 75 Ω coax used in television systems. Requirements for coaxial connectors include low SWR, higher-order-mode-free operation at a high frequency, high repeatability after a connect-disconnect cycle, and mechanical strength. Most connectors are used in pairs, with a male end and a female end (the obvious reason for this nomenclature is left to the imagination of the reader). Below we list some of the characteristics of the most common microwave coaxial connectors.

Type-N connector. This connector was developed in 1942 and named after P. Neill, who worked on its design at Bell Labs. The male and female connectors thread together; the outer diameter of the female connector is about 0.625 in., so this is a relatively large connector. The recommended upper operating frequency ranges from 11 to 18 GHz, depending on cable size. The SWR for a mated connector pair is typically less than 1.07.

SMA connector. The need for a smaller and lighter connector led to the development of the subminiature SMA connector in the early 1960s. The outer diameter of the female end of the SMA connector is about 0.250 in. It can be used up to 25 GHz, and a recent modification, called the K connector, can be used up to 40 GHz. The SMA connector is usually used with teflon-filled braided or semi-rigid cable. It is probably the most frequently used microwave connector today.

SSMA connector. The SSMA (scaled SMA) connector is similar in design to the SMA connector, but smaller in size. The outer diameter of the female end is about 0.192 in., and the maximum operating frequency is about 38 GHz.

APC-7 connector. This is a precision connector (Amphenol precision connector) that can repeatably achieve an SWR less than 1.04 at frequencies up to 18 GHz. The connectors are "sexless," with butt contact between both the inner conductors and the outer conductors.

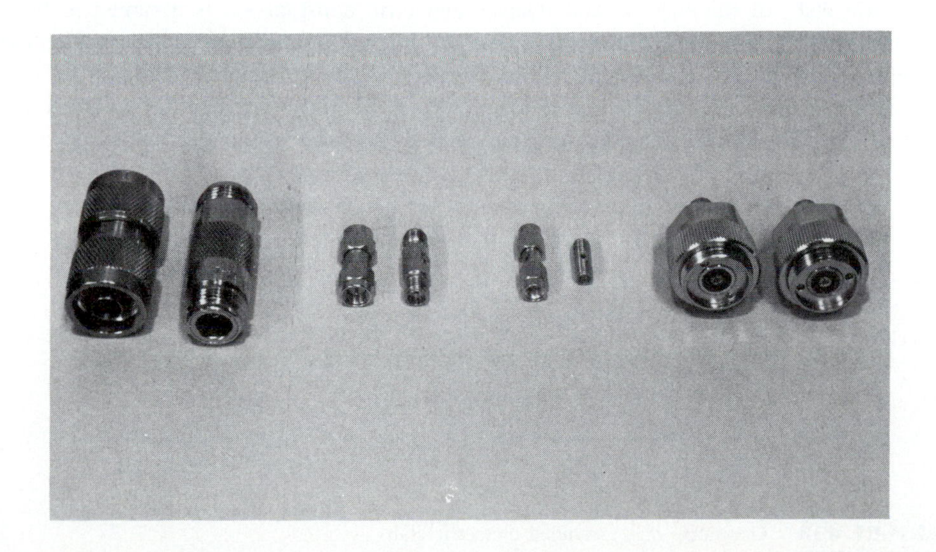

The BNC (baby N connector) and TNC (a threaded BNC connector) are commonly used at RF and IF frequencies, but not for microwave work.

Reference J. H. Bryant, "Coaxial Transmission Lines, Related Two-Conductor Transmission Lines, Connectors, and Components: A U.S. Historical Perspective," *IEEE Trans. Microwave Theory and Techniques*, vol. MTT-32, pp. 970–983, September 1984.

4.6 SURFACE WAVES ON A GROUNDED DIELECTRIC SLAB

We briefly discussed surface waves in Chapter 2, in connection with the field of a plane wave totally reflected from a dielectric interface. In general, surface waves can exist in a variety of geometries involving dielectric interfaces. Here we consider the TM and TE surface waves that can be excited along a grounded dielectric slab. Other geometries that can be used as surface waveguides include an ungrounded dielectric slab, a dielectric rod, a corrugated conductor, or a dielectric coated conducting rod.

Surface waves are typified by a field that decays exponentially away from the dielectric surface, with most of the field contained in or near the dielectric. At higher frequencies the field generally becomes more tightly bound to the dielectric, making such waveguides practical. Because of the presence of the dielectric, the phase velocity of a surface wave is less than the velocity of light in a vacuum. Another reason for studying surface waves is that they may be excited on some types of planar transmission lines, such as microstrip and slotline.

TM Modes

Figure 4.18 shows the geometry of a grounded dielectric slab waveguide. The dielectric slab, of thickness d and relative dielectric constant ϵ_r, is assumed to be of infinite extent in the y and z directions. We will assume propagation in the $+z$ direction with an $e^{-j\beta z}$ propagation factor, and no variation in the y direction ($\partial/\partial y = 0$).

Because there are two distinct regions, with and without a dielectric, we must separately consider the field in these regions, and then match tangential fields across the

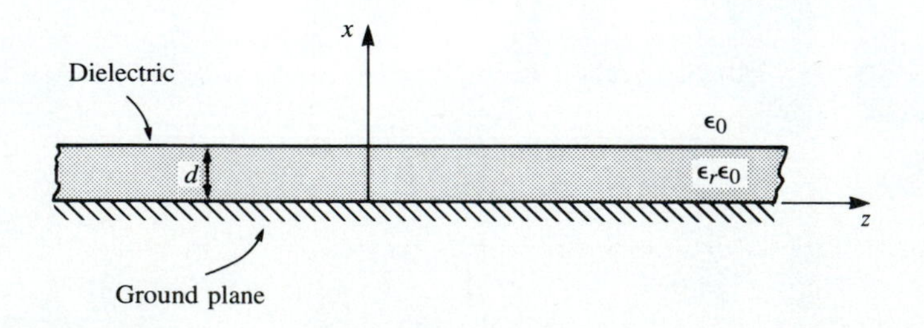

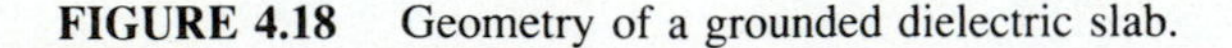

FIGURE 4.18 Geometry of a grounded dielectric slab.

interface. E_z must satisfy the wave equation of (4.25) in each region:

$$\left(\frac{\partial^2}{\partial x^2} + \epsilon_r k_0^2 - \beta^2\right) e_z(x, y) = 0, \quad \text{for } 0 \le x \le d, \tag{4.169a}$$

$$\left(\frac{\partial^2}{\partial x^2} + k_0^2 - \beta^2\right) e_z(x, y) = 0, \quad \text{for } d \le x < \infty, \tag{4.169b}$$

where $E_z(x, y, z) = e_z(x, y)e^{-j\beta z}$.

Now define the cutoff wavenumbers for the two regions as

$$k_c^2 = \epsilon_r k_0^2 - \beta^2, \tag{4.170a}$$

$$h^2 = \beta^2 - k_0^2, \tag{4.170b}$$

where the sign on h^2 has been selected in anticipation of an exponentially decaying result for $x > d$. Observe that the same propagation constant β has been used for both regions. This must be the case to achieve phase matching of the tangential fields at the $x = d$ interface for all values of z.

The general solutions to (4.169) are then

$$e_z(x, y) = A \sin k_c x + B \cos k_c x, \quad \text{for } 0 \le x \le d, \tag{4.171a}$$

$$e_z(x, y) = Ce^{hx} + De^{-hx}, \quad \text{for } d \le x < \infty. \tag{4.171b}$$

Note that these solutions are valid for k_c and h either real or imaginary; it will turn out that both k_c and h are real, because of the choice of definitions in (4.170).

The boundary conditions that must be satisfied are

$$E_z(x, y, z) = 0, \quad \text{at } x = 0, \tag{4.172a}$$

$$E_z(x, y, z) < \infty, \quad \text{as } x \to \infty, \tag{4.172b}$$

$$E_z(x, y, z) \text{ continuous}, \quad \text{at } x = d, \tag{4.172c}$$

$$H_y(x, y, z) \text{ continuous}, \quad \text{at } x = d. \tag{4.172d}$$

From (4.23), $H_x = E_y = H_z = 0$. Condition (4.172a) implies that $B = 0$ in (4.167a). Condition (4.172b) comes about as a requirement for finite fields (and energy) infinitely far away from a source, and implies that $C = 0$. The continuity of E_z leads to

$$A \sin k_c d = De^{-hd}, \tag{4.173a}$$

while (4.23b) must be used to apply continuity to H_y, to obtain

$$\frac{\epsilon_r A}{k_c} \cos k_c d = \frac{D}{h} e^{-hd}. \tag{4.173b}$$

For a nontrivial solution, the determinant of the two equations of (4.173) must vanish, leading to

$$k_c \tan k_c d = \epsilon_r h. \tag{4.174}$$

Eliminating β from (4.170a) and (4.170b) gives

$$k_c^2 + h^2 = (\epsilon_r - 1)k_0^2. \qquad 4.175$$

Equations (4.174) and (4.175) constitute a set of simultaneous transcendental equations which must be solved for the propagation constants k_c and h, given k_o and ϵ_r. These equations are best solved numerically, but Figure 4.19 shows a graphical representation of the solutions. Multiplying both sides of (4.175) by d^2 gives

$$(k_c d)^2 + (hd)^2 = (\epsilon_r - 1)(k_0 d)^2,$$

which is the equation of a circle in the $k_c d$, hd plane, as shown in Figure 4.19. The radius of the circle is $\sqrt{\epsilon_r - 1}k_0 d$, which is proportional to the electrical thickness of the dielectric slab. Multiplying (4.174) by d gives

$$k_c d \tan k_c d = \epsilon_r hd,$$

which is also plotted in Figure 4.19. The intersection of these curves implies a solution to both (4.174) and (4.175). Observe that k_c may be positive or negative; from (4.171a) this is seen to merely change the sign of the constant A. As $\sqrt{\epsilon_r - 1}k_0 d$ becomes larger, the circle may intersect more than one branch of the tangent function, implying that more than one TM mode can propagate. Solutions for negative h, however, must be excluded since we assumed h was positive real when applying boundary condition (4.172b).

For any nonzero thickness slab, with a permittivity greater than unity, there is at least one propagating TM mode, which we will call the TM_0 mode. This is the dominant mode of the dielectric slab waveguide, and has a zero cutoff frequency. (Although for $k_0 = 0$, $k_c = h = 0$ and all fields vanish.) From Figure 4.19, it can be seen that the next TM mode, the TM_1 mode, will not turn on until the radius of the circle becomes greater than π. The cutoff frequency of the TM_n mode can then be derived as

$$f_c = \frac{nc}{2d\sqrt{\epsilon_r - 1}}, \qquad n = 0, 1, 2, \cdots. \qquad 4.176$$

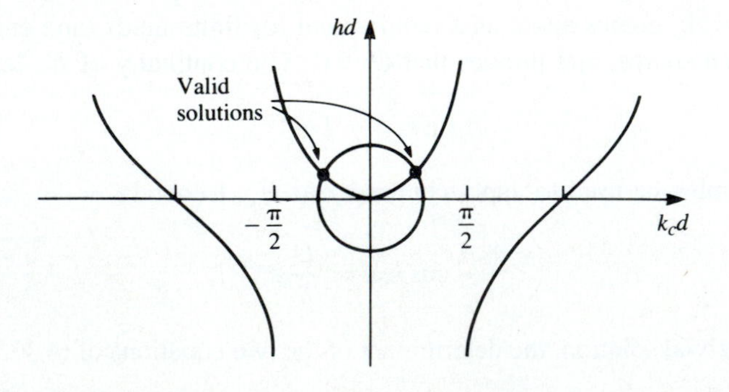

FIGURE 4.19 Graphical solution of the transcendental equation for the cutoff frequency of a TM surface wave mode of the grounded dielectric slab.

Once k_c and h have been found for a particular surface wave mode, the field expressions can be found as,

$$E_z(x,y,z) = \begin{cases} A\sin k_c x e^{-j\beta z} & \text{for } 0 \le x \le d \\ A\sin k_c d e^{-h(x-d)} e^{-j\beta z} & \text{for } d \le x < \infty, \end{cases} \tag{4.177a}$$

$$E_x(x,y,z) = \begin{cases} \dfrac{-j\beta}{k_c} A\cos k_c x e^{-j\beta z} & \text{for } 0 \le x \le d \\ \dfrac{-j\beta}{h} A\sin k_c d e^{-h(x-d)} e^{-j\beta z} & \text{for } d \le x < \infty, \end{cases} \tag{4.177b}$$

$$H_y(x,y,z) = \begin{cases} \dfrac{-j\omega\epsilon_0\epsilon_r}{k_c} A\cos k_c x e^{-j\beta z} & \text{for } 0 \le x \le d \\ \dfrac{-j\omega\epsilon_0}{h} A\sin k_c d e^{-h(x-d)} e^{-j\beta z} & \text{for } d \le x < \infty. \end{cases} \tag{4.177c}$$

TE Modes

TE modes can also be supported by the grounded dielectric slab. The H_z field satisfies the wave equations

$$\left(\frac{\partial^2}{\partial x^2} + k_c^2\right) h_z(x,y) = 0, \qquad \text{for } 0 \le x \le d, \tag{4.178a}$$

$$\left(\frac{\partial^2}{\partial x^2} - h^2\right) h_z(x,y) = 0, \qquad \text{for } d \le x < \infty, \tag{4.178b}$$

with $H_z(x,y,z) = h_z(x,y)e^{-j\beta z}$, and k_c^2 and h^2 defined in (4.170a) and (4.170b). As for the TM modes, the general solutions to (4.178) are

$$h_z(x,y) = A\sin k_c x + B\cos k_c x, \tag{4.179a}$$

$$h_z(x,y) = Ce^{hx} + De^{-hx}. \tag{4.179b}$$

To satisfy the radiation condition, $C = 0$. Using (4.19d) to find E_y from H_z leads to $A = 0$ for $E_y = 0$ at $x = 0$, and to the equation

$$\frac{-B}{k_c}\sin k_c d = \frac{D}{h}e^{-hd}, \tag{4.180a}$$

for continuity of E_y at $x = d$. Continuity of H_z at $x = d$ gives

$$B\cos k_c d = De^{-hd}. \tag{4.180b}$$

Simultaneously solving (4.180a) and (4.180b) leads to the determinantal equation

$$-k_c \cot k_c d = h. \tag{4.181}$$

From (4.170a) and (4.170b) we also have that

$$k_c^2 + h^2 = (\epsilon_r - 1)k_0^2. \tag{4.182}$$

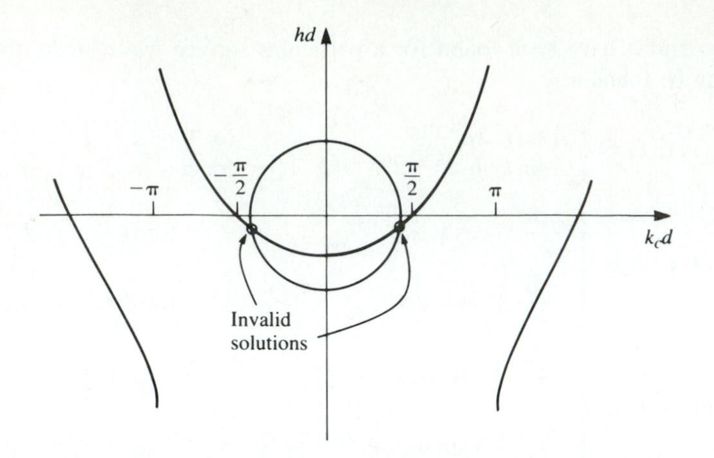

FIGURE 4.20 Graphical solution of the transcendental equation for the cutoff frequency of a TE surface wave mode. Figure depicts a mode below cutoff.

Equations (4.181) and (4.182) must be solved simultaneously for the variables k_c and h. Equation (4.182) again represents circles in the k_cd, hd plane, while (4.181) can be rewritten as

$$-k_cd \cot k_cd = hd,$$

and plotted as a family of curves in the k_cd, hd plane, as shown in Figure 4.20. Since negative values of h must be excluded, we see from Figure 4.20 that the first TE mode does not start to propagate until the radius of the circle, $\sqrt{\epsilon_r - 1}\, k_0d$, becomes greater than $\pi/2$. The cutoff frequency of the TE_n modes can then be found as

$$f_c = \frac{(2n-1)c}{4d\sqrt{\epsilon_r - 1}}, \qquad \text{for } n = 1, 2, 3, \cdots. \tag{4.183}$$

Comparing with (4.176) shows that the order of propagation for the TM_n and TE_n modes is, TM_0, TE_1, TM_1, TE_2, TM_2, $\cdots$.

After finding the constants k_c and h, the field expressions can be derived as

$$H_z(x,y,z) = \begin{cases} B \cos k_cx e^{-j\beta z} & \text{for } 0 \le x \le d \\ B \cos k_cd e^{-h(x-d)} e^{-j\beta z} & \text{for } d \le x < \infty, \end{cases} \tag{4.184a}$$

$$H_x(x,y,z) = \begin{cases} \dfrac{j\beta B}{k_c} \sin k_cx e^{-j\beta z} & \text{for } 0 \le x \le d \\ \dfrac{-j\beta B}{h} \cos k_cd e^{-h(x-d)} e^{-j\beta z} & \text{for } d \le x < \infty, \end{cases} \tag{4.184b}$$

$$E_y(x,y,z) = \begin{cases} \dfrac{-j\omega\mu_o B}{k_c} \sin k_cx e^{-j\beta z} & \text{for } 0 \le x \le d \\ \dfrac{j\omega\mu_o B}{h} \cos k_cd e^{-h(x-d)} e^{-j\beta z} & \text{for } d \le x < \infty. \end{cases} \tag{4.184c}$$

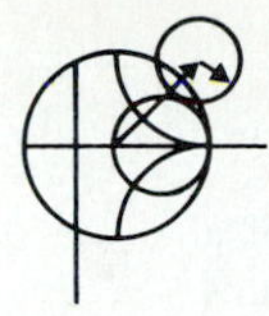

EXAMPLE 4.4

Calculate and plot the propagation constants of the first three propagating surface wave modes of a grounded dielectric sheet with $\epsilon_r = 2.55$, for $d/\lambda_0 = 0$ to 1.2.

Solution

The first three propagating surface wave modes are the TM_0, TE_1, and TM_1 modes. The cutoff frequencies for these modes can be found from (4.176) and (4.183) as

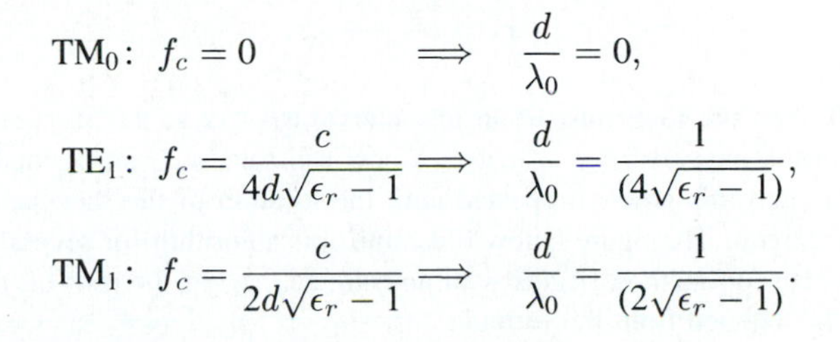

$$TM_0: \ f_c = 0 \quad \Longrightarrow \quad \frac{d}{\lambda_0} = 0,$$

$$TE_1: \ f_c = \frac{c}{4d\sqrt{\epsilon_r - 1}} \Longrightarrow \frac{d}{\lambda_0} = \frac{1}{(4\sqrt{\epsilon_r - 1})},$$

$$TM_1: \ f_c = \frac{c}{2d\sqrt{\epsilon_r - 1}} \Longrightarrow \frac{d}{\lambda_0} = \frac{1}{(2\sqrt{\epsilon_r - 1})}.$$

The propagation constants must be found from the numerical solution of (4.174) and (4.175) for the TM modes, and (4.181) and (4.182) for the TE modes. This can be done with a relatively simple root-finding algorithm (see the Point of Interest on root-finding algorithms); the results are shown in Figure 4.21.

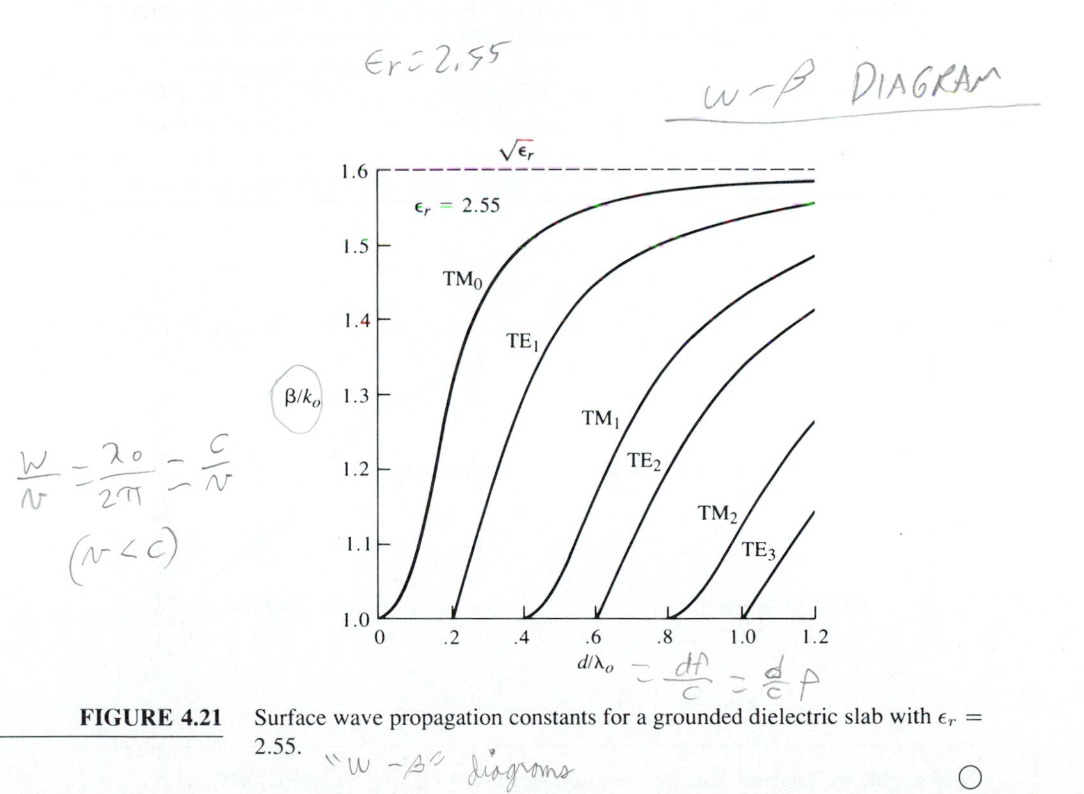

FIGURE 4.21 Surface wave propagation constants for a grounded dielectric slab with $\epsilon_r = 2.55$.

POINT OF INTEREST: Root-Finding Algorithms

In several examples throughout this book we will need to numerically find the root of a transcendental equation, so it may be useful to review two relatively simple but effective algorithms for doing this. Both methods can be easily programmed.

In the interval-halving method the root of $f(x) = 0$ is first bracketed between the values x_1 and x_2. These values can often be estimated from the problem under consideration. If a single root lies between x_1 and x_2, then $f(x_1)f(x_2) < 0$. An estimate, x_3, of the root is made by halving the interval between x_1 and x_2. Thus,

$$x_3 = \frac{x_1 + x_2}{2}.$$

If $f(x_1)f(x_3) < 0$, then the root must lie in the interval $x_1 < x < x_3$; if $f(x_3)f(x_2) < 0$, then the root must be in the interval $x_2 < x < x_3$. A new estimate, x_4, can be made by halving the appropriate interval, and this process repeated until the location of the root has been determined with the desired accuracy. The figure below illustrates this algorithm for several iterations.

The Newton-Rhapson method begins with an estimate, x_1, of the root of $f(x) = 0$. Then a new estimate, x_2, is obtained from the formula

$$x_2 = x_1 - \frac{f(x_1)}{f'(x_1)},$$

where $f'(x_1)$ is the derivative of $f(x)$ at x_1. This result is easily derived from a two-term Taylor series expansion of $f(x)$ near $x = x_1$:$f(x) = f(x_1) + (x - x_1)f'(x_1)$. It can also be interpreted geometrically as fitting a straight line at $x = x_1$ with the same slope as $f(x)$ at this point; this line then intercepts the x-axis at $x = x_2$, as shown in the figure below. Reapplying the above formula gives improved estimates of the root. Convergence is generally much faster than the interval halving method, but a disadvantage is that the derivative of $f(x)$ is required; this can often be computed numerically. The Newton-Rhapson technique can easily be applied to the case where the root is complex (a situation that occurs, for example, when finding the propagation constant of a line or guide with loss).

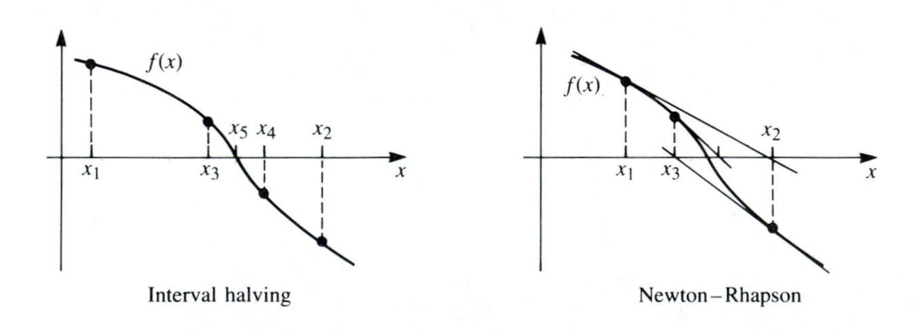

Reference: R. W. Hornbeck, *Numerical Methods*, Quantum Publishers, New York, 1975.

4.7 STRIPLINE

We now consider stripline, a planar-type of transmission line that lends itself well to microwave integrated circuitry and photolithographic fabrication. The geometry of a stripline is shown in Figure 4.22a. A thin conducting strip of width W is centered between two wide conducting ground planes of separation b, and the entire region between the ground planes is filled with a dielectric. In practice, stripline is usually constructed by etching the center conductor on a grounded substrate of thickness $b/2$, and then covering with another grounded substrate of the same thickness. An example of a stripline circuit is shown in Figure 4.23.

Since stripline has two conductors and a homogenous dielectric, it can support a TEM wave, and this is the usual mode of operation. Like the parallel plate guide and coaxial lines, however, the stripline can also support higher-order TM and TE modes, but these are usually avoided in practice (such modes can be suppressed with shorting screws between the ground planes, and by restricting the ground plane spacing to less than $\lambda/4$). Intuitively, one can think of stripline as a sort of "flattened out" coax—both have a center conductor completely enclosed by an outer conductor, and are uniformly filled with a dielectric medium. A sketch of the field lines for stripline is shown in Figure 4.22b. The main difficulty we will have with stripline is that it does not lend itself to a simple analysis, as did the transmission lines and waveguides which we have previously discussed. Since we will be concerned primarily with the TEM mode of the stripline, an electrostatic analysis is sufficient to give the propagation constant and characteristic impedance. An exact solution of Laplace's equation is possible by a conformal mapping approach [1], but the procedure and results are cumbersome. Thus, we will present closed-form expressions that give good approximations to the exact results, and then discuss an approximate numerical technique for solving Laplace's equation for a geometry

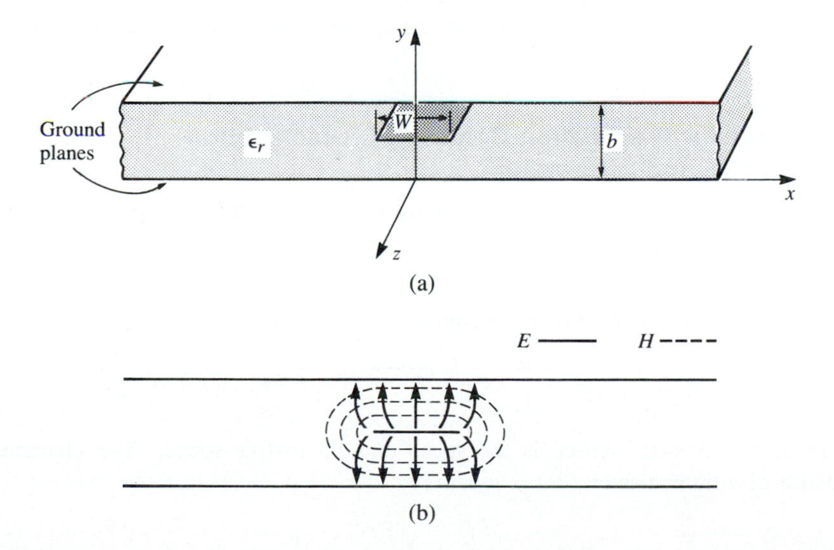

FIGURE 4.22 Stripline transmission line. (a) Geometry. (b) Electric and magnetic field lines.

FIGURE 4.23 Photograph of a stripline circuit assembly, showing two ring hybrids (one used as a power divider and the other in a balanced mixer), and a coupled-line bandpass filter.

Courtesy of Harlan Howe, Jr., M/A-COM Inc.

similar to stripline; this technique will also be applied to microstrip line in the following section.

Formulas for Propagation Constant, Characteristic Impedance, and Attenuation

From Section 4.1 we know that the phase velocity of a TEM mode is given by

$$v_p = 1/\sqrt{\mu_0 \epsilon_0 \epsilon_r} = c/\sqrt{\epsilon_r}, \tag{4.185}$$

thus the propagation constant of the stripline is

$$\beta = \frac{\omega}{v_p} = \omega\sqrt{\mu_0 \epsilon_0 \epsilon_r} = \sqrt{\epsilon_r} k_0. \tag{4.186}$$

In (4.185), $c = 3 \times 10^8$ m/sec is the speed of light in free-space. The characteristic impedance of a transmission line is given by

$$Z_0 = \sqrt{\frac{L}{C}} = \frac{\sqrt{LC}}{C} = \frac{1}{v_p C}, \tag{4.187}$$

where L and C are the inductance and capacitance per unit length of the line. Thus, we can find Z_0 if we know C. As mentioned above, Laplace's equation can be solved by conformal mapping to find the capacitance per unit length of the stripline. The resulting solution, however, involves complicated special functions [1], so for practical computations simple formulas have been developed by curve-fitting to the exact solution [1], [2]. The resulting formula for characteristic impedance is

$$Z_0 = \frac{30\pi}{\sqrt{\epsilon_r}} \frac{b}{W_e + 0.441b}, \tag{4.188a}$$

where W_e is the effective width of the center conductor given by

$$\frac{W_e}{b} = \frac{W}{b} - \begin{cases} 0 & \text{for } \dfrac{W}{b} > 0.35 \\ (0.35 - W/b)^2 & \text{for } \dfrac{W}{b} < 0.35. \end{cases} \tag{4.188b}$$

These formulas assume a zero strip thickness, and are quoted as being accurate to about 1% of the exact results. It is seen from (4.188) that the characteristic impedance decreases as the strip width W increases.

When designing stripline circuits, one usually needs to find the strip width, given the characteristic impedance (and height b and permittivity ϵ_r), which requires the inverse of the formulas in (4.188). Such formulas have been derived as

$$\frac{W}{b} = \begin{cases} x & \text{for } \sqrt{\epsilon_r}\, Z_0 < 120 \\ 0.85 - \sqrt{0.6 - x} & \text{for } \sqrt{\epsilon_r}\, Z_0 > 120, \end{cases} \tag{4.189a}$$

where

$$x = \frac{30\pi}{\sqrt{\epsilon_r} Z_0} - 0.441. \tag{4.189b}$$

Since stripline is a TEM-type of line, the attenuation due to dielectric loss is of the same form as that for other TEM lines, and is given in (4.30). The attenuation due to conductor loss can be found by the perturbation method or Wheeler's incremental inductance rule. An approximate result is

$$\alpha_c = \begin{cases} \dfrac{2.7 \times 10^{-3} R_s \epsilon_r Z_0}{30\pi(b - t)} A & \text{for } \sqrt{\epsilon_r} Z_0 < 120 \\ \dfrac{0.16 R_s}{Z_0 b} B & \text{for } \sqrt{\epsilon_r} Z_0 > 120 \end{cases} \quad \text{Np/m}, \tag{4.190}$$

with

$$A = 1 + \frac{2W}{b - t} + \frac{1}{\pi} \frac{b + t}{b - t} \ln\left(\frac{2b - t}{t}\right),$$

$$B = 1 + \frac{b}{(0.5W + 0.7t)} \left(0.5 + \frac{0.414t}{W} + \frac{1}{2\pi} \ln \frac{4\pi W}{t}\right),$$

where t is the thickness of the strip.

EXAMPLE 4.5

Find the width for a 50 Ω copper stripline conductor, with $b = 0.32$ cm and $\epsilon_r = 2.20$. If the dielectric loss tangent is 0.001 and the operating frequency is 10 GHz, calculate the attenuation in dB/λ. Assume a conductor thickness of $t = 0.01$ mm.

Solution

Since $\sqrt{\epsilon_r} Z_0 = \sqrt{2.2}(50) = 74.2 < 120$, and $x = 30\pi/(\sqrt{\epsilon_r} Z_0) - 0.441 = 0.830$, (4.189) gives the width as $W = bx = (0.32)(0.830) = 0.266$ cm. At 10 GHz, the wavenumber is

$$k = \frac{2\pi f \sqrt{\epsilon_r}}{c} = 310.6 \text{ m}^{-1}.$$

From (4.30) the dielectric attenuation is

$$\alpha_d = \frac{k \tan\delta}{2} = \frac{(310.6)(0.001)}{2} = 0.155 \text{ Np/m}.$$

The surface resistance of copper at 10 GHz is $R_s = 0.026$ Ω. Then from (4.190) the conductor attenuation is

$$\alpha_c = \frac{2.7 \times 10^{-3} R_s \epsilon_r Z_0 A}{30\pi(b - t)} = 0.122 \text{ Np/m},$$

since $A = 4.74$. The total attenuation constant is

$$\alpha = \alpha_d + \alpha_c = 0.277 \text{ Np/m}.$$

In dB,

$$\alpha(\text{dB}) = -20 \log e^{\alpha} = 2.41 \text{ dB/m}.$$

At 10 GHz, the wavelength on the stripline is

$$\lambda = \frac{c}{\sqrt{\epsilon_r} f} = 2.02 \text{ cm},$$

so in terms of wavelength the attenuation is

$$\alpha(\text{dB}) = (2.41)(0.0202) = 0.049 \text{ dB}/\lambda.$$

○

An Approximate Electrostatic Solution

Many practical problems in microwave engineering are very complicated and do not lend themselves to straightforward analytic solutions, but require some sort of numerical approach. Thus it is useful for the student to become aware of such techniques; we will introduce such methods when appropriate throughout this book, beginning with a numerical solution for the characteristic impedance of stripline.

We know that the fields of the TEM mode on a stripline must satisfy Laplace's equation, (4.11), in the region between the two parallel plates. The actual stripline geometry of Figure 4.22a extends to $\pm\infty$, which makes the analysis more difficult.

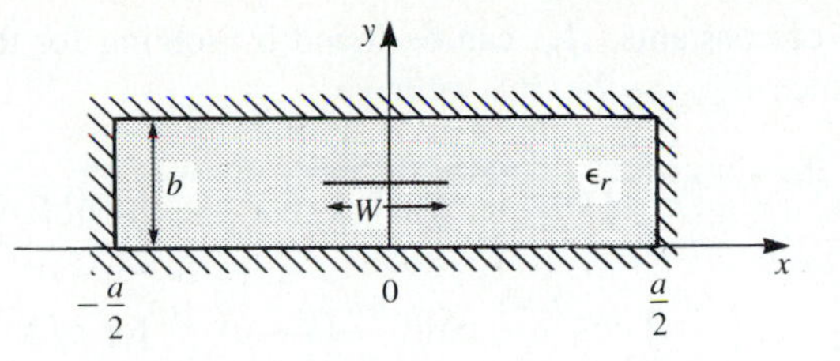

FIGURE 4.24 Geometry of enclosed stripline.

Since we suspect, from the field line drawing of Figure 4.22b, that the field lines do not extend very far away from the center conductor, we can simplify the geometry by truncating the plates beyond some distance, say $|x| > a/2$, and placing metal walls on the sides. Thus, the geometry we will analyze looks like that shown in Figure 4.24, where $a >> b$ so that the fields around the center conductor are not perturbed by the side walls. We then have a closed, finite region in which the potential $\Phi(x, y)$ satisfies Laplace's equation,

$$\nabla_t^2 \Phi(x, y) = 0, \qquad \text{for } |x| \leq a/2, \quad 0 \leq y \leq b, \tag{4.191}$$

with the boundary conditions that

$$\Phi(x, y) = 0, \qquad \text{at } x = \pm a/2, \tag{4.192a}$$

$$\Phi(x, y) = 0, \qquad \text{at } y = 0, b. \tag{4.192b}$$

Laplace's equation can be solved by the method of separation of variables. Since the center conductor at $y = b/2$ will contain a surface charge density, the potential $\Phi(x, y)$ will have a slope discontinuity there, because $\bar{D} = -\epsilon_0 \epsilon_r \nabla_t \Phi$ is discontinuous at $y = b/2$. So separate solutions for $\Phi(x, y)$ must be found for $0 < y < b/2$, and $b/2 < y < b$. The general solutions for $\Phi(x, y)$ in these two regions can be written as

$$\Phi(x, y) = \begin{cases} \displaystyle\sum_{\substack{n=1 \\ \text{odd}}}^{\infty} A_n \cos \frac{n\pi x}{a} \sinh \frac{n\pi y}{a} & \text{for } 0 \leq y \leq b/2 \\ \displaystyle\sum_{\substack{n=1 \\ \text{odd}}}^{\infty} B_n \cos \frac{n\pi x}{a} \sinh \frac{n\pi}{a}(b - y) & \text{for } b/2 \leq y \leq b. \end{cases} \tag{4.193}$$

In this solution, only the odd-n terms are needed because the solution is an even function of x. The reader can verify by substitution that (4.193) satisfies Laplace's equation in the two regions, and satisfies the boundary conditions of (4.192).

Now, the potential must be continuous at $y = b/2$, which from (4.193) leads to

$$A_n = B_n. \tag{4.194}$$

The remaining set of constants, A_n, can be found by solving for the charge density on the center strip. Since $E_y = -\partial\Phi/\partial y$, we have

$$E_y = \begin{cases} -\sum_{\substack{n=1 \\ \text{odd}}}^{\infty} A_n \left(\frac{n\pi}{a}\right) \cos\frac{n\pi x}{a} \cosh\frac{n\pi y}{a} & \text{for } 0 \leq y \leq b/2 \\ \sum_{\substack{n=1 \\ \text{odd}}}^{\infty} A_n \left(\frac{n\pi}{a}\right) \cos\frac{n\pi x}{a} \cosh\frac{n\pi}{a}(b-y) & \text{for } b/2 \leq y \leq b. \end{cases} \quad 4.195$$

The surface charge density on the strip at $y = b/2$ is

$$\begin{aligned} \rho_s &= D_y(x, y = b/2^+) - D_y(x, y = b/2^-) \\ &= \epsilon_0\epsilon_r[E_y(x, y = b/2^+) - E_y(x, y = b/2^-)] \\ &= 2\epsilon_0\epsilon_r \sum_{\substack{n=1 \\ \text{odd}}}^{\infty} A_n \left(\frac{n\pi}{a}\right) \cos\frac{n\pi x}{a} \cosh\frac{n\pi b}{2a}, \end{aligned} \quad 4.196$$

which is seen to be a Fourier series in x for the surface charge density, ρ_s. If we know the surface charge density, we could easily find the unknown constants, A_n, and then the capacitance. We do not know the exact surface charge density, but we can make a good guess by approximating it as a constant over the width of the strip

$$\rho_s(x) = \begin{cases} 1 & \text{for } |x| < W/2 \\ 0 & \text{for } |x| > W/2. \end{cases} \quad 4.197$$

Equating this to (4.196) and using the orthogonality properties of the $\cos(n\pi x/a)$ functions gives the constants A_n as

$$A_n = \frac{2a \sin(n\pi W/2a)}{(n\pi)^2 \epsilon_0\epsilon_r \cosh(n\pi b/2a)}. \quad 4.198$$

The voltage of the center strip relative to the bottom conductor is

$$V = -\int_0^{b/2} E_y(x = 0, y)dy = 2 \sum_{\substack{n=1 \\ \text{odd}}}^{\infty} A_n \sinh\frac{n\pi b}{4a}. \quad 4.199$$

The total charge, per unit length, on the center conductor is

$$Q = \int_{-W/2}^{W/2} \rho_s(x)dx = W \quad \text{C/m}, \quad 4.200$$

so that the capacitance per unit length of the stripline is

$$C = \frac{Q}{V} = \frac{W}{\sum_{\substack{n=1 \\ \text{odd}}}^{\infty} \frac{4a \sin(n\pi W/2a) \sinh(n\pi b/4a)}{(n\pi)^2 \epsilon_0\epsilon_r \cosh(n\pi b/2a)}} \quad \text{Fd/m}. \quad 4.201$$

The characteristic impedance is then found as

$$Z_0 = \sqrt{\frac{L}{C}} = \frac{\sqrt{LC}}{C} = \frac{1}{v_p C} = \frac{\sqrt{\epsilon_r}}{cC},$$

where $c = 3 \times 10^8$ m/sec.

EXAMPLE 4.6

Evaluate the above expressions for a stripline having $\epsilon_r = 2.55$ and $a = 100b$, to find the characteristic impedance for $W/b = 0.25$ to 5.0. Compare with the results from (4.188).

Solution

A short BASIC computer program was written to evaluate (4.201). The series was truncated after 500 terms, and the results are shown below.

W/b	Numerical Eq. (4.201)	Formula Eq. (4.188)
0.25	63.6 Ω	86.6 Ω
0.50	57.8	62.7
1.0	45.0	41.0
2.0	28.0	24.2
3.5	16.8	15.0
5.0	11.8	10.8

We see that the results are in reasonable agreement with the closed form equations of (4.188), particularly for wider strips. Better results could be obtained if more sophisticated estimates were used for the charge density, ρ_s. ○

4.8 MICROSTRIP

Microstrip line is one of the most popular types of planar transmission lines, primarily because it can be fabricated by photolithographic processes and is easily integrated with other passive and active microwave devices. The geometry of a microstrip line is shown in Figure 4.25a. A conductor of width W is printed on a thin, grounded dielectric substrate of thickness d and relative permittivity ϵ_r; a sketch of the field lines is shown in Figure 4.25b.

If the dielectric were not present ($\epsilon_r = 1$), we could think of the line as a two-wire line consisting of two flat strip conductors of width W, separated by a distance $2d$ (the ground plane can be removed via image theory). In this case we would have a simple TEM transmission line, with $v_p = c$ and $\beta = k_0$.

The presence of the dielectric, and particularly the fact that the dielectric does not fill the air region above the strip ($y > d$), complicates the behavior and analysis of microstrip line. Unlike stripline, where all the fields are contained within a homogeneous dielectric

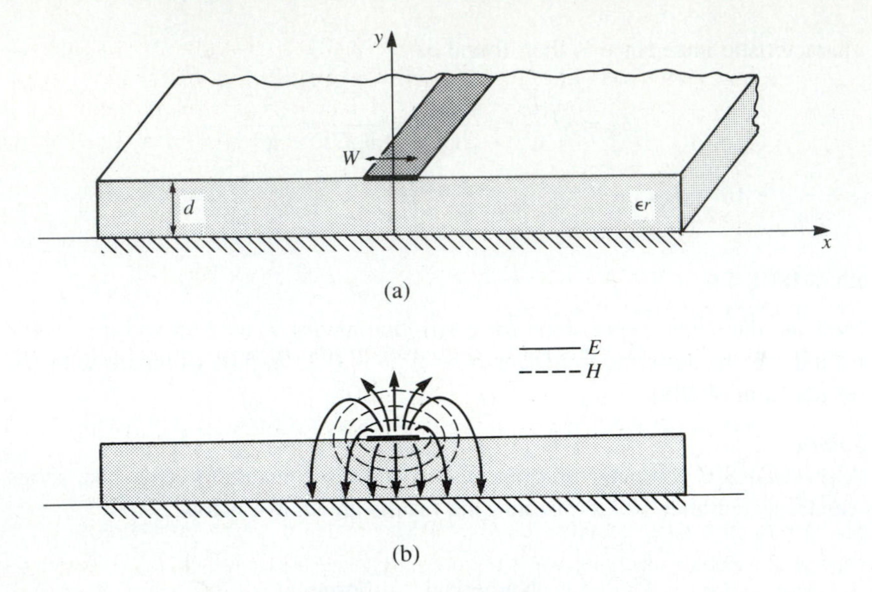

FIGURE 4.25 Microstrip transmission line. (a) Geometry. (b) Electric and magnetic field lines.

region, microstrip has some (usually most) of its field lines in the dielectric region, concentrated between the strip conductor and the ground plane, and some fraction in the air region above the substrate. For this reason the microstrip line cannot support a pure TEM wave, since the phase velocity of TEM fields in the dielectric region would be $c/\sqrt{\epsilon_r}$, but the phase velocity of TEM fields in the air region would be c. Thus, a phase match at the dielectric-air interface would be impossible to attain for a TEM-type wave.

In actuality, the exact fields of a microstrip line constitute a hybrid TM-TE wave, and require more advanced analysis techniques than we are prepared to deal with here. In most practical applications, however, the dielectric substrate is electrically very thin, ($d \ll \lambda$), and so the fields are quasi-TEM. In other words, the fields are essentially the same as those of the static case. Thus, good approximations for the phase velocity, propagation constant, and characteristic impedance can be obtained from static or quasi-static solutions. Then the phase velocity and propagation constant can be expressed as

$$v_p = \frac{c}{\sqrt{\epsilon_e}}, \tag{4.202}$$

$$\beta = k_0\sqrt{\epsilon_e}, \tag{4.203}$$

where ϵ_e is the effective dielectric constant of the microstrip line. Since some of the field lines are in the dielectric region and some are in air, the effective dielectric constant satisfies the relation,

$$1 < \epsilon_e < \epsilon_r,$$

and is dependent on the substrate thickness d, and conductor width, W.

We will first present design formulas for the effective dielectric constant and characteristic impedance of microstrip line; these results are curve-fit approximations to rigorous quasi-static solutions [3], [4]. Then we will outline a numerical method of solution (similar to that used in the previous section for stripline) for the capacitance per unit length of microstrip line.

Formulas for Effective Dielectric Constant, Characteristic Impedance, and Attenuation

The effective dielectric constant of a microstrip line is given approximately by

$$\epsilon_e = \frac{\epsilon_r + 1}{2} + \frac{\epsilon_r - 1}{2} \frac{1}{\sqrt{1 + 12d/W}}. \tag{4.204}$$

The effective dielectric constant can be interpreted as the dielectric constant of a homogeneous medium that replaces the air and dielectric regions of the microstrip, as shown in Figure 4.26. The phase velocity and propagation constant are then given by (4.202) and (4.203).

Given the dimensions of the microstrip line, the characteristic impedance can be calculated as

$$Z_0 = \begin{cases} \dfrac{60}{\sqrt{\epsilon_e}} \ln\left(\dfrac{8d}{W} + \dfrac{W}{4d}\right) & \text{for } W/d \leq 1 \\[2ex] \dfrac{120\pi}{\sqrt{\epsilon_e}\left[W/d + 1.393 + 0.667 \ln\left(W/d + 1.444\right)\right]} & \text{for } W/d \geq 1. \end{cases} \tag{4.205}$$

For a given characteristic impedance Z_0 and dielectric constant ϵ_r, the W/d ratio can be found as

$$\frac{W}{d} = \begin{cases} \dfrac{8e^A}{e^{2A} - 2} & \text{for } W/d < 2 \\[2ex] \dfrac{2}{\pi}\left[B - 1 - \ln(2B - 1) + \dfrac{\epsilon_r - 1}{2\epsilon_r}\left\{\ln(B - 1) + 0.39 - \dfrac{0.61}{\epsilon_r}\right\}\right] & \text{for } W/d > 2, \end{cases} \tag{4.206}$$

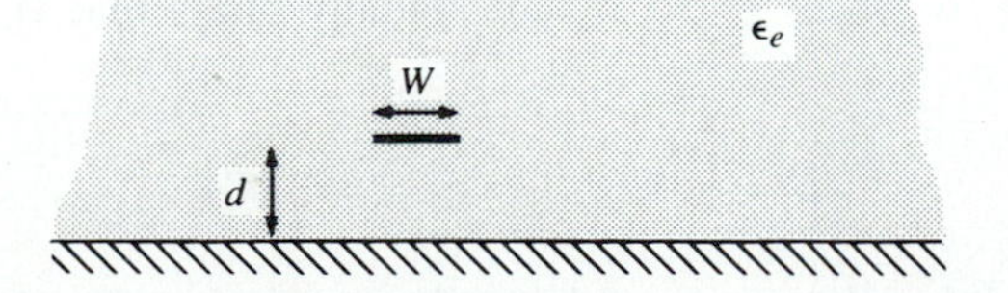

FIGURE 4.26 Equivalent geometry of quasi-TEM microstrip line, where the dielectric slab of thickness d and relative permittivity ϵ_r has been replaced with a homogeneous medium of effective relative permittivity, ϵ_e.

where
$$A = \frac{Z_0}{60}\sqrt{\frac{\epsilon_r + 1}{2}} + \frac{\epsilon_r - 1}{\epsilon_r + 1}\left(0.23 + \frac{0.11}{\epsilon_r}\right)$$

$$B = \frac{377\pi}{2Z_0\sqrt{\epsilon_r}}.$$

Considering microstrip as a quasi-TEM line, the attenuation due to dielectric loss can be determined as

$$\alpha_d = \frac{k_0\epsilon_r(\epsilon_e - 1)\tan\delta}{2\sqrt{\epsilon_e}(\epsilon_r - 1)} \text{ Np/m}, \qquad 4.207$$

where $\tan\delta$ is the loss tangent of the dielectric. This result is derived from (4.30) by multiplying by a "filling factor,"

$$\frac{\epsilon_r(\epsilon_e - 1)}{\epsilon_e(\epsilon_r - 1)},$$

which accounts for the fact that the fields around the microstrip line are partly in air (lossless), and partly in the dielectric. The attenuation due to conductor loss is given approximately by [3]

$$\alpha_c = \frac{R_s}{Z_0 W} \text{ Np/m}, \qquad 4.208$$

where $R_s = \sqrt{\omega\mu_0/2\sigma}$ is the surface resistivity of the conductor. For most microstrip substrates, conductor loss is much more significant than dielectric loss; exceptions may occur with some semiconductor substrates, however.

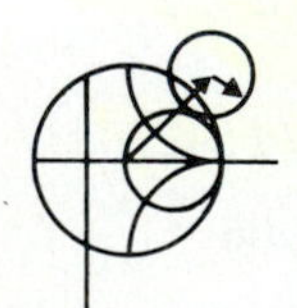

EXAMPLE 4.7

Calculate the width and length of a microstrip line for a 50 Ω characteristic impedance and a 90° phase shift at 2.5 GHz. The substrate thickness is $d = 0.127$ cm, with $\epsilon_r = 2.20$.

Solution

We first find W/d for $Z_0 = 50$ Ω, and initially guess that $W/d > 2$. From (4.206),

$$B = 7.985, \qquad W/d = 3.081.$$

So $W/d > 2$; otherwise we would use the expression for $W/d < 2$. Then $W = 3.081d = 0.391$ cm. From (4.204) the effective dielectric constant is

$$\epsilon_e = 1.87.$$

The line length, ℓ, for a 90° phase shift is found as

$$\phi = 90° = \beta\ell = \sqrt{\epsilon_e}k_0\ell,$$

$$k_0 = \frac{2\pi f}{c} = 52.35 \text{ m}^{-1},$$

$$\ell = \frac{90°(\pi/180°)}{\sqrt{\epsilon_e}k_0} = 2.19 \text{ cm}.$$

An Approximate Electrostatic Solution

ignore

We now look at an approximate quasi-static solution for the microstrip line, so that the appearance of design equations like those of (4.204)–(4.206) is not a complete mystery. This analysis is very similar to that carried out for stripline in the previous section. Like that analysis, it is again convenient to place conducting side walls on the microstrip line, as shown in Figure 4.27. The side walls are placed at $x = \pm a/2$, where $a >> d$, so that the walls should not perturb the field lines localized around the strip conductor. We then can solve Laplace's equation in the region between the side walls:

$$\nabla_t^2\Phi(x, y) = 0, \qquad \text{for } |x| \leq a/2, \;\; 0 \leq y < \infty, \tag{4.209}$$

with boundary conditions,

$$\Phi(x, y) = 0, \qquad \text{at } x = \pm a/2, \tag{4.210a}$$

$$\Phi(x, y) = 0, \qquad \text{at } y = 0, \infty. \tag{4.210b}$$

Since there are two regions defined by the air/dielectric interface, with a charge discontinuity on the strip, we will have separate expressions for $\Phi(x, y)$ in these regions. Solving (4.209) by the method of separation of variables and applying the boundary conditions of (4.210a,b) gives the general solutions as

$$\Phi(x, y) = \begin{cases} \sum\limits_{\substack{n=1\\ \text{odd}}}^{\infty} A_n \cos\dfrac{n\pi x}{a} \sinh\dfrac{n\pi y}{a} & \text{for } 0 \leq y \leq d \\ \sum\limits_{\substack{n=1\\ \text{odd}}}^{\infty} B_n \cos\dfrac{n\pi x}{a} e^{-n\pi y/a} & \text{for } d \leq y < \infty. \end{cases} \tag{4.211}$$

Now the potential must be continuous at $y = d$, so from (4.211) we have that

$$A_n \sinh\frac{n\pi x}{a} = B_n e^{-n\pi d/a}, \tag{4.212}$$

so $\Phi(x, y)$ can be written as

$$\Phi(x, y) = \begin{cases} \sum\limits_{\substack{n=1\\ \text{odd}}}^{\infty} A_n \cos\dfrac{n\pi x}{a} \sinh\dfrac{n\pi y}{a} & \text{for } 0 \leq y \leq d \\ \sum\limits_{\substack{n=1\\ \text{odd}}}^{\infty} A_n \cos\dfrac{n\pi x}{a} \sinh\dfrac{n\pi d}{a} e^{-n\pi(y-d)/a} & \text{for } d \leq y < \infty. \end{cases} \tag{4.213}$$

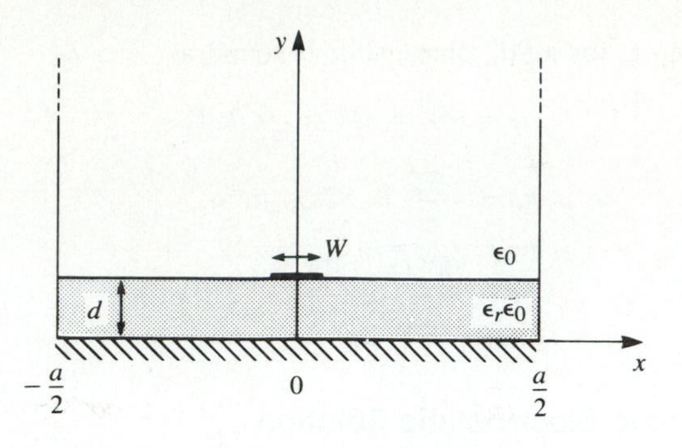

FIGURE 4.27 Geometry of a microstrip line with conducting side walls.

The remaining constants, A_n, can be found by considering the surface charge density on the strip. We first find $E_y = -\partial\Phi/\partial y$:

$$E_y = \begin{cases} -\sum\limits_{\substack{n=1 \\ \text{odd}}}^{\infty} A_n \left(\frac{nx}{a}\right) \cos\frac{n\pi x}{a} \cosh\frac{n\pi y}{a} & \text{for } 0 \leq y < d \\ \sum\limits_{\substack{n=1 \\ \text{odd}}}^{\infty} A_n \left(\frac{n\pi}{a}\right) \cos\frac{n\pi x}{a} \sinh\frac{n\pi d}{a} e^{-n\pi(y-d)/a} & \text{for } d \leq y < \infty. \end{cases} \tag{4.214}$$

Then the surface charge density on the strip at $y = d/2$ is

$$\begin{aligned} \rho_s &= D_y(x, y = d^+) - D_y(x, y = d^-) \\ &= \epsilon_0 E_y(x, y = d^+) - \epsilon_0\epsilon_r E_y(x, y = d^-) \\ &= \epsilon_0 \sum_{\substack{n=1 \\ \text{odd}}}^{\infty} A_n \left(\frac{n\pi}{a}\right) \cos\frac{n\pi x}{a} \left[\sinh\frac{n\pi d}{a} + \epsilon_r \cosh\frac{n\pi d}{a}\right], \end{aligned} \tag{4.215}$$

which is seen to be a Fourier series in x for the surface charge density, ρ_s. As for the stripline case, we can approximate the charge density on the microstrip line by a uniform distribution:

$$\rho_s(x) = \begin{cases} 1 & \text{for } |x| < W/2 \\ 0 & \text{for } |x| > W/2. \end{cases} \tag{4.216}$$

Equating (4.216) to (4.215) and using the orthogonality of the $\cos n\pi x/a$ functions gives the constants A_n as

$$A_n = \frac{4a \sin n\pi W/2a}{(n\pi)^2 \epsilon_0[\sinh(n\pi d/a) + \epsilon_r \cosh(n\pi d/a)]}. \tag{4.217}$$

The voltage of the strip relative to the ground plane is

$$V = -\int_0^d E_y(x = 0, y) dy = \sum_{\substack{n=1 \\ \text{odd}}}^{\infty} A_n \sinh\frac{n\pi d}{a}. \tag{4.218}$$

The total charge, per unit length, on the center strip is

$$Q = \int_{-W/2}^{W/2} \rho_s(x)dx = W \text{ C/m}, \quad 4.219$$

so the static capacitance per unit length of the microstrip line is

$$C = \frac{Q}{V} = \frac{1}{\displaystyle\sum_{\substack{n=1\\ \text{odd}}}^{\infty} \frac{4a \sin(n\pi W/2a) \sinh(n\pi d/a)}{(n\pi)^2 W \epsilon_0[\sinh(n\pi d/a) + \epsilon_r \cosh(n\pi d/a)}}. \quad 4.220$$

Now to find the effective dielectric constant, we consider two cases of capacitance:

Let C = capacitance per unit length of the microstrip line with a dielectric substrate ($\epsilon_r \neq 1$)

Let C_o = capacitance per unit length of the microstrip line with an air dielectric ($\epsilon_r = 1$)

Since capacitance is proportional to the dielectric constant of the material homogeneously filling the region around the conductors, we have that

$$\epsilon_e = \frac{C}{C_o}. \quad 4.221$$

So (4.221) can be evaluated by computing (4.220) twice; once with ϵ_r equal to the dielectric constant of the substrate (for C), and then with $\epsilon_r = 1$ (for C_o). The characteristic impedance is then

$$Z_o = \frac{1}{v_p C} = \frac{\sqrt{\epsilon_e}}{cC}, \quad 4.222$$

where $c = 3 \times 10^8$ m/sec.

EXAMPLE 4.8

Evaluate the above expressions for a microstrip line on a substrate with $\epsilon_r = 2.55$. Calculate the effective dielectric constant and characteristic impedance for $W/d = 0.5$ to 10.0, and compare with the results from (4.204) and (4.205). Let $a = 100d$.

Solution

A computer program was written in BASIC to evaluate (4.220) for $\epsilon = \epsilon_0$ and then $\epsilon = \epsilon_r\epsilon_0$. Then (4.221) was used to evaluate the effective dielectric constant, ϵ_e, and (4.222) to evaluate the characteristic impedance, Z_0. The series was truncated after 50 terms, and the results are shown in the following table.

	Numerical Solutions		Formulas	
W/d	ϵ_e	$Z_o(\Omega)$	ϵ_e	$Z_o(\Omega)$
0.5	1.977	100.9	1.938	119.8
1.0	1.989	94.9	1.990	89.8
2.0	2.036	75.8	2.068	62.2
4.0	2.179	45.0	2.163	39.3
7.0	2.287	29.5	2.245	25.6
10.0	2.351	21.7	2.198	19.1

The comparison is reasonably good, although better results could be obtained from the approximate numerical solution by using a better estimate of the charge density on the strip. ○

4.9 THE TRANSVERSE RESONANCE TECHNIQUE

According to the general solutions to Maxwell's equations for TE or TM waves given in Section 4.1, a uniform waveguide structure always has a propagation constant of the form

$$\beta = \sqrt{k^2 - k_c^2} = \sqrt{k^2 - k_x^2 - k_y^2}, \qquad 4.223$$

where $k_c = \sqrt{k_x^2 + k_y^2}$ is the cutoff wavenumber of the guide and, for a given mode, is a fixed function of the cross-sectional geometry of the guide. Thus, if we know k_c we can determine the propagation constant of the guide. In previous sections we determined k_c by solving the wave equation in the guide, subject to the appropriate boundary conditions; this technique is very powerful and general, but can be complicated for complex waveguides, especially if dielectric layers are present. In addition, the wave equation solution gives a complete field description inside the waveguide, which is much more information than we really need if we are only interested in the propagation constant of the guide. The transverse resonance technique employs a transmission line model of the transverse cross-section of the waveguide, and gives a much simpler and more direct solution for the cutoff frequency. This is another example where circuit and transmission line theory can be used to simplify the field theory solution.

The transverse resonance procedure is based on the fact that in a waveguide at cutoff, the fields form standing waves in the transverse plane of the guide, as can be inferred from the "bouncing plane wave" interpretation of waveguide modes discussed in Section 4.2. This situation can be modelled with an equivalent transmission line circuit operating at resonance. One of the conditions of such a resonant line is the fact that, at any point on the line, the sum of the input impedances seen looking to either side must be zero. That is,

$$Z_{\text{in}}^r(x) + Z_{\text{in}}^\ell(x) = 0, \qquad \text{for all } x, \qquad 4.224$$

where $Z_{\text{in}}^{r}(x)$ and $Z_{\text{in}}^{\ell}(x)$ are the input impedances seen looking to the right and left, respectively, at the point x on the resonant line.

The transverse resonance technique only gives results for the cutoff frequency of the guide. If fields or attenuation due to conductor loss are needed, the complete field theory solution will be required. The procedure will now be illustrated with several examples.

TM Modes for the Parallel Plate Waveguide

We will initially demonstrate the transverse resonance technique by re-solving the problem of Section 4.2 for the TM modes of the parallel plate waveguide. The geometry is shown in Figure 4.28. At cutoff, $k = k_c$, and there is no propagation down the guide in the z direction ($\beta = 0$). The fields thus form a standing wave along the y dimension of the guide. The equivalent circuit is a transmission line of length d (the height of the guide), shorted at both ends (representing the parallel plates at $y = 0, d$), as shown in Figure 4.28. The propagation constant for this line is k_y, and is to be determined. Because of the uniformity in the x direction, $k_x = 0$, so the cutoff wavenumber will be given by $k_c = k_y$. The characteristic impedance of the equivalent transmission line is taken as the wave impedance seen by a TM wave (4.26), with propagation constant k_y:

$$Z_0 = Z_{\text{TM}} = \eta k_y / k, \tag{4.225}$$

where $\eta = \sqrt{\mu/\epsilon}$ is the intrinsic impedance of the material filling the guide, and $k = \omega\sqrt{\mu\epsilon}$ is the wavenumber.

At any point, $0 \leq y \leq d$, along the line we have

$$Z_{\text{in}}^{r}(y) = jZ_{\text{TM}} \tan k_y(d - y), \tag{4.226a}$$

$$Z_{\text{in}}^{\ell}(y) = jZ_{\text{TM}} \tan k_y y. \tag{4.226b}$$

Using these results in (4.224) gives the condition for transverse resonance as

$$jZ_{\text{TM}}[\tan k_y(d - y) + \tan k_y y] = 0,$$

or

$$jZ_{\text{TM}} \frac{\sin k_y d}{\cos k_y(d - y) \cos k_y y} = 0.$$

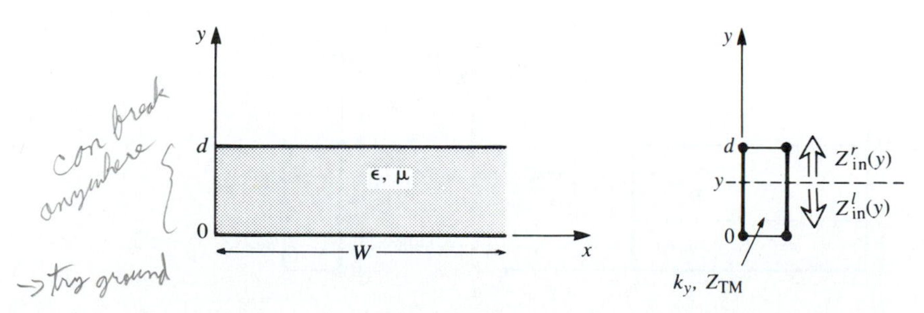

FIGURE 4.28 Transverse resonance equivalent circuit for TM modes of the parallel plate waveguide.

This determines the cutoff wavenumbers as

$$k_c = k_y = \frac{n\pi}{d}, \qquad \text{for } n = 0, 1, 2 \cdots. \tag{4.227}$$

The propagation constant is then found from (4.223). This is the same result as obtained in Section 4.2. For TE modes, we change the characteristic impedance of the line to $Z_{\text{TE}} = \eta k/k_y$, but the same cutoff wavenumber is obtained.

The above procedure can be simplified by noting that condition (4.224) must be valid for any value of x (or y), so we can select a certain point along the transmission line to simplify the evaluation of Z_{in}^r or Z_{in}^ℓ. For example, in the present case we could choose $y = 0$; then $Z_{\text{in}}^\ell(0) = 0$ and $Z_{\text{in}}^r(0) = jZ_{\text{TM}} \tan k_y d$, which yields $k_y = n\pi/d$ more directly.

TE Modes of a Partially Loaded Rectangular Waveguide

The transverse resonance technique is particularly useful when the guide contains dielectric layers because the boundary conditions at the dielectric interfaces, which require the solution of simultaneous algebraic equations in the field theory approach, can be easily handled as junctions of different transmission lines. As an example, consider the rectangular waveguide partially filled with dielectric, as shown in Figure 4.29. To find the cutoff frequencies for the TE modes, the equivalent transverse resonance circuit shown in the figure can be used. The line for $0 < y < t$ represents the dielectric-filled part of the guide, and has a transverse propagation constant k_{yd} and a characteristic impedance for TE modes given by

$$Z_d = \frac{k\eta}{k_{yd}} = \frac{k_0\eta_0}{k_{yd}}, \tag{4.228a}$$

where $k_0 = \omega\sqrt{\mu_0\epsilon_0}, \eta_0 = \sqrt{\mu_0/\epsilon_0}$. For $t < y < b$, the guide is air-filled, and has a transverse propagation constant k_{ya} and an equivalent characteristic impedance given by

$$Z_a = \frac{k_0\eta_0}{k_{ya}}. \tag{4.228b}$$

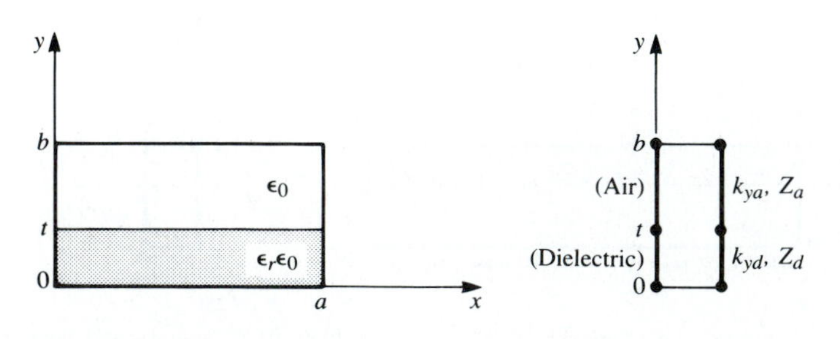

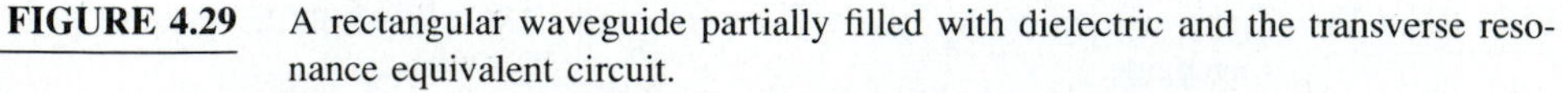

FIGURE 4.29 A rectangular waveguide partially filled with dielectric and the transverse resonance equivalent circuit.

Applying condition (4.224) yields

$$k_{ya} \tan k_{yd} d + k_{yd} \tan k_{ya}(b - t) = 0. \qquad 4.229$$

This equation contains two unknowns, k_{ya} and k_{yd}. An additional equation is obtained from the fact that the propagation constant, β, must be the same in both regions, for phase matching of the tangential fields at the dielectric interface. Thus,

$$\beta = \sqrt{\epsilon_r k_0^2 - k_x^2 - k_{yd}^2} = \sqrt{k_0^2 - k_x^2 - k_{ya}^2},$$

or

$$\epsilon_r k_0^2 - k_{yd}^2 = k_0^2 - k_{ya}^2. \qquad 4.230$$

Equations (4.229) and (4.230) can then be solved (numerically or graphically) to obtain k_{yd} and k_{ya}. There will be an infinite number of solutions, corresponding to the n-dependence (number of variations in y) of the TE_{on} mode. In the x direction, we know that $k_x = m\pi/a$, from Section 4.3 (or from transverse resonance in the x direction). For the TM modes of this structure, we replace the impedances with $k_y\eta/k$.

TM Modes of a Multilayer Surface Waveguide

This example demonstrates how the transverse resonance method can be applied to open waveguide structures, similar to the surface waveguide of Section 4.6. Consider the TM modes for the geometry shown in Figure 4.30; this waveguide can support both TE and TM surface waves propagating along the z direction with no y variation. The equivalent transmission line circuit for the transverse resonance solution is shown in the figure. For $0 < x < t$ the medium is air, and the equivalent transmission line has a transverse propagation constant k_{xa}, and a characteristic impedance (for TM modes) given by

$$Z_a = \frac{k_{xa}\eta_0}{k_0}. \qquad 4.231a$$

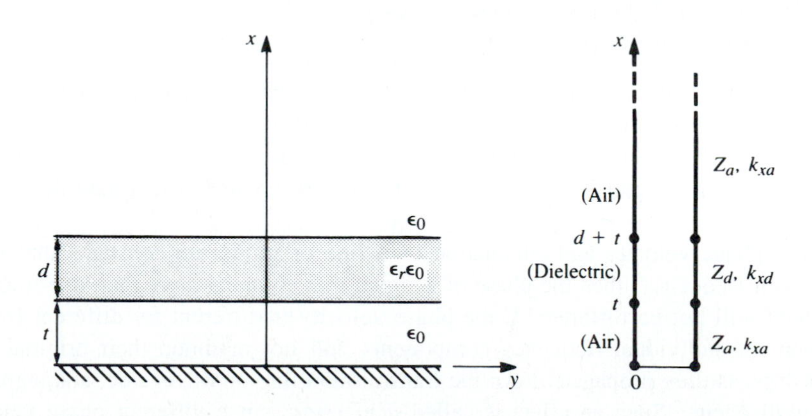

FIGURE 4.30 A surface waveguide consisting of a dielectric sheet above a ground plane and the transverse resonance equivalent circuit.

A dielectric slab exists in the region $t < x < t + d$; the equivalent transmission line has a transverse propagation constant k_{xd}, and a characteristic impedance given by

$$Z_d = \frac{k_{xd}\eta}{k} = \frac{k_{xd}\eta_0}{\epsilon_r k_0}. \tag{4.231b}$$

The air region above the dielectric, for $x > t + d$, is modelled as an infinitely long line with propagation constant k_{xa} and characteristic impedance Z_a. Because this line is infinitely long, the input impedance seen at $x = t + d$ must be Z_a. Then applying (4.224) at $x = t$ gives

$$jZ_a \tan k_{xa}t + Z_d \frac{Z_a + jZ_d \tan k_{xd}d}{Z_d + jZ_a \tan k_{xd}d} = 0. \tag{4.232}$$

From the solution of the surface waveguide of Section 4.6, we anticipate an exponentially decaying field for $x > t + d$; thus we let $k_{xa} = -jh$, where h is real. Then $\tan k_{xa}t = -j \tanh ht$, and $Z_a = -jh\eta_0/k_0$, and (4.232) reduces to

$$\epsilon_r h \tanh t[k_{xd} + h\epsilon_r \tan k_{xd}] + k_{xd}[\epsilon_r h - k_{xd} \tan k_{xd}d] = 0. \tag{4.233}$$

For phase matching at the dielectric-air interfaces, we require that

$$\epsilon_r k_0^2 - k_{xd}^2 = k_0^2 - k_{xa}^2 = k_0^2 + h^2. \tag{4.234}$$

Equations (4.233) and (4.234) can then be solved simultaneously for h and k_{xd}. The reader can verify that (4.233) reduces to the result in (4.174) when the air gap is eliminated ($t \to 0$).

4.10 WAVE VELOCITIES AND DISPERSION

So far, we have encountered two types of velocities related to the propagation of electromagnetic waves:

- The speed of light in a medium ($1/\sqrt{\mu\epsilon}$)
- The phase velocity ($v_p = \omega/\beta$)

The speed of light in a medium is the velocity at which a plane wave would propagate in that medium, while the phase velocity is the speed at which a constant phase point travels. For a TEM plane wave, these two velocities are identical, but for other types of guided wave propagation the phase velocity may be greater or less than the speed of light.

If the phase velocity and attenuation of a line or guide are constants that do not change with frequency, then the phase of a signal that contains more than one frequency component will not be distorted. If the phase velocity is different for different frequencies, then the individual frequency components will not maintain their original phase relationships as they propagate down the transmission line or waveguide, and signal distortion will occur. Such an effect is called *dispersion*, since different phase velocities allow the "faster" waves to lead in phase relative to the "slower" waves, and the original phase relationships will gradually be dispersed as the signal propagates down the line.

In such a case, there is no single phase velocity that can be attributed to the signal as a whole. However, if the bandwidth of the signal is relatively small, or if the dispersion is not too severe, a *group velocity* can be defined in a meaningful way. This velocity then can be used to describe the speed at which the signal propagates.

Group Velocity

As discussed above, the physical interpretation of group velocity is the velocity at which a narrow band signal propagates. We will derive the relation of group velocity to the propagation constant by considering a signal $f(t)$ in the time domain. The Fourier transform of this signal is defined as

$$F(\omega) = \int_{-\infty}^{\infty} f(t)e^{-j\omega t}dt, \tag{4.235a}$$

and the inverse transform is then

$$f(t) = \frac{1}{2\pi}\int_{-\infty}^{\infty} F(\omega)e^{j\omega t}d\omega. \tag{4.235b}$$

Now consider the transmission line or waveguide on which the signal $f(t)$ is propagating as a linear system, with a transfer function $Z(\omega)$ that relates the output, $F_o(\omega)$, of the line to the input, $F(\omega)$, of the line, as shown in Figure 4.31. Thus,

$$F_o(\omega) = Z(\omega)F(\omega). \tag{4.236}$$

For a lossless, matched transmission line or waveguide, the transfer function $Z(\omega)$ can be expressed as

$$Z(\omega) = Ae^{-j\beta z} = |Z(\omega)|e^{-j\psi}, \tag{4.237}$$

where A is a constant and β is the propagation constant of the line or guide. The time-domain representation of the output signal, $f_o(t)$, can then be written as

$$f_o(t) = \frac{1}{2\pi}\int_{-\infty}^{\infty} F(\omega)|Z(\omega)|e^{j(\omega t-\psi)}d\omega. \tag{4.238}$$

Now if $|Z(\omega)| = A$ is a constant, and the phase ψ of $Z(\omega)$ is a linear function of ω, say $\psi = a\omega$, then the output can be expressed as

$$f_o(t) = \frac{1}{2\pi}\int_{-\infty}^{\infty} AF(\omega)e^{j\omega(t-a)}d\omega = Af(t-a), \tag{4.239}$$

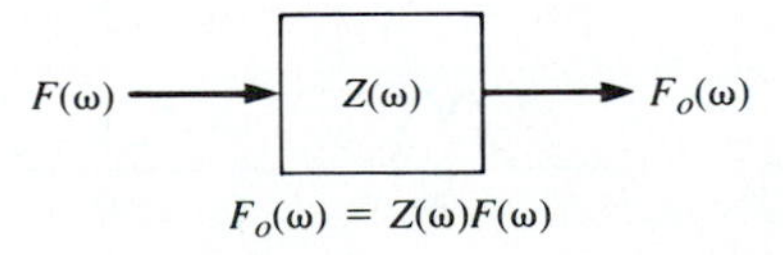

FIGURE 4.31 A transmission line or waveguide represented as a linear system with transfer function $Z(\omega)$.

which is seen to be a replica of $f(t)$, except for an amplitude factor, A, and time shift, a. Thus, a transfer function of the form $Z(\omega) = Ae^{-j\omega a}$ does not distort the input signal. A lossless TEM wave has a propagation constant $\beta = \omega/c$, which is of this form, so a TEM line is dispersionless, and does not lead to signal distortion. If the TEM line is lossy, however, the attenuation may be a function of frequency, which could lead to signal distortion.

Now consider a narrowband input signal of the form

$$s(t) = f(t)\cos\omega_o t = \text{Re}\left\{f(t)e^{j\omega_o t}\right\}, \tag{4.240}$$

which represents an amplitude modulated carrier wave of frequency ω_o. Assume that the highest frequency component of $f(t)$ is ω_m, where $\omega_m << \omega_o$. The Fourier transform, $S(\omega)$, of $s(t)$, is

$$S(\omega) = \int_{-\infty}^{\infty} f(t)e^{-j\omega_o t}e^{j\omega t}dt = F(\omega - \omega_o), \tag{4.241}$$

where we have used the complex form of the input signal as expressed in (4.240). We will then need to take the real part of the output inverse transform to obtain the time-domain output signal. The spectrums of $F(\omega)$ and $S(\omega)$ are depicted in Figure 4.32.

The output signal spectrum is

$$S_o(\omega) = AF(\omega - \omega_o)e^{-j\beta z}, \tag{4.242}$$

and in the time domain,

$$\begin{aligned} s_o(t) &= \frac{1}{2\pi}\text{Re}\int_{-\infty}^{\infty} S_o(\omega)e^{j\omega t}d\omega \\ &= \frac{1}{2\pi}\text{Re}\int_{\omega_o-\omega_m}^{\omega_o+\omega_m} AF(\omega - \omega_o)e^{j(\omega t - \beta z)}d\omega. \end{aligned} \tag{4.243}$$

In general, the propagation constant β may be a complicated function of ω. But if $F(\omega)$ is narrow band ($\omega_m << \omega_o$), then β can be linearized by using a Taylor series expansion about ω_o:

$$\beta(\omega) = \beta(\omega_o) + \left.\frac{d\beta}{d\omega}\right|_{\omega=\omega_o}(\omega - \omega_o) + \frac{1}{2}\left.\frac{d^2\beta}{d\omega^2}\right|_{\omega=\omega_o}(\omega - \omega_o)^2 + \cdots. \tag{4.244}$$

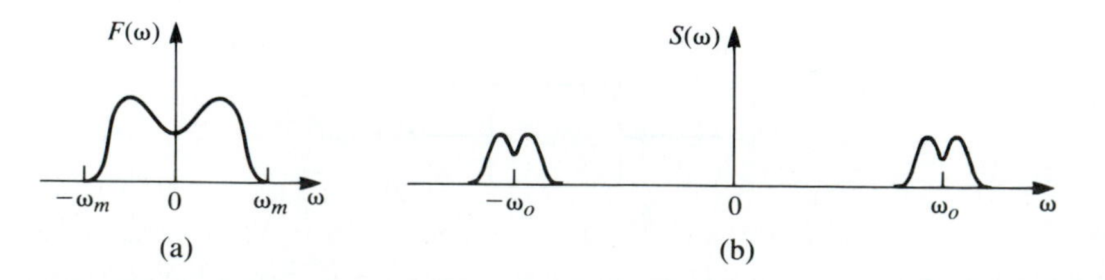

FIGURE 4.32 Fourier spectrums of the signals (a) $f(t)$ and (b) $s(t)$.

Retaining the first two terms gives

$$\beta(\omega) \simeq \beta_o + \beta'_o(\omega - \omega_o), \tag{4.245}$$

where

$$\beta_o = \beta(\omega_o),$$

$$\beta'_o = \left.\frac{d\beta}{d\omega}\right|_{\omega=\omega_o}.$$

Then after a change of variables to $y = \omega - \omega_o$, the expression for $s_o(t)$ becomes

$$s_o(t) = \frac{A}{2\pi}\text{Re}\left\{e^{j(\omega_o t - \beta_o z)} \int_{-\omega_m}^{\omega_m} F(y)e^{j(t-\beta'_o z)y}\,dy\right\}$$

$$= A\text{Re}\left\{f(t-\beta'_o z)e^{j(\omega_o t - \beta_o z)}\right\}$$

$$= Af(t-\beta'_o z)\cos(\omega_o t - \beta_o z), \tag{4.246}$$

which is a time-shifted replica of the original modulation envelope, $f(t)$, of (4.240). The velocity of this envelope is the group velocity, v_g:

$$v_g = \frac{1}{\beta'_o} = \left.\left(\frac{d\beta}{d\omega}\right)^{-1}\right|_{\omega=\omega_o}. \tag{4.247}$$

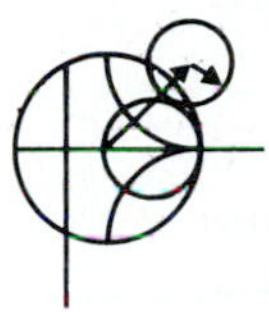

EXAMPLE 4.9

Calculate the group velocity for a waveguide mode propagating in an air-filled guide. Compare this velocity to the phase velocity and speed of light.

Solution

The propagation constant for a mode in an air-filled waveguide is

$$\beta = \sqrt{k_0^2 - k_c^2} = \sqrt{(\omega/c)^2 - k_c^2}.$$

Taking the derivative with respect to frequency gives

$$\frac{d\beta}{d\omega} = \frac{\omega/c^2}{\sqrt{(\omega/c)^2 - k_c^2}} = \frac{k_o}{c\beta},$$

so from (4.247) the group velocity is

$$v_g = \left(\frac{d\beta}{d\omega}\right)^{-1} = \frac{c\beta}{k_0}.$$

The phase velocity is $v_p = \omega/\beta = (k_0 c)/\beta$.

Since $\beta < k_0$, we have that $v_g < c < v_p$, which indicates that the phase velocity of a waveguide mode may be greater than the speed of light, but the group velocity (the velocity of a narrowband signal) will be less than the speed of light. ○

4.11 SUMMARY OF TRANSMISSION LINES AND WAVEGUIDES

In this chapter we have discussed a variety of transmission lines and waveguides; here we will summarize some of the basic properties of these transmission media and their relative advantages in a broader context.

In the beginning of this chapter we made the distinction between TEM, TM, and TE waves, and saw that transmission lines and waveguides can be categorized according to which type of waves they can support. We have seen that TEM waves are nondispersive, with no cutoff frequency, while TM and TE waves exhibit dispersion, and generally have nonzero cutoff frequencies. Other electrical considerations include bandwidth, attenuation, and power handling capacity. Mechanical factors are also very important, however, and include such considerations as physical size (volume and weight), ease of fabrication (cost), and the ability to be integrated with other devices (active or passive). Table 4.6 compares several types of transmission media with regard to the above considerations; this table only gives general guidelines, as specific cases may give better or worse results than those indicated.

Other Types of Lines and Guides

While we have discussed the most common types of waveguides and transmission lines, there are many other guides and lines (and variations) that we have not discussed. A few of the more popular types are briefly mentioned here.

Ridge waveguide. The bandwidth of a rectangular waveguide is, for practical purposes, less than an octave (a 2:1 frequency range). This is because the TE_{20} mode begins to propagate at a frequency equal to twice the cutoff frequency of the TE_{10} mode. The ridge waveguide, shown in Figure 4.33, consists of a rectangular waveguide loaded with conducting ridges on the top and/or bottom walls. This loading tends to lower the cutoff frequency of the dominant mode, leading to increased bandwidth and better impedance characteristics. Such a guide is often used for impedance matching purposes, where the

TABLE 4.6 Comparison of Common Transmission Lines and Waveguides

Characteristic	Coax	Waveguide	Stripline	Microstrip
Modes: Preferred	TEM	TE_{10}	TEM	Quasi-TEM
Other	TM,TE	TM,TE	TM,TE	Hybrid TM,TE
Dispersion	None	Medium	None	Low
Bandwidth	High	Low	High	High
Loss	Medium	Low	High	High
Power capacity	Medium	High	Low	Low
Physical size	Large	Large	Medium	Small
Ease of fabrication	Medium	Medium	Easy	Easy
Integration with other components	Hard	Hard	Fair	Easy

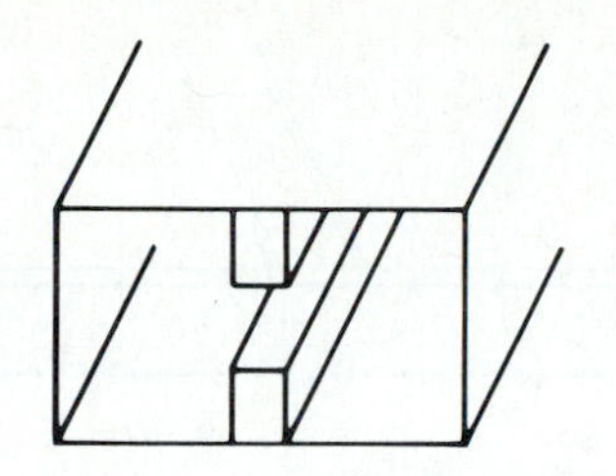

FIGURE 4.33 Cross-section of a ridge waveguide.

ridge may be tapered along the length of the guide. The presence of the ridge, however, reduces the power-handling capacity of the waveguide.

Dielectric waveguide. As we have seen from our study of surface waves, metallic conductors are not necessary to confine and support a propagating electromagnetic field. The dielectric waveguide shown in Figure 4.34 is another example of such a guide, where ϵ_{r2}, the dielectric constant of the ridge, is usually greater than ϵ_{r1}, the dielectric constant of the substrate. The fields are thus mostly confined to the area around the dielectric ridge. This type of guide supports TM and TE modes, and is convenient for integration with active devices. Its small size makes it useful for millimeter wave to optical frequencies, although it can be very lossy at bends or junctions in the ridge line. Many variations in this basic geometry are possible.

Slotline. Of the many types of planar lines that have been proposed, slotline probably ranks next, behind microstrip and stripline, in terms of popularity. The geometry of a slotline is shown in Figure 4.35. It consists of a thin slot in the ground plane on one side of a dielectric substrate. Thus, like microstrip, the two conductors of slotline lead to a quasi-TEM type of mode. Changing the width of the slot changes the characteristic impedance of the line.

Coplanar waveguide. A structure similar to slotline is coplanar waveguide, shown in Figure 4.36. Coplanar waveguide can be thought of as a slotline with a third conductor centered in the slot region. Because of the presence of this additional conductor, this type of line can support even or odd quasi-TEM modes, depending on whether the $\bar{E}$-fields in the two slots are in the opposite direction, or the same direction. Coplanar waveguide is particularly useful for fabricating active circuitry, due to the presence of the center conductor and the close proximity of the ground planes.

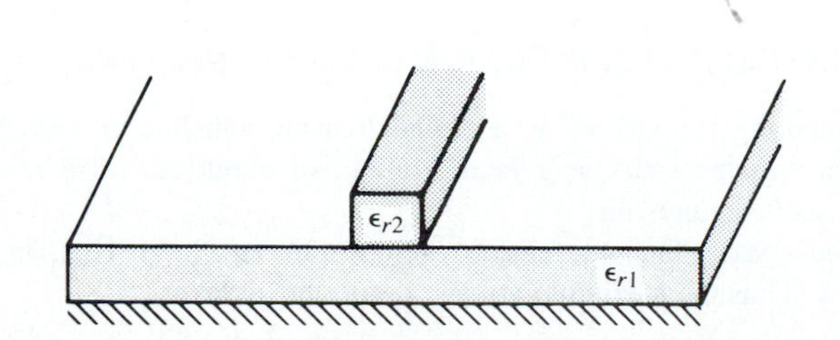

FIGURE 4.34 Dielectric waveguide geometry.

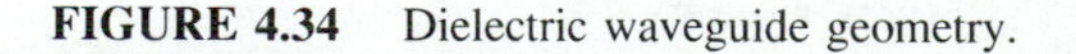

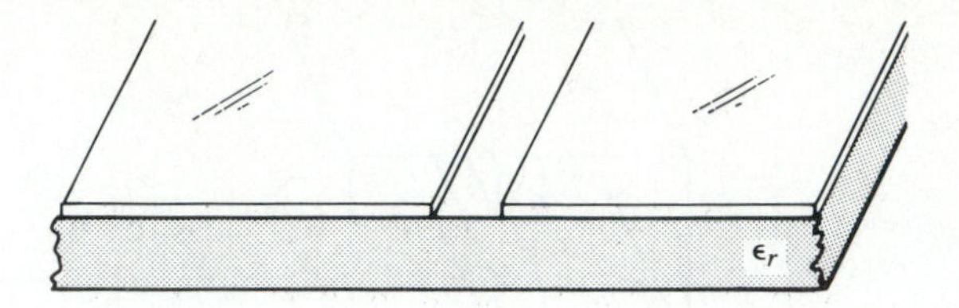

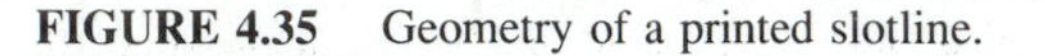
FIGURE 4.35 Geometry of a printed slotline.

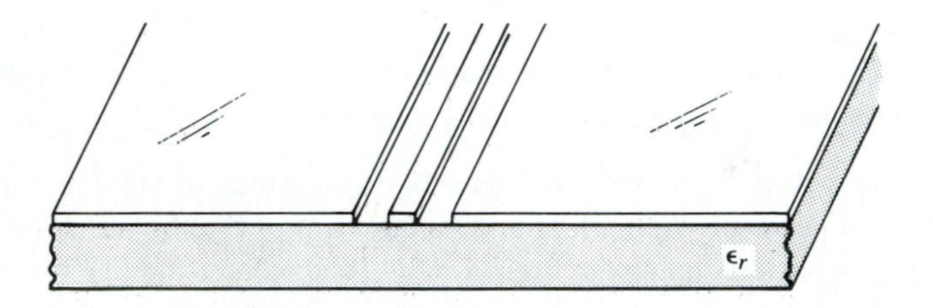

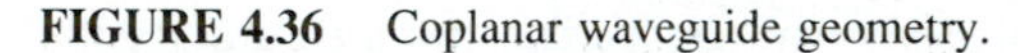
FIGURE 4.36 Coplanar waveguide geometry.

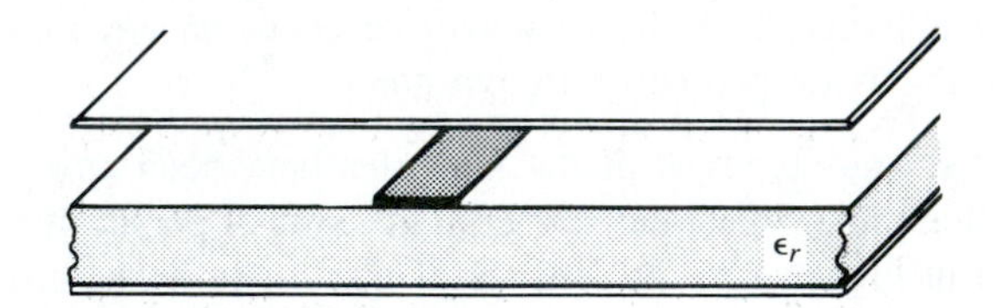

FIGURE 4.37 Covered microstrip line.

Covered microstrip. Many variations of the basic microstrip geometry are possible, but one of the more common is covered microstrip, shown in Figure 4.37. The metallic cover plate is often used for electrical shielding and physical protection of the microstrip circuit, and is usually situated several substrate thicknesses away from the circuit. Its presence can, however, perturb the operation of the circuit enough so that its effect must be taken into account during design.

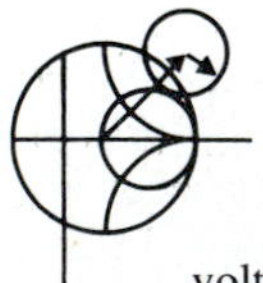

POINT OF INTEREST: Power Capacity of Transmission Lines

The power handling capacity of an air-filled transmission line or waveguide is limited by voltage breakdown, which occurs at a field strength of about $E_d = 3 \times 10^6$ V/m for room temperature air at sea level pressure.

In an air-filled coaxial line, the electric field varies as $E_\rho = V_o/(\rho \ln b/a)$, which has a maximum at $\rho = a$. Thus the maximum voltage before breakdown is

$$V_{\max} = E_d a \ln \frac{b}{a}, \qquad \text{(peak-to-peak)},$$

and the maximum power capacity is then

$$P_{max} = \frac{V_{max}^2}{2Z_0} = \frac{\pi a^2 E_d^2}{\eta_0} \ln \frac{b}{a}.$$

As might be expected, this result shows that power capacity can be increased by using a larger coaxial cable (larger a, b with fixed b/a for the same characteristic impedance). But propagation of higher-order modes limits the maximum operating frequency for a given cable size. Thus, there is an upper limit on the power capacity of a coaxial line for a given maximum operating frequency, f_{max}, which can be shown to be given by

$$P_{max} = \frac{0.025}{\eta_0} \left(\frac{cE_d}{f_{max}} \right)^2 = 5.8 \times 10^{12} \left(\frac{E_d}{f_{max}} \right)^2.$$

As an example, at 10 GHz the maximum peak power capacity of any coaxial line with no higher-order modes is about 520 kW.

In an air-filled rectangular waveguide, the electric field varies as $E_y = E_o \sin(\pi x/a)$, which has a maximum value of E_o at $x = a/2$. Thus the maximum power capacity before breakdown is

$$P_{max} = \frac{abE_o^2}{4Z_w} = \frac{abE_d^2}{4Z_w},$$

which shows that power capacity increases with guide size. For most waveguides, $b \simeq 2a$. To avoid propagation of the TE_{20} mode, we must have $a < c/f_{max}$, where f_{max} is the maximum operating frequency. Then the maximum power capacity of the guide can be shown to be

$$P_{max} = \frac{0.11}{\eta_0} \left(\frac{cE_d}{f_{max}} \right)^2 = 2.6 \times 10^{13} \left(\frac{E_d}{f_{max}} \right)^2.$$

As an example, at 10 GHz the maximum peak power capacity of a rectangular waveguide operating in the TE_{10} mode is about 2300 kW, which is considerably higher than the power capacity of a coaxial cable at the same frequency.

Because arcing and voltage breakdown are very high-speed effects, the above voltage and power limits are peak quantities. In addition, it is good engineering practice to provide a safety factor of at least two, so the maximum powers which can be safely transmitted should be limited to about half of the above values. If there are reflections on the line or guide, the power capacity is further reduced. In the worst case, a reflection coefficient magnitude of unity will double the maximum voltage on the line, so the power capacity will be reduced by a factor of four.

The power capacity of a line can be increased by pressurizing the line with air or an inert gas, or by using a dielectric. The dielectric strength (E_d) of most dielectrics is greater than that of air, but the power capacity may be primarily limited by the heating of the dielectric due to ohmic loss.

Reference: P. A. Rizzi, *Microwave Engineering–Passive Circuits*, Prentice-Hall, New Jersey, 1988.

REFERENCES

[1] H. Howe, Jr., *Stripline Circuit Design*, Artech House, Dedham, Mass., 1974.

[2] I. J. Bahl and R. Garg, "A Designer's Guide to Stripline Circuits," *Microwaves*, January 1978, pp. 90–96.

[3] I. J. Bahl and D. K. Trivedi, "A Designer's Guide to Microstrip Line," *Microwaves*, May 1977, pp. 174–182.

[4] K. C. Gupta, R. Garg, and I. J. Bahl, *Microstrip Lines and Slotlines*, Artech House, Dedham, Mass., 1979.

PROBLEMS

4.1 Derive equations (4.5a)–(4.5d) from equations (4.3) and (4.4).

4.2 Calculate the attenuation due to conductor loss for the TE_n mode of a parallel plate waveguide.

4.3 Consider a section of K-band waveguide. From the dimensions given in Appendix I, determine the cutoff frequencies of the first two propagating modes. From the recommended operating range given in Appendix I for this guide, determine the percentage reduction in bandwidth that this operating range represents, relative to the theoretical bandwidth for one propagating mode.

4.4 Compute the attenuation, in dB/m, for a length of K-band waveguide operating at $f = 20$ GHz. The waveguide is made from brass, and is filled with a dielectric material having $\epsilon_r = 2.6$ and $\tan\delta = 0.01$.

4.5 An attenuator can be made using a section of waveguide operating below cutoff, as shown below. If $a = 2.286$ cm and the operating frequency is 12 GHz, determine the required length of the below-cutoff section of waveguide to achieve an attenuation of 100 dB between the input and output guides. Ignore reflections at the step discontinuities.

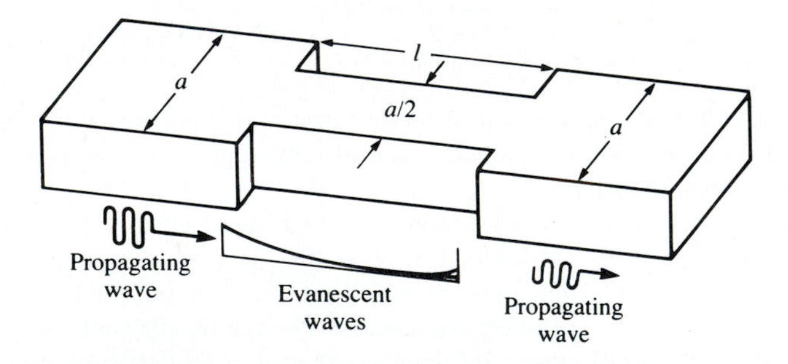

4.6 Find expressions for the electric surface current density on the walls of a rectangular waveguide for a TE_{10} mode. Why can a narrow slot be cut along the centerline of the broad wall of a rectangular waveguide without perturbing the operation of the guide? (Such a slot is often used in a slotted line for a probe to sample the standing wave field inside the guide.)

4.7 Derive the expression for the attenuation of the TM_{mn} mode of a rectangular waveguide, due to imperfectly conducting walls.

4.8 For the partially loaded rectangular waveguide shown below, solve (4.109) with $\beta = 0$ to find the cutoff frequency of the TE_{10} mode. Assume $a = 2.286$ cm, $t = a/2$, and $\epsilon_r = 2.25$.

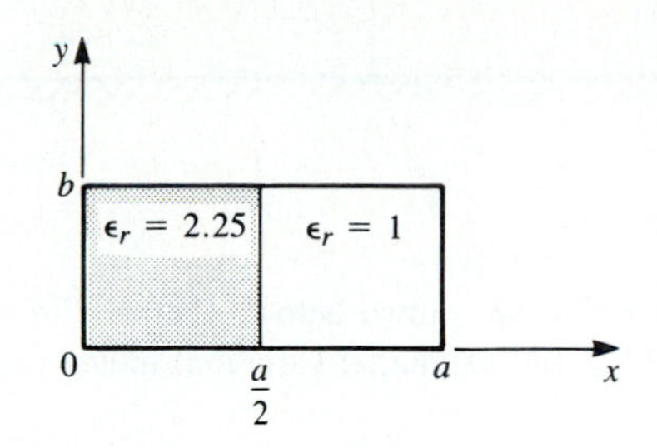

4.9 Consider the partially filled parallel plate waveguide shown below. Derive the solution (fields and cutoff frequency) for the lowest-order TE mode of this structure. Assume the metal plates are infinitely wide. Can a TEM wave propagate on this structure?

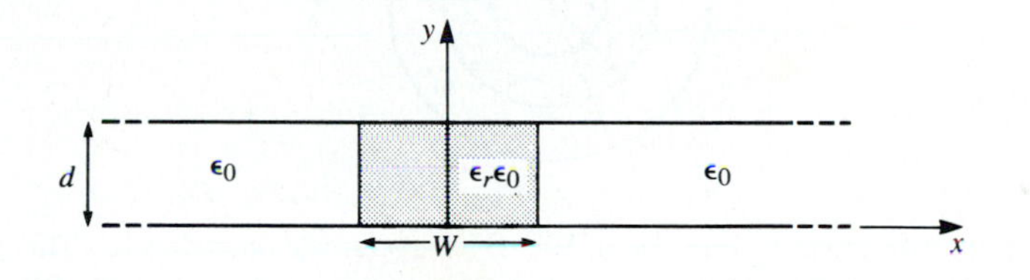

4.10 Consider the partially filled parallel plate waveguide shown below. Derive the solution (fields and cutoff frequency) for the TE modes. Can a TEM wave exist in this structure? Ignore fringing fields at the sides, and assume no x dependence.

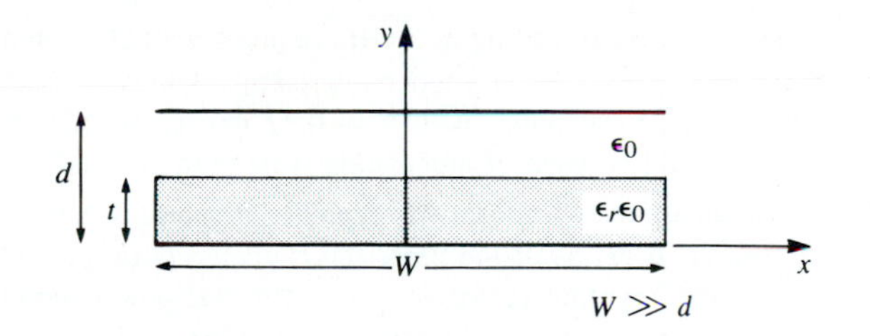

4.11 Derive equations (4.110a–d) for the transverse field components in terms of longitudinal fields, in cylindrical coordinates.

4.12 Derive the expression for the attenuation of the TM_{nm} mode in a circular waveguide with finite conductivity.

4.13 Consider a circular waveguide with $a = 0.8$ cm. Compute the cutoff frequencies and identify the first four propagating modes.

4.14 Derive a transcendental equation for the cutoff frequency of the TM modes of a coaxial waveguide. Using tables, obtain an approximate value of $k_c a$ for the TM_{01} mode, if $b/a = 2$.

4.15 Derive an expression for the attenuation of a TE surface wave on a grounded dielectric slab, when the ground plane has finite conductivity.

4.16 Consider the geometry shown below. Derive a solution for the TM surface waves that can propagate on this structure.

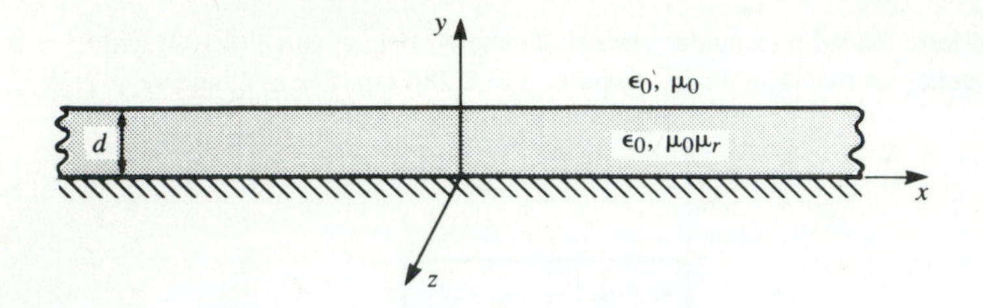

4.17 Consider the partially filled coaxial line shown below. Can a TEM wave propagate on this line? Derive the solution for the TM_{om} (no azimuthal variation) modes of this geometry.

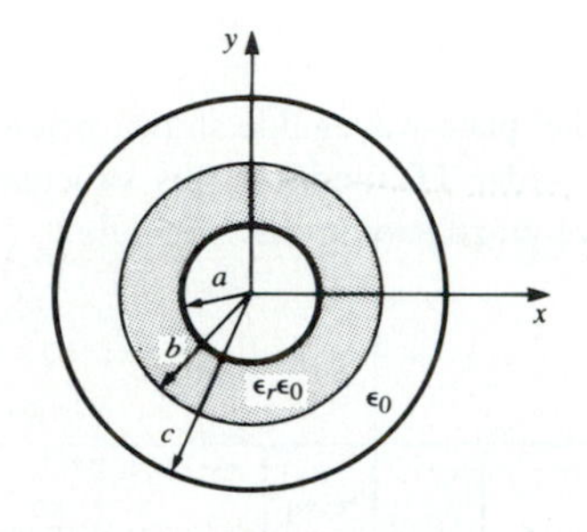

4.18 Design a stripline transmission line for a 100 Ω characteristic impedance. The ground plane separation is 0.316 cm, and the dielectric constant of the filling material is 2.20. What is the guide wavelength on this transmission line if the frequency is 4.0 GHz?

4.19 Design a microstrip transmission line for a 100 Ω characteristic impedance. The substrate thickness is 0.158 cm, with $\epsilon_r = 2.20$. What is the guide wavelength on this transmission line if the frequency is 4.0 GHz?

4.20 A microwave antenna feed network operating at 5 GHz requires a 50 Ω printed transmission line that is 16λ long. Possible choices are (1) copper microstrip, with $d = 0.16$ cm, $\epsilon_r = 2.20$, and $\tan\delta = 0.001$, or (2) copper stripline, with $b = 0.32$ cm, $\epsilon_r = 2.20, t = 0.01$ mm and $\tan\delta = 0.001$. Which line should be used, if attenuation is to be minimized?

4.21 Consider the TE modes of an arbitrary uniform waveguiding structure, where the transverse fields are related to H_z as in (4.19). If H_z is of the form $H_z(x, y, z) = h_z(x, y)e^{-j\beta z}$, where $h_z(x, y)$ is a real function, compute the Poynting vector and show that real power flow occurs only in the z direction. Assume that β is real, corresponding to a propagating mode.

4.22 A piece of rectangular waveguide is air-filled for $z < 0$, and dielectric filled for $z > 0$. Assume that both regions can support only the dominant TE_{10} mode, and that a TE_{10} mode is incident on the interface from $z < 0$. Using a field analysis, write general expressions for the transverse field components of the incident, reflected, and transmitted waves in the two regions, and enforce the boundary conditions at the dielectric interface to find the reflection and transmission coefficients. Compare these results to those obtained with an impedance approach, using Z_{TE} for each region.

4.23 Use the transverse resonance technique to derive a transcendental equation for the propagation constant of the TM modes of a rectangular waveguide that is air-filled for $0 < x < d$, and dielectric-filled for $d < x < a$.

4.24 Apply the transverse resonance technique to find the propagation constants for the TE surface waves that can be supported by the structure of Problem 4.16.

4.25 An X-band waveguide filled with teflon is operating at 9.5 GHz. Calculate the speed of light in this material, and the phase and group velocities in the waveguide.

CHAPTER 5

Microwave Network Analysis

Circuits operating at low frequencies, where the circuit dimensions are small relative to the wavelength, can be treated as an interconnection of lumped passive or active components with unique voltages and currents defined at any point in the circuit. In this situation the circuit dimensions are small enough so that there is negligible phase change from one point in the circuit to another. In addition, the fields can be considered as TEM fields supported by two or more conductors. This leads to a quasi-static type of solution to Maxwell's equations, and to the well-known Kirchhoff voltage and current laws and impedance concepts of circuit theory [1]. As the reader is aware, there exists a powerful and useful set of techniques for analyzing low-frequency circuits. In general, these techniques cannot be directly applied to microwave circuits. It is the purpose of the present chapter, however, to show how circuit and network concepts can be extended to handle many microwave analysis and design problems of practical interest.

The main reason for doing this is that it is usually much easier to apply the simple and intuitive ideas of circuit analysis to a microwave problem than it is to solve Maxwell's equations for the same problem. In a way, field analysis gives us much more information about the particular problem under consideration than we really want or need. That is, because the solution to Maxwell's equations for a given problem is complete, it gives the electric and magnetic fields at all points in space. But usually we are interested in only the voltage or current at a set of terminals, the power flow through a device, or some other type of "global" quantity, as opposed to a minute description of the response at all points in space. Another reason for using circuit or network analysis is because it is then very easy to modify the original problem, or combine several elements together and find the response, without having to analyze in detail the behavior of each element in combination with its neighbors. A field analysis using Maxwell's equations for such problems would be hopelessly difficult. There are situations, however, where such circuit analysis techniques are an oversimplification, leading to erroneous results. In such cases one must resort to a field analysis approach, using Maxwell's equations. It is part of the education of a microwave engineer to be able to determine when circuit analysis concepts apply, and when they should be cast aside.

The basic procedure for microwave network analysis is as follows. We first treat a set of basic, canonical problems rigorously, using field analysis and Maxwell's equations. (As we have done in Chapters 3 and 4, for a variety of transmission line and waveguide problems.) When so doing, we try to obtain quantities that can be directly related to a circuit or transmission line parameter. For example, when we treated various transmission lines and waveguides in Chapter 4 we derived the propagation constant and characteristic impedance of the line. This allowed the transmission line or waveguide to be treated as a distributed component characterized by its length, propagation constant, and characteristic impedance. At this point, we can interconnect various components and use network and/or transmission line theory to analyze the behavior of the entire system of components, including effects such as multiple reflections, loss, impedance transformations, and transitions from one type of transmission media to another (e.g., coax to microstrip). As we will see, transitions between different transmission lines, or discontinuities on a transmission line, generally cannot be treated as a simple junction between two transmission lines, but must be augmented with some type of equivalent circuit to account for reactances associated with the transition or discontinuity.

5.1 IMPEDANCE AND EQUIVALENT VOLTAGES AND CURRENTS

Equivalent Voltages and Currents

At microwave frequencies the measurement of voltage or current is difficult (or impossible), unless a clearly defined terminal pair is available. Such a terminal pair may be present in the case of TEM-type lines (such as coaxial cable, microstrip, or stripline), but does not exist for non-TEM lines (such as rectangular, circular, or surface waveguides).

Figure 5.1 shows the electric and magnetic field lines for an arbitrary two-conductor TEM transmission line. As in Chapter 4, the voltage, V, of the $+$ conductor relative to

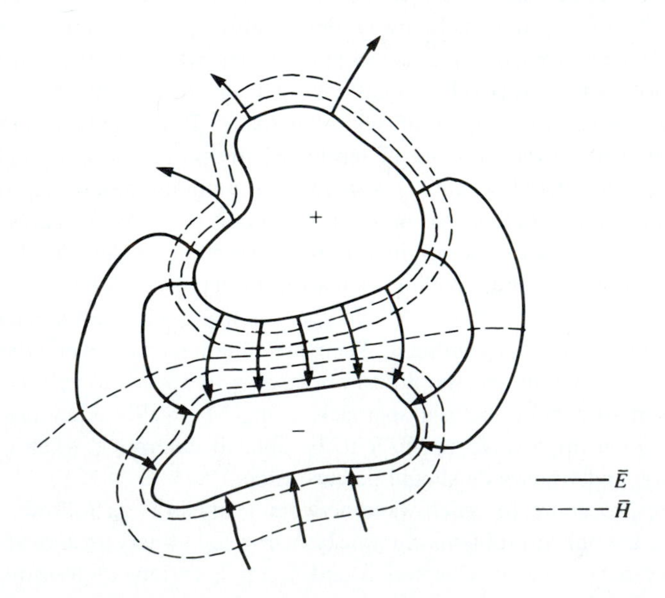

FIGURE 5.1 Electric and magnetic field lines for an arbitrary two-conductor TEM line.

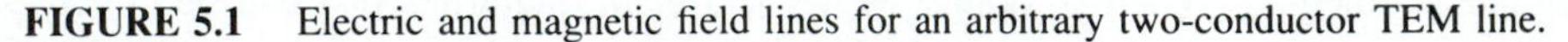

the $-$ conductor can be found as

$$V = \int_{+}^{-} \bar{E} \cdot d\bar{\ell}, \tag{5.1}$$

where the integration path begins on the $+$ conductor and ends on the $-$ conductor. It is important to realize that, because of the electrostatic nature of the transverse fields between the two conductors, the voltage defined in (5.1) is unique and does not depend on the shape of the integration path. The total current flowing on the $+$ conductor can be determined from an application of Ampere's law, as

$$I = \oint_{C+} \bar{H} \cdot d\bar{\ell}, \tag{5.2}$$

where the integration contour is any closed path enclosing the $+$ conductor (but not the $-$ conductor). A characteristic impedance Z_0 can then be defined for traveling waves as

$$Z_0 = \frac{V}{I}. \tag{5.3}$$

At this point, after having defined and determined a voltage, current, and characteristic impedance (and assuming we know the propagation constant for the line) we can proceed to apply the circuit theory for transmission lines developed in Chapter 3 to characterize this line as a circuit element.

The situation is more difficult for waveguides. To see why, we will look at the case of a rectangular waveguide, as shown in Figure 5.2. For the dominant TE_{10} mode, the transverse fields can be written, from Table 4.2, as

$$E_y(x, y, z) = \frac{j\omega\mu a}{\pi} A \sin\frac{\pi x}{a} e^{-j\beta z} = A e_y(x, y) e^{-j\beta z}, \tag{5.4a}$$

$$H_x(x, y, z) = \frac{j\beta a}{\pi} A \sin\frac{\pi x}{a} e^{-j\beta z} = A h_x(x, y) e^{-j\beta z}. \tag{5.4b}$$

Applying (5.1) to the electric field of (5.4a) gives

$$V = \frac{-j\omega\mu a}{\pi} A \sin\frac{\pi x}{a} e^{-j\beta z} \int_y dy. \tag{5.5}$$

Thus it is seen that this voltage depends on the position, x, as well as the length of the integration contour along the y direction. Integrating from $y = 0$ to b for $x = a/2$ gives

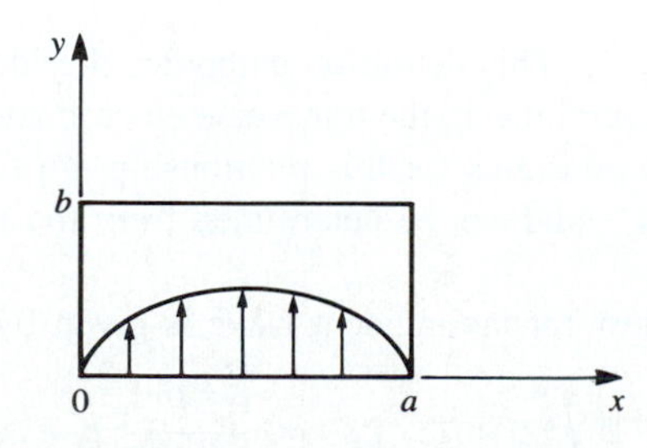

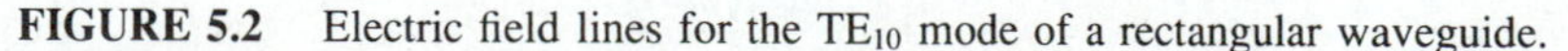

FIGURE 5.2 Electric field lines for the TE_{10} mode of a rectangular waveguide.

a voltage that is quite different from that obtained by integrating from $y = 0$ to b for $x = 0$, for example. What, then, is the correct voltage? The answer is that there is no "correct" voltage in the sense of being unique or pertinent for all applications. A similar problem arises with current, and also impedance. We will now show how we can define voltages, currents, and impedances that are useful for non-TEM lines.

There are many ways to define equivalent voltage, current, and impedance for waveguides, since these quantities are not unique for non-TEM lines, but the following considerations usually lead to the most useful results [1], [2], [3]:

- Voltage and current are only defined for a particular waveguide mode, and are defined so that the voltage is proportional to the transverse electric field, and the current is proportional to the transverse magnetic field.
- In order to be used in a manner similar to voltages and currents of circuit theory, the equivalent voltages and currents should be defined so that their product gives the power flow of the mode.
- The ratio of the voltage to the current for a single traveling wave should be equal to the characteristic impedance of the line. This impedance may be chosen arbitrarily, but is usually selected as equal to the wave impedance of the line, or else normalized to unity.

For an arbitrary waveguide mode with both positively and negatively traveling waves, the transverse fields can be written as

$$\bar{E}_t(x, y, z) = \bar{e}(x, y)(A^+e^{-j\beta z} + A^-e^{j\beta z}) = \frac{\bar{e}(x, y)}{C_1}(V^+e^{-j\beta z} + V^-e^{j\beta z}), \quad 5.6a$$

$$\bar{H}_t(x, y, z) = \bar{h}(x, y)(A^+e^{-j\beta z} - A^-e^{j\beta z}) = \frac{\bar{h}(x, y)}{C_2}(I^+e^{-j\beta z} - I^-e^{j\beta z}), \quad 5.6b$$

where $\bar{e}$ and $\bar{h}$ are the transverse field variations of the mode, and A^+, A^- are the field amplitudes of the traveling waves. Since $\bar{E}_t$ and $\bar{H}_t$ are related by the wave impedance, Z_w, according to (4.22) or (4.26), we also have that

$$\bar{h}(x, y) = \frac{\hat{z} \times \bar{e}(x, y)}{Z_w}. \quad 5.7$$

Equation (5.6) also defines equivalent voltage and current waves as

$$V(z) = V^+e^{-j\beta z} + V^-e^{j\beta z}, \quad 5.8a$$

$$I(z) = I^+e^{-j\beta z} - I^-e^{j\beta z}, \quad 5.8b$$

with $V^+/I^+ = V^-/I^- = Z_0$. This definition embodies the idea of making the equivalent voltage and current proportional to the transverse electric and magnetic fields, respectively. The proportionality constants for this relationship are $C_1 = V^+/A^+ = V^-/A^-$ and $C_2 = I^+/A^+ = I^-/A^-$, and can be determined from the remaining two conditions for power and impedance.

The complex power flow for the incident wave is given by

$$P^+ = \frac{1}{2}|A^+|^2 \iint_S \bar{e} \times \bar{h}^* \cdot \hat{z}\, ds = \frac{V^+I^{+*}}{2C_1C_2^*} \iint_S \bar{e} \times \bar{h}^* \cdot \hat{z}\, ds. \quad 5.9$$

Since we want this power to be equal to $(1/2)V^+I^{+*}$, we have the result that

$$C_1C_2^* = \iint_S \bar{e} \times \bar{h}^* \cdot \hat{z}\, ds, \tag{5.10}$$

where the surface integration is over the cross-section of the waveguide. The characteristic impedance is

$$Z_0 = \frac{V^+}{I^+} = \frac{V^-}{I^-} = \frac{C_1}{C_2}, \tag{5.11}$$

since $V^+ = C_1A$ and $I^+ = C_2A$, from (5.6a,b). If it is desired to have $Z_0 = Z_w$, the wave impedance (Z_{TE} or Z_{TM}) of the mode, then

$$\frac{C_1}{C_2} = Z_w \; (Z_{\text{TE}} \text{ or } Z_{\text{TM}}). \tag{5.12a}$$

Alternatively, it may be desirable to normalize the characteristic impedance to unity ($Z_0 = 1$), in which case we have

$$\frac{C_1}{C_2} = 1. \tag{5.12b}$$

So for a given waveguide mode, (5.10) and (5.12) can be solved for the constants, C_1 and C_2, and equivalent voltages and currents defined. Higher-order modes can be treated in the same way, so that a general field in a waveguide can be expressed in the following form:

$$\bar{E}_t(x,y,z) = \sum_{n=1}^{N} \left(\frac{V_n^+}{C_{1n}} e^{-j\beta_n z} + \frac{V_n^-}{C_{1n}} e^{j\beta_n z} \right) \bar{e}_n(x,y), \tag{5.13a}$$

$$\bar{H}_t(x,y,z) = \sum_{n=1}^{N} \left(\frac{I_n^+}{C_{2n}} e^{-j\beta_n z} - \frac{I_n^-}{C_{2n}} e^{j\beta_n z} \right) \bar{h}_n(x,y), \tag{5.13b}$$

where $V_n^\pm$ and $I_n^\pm$ are the equivalent voltages and currents for the nth mode, and C_{1n} and C_{2n} are the proportionality constants for each mode.

EXAMPLE 5.1

Find the equivalent voltages and currents for a TE_{10} mode in a rectangular waveguide.

Solution

The transverse field components and power flow of the TE_{10} rectangular waveguide mode and the equivalent transmission line model of this mode can be written as follows:

Waveguide Fields	Transmission Line Model
$E_y = (A^+e^{-j\beta z} + A^-e^{j\beta z})\sin(\pi x/a)$	$V(z) = V^+e^{-j\beta z} + V^-e^{j\beta z}$
$H_x = -1/Z_{\text{TE}}(A^+e^{-j\beta z} - A^-e^{j\beta z})\sin(\pi x/a)$	$I(z) = I^+e^{-j\beta z} - I^-e^{j\beta z}$
	$= (V^+/Z_0)e^{-j\beta z} - (V^-/Z_0)e^{j\beta z}$
$P^+ = -(1/2)\int_s E_y H_x^* dx\, dy$	$P = (1/2)V^+I^{+*}$
$= (ab\vert A\vert^2/4Z_{\text{TE}})$	

We now find the constants C_1 and C_2 that relate the equivalent voltage V^+ and current I^+ to the field amplitude, A. Equating incident powers gives

$$\frac{ab|A|^2}{4Z_{\text{TE}}} = \frac{1}{2}V^+I^{+*} = \frac{1}{2}|A|^2C_1C_{2*}.$$

If we choose $Z_0 = Z_{\text{TE}}$, then we also have that

$$\frac{V^+}{I^+} = \frac{C_1}{C_2} = Z_{\text{TE}}.$$

Solving for C_1, C_2 gives

$$C_1 = \sqrt{\frac{ab}{2}},$$

$$C_2 = \frac{1}{Z_{\text{TE}}}\sqrt{\frac{ab}{2}},$$

which completes the transmission line equivalence for the TE_{10} mode. ○

The Concept of Impedance

We have used the idea of impedance in several different applications, so it may be useful at this point to discuss the concept of impedance in more general terms. The term *impedance* was first used by Oliver Heaviside in the nineteenth century to describe the complex ratio V/I in AC circuits consisting of resistors, inductors, and capacitors; the impedance concept quickly became indispensable in the analysis of AC circuits. It was then applied to transmission lines, in terms of lumped-element equivalent circuits and the distributed series impedance and shunt admittance of the line. In the 1930s, Schelkunoff recognized that the impedance concept could be extended to electromagnetic fields in a systematic way, and noted that impedance should be regarded as characteristic of the type of field, as well as the medium [4]. And, as we have seen in Section 3.8 in relation to the analogy between transmission lines and plane wave propagation, impedance may even be dependent on direction. The concept of impedance, then, forms an important link between field theory and transmission line or circuit theory.

Below we summarize the various types of impedance we have used so far and their notation:

- $\eta = \sqrt{\mu/\epsilon}$ = intrinsic impedance of the medium. This impedance is dependent only on the material parameters of the medium, but is equal to the wave impedance for plane waves.

- $Z_w = E_t/H_t = 1/Y_w$ = wave impedance. This impedance is a characteristic of the particular type of wave. TEM, TM, and TE waves each have different wave impedances ($Z_{\text{TEM}}, Z_{\text{TM}}, Z_{\text{TE}}$), which may depend on the type of line or guide, the material, and the operating frequency.
- $Z_0 = 1/Y_0 = \sqrt{L/C}$ = characteristic impedance. Characteristic impedance is the ratio of voltage to current for a traveling wave. Since voltage and current are uniquely defined for TEM waves, the characteristic impedance of a TEM wave is unique. TE and TM waves, however, do not have a uniquely defined voltage and current, so the characteristic impedance for such waves may be defined in various ways.

EXAMPLE 5.2

Consider a rectangular waveguide with $a = 3.485$ cm and $b = 1.580$ cm (C-band guide), air-filled for $z < 0$ and dielectric filled ($\epsilon_r = 2.56$) for $z > 0$, as shown in Figure 5.3. If the operating frequency is 4.5 GHz, use an equivalent transmission line model to compute the reflection coefficient of a TE_{10} wave incident on the interface from $z < 0$.

Solution

The propagation constants in the air ($z < 0$) and the dielectric ($z > 0$) regions are

$$\beta_a = \sqrt{k_0^2 - \left(\frac{\pi}{a}\right)^2} = 27.50 \text{ m}^{-1},$$

$$\beta_d = \sqrt{\epsilon_r k_0^2 - \left(\frac{\pi}{a}\right)^2} = 120.89 \text{ m}^{-1},$$

where $k_0 = 94.25 \text{ m}^{-1}$.

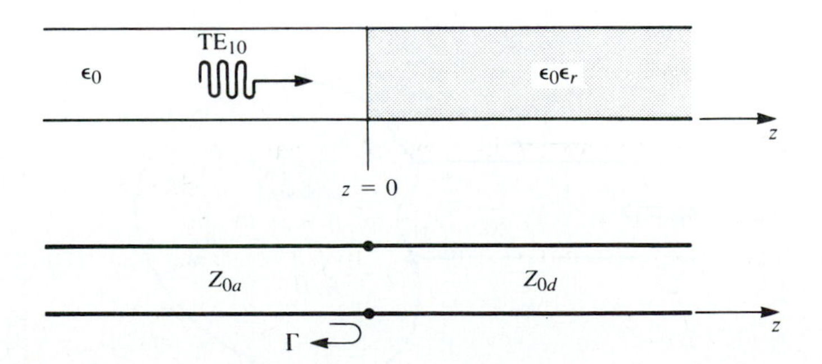

FIGURE 5.3 Geometry of a partially filled waveguide and its transmission line equivalent for Example 5.2 .

The reader may verify that the TE_{10} mode is the only propagating mode in either waveguide region. Now we can set up an equivalent transmission line for the TE_{10} mode in each waveguide, and treat the problem as the reflection of an incident voltage wave at the junction of two infinite transmission lines.

By Example 5.1 and Table 4.2, the equivalent characteristic impedances for the two lines are

$$Z_{0_a} = \frac{k_0\eta_0}{\beta_a} = \frac{(94.25)(377.)}{27.50} = 1292.1\ \Omega,$$

$$Z_{0_d} = \frac{k\eta}{\beta_d} = \frac{k_0\eta_0}{\beta_d} = \frac{(94.25)(377.)}{120.89} = 293.9\ \Omega.$$

The reflection coefficient seen looking into the dielectric filled region is then

$$\Gamma = \frac{Z_{0_d} - Z_{0_a}}{Z_{0_d} + Z_{0_a}} = -0.629.$$

With this result, expressions for the incident, reflected, and transmitted waves can be written in terms of fields, or in terms of equivalent voltages and currents. ○

5.2 IMPEDANCE PROPERTIES OF ONE-PORT NETWORKS

In this section we will discuss some of the basic properties of the driving point impedance for one-port networks. First consider the arbitrary one-port network shown in Figure 5.4. The complex power delivered to this network is given by (2.91):

$$P = \frac{1}{2}\oint_S \bar{E} \times \bar{H}^* \cdot d\bar{s} = P_\ell + 2j\omega(W_m - W_e), \qquad 5.14$$

where P_ℓ is real and represents the average power dissipated by the network, and W_m and W_e represent the stored magnetic and electric energy, respectively. Note that the unit normal vector in Figure 5.4 is pointing into the volume.

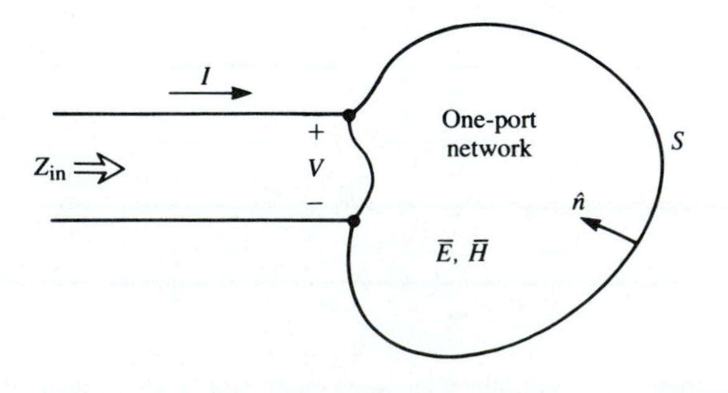

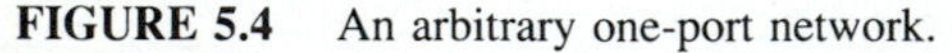
FIGURE 5.4 An arbitrary one-port network.

If we define real transverse modal fields, $\bar{e}$ and $\bar{h}$, over the terminal plane of the network such that

$$\bar{E}_t(x, y, z) = V(z)\bar{e}(x, y)e^{-j\beta z}, \qquad 5.15a$$

$$\bar{H}_t(x, y, z) = I(z)\bar{h}(x, y)e^{-j\beta z}, \qquad 5.15b$$

with a normalization such that

$$\int_S \bar{e} \times \bar{h} \cdot d\bar{s} = 1,$$

then (5.14) can be expressed in terms of the terminal voltage and current:

$$P = \frac{1}{2}\int_S VI^*\bar{e} \times \bar{h} \cdot d\bar{s} = \frac{1}{2}VI^*. \qquad 5.16$$

Then the input impedance is

$$Z_{\text{in}} = R + jX = \frac{V}{I} = \frac{VI^*}{|I|^2} = \frac{P}{1/2|I|^2} = \frac{P_\ell + 2j\omega(W_m - W_e)}{1/2|I|^2}. \qquad 5.17$$

Thus we see that the real part, R, of the input impedance is related to the dissipated power, while the imaginary part, X, is related to the net energy stored in the network. If the network is lossless, then $P_\ell = 0$ and $R = 0$. Then Z_{in} is purely imaginary, with a reactance

$$X = \frac{4\omega(W_m - W_e)}{|I|^2}, \qquad 5.18$$

which is positive for an inductive load ($W_m > W_e$), and negative for a capacitive load ($W_m < W_e$).

Foster's Reactance Theorem

Now let the one-port network of Figure 5.4 be lossless, and consider the effect of a change in frequency. Maxwell's equations are

$$\nabla \times \bar{E} = -j\omega\mu\bar{H},$$

$$\nabla \times \bar{H} = j\omega\epsilon\bar{E}.$$

Differentiating with respect to frequency gives

$$\nabla \times \frac{\partial \bar{E}}{\partial \omega} = -j\omega\mu\frac{\partial \bar{H}}{\partial \omega} - j\mu\bar{H}, \qquad 5.19a$$

$$\nabla \times \frac{\partial \bar{H}}{\partial \omega} = j\omega\epsilon\frac{\partial \bar{E}}{\partial \omega} + j\epsilon\bar{E}. \qquad 5.19b$$

Now use these results, and vector identity (B.8), to expand the quantity:

$$\nabla \cdot \left(\bar{E}^* \times \frac{\partial \bar{H}}{\partial \omega} + \frac{\partial \bar{E}}{\partial \omega} \times \bar{H}^* \right) = \frac{\partial \bar{H}}{\partial \omega} \cdot \nabla \times \bar{E}^* - \bar{E}^* \cdot \nabla \times \frac{\partial \bar{H}}{\partial \omega}$$

$$+ \bar{H}^* \cdot \nabla \times \frac{\partial \bar{E}}{\partial \omega} - \frac{\partial \bar{E}}{\partial \omega} \cdot \nabla \times \bar{H}^*$$

$$= j\omega\mu \bar{H}^* \cdot \frac{\partial \bar{H}}{\partial \omega} - j\omega\epsilon \bar{E}^* \cdot \frac{\partial \bar{E}}{\partial \omega} - j\epsilon |\bar{E}|^2$$

$$- j\omega\mu \bar{H}^* \cdot \frac{\partial \bar{H}}{\partial \omega} - j\mu |\bar{H}|^2 + j\omega\epsilon \frac{\partial \bar{E}}{\partial \omega} \cdot \bar{E}^*$$

$$= -j(\epsilon |\bar{E}|^2 + \mu |\bar{H}|^2). \qquad 5.20$$

Using the divergence theorem on the left-hand side and identifying the terms on the right-hand side with the stored electric and magnetic energies gives

$$\oint_S \left(\bar{E}^* \times \frac{\partial \bar{H}}{\partial \omega} + \frac{\partial \bar{E}}{\partial \omega} \times \bar{H}^* \right) \cdot d\bar{s} = 4j(W_e + W_m), \qquad 5.21$$

where an additional minus sign arises because the divergence theorem assumes an outward pointing normal vector. Using (5.15) puts this result in terms of the terminal voltage and current:

$$\int_S \left(V^* \frac{\partial I}{\partial \omega} \bar{e} \times \bar{h} + V^* I \bar{e} \times \frac{\partial \bar{h}}{\partial \omega} + \frac{\partial V}{\partial \omega} I^* \bar{e} \times \bar{h} + V I^* \frac{\partial \bar{e}}{\partial \omega} \times \bar{h} \right) \cdot d\bar{s} = 4j(W_e + W_m). \qquad 5.22$$

Since $\bar{h} = \hat{z} \times \bar{e}/Z_w$, we have that $\bar{e} \times (\partial \bar{h}/\partial \omega) = (\partial \bar{e}/\partial \omega) \times \bar{h}$, which means that the ω-dependence of $\bar{e}$ and $\bar{h}$ are the same. Also, $V = jXI$, so (5.22) reduces to

$$4j(W_e + W_m) = V^* \frac{\partial I}{\partial \omega} + \frac{\partial V}{\partial \omega} I^*. \qquad 5.23$$

Again using $V = jXI$ gives

$$4j(W_e + W_m) = -jXI^* \frac{\partial I}{\partial \omega} + j\frac{\partial X}{\partial \omega}|I|^2 + jX\frac{\partial I}{\partial \omega} I^* = j|I|^2 \frac{\partial X}{\partial \omega}.$$

Thus,

$$\frac{\partial X}{\partial \omega} = \frac{4(W_e + W_m)}{|I|^2}. \qquad 5.24$$

The right-hand side is always positive, so the slope of reactance versus frequency must always be positive, for a lossless network. Alternatively, we could have used $I = jBV$ in (5.23) to obtain

$$\frac{\partial B}{\partial \omega} = \frac{4(W_e + W_m)}{|V|^2}, \qquad 5.25$$

which shows that the susceptance of a lossless network also has a positive slope with frequency. These results constitute Foster's reactance theorem, and can be used to show that the poles and zeros of a physically realizable reactance or susceptance function must alternate in position.

Even and Odd Properties of $Z(\omega)$ and $\Gamma(\omega)$

Consider the driving point impedance, $Z(\omega)$, at the input port of an electrical network. The voltage and current at this port are related as $V(\omega) = Z(\omega)I(\omega)$. For an arbitrary frequency dependence, we can find the time-domain voltage by taking the inverse Fourier transform of $V(\omega)$:

$$v(t) = \frac{1}{2\pi}\int_{-\infty}^{\infty} V(\omega)e^{j\omega t}d\omega. \tag{5.26}$$

Since $v(t)$ must be real, we have that $v(t) = v^*(t)$, or

$$\int_{-\infty}^{\infty} V(\omega)e^{j\omega t}d\omega = \int_{-\infty}^{\infty} V^*(\omega)e^{-j\omega t}d\omega = \int_{-\infty}^{\infty} V^*(-\omega)e^{j\omega t}d\omega,$$

where the last term was obtained by a change of variable from ω to $-\omega$. This shows that $V(\omega)$ must satisfy the relation

$$V(-\omega) = V^*(\omega), \tag{5.27}$$

which means that $\text{Re}\{V(\omega)\}$ is even in ω, while $\text{Im}\{V(\omega)\}$ is odd in ω. Similar results hold for $I(\omega)$, and for $Z(\omega)$ since

$$V^*(-\omega) = Z^*(-\omega)I^*(-\omega) = Z^*(-\omega)I(\omega) = V(\omega) = Z(\omega)I(\omega).$$

Thus, if $Z(\omega) = R(\omega) + jX(\omega)$, then $R(\omega)$ is even in ω and $X(\omega)$ is odd in (ω). These results can also be inferred from (5.17).

Now consider the reflection coefficient at the input port:

$$\Gamma(\omega) = \frac{Z(\omega) - Z_0}{Z(\omega) + Z_0} = \frac{R(\omega) - Z_0 + jX(\omega)}{R(\omega) + Z_0 + jX(\omega)}. \tag{5.28}$$

Then,

$$\Gamma(-\omega) = \frac{R(\omega) - Z_0 - jX(\omega)}{R(\omega) + Z_0 - jX(\omega)} = \Gamma^*(\omega), \tag{5.29}$$

which shows that the real and imaginary parts of $\Gamma(\omega)$ are even and odd, respectively, in ω. Finally, the magnitude of the reflection coefficient is

$$|\Gamma(\omega)|^2 = \Gamma(\omega)\Gamma^*(\omega) = \Gamma(\omega)\Gamma(-\omega) = |\Gamma(-\omega)|^2, \tag{5.30}$$

which shows that $|\Gamma(\omega)|^2$ and $|\Gamma(\omega)|$ are even functions of ω. This result implies that only even series of the form $a + b\omega^2 + c\omega^4 + \cdots$ can be used to represent $|\Gamma(\omega)|$ or $|\Gamma(\omega)|^2$.

5.3 IMPEDANCE AND ADMITTANCE MATRICES

In the previous section we have seen how equivalent voltages and currents can be defined for TEM and non-TEM waves. Once such voltages and currents have been defined at various points in a microwave network, we can use the impedance and/or admittance matrices of circuit theory to relate these terminal or "port" quantities to each other, and thus to essentially arrive at a matrix description of the network. This type of representation lends itself to the development of equivalent circuits of arbitrary networks, which will be quite useful when we discuss the design of passive components such as couplers and filters.

We begin by considering an arbitrary N-port microwave network, as depicted in Figure 5.5. The ports in Figure 5.5 may be any type of transmission line or transmission line equivalent of a single propagating waveguide mode. (The term *port* was introduced by H. A. Wheeler in the 1950s to replace the less descriptive and more cumbersome phrase, "two-terminal pair" [2], [4].) If one of the physical ports of the network is a waveguide supporting more than one propagating mode, additional electrical ports can be added to account for these modes. At a specific point on the nth port, a terminal plane, t_n, is defined along with equivalent voltages and currents for the incident (V_n^+, I_n^+) and reflected (V_n^-, I_n^-) waves. The terminal planes are important in providing a phase reference for the voltage and current phasors. Now at the nth terminal plane, the total voltage and current is given by

$$V_n = V_n^+ + V_n^-, \quad 5.31a$$

$$I_n = I_n^+ - I_n^-, \quad 5.31b$$

as seen from (5.8) when $z = 0$.

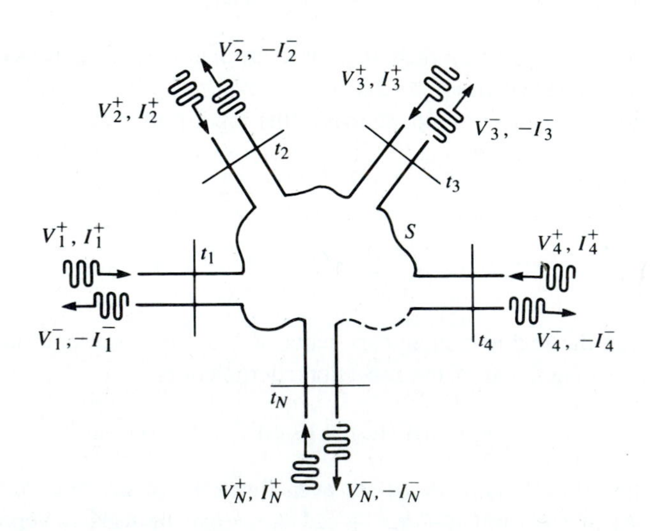

FIGURE 5.5 An arbitrary N-port microwave network.

The impedance matrix $[Z]$ of the microwave network then relates these voltages and currents:

$$\begin{bmatrix} V_1 \\ V_2 \\ \vdots \\ V_N \end{bmatrix} = \begin{bmatrix} Z_{11} & Z_{12} & \cdots & Z_{1N} \\ Z_{21} & & & \vdots \\ \vdots & & & \vdots \\ Z_{N1} & \cdots & \cdots & Z_{NN} \end{bmatrix} \begin{bmatrix} I_1 \\ I_2 \\ \vdots \\ I_N \end{bmatrix},$$

or in matrix form as

$$[V] = [Z][I]. \tag{5.32}$$

Similarly, we can define an admittance matrix [Y] as

$$\begin{bmatrix} I_1 \\ I_2 \\ \vdots \\ I_N \end{bmatrix} = \begin{bmatrix} Y_{11} & Y_{12} & \cdots & Y_{1N} \\ Y_{21} & & & \vdots \\ \vdots & & & \vdots \\ Y_{N1} & \cdots & \cdots & Y_{NN} \end{bmatrix} \begin{bmatrix} V_1 \\ V_2 \\ \vdots \\ V_N \end{bmatrix},$$

or in matrix form as

$$[I] = [Y][V]. \tag{5.33}$$

Of course, the $[Z]$ and $[Y]$ matrices are the inverses of each other:

$$[Y] = [Z]^{-1}. \tag{5.34}$$

Note that both the $[Z]$ and $[Y]$ matrices relate the total port voltages and currents.

From (5.32), we see that Z_{ij} can be found as

$$Z_{ij} = \left.\frac{V_i}{I_j}\right|_{I_k=0 \text{ for } k \neq j}. \tag{5.35}$$

In words, (5.35) states that Z_{ij} can be found by driving port j with the current I_j, open-circuiting all other ports (so $I_k = 0$ for $k \neq j$), and measuring the open-circuit voltage at port i. Thus, Z_{ii} is the input impedance seen looking into port i when all other ports are open-circuited, and Z_{ij} is the transfer impedance between ports i and j when all other ports are open-circuited.

Similarly, from (5.33), Y_{ij} can be found as,

$$Y_{ij} = \left.\frac{I_i}{V_j}\right|_{V_k=0 \text{ for } k \neq j}, \tag{5.36}$$

which states that Y_{ij} can be determined by driving port j with the voltage V_j, short-circuiting all other ports (so $V_k = 0$ for $k \neq j$), and measuring the short-circuit current at port i.

In general, each Z_{ij} or Y_{ij} element may be complex. For an N-port network, the impedance and admittance matrices are $N \times N$ in size, so there are $2N^2$ independent quantities or degrees of freedom for an arbitrary N-port network. In practice, however, many networks are either reciprocal or lossless, or both. If the network is reciprocal (not containing any nonreciprocal media such as ferrites or plasmas, or active devices), we will show that the impedance and admittance matrices are symmetric, so that $Z_{ij} = Z_{ji}$, and $Y_{ij} = Y_{ji}$. If the network is lossless, we can show that all the Z_{ij} or Y_{ij} elements are purely imaginary. Either of these special cases serve to reduce the number of independent quantities or degrees of freedom that an N-port network may have. We now derive the above characteristics for reciprocal and lossless networks.

Reciprocal Networks

Consider the arbitrary network of Figure 5.5 to be reciprocal (no active devices, ferrites, or plasmas), with short circuits placed at all terminal planes except those of ports 1 and 2. Now let $\bar{E}_a, \bar{H}_a$ and $\bar{E}_b, \bar{H}_b$ be the fields anywhere in the network due to two independent sources, a and b, located somewhere in the network. Then the reciprocity theorem of (2.174) states that

$$\oint_S \bar{E}_a \times \bar{H}_b \cdot d\bar{s} = \oint_S \bar{E}_b \times \bar{H}_a \cdot d\bar{s}, \tag{5.37}$$

where we will take S as the closed surface along the boundaries of the network and through the terminal planes of the ports. If the boundary walls of the network and transmission lines are metal, then $\bar{E}_{\text{tan}} = 0$ on these walls (assuming perfect conductors). If the network or the transmission lines are open structures, like microstrip or slotline, the boundaries of the network can be taken arbitrarily far from the lines so that $\bar{E}_{\text{tan}}$ is negligible. Then the only nonzero contribution to the integrals of (5.37) come from the cross-sectional areas of ports 1 and 2.

Now from Section 5.1, the fields due to sources a and b can be evaluated at the terminal planes t_1 and t_2 as

$$\begin{aligned}
\bar{E}_{1a} &= V_{1a}\bar{e}_1 & \bar{H}_{1a} &= I_{1a}\bar{h}_1 \\
\bar{E}_{1b} &= V_{1b}\bar{e}_1 & \bar{H}_{1b} &= I_{1b}\bar{h}_1 \\
\bar{E}_{2a} &= V_{2a}\bar{e}_2 & \bar{H}_{2a} &= I_{2a}\bar{h}_2 \\
\bar{E}_{2b} &= V_{2b}\bar{e}_2 & \bar{H}_{2b} &= I_{2b}\bar{h}_2,
\end{aligned} \tag{5.38}$$

where $\bar{e}_1, \bar{h}_1$ and $\bar{e}_2, \bar{h}_2$ are the transverse modal fields of ports 1 and 2, respectively, and the Vs and Is are the equivalent total voltages and currents. (For instance, $\bar{E}_{1b}$ is the transverse electric field at terminal plane t_1 of port 1 due to source b.) Substituting the fields of (5.38) into (5.37) gives

$$(V_{1a}I_{1b} - V_{1b}I_{1a})\int_{S_1} \bar{e}_1 \times \bar{h}_1 \cdot d\bar{s} + (V_{2a}I_{2b} - V_{2b}I_{2a})\int_{S_2} \bar{e}_2 \times \bar{h}_2 \cdot d\bar{s} = 0, \tag{5.39}$$

where S_1, S_2 are the cross-section areas at the terminal planes of ports 1 and 2.

As in Section 5.1, the equivalent voltages and currents have been defined so that the power through a given port can be expressed as $VI^*/2$; then comparing (5.38) to (5.6) implies that $C_1 = C_2 = 1$ for each port, so that

$$\int_{S_1} \bar{e}_1 \times \bar{h}_1 \cdot d\bar{s} = \int_{S_2} \bar{e}_2 \times \bar{h}_2 \cdot d\bar{s} = 1. \tag{5.40}$$

This reduces (5.39) to

$$V_{1a}I_{1b} - V_{1b}I_{1a} + V_{2a}I_{2b} - V_{2b}I_{2a} = 0 \,. \tag{5.41}$$

Now use the 2×2 admittance matrix of the (effectively) two-port network to eliminate the Is:

$$I_1 = Y_{11}V_1 + Y_{12}V_2,$$
$$I_2 = Y_{21}V_1 + Y_{22}V_2.$$

Substitution into (5.41) gives

$$(V_{1a}V_{2b} - V_{1b}V_{2a})(Y_{12} - Y_{21}) = 0. \tag{5.42}$$

But since the sources a and b are independent, the voltages V_{1a}, V_{1b}, V_{2a}, and V_{2b} can take on arbitrary values. So in order for (5.42) to be satisfied for any choice of sources, we must have $Y_{12} = Y_{21}$, and since the choice of which ports are labeled as 1 and 2 is arbitrary, we have the general result that

$$Y_{ij} = Y_{ji}. \tag{5.43}$$

Then if $[Y]$ is a symmetric matrix, its inverse, $[Z]$, is also symmetric.

Lossless Networks

Now consider a reciprocal lossless N-port junction; we will show that the elements of the impedance and admittance matrices must be pure imaginary. If the network is lossless, then the net real power delivered to the network must be zero. Thus, $\text{Re}\{P_{\text{av}}\} = 0$, where

$$\begin{aligned}
P_{\text{av}} &= \frac{1}{2}[V]^t[I]^* = \frac{1}{2}([Z][I])^t[I]^* = \frac{1}{2}[I]^t[Z][I]^* \\
&= \frac{1}{2}(I_1 Z_{11} I_1^* + I_1 Z_{12} I_2^* + I_2 Z_{21} I_1^* + \cdots) \\
&= \frac{1}{2}\sum_{n=1}^{N}\sum_{m=1}^{N} I_m Z_{mn} I_n^*.
\end{aligned} \tag{5.44}$$

(We have used the result from matrix algebra that $([A][B])^t = [B]^t[A]^t$.) Since the I_ns are independent, we must have the real part of each self term $(I_n Z_{nn} I_n^*)$ equal to zero, since we could set all port currents equal to zero except for the nth current. So,

$$\text{Re}\{I_n Z_{nn} I_n^*\} = |I_n|^2 \text{Re}\{\{Z_{nn}\} = 0,$$

or

$$\text{Re}\{Z_{nn}\} = 0. \tag{5.45}$$

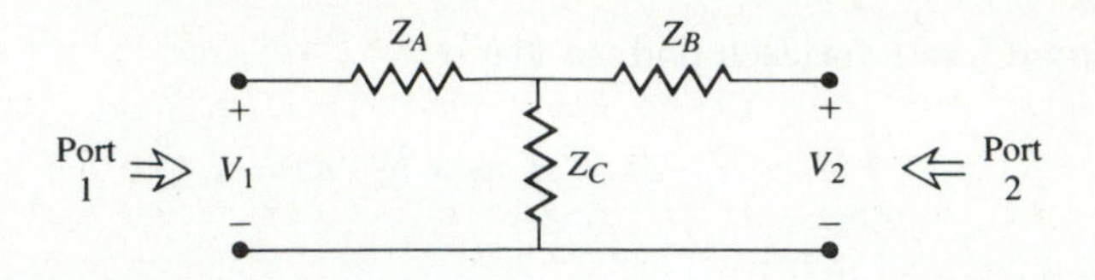

FIGURE 5.6 A two-port T-network.

Now let all port currents be zero except for I_m and I_n. Then (5.44) reduces to

$$\text{Re}\{(I_n I_m^* + I_m I_n^*) Z_{mn}\} = 0,$$

since $Z_{mn} = Z_{nm}$. But $(I_n I_m^* + I_m I_n^*)$ is a purely real quantity which is, in general, nonzero. Thus we must have that

$$\text{Re}\{Z_{mn}\} = 0. \tag{5.46}$$

Then (5.45) and (5.46) imply that $\text{Re}\{Z_{mn}\} = 0$ for any m, n. The reader can verify that this also leads to an imaginary $[Y]$ matrix.

EXAMPLE 5.3

Find the Z parameters of the two-port T-network shown in Figure 5.6.

Solution

From (5.35), Z_{11} can be found as the input impedance of port 1 when port 2 is open-circuited:

$$Z_{11} = \left.\frac{V_1}{I_1}\right|_{I_2=0} = Z_A + Z_C.$$

The transfer impedance Z_{12} can be found measuring the open-circuit voltage at port 1 when a current I_2 is applied at port 2. By voltage division,

$$Z_{12} = \left.\frac{V_1}{I_2}\right|_{I_1=0} = \frac{V_2}{I_2}\frac{Z_C}{Z_B + Z_C} = Z_C.$$

The reader can verify that $Z_{21} = Z_{12}$, indicating that the circuit is reciprocal. Finally, Z_{22} is found as

$$Z_{22} = \left.\frac{V_2}{I_2}\right|_{I_1=0} = Z_B + Z_C.$$

○

5.4 THE SCATTERING MATRIX

We have already discussed the difficulty in defining voltages and currents for non-TEM lines. In addition, a practical problem exists when trying to measure voltages and currents at microwave frequencies because direct measurements usually involve the magnitude (inferred from power) and phase of a wave traveling in a given direction, or of a standing wave. Thus, equivalent voltages and currents, and the related impedance and admittance matrices, become somewhat of an abstraction when dealing with high-frequency networks. A representation more in accord with direct measurements, and with the ideas of incident, reflected, and transmitted waves, is given by the scattering matrix.

Like the impedance or admittance matrix for an N-port network, the scattering matrix provides a complete description of the network as seen at its N ports. While the impedance and admittance matrices relate the total voltages and currents at the ports, the scattering matrix relates the voltage waves incident on the ports to those reflected from the ports. For some components and circuits, the scattering parameters can be calculated using network analysis techniques. Otherwise, the scattering parameters can be measured directly with a vector network analyzer; a photograph of a modern network analyzer is shown in Figure 5.7. Once the scattering parameters of the network are known, conversion to other matrix parameters can be performed, if needed.

Consider the N-port network shown in Figure 5.5, where V_n^+ is the amplitude of the voltage wave incident on port n, and V_n^- is the amplitude of the voltage wave reflected from port n. The scattering matrix, or $[S]$ matrix, is defined in relation to these incident and reflected voltage waves as

$$\begin{bmatrix} V_1^- \\ V_2^- \\ \vdots \\ V_N^- \end{bmatrix} = \begin{bmatrix} S_{11} & S_{12} & \cdots & S_{1N} \\ S_{21} & & & \vdots \\ \vdots & & & \\ S_{N1} & \cdots & & S_{NN} \end{bmatrix} \begin{bmatrix} V_1^+ \\ V_2^+ \\ \vdots \\ V_N^+ \end{bmatrix},$$

or

$$[V^-] = [S][V^+]. \tag{5.47}$$

A specific element of the $[S]$ matrix can be determined as

$$S_{ij} = \left. \frac{V_i^-}{V_j^+} \right|_{V_k^+=0 \text{ for } k\neq j}. \tag{5.48}$$

In words, (5.48) says that S_{ij} is found by driving port j with an incident wave of voltage V_j^+, and measuring the reflected wave amplitude, V_i^-, coming out of port i. The incident waves on all ports except the jth port are set to zero, which means that all ports should be terminated in matched loads to avoid reflections. Thus, S_{ii} is the reflection coefficient seen looking into port i when all other ports are terminated in matched loads, and S_{ij} is the transmission coefficient from port j to port i, when all other ports are terminated in matched loads.

EXAMPLE 5.4

Find the S parameters of the 3 dB attenuator circuit shown in Figure 5.8.

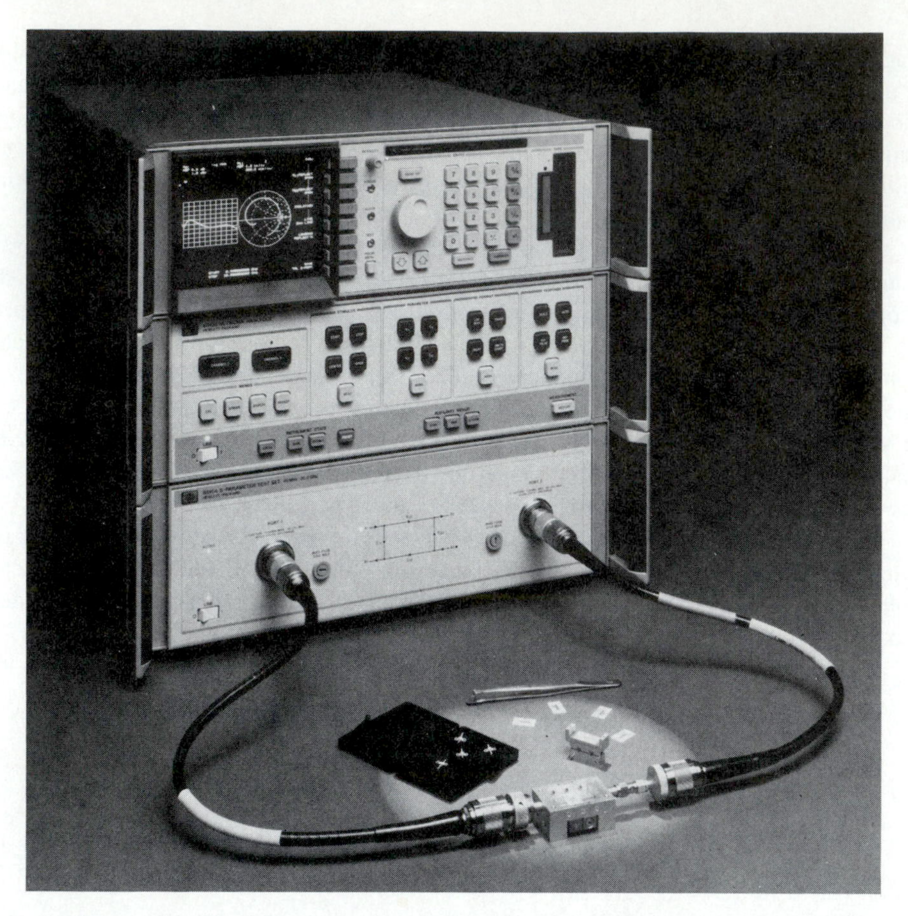

FIGURE 5.7 A photograph of the Hewlett-Packard HP8510B Network Analyzer. This test instrument is used to measure the scattering parameters (magnitude and phase) of a one- or two-port microwave network from 0.05 GHz to 26.5 GHz. Built-in microprocessors provide error correction, a high degree of accuracy, and a wide choice of display formats. This analyzer can also perform a fast Fourier transform of the frequency domain data to provide a time domain response of the network under test.
Courtesy of Hewlett-Packard Company, Santa Rosa, Calif..

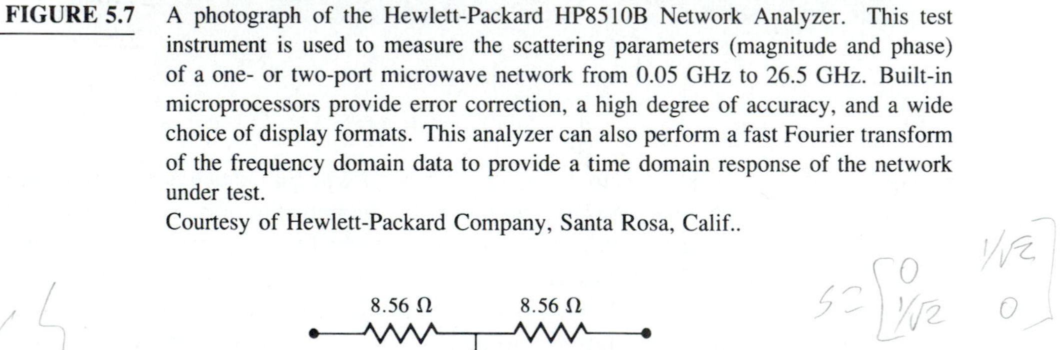

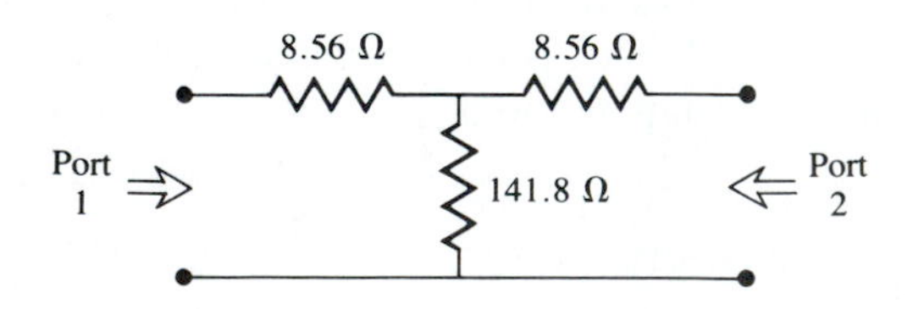

FIGURE 5.8 A matched 3 dB attenuator with a 50 Ω chacteristic impedance (Example 5.4).

Solution

From (5.48), S_{11} can be found as the reflection coefficient seen at port 1 when port 2 is terminated in a matched load ($Z_0 = 50\ \Omega$):

$$S_{11} = \left.\frac{V_1^-}{V_1^+}\right|_{V_2^+=0} = \Gamma^{(1)}|_{V_2^+=0} = \left.\frac{Z_{\text{in}}^{(1)} - Z_0}{Z_{\text{in}}^{(1)} + Z_0}\right|_{Z_0 \text{ on port } 2},$$

but, $Z_{\text{in}}^{(1)} = 8.56 + [141.8(8.56 + 50)]/(141.8 + 8.56 + 50) = 50\ \Omega$, so $S_{11} = 0$. Because of the symmetry of the circuit, $S_{22} = 0$.

S_{21} can be found by applying an incident wave at port 1, V_1^+, and measuring the outcoming wave at port 2, V_2^-. This is equivalent to the transmission coefficient from port 1 to port 2.

$$S_{21} = \left.\frac{V_2^-}{V_1^+}\right|_{V_2^+=0}.$$

From the fact that $S_{11} = S_{22} = 0$, we know that $V_1^- = 0$ when port 2 is terminated in $Z_0 = 50\ \Omega$, and that $V_2^+ = 0$. In this case we then have that $V_1^+ = V_1$ and $V_2^- = V_2$. So by applying a voltage V_1 at port 1 and using voltage division twice we find $V_2^- = V_2$ as the voltage across the 50 Ω load resistor at port 2:

$$V_2^- = V_2 = V_1\left(\frac{41.44}{41.44 + 8.56}\right)\left(\frac{50}{50 + 8.56}\right) = 0.707V_1,$$

where $41.44 = 141.8(58.56)/(141.8 + 58.56)$ is the resistance of the parallel combination of the 50 Ω load and the 8.56 Ω resistor with the 141.8 Ω resistor. Thus, $S_{12} = S_{21} = 0.707$.

If the input power is $|V_1^+|^2/2Z_0$, then the output power is $|V_2^-|^2/2Z_0 = |S_{21}V_1^+|^2/2Z_0 = |S_{21}|^2/2Z_0|V_1^+|^2 = |V_1^+|^2/4Z_0$, which is one-half ($-3$ dB) of the input power. ○

We now show how the $[S]$ matrix can be determined from the $[Z]$ (or $[Y]$) matrix, and vice versa. First, we must assume that the characteristic impedances, Z_{0n}, of all the ports are identical. (This restriction will be removed when we discuss generalized scattering parameters.) Then for convenience, we can set $Z_{0n} = 1$. From (5.31) the total voltage and current at the nth port can be written as

$$V_n = V_n^+ + V_n^-, \tag{5.49a}$$

$$I_n = I_n^+ - I_n^- = V_n^+ - V_n^-. \tag{5.49b}$$

Using the definition of $[Z]$ from (5.32) with (5.49) gives

$$[Z][I] = [Z][V^+] - [Z][V^-] = [V] = [V^+] + [V^-],$$

which can be rewritten as

$$([Z] + [U])[V^-] = ([Z] - [U])[V^+], \tag{5.50}$$

where $[U]$ is the unit, or identity, matrix defined as

$$[U] = \begin{bmatrix} 1 & 0 & \cdots & 0 \\ 0 & 1 & & \vdots \\ \vdots & & \ddots & \\ 0 & & \cdots & 1 \end{bmatrix}.$$

Comparing (5.50) to (5.47) suggests that

$$[S] = ([Z] + [U])^{-1}([Z] - [U]), \tag{5.51}$$

giving the scattering matrix in terms of the impedance matrix. Note that for a one-port network (5.51) reduces to

$$S_{11} = \frac{z_{11} - 1}{z_{11} + 1},$$

in agreement with the result for the reflection coefficient seen looking into a load with a normalized input impedance of z_{11}.

To find $[Z]$ in terms of $[S]$, rewrite (5.51) as $[Z][S] + [U][S] = [Z] - [U]$, and solve for $[Z]$ to give

$$[Z] = ([U] - [S])^{-1}([U] + [S]). \tag{5.52}$$

Reciprocal Networks and Lossless Networks

As we discussed in Section 5.2, the impedance and admittance matrices are symmetric for reciprocal networks, and purely imaginary for lossless networks. Similarly, the scattering matrices for these types of networks have special properties. We will show that the $[S]$ matrix for a reciprocal network is symmetric, and that the $[S]$ matrix for a lossless network is unitary.

By adding (5.49a) and (5.49b) we obtain

$$V_n^+ = \frac{1}{2}(V_n + I_n),$$

or

$$[V^+] = \frac{1}{2}([Z] + [U])[I]. \tag{5.53a}$$

By subtracting (5.49a) and (5.49b) we obtain

$$V_n^- = \frac{1}{2}(V_n - I_n),$$

or

$$[V^-] = \frac{1}{2}([Z] - [U])[I]. \tag{5.53b}$$

Eliminating $[I]$ from (5.53a) and (5.53b) gives

$$[V^-] = ([Z] - [U])([Z] + [U])^{-1}[V^+],$$

so that

$$[S] = ([Z] - [U])([Z] + [U])^{-1}. \tag{5.54}$$

Taking the transpose of (5.54) gives

$$[S]^t = \{([Z] + [U])^{-1}\}^t([Z] - [U])^t.$$

Now $[U]$ is diagonal, so $[U]^t = [U]$, and if the network is reciprocal, $[Z]$ is symmetric so that $[Z]^t = [Z]$. The above then reduces to

$$[S]^t = ([Z] + [U])^{-1}([Z] - [U]),$$

which is equivalent to (5.51). We have thus shown that,

$$[S] = [S]^t, \tag{5.55}$$

for reciprocal networks.

If the network is lossless, then no real power can be delivered to the network. Thus, if the characteristic impedances of all the ports are identical and assumed to be unity, the average power delivered to the network is

$$\begin{aligned} P_{av} &= \frac{1}{2}\mathrm{Re}\{[V]^t[I]^*\} = \frac{1}{2}\mathrm{Re}\{([V^+]^t + [V^-]^t)([V^+]^* - [V^-]^*)\} \\ &= \frac{1}{2}\mathrm{Re}\{[V^+]^t[V^+]^* - [V^+]^t[V^-]^* + [V^-]^t[V^+]^* - [V^-]^t[V^-]^*\} \\ &= \frac{1}{2}[V^+]^t[V^+]^* - \frac{1}{2}[V^-]^t[V^-]^* = 0, \end{aligned} \tag{5.56}$$

since the terms $-[V^+]^t[V^-]^* + [V^-]^t[V^+]^*$ are of the form $A - A^*$, and so are purely imaginary. Of the remaining terms in (5.56), $(1/2)[V^+]^t[V^+]^*$ represents the total incident power, while $(1/2)[V^-]^t[V^-]^*$ represents the total reflected power. So for a lossless junction, we have the intuitive result that the incident and reflected powers are equal:

$$[V^+]^t[V^+]^* = [V^-]^t[V^-]^*. \tag{5.57}$$

Using $[V^-] = [S][V^+]$ in (5.57) gives

$$[V^+]^t[V^+]^* = [V^+]^t[S]^t[S]^*[V^+]^*,$$

so that, for nonzero $[V^+]$,

$$[S]^t[S]^* = [U],$$

or

$$[S]^* = \{[S]^t\}^{-1}. \tag{5.58}$$

A matrix that satisfies the condition of (5.58) is called a *unitary matrix*.

The matrix equation of (5.58) can be written in summation form as

$$\sum_{k=1}^{N} S_{ki}S_{kj}^* = \delta_{ij}, \quad \text{for all } i, j, \tag{5.59}$$

where $\delta_{ij} = 1$ if $i = j$ and $\delta_{ij} = 0$ if $i \neq j$ is the Kronecker delta symbol. Thus, if $i = j$ (5.59) reduces to

$$\sum_{k=1}^{N} S_{ki} S_{ki}^{*} = 1, \tag{5.60a}$$

while if $i \neq j$ (5.59) reduces to

$$\sum_{k=1}^{N} S_{ki} S_{kj}^{*} = 0, \quad \text{for } i \neq j. \tag{5.60b}$$

In words, (5.60a) states that the dot product of any column of $[S]$ with the conjugate of that column gives unity, while (5.60b) states that the dot product of any column with the conjugate of a different column gives zero (orthogonal). If the network is reciprocal, then $[S]$ is symmetric, and the same statements can be made about the rows of the scattering matrix.

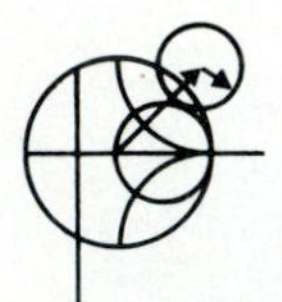

EXAMPLE 5.5

A certain two-port network is measured and the following scattering matrix is obtained:

$$[S] = \begin{bmatrix} 0.1\angle 0 & 0.8\angle 90^\circ \\ 0.8\angle 90^\circ & 0.2\angle 0 \end{bmatrix}.$$

From this data, determine whether the network is reciprocal or lossless. If a shortcircuit is placed on port 2, what will be the resulting return loss at port 1?

Solution

Since $[S]$ is symmetric, the network is reciprocal. To be lossless, the S parameters must satisfy (5.60). Taking the first row (i = 1 in (5.60a)) gives

$$|S_{11}|^2 + |S_{12}|^2 = (0.1)^2 + (0.8)^2 = 0.65 \neq 1.$$

Thus, the network is not lossless.

The reflection coefficient, Γ, at port 1 when port 2 is shorted can be calculated as follows. From the definition of the scattering matrix and the fact that $V_2^+ = -V_2^-$ (for a short circuit at port 2), we can write

$$V_1^- = S_{11}V_1^+ + S_{12}V_2^+ = S_{11}V_1^+ - S_{12}V_2^-,$$

$$V_2^- = S_{21}V_1^+ + S_{22}V_2^+ = S_{21}V_1^+ - S_{22}V_2^-.$$

The second equation gives

$$V_2^- = \frac{S_{21}}{1 + S_{22}} V_1^+.$$

Dividing the first equation by V_1^+, and using the above result, gives the input reflection coefficient as

$$\Gamma = \frac{V_1^-}{V_1^+} = S_{11} - S_{12}\frac{V_2^-}{V_1^+} = S_{11} - \frac{S_{12}S_{21}}{1+S_{22}}$$

$$= 0.1 - \frac{(j0.8)(j0.8)}{1+0.2} = 0.633.$$

So the return loss is

$$RL = -20 \log |\Gamma| = 3.97 \text{ dB}. \qquad \bigcirc$$

An important point to understand about S parameters is that the reflection coefficient looking into port n is not equal to S_{nn}, unless all other ports are matched (this is illustrated in the above example). Similarly, the transmission coefficient from port m to port n is not equal to S_{nm}, unless all other ports are matched. The S parameters of a network are properties only of the network itself (assuming the network is linear), and are defined under the condition that all ports are matched. Changing the terminations or excitations of a network does not change its S parameters, but may change the reflection coefficient seen at a given port, or the transmission coefficient between two ports.

A Shift in Reference Planes

Because the S parameters relate amplitudes (magnitude and phase) of traveling waves incident on and reflected from a microwave network, phase reference planes must be specified for each port of the network. We now show how the S parameters are transformed when the reference planes are moved from their original locations.

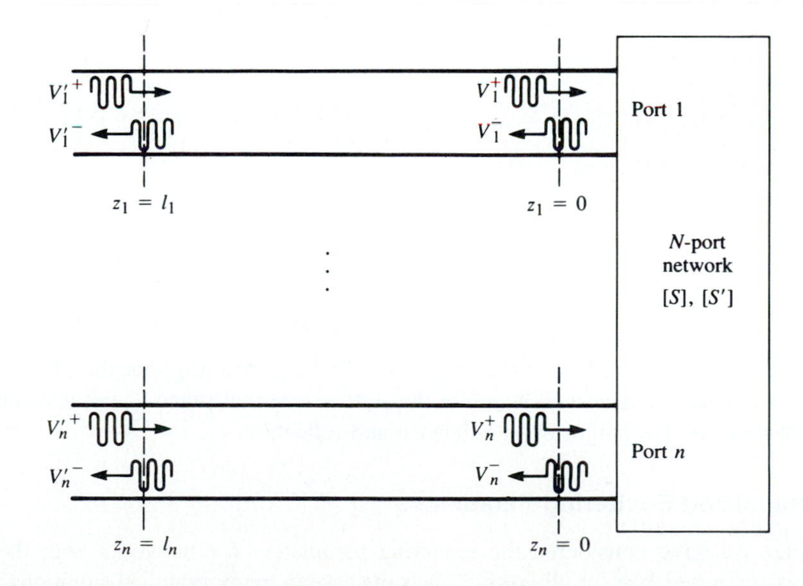

FIGURE 5.9 Shifting reference planes for an N-port network.

for Smith Chart use $2\beta l$ ⟶ double the length

Consider the N-port microwave network shown in Figure 5.9, where the original terminal planes are assumed to be located at $z_n = 0$ for the nth port, and where z_n is an arbitrary coordinate measured along the transmission line feeding the nth port. The scattering matrix for the network with this set of terminal planes is denoted by $[S]$. Now consider a new set of reference planes defined at $z_n = \ell_n$, for the nth port, and let the new scattering matrix be denoted as $[S']$. Then in terms of the incident and reflected port voltages we have that

$$[V^-] = [S][V^+], \tag{5.61a}$$

$$[V'^-] = [S'][V'^+], \tag{5.61b}$$

where the unprimed quantities are referenced to the original terminal planes at $z_n = 0$, and the primed quantities are referenced to the new terminal planes at $z_n = \ell_n$.

Now from the theory of traveling waves on lossless transmission lines we can relate the new wave amplitudes to the original ones as

$$V_n'^+ = V_n^+ e^{j\theta_n}, \tag{5.62a}$$

$$V_n'^- = V_n^- e^{-j\theta_n}, \tag{5.62b}$$

where $\theta_n = \beta_n \ell_n$ is the electrical length of the outward shift of the reference plane of port n. Writing (5.62) in matrix form and substituting into (5.61a) gives

$$\begin{bmatrix} e^{j\theta_1} & & & 0 \\ & e^{j\theta_2} & & \\ & & \ddots & \\ 0 & & & e^{j\theta_N} \end{bmatrix} [V'^-] = [S] \begin{bmatrix} e^{-j\theta_1} & & & 0 \\ & e^{-j\theta_2} & & \\ & & \ddots & \\ 0 & & & e^{-j\theta_N} \end{bmatrix} [V'^+].$$

Multiplying by the inverse of the first matrix on the left gives

$$[V'^-] = \begin{bmatrix} e^{-j\theta_1} & & & 0 \\ & e^{-j\theta_2} & & \\ & & \ddots & \\ 0 & & & e^{-j\theta_N} \end{bmatrix} [S] \begin{bmatrix} e^{-j\theta_1} & & & 0 \\ & e^{-j\theta_2} & & \\ & & \ddots & \\ 0 & & & e^{-j\theta_N} \end{bmatrix} [V'^+].$$

Comparing with (5.61b) shows that

$$[S'] = \begin{bmatrix} e^{-j\theta_1} & & & 0 \\ & e^{-j\theta_2} & & \\ & & \ddots & \\ 0 & & & e^{-j\theta_N} \end{bmatrix} [S] \begin{bmatrix} e^{-j\theta_1} & & & 0 \\ & e^{-j\theta_2} & & \\ & & \ddots & \\ 0 & & & e^{-j\theta_N} \end{bmatrix}, \tag{5.63}$$

which is the desired result. Note that $S'_{nn} = e^{-2j\theta_n} S_{nn}$, meaning that the phase of S_{nn} is shifted by twice the electrical length of the shift in terminal plane n, because the wave travels twice over this length upon incidence and reflection.

Generalized Scattering Parameters

So far we have considered the scattering parameters for networks with the same characteristic impedance for all ports. This is the case in many practical situations, where the characteristic impedance is often 50 Ω. In other cases, however, the characteristic impedances of a multiport network may be different, which requires a generalization of the scattering parameters as defined up to this point.

Consider the N-port network shown in Figure 5.10, where Z_{0n} is the (real) characteristic impedance of the nth port, and V_n^+ and V_n^-, respectively, represent the incident and reflected voltage waves at port n. In order to obtain physically meaningful power relations in terms of wave amplitudes, we must define a new set of wave amplitudes as

$$a_n = V_n^+/\sqrt{Z_{0n}}, \qquad 5.64a$$

$$b_n = V_n^-/\sqrt{Z_{0n}}, \qquad 5.64b$$

where a_n represents an incident wave at the nth port, and b_n represents a reflected wave from that port [1], [5]. Then from (5.49a,b) we have that

$$V_n = V_n^+ + V_n^- = \sqrt{Z_{0n}}(a_n + b_n), \qquad 5.65a$$

$$I_n = \frac{1}{Z_{0n}}(V_n^+ - V_n^-) = \frac{1}{\sqrt{Z_{0n}}}(a_n - b_n). \qquad 5.65b$$

Now the average power delivered to the nth port is

$$P_n = \frac{1}{2}\text{Re}\{V_n I_n^*\} = \frac{1}{2}\text{Re}\{|a_n|^2 - |b_n|^2 + (b_n a_n^* - b_n^* a_n)\} = \frac{1}{2}|a_n|^2 - \frac{1}{2}|b_n|^2, \qquad 5.66$$

since the quantity $(b_n a_n^* - b_n^* a_n)$ is purely imaginary. This is a physically satisfying result, since it says that the average power delivered through port n is equal to the power in the incident wave minus the power in the reflected wave. If expressed in terms of V_n^+ and V_n^-, the corresponding result would be dependent on the characteristic impedance of the nth port.

A generalized scattering matrix can then be used to relate the incident and reflected waves defined in (5.64):

$$[b] = [S][a], \qquad 5.67$$

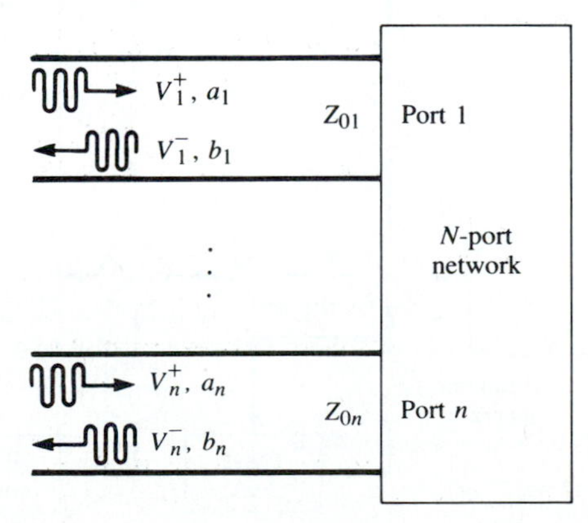

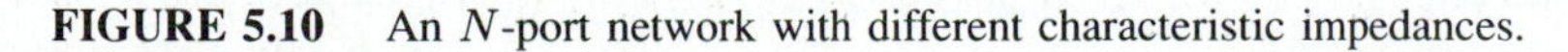
FIGURE 5.10 An N-port network with different characteristic impedances.

where the i, jth element of the scattering matrix is given by

$$S_{ij} = \left.\frac{b_i}{a_j}\right|_{a_k=0 \text{ for } k\neq j}, \tag{5.68}$$

and is analogous to the result of (5.48) for networks with identical characteristic impedance at all ports. Using (5.64) in (5.68) gives

$$S_{ij} = \left.\frac{V_i^- \sqrt{Z_{0j}}}{V_j^+ \sqrt{Z_{0i}}}\right|_{V_k^+=0 \text{ for } k\neq j}, \tag{5.69}$$

which shows how the S parameters of a network with equal characteristic impedance (V_i^-/V_j^+ with $V_k^+ = 0$ for $k \neq j$) can be converted to a network connected to transmission lines with unequal characteristic impedances.

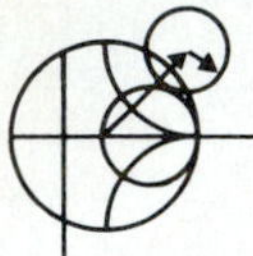

POINT OF INTEREST: The Vector Network Analyzer

The S parameters of passive and active networks can be measured with a vector network analyzer, which is a two- (or four-) channel microwave receiver designed to process the magnitude and phase of the transmitted and reflected waves from the network. A simplified block diagram of a network analyzer similar to the HP8510 system is shown below.

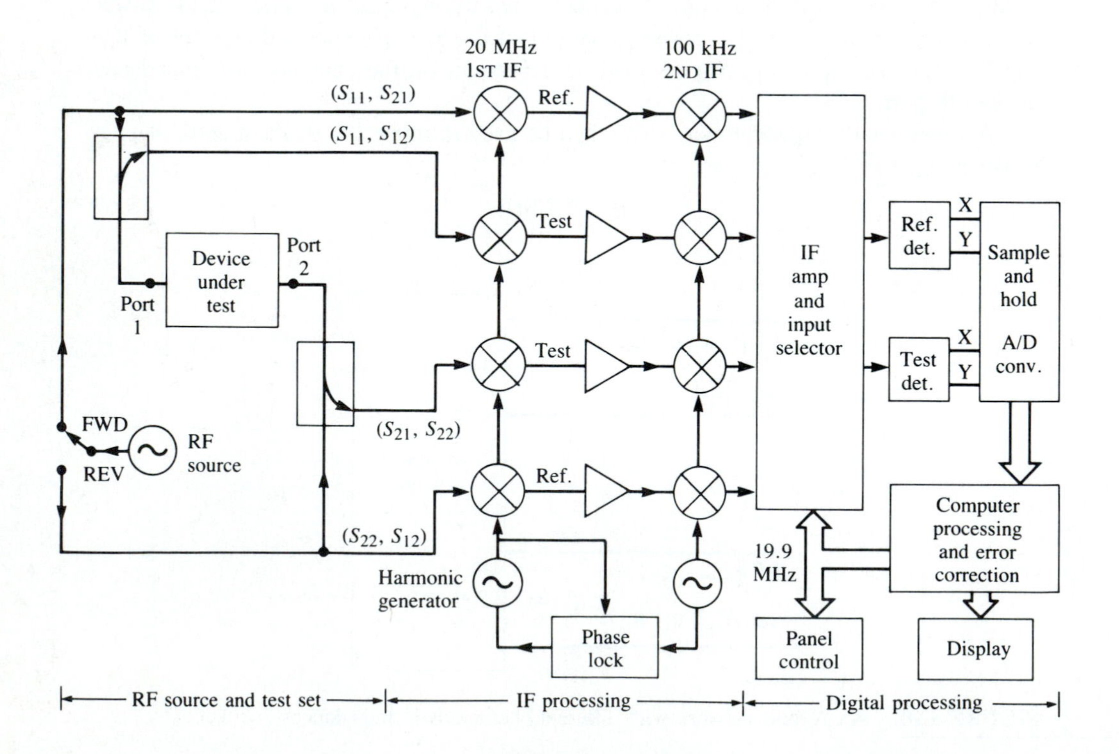

In operation, the RF source is usually set to sweep over a specified bandwidth. A four-port reflectometer samples the incident, reflected, and transmitted RF waves; a switch allows the network to be driven from either port 1 or port 2. Four dual-conversion channels convert these signals to 100 kHz IF frequencies, which are then detected and converted to digital form. A powerful internal computer is used to calculate and display the magnitude and phase of the S parameters, or other quantities that can be derived from the S parameters, such as SWR, return loss, group delay, impedance, etc. An important feature of this network analyzer is the substantial improvement in accuracy made possible with error correcting software. Errors caused by directional coupler mismatch, imperfect directivity, loss, and variations in the frequency response of the analyzer system are accounted for by using a twelve-term error model and a calibration procedure. Another useful feature is the capability to determine the time domain response of the network by calculating the inverse Fourier transform of the frequency domain data.

5.5 THE TRANSMISSION (*ABCD*) MATRIX

The Z, Y, and S parameter representations can be used to characterize a microwave network with an arbitrary number of ports, but in practice many microwave networks consist of a cascade connection of two or more two-port networks. In this case it is convenient to define a 2×2 transmission, or $ABCD$ matrix, for each two-port network. We will then see that the $ABCD$ matrix of the cascade connection of two or more two-port networks can be easily found by multiplying the $ABCD$ matrices of the individual two-ports.

The $ABCD$ matrix is defined for a two-port network in terms of the total voltages and currents as shown in Figure 5.11a and the following:

$$V_1 = AV_2 + BI_2,$$

$$I_1 = CV_2 + DI_2,$$

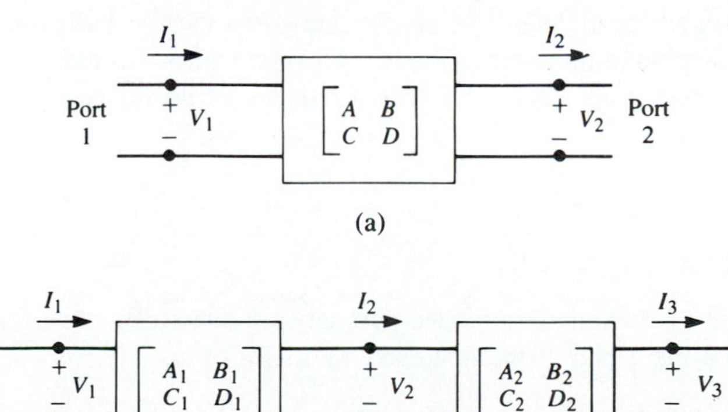

FIGURE 5.11 (a) A two-port network; (b) a cascade connection of two-port networks.

or in matrix form as

$$\begin{bmatrix} V_1 \\ I_1 \end{bmatrix} = \begin{bmatrix} A & B \\ C & D \end{bmatrix} \begin{bmatrix} V_2 \\ I_2 \end{bmatrix}. \tag{5.70}$$

It is important to note from Figure 5.11a that a change in the sign convention of I_2 has been made from our previous definitions, which had I_2 as the current flowing *into* port 2. The convention that I_2 flows *out* of port 2 will be used when dealing with $ABCD$ matrices so that in a cascade network I_2 will be the same current that flows into the adjacent network, as shown in Figure 5.11b. (Alternatively, I_2 in (5.50) could be replaced by $-I_2$, so that the sign convention would not have to be changed [1], [5].) Then the left-hand side of (5.70) represents the voltage and current at port 1 of the network, while the right-hand side of (5.70) represents the voltage and current at port 2.

In the cascade connection of two two-port networks shown in Figure 5.11b, we have that

$$\begin{bmatrix} V_1 \\ I_1 \end{bmatrix} = \begin{bmatrix} A_1 & B_1 \\ C_1 & D_1 \end{bmatrix} \begin{bmatrix} V_2 \\ I_2 \end{bmatrix}, \tag{5.71a}$$

$$\begin{bmatrix} V_2 \\ I_2 \end{bmatrix} = \begin{bmatrix} A_2 & B_2 \\ C_2 & D_2 \end{bmatrix} \begin{bmatrix} V_3 \\ I_3 \end{bmatrix}. \tag{5.71b}$$

Substituting (5.71b) into (5.71a) gives

$$\begin{bmatrix} V_1 \\ I_1 \end{bmatrix} = \begin{bmatrix} A_1 & B_1 \\ C_1 & C_1 \end{bmatrix} \begin{bmatrix} A_2 & B_2 \\ C_2 & D_2 \end{bmatrix} \begin{bmatrix} V_3 \\ I_3 \end{bmatrix}, \tag{5.72}$$

which shows that the $ABCD$ matrix of the cascade connection of the two networks is equal to the product of the $ABCD$ matrices representing the individual two-ports. Note that the order of multiplication of the matrix must be the same as the order in which the networks are arranged, since matrix multiplication is not, in general, commutative.

The usefulness of the $ABCD$ matrix representation lies in the fact that a library of $ABCD$ matrices for elementary two-port networks can be built up, and applied in building-block fashion to more complicated microwave networks that consist of cascades of these simpler two-ports. Table 5.1 lists a number of useful two-port networks, and their $ABCD$ matrices.

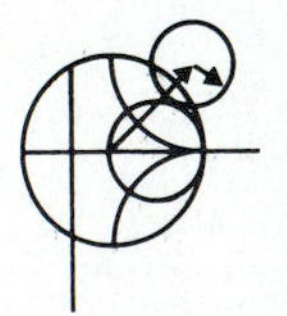

EXAMPLE 5.6

Find the $ABCD$ parameters of a two-port network consisting of a series impedance Z between ports 1 and 2 (the first entry in Table 5.1).

Solution

From the defining relations of (5.70), we have that

$$A = \left.\frac{V_1}{V_2}\right|_{I_2=0},$$

TABLE 5.1 The *ABCD* Parameters of Some Useful Two-Port Circuits

Circuit	*ABCD* Parameters	
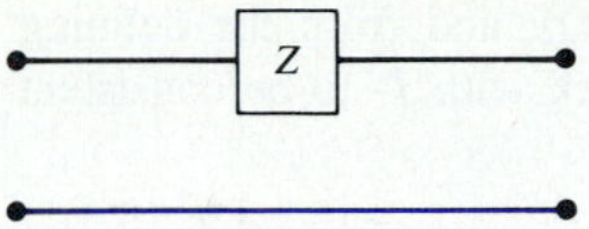 Z	$A = 1$ $C = 0$	$B = Z$ $D = 1$
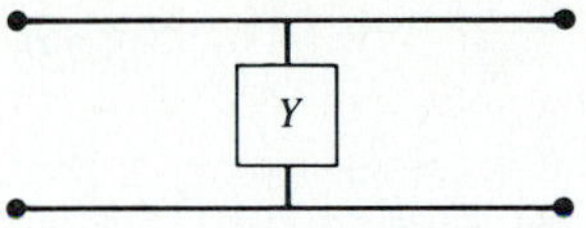 Y	$A = 1$ $C = Y$	$B = 0$ $D = 1$
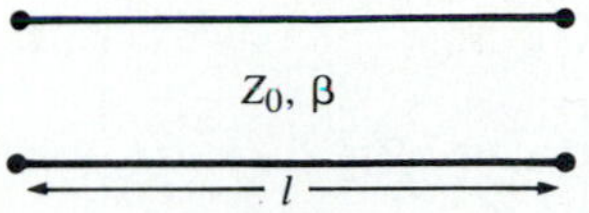 Z_0, β l	$A = \cos\beta\ell$ $C = jY_0 \sin\beta\ell$	$B = jZ_0 \sin\beta\ell$ $D = \cos\beta\ell$
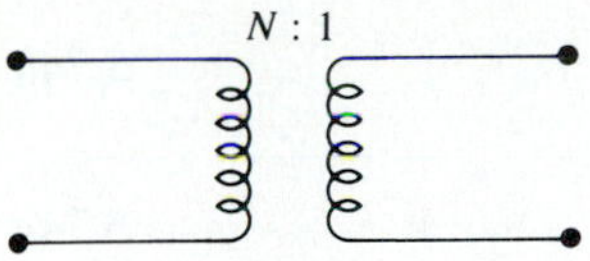 $N : 1$	$A = N$ $C = 0$	$B = 0$ $D = \frac{1}{N}$
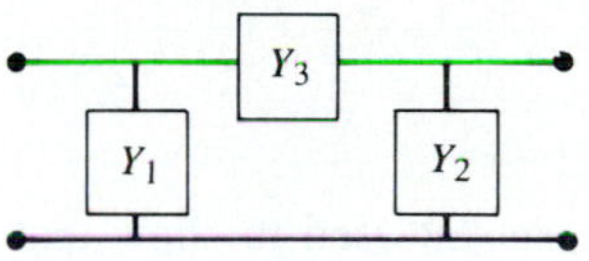 Y_3, Y_1, Y_2	$A = 1 + \frac{Y_2}{Y_3}$ $C = Y_1 + Y_2 + \frac{Y_1 Y_2}{Y_3}$	$B = \frac{1}{Y_3}$ $D = 1 + \frac{Y_1}{Y_3}$
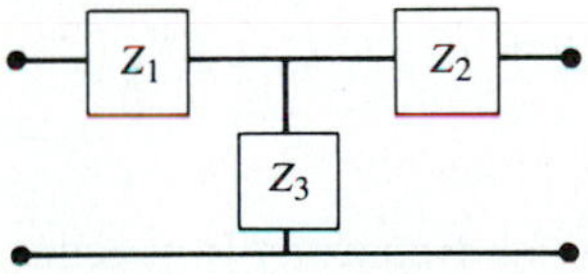 Z_1, Z_2, Z_3	$A = 1 + \frac{Z_1}{Z_3}$ $C = \frac{1}{Z_3}$	$B = Z_1 + Z_2 + \frac{Z_1 Z_2}{Z_3}$ $D = 1 + \frac{Z_2}{Z_3}$

which indicates that A is found by applying a voltage V_1 at port 1, and measuring the open-circuit voltage V_2 at port 2. Thus, $A = 1$. Similarly,

$$B = \left.\frac{V_1}{I_2}\right|_{V_2=0} = \frac{V_1}{V_1/Z} = Z,$$

$$C = \left.\frac{I_1}{V_2}\right|_{I_2=0} = 0,$$

$$D = \left.\frac{I_1}{I_2}\right|_{V_2=0} = \frac{I_1}{I_1} = 1. \qquad \bigcirc$$

Relation to Impedance Matrix

Knowing the Z parameters of a network, one can determine the $ABCD$ parameters. Thus, from the definition of the $ABCD$ parameters in (5.70), and from the defining relations for the Z parameters of (5.32) for a two-port network with I_2 to be consistent with the sign convention used with $ABCD$ parameters,

$$V_1 = I_1 Z_{11} + I_2 Z_{12}, \tag{5.73a}$$

$$V_2 = I_1 Z_{21} + I_2 Z_{22}, \tag{5.73b}$$

we have that

$$A = \left.\frac{V_1}{V_2}\right|_{I_2=0} = \frac{I_1 Z_{11}}{I_1 Z_{21}} = Z_{11}/Z_{21}, \tag{5.74a}$$

$$B = \left.\frac{V_1}{I_2}\right|_{V_2=0} = \left.\frac{I_1 Z_{11} - I_2 Z_{12}}{I_2}\right|_{V_2=0} = Z_{11} \left.\frac{I_1}{I_2}\right|_{V_2=0} - Z_{12}$$

$$= Z_{11} \frac{I_1 Z_{22}}{I_1 Z_{21}} - Z_{12} = \frac{Z_{11} Z_{22} - Z_{12} Z_{21}}{Z_{21}}, \tag{5.74b}$$

$$C = \left.\frac{I_1}{V_2}\right|_{I_2=0} = \frac{I_1}{I_1 Z_{21}} = 1/Z_{21}, \tag{5.74c}$$

$$D = \left.\frac{I_1}{I_2}\right|_{V_2=0} = \frac{I_2 Z_{22}/Z_{21}}{I_2} = Z_{22}/Z_{21}. \tag{5.74d}$$

If the network is reciprocal, then $Z_{12} = Z_{21}$ and (5.74) can be used to show that $AD - BC = 1$.

5.6 TWO-PORT NETWORKS

The special case of a two-port microwave network occurs so frequently in practice that it deserves further attention. We will first discuss the use of equivalent circuits to represent an arbitrary two-port network, and then show how network matrices can be used to treat various interconnections of two-port networks. The terminated two-port network is then discussed, and several types of power gain are defined for such networks. Useful conversions for two-port network parameters are given in Table 5.2.

Equivalent Circuits for Two-Port Networks

Figure 5.12a shows a transition between a coaxial line and a microstrip line, and serves as an example of a two-port network. Terminal planes can be defined at arbitrary points on the two transmission lines; a convenient choice might be as shown in the figure. But because of the physical discontinuity in the transition from a coaxial line to a microstrip line, electric and/or magnetic energy can be stored in the vicinity of the junction, leading to reactive effects. Characterization of such effects can be obtained by measurement or by theoretical analysis (although such an analysis may be quite

TABLE 5.2 **Conversions Between Two-Port Network Parameters**

	S	Z	Y	$ABCD$
S_{11}	S_{11}	$\dfrac{(Z_{11}-Z_0)(Z_{22}+Z_0)-Z_{12}Z_{21}}{\Delta Z}$	$\dfrac{(Y_0-Y_{11})(Y_0+Y_{11})+Y_{12}Y_{21}}{\Delta Y}$	$\dfrac{A+B/Z_0-CZ_0-D}{A+B/Z_0+CZ_0+D}$
S_{12}	S_{12}	$\dfrac{2Z_{12}Z_0}{\Delta Z}$	$\dfrac{-2Y_{12}Y_0}{\Delta Y}$	$\dfrac{2(AD-BC)}{A+B/Z_0+CZ_0+D}$
S_{21}	S_{21}	$\dfrac{2Z_{21}Z_0}{\Delta Z}$	$\dfrac{-2Y_{21}Y_0}{\Delta Y}$	$\dfrac{2}{A+B/Z_0+CZ_0+D}$
S_{22}	S_{22}	$\dfrac{(Z_{11}+Z_0)(Z_{22}-Z_0)-Z_{12}Z_{21}}{\Delta Z}$	$\dfrac{(Y_0+Y_{11})(Y_0-Y_{22})+Y_{12}Y_{21}}{\Delta Y}$	$\dfrac{-A+B/Z_0-CZ_0+D}{A+B/Z_0+CZ_0+D}$
Z_{11}	$Z_0\dfrac{(1+S_{11})(1-S_{22})+S_{12}S_{21}}{(1-S_{11})(1-S_{22})-S_{12}S_{21}}$	Z_{11}	$\dfrac{Y_{22}}{\lvert Y\rvert}$	$\dfrac{A}{C}$
Z_{12}	$Z_0\dfrac{2S_{12}}{(1-S_{11})(1-S_{22})-S_{12}S_{21}}$	Z_{12}	$\dfrac{-Y_{12}}{\lvert Y\rvert}$	$\dfrac{AD-BC}{C}$
Z_{21}	$Z_0\dfrac{2S_{21}}{(1-S_{11})(1-S_{22})-S_{12}S_{21}}$	Z_{21}	$\dfrac{-Y_{21}}{\lvert Y\rvert}$	$\dfrac{1}{C}$
Z_{22}	$Z_0\dfrac{(1-S_{11})(1+S_{22})+S_{12}S_{21}}{(1-S_{11})(1-S_{22})-S_{12}S_{21}}$	Z_{22}	$\dfrac{Y_{11}}{\lvert Y\rvert}$	$\dfrac{D}{C}$
Y_{11}	$Y_0\dfrac{(1-S_{11})(1+S_{22})+S_{12}S_{21}}{(1+S_{11})(1+S_{22})-S_{12}S_{21}}$	$\dfrac{Z_{22}}{\lvert Z\rvert}$	Y_{11}	$\dfrac{D}{B}$
Y_{12}	$Y_0\dfrac{-2S_{12}}{(1+S_{11})(1+S_{22})-S_{12}S_{21}}$	$\dfrac{-Z_{12}}{\lvert Z\rvert}$	Y_{12}	$\dfrac{BC-AD}{B}$
Y_{21}	$Y_0\dfrac{-2S_{21}}{(1+S_{11})(1+S_{22})-S_{12}S_{21}}$	$\dfrac{-Z_{21}}{\lvert Z\rvert}$	Y_{21}	$\dfrac{-1}{B}$
Y_{22}	$Y_0\dfrac{(1+S_{11})(1-S_{22})+S_{12}S_{21}}{(1+S_{11})(1+S_{22})-S_{12}S_{21}}$	$\dfrac{Z_{11}}{\lvert Z\rvert}$	Y_{22}	$\dfrac{A}{B}$
A	$\dfrac{(1+S_{11})(1-S_{22})+S_{12}S_{21}}{2S_{21}}$	$\dfrac{Z_{11}}{Z_{21}}$	$\dfrac{-Y_{22}}{Y_{21}}$	A
B	$Z_0\dfrac{(1+S_{11})(1+S_{22})-S_{12}S_{21}}{2S_{21}}$	$\dfrac{\lvert Z\rvert}{Z_{21}}$	$\dfrac{-1}{Y_{21}}$	B
C	$\dfrac{1}{Z_0}\dfrac{(1-S_{11})(1-S_{22})-S_{12}S_{21}}{2S_{21}}$	$\dfrac{1}{Z_{21}}$	$\dfrac{-\lvert Y\rvert}{Y_{21}}$	C
D	$\dfrac{(1-S_{11})(1+S_{22})-S_{12}S_{21}}{2S_{21}}$	$\dfrac{Z_{22}}{Z_{21}}$	$\dfrac{-Y_{11}}{Y_{21}}$	D

$|Z| = Z_{11}Z_{22} - Z_{12}Z_{21}$; $|Y| = Y_{11}Y_{22} - Y_{12}Y_{21}$; $\Delta Y = (Y_{11} + Y_0)(Y_{22} + Y_0) - Y_{12}Y_{21}$; $\Delta Z = (Z_{11} + Z_0)(Z_{22} + Z_0) - Z_{12}Z_{21}$; $Y_0 = 1/Z_0$

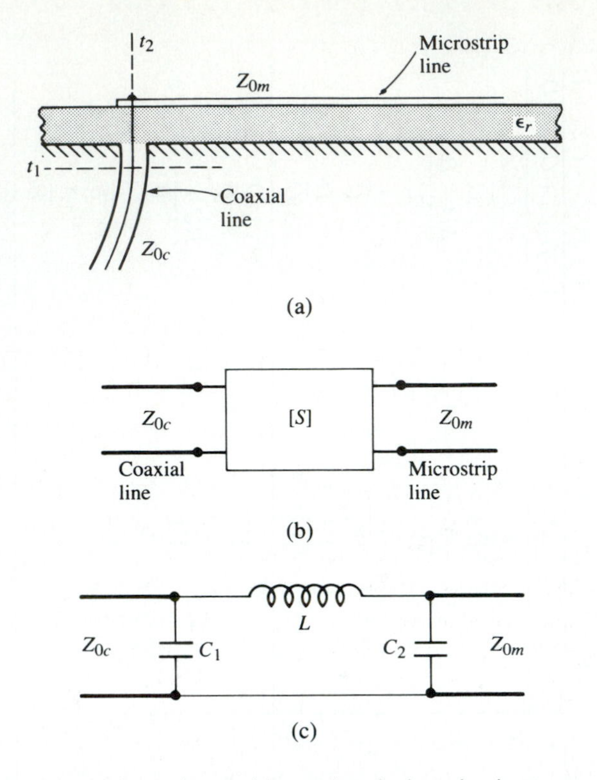

FIGURE 5.12 A coax-to-microstrip transition and equivalent circuit representations. (a) Geometry of the transition. (b) Representation of the transition by a "black box." (c) A possible equivalent circuit for the transition [6].

complicated), and represented by the two-port "black box" shown in Figure 5.12b. The properties of the transition can then be expressed in terms of the network parameters (Z, Y, S, or $ABCD$) of the two-port network. This type of treatment can be applied to a variety of two-port junctions, such as transitions from one type of transmission line to another, transmission line discontinuities such as step changes in width, or bends, etc. When modelling a microwave junction in this way, it is often useful to replace the two-port "black box" with an equivalent circuit containing a few idealized components, as shown in Figure 5.12c. (This is particularly useful if the component values can be related to some physical features of the actual junction.) There are an unlimited number of ways in which such equivalent circuits can be defined; we will discuss some of the most common and useful types below.

As we have seen from the previous sections, an arbitrary two-port network can be described in terms of impedance parameters as

$$V_1 = Z_{11}I_1 + Z_{12}I_2, \qquad 5.75a$$
$$V_2 = Z_{21}I_1 + Z_{22}I_2,$$

or in terms of admittance parameters as

$$I_1 = Y_{11}V_1 + Y_{12}V_2,$$
$$I_2 = Y_{21}V_1 + Y_{22}V_2. \tag{5.75b}$$

If the network is reciprocal, then $Z_{12} = Z_{21}$ and $Y_{12} = Y_{21}$. These representations lead naturally to the T and π equivalent circuits shown in Figure 5.13a and 5.13b. The relations in Table 5.2 can be used to relate the component values to other network parameters.

Other equivalent circuits can also be used to represent a two-port network. If the network is reciprocal, there are six degrees of freedom (the real and imaginary parts of three matrix elements), so the equivalent circuit should have six independent parameters. A nonreciprocal network cannot be represented by a passive equivalent circuit using reciprocal elements.

If the network is lossless, which is a good approximation for many practical two-port junctions, some simplifications can be made in the equivalent circuit. As was shown in Section 5.3, the impedance or admittance matrix elements are purely imaginary for a lossless network. This reduces the degrees of freedom for such a network to three, and implies that the T and π equivalent circuits of Figure 5.13 can be constructed from purely reactive elements. Other possibilities are shown in Figure 5.14.

Interconnected Two-Port Networks

We have seen in Section 5.5 how cascaded two-port networks can be treated using $ABCD$ parameters, but there are other ways in which networks can be interconnected. Consider first the series connection shown in Figure 5.15a. From the figure we see that

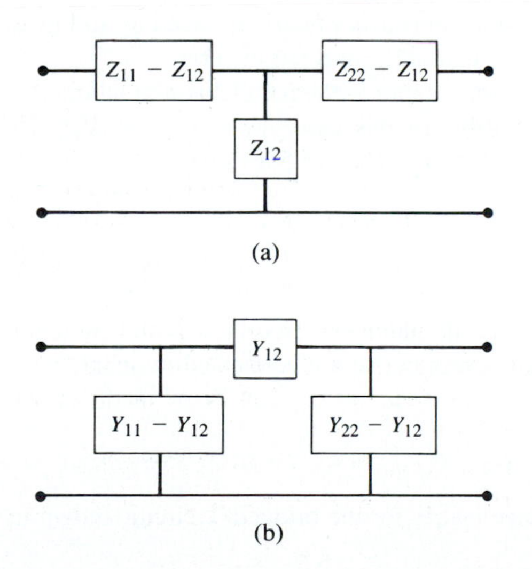

FIGURE 5.13 Equivalent circuits for a reciprocal two-port network. (a) T equivalent. (b) π equivalent.

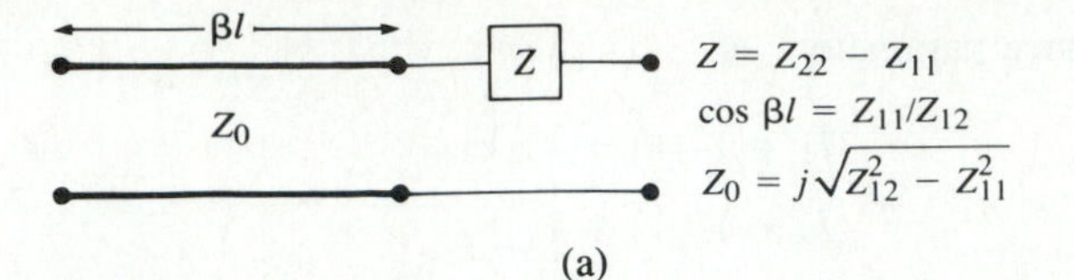

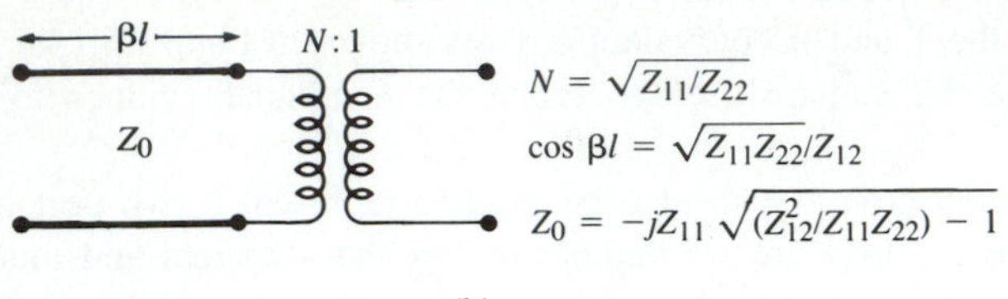

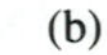

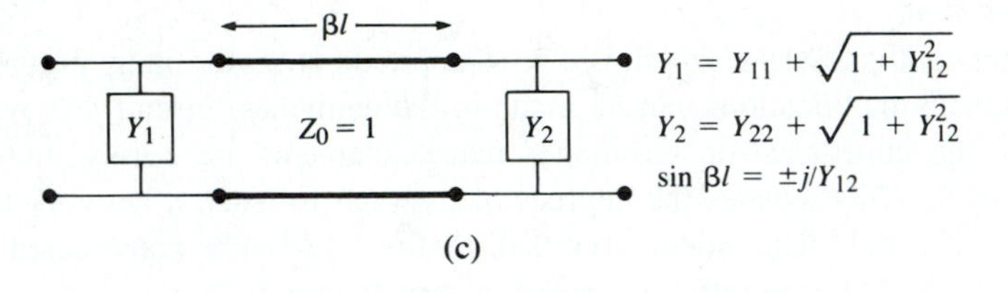

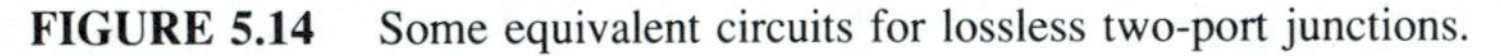
FIGURE 5.14 Some equivalent circuits for lossless two-port junctions.

$V_1 = V_1^a + V_1^b$, $V_2 = V_2^a + V_2^b$, $I_1 = I_1^a = I_1^b$, and $I_2 = I_2^a = I_2^b$. Thus, from (5.75a),

$$V_1 = V_1^a + V_1^b = (Z_{11}^a + Z_{11}^b)I_1 + (Z_{12}^a + Z_{12}^b)I_2,$$
$$V_2 = V_2^a + V_2^b = (Z_{21}^a + Z_{21}^b)I_1 + (Z_{22}^a + Z_{22}^b)I_2, \tag{5.76a}$$

which shows that the overall impedance matrix is found by adding the individual impedance matrices of the networks that are connected in series.

Dual to the above configuration is the parallel combination of two two-port networks, as shown in Figure 5.15b. In this case, $V_1 = V_1^a = V_1^b$, $V_2 = V_2^a = V_2^b$, and $I_1 = I_1^a + I_1^b$, and $I_2 = I_2^a + I_2^b$. Then (5.55b) gives,

$$I_1 = I_1^a + I_1^b = (Y_{11}^a + Y_{11}^b)V_1 + (Y_{12}^a + Y_{12}^b)V_2,$$
$$I_2 = I_2^a + I_2^b = (Y_{21}^a + Y_{21}^b)V_1 + (Y_{22}^a + Y_{22}^b)V_2, \tag{5.76b}$$

which shows that the overall admittance matrix is found by adding the admittance matrices of the individual networks that are connected in shunt.

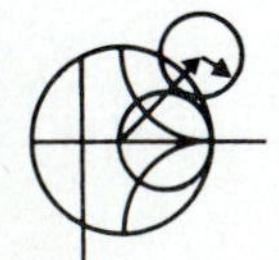

EXAMPLE 5.7

Find the admittance matrix for the bridged-T circuit shown in Figure 5.16a.

Solution

This network can be decomposed into the parallel connection of two simpler networks, as shown in Figure 5.16b. With reference to Figure 5.13 a,b, the

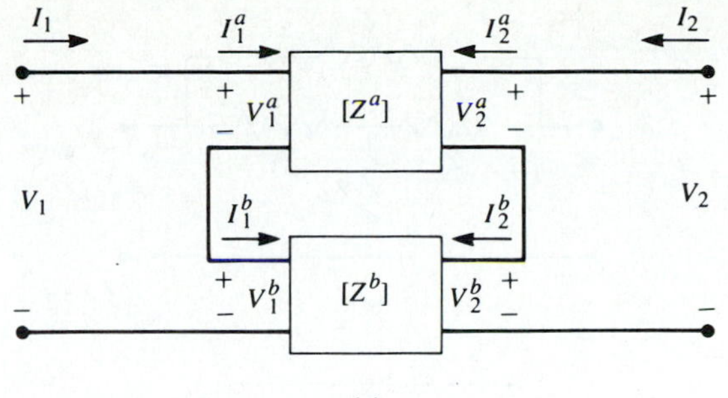

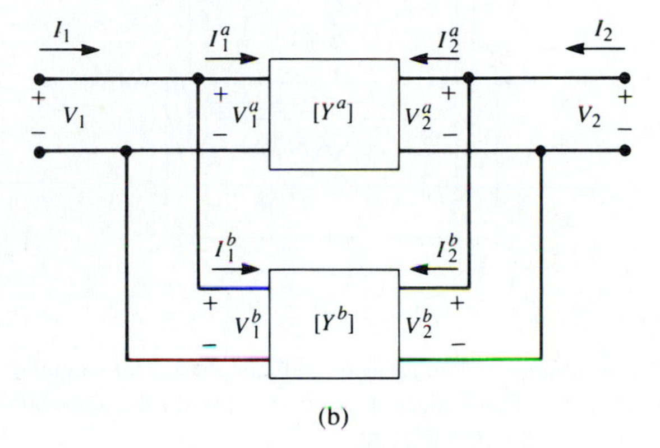

FIGURE 5.15 Series and parallel connections of two-port networks. (a) Series connection. (b) Parallel connection.

impedance and admittance matrices for these two subnetworks can be written as

$$Z_a = \begin{bmatrix} Z_1 + Z_2 & Z_2 \\ Z_2 & Z_1 + Z_2 \end{bmatrix},$$

$$Y_b = \begin{bmatrix} \dfrac{1}{Z_3} & \dfrac{-1}{Z_3} \\ \dfrac{-1}{Z_3} & \dfrac{1}{Z_3} \end{bmatrix}.$$

Inverting Z_a and applying the above result for shunt-connected networks gives the overall admittance matrix as

$$Y = Y_a + Y_b = \begin{bmatrix} \dfrac{1}{Z_3} + \dfrac{Z_1 + Z_2}{D} & -\left(\dfrac{1}{Z_3} + \dfrac{Z_2}{D}\right) \\ -\left(\dfrac{1}{Z_3} + \dfrac{Z_2}{D}\right) & \dfrac{Z_1 + Z_2}{D} + \dfrac{1}{Z_3} \end{bmatrix},$$

where $D = Z_1(Z_1 + 2Z_2)$. ○

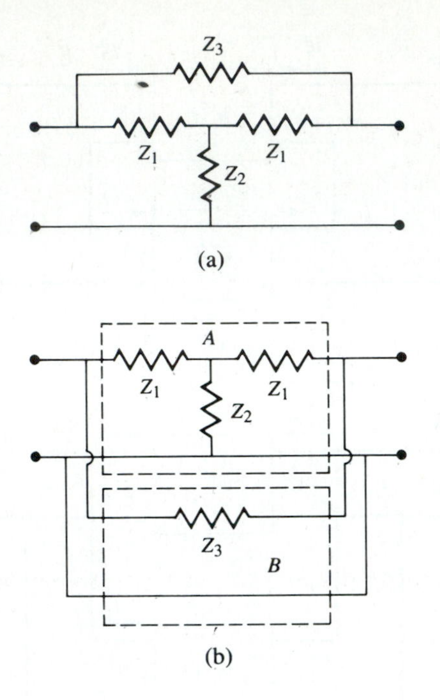

FIGURE 5.16 Decomposition of the network of Example 5.6 into simpler interconnected networks. (a) The bridged-T network. (b) Its decomposition into two simpler networks connected in parallel.

Two-Port Power Gains

We now consider the power transfer characteristics of an arbitrary two-port network with arbitrary sources and load impedances. The general configuration is shown in Figure 5.17, where in practice the two-port network is often a filter or amplifier. We will derive expressions for three types of power gain that are useful for such circuits in terms of the S parameters of the two-port network and the reflection coefficients at the source and load.

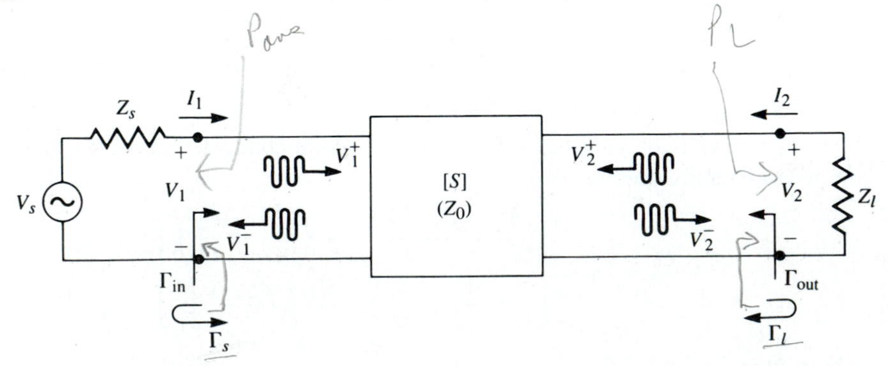

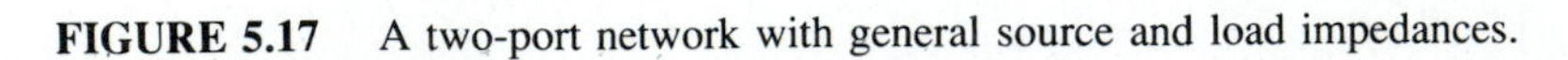

FIGURE 5.17 A two-port network with general source and load impedances.

- Power Gain $= G = P_\ell/P_{\text{in}}$ is the ratio of power dissipated in the load Z_ℓ to the power delivered to the input of the two-port network. This gain is then independent of Z_s, although certain active circuits are strongly dependent on Z_s.
- Available Gain $= G_A = P\text{avn}/P\text{avs}$ is the ratio of the power available from the two-port network to the power available from the source. This gain depends on Z_s, but is independent of Z_ℓ. The characteristics of many active circuits, however, depend on Z_ℓ.
- Transducer Power Gain $= G_T = P_\ell/P\text{avs}$ is the ratio of the power delivered to the load to the power available from the source. It depends on both Z_s and Z_ℓ, and so has advantages over the previous gain definitions.

With reference to Figure 5.17, the reflection coefficient seen looking from the network toward the load is

$$\Gamma_\ell = \frac{Z_\ell - Z_0}{Z_\ell + Z_0}, \qquad 5.77a$$

while the reflection coefficient seen looking from the network toward the source is

$$\Gamma_s = \frac{Z_s - Z_0}{Z_s + Z_0}, \qquad 5.77b$$

where Z_0 is the characteristic impedance reference of the S parameters of the two-port network.

In general, the input of the terminated two-port network will be mismatched with a reflection coefficient given by Γ_{in}, which can be determined as follows. From the definition of the S parameters and that $V_2^+ = \Gamma_\ell V_2^-$, we have

$$V_1^- = S_{11}V_1^+ + S_{12}V_2^+ = S_{11}V_1^+ + S_{12}\Gamma_\ell V_2^-, \qquad 5.78a$$

$$V_2^- = S_{21}V_1^+ + S_{22}V_2^+ = S_{21}V_1^+ + S_{22}\Gamma_\ell V_2^-. \qquad 5.78b$$

Eliminating V_2^- from (5.78a) and solving for V_1^-/V_1^+ gives

$$\Gamma_{\text{in}} = \frac{V_1^-}{V_1^+} = S_{11} + \frac{S_{12}S_{21}\Gamma_\ell}{1 - S_{22}\Gamma_\ell} = \frac{Z_{\text{in}} - Z_0}{Z_{\text{in}} + Z_0}, \qquad 5.79a$$

which is a general result for the input reflection coefficient of a two-port network with an arbitrary load. Z_{in} is the impedance seen looking into port 1 of the terminated network. Similarly, the reflection coefficient seen looking into port 2 of the network when port 1 is terminated by Z_s is

$$\Gamma_{\text{out}} = \frac{V_2^-}{V_2^+} = S_{22} + \frac{S_{12}S_{21}\Gamma_s}{1 - S_{11}\Gamma_s}. \qquad 5.79b$$

By voltage division,

$$V_1 = V_s\frac{Z_{\text{in}}}{Z_s + Z_{\text{in}}} = V_1^+ + V_1^- = V_1^+(1 + \Gamma_{\text{in}}).$$

Using

$$Z_{in} = Z_0 \frac{1+\Gamma_{in}}{1-\Gamma_{in}},$$

from (5.79a) and solving for V_1^+ in terms of V_s gives

$$V_1^+ = \frac{V_s}{2} \frac{(1-\Gamma_s)}{(1-\Gamma_s\Gamma_{in})}. \qquad 5.80$$

If peak values are assumed for all voltages, the average power delivered to the network is

$$P_{in} = \frac{1}{2Z_0}|V_1^+|^2(1-|\Gamma_{in}|^2) = \frac{|V_s|^2}{8Z_0}\frac{|1-\Gamma_s|^2}{|1-\Gamma_s\Gamma_{in}|^2}(1-|\Gamma_{in}|^2), \qquad 5.81$$

where (5.80) was used. The power delivered to the load is

$$P_\ell = \frac{|V_2^-|^2}{2Z_0}(1-|\Gamma_\ell|^2). \qquad 5.82$$

Solving for V_2^- from (5.78b), substituting into (5.82), and using (5.80) gives

$$P_\ell = \frac{|V_1^+|^2}{2Z_0}\frac{|S_{21}|^2(1-|\Gamma_\ell|^2)}{|1-S_{22}\Gamma_\ell|^2}$$

$$= \frac{|V_s|^2}{8Z_0}\frac{|S_{21}|^2(1-|\Gamma_\ell|^2)|1-\Gamma_s|^2}{|1-S_{22}\Gamma_\ell|^2|1-\Gamma_s\Gamma_{in}|^2}. \qquad 5.83$$

The power gain can then be expressed as

$$G = \frac{P_\ell}{P_{in}} = \frac{|S_{21}|^2(1-|\Gamma_\ell|^2)}{|1-S_{22}\Gamma_\ell|^2(1-|\Gamma_{in}|^2)}. \qquad 5.84$$

The power available from the source, P_{avs}, is the maximum power that can be delivered to the network. This occurs when the input impedance of the terminated network is conjugate matched to the source impedance, as discussed in Section 3.6. Thus, from (5.81),

$$P_{avs} = P_{in}\Big|_{\Gamma_{in}=\Gamma_s^*} = \frac{|V_s|^2}{8Z_0}\frac{|1-\Gamma_s|^2}{(1-|\Gamma_s|^2)}. \qquad 5.85$$

Similarly, the power available from the network, P_{avn}, is the maximum power that can be delivered to the load. Thus, from (5.83),

$$P_{avn} = P_\ell\Big|_{\Gamma_\ell=\Gamma_{out}^*} = \frac{|V_s|^2}{8Z_0}\frac{|S_{21}|^2(1-|\Gamma_{out}|^2)|1-\Gamma_s|^2}{|1-S_{22}\Gamma_{out}^*|^2|1-\Gamma_s\Gamma_{in}|^2}\Bigg|_{\Gamma_\ell=\Gamma_{out}^*}. \qquad 5.86$$

In (5.86), Γ_{in} must be evaluated for $\Gamma_\ell = \Gamma_{out}^*$. From (5.79a), it can be shown that

$$|1-\Gamma_s\Gamma_{in}|^2\Big|_{\Gamma_\ell=\Gamma_{out}^*} \quad \frac{|1-S_{11}\Gamma_s|^2(1-|\Gamma_{out}|^2)}{|1-S_{22}\Gamma_{out}^*|^2},$$

which reduces (5.86) to

$$Pavn = \frac{|V_s|^2}{8Z_0} \frac{|S_{21}|^2|1 - \Gamma_s|^2}{|1 - S_{11}\Gamma_s|^2(1 - |\Gamma_{out}|^2)}. \quad 5.87$$

Observe that $Pavs$ and $Pavn$ have been expressed in terms of the source voltage, V_s, which is independent of the input or load impedances. There would be confusion if these quantities were expressed in terms of V_1^+, since V_1^+ is different for each of the calculations of P_ℓ, $Pavs$ and $Pavn$.

Using (5.87) and (5.85), the available power gain is

$$G_A = \frac{Pavn}{Pavs} = \frac{|S_{21}|^2(1 - |\Gamma_s|^2)}{|1 - S_{11}\Gamma_s|^2(1 - |\Gamma_{out}|^2)}. \quad 5.88$$

From (5.83) and (5.85), the transducer power gain is

$$G_T = \frac{P_\ell}{Pavs} = \frac{|S_{21}|^2(1 - |\Gamma_s|^2)(1 - |\Gamma_\ell|^2)}{|1 - S_{22}\Gamma_\ell|^2|1 - \Gamma_s\Gamma_{in}|^2}. \quad 5.89$$

A special case of the transducer power gain is the matched transducer power gain, G_{Tm}, which occurs when both the input and output networks are matched. Then $\Gamma_\ell = \Gamma_s = 0$, and (5.89) reduces to

$$G_{Tm} = |S_{21}|^2. \quad 5.90$$

Another special case is the unilateral transducer power gain, G_{Tu}, where $S_{12} = 0$. This nonreciprocal situation can occur in some amplifier circuits. From (5.79a), $\Gamma_{in} = S_{11}$ when $S_{12} = 0$, so (5.89) gives the unilateral transducer gain as

$$G_{Tu} = \frac{|S_{21}|^2(1 - |\Gamma_s|^2)(1 - |\Gamma_\ell|^2)}{|1 - S_{11}\Gamma_s|^2|1 - S_{22}\Gamma_\ell|^2}. \quad 5.91$$

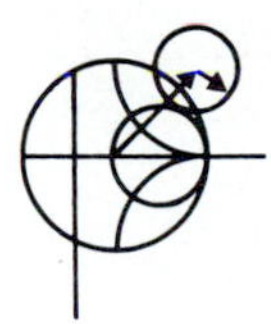

EXAMPLE 5.8

A certain microwave transistor has the following S parameters at 10 GHz with a 50 Ω reference impedance:

$$S_{11} = 0.45\angle 150^\circ,$$
$$S_{12} = 0.01\angle -10^\circ,$$
$$S_{21} = 2.05\angle 10^\circ,$$
$$S_{22} = 0.40\angle -150^\circ.$$

The source impedance is $Z_s = 20\ \Omega$ and the load impedance is $Z_\ell = 30\ \Omega$. Compute the available power gain, the transducer power gain, and the actual power gain.

Solution

From (5.77a,b), the reflection coefficients at the source and load are

$$\Gamma_s = \frac{Z_s - Z_0}{Z_s + Z_0} = \frac{20 - 50}{20 + 50} = -0.429,$$

$$\Gamma_\ell = \frac{Z_\ell - Z_0}{Z_\ell + Z_0} = \frac{30 - 50}{30 + 50} = -0.250.$$

From (5.79a,b) the reflection coefficients seen looking at the input and output of the terminated network are

$$\begin{aligned}\Gamma_{\text{in}} &= S_{11} + \frac{S_{12}S_{21}\Gamma_\ell}{1 - S_{22}\Gamma_\ell} \\ &= 0.45\angle 150^\circ + \frac{(0.01\angle -10^\circ)(2.05\angle 10^\circ)(-0.250)}{1 - (0.40\angle -150^\circ)(-0.250)} \\ &= 0.455\angle 150^\circ,\end{aligned}$$

$$\begin{aligned}\Gamma_{\text{out}} &= S_{22} + \frac{S_{12}S_{21}\Gamma_s}{1 - S_{11}\Gamma_s} \\ &= 0.40\angle -150^\circ + \frac{(0.01\angle -10^\circ)(2.05\angle 10^\circ)(-0.429)}{1 - (0.45\angle 150^\circ)(-0.429)} \\ &= 0.408\angle -151^\circ.\end{aligned}$$

Then from (5.88) the available power gain is

$$\begin{aligned}G_A &= \frac{|S_{21}|^2(1 - |\Gamma_s|^2)}{|1 - S_{11}\Gamma_s|^2(1 - |\Gamma_{\text{out}}|^2)} \\ &= \frac{(2.05)^2(1 - (.429)^2)}{|1 - (0.45\angle 150^\circ)(-.429)|^2(1 - (.408)^2)} \\ &= 5.85.\end{aligned}$$

From (5.89), the transducer power gain is

$$\begin{aligned}G_T &= \frac{|S_{21}|^2(1 - |\Gamma_s|^2)(1 - |\Gamma_\ell|^2)}{|1 - S_{22}\Gamma_\ell|^2|1 - \Gamma_s\Gamma_{\text{in}}|^2} \\ &= \frac{(2.05)^2(1 - (0.429)^2)(1 - (0.250)^2)}{|1 - (0.40\angle -150^\circ)(-0.250)|^2|1 - (-0.429)(0.455\angle 150^\circ)|^2} \\ &= 5.49.\end{aligned}$$

and from (5.84), the actual power gain is

$$\begin{aligned}G &= \frac{|S_{21}|^2(1 - |\Gamma_\ell|^2)}{|1 - S_{22}\Gamma_\ell|^2(1 - |\Gamma_{\text{in}}|^2)} = \frac{(2.05)^2(1 - 0.250)^2}{|1 - (0.40\angle -150^\circ)(-0.250)|^2(1 - (0.455)^2)} \\ &= 5.94.\end{aligned}$$

○

5.7 SIGNAL FLOW GRAPHS

We have seen how transmitted and reflected waves can be represented by scattering parameters, and how the interconnection of sources, networks, and loads can be treated with various matrix representations. In this section we discuss the signal flow graph, which is an additional technique that is very useful for the analysis of microwave networks in terms of transmitted and reflected waves. We first discuss the features and the construction of the flow graph itself, and then present two techniques for the reduction, or solution, of the flow graph.

The primary components of a signal flow graph are nodes and branches:

- Nodes: Each port, i, of the microwave network has two nodes, a_i and b_i. Node a_i is identified with a wave entering port i, while node b_i is identified with a wave reflected from port i.
- Branches: A branch is a directed path between an a-node and a b-node, representing signal flow from node a to node b. Every branch has an associated S parameter or reflection coefficient.

At this point it is useful to consider the flowgraph of an arbitrary two-port network, as shown in Figure 5.18. Figure 5.18a shows a two-port network with incident and reflected waves at each port, and Figure 5.18b shows the corresponding signal flow graph representation. The flow graph gives an intuitive graphical illustration of the network behavior.

For example, a wave of amplitude a_1 incident at port 1 is split, with part going through S_{11} and out port 1 as a reflected wave and part transmitted through S_{21} to node b_2. At node b_2, the wave goes out port 2; if a load with nonzero reflection coefficient is connected at port 2, this wave will be at least partly reflected and re-enter the two-port network at node a_2. Part of the wave can be reflected back out port 2 via S_{22}, and part can be transmitted out port 1 through S_{12}.

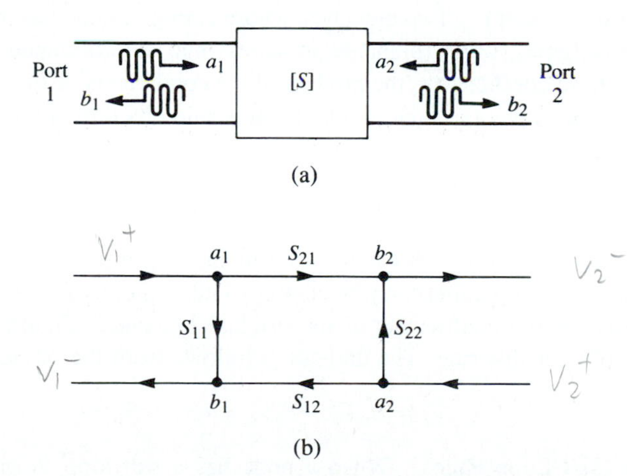

FIGURE 5.18 The signal flow graph representation of a two-port network. (a) Definition of incident and reflected waves. (b) Signal flow graph.

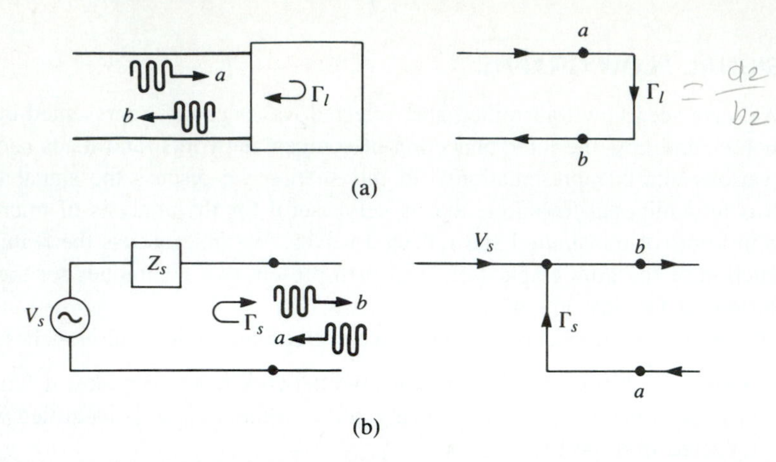

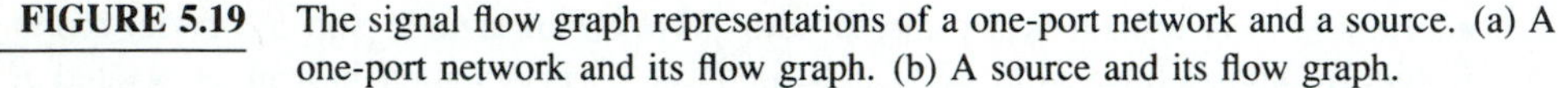
FIGURE 5.19 The signal flow graph representations of a one-port network and a source. (a) A one-port network and its flow graph. (b) A source and its flow graph.

Two other special networks, a one-port network and a voltage source, are shown in Figure 5.19 along with their signal flow graph representations. Once a microwave network has been represented in signal flow graph form, it is a relatively easy matter to solve for the ratio of any combination of wave amplitudes. There are two ways of doing this.

Decomposition of Signal Flow Graphs

In this method, a signal flow graph is reduced to a single branch between two nodes, using the four basic decomposition rules below, to obtain the desired wave amplitude ratio.

- **Rule 1** (Series Rule). Two branches, whose common node has only one incoming and one outgoing wave (branches in series) may be combined to form a single branch whose coefficient is the product of the coefficients of the original branches. Figure 5.20a shows the flow graphs for this rule. Its derivation follows from the basic relation that

$$V_3 = S_{32}V_2 = S_{32}S_{21}V_1. \qquad 5.92$$

- **Rule 2** (Parallel Rule). Two branches from one common node to another common node (branches in parallel) may be combined into a single branch whose coefficient is the sum of the coefficients of the original branches. Figure 5.20b shows the flow graphs for this rule. The derivation follows from the obvious relation that

$$V_2 = S_aV_1 + S_bV_1 = (S_a + S_b)V_1. \qquad 5.93$$

- **Rule 3** (Self-Loop Rule). When a node has a self-loop (a branch that begins and ends on the same node) of coefficient S, the self-loop can be eliminated by multiplying coefficients of the branches feeding that node by $1/(1 - S)$.

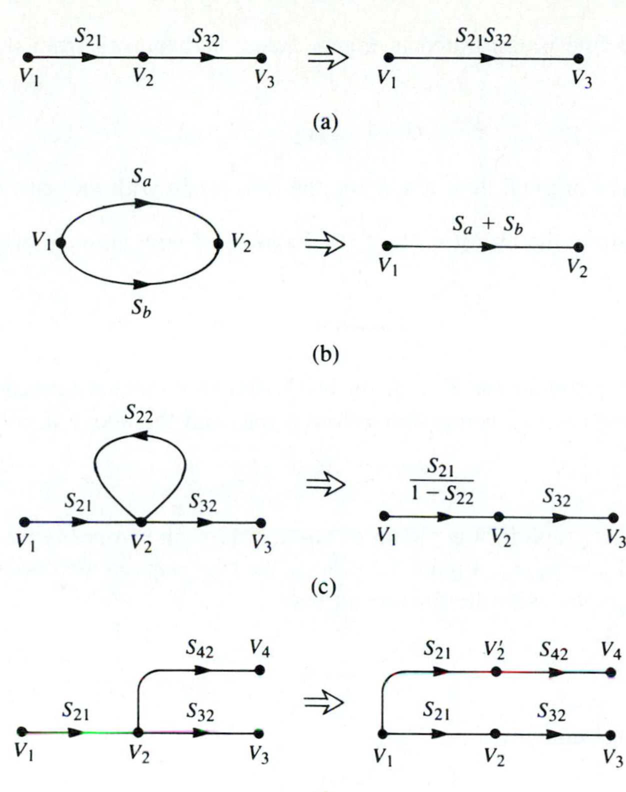

FIGURE 5.20 Decomposition rules. (a) Series rule. (b) Parallel rule. (c) Self-loop rule. (d) Splitting rule.

Figure 5.20c shows the flow graphs for this rule, which can be derived as follows. From the original network we have that

$$V_2 = S_{21}V_1 + S_{22}V_2, \tag{5.94a}$$

$$V_3 = S_{32}V_2. \tag{5.94b}$$

Eliminating V_2 gives

$$V_3 = \frac{S_{32}S_{21}}{1 - S_{22}}V_1, \tag{5.95}$$

which is seen to be the transfer function for the reduced graph of Figure 5.20c.

- **Rule 4** (Splitting Rule). A node may be split into two separate nodes as long as the resulting flow graph contains, once and only once, each combination of separate (not self loops) input and output branches that connect to the original node.

This rule is illustrated in Figure 5.20d, and follows from the observation that

$$V_4 = S_{42}V_2 = S_{21}S_{42}V_1, \tag{5.96}$$

in both the original flow graph and the flow graph with the split node.

We now illustrate the use of each of the above rules with an example.

EXAMPLE 5.9

Derive the expression for Γ_{in}, given in (5.79a), for the terminated two-port network of Figure 5.17 using signal flow graphs and the above decomposition rules.

Solution

The signal flow graph for the circuit of Figure 5.17 is shown in Figure 5.21. We wish to find $\Gamma_{in} = b_1/a_1$. Figure 5.22 shows the four steps in the decomposition of the flow graphs, with the final result that

$$\Gamma_{in} = \frac{b_1}{a_1} = S_{11} + \frac{S_{12}S_{21}\Gamma_\ell}{1 - S_{22}\Gamma_\ell},$$

which is in agreement with (5.79a). ○

Flow Graph Analysis Using Mason's Rule

The decomposition method discussed above is intuitively appealing, and can be applied to any flow graph, but is not very efficient from a computational point of view. Mason's rule, however, can be used to write the desired wave amplitude ratio by inspection. (Mason's rule is also useful in the analysis of feedback control systems.) We first define the following terms:

- *Independent variable node* is the node of an incident wave.
- *Dependent variable node* is the node of a reflected wave.
- *Path* is a series of codirectional branches from an independent node to a dependent node, along which no node is crossed more than once. The value of a path is the

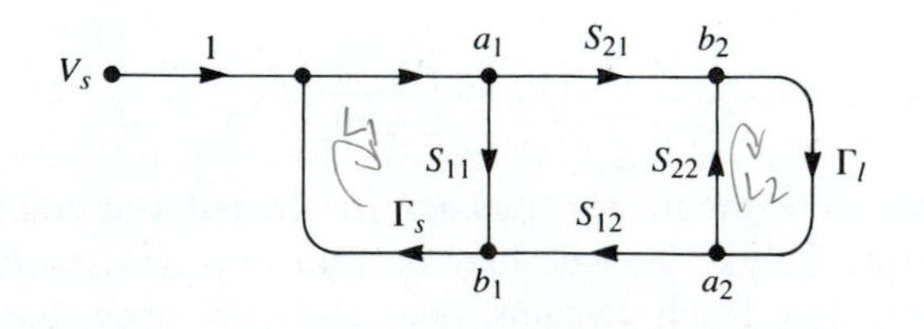

FIGURE 5.21 Signal flow path for the two-port network with general source and load impedances of Figure 5.17.

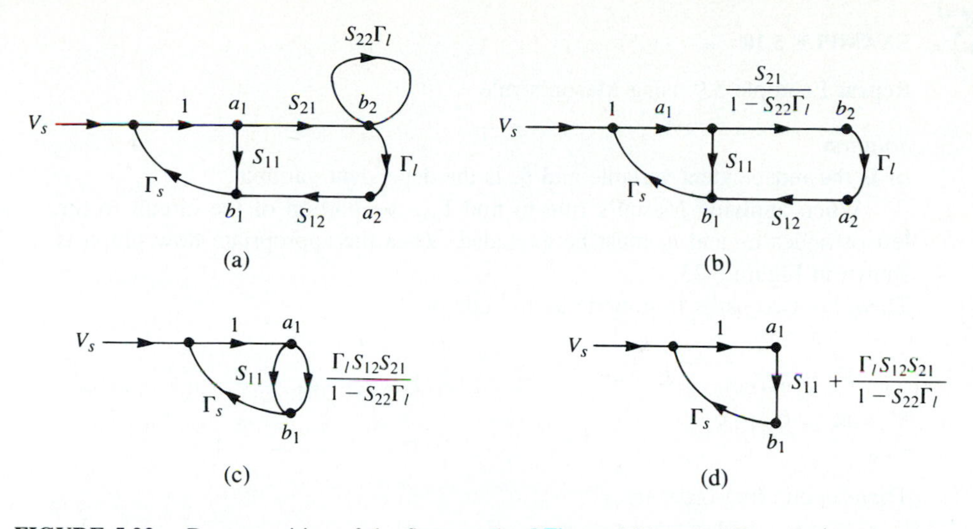

FIGURE 5.22 Decomposition of the flow graph of Figure 5.21 to find $\Gamma_{in} = b_1/a_1$. (a) Using Rule 4 on node a_2. (b) Using Rule 3 for the self-loop. (c) Using Rule 1. (d) Using Rule 2.

product of all the branch coefficients along that path. In the flow graph of Figure 5.21, there are two paths from node a_1 to node b_1, with values S_{11} and $S_{21}\Gamma_\ell S_{12}$.

- *First-order loop* is the product of branch coefficients encountered in a round trip from a node back to that same node, without crossing the same node twice. In the flow graph of Figure 5.21, $S_{22}\Gamma_\ell, S_{11}\Gamma_s$ and $S_{21}\Gamma_\ell S_{12}\Gamma_s$ are first-order loops.
- *Second-order loop* is the product of any two nontouching first-order loops. In the flow graph of Figure 5.21, the first-order loops $S_{11}\Gamma_s$ and $S_{22}\Gamma_\ell$ do not touch, so $S_{11}S_{22}\Gamma_s\Gamma_\ell$ is a second-order loop.
- *Third-order loop* is the product of three nontouching first-order loops. There are no third-order loops in the flow graph of Figure 5.21.

Higher-order loops are similarly defined.

Then Mason's rule for the ratio T of the wave amplitude of a dependent variable to the wave amplitude of an independent variable is given as

$$T = \frac{P_1[1 - \sum L(1)^1 + \sum L(2)^1 - \cdots] + P_2[1 - \sum L(1)^2 + \cdots] + \cdots}{1 - \sum L(1) + \sum L(2) - \sum L(3) + \cdots}, \quad 5.97$$

where $P_1, P_2, \ldots$ are the coefficients of the possible paths connecting the independent and dependent variables, and

$\sum L(1), \sum L(2) \ldots$ are the sums of all first-order, second-order, ... loops,

$\sum L(1)^1, \sum L(2)^1, \ldots$ are the sums of all first-order, second-order, ... loops that do not touch the first path between the variables,

$\sum L(1)^2, \sum L(2)^2, \ldots$ are the sums of all first-order, ... loops that do not touch the second path between the variables,

and so on, for all the paths between the independent and dependent variables.

EXAMPLE 5.10

Repeat Example 5.9 using Mason's rule.

Solution

a_1 is the independent variable and b_1 is the dependent variable.

When applying Mason's rule to find Γ_{in}, the portion of the circuit to the left of nodes a_1 and b_1 must be excluded. Thus the appropriate flow graph is shown in Figure 5.23.

There are two paths from node a_1 to node b_1:

$$P_1 = S_{11},$$

$$P_2 = S_{21}\Gamma_\ell S_{12}.$$

There is one first-order loop:

$$\sum L(1) = \Gamma_\ell S_{22}.$$

There are no second-order (or higher) loops. The sum of all first-order loops that do not touch the S_{11} path is

$$\sum L(1)^1 = \Gamma_\ell S_{22}.$$

The sum of all first-order loops that do not touch the $S_{21}\Gamma_\ell S_{12}$ path is

$$\sum L(1)^2 = 0.$$

Then from (5.97) we have that

$$\Gamma_{in} = \frac{b_1}{a_1} = \frac{S_{11}(1 - \Gamma_\ell S_{22}) + S_{21}\Gamma_\ell S_{12}}{1 - \Gamma_\ell S_{22}} = S_{11} + \frac{S_{21}\Gamma_\ell S_{12}}{1 - \Gamma_\ell S_{22}},$$

which is in agreement with our previous results. Note that one must be careful in forming the flow graph from the proper subset of the network when using this method.

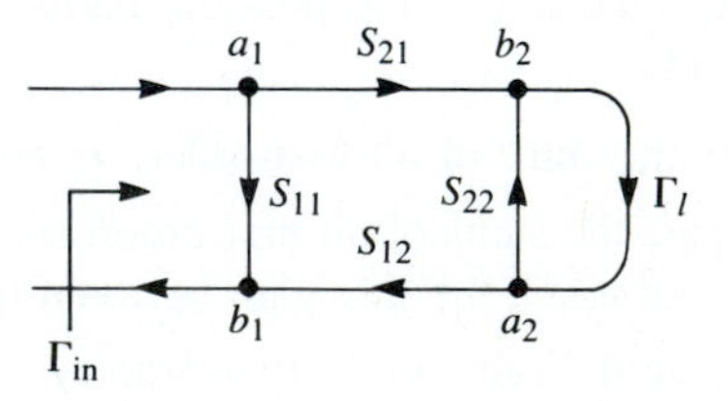

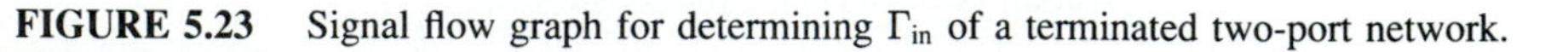
FIGURE 5.23 Signal flow graph for determining Γ_{in} of a terminated two-port network.

POINT OF INTEREST: Computer-Aided Design for Microwave Circuits

A computer-aided design (CAD) software package for microwave circuit analysis and optimization can be a very useful tool for the microwave engineer. Several microwave CAD programs are commercially available, such as SUPERCOMPACT® and TOUCHSTONE®, with the capability of analyzing microwave circuits consisting of transmission lines, lumped elements, active devices, coupled lines, waveguides, and other components. Although such computer programs can be fast, powerful, and accurate, they cannot serve as a substitute for an experienced engineer with a good understanding of microwave design.

A typical design process will usually begin with specifications or design goals for the circuit. Based on previous designs and his own experience, the engineer can develop an initial design, including specific components and a circuit layout. CAD can then be used to model and analyze the design, using data for each of the components and including effects such as loss and discontinuities. The CAD program can be used to optimize the design by adjusting some of the circuit parameters to achieve the best performance. If the specifications are not met, the design may have to be revised. The CAD analysis can also be used to study the effects of component tolerances and errors, to improve circuit reliability and robustness. When the design meets the specifications, an engineering prototype can be built and tested. If the measured results satisfy the specifications, the design process is completed. Otherwise the design will need to be revised, and the procedure repeated.

Without CAD tools, the design process would require the construction and measurement of a laboratory prototype at each iteration, which would be expensive and time consuming. Thus, CAD can greatly decrease the time and cost of a design, while enhancing its quality. The simulation and optimization process is especially important for monolithic microwave integrated circuits (MMICs) because these circuits cannot easily be tuned or trimmed after fabrication.

CAD techniques are not without limitations, however. Of primary importance is the fact that a computer model is only an approximation to a "real-world" circuit, and cannot completely account for the inevitable effects of component and fabricational tolerances, surface roughness, spurious coupling, higher-order modes, and junction discontinuities. These limitations generally become most serious at frequencies above 10 GHz.

END not for class

5.8 DISCONTINUITIES AND MODAL ANALYSIS omit

By either necessity or design, microwave networks often consist of transmission lines with various types of transmission line discontinuities. In some cases discontinuities are an unavoidable result of mechanical or electrical transitions from one medium to another (e.g., a junction between two waveguides, or a coax-to-microstrip transition), and the discontinuity effect is unwanted but may be significant enough to warrant characterization. In other cases discontinuities may be deliberately introduced into the circuit to perform a certain electrical function (e.g., reactive diaphragms in waveguide or stubs in microstrip line for matching or filter circuits). In any event, a transmission line discontinuity can be represented as an equivalent circuit at some point on the transmission line. Depending on the type of discontinuity, the equivalent circuit may be a simple shunt or series element across the line or, in the more general case, a T- or π-equivalent circuit may be required. The component values

® Registered trademarks of Compact Software Corp. and EEsof, Inc., respectively.

of an equivalent circuit depend on the parameters of the line and the discontinuity, as well as the frequency of operation. In some cases the equivalent circuit involves a shift in the phase reference planes on the transmission lines. Once the equivalent circuit of a given discontinuity is known, its effect can be incorporated into the analysis or design of the network using the theory developed previously in this chapter.

The purpose of the present section is to discuss how equivalent circuits are obtained for transmission line discontinuities; we will see that the basic procedure is to start with a field theory solution to a canonical discontinuity problem and develop a circuit model, with component values. This is thus another example of our objective of replacing complicated field analyses with circuit concepts.

Figures 5.24 and 5.25 show some common transmission line discontinuities and their equivalent circuits. As shown in Figures 5.24a–c, thin metallic diaphragms (or "irises") can be placed in the cross-section of a waveguide to yield equivalent shunt inductance, capacitance, or a resonant combination. Similar effects occur with step changes in the height or width of the waveguide, as shown in Figures 5.24d,e. Similar discontinuities can also be made in circular waveguide. The best reference for waveguide discontinuities and their equivalent circuits is *The Waveguide Handbook* [7].

Some typical microstrip discontinuities and transitions are shown in Figure 5.25; similar geometries exist for stripline and other printed transmission lines such as slotline, covered microstrip, coplanar waveguide, etc. Since printed transmission lines are newer, relative to waveguide, and much more difficult to analyze, more research work is needed to accurately characterize printed transmission line discontinuities; some approximate results are given in reference [8].

Modal Analysis of an *H*-Plane Step in Rectangular Waveguide

The field analysis of most discontinuity problems is very difficult, and beyond the scope of this book. The technique of modal analysis, however, is relatively straightforward and similar in principle to the reflection/transmission problems which were discussed in Chapters 2 and 3. In addition, modal analysis is a rigorous and versatile technique that can be applied to many coax, waveguide, and planar transmission line discontinuity problems, and lends itself well to computer implementation. We will present the technique of modal anlaysis by applying it to the problem of finding the equivalent circuit of an H-plane step (change in width) in rectangular waveguide.

The geometry of the H-plane step is shown in Figure 5.26. It is assumed that only the dominant TE_{10} mode is propagating in guide 1 ($z < 0$), and that such a mode is incident on the junction from $z < 0$. It is also assumed that no modes are propagating in guide 2, although the analysis to follow is still valid if propagation can occur in guide 2. From Section 4.3, the transverse components of the incident TE_{10} mode can then be written, for $z < 0$,

$$E_y^i = \sin \frac{\pi x}{a} e^{-j\beta_1^a z}, \tag{5.98a}$$

$$H_x^i = \frac{-1}{Z_1^a} \sin \frac{\pi x}{a} e^{-j\beta_1^a z}, \tag{5.98b}$$

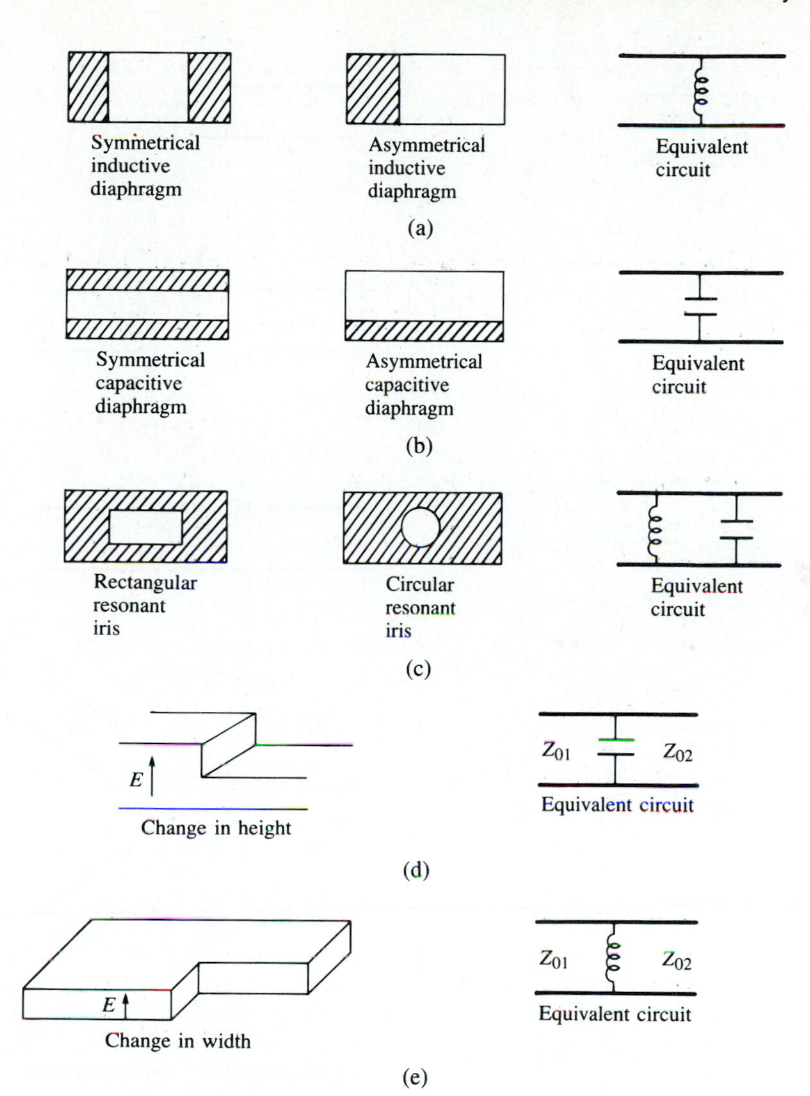

FIGURE 5.24 Rectangular waveguide discontinuities.

where
$$\beta_n^a = \sqrt{k_0^2 - \left(\frac{n\pi}{a}\right)^2} \tag{5.99}$$

is the propagation constant of the TE_{n0} mode in guide 1 (of width a), and

$$Z_n^a = \frac{k_0\eta_0}{\beta_n^a} \tag{5.100}$$

is the wave impedance of the TE_{n0} mode in guide 1. Because of the discontinuity at $z = 0$ there will be reflected and transmitted waves in both guides, consisting of infinite sets of TE_{n0} modes in guides 1 and 2. Only the TE_{10} mode will propagate in guide 1,

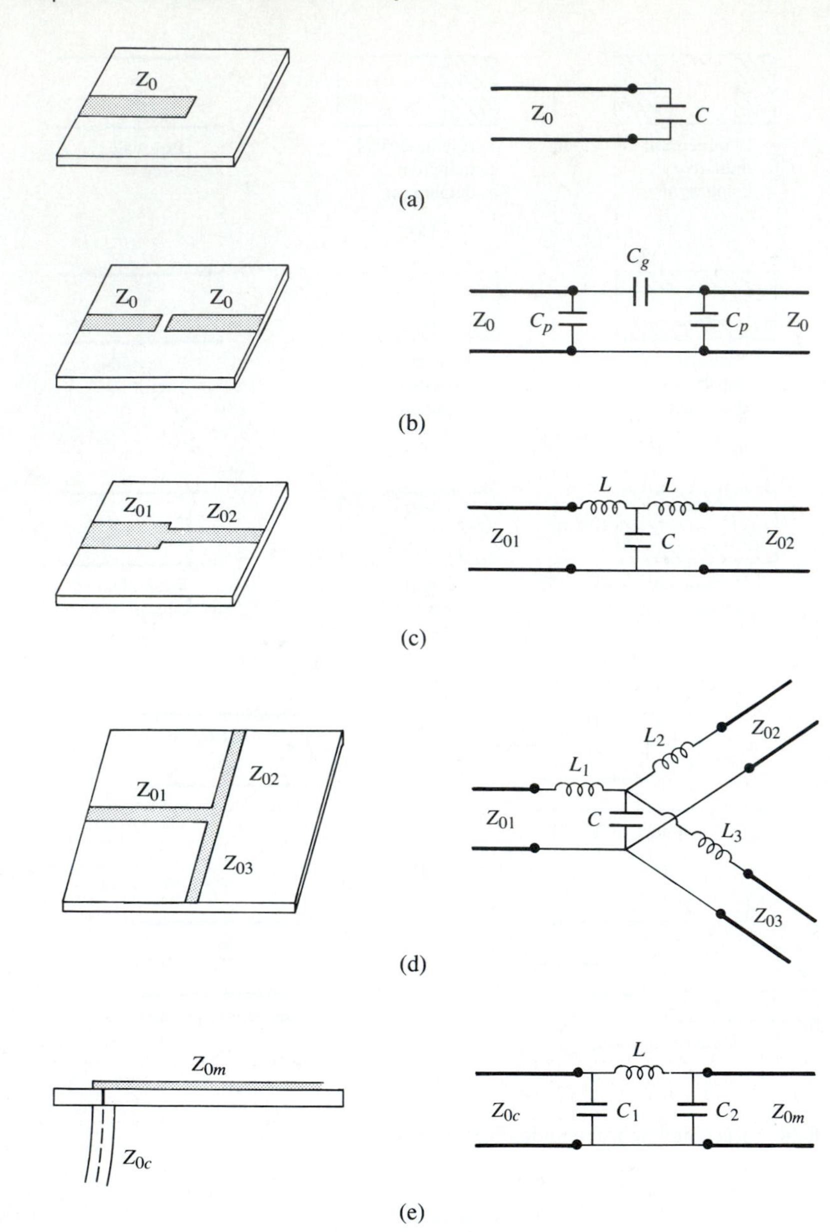

FIGURE 5.25 Some common microstrip discontinuities. (a) Open-ended microstrip. (b) Gap in microstrip. (c) Change in width. (d) T-junction. (e) Coax-to-microstrip junction.

but the higher-order modes are also important in this problem because they account for stored energy, localized near $z = 0$. Because there is no y variation introduced by this discontinuity, TE_{nm} modes for $m \neq 0$ are not excited, nor are any TM modes. A more general discontinuity, however, may excite such modes.

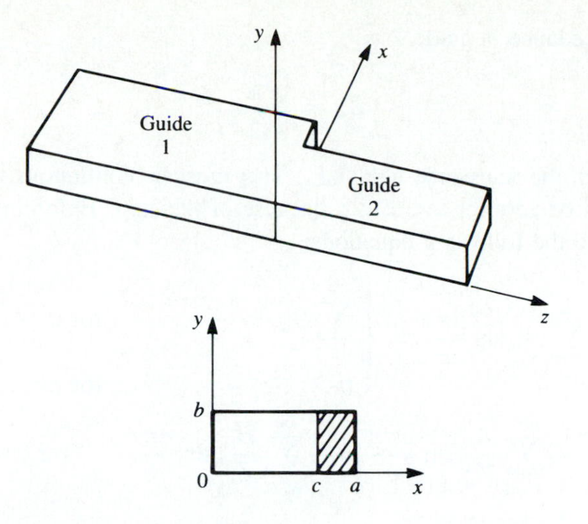

FIGURE 5.26 Geometry of a H-plane step (change in width) in rectangular waveguide.

The reflected modes in guide 1 may then be written, for $z < 0$, as

$$E_y^r = \sum_{n=1}^{\infty} A_n \sin \frac{n\pi x}{a} e^{j\beta_n^a z}, \tag{5.101a}$$

$$H_x^r = \sum_{n=1}^{\infty} \frac{A_n}{Z_n^a} \sin \frac{n\pi x}{a} e^{j\beta_n^a z}, \tag{5.101b}$$

where A_n is the unknown amplitude coefficient of the reflected TE_{n0} mode in guide 1. The reflection coefficient of the incident TE_{10} mode is then A_1. Similarly, the transmitted modes into guide 2 can be written, for $z > 0$, as

$$E_y^t = \sum_{n=1}^{\infty} B_n \sin \frac{n\pi x}{c} e^{-j\beta_n^c z}, \tag{5.102a}$$

$$H_x^t = -\sum_{n=1}^{\infty} \frac{B_n}{Z_n^c} \sin \frac{n\pi x}{c} e^{-j\beta_n^c z}, \tag{5.102b}$$

where the propagation constant in guide 2 is

$$\beta_n^c = \sqrt{k_0^2 - \left(\frac{n\pi}{c}\right)^2}, \tag{5.103}$$

and the wave impedance in guide 2 is

$$Z_n^c = \frac{k_0\eta_0}{\beta_n^c}. \tag{5.104}$$

Now at $z = 0$, the transverse fields (E_y, H_x) must be continuous for $0 < x < c$; in addition, E_y must be zero for $c < x < a$ because of the step. Enforcing these boundary conditions leads to the following equations:

$$E_y = \sin\frac{\pi x}{a} + \sum_{n=1}^{\infty} A_n \sin\frac{n\pi x}{a} = \begin{cases} \displaystyle\sum_{n=1}^{\infty} B_n \sin\frac{n\pi x}{c} & \text{for } 0 < x < c, \\ 0 & \text{for } c < x < a, \end{cases} \tag{5.105a}$$

$$H_x = \frac{-1}{Z_1^a}\sin\frac{\pi x}{a} + \sum_{n=1}^{\infty}\frac{A_n}{Z_n^a}\sin\frac{n\pi x}{a} = -\sum_{n=1}^{\infty}\frac{B_n}{Z_n^c}\sin\frac{n\pi x}{c} \quad \text{for } 0 < x < c. \tag{5.105b}$$

Equations (5.105a) and (5.105b) constitute a doubly infinite set of linear equations for the modal coefficients A_n and B_n. We will first eliminate the B_ns, and then truncate the resulting equation to a finite number of terms and solve for the A_ns.

Multiplying (5.105a) by $\sin(n\pi x/a)$, integrating from $x = 0$ to a, and using the orthogonality relations from Appendix D yields

$$\frac{a}{2}\delta_{m1} + \frac{a}{2}A_m = \sum_{n=1}^{\infty} B_n I_{nm} = \sum_{k=1}^{\infty} B_k I_{km}, \tag{5.106}$$

where

$$I_{mn} = \int_{x=0}^{c} \sin\frac{m\pi x}{c}\sin\frac{n\pi x}{a}\,dx \tag{5.107}$$

is an integral that can be easily evaluated, and

$$\delta_{mn} = \begin{cases} 1 & \text{if } m = n \\ 0 & \text{if } m \neq n \end{cases} \tag{5.108}$$

is the Kronecker delta symbol. Now solve (5.105b) for B_k by multiplying (5.105b) by $\sin(k\pi x/c)$ and integrating from $x = 0$ to c. After using the orthogonality relations, we obtain

$$\frac{-1}{Z_1^a}I_{k1} + \sum_{n=1}^{\infty}\frac{A_n}{Z_n^a}I_{kn} = \frac{-cB_k}{2Z_k^c}. \tag{5.109}$$

Substituting B_k from (5.109) into (5.106) gives an infinite set of linear equations for the A_ns, where $m = 1, 2, \ldots$,

$$\frac{a}{2}A_m + \sum_{n=1}^{\infty}\sum_{k=1}^{\infty}\frac{2Z_k^c I_{km} I_{kn} A_n}{cZ_n^a} = \sum_{k=1}^{\infty}\frac{2Z_k^c I_{km} I_{k1}}{cZ_1^a} - \frac{a}{2}\delta_{m1}. \tag{5.110}$$

For numerical calculation we can truncate the above summations to N terms, which will result in N linear equations for the first N coefficients, A_n. For example, let $N = 1$.

Then (5.110) reduces to

$$\frac{a}{2}A_1 + \frac{2Z_1^c I_{11}^2}{cZ_1^a}A_1 = \frac{2Z_1^c I_{11}^2}{cZ_1^a} - \frac{a}{2}. \tag{5.111}$$

Solving for A_1 (the reflection coefficient of the incident TE_{10} mode) gives

$$A_1 = \frac{Z_\ell - Z_1^a}{Z_\ell + Z_1^a}, \qquad \text{for } N = 1, \tag{5.112}$$

where $Z_\ell = 4Z_1^c I_{11}^2/ac$, which looks like an effective load impedance to guide 1. Accuracy is improved by using larger values of N, and leads to a set of equations which can be written in matrix form as

$$[Q][A] = [P], \tag{5.113}$$

where $[Q]$ is a square $N \times N$ matrix of coefficients,

$$Q_{mn} = \frac{a}{2}\delta_{mn} + \sum_{k=1}^{N} \frac{2Z_k^c I_{km} I_{kn}}{cZ_n^a}, \tag{5.114}$$

$[P]$ is an $N \times 1$ column vector of coefficients given by

$$P_m = \sum_{k=1}^{N} \frac{2Z_k^c I_{km} I_{k1}}{cZ_1^a} - \frac{a}{2}\delta_{m1}, \tag{5.115}$$

and $[A]$ is an $N \times 1$ column vector of the coefficients A_n. After the A_ns are found, the B_ns can be calculated from (5.109), if desired. Equations (5.113)–(5.115) lend themselves well to computer implementation.

Figure 5.27 shows the results of such a calculation. If the width, c, of guide 2 is such that all modes are cutoff (evanescent), then no real power can be transmitted into guide 2, and all the incident power is reflected back into guide 1. The evanescent fields on both sides of the discontinuity store reactive power, however, which implies that the step discontinuity and guide 2 beyond the discontinuity look like a reactance (in this case an inductive reactance) to an incident TE_{10} mode in guide 1. Thus the equivalent circuit of the H-plane step looks like an inductor at the $z = 0$ plane of guide 1, as shown in Figure 5.24e. The equivalent reactance can be found from the reflection coefficient A_1 (after solving (5.113)) as

$$X = -jZ_1^a \frac{1 + A_1}{1 - A_1}. \tag{5.116}$$

Figure 5.27 shows the normalized equivalent inductance versus the ratio of the guide widths, c/a, for a free-space wavelength $\lambda = 1.4a$ and for $N = 1, 2$, and 10 equations. The modal analysis results are compared to calculated data from reference [7]. Note that the solution converges very quickly (because of the fast exponential decay of the higher-order evanescent modes), and that the result using just two modes is very close to the data of reference [7].

The fact that the equivalent circuit of the H-plane step looks inductive is a result of the actual value of the reflection coefficient, A_1, but we can verify this result by

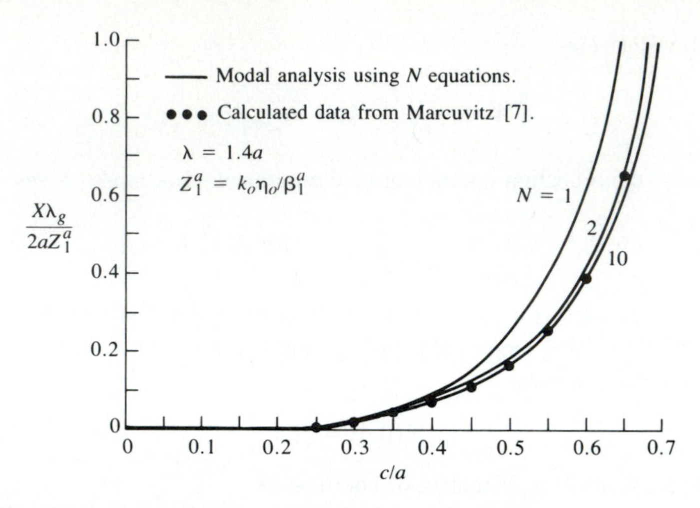

FIGURE 5.27 Equivalent inductance of H-plane asymmetric step.

computing the complex power flow into the evanescent modes on either side of the discontinuity. For example, the complex power flow into guide 2 can be found as

$$\begin{aligned}
P &= \int_{x=0}^{c}\int_{y=0}^{b} \bar{E}\times\bar{H}^*\Big|_{z=0+}\cdot\hat{z}\,dx\,dy \\
&= -b\int_{x=0}^{c} E_y H_x^* dx \\
&= -b\int_{x=0}^{c}\left[\sum_{n=1}^{\infty} B_n \sin\frac{n\pi x}{c}\right]\left[-\sum_{m=1}^{\infty}\frac{B_m^*}{Z_m^{c*}}\sin\frac{m\pi x}{c}\right]dx \\
&= \frac{bc}{2}\sum_{n=1}^{\infty}\frac{|B_n|^2}{Z_n^{c*}} \\
&= \frac{jbc}{2k_0\eta_0}\sum_{n=1}^{\infty}|B_n|^2|\beta_n^c|,
\end{aligned} \qquad 5.117$$

where the orthogonality property of the sine functions was used, as well as (5.102)–(5.104). Equation (5.117) shows that the complex power flow into guide 2 is purely inductive. A similar result can be derived for the evanescent modes in guide 1; this is left as a problem.

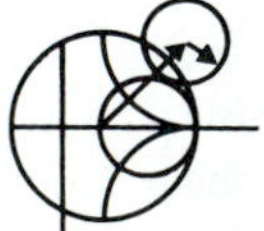

POINT OF INTEREST: Microstrip Discontinuity Compensation

Because a microstrip circuit is easy to fabricate and allows the convenient integration of passive and active components, many types of microwave circuits and subsystems are made in microstrip form. One problem with microstrip circuits (and other planar circuits), however, is that the inevitable discontinuities at bends, step changes in widths, and junctions can cause a degradation in circuit performance. This is because such discontinuities introduce parasitic reactances that can lead to phase and amplitude errors, input and output mismatch, and possibly spurious coupling. One approach for eliminating such effects is to construct an equivalent circuit for the discontinuity (perhaps by measurement), including it in the design of the circuit, and compensating for its effect by adjusting other circuit parameters (such as line lengths and characteristic impedances, or tuning stubs). Another approach is to minimize the effect of a discontinuity by compensating the discontinuity directly, often by chamfering or mitering the conductor.

Consider the case of a bend in a microstrip line. The straightforward right-angle bend shown below has a parasitic discontinuity capacitance caused by the increased conductor area near the bend. This effect could be eliminated by making a smooth, "swept" bend with a radius $r \geq 3W$, but this takes up more space. Alternatively, the right-angle bend can be compensated by mitering the corner, which has the effect of reducing the excess capacitance at the bend. As shown below, this technique can be applied to bends of arbitrary angle. The optimum value of the miter length, a, depends on the characteristic impedance and the bend angle, but a value of $a = 1.8W$ is often used in practice.

The technique of mitering can also be used to compensate step and T-junction discontinuities, as shown below.

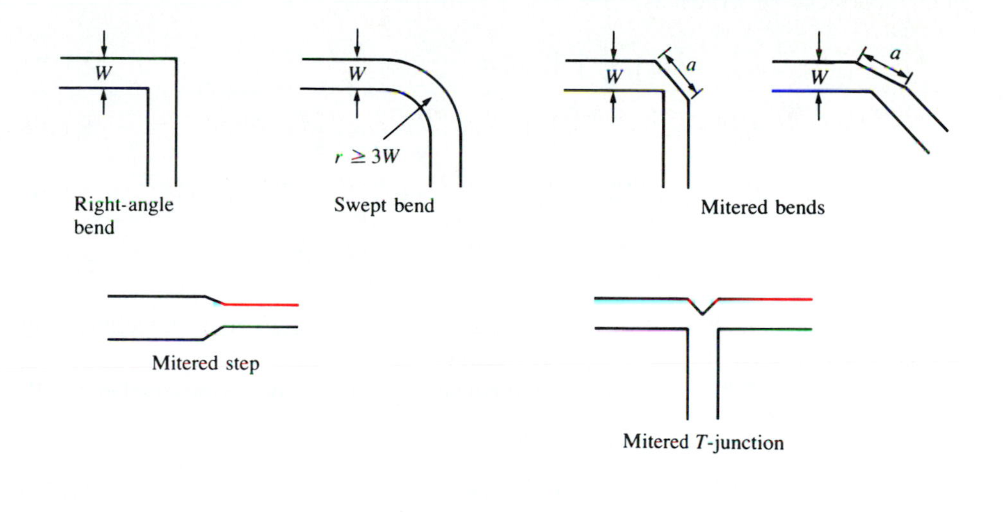

Reference: T. C. Edwards, *Foundations for Microwave Circuit Design*, Wiley, New York, 1981.

5.9 EXCITATION OF WAVEGUIDES—ELECTRIC AND MAGNETIC CURRENTS

So far we have considered the propagation, reflection, and transmission of guided waves in the absence of sources, but obviously the waveguide or transmission line must be coupled to a generator or some other source of power. For TEM or quasi-TEM lines, there is usually only one propagating mode that can be excited by a given source, although there may be reactance (stored energy) associated with a given feed. In the waveguide case, it may be possible for several propagating modes to be excited, along with evanescent modes that store energy. In this section we will develop a formalism for determining the excitation of a given waveguide mode due to an arbitrary electric or magnetic current source. This theory can then be used to find the excitation and input impedance of probe and loop feeds and, in the next section, to determine the excitation of waveguides by apertures.

Current Sheets That Excite Only One Waveguide Mode

Consider an infinitely long rectangular waveguide with a transverse sheet of electric surface current density at $z = 0$, as shown in Figure 5.28. First assume that this current has $\hat{x}$ and $\hat{y}$ components given as

$$\bar{J}_s^{\text{TE}}(x, y) = -\hat{x}\frac{2A_{mn}^{+}n\pi}{b}\cos\frac{m\pi x}{a}\sin\frac{n\pi y}{b} + \hat{y}\frac{2A_{mn}^{+}m\pi}{a}\sin\frac{m\pi x}{a}\cos\frac{n\pi y}{b}. \quad 5.118$$

We will show that such a current excites a TE_{mn} waveguide mode traveling away from the current source in both the $+z$ and $-z$ directions.

From Table 4.2, the transverse fields for positive and negative traveling TE_{mn} waveguide modes can be written as

$$E_x^{\pm} = Z_{\text{TE}}\left(\frac{n\pi}{b}\right)A_{mn}^{\pm}\cos\frac{m\pi x}{a}\sin\frac{n\pi y}{b}e^{\mp j\beta z}, \quad 5.119a$$

$$E_y^{\pm} = -Z_{\text{TE}}\left(\frac{m\pi}{a}\right)A_{mn}^{\pm}\sin\frac{m\pi x}{a}\cos\frac{n\pi y}{b}e^{\mp j\beta z}, \quad 5.119b$$

$$H_x^{\pm} = \pm\left(\frac{m\pi}{a}\right)A_{mn}^{\pm}\sin\frac{m\pi x}{a}\cos\frac{n\pi y}{b}e^{\mp j\beta z}, \quad 5.119c$$

$$H_y^{\pm} = \pm\left(\frac{n\pi}{b}\right)A_{mn}^{\pm}\cos\frac{m\pi x}{a}\sin\frac{n\pi y}{b}e^{\mp j\beta z}, \quad 5.119d$$

where the $\pm$ notation refers to waves traveling in the $+z$ direction or $-z$ direction, with amplitude coefficients A_{mn}^{+} and A_{mn}^{-}, respectively.

From (2.36) and (2.37), the following boundary conditions must be satisfied at $z = 0$:

$$(\bar{E}^{+} - \bar{E}^{-}) \times \hat{z} = 0, \quad 5.120a$$

$$\hat{z} \times (\bar{H}^{+} - \bar{H}^{-}) = \bar{J}_s. \quad 5.120b$$

Equation (5.120a) states that the transverse components of the electric field must be continuous at $z = 0$, which when applied to (5.119a) and (5.119b) gives

$$A_{mn}^{+} = A_{mn}^{-}. \quad 5.121$$

Equation (5.120b) states that the discontinuity in the transverse magnetic field is equal to the electric surface current density. Thus, the surface current density at $z = 0$ must be

$$\begin{aligned}\bar{J}_s &= \hat{y}(H_x^{+} - H_x^{-}) - \hat{x}(H_y^{+} - H_y^{-}) \\ &= -\hat{x}\frac{2A_{mn}^{+}n\pi}{b}\cos\frac{m\pi x}{a}\sin\frac{n\pi y}{b} + \hat{y}\frac{2A_{mn}^{+}m\pi}{a}\sin\frac{m\pi x}{a}\cos\frac{n\pi y}{b},\end{aligned} \quad 5.122$$

where (5.121) was used. This current is seen to be the same as the current of (5.118), which shows, by the uniqueness theorem, that such a current will excite only the TE_{mn}

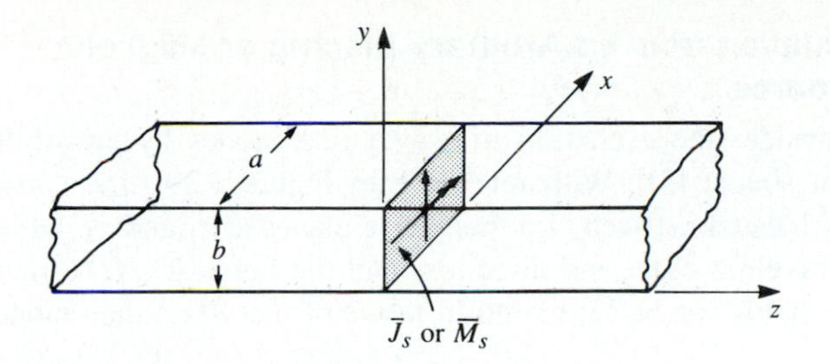

FIGURE 5.28 An infinitely long rectangular waveguide with surface current densities at $z = 0$.

mode propagating in each direction, since Maxwell's equations and all boundary conditions are satisfied.

The analogous electric current that excites only the TM_{mn} mode can be shown to be

$$\bar{J}_s^{\text{TM}}(x, y) = \hat{x}\frac{2B_{mn}^{+}m\pi}{a}\cos\frac{m\pi x}{a}\sin\frac{n\pi y}{b} + \hat{y}\frac{2B_{mn}^{+}n\pi}{b}\sin\frac{m\pi x}{a}\cos\frac{n\pi y}{b}. \qquad 5.123$$

It is left as a problem to verify that this current excites TM_{mn} modes that satisfy the appropriate boundary conditions.

Similar results can be derived for magnetic surface current sheets. From (2.36) and (2.37) the appropriate boundary conditions are

$$(\bar{E}^{+} - \bar{E}^{-}) \times \hat{z} = \bar{M}_s, \qquad 5.124a$$

$$\hat{z} \times (\bar{H}^{+} - \bar{H}^{-}) = 0. \qquad 5.124b$$

For a magnetic current sheet at $z = 0$, the TE_{mn} waveguide mode fields of (5.119) must now have continuous H_x and H_y field components, due to (5.124b). This results in the condition that

$$A_{mn}^{+} = -A_{mn}^{-}. \qquad 5.125$$

Then applying (5.124a) gives the source current as

$$\bar{M}_s^{\text{TE}} = \frac{-\hat{x}2Z_{\text{TE}}A_{mn}^{+}m\pi}{a}\sin\frac{m\pi x}{a}\cos\frac{n\pi y}{b} - \hat{y}\frac{2Z_{\text{TE}}A_{mn}^{+}n\pi}{b}\cos\frac{m\pi x}{a}\sin\frac{n\pi y}{b}. \qquad 5.126$$

The corresponding magnetic surface current that excites only the TM_{mn} mode can be shown to be

$$\bar{M}_s^{\text{TM}} = \frac{-\hat{x}2B_{mn}^{+}n\pi}{b}\sin\frac{m\pi x}{a}\cos\frac{n\pi y}{b} + \frac{\hat{y}2B_{mn}^{+}m\pi}{a}\cos\frac{m\pi x}{a}\sin\frac{n\pi y}{b}. \qquad 5.127$$

These results show that a single waveguide mode can be selectively excited, to the exclusion of all other modes, by either an electric or magnetic current sheet of the appropriate form. In practice, however, such currents are very difficult to generate, and are usually only approximated with one or two probes or loops. In this case many modes may be excited, but usually most of these modes are evanescent.

Mode Excitation from an Arbitrary Electric or Magnetic Current Source

We now consider the excitation of waveguide modes by an arbitrary electric or magnetic current source [3]. With reference to Figure 5.29, first consider an electric current source $\bar{J}$ located between two transverse planes at z_1 and z_2, which generates the fields $\bar{E}^+, \bar{H}^+$ traveling in the $+z$ direction, and the fields $\bar{E}^-, \bar{H}^-$ traveling in the $-z$ direction. These fields can be expressed in terms of the waveguide modes as follows:

$$\bar{E}^+ = \sum_n A_n^+ \bar{E}_n^+ = \sum_n A_n^+ (\bar{e}_n + \hat{z}e_{zn})e^{-j\beta_n z}, \qquad z > z_2, \tag{5.128a}$$

$$\bar{H}^+ = \sum_n A_n^+ \bar{H}_n^+ = \sum_n A_n^+ (\bar{h}_n + \hat{z}h_{zn})e^{-j\beta_n z}, \qquad z > z_2, \tag{5.128b}$$

$$\bar{E}^- = \sum_n A_n^- \bar{E}_n^- = \sum_n A_n^- (\bar{e}_n - \hat{z}e_{zn})e^{j\beta_n z}, \qquad z < z_1, \tag{5.128c}$$

$$\bar{H}^- = \sum_n A_n^- \bar{H}_n^- = \sum_n A_n^- (-\bar{h}_n + \hat{z}h_{zn})e^{j\beta_n z}, \qquad z < z_1, \tag{5.128d}$$

where the single index n is used to represent any possible TE or TM mode. For a given current $\bar{J}$, we can determine the unknown amplitude A_n^+ by using the Lorentz reciprocity theorem of (2.173) with $\bar{M}_1 = \bar{M}_2 = 0$ (since here we are only considering an electric current source),

$$\oint_S (\bar{E}_1 \times \bar{H}_2 - \bar{E}_2 \times \bar{H}_1) \cdot d\bar{s} = \int_V (\bar{E}_2 \cdot \bar{J}_1 - \bar{E}_1 \cdot \bar{J}_2) dv,$$

where S is a closed surface enclosing the volume V, and $\bar{E}_i, \bar{H}_i$ are the fields due to the current source $\bar{J}_i$ (for $i = 1$ or 2).

To apply the reciprocity theorem to the present problem, we let the volume V be the region between the waveguide walls and the transverse cross-section planes at z_1 and z_2. Then let $\bar{E}_1 = \bar{E}^\pm$ and $\bar{H}_1 = \bar{H}^\pm$, depending on whether $z \geq z_2$, or $z \leq z_1$, and let $\bar{E}_2, \bar{H}_2$ be the nth waveguide mode traveling in the negative z direction:

$$\bar{E}_2 = \bar{E}_n^- = (\bar{e}_n - \hat{z}e_{zn})e^{j\beta_n z},$$

$$\bar{H}_2 = \bar{H}_n^- = (-\bar{h}_n + \hat{z}h_{zn})e^{j\beta_n z}.$$

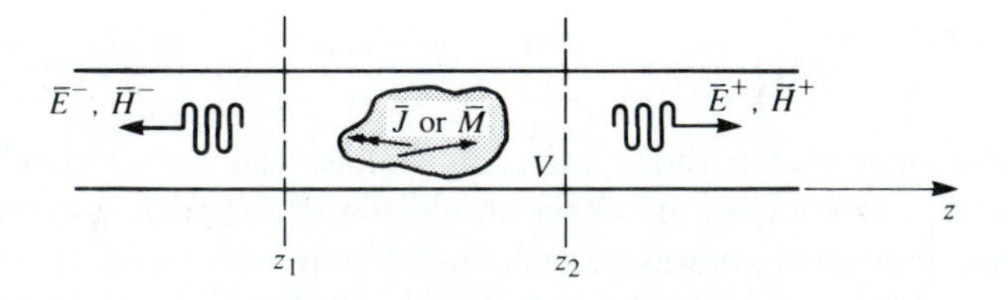

FIGURE 5.29 An arbitrary electric or magnetic current source in an infinitely long waveguide.

Substitution into the above form of the reciprocity theorem gives, with $\bar{J}_1 = \bar{J}$ and $\bar{J}_2 = 0$,

$$\oint_S (\bar{E}^\pm \times \bar{H}_n^- - \bar{E}_n^- \times \bar{H}^\pm) \cdot d\bar{s} = \int_V \bar{E}_n^- \cdot \bar{J} dv. \quad 5.129$$

The portion of the surface integral over the waveguide walls vanishes because the tangential electric field is zero there; that is, $\bar{E} \times \bar{H} \cdot \hat{z} = \bar{H} \cdot (\hat{z} \times \bar{E}) = 0$ on the waveguide walls. This reduces the integration to the guide cross-section, S_0, at the planes z_1 and z_2. In addition, the waveguide modes are orthogonal over the guide cross-section:

$$\int_{S_0} \bar{E}_m^\pm \times \bar{H}_n^\pm \cdot d\bar{s} = \int_{S_0} (\bar{e}_m \pm \hat{z} e_{zn}) \times (\pm \bar{h}_n + \hat{z} h_{zn}) \cdot \hat{z}\, ds$$

$$= \pm \int_{S_0} \bar{e}_m \times \bar{h}_n \cdot \hat{z}\, ds = 0, \qquad \text{for } m \neq n. \quad 5.130$$

Using (5.128) and (5.130) then reduces (5.129) to

$$A_n^+ \int_{z_2} (\bar{E}_n^+ \times \bar{H}_n^- - \bar{E}_n^- \times \bar{H}_n^+) \cdot d\bar{s} + A_n^- \int_{z_1} (\bar{E}_n^- \times \bar{H}_n^- - \bar{E}_n^- \times \bar{H}_n^-) \cdot d\bar{s}$$

$$= \int_V \bar{E}_n^- \cdot \bar{J} dv.$$

Since the second integral vanishes, this further reduces to

$$A_n^+ \int_{z_2} [(\bar{e}_n + \hat{z} e_{zn}) \times (-\bar{h}_n + \hat{z} h_{zn}) - (\bar{e}_n - \hat{z} e_{zn}) \times (\bar{h}_n + \hat{z} h_{zn})] \cdot \hat{z}\, ds$$

$$= -2A_n^+ \int_{z_2} \bar{e}_n \times \bar{h}_n \cdot \hat{z}\, ds = \int_V \bar{E}_n^- \cdot \bar{J}\, dv,$$

or

$$A_n^+ = \frac{-1}{P_n} \int_V \bar{E}_n^- \cdot \bar{J} dv = \frac{-1}{P_n} \int_V (\bar{e}_n - \hat{z} e_{zn}) \cdot \bar{J} e^{j\beta_n z} dv, \quad 5.131$$

where

$$P_n = 2 \int_{S_0} \bar{e}_n \times \bar{h}_n \cdot \hat{z}\, ds. \quad 5.132$$

is a normalization constant proportional to the power flow of the nth mode.

By repeating the above procedure with $\bar{E}_2 = \bar{E}_n^+$ and $\bar{H}_2 = \bar{H}_n^+$, the amplitude of the negatively traveling waves can be derived as

$$A_n^- = \frac{-1}{P_n} \int_V \bar{E}_n^+ \cdot \bar{J}\, dv = \frac{-1}{P_n} \int_V (\bar{e}_n + \hat{z} e_{zn}) \cdot \bar{J} e^{-j\beta_n z} dv. \quad 5.133$$

The above results are quite general, being applicable to any type of waveguide (including planar lines such as stripline and microstrip), where modal fields can be defined. Example 5.11 applies this theory to the problem of a probe-fed rectangular waveguide.

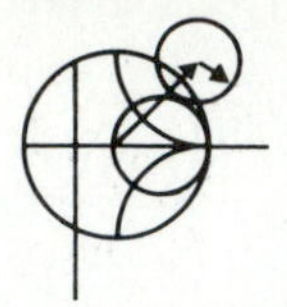

EXAMPLE 5.11

For the probe-fed rectangular waveguide shown in Figure 5.30, determine the amplitudes of the forward and backward traveling TE_{10} modes, and the input resistance seen by the probe. Assume that the TE_{10} mode is the only propagating mode.

Solution

If the current probe is assumed to have an infinitesimal diameter, the source volume current density $\bar{J}$ can be written as

$$\bar{J}(x, y, z) = I_0 \delta\left(x - \frac{a}{2}\right) \delta(z)\hat{y}, \qquad \text{for } 0 \leq y \leq b.$$

From Chapter 4 the TE_{10} modal fields can be written as

$$\bar{e}_1 = \hat{y} \sin \frac{\pi x}{a},$$

$$\bar{h}_1 = \frac{-\hat{x}}{Z_1} \sin \frac{\pi x}{a},$$

where $Z_1 = k_0\eta_0/\beta_1$ is the TE_{10} wave impedance. From (5.132) the normalization constant P_1 is,

$$P_1 = \frac{2}{Z_1} \int_{x=0}^{a} \int_{y=0}^{b} \sin^2 \frac{\pi x}{a} dx\, dy = \frac{ab}{Z_1}.$$

Then from (5.131) the amplitude A_1^+ is

$$A_1^+ = \frac{-1}{P_1} \int_V \sin \frac{\pi x}{a} e^{j\beta_1 z} I_0 \delta\left(x - \frac{a}{2}\right) \delta(z) dx\, dy\, dz = \frac{-I_0 b}{P_1} = \frac{-Z_1 I_0}{a}.$$

Similarly,

$$A_1^- = \frac{-Z_1 I_0}{a}.$$

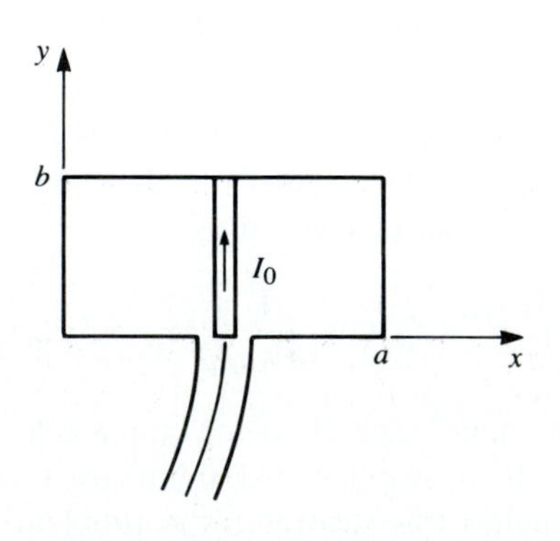

FIGURE 5.30 A uniform current probe in a rectangular waveguide.

If the TE_{10} mode is the only propagating mode in the waveguide, then this mode carries all of the average power, which can be calculated for real Z_1 as

$$\begin{aligned} P &= \frac{1}{2}\int_{S_0} \bar{E}^+ \times \bar{H}^{+*} \cdot d\bar{s} + \frac{1}{2}\int_{S_0} \bar{E}^- \times \bar{H}^{-*} \cdot d\bar{s} \\ &= \int_{S_0} \bar{E}^+ \times \bar{H}^{+*} \cdot d\bar{s} \\ &= \int_{x=0}^{a}\int_{y=0}^{b} \frac{|A_1^+|^2}{Z_1} \sin^2 \frac{\pi x}{a} dx\, dy \\ &= \frac{ab|A_1^+|^2}{2Z_1}. \end{aligned}$$

If the input resistance seen looking into the probe is R_{in}, and the terminal current is I_0, then $P = I_0^2 R_{\mathrm{in}}/2$, so that the input resistance is

$$R_{\mathrm{in}} = \frac{2P}{I_0^2} = \frac{ab|A_1^+|^2}{I_0^2 Z_1} = \frac{bZ_1}{a},$$

which is real for real Z_1 (corresponding to a propagating TE_{10} mode). ○

A similar derivation can be carried out for a magnetic current source $\bar{M}$. This source will also generate positively and negatively traveling waves which can be expressed as a superposition of waveguide modes, as in (5.128). For $\bar{J}_1 = \bar{J}_2 = 0$, the reciprocity theorem of (2.173) reduces to

$$\oint_S (\bar{E}_1 \times \bar{H}_2 - \bar{E}_2 \times \bar{H}_1) \cdot d\bar{s} = \int_V (\bar{H}_1 \cdot \bar{M}_2 - \bar{H}_2 \cdot \bar{M}_1) dv. \quad 5.134$$

By following the same procedure as for the electric current case, the excitation coefficients of the nth waveguide mode can be derived as

$$A_n^+ = \frac{1}{P_n}\int_V \bar{H}_n^- \cdot \bar{M} dv = \frac{1}{P_n}\int_V (-\bar{h}_n + \hat{z}h_{zn}) \cdot \bar{M} e^{j\beta_n z} dv, \quad 5.135$$

$$A_n^- = \frac{1}{P_n}\int_V \bar{H}_n^+ \cdot \bar{M} dv = \frac{1}{P_n}\int_V (\bar{h}_n + \hat{z}h_{zn}) \cdot \bar{M} e^{-j\beta_n z} dv, \quad 5.136$$

where P_n is defined in (5.132).

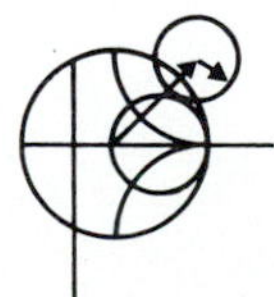

EXAMPLE 5.12

Find the excitation coefficient of the forward traveling TE_{10} mode generated by the loop in the end wall of the waveguide shown in Figure 5.31a.

Solution
By image theory, the half-loop of current I_0 on the end wall of the waveguide can be replaced by a full loop of current I_0, without the end wall, as shown in

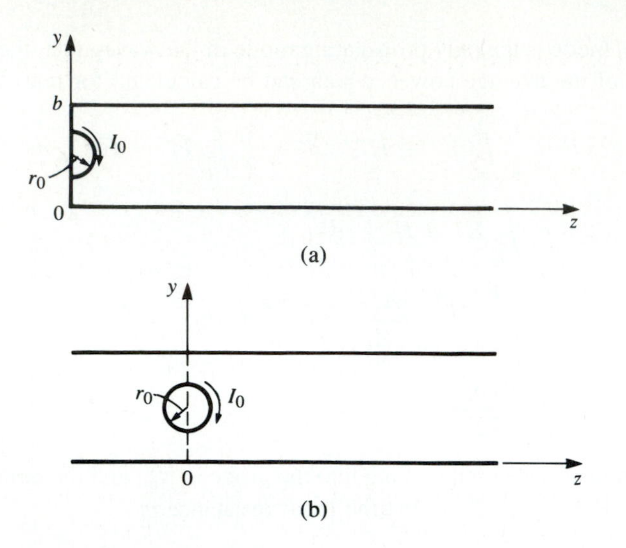

FIGURE 5.31 Application of image theory to a loop in the end wall of a rectangular waveguide. (a) Original geometry. (b) Using image theory to replace the end wall with the image of the half-loop.

Figure 5.31b. Assuming that the current loop is very small, it is equivalent to a magnetic dipole moment,

$$\bar{P}_m = \hat{x} I_0 \pi r_0^2 \delta \left(x - \frac{a}{2} \right) \delta \left(y - \frac{b}{2} \right) \delta(z).$$

Now since $\nabla \times \bar{E} = -j\omega \bar{B} - \bar{M} = -j\omega\mu_0 \bar{H} - j\omega\mu_0 \bar{P}_m - \bar{M}$, a magnetic polarization current $\bar{P}_m$ can be related to an equivalent magnetic current density $\bar{M}$ as

$$\bar{M} = j\omega\mu_0 \bar{P}_m.$$

Thus, the loop can be represented as a magnetic current density:

$$\bar{M} = \hat{x} j\omega\mu_0 I_0 \pi r_0^2 \delta \left(x - \frac{a}{2} \right) \delta \left(y - \frac{b}{2} \right) \delta(z) \text{ V/m}^2.$$

If we define the modal $\bar{h}_1$ field as

$$\bar{h}_1 = \frac{-\hat{x}}{Z_1} \sin \frac{\pi x}{a},$$

then (5.135) gives the forward wave excitation coefficient A_1^+ as

$$A_1^+ = \frac{1}{P_1} \int_V -\bar{h}_1 \cdot \bar{M} \, dv = \frac{j k_0 \eta_0 I_0 \pi r_0^2}{ab}.$$ ○

5.10 EXCITATION OF WAVEGUIDES—APERTURE COUPLING

Besides the probe and loop feeds of the previous section, waveguides and other transmission lines can also be coupled through small apertures. One common application of such coupling is in directional couplers and power dividers, where power from one guide is coupled to another guide through small apertures in a common wall. Figure 5.32 shows a variety of waveguide and other transmission line configurations where aperture coupling can be employed. We will first develop an intuitive explanation for the fact that a small aperture can be represented as an infinitesimal electric and/or an infinitesimal magnetic dipole, then we will use the results of Section 5.9 to find the fields generated by these equivalent currents. Our analysis will be somewhat phenomenological [3], [9]; a more advanced theory of aperture coupling based on the equivalence theorem can be found in reference [10].

Consider Figure 5.33a, which shows the normal electric field lines near a conducting wall (the tangential electric field is zero near the wall). If a small aperture is cut into the conductor the electric field lines will fringe through and around the aperture as shown in Figure 5.33b. Now consider Figure 5.33c, which shows the fringing field lines around two infinitesimal electric polarization currents, $\bar{P}_e$, normal to a conducting wall (without an aperture). The similarity of the field lines of Figures 5.33c and 5.33b suggests that an aperture excited by a normal electric field can be represented by two oppositely directed infinitesimal electric polarization currents, $\bar{P}_e$, normal to the closed conducting wall.

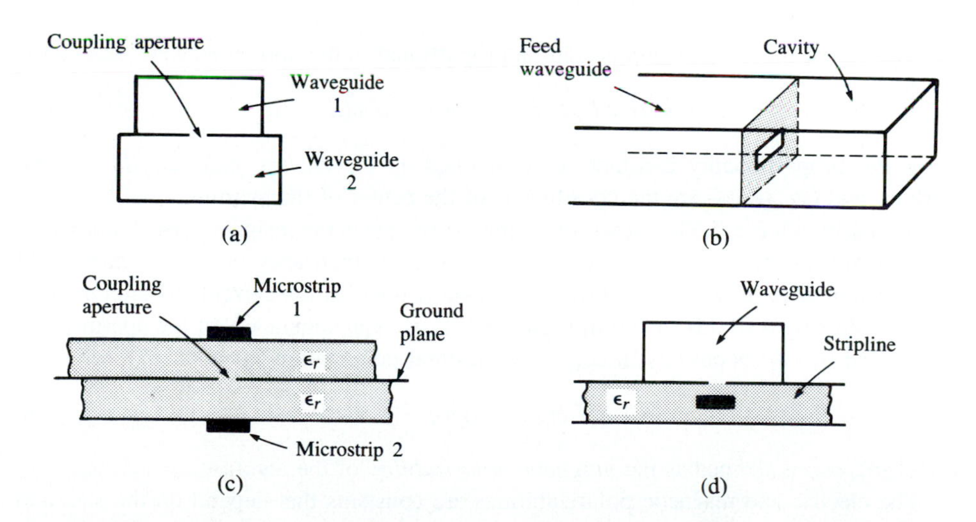

FIGURE 5.32 Various waveguide and other transmission line configurations using aperture coupling. (a) Coupling between two waveguides via an aperture in the common broad wall. (b) Coupling to a waveguide cavity via an aperture in a transverse wall. (c) Coupling between two microstrip lines via an aperture in the common ground plane. (d) Coupling from a waveguide to a stripline via an aperture.

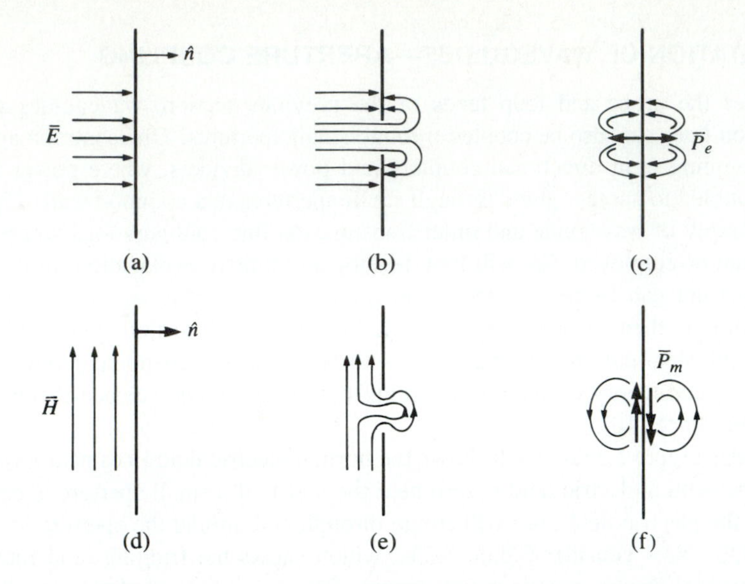

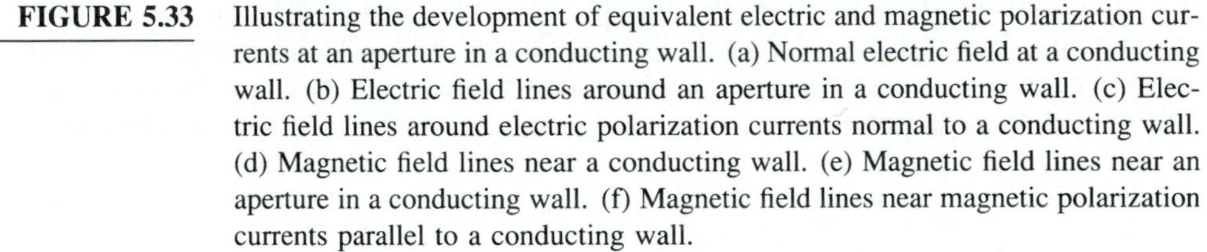
FIGURE 5.33 Illustrating the development of equivalent electric and magnetic polarization currents at an aperture in a conducting wall. (a) Normal electric field at a conducting wall. (b) Electric field lines around an aperture in a conducting wall. (c) Electric field lines around electric polarization currents normal to a conducting wall. (d) Magnetic field lines near a conducting wall. (e) Magnetic field lines near an aperture in a conducting wall. (f) Magnetic field lines near magnetic polarization currents parallel to a conducting wall.

The strength of this polarization current is proportional to the normal electric field, thus,

$$\bar{P}_e = \epsilon_0 \alpha_e \hat{n} E_n \delta(x - x_0)\delta(y - y_0)\delta(z - z_0), \tag{5.137}$$

where the proportionality constant α_e is defined as the *electric polarizability* of the aperture, and (x_0, y_0, z_0) are the coordinates of the center of the aperture.

Similarly, Figure 5.33e shows the fringing of tangential magnetic field lines (the normal magnetic field is zero at the conductor) near a small aperture. Since these field lines are similar to those produced by two magnetic polarization currents located parallel to the conducting wall, (as shown in Figure 5.33f), we can conclude that the aperture can be replaced by two oppositely directed infinitesimal polarization currents, $\bar{P}_m$, where

$$\bar{P}_m = -\alpha_m \bar{H}_t \delta(x - x_0)\delta(y - y_0)\delta(z - z_0). \tag{5.138}$$

In (5.138), α_m is defined as the *magnetic polarizability* of the aperture.

The electric and magnetic polarizabilities are constants that depend on the size and shape of the aperture, and have been derived for a variety of simple shapes [2], [9], [10]. The polarizabilities for circular and rectangular apertures, which are probably the most commonly used shapes, are given in Table 5.3.

We now show that the electric and magnetic polarization currents, $\bar{P}_e$ and $\bar{P}_m$, can be related to electric and magnetic current sources, $\bar{J}$ and $\bar{M}$, respectively. From Maxwell's

TABLE 5.3 Electric and Magnetic Polarizations

Aperture Shape	α_e	α_m
Round hole	$\frac{2r_0^3}{3}$	$\frac{4r_0^3}{3}$
Rectangular slot ($\bar{H}$ across slot)	$\frac{\pi \ell d^2}{16}$	$\frac{\pi \ell d^2}{16}$

equations (2.27a) and (2.27b) we have

$$\nabla \times \bar{E} = -j\omega\mu\bar{H} - \bar{M}, \tag{5.139a}$$

$$\nabla \times \bar{H} = j\omega\epsilon\bar{E} + \bar{J}. \tag{5.139b}$$

Then using (2.15) and (2.23), which define $\bar{P}_e$ and $\bar{P}_m$, we obtain

$$\nabla \times \bar{E} = -j\omega\mu_0\bar{H} - j\omega\mu_0\bar{P}_m - \bar{M}, \tag{5.140a}$$

$$\nabla \times \bar{H} = j\omega\epsilon_0\bar{E} + j\omega\bar{P}_e + \bar{J}. \tag{5.140b}$$

Thus, since $\bar{M}$ has the same role in these equations as $j\omega\mu_0\bar{P}_m$, and $\bar{J}$ has the same role as $j\omega\bar{P}_e$, we can define equivalent currents as

$$\bar{J} = j\omega\bar{P}_e, \tag{5.141a}$$

$$\bar{M} = j\omega\mu_0\bar{P}_m. \tag{5.141b}$$

These results then allow us to use the formulas of (5.131), (5.133), (5.135), and (5.136) to compute the fields from these currents.

The above theory is approximate because of various assumptions involved in the evaluation of the polarizabilities, but generally gives reasonable results for apertures which are small (where the term *small* implies small relative to an electrical wavelength), and not located too close to edges or corners of the guide. In addition, it is important to realize that the equivalent dipoles given by (5.137) and (5.138) radiate in the presence of the conducting wall to give the fields transmitted through the aperture. The fields on the input side of the conducting wall are also affected by the presence of the aperture, and this effect is accounted for by the equivalent dipoles on the incident side of the conductor (which are the negative of those on the output side). In this way, continuity of tangential fields is preserved across the aperture. In both cases, the presence of the (closed) conducting wall can be accounted for by using image theory to remove the wall and double the strength of the dipoles. These details will be clarified by applying this theory to apertures in transverse and broad walls of waveguides.

Coupling Through an Aperture in a Transverse Waveguide Wall

Consider a small circular aperture centered in the transverse wall of a waveguide, as shown in Figure 5.34a. Assume that only the TE_{10} mode propagates in the guide, and that such a mode is incident on the transverse wall from $z < 0$. Then, if the aperture is assumed to be closed, as in Figure 5.34b, the standing wave fields in the region $z < 0$ can be written as

$$E_y = A(e^{-j\beta z} - e^{j\beta z}) \sin \frac{\pi x}{a}, \tag{5.142a}$$

$$H_x = \frac{-A}{Z_{10}}(e^{-j\beta z} + e^{j\beta z}) \sin \frac{\pi x}{a}, \tag{5.142b}$$

where β and Z_{10} are the propagation constant and wave impedance of the TE_{10} mode. From (5.137) and (5.138) we can determine the equivalent electric and magnetic polarization currents from the above fields as

$$\bar{P}_e = \hat{z}\epsilon_0 \alpha_e E_z \delta\left(x - \frac{a}{2}\right)\delta\left(y - \frac{b}{2}\right)\delta(z) = 0, \tag{5.143a}$$

$$\begin{aligned}\bar{P}_m &= -\hat{x}\alpha_m H_x \delta(x - \frac{a}{2})\delta\left(y - \frac{b}{2}\right)\delta(z) \\ &= \hat{x}\frac{2A\alpha_m}{Z_{10}}\delta\left(x - \frac{a}{2}\right)\delta\left(y - \frac{b}{2}\right)\delta(z),\end{aligned} \tag{5.143b}$$

since $E_z = 0$ for a TE mode. Now, by (5.141b), the magnetic polarization current $\bar{P}_m$ is equivalent to a magnetic current density,

$$\bar{M} = j\omega\mu_0 \bar{P}_m = \hat{x}\frac{2j\omega\mu_0 A\alpha_m}{Z_{10}}\delta\left(x - \frac{a}{2}\right)\delta\left(y - \frac{b}{2}\right)\delta(z). \tag{5.144}$$

As shown in Figure 5.34d, the fields scattered by the aperture are considered as being produced by the equivalent currents $\bar{P}_m$ and $-\bar{P}_m$ on either side of the closed wall. The presence of the conducting wall is easily accounted for using image theory, which has the effect of doubling the dipole strengths and removing the wall, as depicted in Figure 5.34e (for $z < 0$) and Figure 5.34f (for $z > 0$). Thus the coefficients of the transmitted and reflected waves caused by the equivalent aperture currents can be found by using (5.144) in (5.135) and (5.136) to give

$$A_{10}^{+} = \frac{-1}{P_{10}}\int \bar{h}_{10} \cdot (2j\omega\mu_0 \bar{P}_m) dv = \frac{4jA\omega\mu_0\alpha_m}{abZ_{10}} = \frac{4jA\beta\alpha_m}{ab}, \tag{5.145a}$$

$$A_{10}^{-} = \frac{-1}{P_{10}}\int \bar{h}_{10} \cdot (-2j\omega\mu_0 \bar{P}_m) dv = \frac{4jA\omega\mu_0\alpha_m}{abZ_{10}} = \frac{4jA\beta\alpha_m}{ab}, \tag{5.145b}$$

since $\bar{h}_{10} = (-\hat{x}/Z_{10}) \sin(\pi x/a)$, and $P_{10} = ab/Z_{10}$. The magnetic polarizability α_m is given in Table 5.3. The complete fields can now be written as

$$E_y = [Ae^{-j\beta z} + (A_{10}^{-} - A)e^{j\beta z}] \sin \frac{\pi x}{a}, \quad \text{for } z < 0, \tag{5.146a}$$

$$H_x = \frac{1}{Z_{10}}[-Ae^{-j\beta z} + (A_{10}^{-} - A)e^{j\beta z}] \sin \frac{\pi x}{a}, \quad \text{for } z < 0, \tag{5.146b}$$

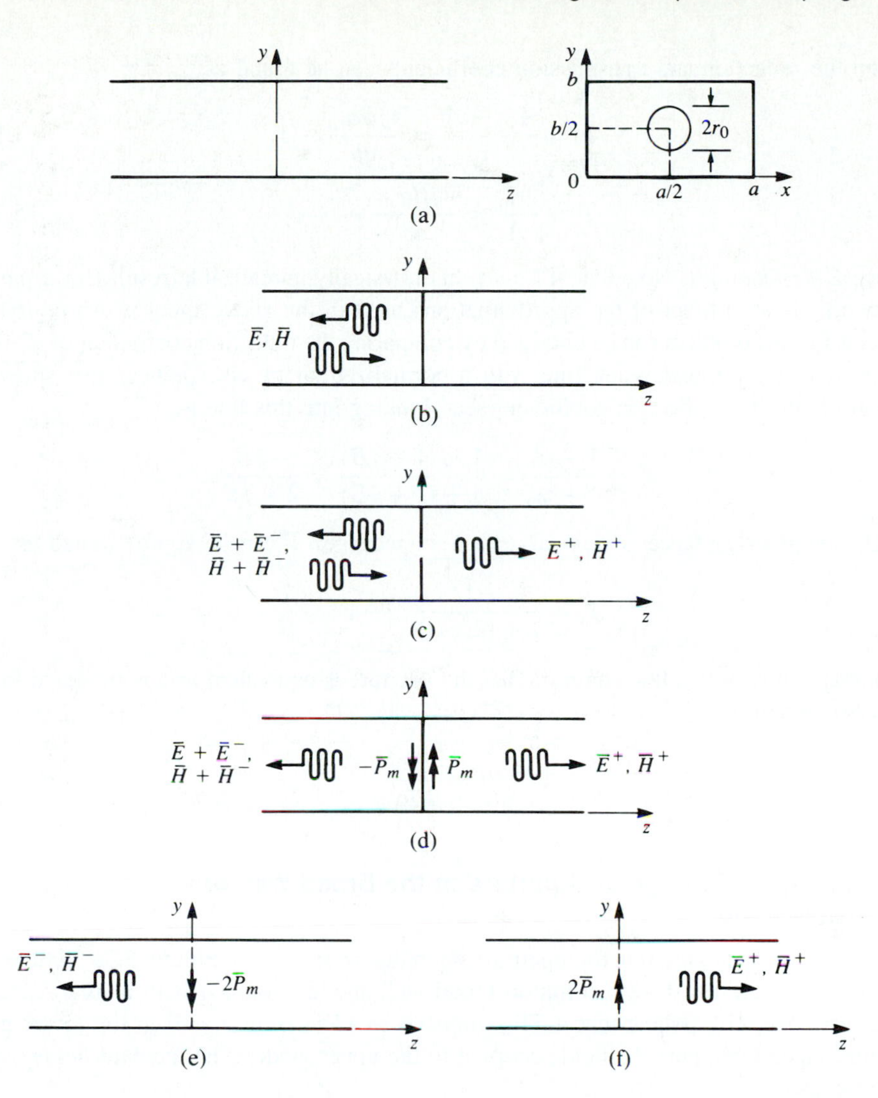

FIGURE 5.34 Applying small-hole coupling theory and image theory to the problem of an aperture in the transverse wall of a waveguide. (a) Geometry of a circular aperture in the transverse wall of a waveguide. (b) Fields with aperture closed. (c) Fields with aperture open. (d) Fields with aperture closed and replaced with equivalent dipoles. (e) Fields radiated by equivalent dipoles for $z < 0$; wall removed by image theory. (f) Fields radiated by equivalent dipoles for $z > 0$; wall removed by image theory.

and

$$E_y = A_{10}^{+} e^{-j\beta z} \sin \frac{\pi x}{a}, \qquad \text{for } z > 0, \tag{5.147a}$$

$$H_x = \frac{-A_{10}^{+}}{Z_{10}} e^{-j\beta z} \sin \frac{\pi x}{a}, \qquad \text{for } z > 0. \tag{5.147b}$$

Then the reflection and transmission coefficients can be found as

$$\Gamma = \frac{A_{10}^{-} - A}{A} = \frac{4j\beta\alpha_m}{ab} - 1, \tag{5.148a}$$

$$T = \frac{A_{10}^{+}}{A} = \frac{4j\beta\alpha_m}{ab}, \tag{5.148b}$$

since $Z_{10} = k_0\eta_0/\beta$. Note that $|\Gamma| > 1$; this physically unrealizable result (for a passive network) is an artifact of the approximations used in the above theory. An equivalent circuit for this problem can be obtained by comparing the reflection coefficient of (5.148a) with that of the transmission line with a normalized shunt susceptance, jB, shown in Figure 5.35. The reflection coefficient seen looking into this line is

$$\Gamma = \frac{1 - y_{\text{in}}}{1 + y_{\text{in}}} = \frac{1 - (1 + jB)}{1 + (1 + jB)} = \frac{-jB}{2 + jB}.$$

If the shunt susceptance is very large (low impedance), Γ can be approximated as

$$\Gamma = \frac{-1}{1 + (2/jB)} \simeq -1 - j\frac{2}{B}.$$

Comparison with (5.148a) suggests that the aperture is equivalent to a normalized inductive susceptance,

$$B = \frac{-ab}{2\beta\alpha_m}.$$

Coupling Through an Aperture in the Broad Wall of a Waveguide

Another configuration for aperture coupling is shown in Figure 5.36, where two parallel waveguides share a common broad wall and are coupled with a small centered aperture. We will assume that a TE_{10} mode is incident from $z < 0$ in the lower guide (guide 1), and compute the fields coupled to the upper guide. The incident fields can be written as

$$E_y = A \sin\frac{\pi x}{a} e^{-j\beta z}, \tag{5.149a}$$

$$H_x = \frac{-A}{Z_{10}} \sin\frac{\pi x}{a} e^{-j\beta z}. \tag{5.149b}$$

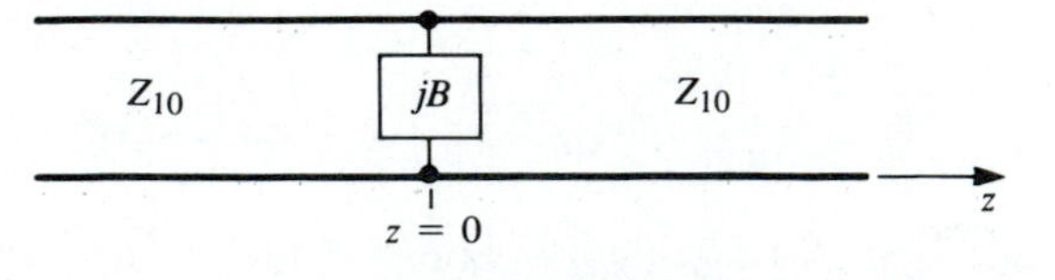

FIGURE 5.35 Equivalent circuit of the aperture in a transverse waveguide wall.

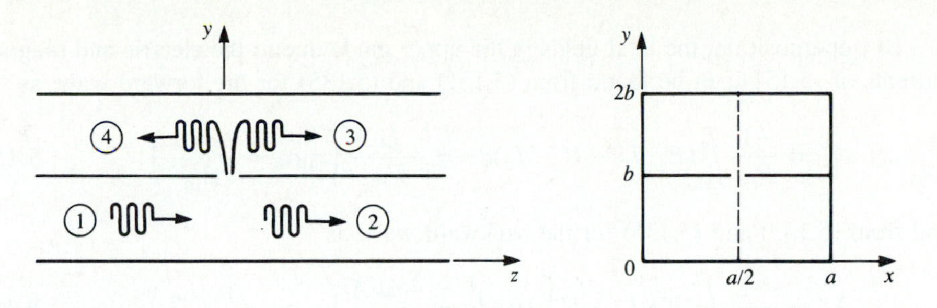

FIGURE 5.36 Two parallel waveguides coupled through an aperture in a common broad wall.

The excitation field at the center of the aperture at $(x = a/2, y = b, z = 0)$ is then

$$E_y = A, \tag{5.150a}$$

$$H_x = \frac{-A}{Z_{10}}. \tag{5.150b}$$

(If the aperture were not centered at $x = a/2$, the H_z field would be nonzero and would have to be included.)

Now from (5.137), (5.138), and (5.141), the equivalent electric and magnetic dipoles for coupling to the fields in the upper guide are

$$J_y = j\omega\epsilon_0\alpha_e A\delta\left(x - \frac{a}{2}\right)\delta(y - b)\delta(z), \tag{5.151a}$$

$$M_x = \frac{j\omega\mu_0\alpha_m A}{Z_{10}}\delta\left(x - \frac{a}{2}\right)\delta(y - b)\delta(z). \tag{5.151b}$$

Note that in this case we have excited both an electric and a magnetic dipole. Now let the fields in the upper guide be expressed as

$$E_y^- = A^- \sin\frac{\pi x}{a}e^{+j\beta z}, \qquad \text{for } z < 0, \tag{5.152a}$$

$$H_x^- = \frac{A^-}{Z_{10}}\sin\frac{\pi x}{a}e^{+j\beta z}, \qquad \text{for } z < 0, \tag{5.152b}$$

$$E_y^+ = A^+ \sin\frac{\pi x}{a}e^{-j\beta z}, \qquad \text{for } z > 0, \tag{5.153a}$$

$$H_x^+ = \frac{-A^+}{Z_{10}}\sin\frac{\pi x}{a}e^{-j\beta z}, \qquad \text{for } z > 0, \tag{5.153b}$$

where A^+, A^- are the unknown amplitudes of the forward and backward traveling waves in the upper guide, respectively.

By superposition, the total fields in the upper guide due to the electric and magnetic currents of (5.151) can be found from (5.131) and (5.135) for the forward wave as

$$A^{+} = \frac{-1}{P_{10}} \int_V (E_y^- J_y - H_x^- M_x) dv = \frac{-j\omega A}{P_{10}} \left(\epsilon_0 \alpha_e - \frac{\mu_0 \alpha_m}{Z_{10}^2} \right), \tag{5.154a}$$

and from (5.133) and (5.136) for the backward wave as

$$A^{-} = \frac{-1}{P_{10}} \int_V (E_y^+ J_y - H_x^+ M_x) dv = \frac{-j\omega A}{P_{10}} \left(\epsilon_0 \alpha_e + \frac{\mu_0 \alpha_m}{Z_{10}^2} \right), \tag{5.154b}$$

where $P_{10} = ab/Z_{10}$. Note that the electric dipole excites the same fields in both directions, but the magnetic dipole excites oppositely polarized fields in the forward and backward directions.

REFERENCES

[1] S. Ramo, T. R. Whinnery, and T. van Duzer, *Fields and Waves in Communication Electronics*, John Wiley & Sons, N.Y., 1965.

[2] C. G. Montgomery, R. H. Dicke, and E. M. Purcell, *Principles of Microwave Circuits*, vol. 8 of MIT Rad. Lab. Series, McGraw-Hill, N.Y., 1948.

[3] R. E. Collin, *Foundations for Microwave Engineering*, McGraw-Hill, N.Y., 1966.

[4] A. A. Oliner, "Historical Perspectives on Microwave Field Theory," *IEEE Trans. Microwave Theory and Techniques*, vol. MTT-32, pp. 1022–1045, September 1984.

[5] G. Gonzalez, *Microwave Transistor Amplifiers*, Prentice-Hall, N.J., 1984.

[6] J. S. Wright, O. P. Jain, W. J. Chudobiak, and V. Makios, "Equivalent Circuits of Microstrip Impedance Discontinuities and Launchers," *IEEE Trans. Microwave Theory and Techniques*, vol. MTT-22, pp. 48–52, January 1974.

[7] N. Marcuvitz, *Waveguide Handbook*, vol. 10 of MIT Rad. Lab. Series, McGraw-Hill, N.Y., 1948.

[8] K. C. Gupta, R. Garg, and I. J. Bahl, *Microstrip Lines and Slotlines*, Artech House, Dedham, Mass., 1979.

[9] G. Matthaei, L. Young, and E. M. T. Jones, *Microwave Filters, Impedance-Matching Networks, and Coupling Structures*, Chapter 5. Artech House, Dedham, Mass., 1980.

[10] R. E. Collin, *Field Theory of Guided Waves*, McGraw-Hill, N.Y., 1960.

PROBLEMS

5.1 Solve the problem of Example 5.2 by writing expressions for the incident, reflected, and transmitted E_y and H_x fields for the regions $z < 0$ and $z > 0$, and applying the boundary conditions for these fields at the dielectric interface at $z = 0$.

5.2 Consider the reflection of a TE_{10} mode, incident from $z < 0$, at a step change in the height of a rectangular waveguide, as shown below. Show that if the method of Example 5.2 is used, the result $\Gamma = 0$ is obtained. Do you think this is the correct solution? Why? (This problem shows that the one-mode impedance viewpoint does not always provide a correct analysis.)

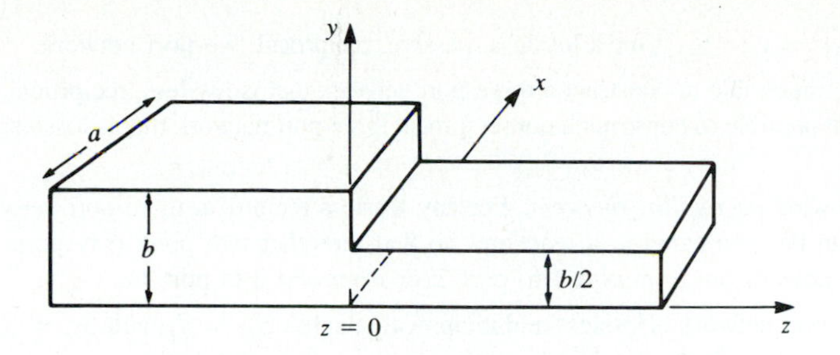

5.3 Consider a series RLC circuit with a current, I. Calculate the power lost and the stored electric and magnetic energies, and show that the input impedance can be expressed as in (5.17).

5.4 Verify that the slope of the reactance of a series LC circuit is given by (5.24).

5.5 Show that the input impedance, Z, of a parallel RLC circuit satisfies the condition that $Z(-\omega) = Z^*(\omega)$.

5.6 Show that the admittance matrix of a lossless N-port network has purely imaginary elements.

5.7 Does a nonreciprocal lossless network always have a purely imaginary impedance matrix?

5.8 Derive the $[Z]$ and $[Y]$ matrices for the two-port networks shown below.

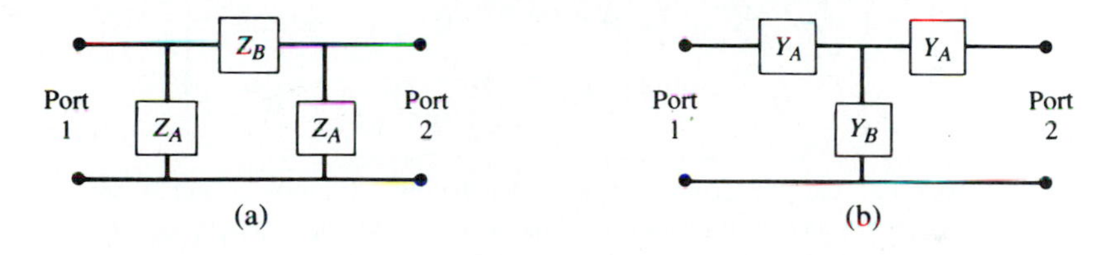

5.9 Consider a two-port network, and let $Z_{SC}^{(1)}$, $Z_{SC}^{(2)}$, $Z_{OC}^{(1)}$, $Z_{OC}^{(2)}$ be the input impedance seen when port 2 is short-circuited, when port 1 is short-circuited, when port 2 is open-circuited, and when port 1 is open-circuited, respectively. Show that the impedance matrix elements are given by

$$Z_{11} = Z_{OC}^{(1)}, \qquad Z_{22} = Z_{OC}^{(2)}, \qquad Z_{12}^2 = Z_{21}^2 = (Z_{OC}^{(1)} - Z_{SC}^{(1)})Z_{OC}^{(2)}$$

5.10 A two-port network is driven at both ports such that the port voltages and currents have the following values:

$$V_1 = 10\angle 0, \qquad I_1 = 0.1\angle 40^\circ,$$

$$V_2 = 12\angle 30^\circ, \qquad I_2 = 0.15\angle 100^\circ.$$

Determine the incident and reflected voltages at both ports, if the characteristic impedance is 50 Ω.

5.11 Derive the scattering matrix for a two-port network consisting of a length ℓ of transmission line with characteristic impedance Z_0.

5.12 Consider two two-port networks with individual scattering matrices, $[S^A]$ and $[S^B]$. Show that the overall S_{21} parameter of the cascade of these two networks is given by

$$S_{21} = \frac{S_{21}^A S_{21}^B}{1 - S_{22}^A S_{11}^B}.$$

5.13 Prove that $|S_{21}|^2 = 1 - |S_{11}|^2$ for a lossless, passive, reciprocal two-port network.

5.14 Show that it is impossible to construct a three-port network that is lossless, reciprocal, and matched at all ports. Is it possible to construct a nonreciprocal three-port network that is lossless and matched at all ports?

5.15 Prove the following *decoupling theorem*: For any lossless reciprocal three-port network, one port (say port 3) can be terminated in a reactance so that the other two ports (say ports 1 and 2) are decoupled (no power flow from port 1 to port 2, or from port 2 to port 1).

5.16 A certain three-port network is lossless and reciprocal, and has $S_{13} = S_{23}$ and $S_{11} = S_{22}$. Show that if port 2 is terminated with a matched load, then port 1 can be matched by placing an appropriate reactance at port 3.

5.17 A four-port network has the scattering matrix shown below. (a) Is this network lossless? (b) Is this network reciprocal? (c) What is the return loss at port 1 when all other ports are matched? (d) What is the insertion loss and phase between ports 2 and 4, when all other ports are matched? (e) What is the reflection coefficient seen at port 1 if a short circuit is placed at the terminal plane of port 3, and all other ports are matched?

$$[S] = \begin{bmatrix} 0.1\angle 90^\circ & \frac{1}{\sqrt{2}}\angle -45^\circ & \frac{1}{\sqrt{2}}\angle 45^\circ & 0 \\ \frac{1}{\sqrt{2}}\angle -45^\circ & 0 & 0 & \frac{1}{\sqrt{2}}\angle 45^\circ \\ \frac{1}{\sqrt{2}}\angle -45^\circ & 0 & 0 & \frac{1}{\sqrt{2}}\angle -45^\circ \\ 0 & \frac{1}{\sqrt{2}}\angle 45^\circ & \frac{1}{\sqrt{2}}\angle -45^\circ & 0 \end{bmatrix}.$$

5.18 A four-port network has the scattering matrix shown below. If ports 3 and 4 are connected together with a lossless matched transmission line with an electrical length of 100°, find the resulting insertion loss and phase between ports 1 and 2.

$$[S] = \begin{bmatrix} 0.6\angle 90^\circ & 0 & 0 & 0.8\angle 0 \\ 0 & 0.707\angle 45^\circ & 0.707\angle -45^\circ & 0 \\ 0 & 0.707\angle -45^\circ & 0.707\angle 45^\circ & 0 \\ 0.8\angle 0 & 0 & 0 & 0.6\angle 90^\circ \end{bmatrix}.$$

5.19 Consider a two-port network consisting of a junction of two transmission lines with characteristic impedances Z_{01} and Z_{02}, as shown below. Find the generalized scattering parameters of this network.

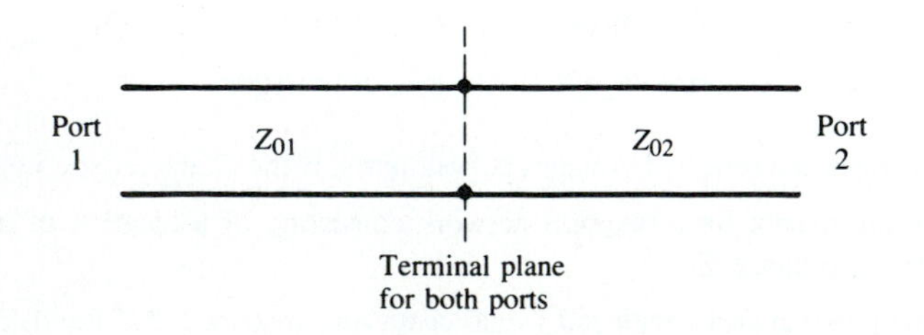

5.20 The scattering parameters of a certain two-port network were measured to be

$$S_{11} = 0.3 + j0.7, \qquad S_{12} = S_{21} = j0.6, \qquad S_{22} = 0.3 - j0.7.$$

Find the equivalent impedance parameters for this network, if the characteristic impedance is 50 Ω.

5.21 When normalized to a characteristic impedance Z_0, a certain two-port network has scattering parameters S_{ij}. This network is now placed in a circuit as shown below. Find the new (generalized) scattering parameters S'_{ij}, relative to the characteristic impedances Z_{01} and Z_{02}, in terms of S_{ij}.

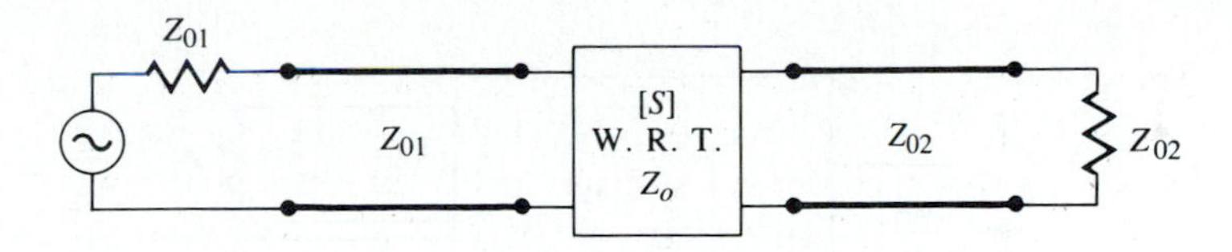

5.22 Find the impedance parameters of a section of transmission line with length ℓ, characteristic impedance Z_0, and propagation constant β.

5.23 The $ABCD$ parameters of the first entry in Table 5.1 were derived in Example 5.6. Verify the $ABCD$ parameters for the second, third, and fourth entries.

5.24 Derive expressions that give the impedance parameters in terms of the $ABCD$ parameters.

5.25 Use $ABCD$ matrices to find the voltage V_L across the load resistor in the circuit shown below.

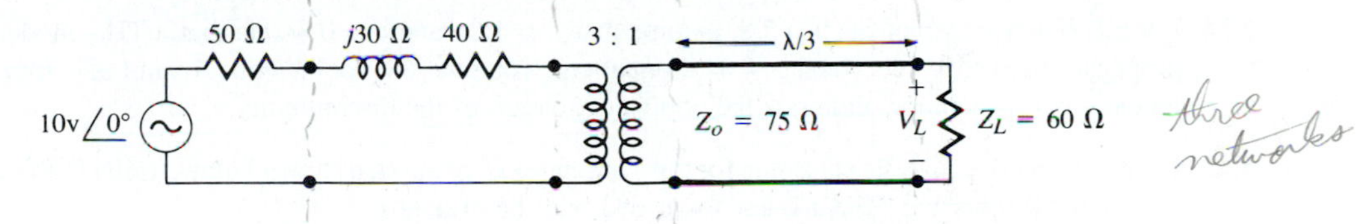

5.26 Find the $ABCD$ matrix for the shunt-connected transmission lines shown below, and the length and characteristic impedance of a single equivalent transmission line. Simplify the result for the case where $\theta_1 = \theta_2$.

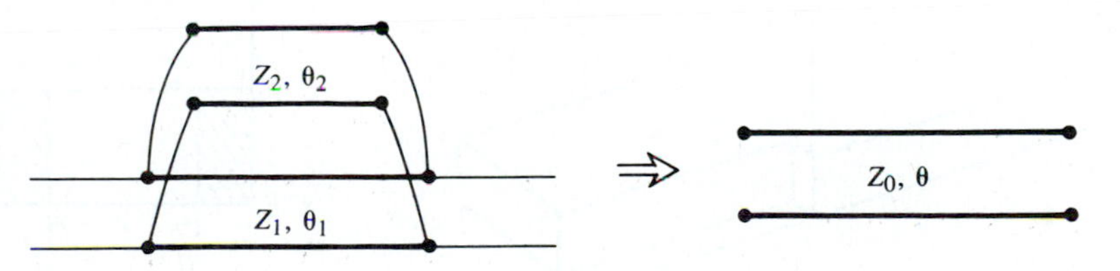

5.27 Derive the expressions for the equivalent circuit parameters given in Figure 5.13a.

5.28 Derive the expressions for S parameters in terms of the $ABCD$ parameters, as given in Table 5.2.

5.29 Derive the expressions for Z parameters in terms of the S parameters, as given in Table 5.2.

5.30 Consider the microwave network shown below, consisting of a 50 Ω source, a 50 Ω, 3 dB matched attenuator, and a 50 Ω load. Compute the available power gain, the transducer power gain, and the actual power gain. How do these gains change if the load is changed to 25 Ω?

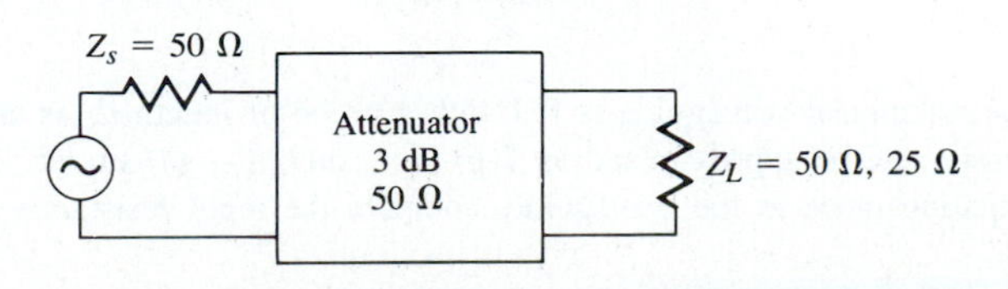

5.31 Find the signal flow graph for a matched length of lossless transmission line, with an electrical length $\beta\ell$.

5.32 Use signal flow graphs to find the power ratios P_2/P_1 and P_3/P_1 for the mismatched three-port network shown below.

P_1 → Port 1

$$[S] = \begin{bmatrix} 0 & S_{12} & 0 \\ S_{12} & 0 & S_{23} \\ 0 & S_{23} & 0 \end{bmatrix}$$

Port 2 Γ_2 P_2

Port 3 Γ_3 P_3

5.33 For the H-plane step analysis of Section 5.8, compute the complex power flow in the reflected modes in guide 1, and show that the reactive power is inductive.

5.34 For the H-plane step of Section 5.8, assume that $\lambda = 1.2a$ and $c = 0.8a$, so that a TE_{10} mode can propagate in each guide. Using $N = 2$ equations, compute the coefficients A_1 and A_2 from the modal analysis solution and draw the equivalent circuit of the discontinuity.

5.35 Derive the modal analysis equations for the symmetric H-plane step shown below. (HINT: Because of symmetry, only the TE_{n0} modes, for n odd, will be excited.)

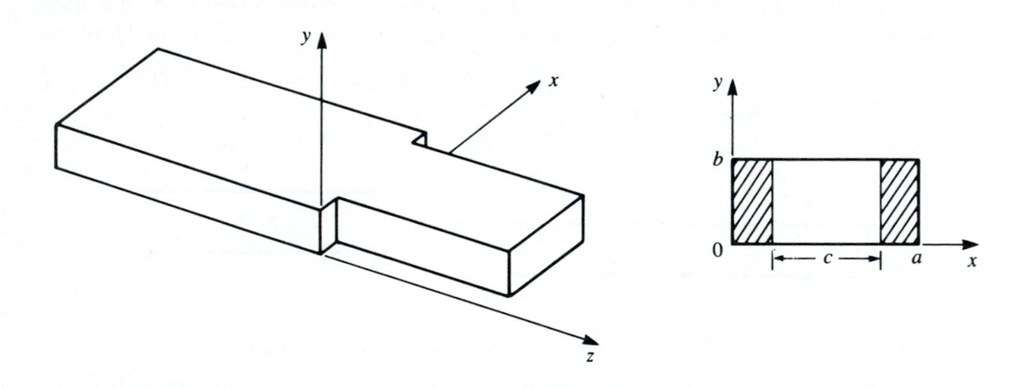

5.36 Find the transverse $\bar{E}$ and $\bar{H}$ fields excited by the current of (5.123) by postulating traveling TM_{mn} modes on either side of the source at $z = 0$, and applying the appropriate boundary conditions.

5.37 Show that the magnetic surface current density of (5.127) excites TM_{mn} waves traveling away from the source.

5.38 An infinitely long rectangular waveguide is fed with a probe of length d, as shown below. The current on this probe can be approximated as $I(y) = I_0 \sin k(d - y)/\sin kd$. If the TE_{10} mode is the only propagating mode in the waveguide, compute the input resistance seen at the probe terminals.

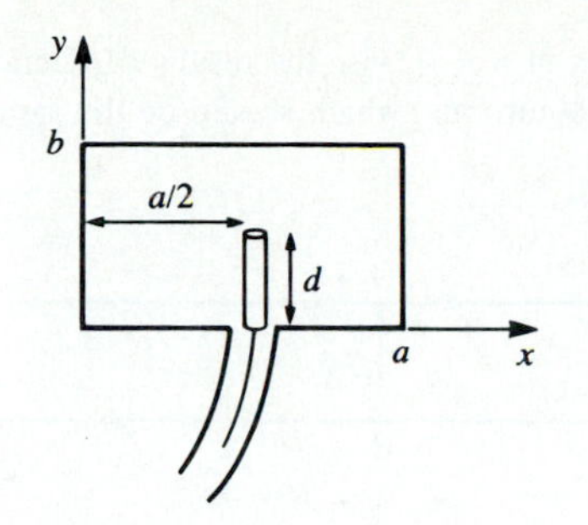

5.39 Consider the infinitely long waveguide fed with two probes driven 180° out of phase, as shown below. What are the resulting excitation coefficients for the TE_{10} and TE_{20} modes? What other modes can be excited by this feeding arrangement?

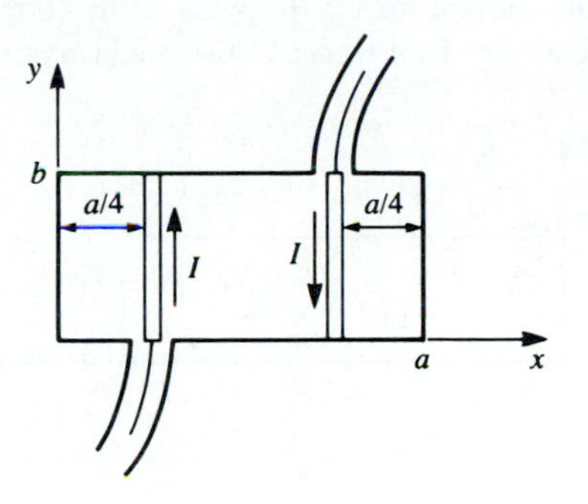

5.40 Beginning with equation (5.134), derive the modal excitation coefficients for a magnetic current source, as given by (5.135) and (5.136).

5.41 Consider a small current loop on the side wall of a rectangular waveguide, as shown below. Find the TE_{10} fields excited by this loop, if the loop is of radius r_0.

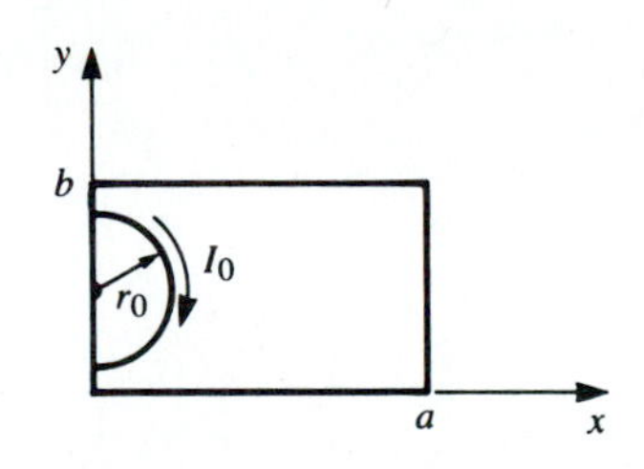

5.42 A rectangular waveguide is shorted at $z = 0$, and has an electric current sheet, J_{sy}, located at $z = d$, where

$$J_{sy} = \frac{2\pi A}{a} \sin \frac{\pi x}{a}.$$

Find expressions for the fields generated by this current by assuming standing wave fields for $0 < z < d$, and traveling wave fields for $z > d$, and applying boundary conditions at $z = 0$ and $z = d$. Now solve the problem using image theory, by placing a current sheet $-J_{sy}$ at $z = -d$, and removing the shorting wall at $z = 0$. Use the results of Section 5.9 and superposition to find the fields radiated by these two currents, which should be the same as the first results for $z > 0$.

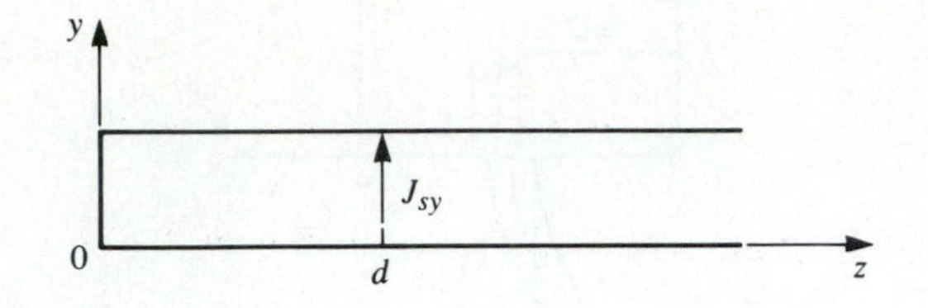

5.43 A rectangular waveguide is shorted at $z = 0$, and has a magnetic current sheet, M_{sx}, located at $z = d$, where

$$M_{sx} = \frac{2\pi B}{a} \sin \frac{\pi x}{a}.$$

Find the fields radiated by this current for $z > 0$. What is the correct image current that should be placed at $z = -d$, to account for the presence of the conducting wall at $z = 0$?

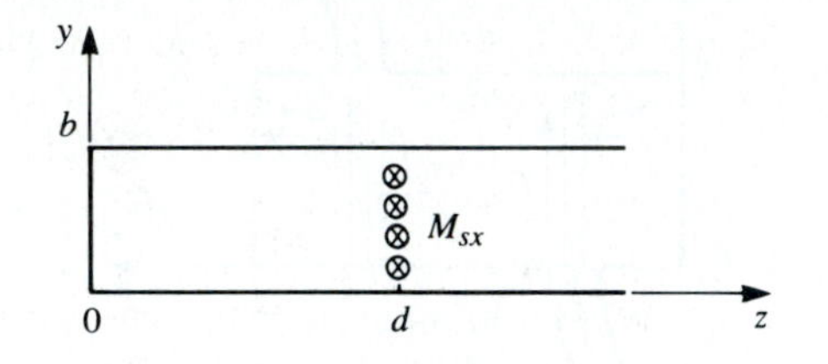

CHAPTER 6

Impedance Matching and Tuning

This chapter marks a turning point in that we now begin to apply the theory and techniques of the previous chapters to practical problems in microwave engineering. We begin with the topic of impedance matching, which is often a part of the larger design process for a microwave component or system. The basic idea of impedance matching is illustrated in Figure 6.1, which shows an impedance matching network placed between a load impedance and a transmission line. The matching network is ideally lossless, to avoid unnecessary loss of power, and is usually designed so that the impedance seen looking into the matching network is Z_0. Then reflections are eliminated on the transmission line to the left of the matching network, although there will be multiple reflections between the matching network and the load. This procedure is also referred to as tuning. Impedance matching or tuning is important for the following reasons:

- Maximum power is delivered when the load is matched to the line (assuming the generator is matched), and power loss in the feed line is minimized.
- Impedance matching sensitive receiver components (antenna, low-noise amplifier, etc.) improves the signal-to-noise ratio of the system.
- Impedance matching in a power distribution network (such as an antenna array feed network) will reduce amplitude and phase errors.

As long as the load impedance, Z_L, has some nonzero real part, a matching network can always be found. Many choices are available, however, and we will discuss the design and performance of several types of practical matching networks. Factors that may be important in the selection of a particular matching network include the following:

- *Complexity*—As with most engineering solutions, the simplest design that satisfies the required specifications is generally the most preferable. A simpler matching

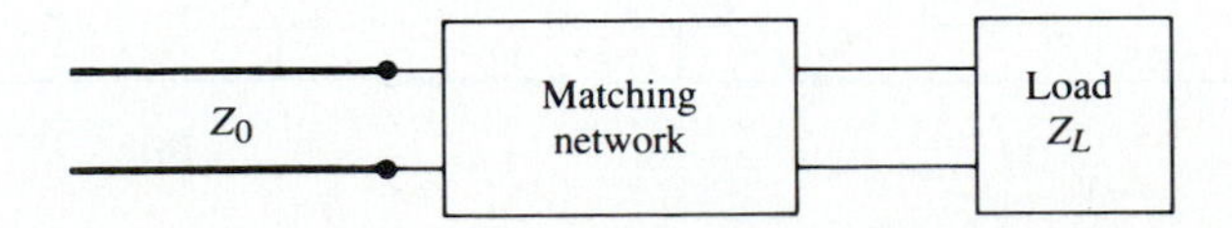

FIGURE 6.1 A lossless network matching an arbitrary load impedance to a transmission line.

network is usually cheaper, more reliable, and less lossy than a more complex design.

- *Bandwidth*—Any type of matching network can ideally give a perfect match (zero reflection) at a single frequency. In many applications, however, it is desirable to match a load over a band of frequencies. There are several ways of doing this with, of course, a corresponding increase in complexity.
- *Implementation*—Depending on the type of transmission line or waveguide being used, one type of matching network may be preferable compared to another. For example, tuning stubs are much easier to implement in waveguide than are multisection quarter-wave transformers.
- *Adjustability*—In some applications the matching network may require adjustment to match a variable load impedance. Some types of matching networks are more amenable than others in this regard.

6.1 MATCHING WITH LUMPED ELEMENTS (*L NETWORKS*)

Probably the simplest type of matching network is the L section, which uses two reactive elements to match an arbitrary load impedance to a transmission line. There are two possible configurations for this network, as shown in Figure 6.2. If the normalized load impedance, $z_L = Z_L/Z_0$, is inside the $1 + jx$ circle on the Smith chart, then the circuit of Figure 6.2a should be used. If the normalized load impedance is outside the $1 + jx$ circle on the Smith chart, the circuit of Figure 6.2b should be used. The $1 + jx$ circle is the resistance circle on the impedance Smith chart for which $r = 1$.

In either of the configurations of Figure 6.2, the reactive elements may be either inductors or capacitors, depending on the load impedance. Thus, there are eight distinct possibilities for the matching circuit for various load impedances. If the frequency is low enough and/or the circuit size is small enough, actual lumped-element capacitors and inductors can be used. This may be feasible for frequencies up to about 1 GHz or so, although modern microwave integrated circuits may be small enough so that lumped elements can be used at higher frequencies as well. There is, however, a large range of frequencies and circuit sizes where lumped elements may not be realizable. This is a limitation of the L section matching technique.

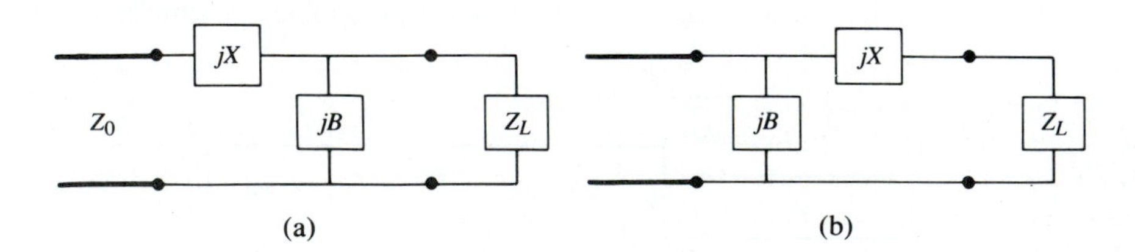

FIGURE 6.2 L section matching networks. (a) Network for z_L inside the $1 + jx$ circle. (b) Network for z_L outside the $1 + jx$ circle.

We will now derive the analytic expressions for the matching network elements of the two cases in Figure 6.2, then illustrate an alternative design procedure using the Smith chart.

Analytic Solutions

Although we will discuss a simple graphical solution using the Smith chart, it may be useful to derive expressions for the L section matching network components. Such expressions would be useful in a computer-aided design program for L section matching, or when it is necessary to have more accuracy than the Smith chart can provide.

Consider first the circuit of Figure 6.2a, and let $Z_L = R_L + jX_L$. We stated that this circuit would be used when $z_L = Z_L/Z_0$ is inside the $1 + jx$ circle on the Smith chart, which implies that $R_L > Z_0$ for this case.

The impedance seen looking into the matching network followed by the load impedance must be equal to Z_0, for a match:

$$Z_0 = jX + \frac{1}{jB + 1/(R_L + jX_L)}. \tag{6.1}$$

Rearranging and separating into real and imaginary parts gives two equations for the two unknowns, X and B:

$$B(XR_L - X_L Z_0) = R_L - Z_0, \tag{6.2a}$$

$$X(1 - BX_L) = BZ_0 R_L - X_L. \tag{6.2b}$$

Solving (6.2a) for X and substituting into (6.2b) gives a quadratic equation for B. The solution is

$$B = \frac{X_L \pm \sqrt{R_L/Z_0}\sqrt{R_L^2 + X_L^2 - Z_0 R_L}}{R_L^2 + X_L^2}. \tag{6.3a}$$

Note that since $R_L > Z_0$, the argument of the second square root is always positive. Then the series reactance can be found as

$$X = \frac{1}{B} + \frac{X_L Z_0}{R_L} - \frac{Z_0}{BR_L}. \tag{6.3b}$$

Equation (6.3a) indicates that two solutions are possible for B and X. Both of these solutions are physically realizable, since both positive and negative values of B and X are possible (positive X implies an inductor, negative X implies a capacitor, while positive B implies a capacitor and negative B implies an inductor.) One solution, however, may result in significantly smaller values for the reactive components, and may be the preferred solution if the bandwidth of the match is better, or the SWR on the line between the matching network and the load is smaller.

Now consider the circuit of Figure 6.2b. This circuit is to be used when z_L is outside the $1 + jx$ circle on the Smith chart, which implies that $R_L < Z_0$. The admittance seen looking into the matching network followed by the load impedance $Z_L = R_L + jX_L$

must be equal to $1/Z_0$, for a match:

$$\frac{1}{Z_0} = jB + \frac{1}{R_L + j(X + X_L)}. \tag{6.4}$$

Rearranging and separating into real and imaginary parts gives two equations for the two unknowns, X and B:

$$BZ_0(X + X_L) = Z_0 - R_L, \tag{6.5a}$$

$$(X + X_L) = BZ_0 R_L. \tag{6.5b}$$

Solving for X and B gives

$$X = \pm\sqrt{R_L(Z_0 - R_L)} - X_L, \tag{6.6a}$$

$$B = \pm\frac{\sqrt{(Z_0 - R_L)/R_L}}{Z_0}. \tag{6.6b}$$

Since $R_L < Z_0$, the arguments of the square roots are always positive. Again, note that two solutions are possible.

In order to match an arbitrary complex load to a line of characteristic impedance Z_0, the real part of the input impedance to the matching network must be Z_0, while the imaginary part must be zero. This implies that a general matching network must have at least two degrees of freedom; in the L section matching circuit these two degrees of freedom are provided by the values of the two reactive components.

Smith Chart Solutions

Instead of the above formulas, the Smith chart can be used to quickly and accurately design L section matching networks, a procedure best illustrated by an example.

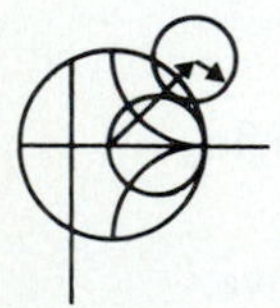

EXAMPLE 6.1

Design an L section matching network to match a series RC load with an impedance $Z_L = 200 - j100\ \Omega$, to a $100\ \Omega$ line, at a frequency of 500 MHz.

Solution

The normalized load impedance is $z_L = 2 - j1$, which is plotted on the Smith chart of Figure 6.3a. This point is inside the $1 + jx$ circle, so we will use the matching circuit of Figure 6.2a. Since the first element from the load is a shunt susceptance, it makes sense to convert to admittance by drawing the SWR circle through the load, and a straight line from the load through the center of the chart, as shown in Figure 6.3a. Now, after we add the shunt susceptance and convert back to impedance, we want to be on the $1 + jx$ circle, so that we can add a series reactance to cancel the jx and match the load. This means that the shunt susceptance must move us from y_L to the $1 + jx$ circle on the *admittance* Smith chart. Thus, we construct the rotated $1 + jx$ circle as shown in Figure 6.3a (center at 0.333). (A combined ZY chart is convenient to use here, if it is not too confusing.) Then we see that adding a susceptance of $jb = j0.3$ will move us along a constant conductance circle to $y = 0.4 + j0.5$ (this choice is

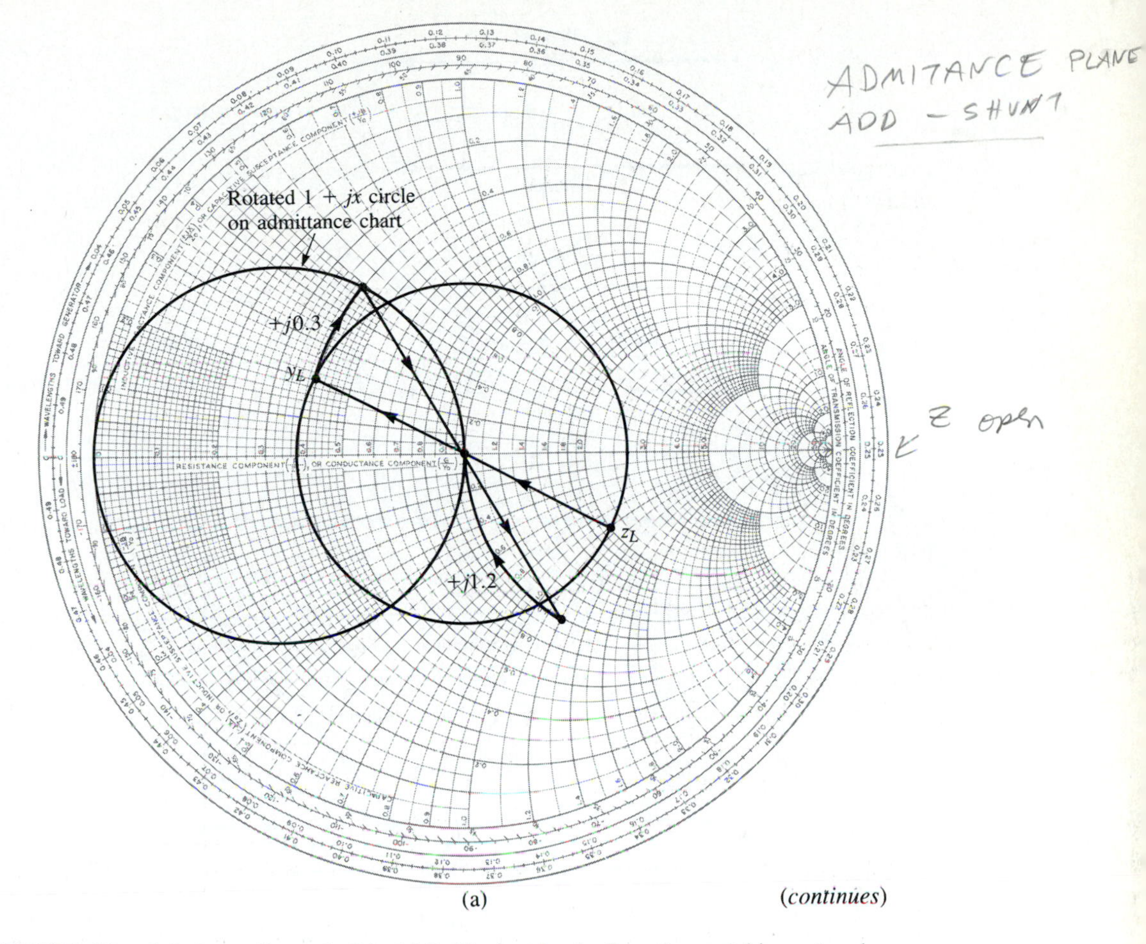

(a) *(continues)*

FIGURE 6.3 Solution to Example 6.1. (a) Smith chart for the L section matching networks.

the shortest distance from y_L to the shifted $1 + jx$ circle). Converting back to impedance leaves us at $z = 1 - j1.2$, indicating that a series reactance $x = j1.2$ will bring us to the center of the chart. For comparison, the formulas of (6.3a,b) give the solution as $b = 0.29, x = 1.22$.

This matching circuit consists of a shunt capacitor and a series inductor, as shown in Figure 6.3b. For a frequency of $f = 500$ MHz, the capacitor has a value of

$$C = \frac{b}{2\pi f Z_0} = 0.92 \text{ pF},$$

and the inductor has a value of

$$L = \frac{x Z_0}{2\pi f} = 38.8 \text{ nH}.$$

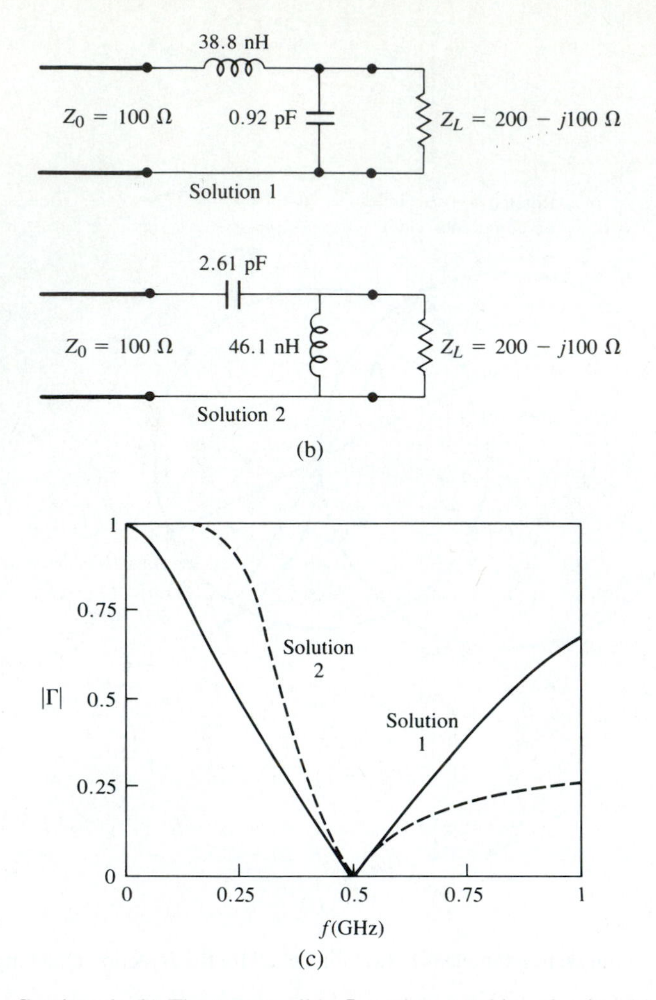

FIGURE 6.3 Continued. (b) The two possible L section matching circuits. (c) Reflection coefficient magnitudes versus frequency for the matching circuits of (b).

It may also be interesting to look at the second solution to this matching problem. If instead of adding a shunt susceptance of $b = 0.3$, we use a shunt susceptance of $b = -0.7$, we will move to a point on the lower half of the shifted $1 + jx$ circle, to $y = 0.4 - j0.5$. Then converting to impedance and adding a series reactance of $x = -1.2$ leads to a match as well. The formulas of (6.3a,b) give this solution as $b = -0.69, x = -1.22$. This matching circuit is also shown in Figure 6.3b, and is seen to have the positions of the inductor and capacitor reversed from the first matching network. At a frequency of $f = 500$ MHz, the capacitor has a value of

$$C = \frac{-1}{2\pi f x Z_0} = 2.61 \text{ pF},$$

while the inductor has a value of

$$L = \frac{-Z_0}{2\pi f b} = 46.1 \text{ nH}.$$

Figure 6.3c shows the reflection coefficient magnitude versus frequency for these two matching networks, assuming that the load impedance of $Z_L = 200 - j100\ \Omega$ at 500 MHz consists of a 200 Ω resistor and a 3.18 pF capacitor in series. There is not a substantial difference in bandwidth for these two solutions. ○

POINT OF INTEREST: Lumped Elements for Microwave Integrated Circuits

Lumped R, L, and C elements can be practically realized at microwave frequencies if the length, ℓ, of the component is very small relative to the operating wavelength. Over a limited range of values, such components can be used in hybrid and monolithic microwave integrated circuits (MICs) at frequencies up to 60 GHz, if the condition that $\ell < \lambda/10$ is satisfied. Usually, however, the characteristics of such an element are far from ideal, requiring that undesirable effects such as parasitic capacitance and/or inductance, spurious resonances, fringing fields, loss, and perturbations caused by a ground plane be incorporated in the design via a CAD model (see the Point of Interest concerning CAD).

Resistors are fabricated with thin films of lossy material such as nichrome, tantalum nitride, or doped semiconductor material. In monolithic circuits such films can be deposited or grown, while chip resistors made from a lossy film deposited on a ceramic chip can be bonded or soldered in a hybrid circuit. Low resistances are hard to obtain.

Small values of inductance can be realized with a short length or loop of transmission line, and larger values (up to about 10 nH) can be obtained with a spiral inductor, as shown in the following figures. Larger inductance values generally incur more loss, and more shunt capacitance; this leads to a resonance that limits the maximum operating frequency.

Capacitors can be fabricated in several ways. A short transmission line stub can provide a shunt capacitance in the range of 0 to 0.1 pF. A single gap or interdigital set of gaps in a transmission

Lossy film

Planar resistor

Chip resistor

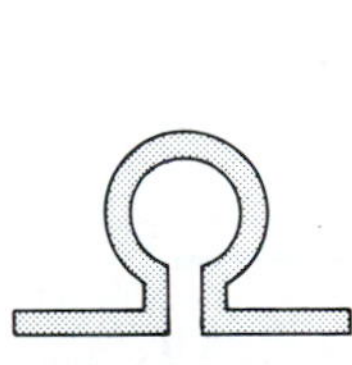

Loop inductor

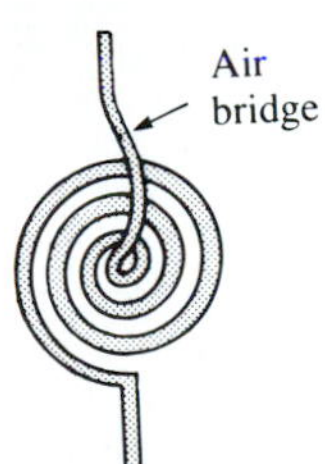

Spiral inductor

Interdigital
gap capacitor

Dielectric

ϵ_r

Metal-insulator-
metal capacitor

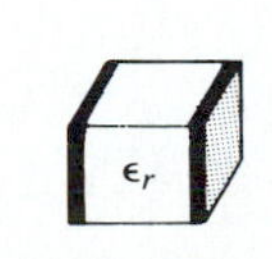

Chip capacitor

line can provide a series capacitance up to about 0.5 pF. Greater values (up to about 25 pF) can be obtained using a metal-insulator-metal (MIM) sandwich, either in monolithic or chip (hybrid) form.

6.2 SINGLE-STUB TUNING

We next consider a matching technique that uses a single open-circuited or short-circuited length of transmission line (a "stub"), connected either in parallel or in series with the transmission feed line at a certain distance from the load, as shown in Figure 6.4. Such a tuning circuit is convenient from a microwave fabrication aspect, since lumped elements are not required. The shunt tuning stub is especially easy to fabricate in microstrip or stripline form.

In single-stub tuning, the two adjustable parameters are the distance, d, from the load to the stub position, and the value of susceptance or reactance provided by the shunt or series stub. For the shunt-stub case, the basic idea is to select d so that the admittance,

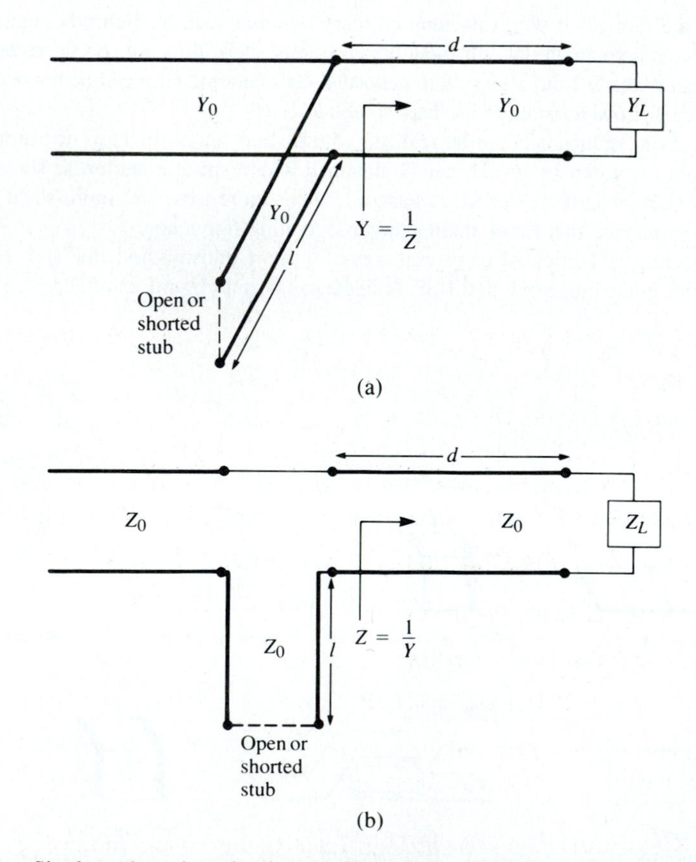

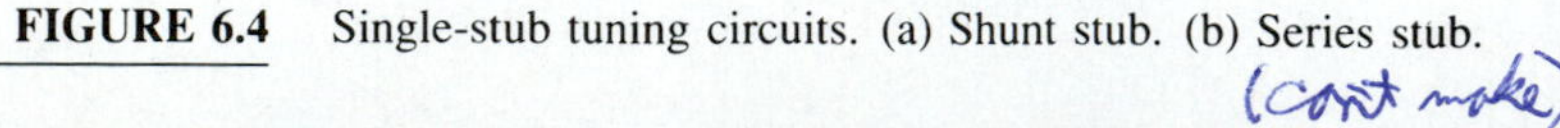

FIGURE 6.4 Single-stub tuning circuits. (a) Shunt stub. (b) Series stub.

Y, seen looking into the line at distance d from the load is of the form $Y_0 + jB$. Then the stub susceptance is chosen as $-jB$, resulting in a matched condition. For the series stub case, the distance d is selected so that the impedance, Z, seen looking into the line at a distance d from the load is of the form $Z_0 + jX$. Then the stub reactance is chosen as $-jX$, resulting in a matched condition.

As discussed in Chapter 3, the proper length of open or shorted transmission line can provide any desired value of reactance or susceptance. For a given susceptance or reactance, the difference in lengths of an open- or short-circuited stub is $\lambda/4$. For transmission line media such as microstrip or stripline, open-circuited stubs are easier to fabricate since a via hole through the substrate to the ground plane is not needed. For lines like coax or waveguide, however, short-circuited stubs are usually preferred, because the cross-sectional area of such an open-circuited line may be large enough (electrically) to radiate, in which case the stub is no longer purely reactive.

Below we discuss both Smith chart and analytic solutions for shunt and series stub tuning. The Smith chart solutions are fast, intuitive, and usually accurate enough in practice. The analytic expressions are more accurate, and useful for computer analysis.

Shunt Stubs

The single–stub shunt tuning circuit is shown in Figure 6.4a. We will first discuss an example illustrating the Smith chart solution, and then derive formulas for d and ℓ.

EXAMPLE 6.2

For a load impedance $Z_L = 15 + j10\ \Omega$, design two single-stub shunt tuning networks to match this load to a 50 Ω line. Assuming that the load is matched at 2 GHz, and that the load consists of a resistor and inductor in series, plot the reflection coefficient magnitude from 1 GHz to 3 GHz for each solution.

Solution

The first step is to plot the normalized load impedance $z_L = 0.3 + j0.2$, construct the appropriate SWR circle, and convert to the load admittance, y_L, as shown on the Smith chart in Figure 6.5a. For the remaining steps we consider the Smith chart as an admittance chart. Now notice that the SWR circle intersects the $1 + jb$ circle at two points, denoted as y_1 and y_2 in Figure 6.5a. Thus the distance d, from the load to the stub, is given by either of these two intersections. Reading the WTG scale, we obtain

$$d_1 = 0.328 - 0.284 = 0.044\lambda,$$
$$d_2 = (0.5 - 0.284) + 0.171 = 0.387\lambda.$$

Actually, there are an infinite number of distances, d, on the SWR circle that intersect the $1 + jb$ circle. Usually, it is desired to keep the matching stub as close as possible to the load, to improve the bandwidth of the match and to reduce losses caused by a possibly large standing wave ratio on the line between the stub and the load.

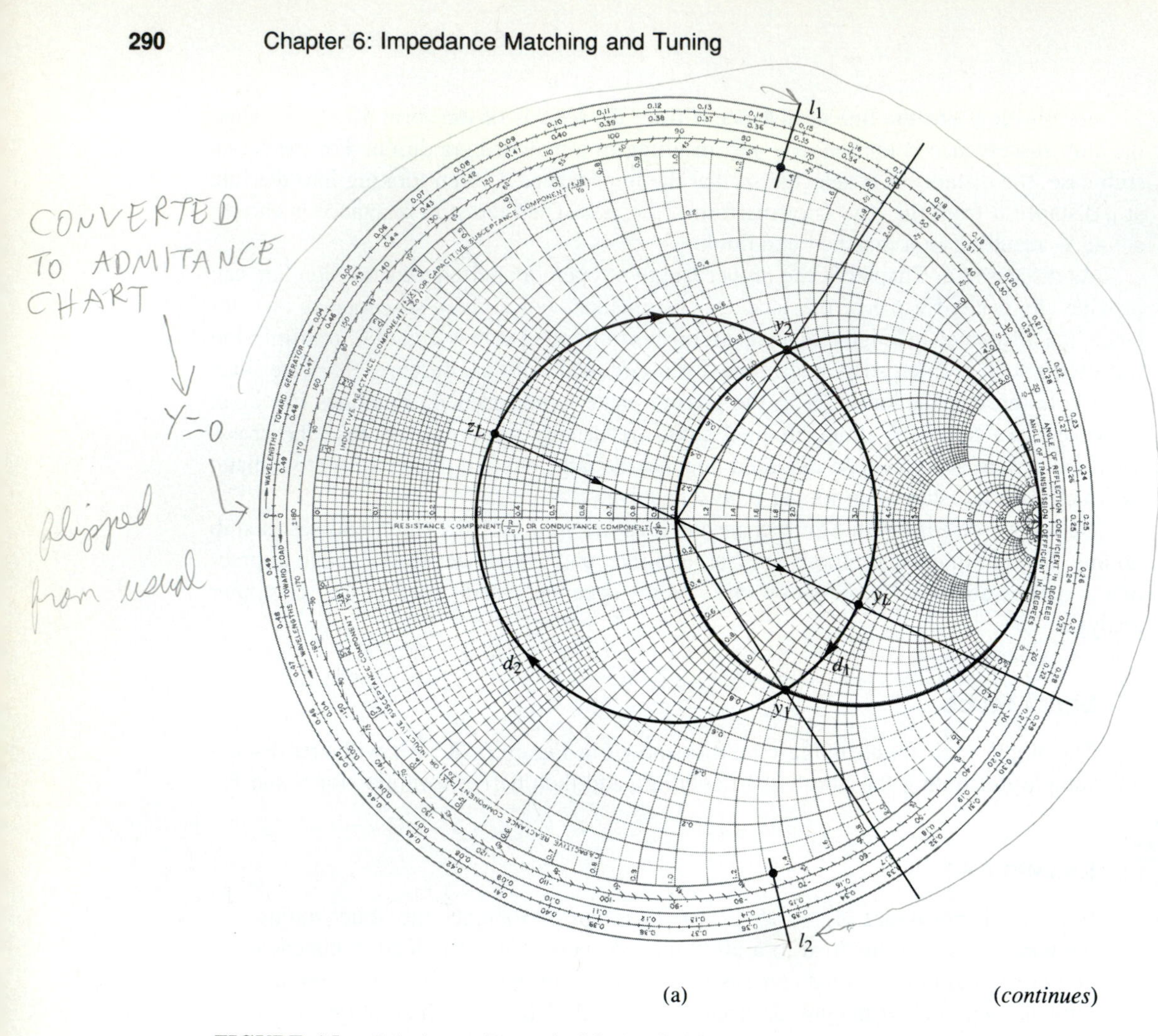

(a) *(continues)*

FIGURE 6.5 Solution to Example 6.2. (a) Smith chart for the shunt-stub tuners.

At the two intersection points, the normalized admittances are

$$y_1 = 1 - j1.33,$$

$$y_2 = 1 + j1.33.$$

Thus, the first tuning solution requires a stub with a susceptance of $j1.33$. The length of an open-circuited stub that gives this susceptance can be found on the Smith chart by starting at $y = 0$ (the open circuit) and moving along the outer edge of the chart ($g = 0$) towards the generator to the $j1.33$ point. The length is then

$$\ell_1 = 0.147\lambda.$$

Similarly, the required open-circuit stub length for the second solution is

$$\ell_2 = 0.353\lambda.$$

This completes the tuner designs.

To analyze the frequency dependence of these two designs, we need to know the load impedance as a function of frequency. The series-RL load impedance is $Z_L = 15 + j10\ \Omega$ at 2 GHz, so $R = 15\ \Omega$ and $L = 0.796$ nH. The two tuning circuits are shown in Figure 6.5b. Figure 6.5c shows the calculated reflection coefficient magnitudes for these two solutions. Observe that solution 1 has a significantly better bandwidth than solution 2; this is because both d and ℓ are shorter for solution 1, which reduces the frequency variation of the match. ○

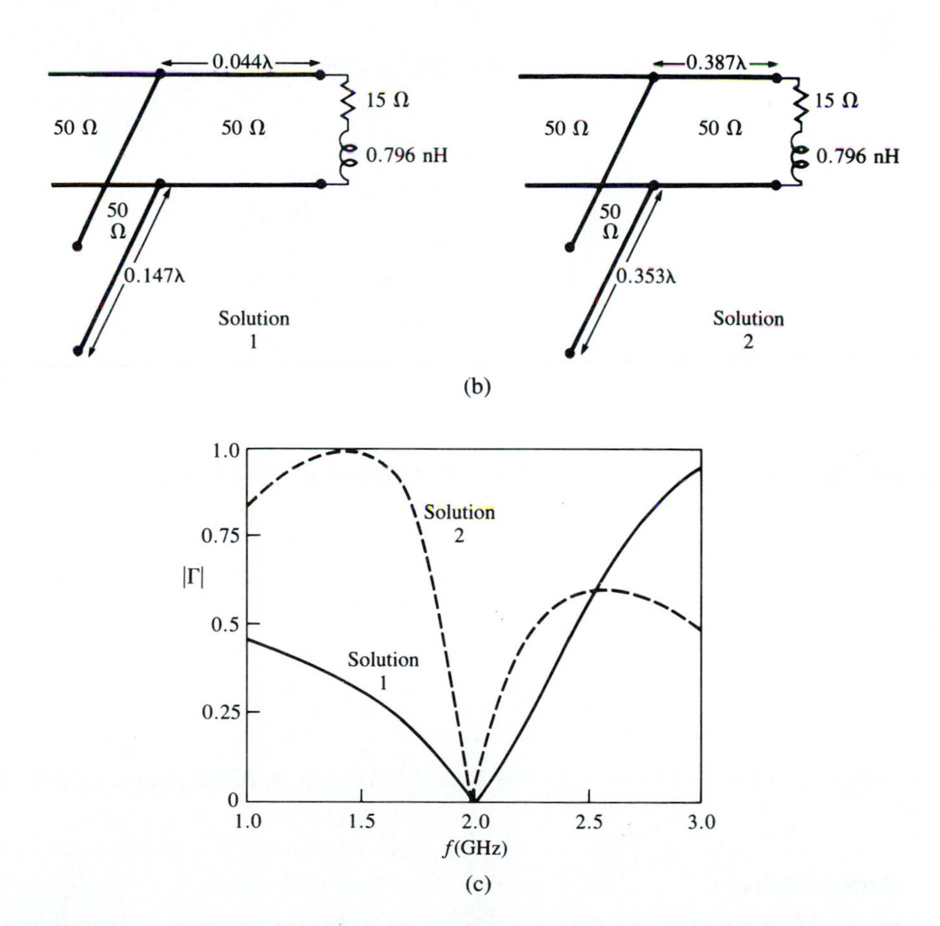

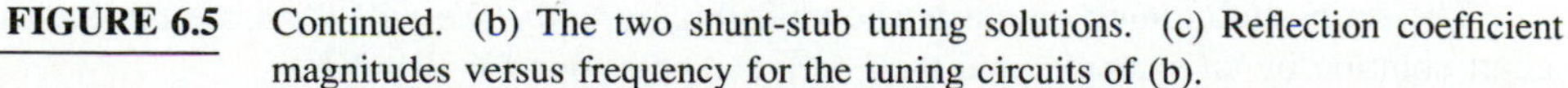

FIGURE 6.5 Continued. (b) The two shunt-stub tuning solutions. (c) Reflection coefficient magnitudes versus frequency for the tuning circuits of (b).

To derive formulas for d and ℓ, let the load impedance be written as $Z_L = 1/Y_L = R_L + jX_L$. Then the impedance Z down a length, d, of line from the load is

$$Z = Z_0 \frac{(R_L + jX_L) + jZ_0 t}{Z_0 + j(R_L + jX_L)t}, \quad 6.7$$

where $t = \tan \beta d$. The admittance at this point is

$$Y = G + jB = \frac{1}{Z},$$

where

$$G = \frac{R_L(1+t^2)}{R_L^2 + (X_L + Z_0 t)^2}, \quad 6.8a$$

$$B = \frac{R_L^2 t - (Z_0 - X_L t)(X_L + Z_0 t)}{Z_0[R_L^2 + (X_L + Z_0 t)^2]}. \quad 6.8b$$

Now d (which implies t) is chosen so that $G = Y_0 = 1/Z_0$. From (6.8a), this results in a quadratic equation for t:

$$Z_0(R_L - Z_0)t^2 - 2X_L Z_0 t + (R_L Z_0 - R_L^2 - X_L^2) = 0.$$

Solving for t gives

$$t = \frac{X_L \pm \sqrt{R_L[(Z_0 - R_L)^2 + X_L^2]/Z_0}}{R_L - Z_0}, \quad \text{for } R_L \neq Z_0. \quad 6.9$$

If $R_L = Z_0$, then $t = -X_L/2Z_0$. Thus, the two principal solutions for d are

$$\frac{d}{\lambda} = \begin{cases} \frac{1}{2\pi} \tan^{-1} t & \text{for } t \geq 0 \\ \frac{1}{2\pi}(\pi + \tan^{-1} t) & \text{for } t < 0. \end{cases} \quad 6.10$$

To find the required stub lengths, first use t in (6.8b) to find the stub susceptance, $B_s = -B$. Then, for an open-circuited stub,

$$\frac{\ell_o}{\lambda} = \frac{1}{2\pi} \tan^{-1}\left(\frac{B_s}{Y_0}\right) = \frac{-1}{2\pi} \tan^{-1}\left(\frac{B}{Y_0}\right), \quad 6.11a$$

while for a short-circuited stub,

$$\frac{\ell_s}{\lambda} = \frac{-1}{2\pi} \tan^{-1}\left(\frac{Y_0}{B_s}\right) = \frac{1}{2\pi} \tan^{-1}\left(\frac{Y_0}{B}\right). \quad 6.11b$$

If the length given by (6.11a) or (6.11b) is negative, $\lambda/2$ can be added to give a positive result.

Series Stubs

The series stub tuning circuit is shown in Figure 6.4b. We will illustrate the Smith chart solution by an example, and then derive expressions for d and ℓ.

EXAMPLE 6.3

Match a load impedance of $Z_L = 100 + j80$ to a 50 Ω line using a single series open-circuit stub. Assuming that the load is matched at 2 GHz, and that the load consists of a resistor and inductor in series, plot the reflection coefficient magnitude from 1 GHz to 3 GHz.

Solution

The first step is to plot the normalized load impedance, $z_L = 2 + j1.6$, and draw the SWR circle. For the series-stub design, the chart is an impedance chart. Note that the SWR circle intersects the $1 + jx$ circle at two points, denoted as z_1 and z_2 in Figure 6.6a. The shortest distance, d_1, from the load to

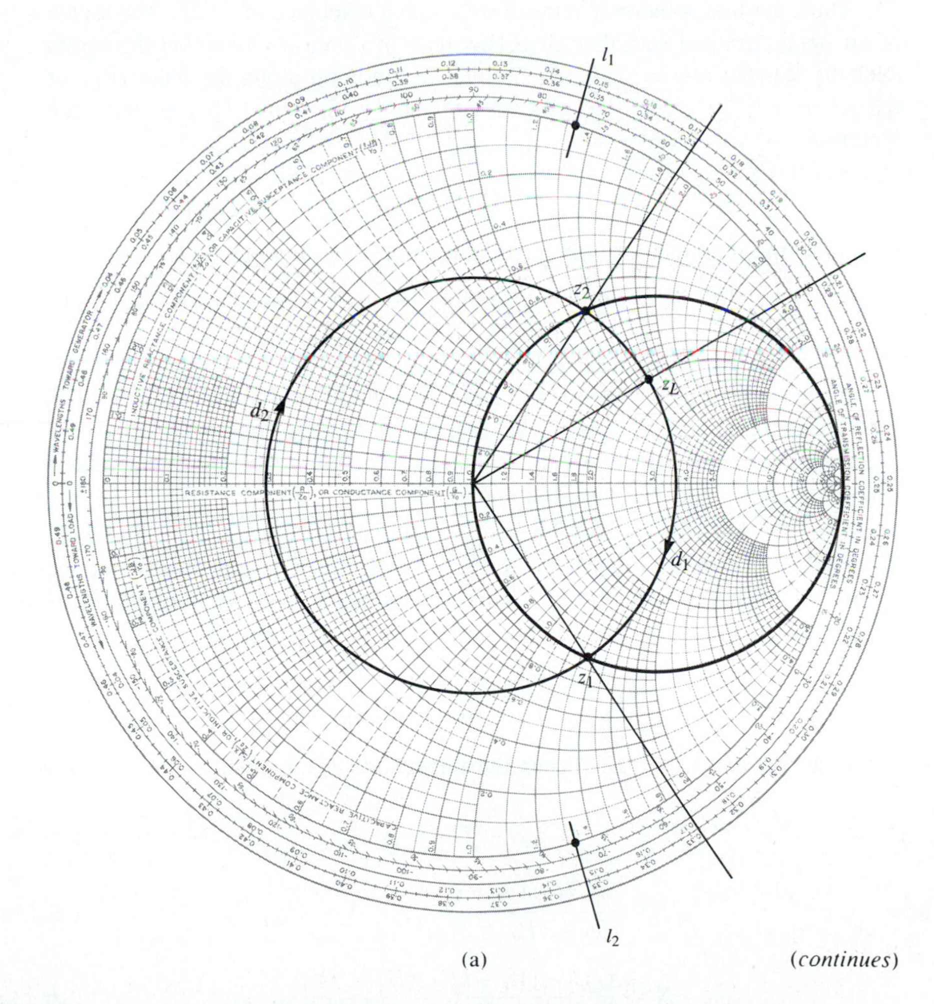

(a) *(continues)*

FIGURE 6.6 Solution to Example 6.3. (a) Smith chart for the series-stub tuners.

the stub is, from the WTG scale,

$$d_1 = 0.328 - 0.208 = 0.120\lambda,$$

while the second distance is

$$d_2 = (0.5 - 0.208) + 0.172 = 0.463\lambda.$$

As in the shunt-stub case, additional rotations around the SWR circle lead to additional solutions, but these are usually not of practical interest.

The normalized impedances at the two intersection points are

$$z_1 = 1 - j1.33,$$
$$z_2 = 1 + j1.33.$$

Thus, the first solution requires a stub with a reactance of $j1.33$. The length of an open-circuited stub that gives this reactance can be found on the Smith chart by starting at $z = \infty$ (open circuit), and moving along the outer edge of the chart ($r = 0$) toward the generator to the $j1.33$ point. This gives a stub length of

$$\ell_1 = 0.397\lambda.$$

Similarly, the required open-circuited stub length for the second solution is

$$\ell_2 = 0.103\lambda.$$

This completes the tuner designs.

If the load is a series resistor and inductor with $Z_L = 100+j80\ \Omega$ at 2 GHz, then $R = 100\ \Omega$ and $L = 6.37$ nH. The two matching circuits are shown in Figure 6.6b. Figure 6.6c shows the calculated reflection coefficient magnitudes versus frequency for the two solutions. ○

To derive formulas for d and ℓ for the series-stub tuner, let the load admittance be written as $Y_L = 1/Z_L = G_L + jB_L$. Then the admittance Y down a length, d, of line from the load is

$$Y = Y_0 \frac{(G_L + jB_L) + jtY_0}{Y_0 + jt(G_L + jB_L)}, \quad (6.12)$$

where $t = \tan \beta d$, and $Y_0 = 1/Z_0$. Then the impedance at this point is

$$Z = R + jX = \frac{1}{Y},$$

where

$$R = \frac{G_L(1 + t^2)}{G_L^2 + (B_L + Y_0 t)^2}, \quad (6.13a)$$

$$X = \frac{G_L^2 t - (Y_0 - tB_L)(B_L + tY_0)}{Y_0[G_L^2 + (B_L + Y_0 t)^2]}. \quad (6.13b)$$

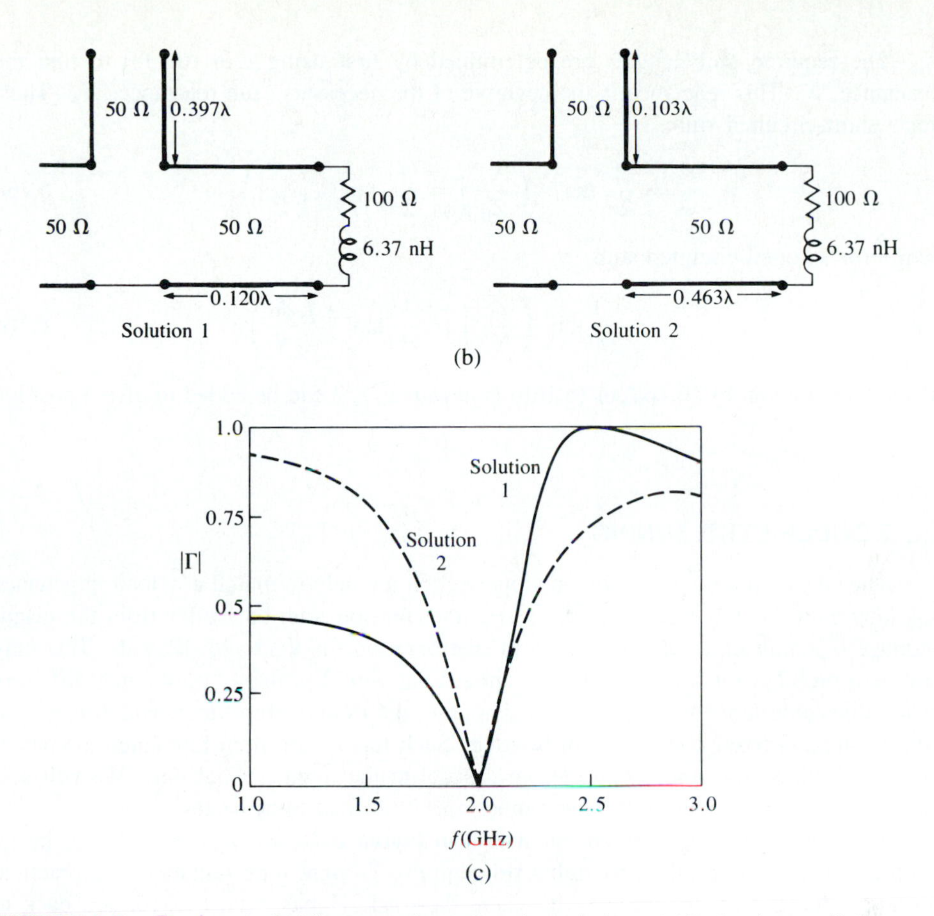

FIGURE 6.6 Continued. (b) The two series-stub tuning solutions. (c) Reflection coefficient magnitudes versus frequency for the tuning curcuits of (b).

Now d (which implies t) is chosen so that $R = Z_0 = 1/Y_0$. From (6.13a), this results in a quadratic equation for t:

$$Y_0(G_L - Y_0)t^2 - 2B_L Y_0 t + (G_L Y_0 - G_L^2 - B_L^2) = 0.$$

Solving for t gives

$$t = \frac{B_L \pm \sqrt{G_L[(Y_0 - G_L)^2 + B_L^2]/Y_0}}{G_L - Y_0}, \quad \text{for } G_L \neq Y_0. \tag{6.14}$$

If $G_L = Y_0$, then $t = -B_L/2Y_0$. Then the two principal solutions for d are

$$d/\lambda = \begin{cases} \dfrac{1}{2\pi}\tan^{-1} t & \text{for } t \geq 0 \\ \dfrac{1}{2\pi}(\pi + \tan^{-1} t) & \text{for } t < 0. \end{cases} \tag{6.15}$$

The required stub lengths are determined by first using t in (6.13b) to find the reactance, X. This reactance is the negative of the necessary stub reactance, X_s. Thus, for a short-circuited stub,

$$\frac{\ell_s}{\lambda} = \frac{1}{2\pi} \tan^{-1}\left(\frac{X_s}{Z_0}\right) = \frac{-1}{2\pi} \tan^{-1}\left(\frac{X}{Z_0}\right), \tag{6.16a}$$

while for an open-circuited stub,

$$\frac{\ell_o}{\lambda} = \frac{-1}{2\pi} \tan^{1}\left(\frac{Z_0}{X_s}\right) = \frac{1}{2\pi} \tan^{-1}\left(\frac{Z_0}{X}\right). \tag{6.16b}$$

If the length given by (6.16a) or (6.16b) is negative, $\lambda/2$ can be added to give a positive result.

6.3 DOUBLE-STUB TUNING

The single-stub tuners of the previous section are able to match any load impedance (as long as it has a nonzero real part) to a transmission line, but suffer from the disadvantage of requiring a variable length of line between the load and the stub. This may not be a problem for a fixed matching circuit, but would probably pose some difficulty if an adjustable tuner was desired. In this case, the double-stub tuner, which uses two tuning stubs in fixed positions, can be used. Such tuners are often fabricated in coaxial line, with adjustable stubs connected in parallel to the main coaxial line. We will see, however, that the double-stub tuner cannot match all load impedances.

The double-stub tuner circuit is shown in Figure 6.7a, where the load may be an arbitrary distance from the first stub. Although this is more representative of a practical situation, the circuit of Figure 6.7b, where the load Y_L' has been transformed back to the position of the first stub, is easier to deal with and does not lose any generality. The stubs shown in Figure 6.7 are shunt stubs, which are usually easier to implement in practice than are series stubs; the latter could be used just as well, in principle. In either case, the stubs can be open-circuited or short-circuited.

Smith Chart Solution

The Smith chart of Figure 6.8 illustrates the basic operation of the double-stub tuner. As in the case of the single-stub tuners, two solutions are possible. The susceptance of the first stub, b_1 (or b_1', for the second solution), moves the load admittance to y_1 (or y_1'). These points lie on the rotated $1 + jb$ circle; the amount of rotation is d wavelengths toward the load, where d is the electrical distance between the two stubs. Then transforming y_1 (or y_1') toward the generator through a length, d, of line leaves us at the point y_2 (or y_2'), which must be on the $1 + jb$ circle. The second stub then adds a susceptance b_2 (or b_2'), which brings us to the center of the chart, and completes the match.

Notice from Figure 6.8 that if the load admittance, y_L, were inside the shaded region of the $g_0 + jb$ circle, no value of stub susceptance b_1 could ever bring the load point to intersect the rotated $1 + jb$ circle. This shaded region thus forms a forbidden range of load admittances, which cannot be matched with this particular double-stub tuner. A

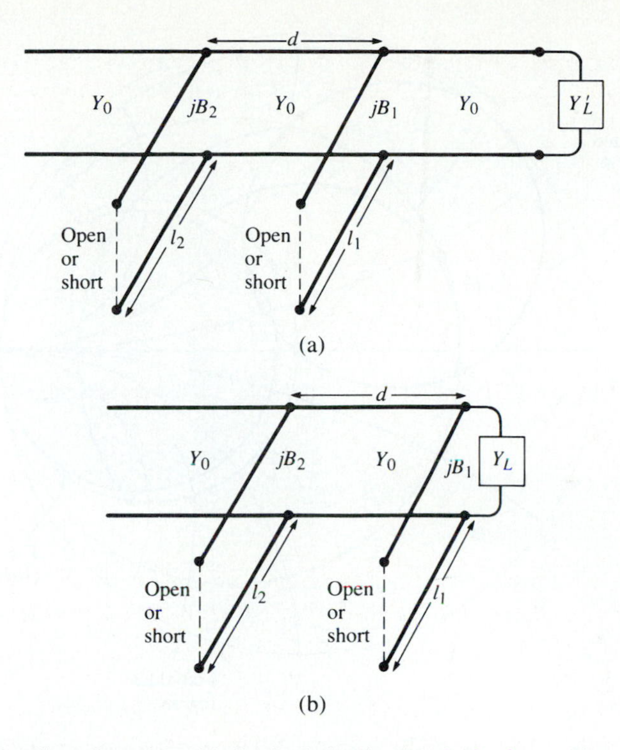

FIGURE 6.7 Double-stub tuning. (a) Original circuit with the load an arbitrary distance from the first stub. (b) Equivalent circuit with load at the first stub.

simple way of reducing the forbidden range is to reduce the distance, d, between the stubs. This has the effect of swinging the rotated $1 + jb$ circle back towards the $y = \infty$ point, but d must be kept large enough for the practical purpose of fabricating the two separate stubs. In addition, stub spacings near 0 or $\lambda/2$ lead to matching networks that are very frequency sensitive. In practice, stub spacings are usually chosen as $\lambda/8$ or $3\lambda/8$. If the length of line between the load and the first stub can be adjusted, then the load admittance y_L can always be moved out of the forbidden region.

EXAMPLE 6.4

Design a double-stub shunt tuner to match a load impedance $Z_L = 60 - j80\ \Omega$ to a 50 Ω line. The stubs are to be short-circuited stubs, and are spaced $\lambda/8$ apart. Assuming that this load consists of a series resistor and capacitor, and that the match frequency is 2 GHz, plot the reflection coefficient magnitude versus frequency from 1 GHz to 3 GHz.

Solution

The normalized load admittance is $y_L = 0.3 + j0.4$, which is plotted on the Smith chart of Figure 6.9a. Next we construct the rotated $1 + jb$ conductance circle, by moving every point on the $g = 1$ circle $\lambda/8$ toward the load. We then

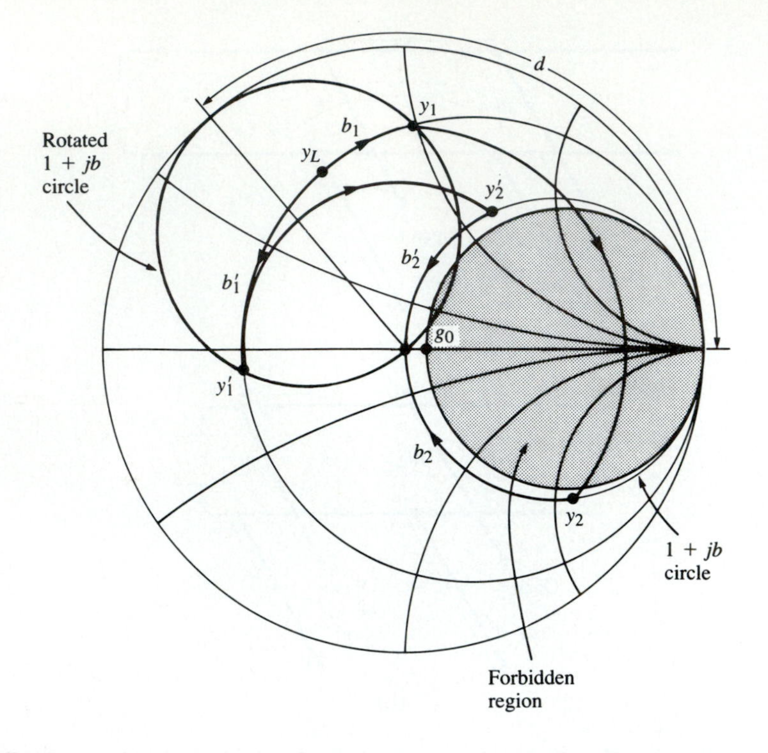

FIGURE 6.8 Smith chart diagram for the operation of a double-stub tuner.

find the susceptance of the first stub, which can be one of two possible values:

$$b_1 = 1.314,$$

or

$$b_1' = -0.114.$$

We now transform through the $\lambda/8$ section of line by rotating along a constant radius (SWR) circle $\lambda/8$ toward the generator. This brings the two solutions to the following points:

$$y_2 = 1 - j3.38,$$

or

$$y_2' = 1 + j1.38.$$

Then the susceptance of the second stub should be

$$b_2 = 3.38,$$

or

$$b_2' = -1.38.$$

The lengths of the short-circuited stubs are then found as

$$\ell_1 = 0.396\lambda, \qquad \ell_2 = 0.454\lambda,$$

or

$$\ell_1' = 0.232\lambda, \qquad \ell_2' = 0.100\lambda.$$

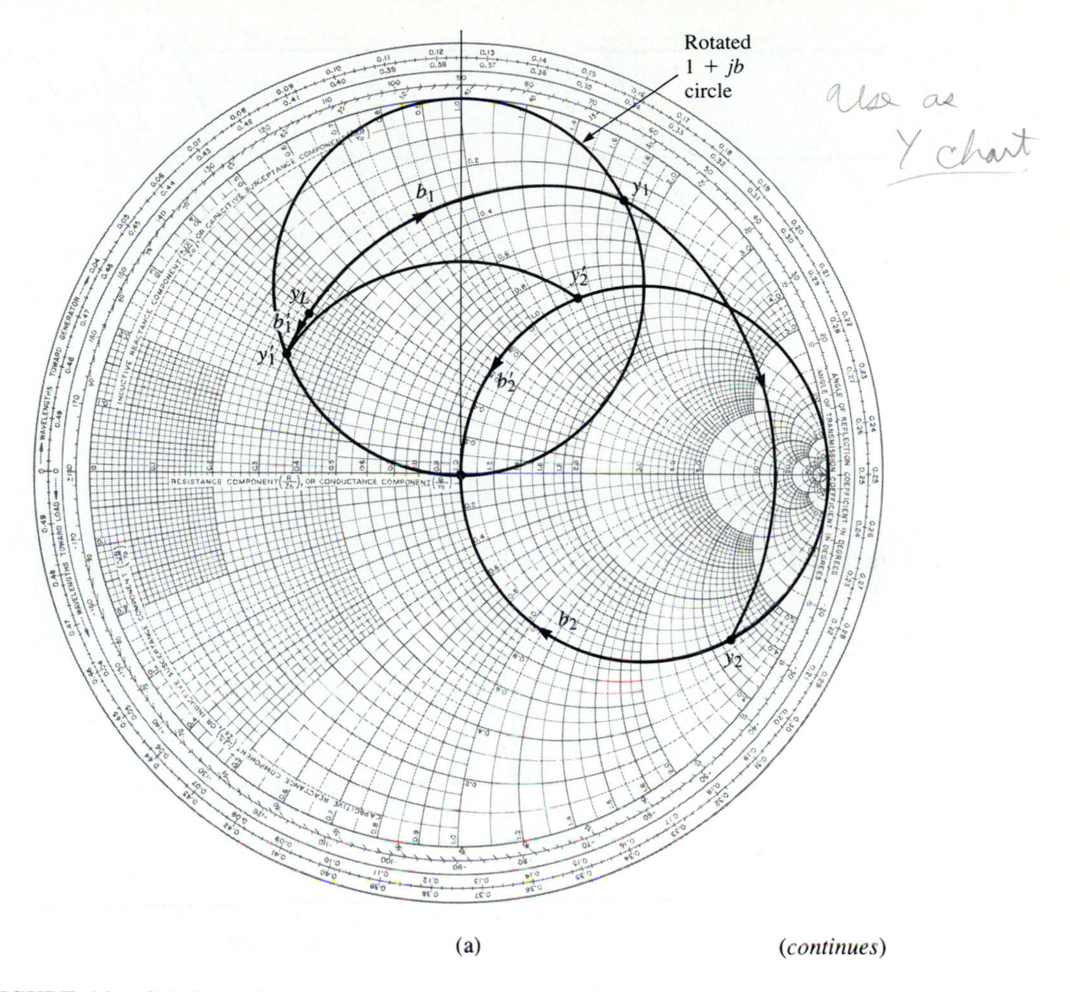

(a) *(continues)*

FIGURE 6.9 Solution to Example 6.4. (a) Smith chart for the double-stub tuners.

This completes both solutions for the double-stub tuner design.

Now if the resistor-capacitor load $Z_L = 60 - j80\ \Omega$ at $f = 2$ GHz, then $R = 60\ \Omega$ and $C = 0.995$ pF. The two tuning circuits are then shown in Figure 6.9b, and the reflection coefficient magnitudes are plotted versus frequency in Figure 6.9c. Note that the first solution has a much narrower bandwidth than the second (primed) solution, due to the fact that both stubs for the first solution are somewhat longer (and closer to $\lambda/2$) than the stubs of the second solution. ○

Analytic Solution

Just to the left of the first stub in Figure 6.7b, the admittance is

$$Y_1 = G_L + j(B_L + B_1), \tag{6.17}$$

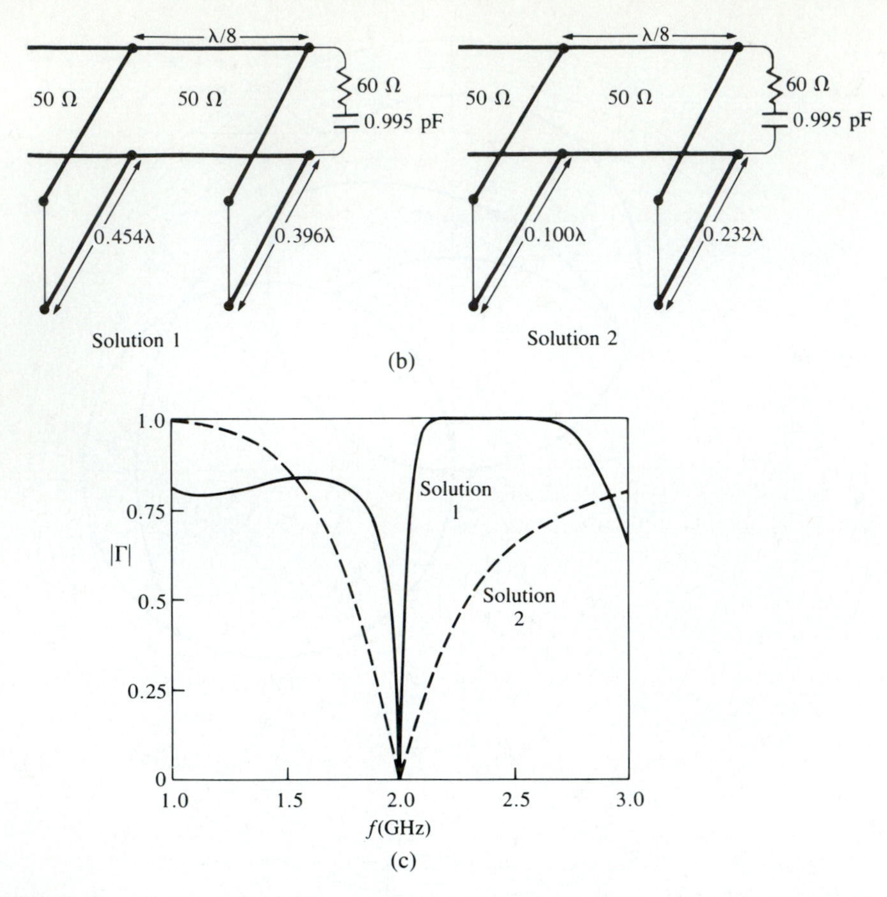

FIGURE 6.9 Continued. (b) The two double-stub tuning solutions. (c) Reflection coefficient magnitudes versus frequency for the tuning circuits of (b).

where $Y_L = G_L + jB_L$ is the load admittance and B_1 is the susceptance of the first stub. After transforming through a length d of transmission line, the admittance just to the right of the second stub is

$$Y_2 = Y_0 \frac{G_L + j(B_L + B_1 + Y_0 t)}{Y_0 + jt(G_L + jB_L + jB_1)}, \tag{6.18}$$

where $t = \tan \beta d$ and $Y_0 = 1/Z_0$. At this point, the real part of Y_2 must equal Y_0, which leads to the equation

$$G_L^2 - G_L Y_0 \frac{1 + t^2}{t^2} + \frac{(Y_0 - B_L t - B_1 t) Y_0}{t^2} = 0. \tag{6.19}$$

Solving for G_L gives

$$G_L = Y_0 \frac{1+t^2}{2t^2}\left[1 \pm \sqrt{1 - \frac{4t^2(Y_0 - B_L t - B_1 t)^2}{Y_0(1+t^2)^2}}\right]. \tag{6.20}$$

Since G_L is real, the quantity within the square root must be nonnegative, and so

$$0 \leq \frac{4t^2(Y_0 - B_L t - B_1 t)^2}{Y_0(1+t^2)^2} \leq 1.$$

This implies that

$$0 \leq G_L \leq Y_0 \frac{1+t^2}{2t^2} = \frac{Y_0}{\sin^2 \beta d}, \tag{6.21}$$

which gives the range on G_L that can be matched for a given stub spacing, d. After d has been fixed, the first stub susceptance can be determined from (6.19) as

$$B_1 = -B_L + \frac{Y_0 \pm \sqrt{(1+t^2)G_L Y_0 - G_L^2 t^2}}{t}. \tag{6.22}$$

Then the second stub susceptance can be found from the negative of the imaginary part of (6.18) to be

$$B_2 = \frac{\pm Y_0 \sqrt{Y_0 G_L (1+t^2) - G_L^2 t^2} + G_L Y_0}{G_L t}. \tag{6.23}$$

The upper and lower signs in (6.22) and (6.23) correspond to the same solutions. The open-circuited stub length is found as

$$\frac{\ell_o}{\lambda} = \frac{1}{2\pi} \tan^{-1}\left(\frac{B}{Y_0}\right), \tag{6.24a}$$

while the short-circuited stub length is found as

$$\frac{\ell_s}{\lambda} = \frac{-1}{2\pi} \tan^{-1}\left(\frac{Y_0}{B}\right), \tag{6.24b}$$

where $B = B_1, B_2$.

6.4 THE QUARTER-WAVE TRANSFORMER

As discussed in Section 3.5, the quarter-wave transformer is a simple and useful circuit for matching a real load impedance to a transmission line. An additional feature of the quarter-wave transformer is that it can be extended to multisection designs in a methodical manner, for broader bandwidth. If only a narrow band impedance match is required, a single-section transformer may suffice. But, as we will see in the next few sections, multisection quarter-wave transformer designs can be synthesized to yield optimum matching characteristics over a desired frequency band. We will see in Chapter 9 that such networks are closely related to bandpass filters.

One drawback of the quarter-wave transformer is that it can only match a real load impedance. A complex load impedance can always be transformed to a real impedance, however, by using an appropriate length of transmission line between the load and the transformer, or an appropriate series or shunt reactive stub. These techniques will usually alter the frequency dependence of the equivalent load, which often has the effect of reducing the bandwidth of the match.

In Section 3.5 we analyzed the operation of the quarter-wave transformer from an impedance viewpoint and a multiple reflection viewpoint. Here we will concentrate on the bandwidth performance of the transformer, as a function of the load mismatch; this discussion will also serve as a prelude to the more general case of multisection transformers in the sections to follow.

The single-section quarter wave matching transformer circuit is shown in Figure 6.10. The characteristic impedance of the matching section is

$$Z_1 = \sqrt{Z_0 Z_L}. \tag{6.25}$$

At the design frequency, f_0, the electrical length of the matching section is $\lambda_0/4$, but at other frequencies the length is different, so a perfect match is no longer obtained. We will now derive an approximate expression for the mismatch versus frequency.

The input impedance seen looking into the matching section is

$$Z_{\text{in}} = Z_1 \frac{Z_L + jZ_1 t}{Z_1 + jZ_L t}, \tag{6.26}$$

where $t = \tan \beta\ell = \tan \theta$, and $\beta\ell = \theta = \pi/2$ at the design frequency, f_0. The reflection coefficient is then

$$\Gamma = \frac{Z_{\text{in}} - Z_0}{Z_{\text{in}} + Z_0} = \frac{Z_1(Z_L - Z_0) + jt(Z_1^2 - Z_0 Z_L)}{Z_1(Z_L + Z_0) + jt(Z_1^2 + Z_0 Z_L)}. \tag{6.27}$$

Since $Z_1^2 = Z_0 Z_L$, this reduces to

$$\Gamma = \frac{Z_L - Z_0}{Z_L + Z_0 + j2t\sqrt{Z_0 Z_L}}. \tag{6.28}$$

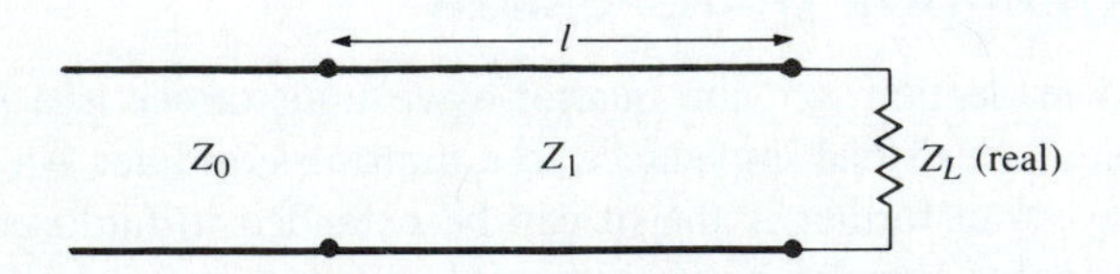

FIGURE 6.10 A single-section quarter-wave matching transformer. $\ell = \lambda_0/4$ at the design frequency f_0.

The reflection coefficient magnitude is

$$|\Gamma| = \frac{|Z_L - Z_0|}{\left[(Z_L + Z_0)^2 + 4t^2 Z_0 Z_L\right]^{1/2}}$$

$$= \frac{1}{\left\{\left(Z_L + Z_0/Z_L - Z_0\right)^2 + [4t^2 Z_0 Z_L/(Z_L - Z_0)^2]\right\}^{1/2}}$$

$$= \frac{1}{\left\{1 + [4Z_0 Z_L/(Z_L - Z_0)^2] + [4Z_0 Z_L t^2/(Z_L - Z_0)^2]\right\}^{1/2}}$$

$$= \frac{1}{\left\{1 + [4Z_0 Z_L/(Z_L - Z_0)^2] \sec^2\theta\right\}^{1/2}}, \qquad 6.29$$

since $1 + t^2 = 1 + \tan^2\theta = \sec^2\theta$.

Now if we assume that the frequency is near the design frequency, f_0, then $\ell \simeq \lambda_0/4$ and $\theta \simeq \pi/2$. Then $\sec^2\theta \gg 1$, and (6.29) simplifies to

$$|\Gamma| \simeq \frac{|Z_L - Z_0|}{2\sqrt{Z_0 Z_L}} |\cos\theta|, \qquad \text{for } \theta \text{ near } \pi/2. \qquad 6.30$$

This result gives the approximate mismatch of the quarter–wave transformer near the design frequency, as sketched in Figure 6.11.

If we set a maximum value, Γ_m, of the reflection coefficient magnitude that can be tolerated, then we can define the bandwidth of the matching transformer as

$$\Delta\theta = 2\left(\frac{\pi}{2} - \theta_m\right), \qquad 6.31$$

since the response of (6.29) is symmetric about $\theta = \pi/2$, and $\Gamma = \Gamma_m$ at $\theta = \theta_m$ and at $\theta = \pi - \theta_m$. Equating Γ_m to the exact expression for reflection coefficient magnitude in (6.29) allows us to solve for θ_m:

$$\frac{1}{\Gamma_m^2} = 1 + \left(\frac{2\sqrt{Z_0 Z_L}}{Z_L - Z_0} \sec\theta\right)^2,$$

or

$$\cos\theta_m = \frac{\Gamma_m}{\sqrt{1 - \Gamma_m^2}} \frac{2\sqrt{Z_0 Z_L}}{|Z_L - Z_0|}. \qquad 6.32$$

If we assume TEM lines, then

$$\theta = \beta\ell = \frac{2\pi f}{v_p} \frac{v_p}{4 f_0} = \frac{\pi f}{2 f_0},$$

therefore the frequency of the lower band edge at $\theta = \theta_m$ is

$$f_m = \frac{2\theta_m f_0}{\pi},$$

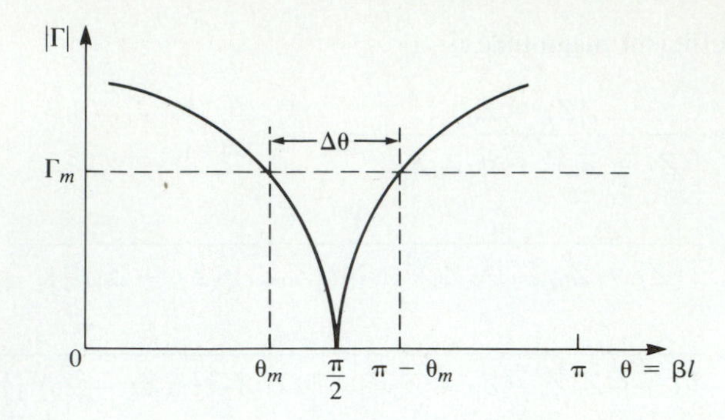

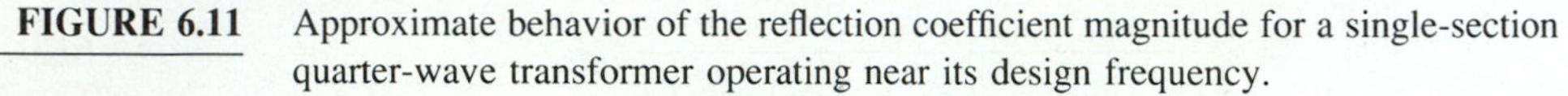

FIGURE 6.11 Approximate behavior of the reflection coefficient magnitude for a single-section quarter-wave transformer operating near its design frequency.

and the fractional bandwidth is, using (6.32),

$$\frac{\Delta f}{f_0} = \frac{2(f_0 - f_m)}{f_0} = 2 - \frac{2f_m}{f_0} = 2 - \frac{4\theta_m}{\pi}$$
$$= 2 - \frac{4}{\pi}\cos^{-1}\left[\frac{\Gamma_m}{\sqrt{1-\Gamma_m^2}}\frac{2\sqrt{Z_0 Z_L}}{|Z_L - Z_0|}\right]. \quad 6.33$$

The fractional bandwidth is usually expressed as a percentage, $100\Delta f/f_0$ %. Note that the bandwidth of the transformer increases as Z_L becomes closer to Z_0 (a less mismatched load).

The above results are strictly valid only for TEM lines. When non-TEM lines (such as waveguides) are used, the propagation constant is no longer a linear function of frequency, and the wave impedance will be frequency dependent. These factors serve to complicate the general behavior of quarter-wave transformers for non-TEM lines, but in practice the bandwidth of the transformer is often small enough so that these complications do not substantially affect the result. Another factor ignored in the above analysis is the effect of reactances associated with discontinuities when there is a step change in the dimensions of a transmission line. This can often be compensated for by making a small adjustment in the length of the matching section.

Figure 6.12 shows a plot of the reflection coefficient magnitude versus normalized frequency for various mismatched loads. Note the trend of increased bandwidth for smaller load mismatches.

EXAMPLE 6.5

Design a single-section quarter-wave matching transformer to match a 10 Ω load to a 50 Ω line, at $f_0 = 3$ GHz. Determine the percent bandwidth for which the SWR ≤ 1.5.

Solution

From (6.25), the characteristic impedance of the matching section is

$$Z_1 = \sqrt{Z_0 Z_L} = \sqrt{(50)(10)} = 22.36\ \Omega,$$

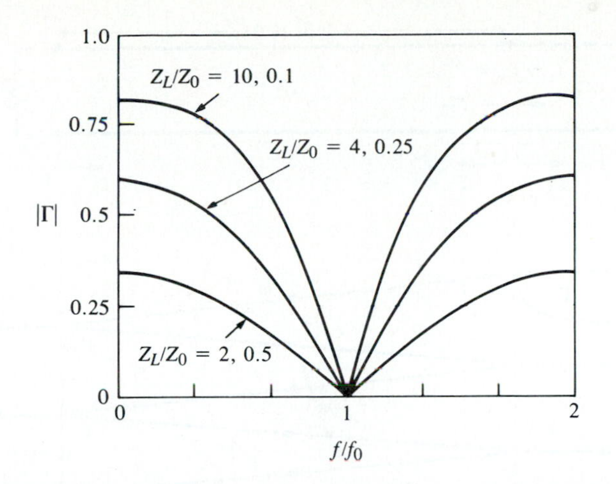

FIGURE 6.12 Reflection coefficient magnitude versus frequency for a single-section quarter-wave matching transformer with various load mismatches.

and the length of the matching section is $\lambda/4$ at 3 GHz. An SWR of 1.5 corresponds to a reflection coefficient magnitude of

$$\Gamma_m = \frac{\text{SWR} - 1}{\text{SWR} + 1} = \frac{1.5 - 1}{1.5 + 1} = 0.2.$$

The fractional bandwidth is computed from (6.33) as

$$\begin{aligned}\frac{\Delta f}{f_0} &= 2 - \frac{4}{\pi} \cos^{-1} \left[\frac{\Gamma_m}{\sqrt{1 - \Gamma_m^2}} \frac{2\sqrt{Z_0 Z_L}}{|Z_L - Z_0|} \right] \\ &= 2 - \frac{4}{\pi} \cos^{-1} \left[\frac{0.2}{\sqrt{1 - (0.2)^2}} \frac{2\sqrt{(50)(10)}}{|10 - 50|} \right] \\ &= 0.29, \text{ or } 29\%.\end{aligned}$$

○

6.5 THE THEORY OF SMALL REFLECTIONS

The quarter-wave transformer provides a simple means of matching any real load impedance to any line impedance. For applications requiring more bandwidth than a single quarter-wave section can provide, multisection transformers can be used. The design of such transformers is the subject of the next two sections, but prior to that material we need to derive some approximate results for the total reflection coefficient caused by the partial reflections from several small discontinuities. This topic is generally referred to as the theory of small reflections [1].

Single-Section Transformer

Consider the single-section transformer shown in Figure 6.13; we will derive an approximate expression for the overall reflection coefficient Γ. The partial reflection and

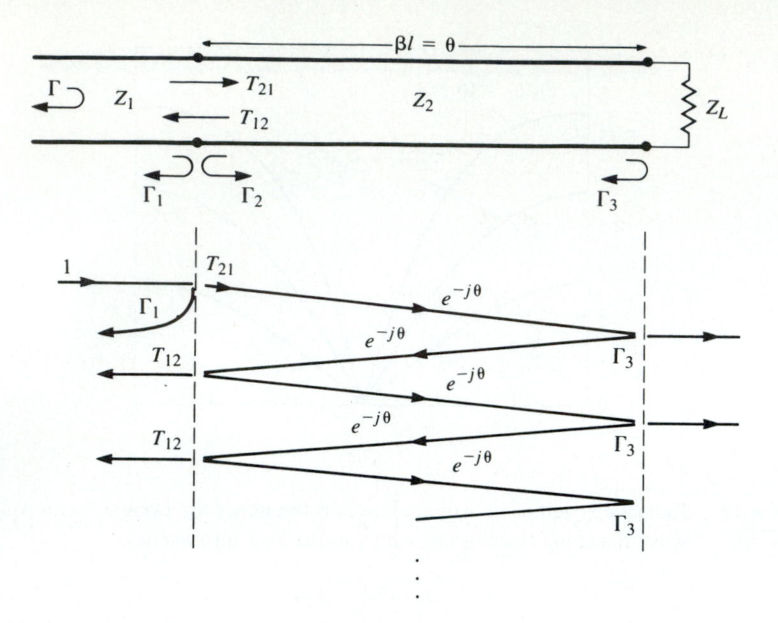

FIGURE 6.13 Partial reflections and transmissions on a single-section matching transformer.

transmission coefficients are

$$\Gamma_1 = \frac{Z_2 - Z_1}{Z_2 + Z_1}, \tag{6.34}$$

$$\Gamma_2 = -\Gamma_1, \tag{6.35}$$

$$\Gamma_3 = \frac{Z_L - Z_2}{Z_L + Z_2}, \tag{6.36}$$

$$T_{21} = 1 + \Gamma_1 = \frac{2Z_2}{Z_1 + Z_2}, \tag{6.37}$$

$$T_{12} = 1 + \Gamma_2 = \frac{2Z_1}{Z_1 + Z_2}. \tag{6.38}$$

We can compute the total reflection, Γ, seen by the feed line by the impedance method or by the multiple reflection method, as discussed in Section 3.5. For our present purpose the latter technique is preferred, so we can express the total reflection as an infinite sum of partial reflections and transmissions as follows:

$$\begin{aligned}\Gamma &= \Gamma_1 + T_{12}T_{21}\Gamma_3 e^{-2j\theta} + T_{12}T_{21}\Gamma_3^2\Gamma_2 e^{-4j\theta} + \cdots \\ &= \Gamma_1 + T_{12}T_{21}\Gamma_3 e^{-2j\theta}\sum_{n=0}^{\infty}\Gamma_2^n\Gamma_3^n e^{-2jn\theta}.\end{aligned} \tag{6.39}$$

Using the geometric series,

$$\sum_{n=0}^{\infty} x^n = \frac{1}{1-x}, \qquad \text{for } |x| < 1,$$

(6.39) can be expressed in closed form as

$$\Gamma = \Gamma_1 + \frac{T_{12}T_{21}\Gamma_3 e^{-2j\theta}}{1 - \Gamma_2\Gamma_3 e^{-2j\theta}}. \tag{6.40}$$

From (6.35), (6.37), and (6.38), we use $\Gamma_2 = -\Gamma_1, T_{21} = 1 + \Gamma_1$, and $T_{12} = 1 - \Gamma_1$ in (6.40) to give

$$\Gamma = \frac{\Gamma_1 + \Gamma_3 e^{-2j\theta}}{1 + \Gamma_1\Gamma_3 e^{-2j\theta}}. \tag{6.41}$$

Now if the discontinuities between the impedances Z_1, Z_2 and Z_2, Z_L are small, then $|\Gamma_1\Gamma_3| \ll 1$, so we can approximate (6.41) as

$$\Gamma \simeq \Gamma_1 + \Gamma_3 e^{-2j\theta}. \tag{6.42}$$

This result states the intuitive idea that the total reflection is dominated by the reflection from the initial discontinuity between Z_1 and Z_2 (Γ_1), and the first reflection from the discontinuity between Z_2 and Z_L ($\Gamma_3 e^{-2j\theta}$). The $e^{-2j\theta}$ term accounts for the phase delay when the incident wave travels up and down the line. The following example demonstrates the accuracy of this approximation.

EXAMPLE 6.6

Consider the quarter-wave transformer of Figure 6.13, with $Z_1 = 100\ \Omega$, $Z_2 = 150\ \Omega$, and $Z_L = 225\ \Omega$. Evaluate the worst-case percent error in computing $|\Gamma|$ from the approximate expression of (6.42).

Solution

The partial reflection coefficients from (6.34) and (6.36) are

$$\Gamma_1 = \frac{Z_2 - Z_1}{Z_2 + Z_1} = \frac{150 - 100}{150 + 100} = 0.2,$$

$$\Gamma_3 = \frac{Z_L - Z_2}{Z_L + Z_2} = \frac{225 - 150}{225 + 150} = 0.2.$$

Since the approximate expression for Γ in (6.42) is identical to the numerator for the exact expression in (6.41), the greatest error will occur when the denominator of (6.41) departs from unity to the greatest extent. This occurs for $\theta = 0$ or $180°$, since for $\theta = 90°$ both results are zero. Then (6.41) gives the exact result as $\Gamma = 0.384$, while (6.42) gives the approximate result as $\Gamma = 0.4$. The error is about 4%. ○

Multisection Transformer

Now consider the multisection transformer shown in Figure 6.14. This transformer consists of N equal-length (*commensurate*) sections of transmission lines. We will derive an approximate expression for the total reflection coefficient Γ.

Partial reflection coefficients can be defined at each junction, as follows:

$$\Gamma_0 = \frac{Z_1 - Z_0}{Z_1 + Z_0}, \tag{6.43a}$$

$$\Gamma_n = \frac{Z_{n+1} - Z_n}{Z_{n+1} + Z_n}, \tag{6.43b}$$

$$\Gamma_N = \frac{Z_L - Z_N}{Z_L + Z_N}. \tag{6.43c}$$

We also assume that all Z_n increase or decrease monotonically across the transformer, and that Z_L is real. This implies that all Γ_n will be real, and of the same sign ($\Gamma_n > 0$ if $Z_L > Z_0$; $\Gamma_n < 0$ if $Z_L < Z_0$). Then using the results of the previous section, the overall reflection coefficient can be approximated as

$$\Gamma(\theta) = \Gamma_0 + \Gamma_1 e^{-2j\theta} + \Gamma_2 e^{-4j\theta} + \cdots + \Gamma_N e^{-2jN\theta}. \tag{6.44}$$

Further assume that the transformer can be made symmetrical, so that $\Gamma_0 = \Gamma_N, \Gamma_1 = \Gamma_{N-1}, \Gamma_2 = \Gamma_{N-2}$, etc. (Note that this does *not* imply that the Z_ns are symmetrical). Then (6.44) can be written as

$$\Gamma(\theta) = e^{-jN\theta}\{\Gamma_0[e^{jN\theta} + e^{-jN\theta}] + \Gamma_1[e^{j(N-2)\theta} + e^{-j(N-2)\theta}] + \cdots\}. \tag{6.45}$$

If N is odd, the last term is $\Gamma_{(N-1)/2}(e^{j\theta} - e^{-j\theta})$, while if N is even the last term is $\Gamma_{N/2}$. Equation (6.45) is then seen to be of the form of a finite Fourier cosine series in θ, which can be written as

$$\Gamma(\theta) = 2e^{-jN\theta}\Big[\Gamma_0 \cos N\theta + \Gamma_1 \cos(N-2)\theta + \cdots + \Gamma_n \cos(N-2n)\theta + \cdots + \frac{1}{2}\Gamma_{N/2}\Big], \quad \text{for } N \text{ even}, \tag{6.46a}$$

$$\Gamma(\theta) = 2e^{-jN\theta}[\Gamma_0 \cos N\theta + \Gamma_1 \cos(N-2)\theta + \cdots + \Gamma_n \cos(N-2n)\theta + \cdots + \Gamma_{(N-1)/2} \cos\theta], \quad \text{for } N \text{ odd}. \tag{6.46b}$$

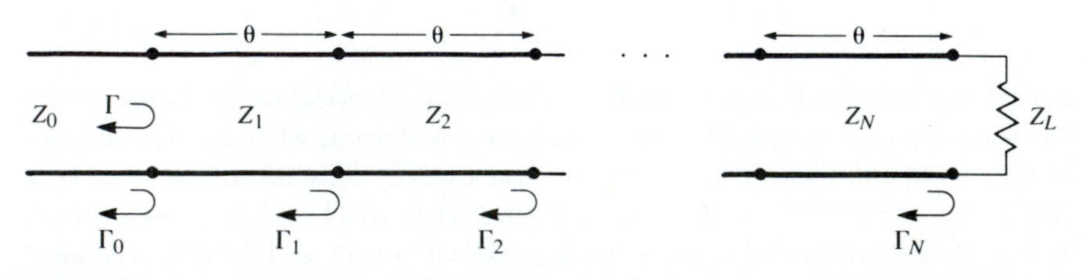

FIGURE 6.14 Partial reflection coefficients for a multisection matching transformer.

The importance of these results lies in the fact that we can synthesize any desired reflection coefficient response as a function of frequency (θ), by properly choosing the Γ_ns and using enough sections (N). This should be clear from the realization that a Fourier series can represent an arbitrary smooth function, if enough terms are used. In the next two sections we will show how to use this theory to design multisection transformers for two of the most commonly used passband responses: the binomial (maximally flat) response, and the Chebyshev (equal ripple) response.

6.6 BINOMIAL MULTISECTION MATCHING TRANSFORMERS

The passband response of a binomial matching transformer is optimum in the sense that, for a given number of sections, the response is as flat as possible near the design frequency. Thus, such a transformer is also known as maximally flat. This type of response is designed, for an N-section transformer, by setting the first $N-1$ derivatives of $|\Gamma(\theta)|$ to zero, at the center frequency f_0. Such a response can be obtained if we let

$$\Gamma(\theta) = A(1 + e^{-2j\theta})^N. \tag{6.47}$$

Then the magnitude $|\Gamma(\theta)|$ is

$$\begin{aligned} |\Gamma(\theta)| &= |A||e^{-j\theta}|^N |e^{j\theta} + e^{-j\theta}|^N \\ &= 2^N |A| |\cos\theta|^N \end{aligned} \tag{6.48}$$

Note that $|\Gamma(\theta)| = 0$ for $\theta = \pi/2$, and that $(d^n|\Gamma(\theta)|)/d\theta^n = 0$ at $\theta = \pi/2$ for $n = 1, 2, ..., N-1$. ($\theta = \pi/2$ corresponds to the center frequency f_0, for which $\ell = \lambda/4$ and $\theta = \beta\ell = \pi/2$.)

We can determine the constant A by letting $f \rightarrow 0$. Then $\theta = \beta\ell = 0$, and (6.48) reduces to

$$|\Gamma(0)| = 2^N|A| = \left|\frac{Z_L - Z_0}{Z_L + Z_0}\right|,$$

since for $f = 0$ all sections are of zero electrical length. Thus the constant A can be written as

$$A = 2^{-N}\left|\frac{Z_L - Z_0}{Z_L + Z_0}\right|. \tag{6.49}$$

Now expand $\Gamma(\theta)$ in (6.47) according to the binomial expansion:

$$\Gamma(\theta) = A(1 + e^{-2j\theta})^N = A\sum_{n=0}^{N} C_n^N e^{-2jn\theta}, \tag{6.50}$$

where

$$C_n^N = \frac{N!}{(N-n)!n!}, \tag{6.51}$$

are the binomial coefficients. Note that $C_n^N = C_{N-n}^N$, $C_0^N = 1$, and $C_1^N = N = C_{N-1}^N$. The key step is now to equate the desired passband response as given in (6.50), to the actual response as given (approximately) by (6.44):

$$\Gamma(\theta) = A \sum_{n=0}^{N} C_n^N e^{-2jn\theta} = \Gamma_0 + \Gamma_1 e^{-2j\theta} + \Gamma_2 e^{-4j\theta} + \cdots + \Gamma_N e^{-2jN\theta}.$$

This shows that the Γ_n must be chosen as

$$\Gamma_n = AC_n^N. \qquad 6.52$$

where A is given by (6.49), and C_n^N is a binomial coefficient.

At this point, the characteristic impedances Z_n can be found via (6.43), but a simpler solution can be obtained using the following approximation [1]. Since we assumed that the Γ_n are small, we can write

$$\Gamma_n = \frac{Z_{n+1} - Z_n}{Z_{n+1} + Z_n} \simeq \frac{1}{2} \ln \frac{Z_{n+1}}{Z_n},$$

since $\ln x \simeq 2(x-1)/(x+1)$. Then, using (6.52) and (6.49) gives

$$\ln \frac{Z_{n+1}}{Z_n} = 2\Gamma_n = 2AC_n^N = 2(2^{-N}) \frac{Z_L - Z_0}{Z_L + Z_0} C_n^N = 2^{-N} C_n^N \ln \frac{Z_L}{Z_0}, \qquad 6.53$$

which can be used to find Z_{n+1}, starting with $n = 0$. These results are approximate, but generally give usable results for $0.5Z_0 < Z_L < 2Z_0$.

Exact results can be found by using the transmission line equations for each section and numerically solving for the characteristic impedances [2]. The results of such calculations are listed in Table 6.1, which give the exact line impedances for $N = 2, 3, 4, 5,$ and 6 section binomial matching transformers, for various ratios of load impedance, Z_L, to feed line impedance, Z_0. The table gives results only for $Z_L/Z_0 > 1$; if $Z_L/Z_0 < 1$, the results for Z_0/Z_L should be used, but with Z_1 starting at the load end. This is because the solution is symmetric about $Z_L/Z_0 = 1$; the same transformer that matches Z_L to Z_0 can be reversed and used to match Z_0 to Z_L. More extensive tables can be found in reference [2].

The bandwidth of the binomial transformer can be evaluated as follows. As in Section 6.4, let Γ_m be the maximum value of reflection coefficient that can be tolerated over the passband. Then from (6.48),

$$\Gamma_m = 2^N |A| \cos^N \theta_m,$$

where $\theta_m < \pi/2$ is the lower edge of the passband, as shown in Figure 6.11. Thus,

$$\theta_m = \cos^{-1} \left[\frac{1}{2} \left(\frac{\Gamma_m}{A} \right)^{1/N} \right], \qquad 6.54$$

and using (6.33) gives the fractional bandwidth as

$$\frac{\Delta f}{f_0} = \frac{2(f_0 - f_m)}{f_0} = 2 - \frac{4\theta_m}{\pi}$$
$$= 2 - \frac{4}{\pi} \cos^{-1} \left[\frac{1}{2} \left(\frac{\Gamma_m}{A} \right)^{1/N} \right]. \qquad 6.55$$

TABLE 6.1 Binomial Transformer Design

Z_L/Z_0	$N=2$ Z_1/Z_0	Z_2/Z_0	$N=3$ Z_1/Z_0	Z_2/Z_0	Z_3/Z_0	$N=4$ Z_1/Z_0	Z_2/Z_0	Z_3/Z_0	Z_4/Z_0
1.0	1.0000	1.0000	1.0000	1.0000	1.0000	1.0000	1.0000	1.0000	1.0000
1.5	1.1067	1.3554	1.0520	1.2247	1.4259	1.0257	1.1351	1.3215	1.4624
2.0	1.1892	1.6818	1.0907	1.4142	1.8337	1.0444	1.2421	1.6102	1.9150
3.0	1.3161	2.2795	1.1479	1.7321	2.6135	1.0718	1.4105	2.1269	2.7990
4.0	1.4142	2.8285	1.1907	2.0000	3.3594	1.0919	1.5442	2.5903	3.6633
6.0	1.5651	3.8336	1.2544	2.4495	4.7832	1.1215	1.7553	3.4182	5.3500
8.0	1.6818	4.7568	1.3022	2.8284	6.1434	1.1436	1.9232	4.1597	6.9955
10.0	1.7783	5.6233	1.3409	3.1623	7.4577	1.1613	2.0651	4.8424	8.6110

Z_L/Z_0	$N=5$ Z_1/Z_0	Z_2/Z_0	Z_3/Z_0	Z_4/Z_0	Z_5/Z_0	$N=6$ Z_1/Z_0	Z_2/Z_0	Z_3/Z_0	Z_4/Z_0	Z_5/Z_0	Z_6/Z_0
1.0	1.0000	1.0000	1.0000	1.0000	1.0000	1.0000	1.0000	1.0000	1.0000	1.0000	1.0000
1.5	1.0128	1.0790	1.2247	1.3902	1.4810	1.0064	1.0454	1.1496	1.3048	1.4349	1.4905
2.0	1.0220	1.1391	1.4142	1.7558	1.9569	1.0110	1.0790	1.2693	1.5757	1.8536	1.9782
3.0	1.0354	1.2300	1.7321	2.4390	2.8974	1.0176	1.1288	1.4599	2.0549	2.6577	2.9481
4.0	1.0452	1.2995	2.0000	3.0781	3.8270	1.0225	1.1661	1.6129	2.4800	3.4302	3.9120
6.0	1.0596	1.4055	2.4495	4.2689	5.6625	1.0296	1.2219	1.8573	3.2305	4.9104	5.8275
8.0	1.0703	1.4870	2.8284	5.3800	7.4745	1.0349	1.2640	2.0539	3.8950	6.3291	7.7302
10.0	1.0789	1.5541	3.1623	6.4346	9.2687	1.0392	1.2982	2.2215	4.5015	7.7030	9.6228

EXAMPLE 6.7

Design a three-section binomial transformer to match a 50 Ω load to a 100 Ω line, and calculate the bandwidth for $\Gamma_m = 0.05$. Plot the reflection coefficient magnitude versus normalized frequency for the exact designs using 1, 2, 3, 4, and 5 sections.

Solution

For $N = 3, Z_L = 50\ \Omega, Z_0 = 100\ \Omega$ we have, from (6.49),

$$A = 2^{-N}\left|\frac{Z_L - Z_0}{Z_L + Z_0}\right| = 2^{-3}\left|\frac{50-100}{50+100}\right| = 0.0417.$$

From (6.55) the bandwidth is

$$\frac{\Delta f}{f_0} = 2 - \frac{4}{\pi}\cos^{-1}\left[\frac{1}{2}\left(\frac{\Gamma_m}{A}\right)^{1/N}\right] = 2 - \frac{4}{\pi}\cos^{-1}\left[\frac{1}{2}\left(\frac{0.05}{0.0417}\right)^{1/3}\right] = 0.71, \text{ or } 71\%.$$

The necessary binomial coefficients are

$$C_0^3 = \frac{3!}{3!0!} = 1,$$

$$C_1^3 = \frac{3!}{2!1!} = 3,$$

$$C_2^3 = \frac{3!}{1!2!} = 3.$$

Then using (6.53) gives the required characteristic impedances as

$$n = 0:\ \ln Z_1 = \ln Z_0 + 2^{-N}C_0^3 \ln\frac{Z_L}{Z_0}$$

$$= \ln 100 + 2^{-3}(1)\ln\frac{50}{100} = 4.518,$$

$$Z_1 = 91.7\ \Omega;$$

$$n = 1:\ \ln Z_2 = \ln Z_1 + 2^{-N}C_1^3 \ln\frac{Z_L}{Z_0}$$

$$= \ln 91.7 + 2^{-3}(3)\ln\frac{50}{100} = 4.26,$$

$$Z_2 = 70.7\ \Omega;$$

$$n = 2:\ \ln Z_3 = \ln Z_2 + 2^{-N}C_2^3 \ln\frac{Z_L}{Z_0}$$

$$= \ln 70.7 + 2^{-3}(3)\ln\frac{50}{100} = 4.00,$$

$$Z_3 = 54.5\ \Omega.$$

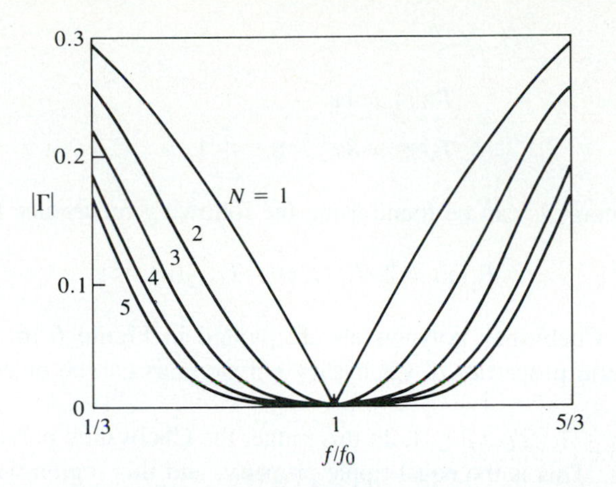

FIGURE 6.15 Reflection coefficient magnitude versus frequency for multisection binomial matching transformers of Example 6.7. $Z_L = 50\ \Omega$ and $Z_0 = 100\ \Omega$.

To use the data in Table 6.1, we reverse the source and load impedances and consider the problem of matching a 100 Ω load to a 50 Ω line. Then $Z_L/Z_0 = 2.0$, and we obtain the exact characteristic impedances as $Z_1 = 91.7\ \Omega$, $Z_2 = 70.7\ \Omega$, and $Z_3 = 54.5\ \Omega$, which agree with the approximate results to three significant digits. Figure 6.15 shows the reflection coefficient magnitude versus frequency for exact designs using $N = 1, 2, 3, 4$, and 5 sections. Observe that greater bandwidth is obtained for transformers using more sections. ○

6.7 CHEBYSHEV MULTISECTION MATCHING TRANSFORMERS

In contrast with the binomial matching transformer, the Chebyshev transformer optimizes bandwidth at the expense of passband ripple. If such a passband characteristic can be tolerated, the bandwidth of the Chebyshev transformer will be substantially better than that of the binomial transformer, for a given number of sections. The Chebyshev transformer is designed by equating $\Gamma(\theta)$ to a Chebyshev polynomial, which has the optimum characteristics needed for this type of transformer. Thus we will first discuss the properties of the Chebyshev polynomials, and then derive a design procedure for Chebyshev matching transformers using the small reflection theory of Section 6.5.

Chebyshev Polynomials

The nth order Chebyshev polynomial is a polynomial of degree n, and is denoted by $T_n(x)$. The first four Chebyshev polynomials are

$$T_1(x) = x, \qquad 6.56a$$

$$T_2(x) = 2x^2 - 1, \qquad 6.56b$$

$$T_3(x) = 4x^3 - 3x, \tag{6.56c}$$

$$T_4(x) = 8x^4 - 8x^2 + 1. \tag{6.56d}$$

Higher-order polynomials can be found using the following recurrence formula:

$$T_n(x) = 2xT_{n-1}(x) - T_{n-2}(x). \tag{6.57}$$

The first four Chebyshev polynomials are plotted in Figure 6.16, from which the following very useful properties of Chebyshev polynomials can be noted:

- For $-1 \leq x \leq 1$, $|T_n(x)| \leq 1$. In this range, the Chebyshev polynomials oscillate between ± 1. This is the equal ripple property, and this region will be mapped to the passband of the matching transformer.
- For $|x| > 1$, $|T_n(x)| > 1$. This region will map to the frequency range outside the passband.
- For $|x| > 1$, the $|T_n(x)|$ increases faster with x as n increases.

Now let $x = \cos\theta$ for $|x| < 1$. Then it can be shown that the Chebyshev polynomials can be expressed as

$$T_n(\cos\theta) = \cos n\theta,$$

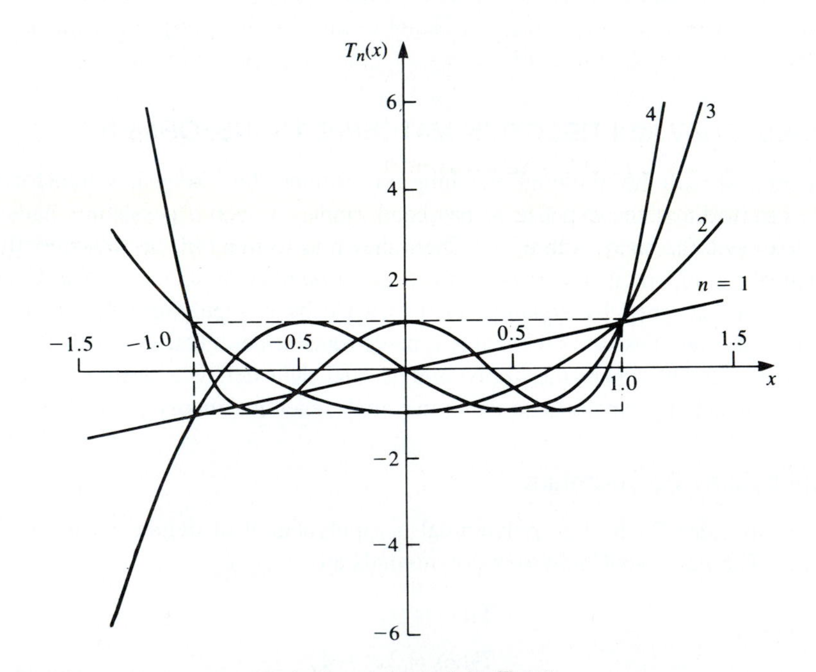

FIGURE 6.16 The first four Chebyshev polynomials, $T_n(x)$.

or more generally as

$$T_n(x) = \cos(n \cos^{-1} x), \quad \text{for } |x| < 1, \tag{6.58a}$$

$$T_n(x) = \cosh(n \cosh^{-1} x), \quad \text{for } |x| > 1. \tag{6.58b}$$

We desire equal ripple in the passband of the transformer, so it is necessary to map θ_m to $x = 1$ and $\pi - \theta_m$ to $x = -1$, where θ_m and $\pi - \theta_m$ are the lower and upper edges of the passband, as shown in Figure 6.11. This can be accomplished by replacing $\cos\theta$ in (6.58a) with $\cos\theta/\cos\theta_m$:

$$T_n\left(\frac{\cos\theta}{\cos\theta_m}\right) = T_n(\sec\theta_m \cos\theta) = \cos n\left[\cos^{-1}\left(\frac{\cos\theta}{\cos\theta_m}\right)\right]. \tag{6.59}$$

Then $|\sec\theta_m \cos\theta| \leq 1$ for $\theta_m < \theta < \pi - \theta_m$, so $|T_n(\sec\theta_m \cos\theta)| \leq 1$ over this same range.

Since $\cos^n\theta$ can be expanded into a sum of terms of the form $\cos(n - 2m)\theta$, the Chebyshev polynomials of (6.56) can be rewritten in the following useful form:

$$T_1(\sec\theta_m \cos\theta) = \sec\theta_m \cos\theta, \tag{6.60a}$$

$$T_2(\sec\theta_m \cos\theta) = \sec^2\theta_m(1 + \cos 2\theta) - 1, \tag{6.60b}$$

$$T_3(\sec\theta_m \cos\theta) = \sec^3\theta_m(\cos 3\theta + 3\cos\theta) - 3\sec\theta_m \cos\theta, \tag{6.60c}$$

$$T_4(\sec\theta_m \cos\theta) = \sec^4\theta_m(\cos 4\theta + 4\cos 2\theta + 3), \\ - 4\sec^2\theta_m(\cos 2\theta + 1) + 1. \tag{6.60d}$$

The above results can be used to design matching transformers with up to four sections, and will also be used in later chapters for the design of directional couplers and filters.

Design of Chebyshev Transformers

We can now synthesize a Chebyshev equal-ripple passband by making $\Gamma(\theta)$ proportional to $T_N(\sec\theta_m \cos\theta)$, where N is the number of sections in the transformer. Thus, using (6.46),

$$\Gamma(\theta) = 2e^{-jN\theta}[\Gamma_0 \cos N\theta + \Gamma_1 \cos(N-2)\theta + \cdots + \Gamma_n \cos(N-2n)\theta + \cdots] \\ = Ae^{-jN\theta} T_N(\sec\theta_m \cos\theta), \tag{6.61}$$

where the last term in the series of (6.61) is $(1/2)\Gamma_{N/2}$ for N even and $\Gamma_{(N-1)/2}\cos\theta$ for N odd. As in the binomial transformer case, we can find the constant A by letting $\theta = 0$, corresponding to zero frequency. Thus,

$$\Gamma(0) = \frac{Z_L - Z_0}{Z_L + Z_0} = AT_N(\sec\theta_m),$$

so we have

$$A = \frac{Z_L - Z_0}{Z_L + Z_0} \frac{1}{T_N(\sec\theta_m)}. \tag{6.62}$$

Now if the maximum allowable reflection coefficient magnitude in the passband is Γ_m, then from (6.61) $\Gamma_m = A$, since the maximum value of $T_n(\sec\theta_m \cos\theta)$ in the passband is unity. Then from (6.62) θ_m is determined as

$$T_N(\sec\theta_m) = \frac{1}{\Gamma_m}\left|\frac{Z_L - Z_0}{Z_L + Z_0}\right|,$$

or, using (6.58b),

$$\sec\theta_m = \cosh\left[\frac{1}{N}\cosh^{-1}\left(\frac{1}{\Gamma_m}\left|\frac{Z_L - Z_0}{Z_L + Z_0}\right|\right)\right]. \tag{6.63}$$

Once θ_m is known, the fractional bandwidth can be calculated from (6.33) as

$$\frac{\Delta f}{f_0} = 2 - \frac{4\theta_m}{\pi}. \tag{6.64}$$

From (6.61), the Γ_n can be determined using the results of (6.60) to expand $T_N(\sec\theta_m \cos\theta)$ and equating similar terms of the form $\cos(N-2n)\theta$. The characteristic impedances Z_n can then be found from (6.43). This procedure will be illustrated in Example 6.8.

The above results are approximate because of the reliance on small reflection theory, but are general enough to design transformers with an arbitrary ripple level, Γ_m. Table 6.2 gives exact results [2] for a few specific values of Γ_m, for $N = 2, 3$, and 4 sections; more extensive tables can be found in reference [2].

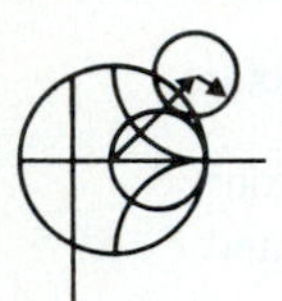

EXAMPLE 6.8

Design a three-section Chebyshev transformer to match a 100 Ω load to a 50 Ω line, with $\Gamma_m = 0.05$, using the above theory. Plot the reflection coefficient magnitude versus normalized frequency for exact designs using 1, 2, 3, and 4 sections.

Solution
From (6.61) with $N = 3$,

$$\Gamma(\theta) = 2e^{-j3\theta}[\Gamma_0 \cos 3\theta + \Gamma_1 \cos\theta] = Ae^{-j3\theta}T_3(\sec\theta_m \cos\theta).$$

Then, $A = \Gamma_m = 0.05$, and from (6.63),

$$\begin{aligned}\sec\theta_m &= \cosh\left[\frac{1}{N}\cosh^{-1}\left(\frac{1}{\Gamma_m}\left|\frac{Z_L - Z_0}{Z_L + Z_0}\right|\right)\right]\\ &= \cosh\left[\frac{1}{3}\cosh^{-1}\left(\frac{1}{0.05}\left|\frac{100 - 50}{100 + 50}\right|\right)\right]\\ &= 1.395,\end{aligned}$$

so, $\theta_m = 44.2°$.

Using (6.60c) for T_3 gives

$$2[\Gamma_0 \cos 3\theta + \Gamma_1 \cos\theta] = A\sec^3\theta_m(\cos 3\theta + 3\cos\theta) - 3A\sec\theta_m \cos\theta.$$

TABLE 6.2 Chebyshev Transformer Design

	$N=2$				$N=3$					
	$\Gamma_m = 0.05$		$\Gamma_m = 0.20$		$\Gamma_m = 0.05$			$\Gamma_m = 0.20$		
Z_L/Z_0	Z_1/Z_0	Z_2/Z_0	Z_1/Z_0	Z_2/Z_0	Z_1/Z_0	Z_2/Z_0	Z_3/Z_0	Z_1/Z_0	Z_2/Z_0	Z_3/Z_0
1.0	1.0000	1.0000	1.0000	1.0000	1.0000	1.0000	1.0000	1.0000	1.0000	1.0000
1.5	1.1347	1.3219	1.2247	1.2247	1.1029	1.2247	1.3601	1.2247	1.2247	1.2247
2.0	1.2193	1.6402	1.3161	1.5197	1.1475	1.4142	1.7429	1.2855	1.4142	1.5558
3.0	1.3494	2.2232	1.4565	2.0598	1.2171	1.7321	2.4649	1.3743	1.7321	2.1829
4.0	1.4500	2.7585	1.5651	2.5558	1.2662	2.0000	3.1591	1.4333	2.0000	2.7908
6.0	1.6047	3.7389	1.7321	3.4641	1.3383	2.4495	4.4833	1.5193	2.4495	3.9492
8.0	1.7244	4.6393	1.8612	4.2983	1.3944	2.8284	5.7372	1.5766	2.8284	5.0742
10.0	1.8233	5.4845	1.9680	5.0813	1.4385	3.1623	6.9517	1.6415	3.1623	6.0920

	$N=4$							
	$\Gamma_m = 0.05$				$\Gamma_m = 0.20$			
Z_L/Z_0	Z_1/Z_0	Z_2/Z_0	Z_3/Z_0	Z_4/Z_0	Z_1/Z_0	Z_2/Z_0	Z_3/Z_0	Z_4/Z_0
1.0	1.0000	1.0000	1.0000	1.0000	1.0000	1.0000	1.0000	1.0000
1.5	1.0892	1.1742	1.2775	1.3772	1.2247	1.2247	1.2247	1.2247
2.0	1.1201	1.2979	1.5409	1.7855	1.2727	1.3634	1.4669	1.5715
3.0	1.1586	1.4876	2.0167	2.5893	1.4879	1.5819	1.8965	2.0163
4.0	1.1906	1.6414	2.4369	3.3597	1.3692	1.7490	2.2870	2.9214
6.0	1.2290	1.8773	3.1961	4.8820	1.4415	2.0231	2.9657	4.1623
8.0	1.2583	2.0657	3.8728	6.3578	1.4914	2.2428	3.5670	5.3641
10.0	1.2832	2.2268	4.4907	7.7930	1.5163	2.4210	4.1305	6.5950

Equating similar terms in $\cos n\theta$ gives the following results:

$$\cos 3\theta\colon\ 2\Gamma_0 = A \sec^3 \theta_m,$$

$$\Gamma_0 = 0.0678;$$

$$\cos\theta\colon\ 2\Gamma_1 = 3A(\sec^3\theta_m - \sec\theta_m),$$

$$\Gamma_1 = 0.099.$$

From symmetry we also have that

$$\Gamma_3 = \Gamma_0 = 0.0678,$$

and

$$\Gamma_2 = \Gamma_1 = 0.099.$$

Then the characteristic impedances are

$$Z_1 = Z_0 \frac{1+\Gamma_0}{1-\Gamma_0} = 50\frac{1+.0678}{1-.0678} = 57.27\ \Omega,$$

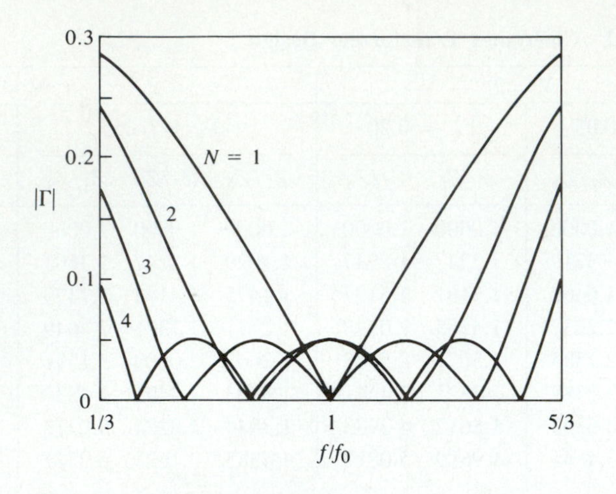

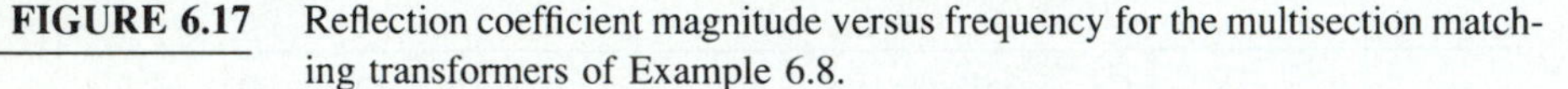

FIGURE 6.17 Reflection coefficient magnitude versus frequency for the multisection matching transformers of Example 6.8.

$$Z_2 = Z_1 \frac{1+\Gamma_1}{1-\Gamma_1} = 57.27 \frac{1+.099}{1-.099} = 69.86\ \Omega,$$

$$Z_3 = Z_L \frac{1-\Gamma_3}{1+\Gamma_3} = 100 \frac{1-.0678}{1+.0678} = 87.30\ \Omega.$$

Note that Z_1 and Z_2 were calculated using (6.43b), starting at the input side of the transformer, but Z_3 was calculated from the load side, using (6.43c). This avoids the cumulative error which would occur if we calculated all impedances in a progression from one side. These values can be compared to the exact values from Table 6.2 of $Z_1 = 57.37\ \Omega$, $Z_2 = 70.71\ \Omega$, and $Z_3 = 87.15\ \Omega$. The bandwidth, from (6.64), is

$$\frac{\Delta f}{f_0} = 2 - \frac{4\theta_m}{\pi} = 2 - 4\left(\frac{44.2^\circ}{180^\circ}\right) = 1.02,$$

or 102%. This is significantly greater than the bandwidth of the binomial transformer of Example 6.7 (71%), which was for the same type of mismatch. The trade-off, of course, is a nonzero ripple in the passband of the Chebyshev transformer.

Figure 6.17 shows reflection coefficient magnitudes versus frequency for the exact designs from Table 6.2 for $N = 1, 2, 3$, and 4 sections. ○

6.8 TAPERED LINES

In the preceding sections we discussed how an arbitrary real load impedance could be matched to a line over a desired bandwidth by using multisection matching transformers.

As the number, N, of discrete sections increases, the step changes in characteristic impedance between the sections become smaller. Thus, in the limit of an infinite number of sections, we approach a continuously tapered line. In practice, of course, a matching transformer must be of finite length, often no more than a few sections long. But instead of discrete sections, the line can be continuously tapered, as suggested in Figure 6.18a. Then by changing the type of taper, we can obtain different passband characteristics.

In this section we will derive an approximate theory, based on the theory of small reflections, to predict the reflection coefficient response as a function of the impedance taper, $Z(z)$. We will then apply these results to a few common types of tapers.

Consider the continuously tapered line of Figure 6.18a as being made up of a number of incremental sections of length Δz, with an impedance change $\Delta Z(z)$ from one section to the next, as shown in Figure 6.18b. Then the incremental reflection coefficient from the step at z is given by

$$\Delta\Gamma = \frac{(Z+\Delta Z)-Z}{(Z+\Delta Z)+Z} \simeq \frac{\Delta Z}{2Z}. \tag{6.65}$$

In the limit as $\Delta z \to 0$, we have an exact differential:

$$d\Gamma = \frac{dZ}{2Z} = \frac{1}{2}\frac{d(\ln Z/Z_0)}{dz}dz, \tag{6.66}$$

since

$$\frac{d(\ln f(z))}{dz} = \frac{1}{f}\frac{df(z)}{dz}.$$

Then, by using the theory of small reflections, the total reflection coefficient at $z = 0$

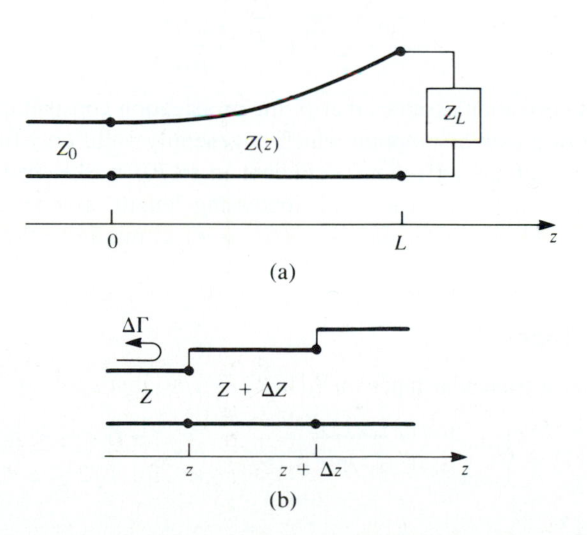

FIGURE 6.18 A tapered transmission line matching section and the model for an incremental length of tapered line. (a) The tapered transmission line matching section. (b) Model for an incremental step change in impedance of the tapered line.

can be found by summing all the partial reflections with their appropriate phase shifts:

$$\Gamma(\theta) = \frac{1}{2}\int_{z=0}^{L} e^{-2j\beta z}\frac{d}{dz}\ln\left(\frac{Z}{Z_0}\right)dz, \qquad 6.67$$

where $\theta = 2\beta z$. So if $Z(z)$ is known, $\Gamma(\theta)$ can be found as a function of frequency. Alternatively, if $\Gamma(\theta)$ is specified, then in principle $Z(z)$ can be found. This latter procedure is difficult, and is generally avoided in practice; the reader is referred to references [1], [4] for further discussion along these lines. Here we will consider three special cases of $Z(z)$ impedance tapers, and evaluate the resulting responses.

Exponential Taper

Consider first an exponential taper, where

$$Z(z) = Z_0 e^{az}, \quad \text{for } 0 < z < L, \qquad 6.68$$

as indicated in Figure 6.19a. At $z = 0$, $Z(0) = Z_0$, as desired. At $z = L$, we wish to have $Z(L) = Z_L = Z_0 e^{aL}$, which determines the constant a as

$$a = \frac{1}{L}\ln\left(\frac{Z_L}{Z_0}\right). \qquad 6.69$$

We now find $\Gamma(\theta)$ by using (6.68) and (6.69) in (6.67):

$$\begin{aligned}\Gamma &= \frac{1}{2}\int_0^L e^{-2j\beta z}\frac{d}{dz}(\ln e^{az})dz \\ &= \frac{\ln Z_L/Z_0}{2L}\int_0^L e^{-2j\beta z}dz \\ &= \frac{\ln Z_L/Z_0}{2}e^{-j\beta L}\frac{\sin\beta L}{\beta L}.\end{aligned} \qquad 6.70$$

Observe that this derivation assumes that β, the propagation constant of the tapered line, is not a function of z—an assumption which is generally valid only for TEM lines.

The magnitude of the reflection coefficient in (6.70) is sketched in Figure 6.19b; note that the peaks in $|\Gamma|$ decrease with increasing length, as one might expect, and that the length should be greater than $\lambda/2$ ($\beta L > \pi$) to minimize the mismatch at low frequencies.

Triangular Taper

Next consider a triangular taper for $(d\ln Z/Z_0)/dz$, that is,

$$Z(z) = \begin{cases} Z_0 e^{2(z/L)^2 \ln Z_L/Z_0} & \text{for } 0 \le z \le L/2 \\ Z_0 e^{(4z/L - 2z^2/L^2 - 1)\ln Z_L/Z_0} & \text{for } L/2 \le z \le L. \end{cases} \qquad 6.71$$

Then,

$$\frac{d(\ln Z/Z_0)}{dz} = \begin{cases} 4z/L^2 \ln Z_L/Z_0 & \text{for } 0 \le z \le L/2 \\ (4/L - 4z/L^2)\ln Z_L/Z_0 & \text{for } L/2 \le z \le L. \end{cases} \qquad 6.72$$

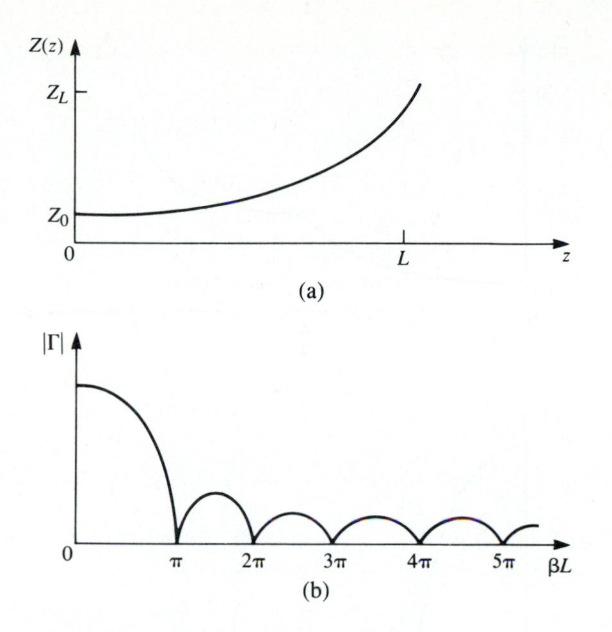

FIGURE 6.19 A matching section with an exponential impedance taper. (a) Variation of impedance. (b) Resulting reflection coefficient magnitude response.

$Z(z)$ is plotted in Figure 6.20a. Evaluating Γ from (6.67) gives

$$\Gamma(\theta) = \frac{1}{2} e^{-j\beta L} \ln\left(\frac{Z_L}{Z_0}\right) \left[\frac{\sin(\beta L/2)}{\beta L/2}\right]^2. \qquad 6.73$$

The magnitude of this result is sketched in Figure 6.20b. Note that, for $\beta L > 2\pi$, the peaks of the triangular taper are lower than the corresponding peaks of the exponential case. But the first null for the triangular taper occurs at $\beta L = 2\pi$, whereas for the exponential taper it occurs at $\beta L = \pi$.

Klopfenstein Taper

Considering the fact that there are an infinite number of possibilities for choosing an impedance matching taper, it is logical to ask if there is a design which is "best." For a given taper length (greater than a critical value), the Klopfenstein impedance taper [4], [5] has been shown to be optimum in the sense that the reflection coefficient is minimum over the passband. Alternatively, for a maximum reflection coefficient specification in the passband, the Klopfenstein taper yields the shortest matching section.

The Klopfenstein taper is derived from a stepped Chebyshev transformer as the number of sections increases to infinity, and is analogous to the Taylor distribution of antenna array theory. We will not present the details of this derivation, which can be found in references [1], [4]; only the necessary results for the design of Klopfenstein tapers are given below.

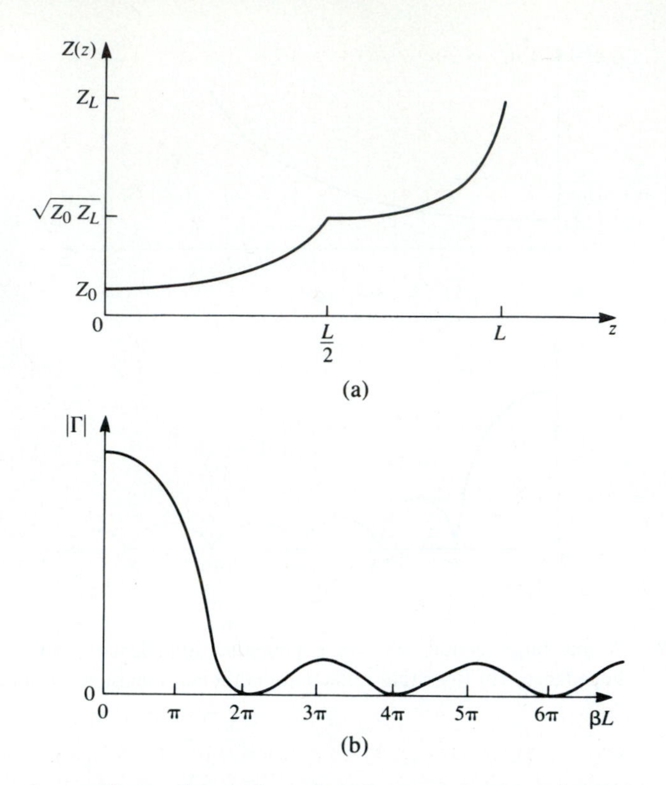

FIGURE 6.20 A matching section with a triangular taper for $d(\ln Z/Z_0)/dz$. (a) Variation of impedance. (b) Resulting reflection coefficient magnitude response.

The logarithm of the characteristic impedance variation for the Klopfenstein taper is given by

$$\ln Z(z) = \frac{1}{2} \ln Z_0 Z_L + \frac{\Gamma_0}{\cosh A} A^2 \phi(2z/L - 1, A), \quad \text{for } 0 \leq z \leq L, \qquad 6.74$$

where the function $\phi(x, A)$ is defined as

$$\phi(x, A) = -\phi(-x, A) = \int_0^x \frac{I_1(A\sqrt{1-y^2})}{A\sqrt{1-y^2}} dy, \quad \text{for } |x| \leq 1, \qquad 6.75$$

where $I_1(x)$ is the modified Bessel function. This function takes the following special values:

$$\phi(0, A) = 0$$

$$\phi(x, 0) = \frac{x}{2}$$

$$\phi(1, A) = \frac{\cosh A - 1}{A^2},$$

but otherwise must be calculated numerically. A very simple and efficient method for doing this is available [6].

The resulting reflection coefficient is given by

$$\Gamma(\theta) = \Gamma_0 e^{-j\beta L} \frac{\cos\sqrt{(\beta L)^2 - A^2}}{\cosh A}, \qquad \text{for } \beta L > A. \tag{6.76}$$

If $\beta L < A$, the $\cos\sqrt{(\beta L)^2 - A^2}$ term becomes $\cosh\sqrt{A^2 - (\beta L)^2}$.

In (6.74) and (6.76), Γ_0 is the reflection coefficient at zero frequency, given as

$$\Gamma_0 = \frac{Z_L - Z_0}{Z_L + Z_0} \simeq \frac{1}{2} \ln\left(\frac{Z_L}{Z_0}\right). \tag{6.77}$$

The passband is defined as $\beta L \geq A$, and so the maximum ripple in the passband is

$$\Gamma_m = \frac{\Gamma_0}{\cosh A}, \tag{6.78}$$

because $\Gamma(\theta)$ oscillates between $\pm\Gamma_0/\cosh A$ for $\beta L > A$.

It is interesting to note that the impedance taper of (6.74) has steps at $z = 0$ and L (the ends of the tapered section), and so does not smoothly join the source and load impedances. A typical Klopfenstein impedance taper and its response are given in the following example.

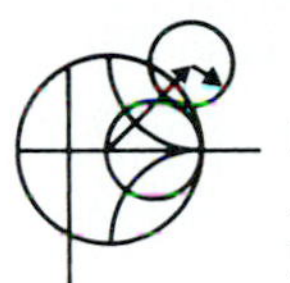

EXAMPLE 6.9

Design a triangular taper, an exponential taper, and a Klopfenstein taper (with $\Gamma_m = 0.02$) to match a 50 Ω load to a 100 Ω line. Plot the impedance variations and resulting reflection coefficient magnitudes versus βL.

Solution

Triangular taper: From (6.71) the impedance variation is

$$Z(z) = Z_0 \begin{cases} e^{2(z/L)^2 \ln Z_L/Z_0} & \text{for } 0 \leq z \leq L/2 \\ e^{(4z/L - 2z^2/L^2 - 1)\ln Z_L/Z_0} & \text{for } L/2 \leq z \leq L, \end{cases}$$

with $Z_0 = 100$ Ω and $Z_L = 50$ Ω. The resulting reflection coefficient response is given by (6.73):

$$|\Gamma(\theta)| = \frac{1}{2} \ln\left(\frac{Z_L}{Z_0}\right) \left[\frac{\sin(\beta L/2)}{\beta L/2}\right]^2.$$

Exponential taper: From (6.68) the impedance variation is

$$Z(z) = Z_0 e^{az}, \qquad \text{for } 0 < z < L,$$

with $a = (1/L) \ln Z_L/Z_0 = 0.693/L$. The reflection coefficient response is, from (6.70),

$$|\Gamma(\theta)| = \frac{1}{2} \ln\left(\frac{Z_L}{Z_0}\right) \frac{\sin \beta L}{\beta L}.$$

Klopfenstein taper: Using (6.77) gives Γ_0 as

$$\Gamma_0 = \frac{1}{2} \ln\left(\frac{Z_L}{Z_0}\right) = 0.346,$$

and (6.78) gives A as

$$A = \cosh^{-1}\left(\frac{\Gamma_0}{\Gamma_m}\right) = \cosh^{-1}\left(\frac{0.346}{0.002}\right) = 3.543.$$

The impedance taper must be numerically evaluated from (6.74). The reflection coefficient magnitude is given by (6.76):

$$|\Gamma(\theta)| = \Gamma_0 \frac{\cos\sqrt{(\beta L)^2 - A^2}}{\cosh A}.$$

The passband for the Klopfenstein taper is defined as $\beta L > A = 3.543 = 1.13\pi$.

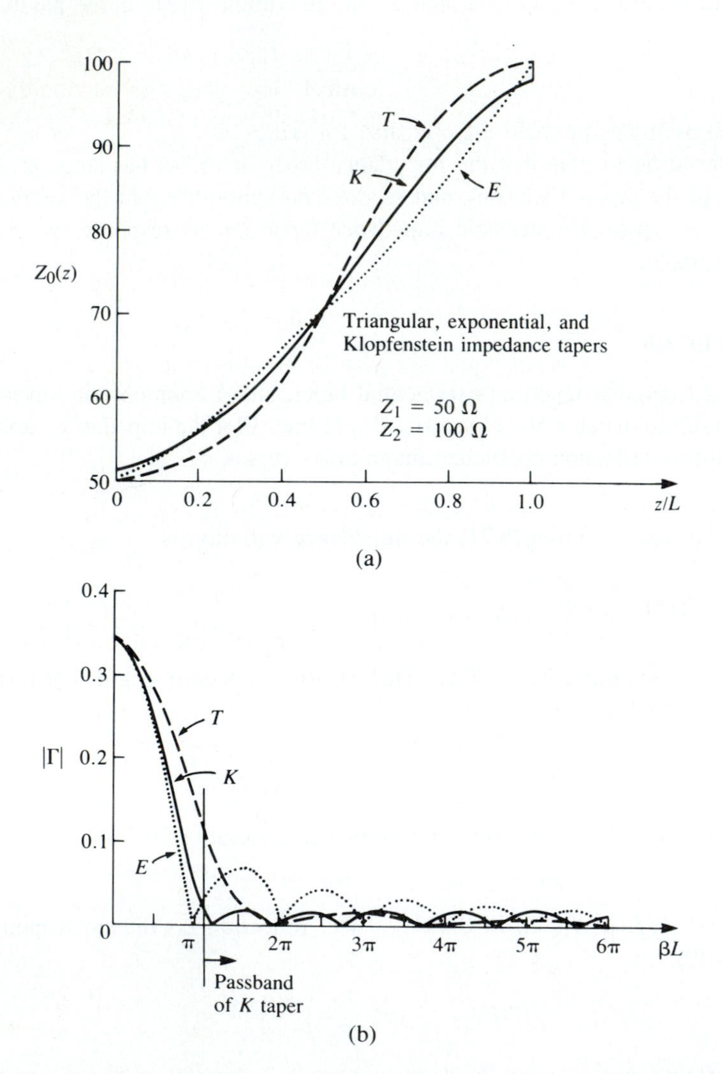

FIGURE 6.21 Solution to Example 6.9. (a) Impedance variations for the triangular, exponential, and Klopfenstein tapers. (b) Resulting reflection coefficient magnitude versus frequency for the tapers of (a).

Figure 6.21a,b shows the impedance variations (versus z/L), and the resulting reflection coefficient magnitude (versus βL) for the three types of tapers. The Klopfenstein taper is seen to give the desired response of $|\Gamma| \leq \Gamma_m = 0.02$ for $\beta L \geq 1.13\pi$, which is lower than either the triangular or exponential taper responses. Also note that, like the stepped-Chebyshev matching transformer, the response of the Klopfenstein taper has equal-ripple lobes versus frequency in its passband. ○

6.9 THE BODE-FANO CRITERIA

In this chapter we discussed several techniques for matching an arbitrary load at a single frequency, using lumped elements, tuning stubs, and single-section quarter-wave transformers. We then presented multisection matching transformers and tapered lines as a means of obtaining broader bandwidths, with various passband characteristics. We will now close our study of impedance matching with a somewhat qualitative discussion of the theoretical limits that constrain the performance of an impedance matching network.

We limit our discussion to the circuit of Figure 6.1, where a lossless network is used to match an arbitrary complex load, generally over a nonzero bandwidth. From a very general perspective, we might raise the following questions in regard to this problem:

- Can we achieve a perfect match (zero reflection) over a specified bandwidth?
- If not, how good can we do? What is the trade-off between Γ_m, the maximum allowable reflection in the passband, and the bandwidth?
- How complex must the matching network be for a given specification?

These questions can be answered by the Bode-Fano criteria [7], [8] which gives, for certain canonical types of load impedances, a theoretical limit on the minimum reflection coefficient magnitude that can be obtained with an arbitrary matching network. The Bode-Fano criteria thus represents the optimum result that can be ideally achieved, even though such a result may only be approximated in practice. Such optimal results are always important, however, because they give us the upper limit of performance, and provide a benchmark against which a practical design can be compared.

Figure 6.22a shows a lossless network used to match a parallel *RC* load impedance. The Bode-Fano criteria states that

$$\int_0^\infty \ln \frac{1}{|\Gamma(\omega)|} d\omega \leq \frac{\pi}{RC}, \tag{6.79}$$

where $\Gamma(\omega)$ is the reflection coefficient seen looking into the arbitrary lossless matching network. The derivation of this result is beyond the scope of this text (the interested reader is referred to references [7] and [8]), but our goal here is to discuss the implications of the above result.

Assume that we desire to synthesize a matching network with a reflection coefficient response like that shown in Figure 6.23a. Applying (6.79) to this function gives

$$\int_0^\infty \ln \frac{1}{|\Gamma|} d\omega = \int_{\Delta\omega} \ln \frac{1}{\Gamma_m} d\omega = \Delta\omega \ln \frac{1}{\Gamma_m} \leq \frac{\pi}{RC}, \tag{6.80}$$

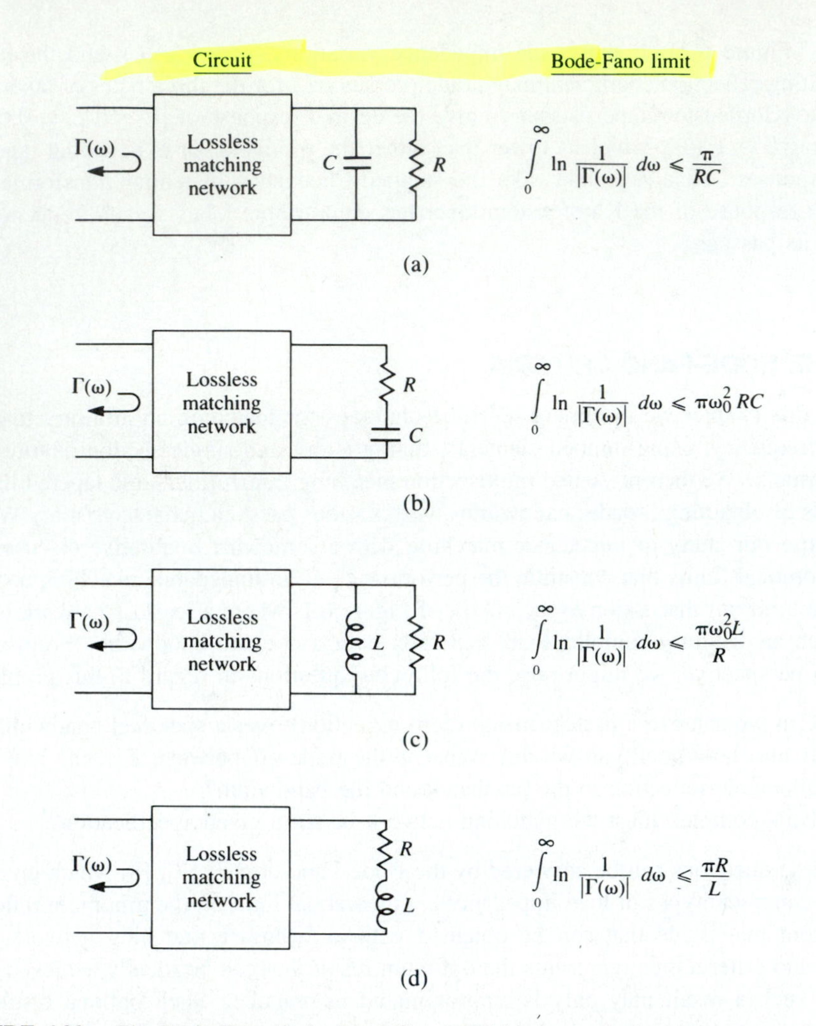

FIGURE 6.22 The Bode-Fano limits for RC and RL loads matched with passive and lossless networks (ω_0 is the center frequency of the matching bandwidth). (a) Parallel RC. (b) Series RC. (c) Parallel RL. (d) Series RL.

which leads to the following conclusions:

- For a given load (fixed RC product), a broader bandwidth ($\Delta\omega$) can only be achieved at the expense of a higher reflection coefficient in the passband (Γ_m).
- The passband reflection coefficient Γ_m cannot be zero unless $\Delta\omega = 0$. Thus a perfect match can only be achieved at a finite number of frequencies, as illustrated in Figure 6.23b.
- As R and/or C increase, the quality of the match ($\Delta\omega$ and/or $1/\Gamma_m$) must decrease. Thus, higher-Q circuits are intrinsically harder to match than are lower-Q circuits.

Since $\ln 1/|\Gamma|$ is proportional to the return loss (in dB) at the input of the matching network, (6.79) can be interpreted as requiring that the area between the return loss curve

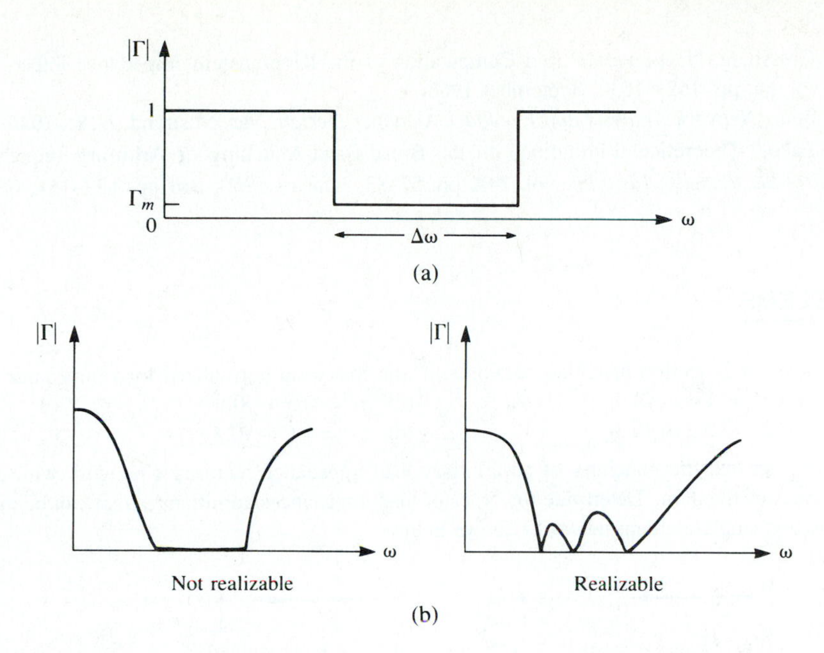

FIGURE 6.23 Illustrating the Bode-Fano criteria. (a) A possible reflection coefficient response. (b) Nonrealizable and realizable reflection coefficient responses.

and the $|\Gamma| = 1$ ($RL = 0$ dB) axis must be less than or equal to a constant. Optimization then implies that the return loss curve be adjusted so that $|\Gamma| = \Gamma_m$ over the passband and $|\Gamma| = 1$ elsewhere, as in Figure 6.23a. In this way, no area under the return loss curve is wasted outside the passband, or lost in regions within the passband for which $|\Gamma| < \Gamma_m$. The square-shaped response of Figure 6.23a is thus the optimum response, but cannot be realized in practice because it would require an infinite number of elements in the matching network. It can be approximated, however, with a reasonably small number of elements, as described in reference [8]. Finally, note that the Chebyshev matching transformer can be considered as a close approximation to the ideal passband of Figure 6.23a, when the ripple of the Chebyshev response is made equal to Γ_m. Figure 6.22 lists the Bode-Fano limits for other types of RC and RL loads.

REFERENCES

[1] R. E. Collin, *Foundations for Microwave Engineering*, McGraw-Hill, N.Y. 1966.

[2] G. L. Matthaei, L. Young, and E. M. T. Jones, *Microwave Filters, Impedance-Matching Networks, and Coupling Structures,* Artech House Books, Dedham, Mass. 1980.

[3] P. Bhartia and I. J. Bahl, *Millimeter Wave Engineering and Applications*, Wiley Interscience, N.Y., 1984.

[4] R. E. Collin, "The Optimum Tapered Transmission Line Matching Section," *Proc. IRE,* vol. 44, pp. 539-548, April 1956.

[5] R. W. Klopfenstein, "A Transmission Line Taper of Improved Design," *Proc. IRE,* vol. 44, pp. 31-15, January 1956.

[6] M. A. Grossberg, "Extremely Rapid Computation of the Klopfenstein Impedance Taper," *Proc. IEEE,* vol. 56, pp. 1629-1630, September 1968.

[7] H. W. Bode, *Network Analysis and Feedback Amplifier Design,* Van Nostrand, N.Y., 1945.

[8] R. M. Fano, "Theoretical Limitations on the Broad-Band Matching of Arbitrary Impedances," *Journal of the Franklin Institute,* vol. 249, pp. 57–83, January 1950, and pp. 139–154, February 1950.

PROBLEMS

6.1 Design lossless L section matching networks for the following normalized load impedances:

(a) $z_L = 1.4 + j2.0$ (c) $z_L = 0.5 + j0.9$

(b) $z_L = 0.2 + j0.3$ (d) $z_L = 1.6 - j0.3$

6.2 We have seen that the matching of an arbitrary load impedance requires a network with at least two degrees of freedom. Determine the types of load impedances/admittances that can be matched with the two single-element networks shown below.

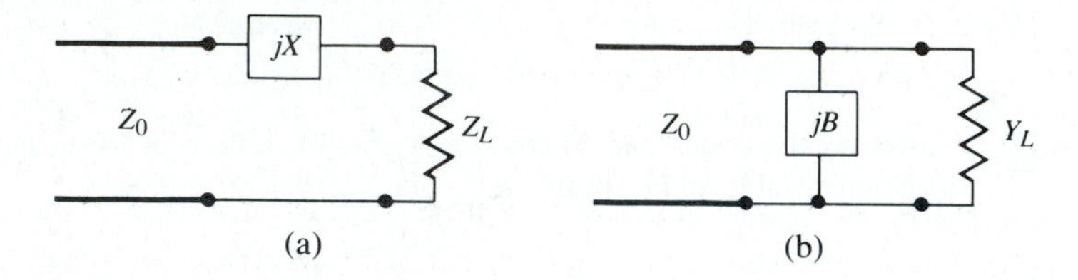

6.3 A load impedance $Z_L = 200 + j160\ \Omega$ is to be matched to a 100 Ω line using a single shunt-stub tuner. Find two solutions using open-circuited stubs.

6.4 Repeat Problem 6.3 using short-circuited stubs.

6.5 A load impedance $Z_L = 20 - j60\ \Omega$ is to be matched to a 50 Ω line using a single series stub tuner. Find two solutions using open-circuited stubs.

6.6 Repeat Problem 6.5 using short-circuited stubs.

6.7 In the circuit shown below a $Z_L = 200 + j100\ \Omega$ load is to be matched to a 40 Ω line, using a length, ℓ, of lossless transmission line of characteristic impedance, Z_1. Find ℓ and Z_1. Determine, in general, what type of load impedances can be matched using such a circuit.

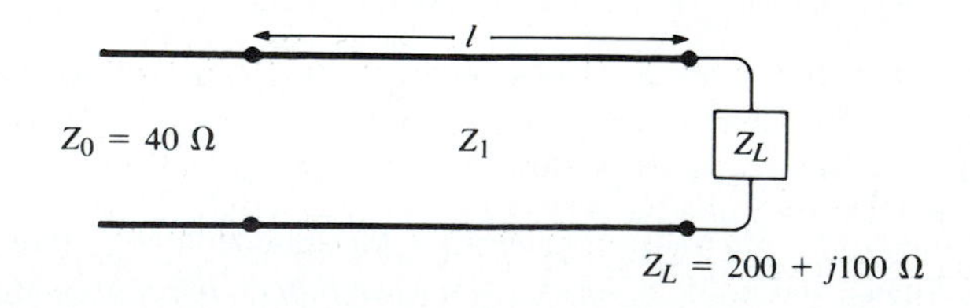

6.8 An open-circuit tuning stub is to be made from a lossy transmission line with an attenuation constant $\alpha = 0.01$ Np/λ. What is the maximum value of normalized reactance that can be obtained with this stub? What is the maximum value of normalized reactance that can be obtained with a shorted stub of the same type of transmission line?

6.9 Design a double-stub tuner using open-circuited stubs with a $\lambda/8$ spacing to match a load admittance $Y_L = (1.4 + j2)Y_0$.

6.10 Repeat Problem 6.9 using a double-stub tuner with short-circuited stubs and a $3\lambda/8$ spacing.

6.11 Derive the design equations for a double-stub tuner using two series stubs, spaced a distance d apart. Assume the load impedance is $Z_L = R_L + jX_L$.

6.12 Consider matching a load $Z_L = 200\ \Omega$ to a $100\ \Omega$ line, using single shunt-stub, single series stub, and double shunt-stub tuners, with short-circuited stubs. Which tuner will give the best bandwidth? Justify your answer by calculating the reflection coefficient for all *six* solutions at $1.1f_0$, where f_0 is the match frequency.

6.13 Design a single-section quarter-wave matching transformer to match a $350\ \Omega$ load to a $100\ \Omega$ line. What is the percent bandwidth of this transformer, for SWR≤ 2? If the design frequency is 4 GHz, sketch the layout of a microstrip circuit, including dimensions, to implement this matching transformer. Assume the substrate is 0.159 cm thick, with a dielectric constant of 2.2.

6.14 A complex load impedance of $Z_L = 80 + j40\ \Omega$ is to be matched to a $50\ \Omega$ line. Find the shortest length of an open-circuited shunt stub at the load required to produce a purely real impedance, then design an appropriate single-section matching transformer.

6.15 A waveguide load with an equivalent TE_{10} wave impedance of $377\ \Omega$ must be matched to an air-filled X-band rectangular guide at 10 GHz. A quarter-wave matching transformer is to be used, and is to consist of a section of guide filled with dielectric. Find the required dielectric constant and physical length of the matching section.

6.16 Design a four-section binomial matching transformer to match a $10\ \Omega$ load to a $50\ \Omega$ line. What is the bandwidth of this transformer, for $\Gamma_m = 0.05$?

6.17 Derive the exact characteristic impedance for a two-section binomial matching transformer, for a normalized load impedance $Z_L/Z_0 = 1.5$. Check your results with Table 6.1.

6.18 Calculate and plot the percent bandwidth for a $N = 1, 2$, and 4 section binomial matching transformer, versus $Z_L/Z_0 = 1.5$ to 6 for $\Gamma_m = 0.2$.

6.19 Using (6.56) and trigonometric identities, verify the results of (6.60).

6.20 Design a four-section Chebyshev matching transformer to match a $40\ \Omega$ line to a $60\ \Omega$ load. The maximum permissible SWR over the passband is 1.2. What is the resulting bandwidth? Use the approximate theory developed in the text, as opposed to the tables.

6.21 Derive the exact characteristic impedances for a two-section Chebyshev matching transformer, for a normalized load impedance $Z_L/Z_0 = 1.5$. Check your results with Table 6.2 for $\Gamma_m = 0.05$.

6.22 A load of $Z_L/Z_0 = 1.5$ is to be matched to a feed line using a multisection transformer, and it is desired to have a passband response with $|\Gamma(\theta)| = A(0.1 + \cos^2\theta)$, for $0 \leq \theta \leq \pi$. Use the approximate theory for multisection transformers to design a two–section transformer.

6.23 A tapered matching section has $d(\ln Z/Z_0)/dz = A \sin \pi z/L$. Find the constant A so that $Z(0) = Z_0$ and $Z(L) = Z_L$. Compute Γ, and plot $|\Gamma|$ versus βL.

6.24 Design an exponentially tapered matching transformer to match a $100\ \Omega$ load to a $50\ \Omega$ line. Plot $|\Gamma|$ versus βL, and find the length of the matching section (at the center frequency) required to obtain $|\Gamma| \leq 0.05$ over a 100% bandwidth. How many sections would be required if a Chebyshev matching transformer were used to achieve the same specifications?

6.25 A parallel RC load with $R = 100\ \Omega$ and $C = 1.5$ pF is to be matched to a $50\ \Omega$ line over a frequency band from 2.0 to 10.0 GHz. What is the best return loss over this band that can be obtained with an optimum matching network?

6.26 Consider a series RL load with $R = 80\ \Omega$ and $L = 5$ nH. Design a lumped element L section matching network to match this load to a $50\ \Omega$ line at 2 GHz. Plot $|\Gamma|$ versus frequency for this network to determine the bandwidth for which $|\Gamma| \leq \Gamma_m = 0.1$. Compare this with the maximum possible bandwidth for this load, as given by the Bode-Fano criteria. (Assume a square reflection coefficient response like that of Figure 6.23a.)

CHAPTER 7

Microwave Resonators

Microwave resonators are used in a variety of applications, including filters, oscillators, frequency meters, and tuned amplifiers. Since the operation of microwave resonators is very similar to that of the lumped-element resonators of circuit theory, we will begin by reviewing the basic characteristics of series and parallel *RLC* resonant circuits. We will then discuss various implementations of resonators at microwave frequencies using distributed elements such as transmission lines, rectangular and circular waveguide, and dielectric cavities. We will also discuss open resonators of the Fabry-Perot type, and the excitation of resonators using apertures and current sheets.

7.1 SERIES AND PARALLEL RESONANT CIRCUITS

Near resonance, a microwave resonator can usually be modeled by either a series or parallel *RLC* lumped-element equivalent circuit, and so we will derive some of the basic properties of such circuits below.

Series Resonant Circuit

A series *RLC* lumped-element resonant circuit is shown in Figure 7.1a. The input impedance is

$$Z_{\text{in}} = R + j\omega L - j\frac{1}{\omega C}, \tag{7.1}$$

and the complex power delivered to the resonator is

$$\begin{aligned} P_{\text{in}} &= \frac{1}{2}VI^* = \frac{1}{2}Z_{\text{in}}|I|^2 = \frac{1}{2}Z_{\text{in}}\left|\frac{V}{Z_{\text{in}}}\right|^2 \\ &= \frac{1}{2}|I|^2\left(R + j\omega L - j\frac{1}{\omega C}\right). \end{aligned} \tag{7.2}$$

The power dissipated by the resistor, R, is

$$P_{\text{loss}} = \frac{1}{2}|I|^2 R, \tag{7.3a}$$

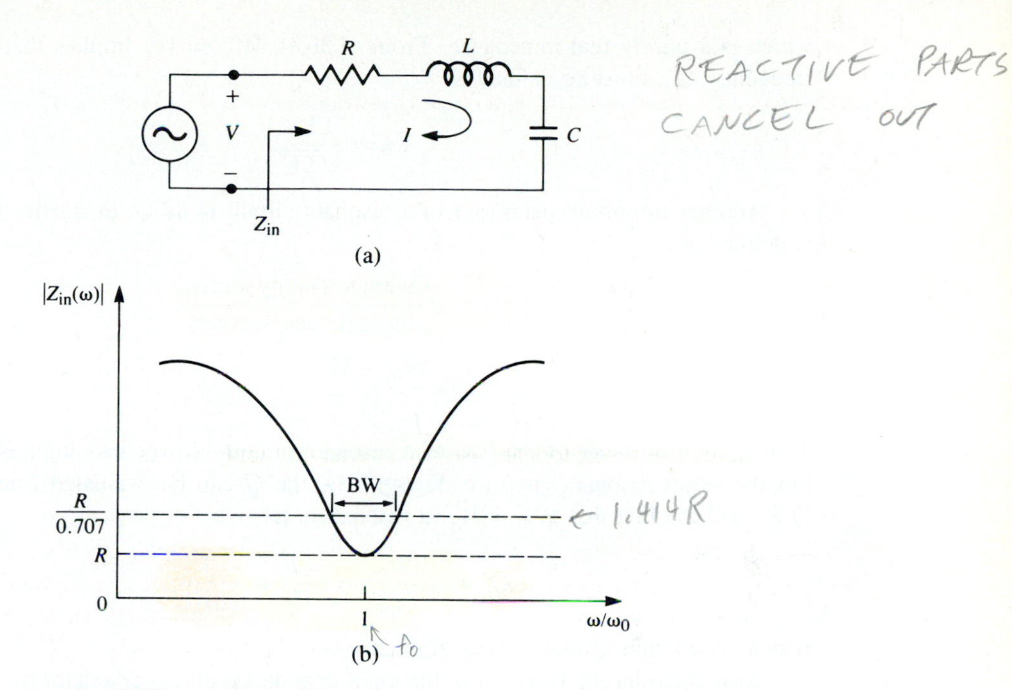

FIGURE 7.1 A series RLC resonator and its response. (a) The series RLC curcuit. (b) The input impedance magnitude versus frequency.

the average magnetic energy stored in the inductor, L, is

$$W_m = \frac{1}{4}|I|^2 L, \tag{7.3b}$$

and the average electric energy stored in the capacitor, C, is

$$W_e = \frac{1}{4}|V_c|^2 C = \frac{1}{4}|I|^2 \frac{1}{\omega^2 C}, \tag{7.3c}$$

where V_c is the voltage across the capacitor. Then the complex power of (7.2) can be rewritten as

$$P_{in} = P_{loss} + 2j\omega(W_m - W_e), \tag{7.4}$$

and the input impedance of (7.1) can be rewritten as

$$Z_{in} = \frac{2P_{in}}{|I|^2} = \frac{P_{loss} + 2j\omega(W_m - W_e)}{|I|^2/2}. \tag{7.5}$$

Resonance occurs when the average stored magnetic and electric energies are equal, or $W_m = W_e$. Then from (7.5) and (7.3a), the input impedance at resonance is

$$Z_{in} = \frac{P_{loss}}{|I|^2/2} = R,$$

which is a purely real impedance. From (7.3b,c), $W_m = W_e$ implies that the resonant frequency, ω_0, must be defined as

$$\omega_0 = \frac{1}{\sqrt{LC}}. \tag{7.6}$$

Another important parameter of a resonant circuit is its Q, or quality factor, which is defined as

$$\begin{aligned} Q &= \omega \frac{\text{(average energy stored)}}{\text{(energy loss/second)}} \\ &= \omega \frac{W_m + W_e}{P_\ell}. \end{aligned} \tag{7.7}$$

Thus Q is a measure of the loss of a resonant circuit—lower loss implies a higher Q. For the series resonant circuit of Figure 7.1a, the Q can be evaluated from (7.7) using (7.3), and the fact that $W_m = W_e$ at resonance, to give

SERIES →

$$Q = \omega_0 \frac{2W_m}{P_{\text{loss}}} = \frac{\omega_0 L}{R} = \frac{1}{\omega_0 RC}, \tag{7.8}$$

which shows that Q increases as R decreases.

Now consider the behavior of the input impedance of this resonator near its resonant frequency [1]. We let $\omega = \omega_0 + \Delta\omega$, where $\Delta\omega$ is small. The input impedance can then be rewritten from (7.1) as

$$\begin{aligned} Z_{\text{in}} &= R + j\omega L \left(1 - \frac{1}{\omega^2 LC}\right) \\ &= R + j\omega L \left(\frac{\omega^2 - \omega_0^2}{\omega^2}\right), \end{aligned}$$

since $\omega_0^2 = 1/LC$. Now $\omega^2 - \omega_0^2 = (\omega - \omega_0)(\omega + \omega_0) = \Delta\omega(2\omega - \Delta\omega) \simeq 2\omega\Delta\omega$ for small $\Delta\omega$. Thus,

$$\begin{aligned} Z_{\text{in}} &\simeq R + j2L\Delta\omega \\ &\simeq R + j\frac{2RQ\Delta\omega}{\omega_0}. \end{aligned} \tag{7.9}$$

This form will be useful for identifying equivalent circuits with distributed element resonators.

Alternatively, a resonator with loss can be treated as a lossless resonator whose resonant frequency ω_0 has been replaced with a complex effective resonant frequency:

$$\omega_0 \longleftarrow \omega_0 \left(1 + \frac{j}{2Q}\right). \tag{7.10}$$

This can be seen by considering the input impedance of a series resonator with no loss, as given by (7.9) with $R = 0$:

$$Z_{\text{in}} = j2L(\omega - \omega_0).$$

Then substituting the complex frequency of (7.10) for ω_0 to give

$$Z_{\text{in}} = j2L\left(\omega - \omega_0 - j\frac{\omega_0}{2Q}\right)$$

$$= \frac{\omega_0 L}{Q} + j2L(\omega - \omega_0) = R + j2L\Delta\omega,$$

which is identical to (7.9). This is a useful procedure because for most practical resonators the loss is very small, so the Q can be found using the perturbation method, beginning with the solution for the lossless case. Then the effect of loss can be added to the input impedance by replacing ω_0 with the complex resonant frequency given in (7.10).

Finally, consider the half-power fractional bandwidth of the resonator. Figure 7.1b shows the variation of the magnitude of the input impedance versus frequency. When the frequency is such that $|Z_{\text{in}}|^2 = 2R^2$, then by (7.2) the average (real) power delivered to the circuit is one-half that delivered at resonance. If BW is the fractional bandwidth, then $\Delta\omega/\omega_0 = \text{BW}/2$ at the upper band edge. Then using (7.9) gives

$$|R + jRQ(\text{BW})|^2 = 2R^2,$$

or

$$\text{BW} = \frac{1}{Q}. \tag{7.11}$$

Parallel Resonant Circuit

The parallel RLC resonant circuit, shown in Figure 7.2a, is the dual of the series RLC circuit. The input impedance is

$$Z_{\text{in}} = \left(\frac{1}{R} + \frac{1}{j\omega L} + j\omega C\right)^{-1}, \tag{7.12}$$

and the complex power delivered to the resonator is

$$P_{\text{in}} = \frac{1}{2}VI^* = \frac{1}{2}Z_{\text{in}}|I|^2 = \frac{1}{2}|V|^2\frac{1}{Z_{\text{in}}^*}$$

$$= \frac{1}{2}|V|^2\left(\frac{1}{R} + \frac{j}{\omega L} - j\omega C\right). \tag{7.13}$$

The power dissipated by the resistor, R, is

$$P_{\text{loss}} = \frac{1}{2}\frac{|V|^2}{R}, \tag{7.14a}$$

the average electric energy stored in the capacitor, C, is

$$W_e = \frac{1}{4}|V|^2 C, \tag{7.14b}$$

and the average magnetic energy stored in the inductor, L, is

$$W_m = \frac{1}{4}|I_L|^2 L = \frac{1}{4}|V|^2\frac{1}{\omega^2 L}, \tag{7.14c}$$

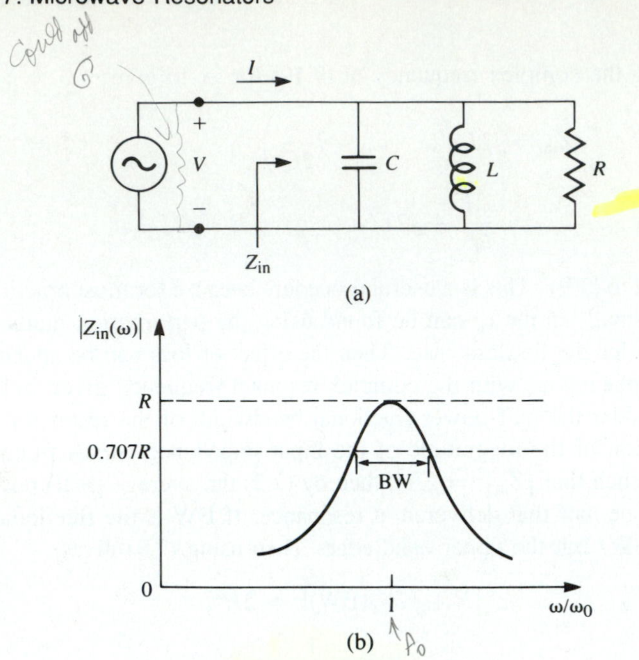

FIGURE 7.2 A parallel RLC resonator and its response. (a) The parallel RLC circuit. (b) The input impedance magnitude versus frequency.

where I_L is the current through the inductor. Then the complex power of (7.13) can be rewritten as

$$P_{\text{in}} = P_{\text{loss}} + 2j\omega(W_m - W_e), \quad 7.15$$

which is identical to (7.4). Similarly, the input impedance can be expressed as

$$Z_{\text{in}} = \frac{2P_{\text{in}}}{|I|^2} = \frac{P_{\text{loss}} + 2j\omega(W_m - W_e)}{|I|^2/2}, \quad 7.16$$

which is identical to (7.5).

As in the series case, resonance occurs when $W_m = W_e$. Then from (7.16) and (7.14a) the input impedance at resonance is

$$Z_{\text{in}} = \frac{P_{\text{loss}}}{|I|^2/2} = R,$$

which is a purely real impedance. From (7.14b, c), $W_m = W_e$ implies that the resonant frequency, ω_0, should be defined as

$$\omega_0 = \frac{1}{\sqrt{LC}}, \quad 7.17$$

which again is identical to the series resonant circuit case.

From the definition of (7.7), and the results in (7.14), the Q of the parallel resonant circuit can be expressed as

$$Q = \omega_0 \frac{2W_m}{P_{\text{loss}}} = \frac{R}{\omega_0 L} = \omega_0 RC, \tag{7.18}$$

since $W_m = W_e$ at resonance. This result shows that the Q of the parallel resonant circuit increases as R increases.

Near resonance, the input impedance of (7.12) can be simplified using the result that

$$\frac{1}{1+x} \simeq 1 - x + \cdots.$$

Letting $\omega = \omega_0 + \Delta\omega$, where $\Delta\omega$ is small, (7.12) can be rewritten as [1]

$$\begin{aligned} Z_{\text{in}} &\simeq \left(\frac{1}{R} + \frac{1 - \Delta\omega/\omega_0}{j\omega_0 L} + j\omega_0 C + j\Delta\omega C\right)^{-1} \\ &\simeq \left(\frac{1}{R} + j\frac{\Delta\omega}{\omega_0^2 L} + j\Delta\omega C\right)^{-1} \\ &\simeq \left(\frac{1}{R} + 2j\Delta\omega C\right)^{-1} \\ &\simeq \frac{R}{1 + 2j\Delta\omega RC} = \frac{R}{1 + 2jQ\Delta\omega/\omega_0}, \end{aligned} \tag{7.19}$$

since $\omega_0^2 = 1/LC$. When $R = 0$ (7.19) reduces to

$$Z_{\text{in}} = \frac{1}{j2C(\omega - \omega_0)}.$$

As in the series resonator case, the effect of loss can be accounted for by replacing ω_0 in this expression with a complex effective resonant frequency:

$$\omega_0 \longleftarrow \omega_0 \left(1 + \frac{j}{2Q}\right) \tag{7.20}$$

Figure 7.2b shows the behavior of the magnitude of the input impedance versus frequency. The half-power bandwidth edges occur at frequencies ($\Delta\omega/\omega_0 = \text{BW}/2$), such that

$$|Z_{\text{in}}|^2 = \frac{R^2}{2},$$

which, from (7.19), implies that

$$\text{BW} = \frac{1}{Q}, \tag{7.21}$$

as in the series resonance case.

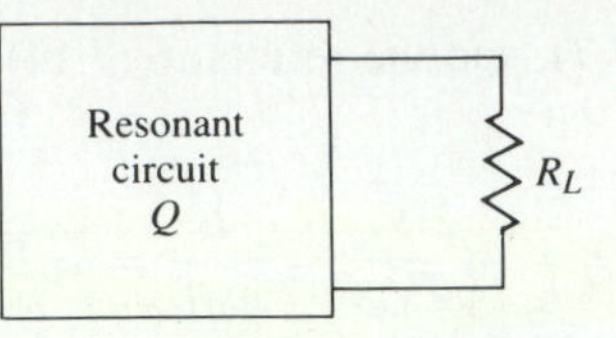

FIGURE 7.3 A resonant circuit connected to an external load, R_L.

Loaded and Unloaded Q

The Q defined in the preceding sections is a characteristic of the resonant circuit itself, in the absence of any loading effects caused by external circuitry, and so is called the unloaded Q. In practice, however, a resonant circuit is invariably coupled to other circuitry, which will always have the effect of lowering the overall, or loaded Q, Q_L, of the circuit. Figure 7.3 depicts a resonator coupled to an external load resistor, R_L. If the resonator is a series RLC circuit, the load resistor R_L adds in series with R so that the effective resistance in (7.8) is $R + R_L$. If the resonator is a parallel RLC circuit, the load resistor R_L combines in parallel with R so that the effective resistance in (7.18) is $RR_L/(R + R_L)$. If we define an external Q, Q_e, as

$$Q_e = \begin{cases} \dfrac{\omega_0 L}{R_L} & \text{for series circuits} \\ \dfrac{R_L}{\omega_0 L} & \text{for parallel circuits,} \end{cases} \tag{7.22}$$

then the loaded Q can be expressed as

$$\frac{1}{Q_L} = \frac{1}{Q_e} + \frac{1}{Q}. \tag{7.23}$$

Table 7.1 summarizes the above results for series and parallel resonant circuits.

7.2 TRANSMISSION LINE RESONATORS

As we have seen, ideal lumped elements are usually unattainable at microwave frequencies, so distributed elements are more commonly used. In this section we will study the use of transmission line sections with various lengths and terminations (usually open or short circuited) to form resonators. Since we will be interested in the Q of these resonators, we must consider lossy transmission lines.

Short-Circuited $\lambda/2$ Line (SERIES)

Consider a length of lossy transmission line, short circuited at one end, as shown in Figure 7.4. The line has a characteristic impedance Z_0, propagation constant β, and attenuation constant α. At the frequency $\omega = \omega_0$, the length of the line is $\ell = \lambda/2$,

TABLE 7.1 Summary of Results for Series and Parallel Resonators

Quantity	Series Resonator	Parallel Resonator
Input Impedance/admittance	$Z_{in} = R + j\omega L - j\frac{1}{\omega C}$	$Y_{in} = \frac{1}{R} + j\omega C - j\frac{1}{\omega L}$
	$\simeq R + j\frac{2RQ\Delta\omega}{\omega_0}$	$\simeq \frac{1}{R} + j\frac{2Q\Delta\omega}{R\omega_0}$
Power loss	$P_{loss} = \frac{1}{2}\|I\|^2 R$	$P_{loss} = \frac{1}{2}\frac{\|V\|^2}{R}$
Stored magnetic energy	$W_m = \frac{1}{4}\|I\|^2 L$	$W_m = \frac{1}{4}\|V\|^2 \frac{1}{\omega^2 L}$
Stored electric energy	$W_e = \frac{1}{4}\|I\|^2 \frac{1}{\omega^2 C}$	$W_e = \frac{1}{4}\|V\|^2 C$
Resonant frequency	$\omega_0 = \frac{1}{\sqrt{LC}}$	$\omega_0 = \frac{1}{\sqrt{LC}}$
Unloaded Q	$Q = \frac{\omega_0 L}{R} = \frac{1}{\omega_0 RC}$	$Q = \omega_0 RC = \frac{R}{\omega_0 L}$
External Q	$Q_e = \frac{\omega_0 L}{R_L}$	$Q_e = \frac{R_L}{\omega_0 L}$

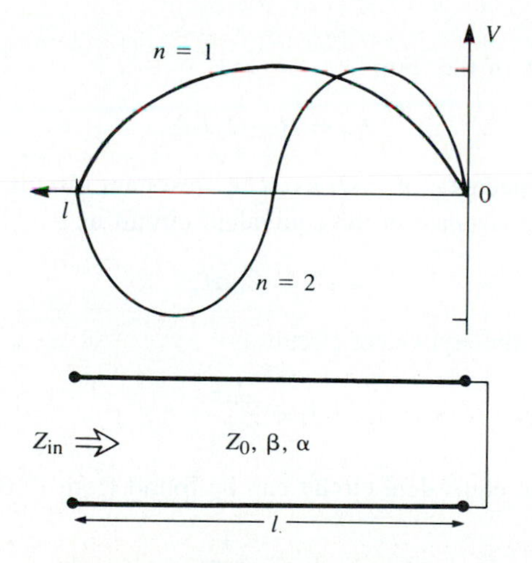

FIGURE 7.4 A short-circuited length of lossy transmission line, and the voltage distributions for $n = 1$ ($\ell = \lambda/2$) and $n = 2$ ($\ell = \lambda$) resonators.

where $\lambda = 2\pi/\beta$. From (3.99), the input impedance is,

$$Z_{in} = Z_0 \tanh(\alpha + j\beta)\ell.$$

Using an identity for the hyperbolic tangent gives

$$Z_{\text{in}} = Z_0 \frac{\tanh \alpha \ell + j \tan \beta \ell}{1 + j \tan \beta \ell \tanh \alpha \ell}. \tag{7.24}$$

Observe that $Z_{\text{in}} = jZ_0 \tan \beta \ell$ if $\alpha = 0$ (no loss).

In practice, most transmission lines have small loss, so we can assume that $\alpha \ell \ll 1$, and so $\tanh \alpha \ell \simeq \alpha \ell$. Now let $\omega = \omega_0 + \Delta\omega$, where $\Delta\omega$ is small. Then, assuming a TEM line,

$$\beta \ell = \frac{\omega \ell}{v_p} = \frac{\omega_0 \ell}{v_p} + \frac{\Delta\omega \ell}{v_p},$$

where v_p is the phase velocity of the transmission line. Since $\ell = \lambda/2 = \pi v_p/\omega_0$ for $\omega = \omega_0$, we have

$$\beta \ell = \pi + \frac{\Delta\omega \pi}{\omega_0},$$

and then

$$\tan \beta \ell = \tan \left(\pi + \frac{\Delta\omega \pi}{\omega_0} \right) = \tan \frac{\Delta\omega \pi}{\omega_0} \simeq \frac{\Delta\omega \pi}{\omega_0}.$$

Using these results in (7.24) gives

$$Z_{\text{in}} \simeq Z_0 \frac{\alpha \ell + j(\Delta\omega \pi/\omega_0)}{1 + j(\Delta\omega \pi/\omega_0)\alpha \ell} \simeq Z_0 \left(\alpha \ell + j \frac{\Delta\omega \pi}{\omega_0} \right), \tag{7.25}$$

since $\Delta\omega \alpha \ell/\omega_0 \ll 1$.

Equation (7.25) is of the form

$$Z_{\text{in}} = R + 2jL\Delta\omega,$$

which is the input impedance of a series RLC resonant circuit, as given by (7.9). We can then identify the resistance of the equivalent circuit as

$$R = Z_0 \alpha \ell, \tag{7.26a}$$

and the inductance of the equivalent circuit as

$$L = \frac{Z_0 \pi}{2\omega_0}. \tag{7.26b}$$

The capacitance of the equivalent circuit can be found from (7.6) as

$$C = \frac{1}{\omega_0^2 L} \tag{7.26c}$$

The resonator of Figure 7.4 thus resonates for $\Delta\omega = 0$ ($\ell = \lambda/2$), and its input impedance at this frequency is $Z_{\text{in}} = R = Z_0 \alpha \ell$. Resonance also occurs for $\ell = n\lambda/2$, $n = 1, 2, 3, \ldots$. The voltage distributions for the $n = 1$ and $n = 2$ resonant modes are shown in Figure 7.4.

The Q of this resonator can be found from (7.8) and (7.26) as

$$Q = \frac{\omega_0 L}{R} = \frac{\pi}{2\alpha \ell} = \frac{\beta}{2\alpha}, \tag{7.27}$$

since $\beta\ell = \pi$ at the first resonance. This result shows that the Q decreases as the attenuation of the line increases, as expected.

EXAMPLE 7.1

A $\lambda/2$ resonator is to be made from a piece of copper coaxial line, with an inner conductor radius of 1 mm and an outer conductor radius of 4 mm. If the resonant frequency is 5 GHz, compare the Q of an air-filled coaxial line resonator to that of a teflon-filled coaxial line resonator.

Solution

We must first compute the attenuation of the coaxial line, which can be done using the results of Example 3.8 or 3.9. From Appendix F, the conductivity of copper is $\sigma = 5.813 \times 10^7$ S/m. Then the surface resistivity is

$$R_s = \sqrt{\frac{\omega\mu_0}{2\sigma}} = 1.84 \times 10^{-2}\ \Omega,$$

and the attenuation due to conductor loss for the air-filled line is

$$\alpha_c = \frac{R_s}{2\eta \ln b/a}\left(\frac{1}{a} + \frac{1}{b}\right)$$
$$= \frac{1.84 \times 10^{-2}}{2(377)\ln(0.004/0.001)}\left(\frac{1}{0.001} + \frac{1}{0.004}\right) = 0.022\ \text{Np/m}.$$

For teflon, $\epsilon_r = 2.08$ and $\tan\delta = 0.0004$, so the attenuation due to conductor loss for the teflon-filled line is

$$\alpha_c = \frac{1.84 \times 10^{-2}\sqrt{2.08}}{2(377)\ln(0.004/0.001)}\left(\frac{1}{0.001} + \frac{1}{0.004}\right) = 0.032\ \text{Np/m}.$$

The dielectric loss of the air-filled line is zero, but the dielectric loss of the teflon-filled line is

$$\alpha_d = k_0 \frac{\sqrt{\epsilon_r}}{2} \tan\delta$$
$$= \frac{(104.7)\sqrt{2.08}(0.0004)}{2} = 0.030\ \text{Np/m}.$$

Finally, from (7.27), the Qs can be computed as

$$Q_{\text{air}} = \frac{\beta}{2\alpha} = \frac{104.7}{2(0.022)} = 2380,$$

$$Q_{\text{teflon}} = \frac{\beta}{2\alpha} = \frac{104.7\sqrt{2.08}}{2(0.032 + 0.030)} = 1218.$$

Thus it is seen that the Q of the air-filled line is almost twice that of the teflon-filled line. The Q can be further increased by using silver-plated conductors. ○

Short-Circuited $\lambda/4$ Line (SERIES)

A parallel type of resonance (antiresonance) can be achieved using a short-circuited transmission line of length $\lambda/4$. The input impedance of the shorted line of length ℓ is

$$\begin{aligned} Z_{\text{in}} &= Z_0 \tanh(\alpha + j\beta)\ell \\ &= Z_0 \frac{\tanh \alpha\ell + j \tan \beta\ell}{1 + j \tan \beta\ell \tanh \alpha\ell} \\ &= Z_0 \frac{1 - j \tanh \alpha\ell \cot \beta\ell}{\tanh \alpha\ell - j \cot \beta\ell}, \end{aligned} \qquad 7.28$$

where the last result was obtained by multiplying both numerator and denominator by $-j \cot \beta\ell$. Now assume that $\ell = \lambda/4$ at $\omega = \omega_0$, and let $\omega = \omega_0 + \Delta\omega$. Then, for a TEM line,

$$\beta\ell = \frac{\omega_0 \ell}{v_p} + \frac{\Delta\omega\ell}{v_p} = \frac{\pi}{2} + \frac{\pi\Delta\omega}{2\omega_0},$$

and so $$\cot \beta\ell = \cot\left(\frac{\pi}{2} + \frac{\pi\Delta\omega}{2\omega_0}\right) = -\tan \frac{\pi\Delta\omega}{2\omega_0} \simeq \frac{-\pi\Delta\omega}{2\omega_0}.$$

Also, as before, $\tanh \alpha\ell \simeq \alpha\ell$ for small loss. Using these results in (7.28) gives

$$Z_{\text{in}} = Z_0 \frac{1 + j\alpha\ell\pi\Delta\omega/2\omega_0}{\alpha\ell + j\pi\Delta\omega/2\omega_0} \simeq \frac{Z_0}{\alpha\ell + j\pi\Delta\omega/2\omega_0}, \qquad 7.29$$

since $\alpha\ell\pi\Delta\omega/2\omega_0 << 1$. This result is of the same form as the impedance of a parallel RLC circuit, as given in (7.19):

$$Z_{in} = \frac{1}{(1/R) + 2j\Delta\omega C}.$$

Then we can identify the resistance of the equivalent circuit as

$$R = \frac{Z_0}{\alpha\ell} \qquad 7.30a$$

and the capacitance of the equivalent circuit as

$$C = \frac{\pi}{4\omega_0 Z_0}. \qquad 7.30b$$

The inductance of the equivalent circuit can be found as

$$L = \frac{1}{\omega_0^2 C}. \qquad 7.30c$$

The resonator of Figure 7.4 thus has a parallel type resonance for $\ell = \lambda/4$, with an input impedance at resonance of $Z_{in} = R = Z_0/\alpha\ell$. From (7.18) and (7.30) the Q of this resonator is

$$Q = \omega_0 RC = \frac{\pi}{4\alpha\ell} = \frac{\beta}{2\alpha}, \tag{7.31}$$

since $\ell = \pi/2\beta$ at resonance.

Open-Circuited $\lambda/2$ Line (PARALLEL)

A practical resonator that is often used in microstrip circuits consists of an open-circuited length of transmission line, as shown in Figure 7.5. Such a resonator will behave as a parallel resonant circuit when the length is $\lambda/2$, or multiples of $\lambda/2$.

The input impedance of an open-circuited line of length ℓ is

$$Z_{in} = Z_0 \coth(\alpha + j\beta)\ell = Z_0 \frac{1 + j \tan\beta\ell \tanh\alpha\ell}{\tanh\alpha\ell + j\tan\beta\ell}. \tag{7.32}$$

As before, assume that $\ell = \lambda/2$ at $\omega = \omega_0$, and let $\omega = \omega_0 + \Delta\omega$. Then,

$$\beta\ell = \pi + \frac{\pi\Delta\omega}{\omega_0},$$

and so

$$\tan\beta\ell = \tan\frac{\Delta\omega\pi}{\omega} \simeq \frac{\Delta\omega\pi}{\omega_0},$$

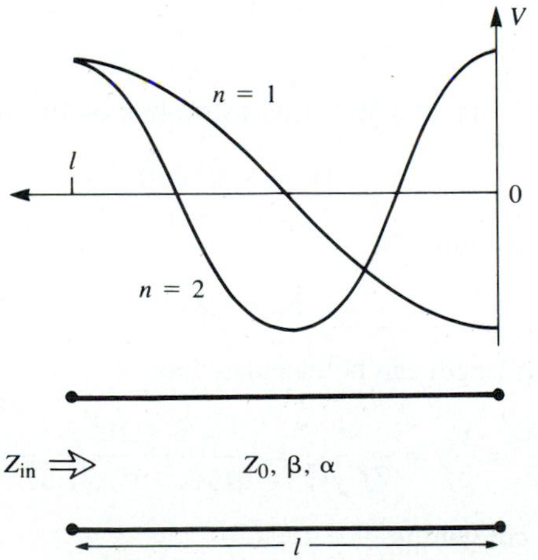

FIGURE 7.5 An open-circuited length of lossy transmission line, and the voltage distributions for $n = 1$ ($\ell = \lambda/2$) and $n = 2$ ($\ell = \lambda$) resonators.

and $\tanh \alpha\ell \simeq \alpha\ell$. Using these results in (7.32) gives

$$Z_{\rm in} = \frac{Z_0}{\alpha\ell + j(\Delta\omega\pi/\omega_0)}. \tag{7.33}$$

Comparing with the input impedance of a parallel resonant circuit as given by (7.19) suggests that the resistance of the equivalent RLC circuit is,

$$R = \frac{Z_0}{\alpha\ell}, \tag{7.34a}$$

and the capacitance of the equivalent circuit is

$$C = \frac{\pi}{2\omega_0 Z_0}. \tag{7.34b}$$

The inductance of the equivalent circuits is

$$L = \frac{1}{\omega_0^2 C}. \tag{7.34c}$$

From (7.18) and (7.34) the Q is

$$Q = \omega_0 RC = \frac{\pi}{2\alpha\ell} = \frac{\beta}{2\alpha}, \tag{7.35}$$

since $\ell = \pi/\beta$ at resonance.

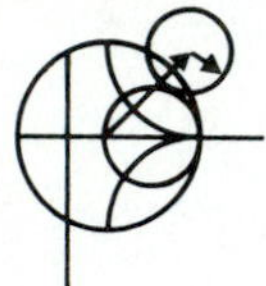

EXAMPLE 7.2

Consider a microstrip resonator constructed from a $\lambda/2$ length of 50 Ω open-circuited microstrip line. The substrate thickness is 0.159 cm, with $\epsilon_r = 2.2$ and $\tan\delta = 0.001$. The conductors are copper. Compute the length of the line for resonance at 5 GHz, and the Q of the resonator. Ignore fringing fields at the end of the line.

Solution
From (4.206), the width of a 50 Ω microstrip line on this substrate is found to be

$$W = 0.49 \text{ cm},$$

and the effective permittivity is

$$\epsilon_e = 1.87.$$

Then the resonant length can be calculated as

$$\ell = \frac{\lambda}{2} = \frac{v_p}{2f} = \frac{c}{2f\sqrt{\epsilon_e}} = \frac{3 \times 10^8}{2(5 \times 10^9)\sqrt{1.87}} = 2.19 \text{ cm}.$$

The propagation constant is

$$\beta = \frac{2\pi f}{v_p} = \frac{2\pi f\sqrt{\epsilon_e}}{c} = \frac{2\pi(5 \times 10^9)\sqrt{1.87}}{3 \times 10^8} = 143.2 \text{ rad/m}.$$

From (4.208), the attenuation due to conductor loss is

$$\alpha_c = \frac{R_s}{Z_0 W} = \frac{1.84 \times 10^{-2}}{50(0.0049)} = 0.075 \text{ Np/m},$$

where we used R_s from Example 7.1. From (4.207), the attenuation due to dielectric loss is

$$\alpha_d = \frac{k_0 \epsilon_r (\epsilon_e - 1) \tan \delta}{2\sqrt{\epsilon_e}(\epsilon_r - 1)} = \frac{(104.7)(2.2)(0.87)(0.001)}{2\sqrt{1.87}(1.2)} = 0.0611 \text{ Np/m}.$$

Then from (7.35) the Q is

$$Q = \frac{\beta}{2\alpha} = \frac{143.2}{2(0.075 + 0.0611)} = 526.$$

○

7.3 RECTANGULAR WAVEGUIDE CAVITIES

Resonators can also be constructed from closed sections of waveguide, which should not be surprising since waveguides are a type of transmission line. Because of radiation loss from open-ended waveguide, waveguide resonators are usually short circuited at both ends, thus forming a closed box or cavity. Electric and magnetic energy is stored within the cavity, and power can be dissipated in the metallic walls of the cavity as well as in the dielectric filling the cavity. Coupling to the resonator can be by a small aperture or a small probe or loop.

We will first derive the resonant frequencies for a general TE or TM resonant mode, and then derive an expression for the Q of the $\text{TE}_{10\ell}$ mode. A complete treatment of the Q for arbitrary TE and TM modes can be made using the same procedure, but is not included here because of its length and complexity.

Resonant Frequencies

The geometry of a rectangular cavity is shown in Figure 7.6. It consists of a length d of rectangular waveguide shorted at both ends ($z = 0, d$). We first find the resonant frequencies of this cavity under the assumption that the cavity is lossless, then we determine the Q using the perturbation method outlined in Section 3.9. We could begin with the wave equations and use the method of separation of variables to solve for the electric and magnetic fields that satisfy the boundary conditions of the cavity, but it is easier to start with the TE and TM waveguide fields, which already satisfy the necessary boundary conditions on the side walls ($x = 0, a$ and $y = 0, b$) of the cavity. Then it is only necessary to enforce the boundary conditions that $E_x = E_y = 0$ on the end walls at $z = 0, d$.

From Table 4.2 the transverse electric fields (E_x, E_y) of the TE_{mn} or TM_{mn} rectangular waveguide mode can be written as

$$\bar{E}_t(x, y, z) = \bar{e}(x, y)[A^+ e^{-j\beta_{mn} z} + A^- e^{j\beta_{mn} z}], \quad 7.36$$

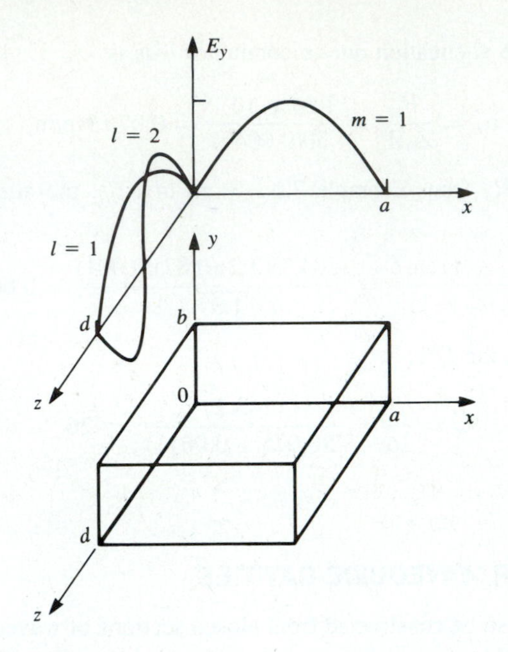

FIGURE 7.6 A rectangular resonant cavity, and the electric field distributions for the TE_{101} and TE_{102} resonant modes.

where $\bar{e}(x, y)$ is the transverse variation of the mode, and A^+, A^- are arbitrary amplitudes of the forward and backward traveling waves. The propagation constant of the m, nth TE or TM mode is

$$\beta_{mn} = \sqrt{k^2 - \left(\frac{m\pi}{a}\right)^2 - \left(\frac{n\pi}{b}\right)^2}, \tag{7.37}$$

where $k = \omega\sqrt{\mu\epsilon}$, and μ, ϵ are the permeability of permittivity of the material filling the cavity.

Applying the condition that $\bar{E}_t = 0$ at $z = 0$ to (7.36) implies that $A^+ = -A^-$ (as we should expect for reflection from a perfectly conducting wall). Then the condition that $\bar{E}_t = 0$ at $z = d$ leads to the equation

$$\bar{E}_t(x, y, d) = -\bar{e}(x, y)A^+ 2j \sin \beta_{mn} d = 0.$$

The only nontrivial ($A^+ \neq 0$) solution thus occurs for

$$\beta_{mn} d = \ell\pi, \quad \ell = 1, 2, 3, \cdots, \tag{7.38}$$

which implies that the cavity must be an integer multiple of a half-guide wavelength long at the resonant frequency. No nontrivial solutions are possible for other lengths, or for frequencies other than the resonant frequencies. The rectangular cavity is thus a waveguide version of the short-circuited $\lambda/2$ transmission line resonator.

A cutoff wavenumber for the rectangular cavity can be defined as

$$k_{mn\ell} = \sqrt{\left(\frac{m\pi}{a}\right)^2 + \left(\frac{n\pi}{b}\right)^2 + \left(\frac{\ell\pi}{d}\right)^2}. \tag{7.39}$$

Then we can refer to the $\text{TE}_{mn\ell}$ or $\text{TM}_{mn\ell}$ resonant mode of the cavity, where the indices m, n, ℓ refer to the number of variations in the standing wave pattern in the x, y, z directions, respectively. The resonant frequency of the $\text{TE}_{mn\ell}$ or $\text{TM}_{mn\ell}$ mode is then given by

$$f_{mn\ell} = \frac{ck_{mn\ell}}{2\pi\sqrt{\mu_r\epsilon_r}} = \frac{c}{2\pi\sqrt{\mu_r\epsilon_r}}\sqrt{\left(\frac{m\pi}{a}\right)^2 + \left(\frac{n\pi}{b}\right)^2 + \left(\frac{\ell\pi}{d}\right)^2}. \tag{7.40}$$

If $b < a < d$, the dominant resonant mode (lowest resonant frequency) will be the TE_{101} mode, corresponding to the TE_{10} dominant waveguide mode in a shorted guide of length $\lambda_g/2$. The dominant TM mode is the TM_{110} mode.

Q of the $\text{TE}_{10\ell}$ Mode

From Table 4.2, (7.36), and the fact that $A^- = -A^+$, the total fields for the $\text{TE}_{10\ell}$ resonant mode can be written as

$$E_y = A^+ \sin\frac{\pi x}{a}[e^{-j\beta z} - e^{j\beta z}], \tag{7.41a}$$

$$H_x = \frac{-A^+}{Z_{\text{TE}}}\sin\frac{\pi x}{a}[e^{-j\beta z} + e^{j\beta z}], \tag{7.41b}$$

$$H_z = \frac{j\pi A^+}{k\eta a}\cos\frac{\pi x}{a}[e^{-j\beta z} - e^{j\beta z}]. \tag{7.41c}$$

Letting $E_0 = -2jA^+$ and using (7.38) allows these expressions to be reduced to

$$E_y = E_0 \sin\frac{\pi x}{a}\sin\frac{\ell\pi z}{d}, \tag{7.42a}$$

$$H_x = \frac{-jE_0}{Z_{\text{TE}}}\sin\frac{\pi x}{a}\cos\frac{\ell\pi z}{d}, \tag{7.42b}$$

$$H_z = \frac{j\pi E_0}{k\eta a}\cos\frac{\pi x}{a}\sin\frac{\ell\pi z}{d}, \tag{7.42c}$$

which clearly show that the fields form standing waves inside the cavity. We can now compute the Q of this mode by finding the stored electric and magnetic energies, and the power lost in the conducting walls and the dielectric.

The stored electric energy is, from (2.84),

$$W_e = \frac{\epsilon}{4}\int_V E_y E_y^* dv = \frac{\epsilon abd}{16}E_0^2, \tag{7.43a}$$

while the stored magnetic energy is, from (2.86),

$$W_m = \frac{\mu}{4}\int_V (H_x H_x^* + H_z H_z^*) dv$$
$$= \frac{\mu abd}{16} E_0^2 \left(\frac{1}{Z_{\text{TE}}^2} + \frac{\pi^2}{k^2\eta^2 a^2} \right). \qquad 7.43b$$

Since $Z_{\text{TE}} = k\eta/\beta$, and $\beta = \beta_{10} = \sqrt{k^2 - (\pi/a)^2}$, the quantity in parentheses in (7.43b) can be reduced to

$$\left(\frac{1}{Z_{\text{TE}}^2} + \frac{\pi^2}{k^2\eta^2 a^2} \right) = \frac{\beta^2 + (\pi/a)^2}{k^2\eta^2} = \frac{1}{\eta^2} = \frac{\epsilon}{\mu},$$

which shows that $W_e = W_m$. Thus, the stored electric and magnetic energies are equal at resonance, analogous to the *RLC* resonant circuits of Section 7.1.

For small losses we can find the power dissipated in the cavity walls using the perturbation method of Section 3.9. Thus, the power lost in the conducting walls is given by (2.131) as

$$P_c = \frac{R_s}{2}\int_{\text{walls}} |H_t|^2 ds, \qquad 7.44$$

where $R_s = \sqrt{\omega\mu_o/2\sigma}$ is the surface resistivity of the metallic walls, and H_t is the tangential magnetic field at the surface of the walls. Using (7.42b,c) in (7.44) gives,

$$P_c = \frac{R_s}{2}\left\{ 2\int_{y=0}^{b}\int_{x=0}^{a} |H_x(z=0)|^2 dx\, dy + 2\int_{z=0}^{d}\int_{y=0}^{b} |H_z(x=0)|^2 dy\, dz \right.$$
$$\left. + 2\int_{z=0}^{d}\int_{x=0}^{a}\left[|H_x(y=0)|^2 + |H_z(y=0)|^2 \right] dx\, dz \right\}$$
$$= \frac{R_s E_0^2 \lambda^2}{8\eta^2}\left(\frac{\ell^2 ab}{d^2} + \frac{bd}{a^2} + \frac{\ell^2 a}{2d} + \frac{d}{2a} \right), \qquad 7.45$$

where use has been made of the symmetry of the cavity in doubling the contributions from the walls at $x = 0, y = 0$, and $z = 0$ to account for the contributions from the walls at $x = a, y = b$, and $z = d$, respectively. The relations that $k = 2\pi/\lambda$ and $Z_{\text{TE}} = k\eta/\beta = 2d\eta/\ell\lambda$ were also used in simplifying (7.45). Then, from (7.7), the Q of the cavity with lossy conducting walls but lossless dielectric can be found as

$$Q_c = \frac{2\omega_0 W_e}{P_c}$$
$$= \frac{k^3 ab\, d\eta}{4\pi^2 R_s} \frac{1}{[(\ell^2 ab/d^2) + (bd/a^2) + (\ell^2 a/2d) + (d/2a)]}$$
$$= \frac{(kad)^3 b\eta}{2\pi^2 R_s} \frac{1}{(2\ell^2 a^3 b + 2bd^3 + \ell^2 a^3 d + ad^3)}. \qquad 7.46$$

We now compute the power lost in the dielectric. As discussed in Chapter 2, a lossy dielectric has an effective conductivity $\sigma = \omega\epsilon'' = \omega\epsilon_r\epsilon_0 \tan\delta$, where $\epsilon = \epsilon' - j\epsilon'' =$

$\epsilon_r \epsilon_0(1 - j \tan\delta)$, and $\tan\delta$ is the loss tangent of the material. Then the power dissipated in the dielectric is, from (2.92),

$$P_d = \frac{1}{2}\int_V \bar{J} \cdot \bar{E}^* dv = \frac{\omega\epsilon''}{2}\int_V |\bar{E}|^2 dv = \frac{abd\omega\epsilon''|E_0|^2}{8}, \tag{7.47}$$

where $\bar{E}$ is given by (7.42a). Then from (7.7) the Q of the cavity with a lossy dielectric filling, but with perfectly conducting walls, is

$$Q_d = \frac{2\omega W_e}{P_d} = \frac{\epsilon'}{\epsilon''} = \frac{1}{\tan\delta}. \tag{7.48}$$

The simplicity of this result is due to the fact that the integral in (7.43a) for W_e cancels with the identical integral in (7.47) for P_d. This result thus applies to Q_d for an arbitrary resonant cavity mode. When both wall losses and dielectric losses are present, the total power loss is $P_c + P_d$, so (7.7) gives the total Q as

$$Q = \left(\frac{1}{Q_c} + \frac{1}{Q_d}\right)^{-1}. \tag{7.49}$$

EXAMPLE 7.3

A rectangular waveguide cavity is made from a piece of copper WR-187 H-band waveguide, with $a = 4.755$ cm and $b = 2.215$ cm. The cavity is filled with polyethylene ($\epsilon_r = 2.25$, $\tan\delta = 0.0004$). If resonance is to occur at $f = 5$ GHz, find the required length, d, and the resulting Q for the $\ell = 1$ and $\ell = 2$ resonant modes.

Solution

The wavenumber k is

$$k = \frac{2\pi f\sqrt{\epsilon_r}}{c} = 157.08 \text{ m}^{-1}.$$

From (7.40) the required length for resonance can be found as $(m = 1,\ n = 0)$

$$d = \frac{\ell\pi}{\sqrt{k^2 - (\pi/a)^2}},$$

$$\text{for } \ell = 1, \quad d = \frac{\pi}{\sqrt{(157.08)^2 - (\pi/0.02215)^2}} = 4.65 \text{ cm},$$

$$\text{for } \ell = 2, \quad d = 2(4.65) = 9.30 \text{ cm}.$$

From Example 7.1, the surface resistivity of copper at 5 GHz is $R_s = 1.84 \times 10^{-2}\ \Omega$. The intrinsic impedance is

$$\eta = \frac{377}{\sqrt{\epsilon_r}} = 251.3\ \Omega.$$

Then from (7.46) the Q due to conductor loss only is

$$\text{for } \ell = 1, \quad Q_c = 3380,$$

$$\text{for } \ell = 2, \quad Q_c = 3864.$$

From (7.48) the Q due to dielectric loss only is, for both $\ell = 1$ and $\ell = 2$,

$$Q_d = \frac{1}{\tan\delta} = \frac{1}{0.0004} = 2500.$$

So the total Qs are, from (7.49)

$$\text{for } \ell = 1, \quad Q = \left(\frac{1}{3380} + \frac{1}{2500}\right)^{-1} = 1437,$$

$$\text{for } \ell = 2, \quad Q = \left(\frac{1}{3864} + \frac{1}{2500}\right)^{-1} = 1518.$$

Note that the dielectric loss has the dominant effect on the Q; higher Q could thus be obtained using an air-filled cavity. These results can be compared to those of Examples 7.1 and 7.2, which used similar types of materials at the same frequency. ○

7.4 CIRCULAR WAVEGUIDE CAVITIES

A cylindrical cavity resonator can be constructed from a section of circular waveguide shorted at both ends, similar to rectangular cavities. Since the dominant circular waveguide mode is the TE_{11} mode, the dominant cylindrical cavity mode is the TE_{111} mode. We will derive the resonant frequencies for the $TE_{nm\ell}$ and $TM_{nm\ell}$ circular cavity modes, and the expression for the Q of the $TE_{nm\ell}$ mode.

Circular cavities are often used for microwave frequency meters. The cavity is constructed with a movable top wall to allow mechanical tuning of the resonant frequency, and the cavity is loosely coupled to a waveguide with a small aperture. In operation, power will be absorbed by the cavity as it is tuned to the operating frequency of the system; this absorption can be monitored with a power meter elsewhere in the system. The tuning dial is usually directly calibrated in frequency, as in the model shown in Figure 7.7. Since frequency resolution is determined by the Q of the resonator, the TE_{011} mode is often used for frequency meters because its Q is much higher than the Q of the dominant circular cavity mode. This is also the reason for a loose coupling to the cavity.

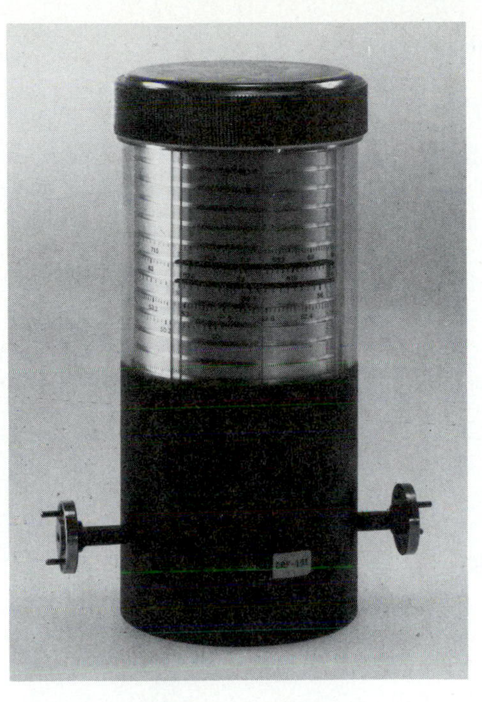

FIGURE 7.7 Photograph of a V-band waveguide frequency meter. The knob on top of the device rotates to change the length of the circular cavity resonator; the scale gives a readout of the frequency.

Photograph courtesy of Millitech Corporation, S. Deerfield, Mass.

Resonant Frequencies

The geometry of a cylindrical cavity is shown in Figure 7.8. As in the case of the rectangular cavity, the solution is simplified by beginning with the circular waveguide modes, which already satisfy the necessary boundary conditions on the circular waveguide wall. From Table 4.5, the transverse electric fields (E_ρ, E_ϕ) of the TE_{nm} or TM_{nm}

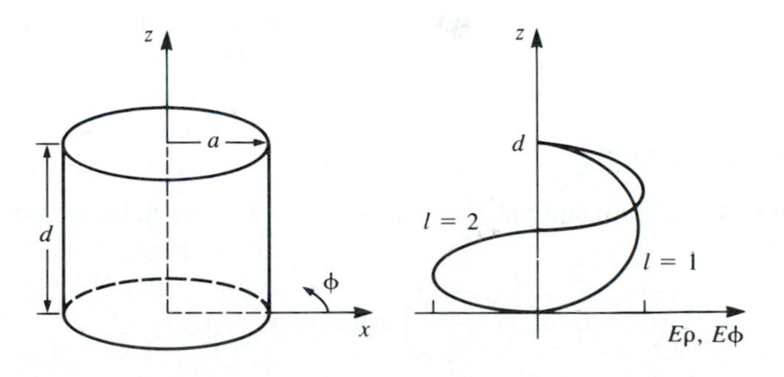

FIGURE 7.8 A cylindrical resonant cavity, and the electric field distribution for resonant modes with $\ell = 1$ or $\ell = 2$.

circular waveguide mode can be written as

$$\bar{E}_t(\rho, \phi, z) = \bar{e}(\rho, \phi)\left[A^+ e^{-j\beta_{nm}z} + A^- e^{j\beta_{nm}z}\right], \tag{7.50}$$

where $\bar{e}(\rho, \phi)$ represents the transverse variation of the mode, and A^+ and A^- are arbitrary amplitudes of the forward and backward traveling waves. The propagation constant of the TE_{nm} mode is, from (4.126),

$$\beta_{nm} = \sqrt{k^2 - \left(\frac{p'_{nm}}{a}\right)^2}, \tag{7.51a}$$

while the propagation constant of the TM_{nm} mode is, from (4.139),

$$\beta_{nm} = \sqrt{k^2 - \left(\frac{p_{nm}}{a}\right)^2}, \tag{7.51b}$$

where $k = \omega\sqrt{\mu\epsilon}$.

Now in order to have $\bar{E}_t = 0$ at $z = 0, d$, we must have $A^+ = -A^-$, and

$$A^+ \sin \beta_{nm} d = 0,$$

or

$$\beta_{nm} d = \ell\pi, \qquad \text{for } \ell = 0, 1, 2, 3, \cdots, \tag{7.52}$$

which implies that the waveguide must be an integer number of half-guide wavelengths long. Thus, the resonant frequency of the $\mathrm{TE}_{nm\ell}$ mode is

$$f_{nm\ell} = \frac{c}{2\pi\sqrt{\mu_r\epsilon_r}}\sqrt{\left(\frac{p'_{nm}}{a}\right)^2 + \left(\frac{\ell\pi}{d}\right)^2}, \tag{7.53a}$$

and the resonant frequency of the $\mathrm{TM}_{nm\ell}$ mode is

$$f_{nm\ell} = \frac{c}{2\pi\sqrt{\mu_r\epsilon_r}}\sqrt{\left(\frac{p_{nm}}{a}\right)^2 + \left(\frac{\ell\pi}{d}\right)^2}. \tag{7.53b}$$

Then the dominant TE mode is the TE_{111} mode, while the dominant TM mode is the TM_{110} mode. Figure 7.9 shows a *mode chart* for the lower order resonant modes of a cylindrical cavity. Such a chart is very useful for the design of cavity resonators, as it shows what modes can be excited at a given frequency, for a given cavity size.

Q of the $\mathrm{TE}_{nm\ell}$ Mode

From Table 4.5, (7.50), and the fact that $A^+ = -A^-$, the fields of the $\mathrm{TE}_{nm\ell}$ mode can be written as

$$H_z = H_0 J_n\left(\frac{p'_{nm}\rho}{a}\right)\cos n\phi \sin\frac{\ell\pi z}{d}, \tag{7.54a}$$

$$H_\rho = \frac{\beta a H_0}{p'_{nm}} J'_n\left(\frac{p'_{nm}\rho}{a}\right)\cos n\phi \cos\frac{\ell\pi z}{d}, \tag{7.54b}$$

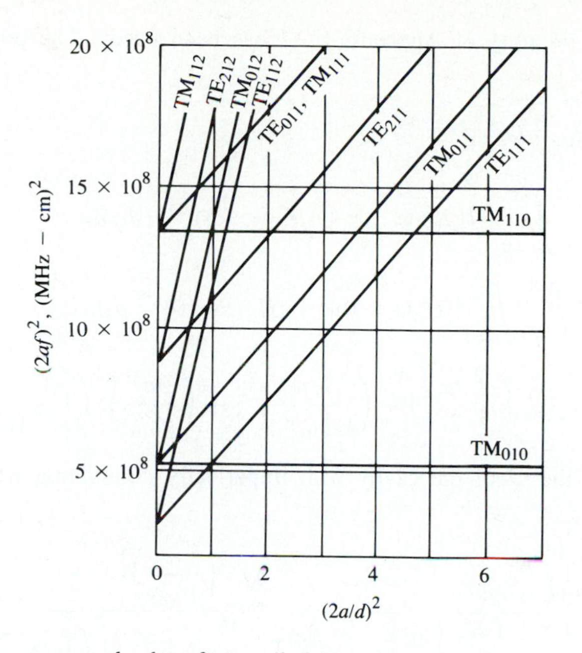

FIGURE 7.9 Resonant mode chart for a cylindrical cavity.

Adapted from data from R. E. Collin, *Foundations for Microwave Engineering* (New York: McGraw-Hill, 1966). Used with permission.

$$H_\phi = \frac{-\beta a^2 n H_0}{(p'_{nm})^2 \rho} J_n \left(\frac{p'_{nm}\rho}{a}\right) \sin n\phi \cos \frac{\ell \pi z}{d}, \tag{7.54c}$$

$$E_\rho = \frac{jk\eta a^2 n H_0}{(p'_{nm})^2 \rho} J_n \left(\frac{p'_{nm}\rho}{a}\right) \sin n\phi \sin \frac{\ell \pi z}{d}, \tag{7.54d}$$

$$E_\phi = \frac{jk\eta a H_0}{p'_{nm}} J'_n \left(\frac{p'_{nm}\rho}{a}\right) \cos n\phi \sin \frac{\ell \pi z}{d}, \tag{7.54e}$$

$$E_z = 0, \tag{7.54f}$$

where $\eta = \sqrt{\mu/\epsilon}$, and $H_0 = -2jA^+$.

Since the time-average stored electric and magnetic energies are equal, the total stored energy is

$$\begin{aligned} W = 2W_e &= \frac{\epsilon}{2} \int_{z=0}^{d} \int_{\phi=0}^{2\pi} \int_{\rho=0}^{a} (|E_\rho|^2 + |E_\phi|^2) \rho d\rho d\phi \, dz \\ &= \frac{\epsilon k^2 \eta^2 a^2 \pi d H_0^2}{4(p'_{nm})^2} \int_{\rho=0}^{a} \left[J_n'^2 \left(\frac{p'_{nm}\rho}{a}\right) + \left(\frac{na}{p'_{nm}\rho}\right)^2 J_n^2 \left(\frac{p'_{nm}\rho}{a}\right)\right] \rho \, d\rho \\ &= \frac{\epsilon k^2 \eta^2 a^4 H_0^2 \pi d}{8(p'_{nm})^2} \left[1 - \left(\frac{n}{p'_{nm}}\right)^2\right] J_n^2(p'_{nm}), \end{aligned} \tag{7.55}$$

where the integral identity of Appendix C.17 has been used. The power loss in the conducting walls is

$$
\begin{aligned}
P_c &= \frac{R_s}{2}\int_S |\bar{H}_{\tan}|^2 ds \\
&= \frac{R_s}{2}\left\{\int_{z=0}^{d}\int_{\phi=0}^{2\pi}\left[|H_\phi(\rho=a)|^2+|H_z(\rho=a)|^2\right] a\, d\phi\, dz \right. \\
&\qquad \left. + 2\int_{\phi=0}^{2\pi}\int_{\rho=0}^{a}\left[|H_\rho(z=0)|^2+|H_\phi(z=0)|^2\right]\rho\, d\rho\, d\phi\right\} \\
&= \frac{R_s}{2}\pi H_0^2 J_n^2(p'_{nm})\left\{\frac{da}{2}\left[1+\left(\frac{\beta an}{(p'_{nm})^2}\right)^2\right]+\left(\frac{\beta a^2}{p'_{nm}}\right)^2\left(1-\frac{n^2}{(p'_{nm})^2}\right)\right\}.
\end{aligned}
\quad 7.56
$$

Then, from (7.8), the Q of the cavity with imperfectly conducting walls but lossless dielectric is

$$
Q_c = \frac{\omega_0 W}{P_c} = \frac{(ka)^3 \eta a d}{4(p'_{nm})^2 R_s}\frac{1-\left(\frac{n}{p'_{nm}}\right)^2}{\left\{\frac{ad}{2}\left[1+\left(\frac{\beta an}{(p'_{nm})^2}\right)^2\right]+\left(\frac{\beta a^2}{p'_{nm}}\right)^2\left(1-\frac{n^2}{(p'_{nm})^2}\right)\right\}}. \quad 7.57
$$

From (7.52) and (7.51) we see that $\beta = \ell\pi/d$ and $(ka)^2$ are constants that do not vary with frequency, for a cavity with fixed dimensions. Thus, the frequency dependence of Q_c is given by k/R_s, which varies as $1/\sqrt{f}$; this gives the variation in Q_c for a given resonant mode and cavity shape (fixed n, m, ℓ, and a/d).

Figure 7.10 shows the normalized Q due to conductor loss for various resonant modes of a cylindrical cavity. Observe that the TE_{011} mode has a Q significantly higher than the lower-order TE_{111}, TM_{010}, or TM_{111} modes.

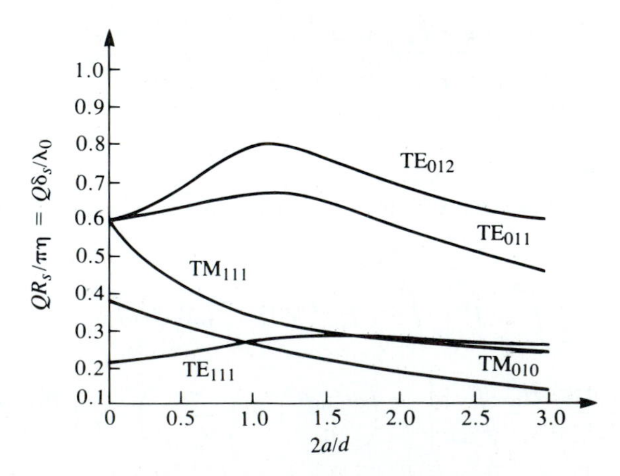

FIGURE 7.10 Normalized Q for various cylindrical cavity modes.

Adapted from data from R. E. Collin, *Foundations for Microwave Engineering* (New York: McGraw-Hill, 1966). Used with permission.

To compute the Q due to dielectric loss, we must compute the power dissipated in the dielectric. Thus,

$$P_d = \frac{1}{2}\int_V \bar{J}\cdot\bar{E}^* dv = \frac{\omega\epsilon''}{2}\int_V \left[|E_p|^2 + |E_\phi|^2\right] dv$$
$$= \frac{\omega\epsilon'' k^2\eta^2 a^2 H_0^2 \pi d}{4(p'_{nm})^2}\int_{\rho=0}^{a}\left[\left(\frac{na}{p'_{nm}\rho}\right)^2 J_n^2\left(\frac{p'_{nm}\rho}{a}\right) + J_n'^2\left(\frac{p'_{nm}\rho}{a}\right)\right]\rho\, d\rho$$
$$= \frac{\omega\epsilon'' k^2\eta^2 a^4 H_0^2}{8(p'_{nm})^2}\left[1-\left(\frac{n}{p'_{nm}}\right)^2\right] J_n^2(p'_{nm}). \qquad 7.58$$

Then (7.8) gives the Q as

$$Q_d = \frac{\omega W}{P_d} = \frac{\epsilon}{\epsilon''} = \frac{1}{\tan\delta}, \qquad 7.59$$

where $\tan\delta$ is the loss tangent of the dielectric. This is the same as the result for Q_d of (7.48) for the rectangular cavity. When both conductor and dielectric losses are present, the total cavity Q can be found from (7.49).

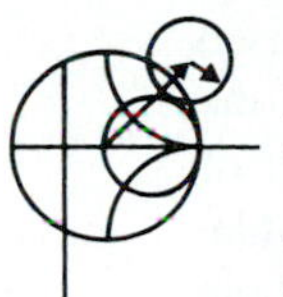

EXAMPLE 7.4

A circular cavity resonator with $d = 2a$ is to be designed to resonate at 5.0 GHz in the TE_{011} mode. If the cavity is made from copper and is air-filled, find its dimensions and Q.

Solution

$$\lambda = \frac{300}{5000} = 0.06 \text{ m},$$
$$k = \frac{2\pi}{\lambda} = 104.7 \text{ m}^{-1}.$$

From (7.53a), the resonant frequency of the TE_{011} mode is

$$f_{011} = \frac{c}{2\pi}\sqrt{\left(\frac{p'_{01}}{a}\right)^2 + \left(\frac{\pi}{d}\right)^2}.$$

Thus, since $d = 2a$

$$\frac{2\pi f_{011}}{c} = k = \sqrt{\left(\frac{p'_{01}}{a}\right)^2 + \left(\frac{\pi}{2a}\right)^2}.$$

Solving for a gives

$$a = \frac{\sqrt{(p'_{01})^2 + (\pi/2)^2}}{k} = \frac{\sqrt{(3.832)^2 + (\pi/2)^2}}{104.7} = 3.96 \text{ cm}.$$

Then $d = 2a = 7.91$ cm.

The surface resistivity of copper ($\sigma = 5.813 \times 10^7$ S/m) at 5 GHz is

$$R_s = \sqrt{\frac{\omega\mu}{2\sigma}} = 0.0184\ \Omega.$$

Then from (7.57), with $n = 0$, $m = \ell = 1$, and $d = 2a$, the Q is

$$Q_c = \frac{(ka)^3 \eta a d}{4(p'_{01})^2 R_s} \frac{1}{[(ad/2) + (\beta a^2/p'_{01})]^2} = \frac{ka\eta}{2R_s} = 42,400.$$

This can be compared with the rectangular cavity of Example 7.3, which had $Q_c = 3,380$ for the TE_{101} mode and $Q_c = 3,864$ for the TE_{102} mode. Note from Figure 7.10 that $d = 2a$ is close to the optimum size for maximizing the Q of the TE_{011} mode. ○

7.5 DIELECTRIC RESONATORS

A small disc or cube of low-loss high dielectric constant material can also be used as a microwave resonator. Such dielectric resonators are similar in principle to the rectangular or cylindrical cavities previously discussed; the high-dielectric constant of the resonator ensures that most of the fields are contained within the dielectric but, unlike metallic cavities, there is some field fringing or leakage from the sides and ends of the dielectric resonator. Such a resonator is generally smaller in cost, size, and weight than an equivalent metallic cavity, and can very easily be incorporated into microwave integrated circuits and coupled to planar transmission lines. Materials with dielectric constants $10 \leq \epsilon_r \leq 100$ are generally used, with barium tetratitanate and titanium dioxide being typical examples. Conductor losses are absent, but dielectric loss usually increases with dielectric constant; Qs of up to several thousand can be achieved, however. By using an adjustable metal plate above the resonator, the resonant frequency can be mechanically tuned.

Below we will present an approximate analysis for the resonant frequencies of the $\text{TE}_{01\delta}$ mode of a cylindrical dielectric resonator; this mode is the one most commonly used in practice, and is analogous to the TE_{011} mode of a circular metallic cavity.

$\text{TE}_{01\delta}$ Resonant Frequencies

The geometry of a cylindrical dielectric resonator is shown in Figure 7.11. The basic operation of the $\text{TE}_{01\delta}$ mode can be explained as follows. The dielectric resonator is considered as a short length, L, of dielectric waveguide open at both ends. The lowest order TE mode of this guide is the TE_{01} mode, and is the dual of the TM_{01} mode of a circular metallic waveguide. Because of the high permittivity of the resonator, propagation along the z-axis can occur inside the dielectric at the resonant frequency, but the fields will be cut off (evanescent) in the air regions around the dielectric. Thus

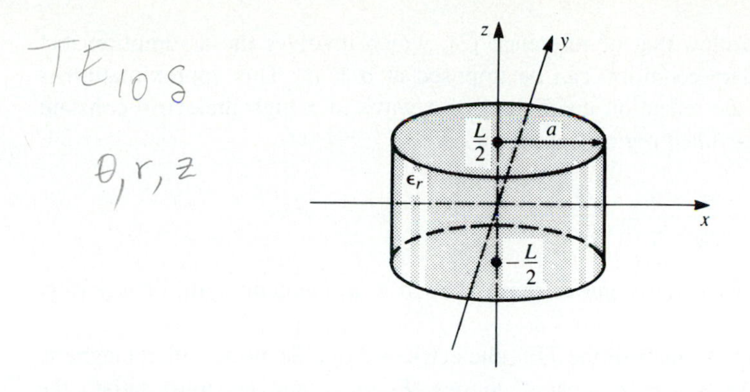

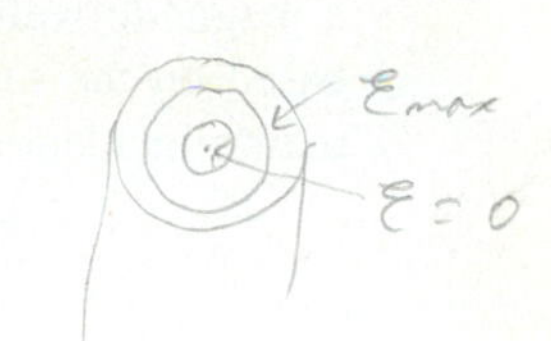

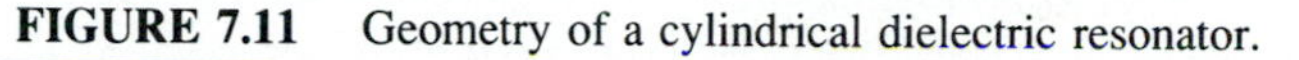

FIGURE 7.11 Geometry of a cylindrical dielectric resonator.

the H_z field will look like that sketched in Figure 7.12; higher-order resonant modes will have more variations in the z direction inside the resonator. Since the resonant length, L, for the $\text{TE}_{01\delta}$ mode is less than $\lambda_g/2$, (where λ_g is the guide wavelength of the TE_{01} dielectric waveguide mode), the symbol $\delta = 2L/\lambda_g < 1$ is used to denote the z variation of the resonant mode. Thus the equivalent circuit of the resonator looks like a length of transmission line terminated in purely reactive loads at both ends.

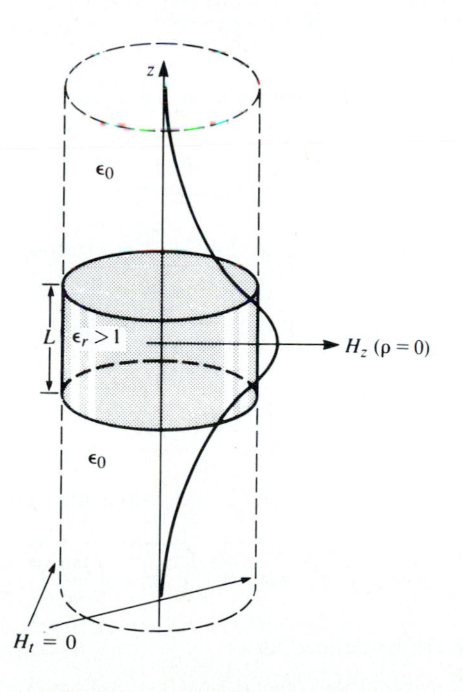

FIGURE 7.12 Magnetic wall boundary condition approximation and distribution of H_z versus z for $\rho = 0$ of the first mode of the cylindrical dielectric resonator.

Our analysis will follow that of reference [2], which involves the assumption that a magnetic wall boundary condition can be imposed at $\rho = a$. This approximation is based on the fact that the reflection coefficient of a wave in a high dielectric constant region incident on an air-filled region approaches +1:

$$\Gamma = \frac{\eta_0 - \eta}{\eta_0 + \eta} = \frac{\sqrt{\epsilon_r} - 1}{\sqrt{\epsilon_r} + 1} \rightarrow 1, \qquad \text{as } \epsilon_r \rightarrow \infty.$$

This reflection coefficient is the same as that obtained at a magnetic wall, or a perfect open circuit.

We begin by finding the fields of the TE_{01} dielectric waveguide mode with a magnetic wall boundary condition at $\rho = a$. For TE modes, $E_z = 0$, and H_z must satisfy the wave equation

$$(\nabla^2 + k^2)H_z = 0, \tag{7.60}$$

where

$$k = \begin{cases} \sqrt{\epsilon_r} k_0 & \text{for } |z| < L/2 \\ k_0 & \text{for } |z| > L/2. \end{cases} \tag{7.61}$$

Since $\partial/\partial\phi = 0$, the transverse fields are given by (4.110) as follows:

$$E_\phi = \frac{j\omega\mu_0}{k_c^2} \frac{\partial H_z}{\partial \rho}, \tag{7.62a}$$

$$H_\rho = \frac{-j\beta}{k_c^2} \frac{\partial H_z}{\partial \rho}, \tag{7.62b}$$

where $k_c^2 = k^2 - \beta^2$. Since H_z must be finite at $\rho = 0$ and zero at $\rho = a$ (the magnetic wall), we have

$$H_z = H_0 J_0(k_c \rho) e^{\pm j\beta z}, \tag{7.63}$$

where $k_c = p_{01}/a$, and $J_0(p_{01}) = 0$ ($p_{01} = 2.405$). Then from (7.62) the transverse fields are

$$E_\phi = \frac{j\omega\mu_0 H_0}{k_c} J_0'(k_c \rho) e^{\pm j\beta z}, \tag{7.64a}$$

$$H_\rho = \frac{\mp j\beta H_0}{k_c} J_0'(k_c \rho) e^{\pm j\beta z}. \tag{7.64b}$$

Now in the dielectric region, $|z| < L/2$, the propagation constant is real:

$$\beta = \sqrt{\epsilon_r k_0^2 - k_c^2} = \sqrt{\epsilon_r k_0^2 - \left(\frac{p_{01}}{a}\right)^2}, \tag{7.65a}$$

and a wave impedance can be defined as

$$Z_d = \frac{E_\phi}{H_\rho} = \frac{\omega\mu_0}{\beta}. \tag{7.65b}$$

In the air region, $|z| > L/2$, the propagation constant will be imaginary, so it is convenient to write

$$\alpha = \sqrt{k_c^2 - k_0^2} = \sqrt{\left(\frac{p_{01}}{a}\right)^2 - k_0^2}, \qquad 7.66a$$

and to define a wave impedance in the air region as

$$Z_a = \frac{j\omega\mu_0}{\alpha}, \qquad 7.66b$$

which is seen to be imaginary.

From symmetry, the H_z and E_ϕ field distributions for the lowest-order mode will be even functions about $z = 0$. Thus the transverse fields for the $TE_{01\delta}$ mode can be written for $|z| < L/2$ as

$$E_\phi = AJ_0'(k_c\rho)\cos\beta z, \qquad 7.67a$$

$$H_\rho = \frac{-jA}{Z_d} J_0'(k_c\rho)\sin\beta z, \qquad 7.67b$$

and for $|z| > L/2$ as

$$E_\phi = BJ_0'(k_c\rho)e^{-\alpha|z|}, \qquad 7.68a$$

$$H_\rho = \frac{\pm B}{Z_a} J_0'(k_c\rho)e^{-\alpha|z|}, \qquad 7.68b$$

where A and B are unknown amplitude coefficients. In (7.68b), the $\pm$ sign is used for $z > L/2$ or $z < -L/2$, respectively.

Matching tangential fields at $z = L/2$ (or $z = -L/2$) leads to the following two equations:

$$A\cos\frac{\beta L}{2} = Be^{-\alpha L/2}, \qquad 7.69a$$

$$\frac{-jA}{Z_d}\sin\frac{\beta L}{2} = \frac{B}{Z_a}e^{-\alpha L/2}, \qquad 7.69b$$

which can be reduced to a single transcendental equation:

$$-jZ_a\sin\frac{\beta L}{2} = Z_d\cos\frac{\beta L}{2}.$$

Using (7.65b) and (7.66b) allows this to be written as

$$\tan\frac{\beta L}{2} = \frac{\alpha}{\beta}, \qquad 7.70$$

where β is given by (7.65a) and α is given by (7.66a). This equation can be solved numerically for k_0, which determines the resonant frequency.

This solution is relatively crude, since it ignores fringing fields at the sides of the resonator, and yields accuracies only on the order of 10% (not accurate enough for most practical purposes), but it serves to illustrate the basic behavior of dielectric resonators. More accurate solutions are available in the literature [3].

The Q of the resonator can be calculated by determining the stored energy (inside and outside the dielectric cylinder), and the power dissipated in the dielectric and possibly lost to radiation. If the latter is small, the Q can be approximated as $1/\tan\delta$, as in the case of the metallic cavity resonators.

EXAMPLE 7.5

Find the resonant frequency and approximate Q for the $TE_{01\delta}$ mode of a dielectric resonator made from titania, with $\epsilon_r = 95$, and $\tan\delta = 0.001$. The resonator dimensions are $a = 0.413$ cm, and $L = 0.8255$ cm.

Solution
The transcendental equation of (7.70) must be solved for k_0, with β and α given by (7.65a) and (7.66a). Thus,

$$\tan\frac{\beta L}{2} = \frac{\alpha}{\beta},$$

where

$$\alpha = \sqrt{(2.405/a)^2 - k_0^2},$$

$$\beta = \sqrt{\epsilon_r k_0^2 - (2.405/a)^2},$$

and

$$k_0 = \frac{2\pi f}{c}.$$

Since α and β must both be real, the possible frequency range is from f_1 to f_2, where

$$f_1 = \frac{ck_0}{2\pi} = \frac{c(2.405)}{2\pi\sqrt{\epsilon_r}a} = 2.853 \text{ GHz},$$

$$f_2 = \frac{ck_0}{2\pi} = \frac{c(2.405)}{2\pi a} = 27.804 \text{ GHz}.$$

Using the interval-halving method (see the Point of Interest on root-finding algorithms in Chapter 4) to find the root of the above equation gives a resonant frequency of about 3.152 GHz. This compares with a measured value of about 3.4 GHz from reference [2], indicating a 10% error. The approximate Q, due to dielectric loss, is

$$Q_d = \frac{1}{\tan\delta} = 1000.$$

○

7.6 FABRY-PEROT RESONATORS

In principle, the previously described resonators can be used at arbitrarily high frequencies. But examination of the expressions for Q_c due to conductor losses shows that the Q will decrease as $1/\sqrt{f}$ for a given cavity or transmission line resonator. Thus, at very high frequencies the Q of such resonators may be too small to be useful.

In addition, at high frequencies the physical size of a cavity operating in a low-order mode may be too small to be practical. If a high-order mode is used, the resonances of nearby modes will be very close in frequency, and because of the finite bandwidth of these modes there may be little or no separation between them, making such resonators unusable.

A conceptual way of avoiding these difficulties is to remove the side walls of a cavity resonator, which has the effect of reducing conductor losses and the number of possible resonant modes. Such an *open* resonator thus consists of two parallel metal plates, as shown in Figure 7.13; this is also known as a Fabry-Perot resonator, since it is similar in principle to the optical Fabry-Perot interferometer. In order for the device to be useful, the plates must be very parallel, and large enough in extent so that no significant radiation leaves the region between the plates. These constraints can be relaxed by using spherical or parabolic-shaped reflecting mirrors to focus and confine the energy to a stable mode pattern. Such *quasi-optical* resonators are very useful at millimeter wave and sub-millimeter wave frequencies, and are similar to those used in laser applications at infrared and visible wavelengths. Figure 7.14 shows a photograph of such a resonator. These types of resonators also find applications in the measurement of dielectric constants at millimeter wave frequencies [4]. We will first consider the operation of the idealized parallel plate resonator of Figure 7.13, and then discuss the stability of resonators with plane and spherical mirrors.

If we assume the parallel plates of Figure 7.13 to be infinite in extent, then a TEM (plane wave) standing wave field can exist between the plates with the following form:

$$E_x = E_0 \sin k_0 z, \qquad 7.71a$$

$$H_y = \frac{jE_0}{\eta_0} \cos k_0 z, \qquad 7.71b$$

where E_0 is an arbitrary amplitude constant and $\eta_0 = 377\ \Omega$ is the intrinsic impedance of free-space. These fields already satisfy the boundary condition that $E_x = 0$ at $z = 0$; to satisfy the boundary conditions that $E_x = 0$ at $z = d$, we must have that

$$k_0 d = \ell\pi, \qquad \ell = 1, 2, 3, \ldots,$$

or

$$f_0 = \frac{ck_0}{2\pi} = \frac{c\ell}{2d}, \qquad \text{for } \ell = 1, 2, 3, \ldots, \qquad 7.72$$

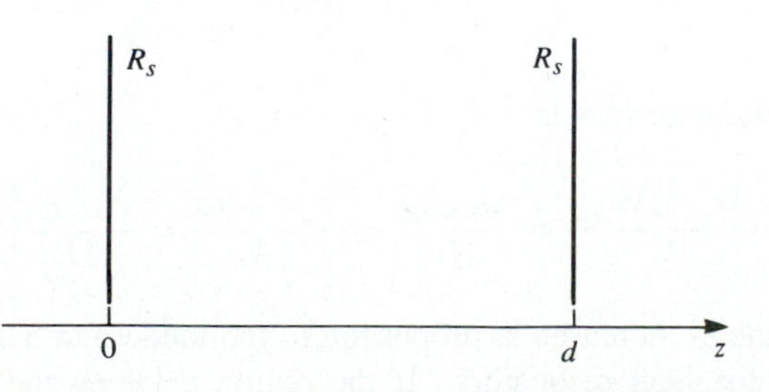

FIGURE 7.13 A Fabry-Perot resonator consisting of two parallel conducting plates.

FIGURE 7.14 Photograph of a Fabry-Perot resonator operating at 183 GHz with a mode number of 244 (nominally). One reflector is a flat plate, and the other is a movable spherical reflector.
Photograph courtesy of Millitech Corporation, S. Deerfield, Mass.

which determines the resonant frequencies. Note that there is only a single index, ℓ for these modes, as opposed to three indices for rectangular or cylindrical cavities; this is a consequence of the absence of conducting sidewalls.

We can determine the Q of the resonator as follows. The stored electric energy is, for 1 m^2 of cross-section,

$$W_e = \frac{\epsilon_0}{4}\int_{z=0}^{d} |E_x|^2 dz = \frac{\epsilon_0 |E_0|^2}{4}\int_{z=0}^{d} \sin^2 \frac{\ell\pi z}{d} dz = \frac{\epsilon_0 |E_0|^2 d}{8}. \quad 7.73a$$

The stored magnetic energy per square meter of cross-section is

$$W_m = \frac{\mu_0}{4}\int_{z=0}^{d} |H_y|^2 dz = \frac{\mu_0 |E_0|^2}{4\eta_0^2}\int_{z=0}^{d} \cos^2 \frac{\ell\pi z}{d} dz = \frac{\mu_0 |E_0|^2 d}{8\eta_0^2} = \frac{\epsilon_0 |E_0|^2 d}{8}, \quad 7.73b$$

which is seen to be equal to the stored electric energy. The power lost per square meter in both conducting plates is

$$P_c = 2\left(\frac{R_s}{2}\right)|H_y(z=0)|^2 = \frac{R_s |E_0|^2}{\eta_0^2}, \quad 7.74$$

so the Q due to conductor loss is

$$Q_c = \frac{\omega(W_e + W_m)}{P_c} = \frac{\omega\epsilon_0 d\eta_0^2}{4R_s} = \frac{\pi f_0 \epsilon_0 d\eta_0^2}{2R_s} = \frac{c\pi\ell\epsilon_0\eta_0^2}{4R_s} = \frac{\pi\ell\eta_0}{4R_s}, \quad 7.75$$

which shows that the Q increases in proportion to the mode number ℓ; ℓ is often several thousand or more for such resonators. If the region between the plates is filled with a dielectric material with tangent δ, it is easy to show that the Q due to dielectric loss is

$$Q_d = \frac{1}{\tan \delta}. \qquad 7.76$$

Dielectric is seldom used in such resonators, however, because of its limiting effect on the Q.

EXAMPLE 7.6

Consider a parallel-plate Fabry-Perot resonator with large copper plates spaced 4 cm apart. If the frequency is 94 GHz, find the mode number and Q of this resonator.

Solution
From (7.72), the mode number is

$$\ell = \frac{2df_0}{c} = \frac{2(0.04)(94 \times 10^9)}{3 \times 10^8} = 25.$$

At 94 GHz the surface resistivity of copper is

$$R_s = \sqrt{\frac{\omega\mu_0}{2\sigma}} = 8.0 \times 10^{-2}\ \Omega.$$

Then from (7.75) the Q is

$$Q = \frac{\pi \ell \eta_0}{4R_s} = \frac{\pi(25)(377)}{4(8 \times 10^{-2})} = 92,530.$$

○

Stability of Open Resonators

We now qualitatively discuss some of the properties of open resonators using curved mirrors. The general geometry is shown in Figure 7.15, which shows two spherical mirrors having radii of curvature R_1 and R_2, and separated by a distance d. Depending on the focusing properties of these mirrors, the energy in the resonator may be confined to a narrow region about the axis of the mirrors (stable), or it may spread out beyond the edges of the mirrors (unstable), resulting in a high degree of loss.

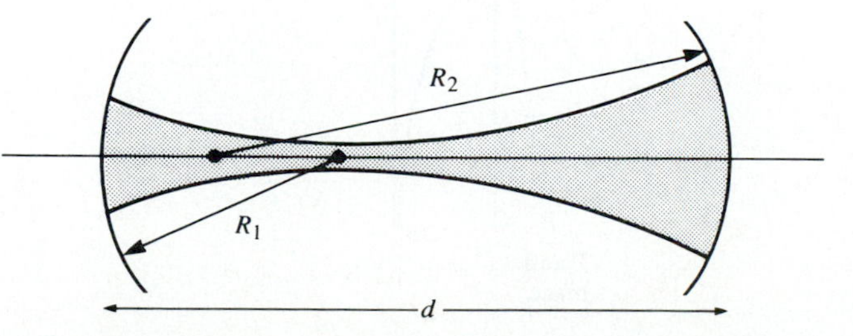

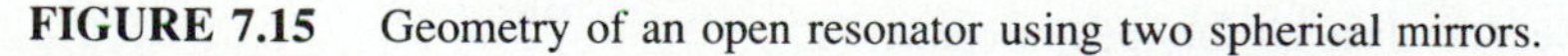

FIGURE 7.15 Geometry of an open resonator using two spherical mirrors.

Using ray optics [5], it can be shown that the open resonator geometry of Figure 7.15 will support a stable mode if the following condition is met:

$$0 \leq \left(1 - \frac{d}{R_1}\right)\left(1 - \frac{d}{R_2}\right) \leq 1. \tag{7.77}$$

This stability criterion can be presented in graphical form, as shown in Figure 7.16. The boundaries associated with the inequality on the left-hand side of (7.77) are straight lines at $d/R_1 = 1$ and $d/R - 2 = 1$, while the boundaries associated with the inequality on the right-hand side of (7.77) are hyperbolas with a focus at $d/R_1 = d/R - 2 = 1$. We can now consider some special configurations.

Parallel-plane resonator. This is essentially the idealized resonator of Figure 7.13. The radii of curvature are $R_1 = R_2 = \infty$, so this configuration corresponds to the point $d/R_1 = d/R_2 = 0$ in Figure 7.16, which is seen to be right on the boundary between a stable and unstable region. Thus, any irregularities, such as a lack of parallelism in the mirrors, will result in an unstable system.

Confocal resonator. In this case, $R_1 = R_2 = d$, corresponding to a symmetrical configuration. This resonator is represented by a point between a stable region and unstable region, and so is very sensitive to irregularities.

Concentric resonators. Here $R_1 = R_2 = d/2$, and the two mirrors have the same center, hence the term concentric. This configuration also lies at the edge of a stable and unstable region.

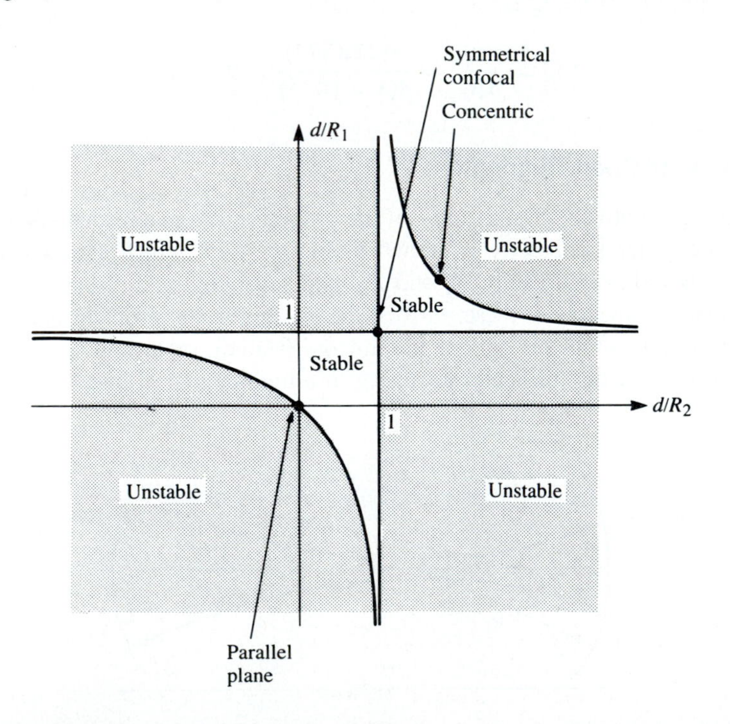

FIGURE 7.16 Stability diagram for open resonators.

Stable resonators. Symmetrical spherical resonators can be made stable by choosing $d/R_1 = d/R_2$ around 0.6, in which case the resonator is between the confocal and parallel plane designs, or around 1.4, in which case the resonator is between the confocal and concentric designs.

7.7 EXCITATION OF RESONATORS

We now discuss how the resonators of the previous sections can be coupled to external circuitry. In general, the way in which this is done depends on the type of resonator under consideration; some typical coupling techniques are shown for various resonators in Figure 7.17. In this section we will discuss the operation of some of the more common coupling techniques, notably gap coupling and aperture coupling. First we will illustrate the concept of critical coupling, whereby a resonator can be matched to a feedline, using a lumped-element resonant circuit.

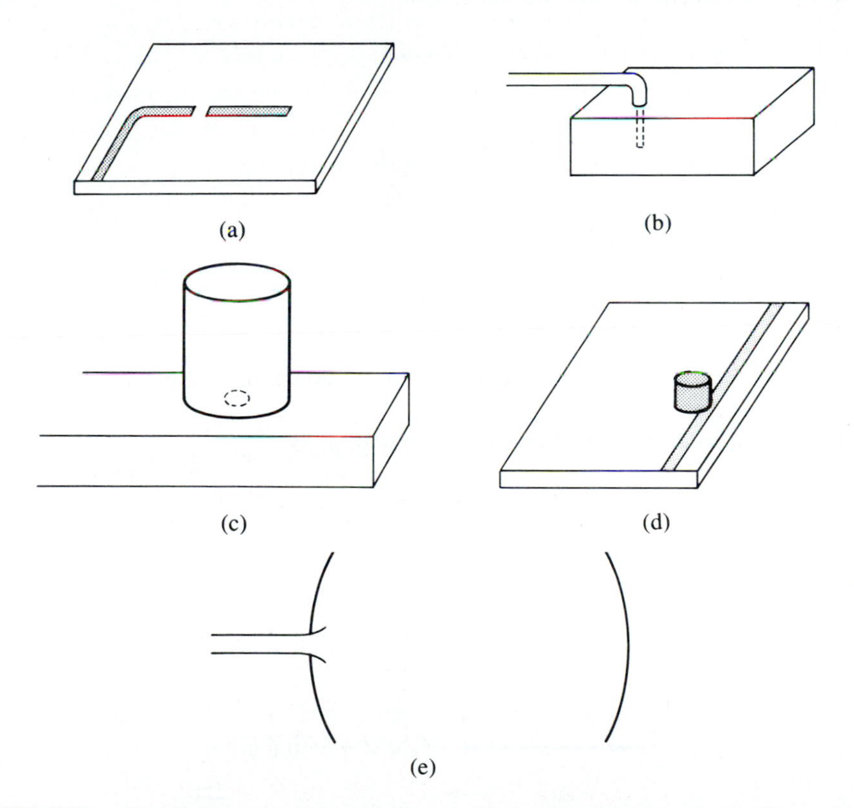

FIGURE 7.17 Coupling to microwave resonators. (a) A microstrip transmission line resonator gap coupled to a microstrip feed line. (b) A rectangular cavity resonator fed by a coaxial probe. (c) A circular cavity resonator aperture coupled to a rectangular waveguide. (d) A dielectric resonator coupled to a microstrip feed line. (e) A Fabry-Perot resonator fed by a waveguide horn antenna.

Critical Coupling

To obtain maximum power transfer between a resonator and a feed line, the resonator must be matched to the feed at the resonant frequency. The resonator is then said to be critically coupled to the feed. We will first illustrate the basic concept of critical coupling by considering the series resonant circuit shown in Figure 7.18.

From (7.9), the input impedance near resonance of the series resonant circuit of Figure 7.18 is given by

$$Z_{\text{in}} = R + j2L\Delta\omega = R + j\frac{2RQ\Delta\omega}{\omega_0}, \tag{7.78}$$

and the unloaded Q is, from (7.8),

$$Q = \frac{\omega_0 L}{R}. \tag{7.79}$$

At resonance, $\Delta\omega = 0$, so from (7.78) the input impedance is $Z_{\text{in}} = R$. In order to match the resonator to the line we must have,

$$R = Z_0. \tag{7.80}$$

Then the unloaded Q is

$$Q = \frac{\omega_0 L}{Z_0}. \tag{7.81}$$

From (7.22), the external Q is

$$Q_e = \frac{\omega_0 L}{Z_0} = Q, \tag{7.82}$$

which shows that the external and unloaded Qs are equal under the condition of critical coupling.

It is useful to define a coefficient of coupling, g, as

$$g = \frac{Q}{Q_e}, \tag{7.83}$$

which can be applied to both series ($g = Z_0/R$) and parallel ($g = R/Z_0$) resonant circuits. Then, three cases can be distinguished.

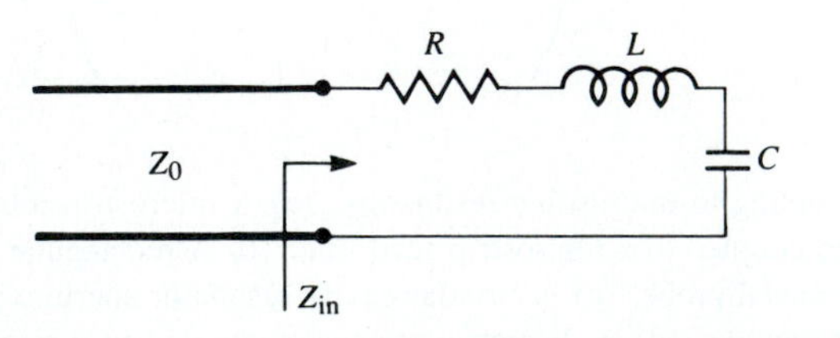

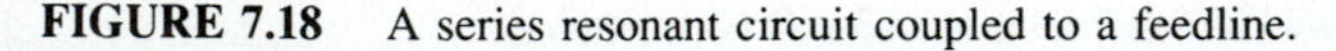

FIGURE 7.18 A series resonant circuit coupled to a feedline.

1. $g < 1$ The resonator is said to be undercoupled to the feed line.
2. $g = 1$ The resonator is critically coupled to the feed line.
3. $g > 1$ The resonator is said to be overcoupled to the feed line.

Figure 7.19 shows a Smith chart sketch of the impedance loci for the series resonant circuit, as given by (7.78), for various values of R corresponding to the above cases.

A Gap-Coupled Microstrip Resonator

Next we consider a $\lambda/2$ open-circuited microstrip resonator coupled to a microstrip feed line, as shown in Figure 7.17a. The gap in the microstrip line can be approximated as a series capacitor, so the equivalent circuit of this resonator and feed can be constructed as shown in Figure 7.20. The normalized input impedance seen by the feed line is

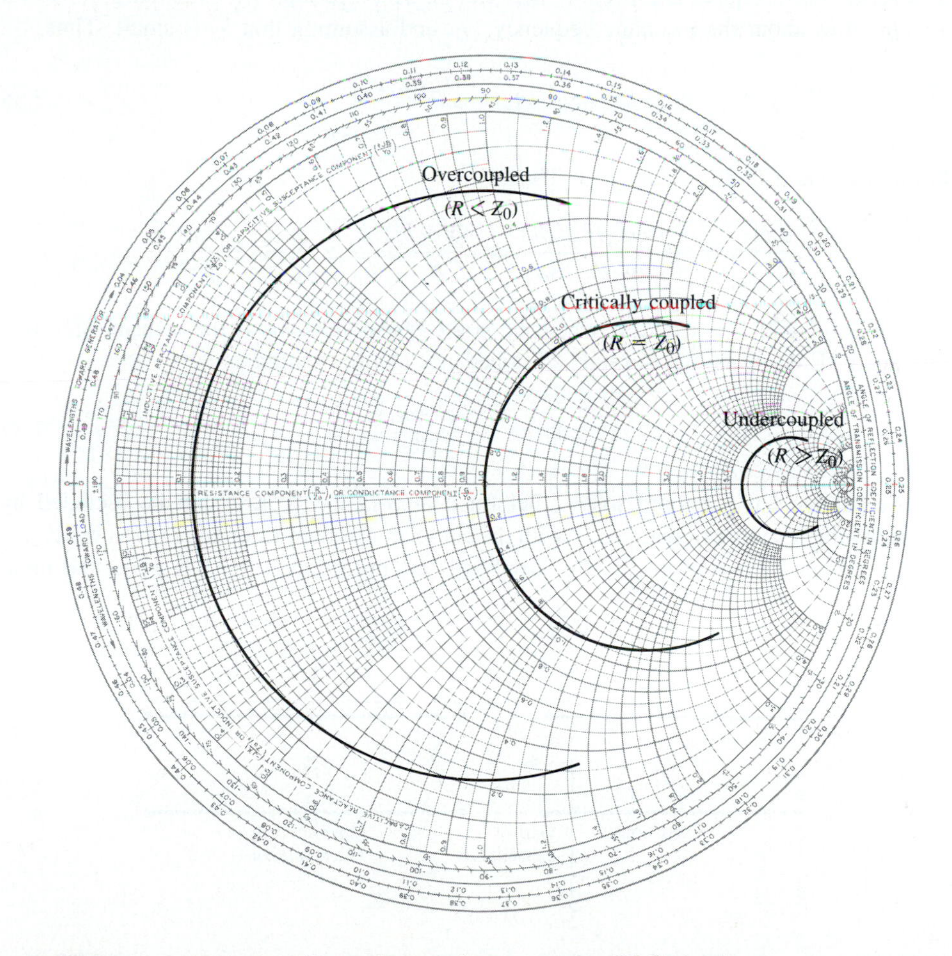

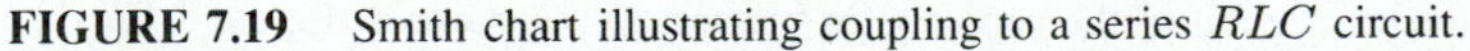
FIGURE 7.19 Smith chart illustrating coupling to a series RLC circuit.

then

$$z = \frac{Z}{Z_0} = -j\frac{[(1/\omega C) + Z_0 \cot \beta \ell]}{Z_0} = -j\left(\frac{\tan \beta \ell + b_c}{b_c \tan \beta \ell}\right), \tag{7.84}$$

where $b_c = Z_0 \omega C$ is the normalized susceptance of the coupling capacitor, C. Resonance occurs with $z = 0$, or when

$$\tan \beta \ell + b_c = 0. \tag{7.85}$$

The solutions to this transcendental equation are sketched in Figure 7.21. In practice, $b_c \ll 1$, so that the first resonant frequency, ω_1, will be close to the frequency for which $\beta \ell = \pi$ (the first resonant frequency of the unloaded resonator). In this case the coupling of the feedline to the resonator has the effect of lowering its resonant frequency.

We now wish to simplify the driving point impedance of (7.84) to relate this resonator to a series RLC equivalent circuit. This can be accomplished by expanding $z(\omega)$ in a Taylor series about the resonant frequency, ω_1, and assuming that b_c is small. Thus,

$$z(\omega) = z(\omega_1) + (\omega - \omega_1)\left.\frac{dz(\omega)}{d\omega}\right|_{\omega_1} + \cdots. \tag{7.86}$$

From (7.84) and (7.85), $z(\omega_1) = 0$. Then,

$$\left.\frac{dz}{d\omega}\right|_{\omega_1} = \frac{-j\sec^2 \beta \ell}{b_c \tan \beta \ell}\frac{d(\beta \ell)}{d\omega} = \frac{j(1 + b_c^2)}{b_c^2}\frac{\ell}{v_p} \simeq \frac{j}{b_c^2}\frac{\ell}{v_p} \simeq \frac{j\pi}{\omega_1 b_c^2},$$

since $b_c \ll 1$ and $\ell \simeq \pi v_p/\omega_1$, where v_p is the phase velocity of the transmission line (assumed TEM). Then the normalized impedance can be written as

$$z(\omega) = \frac{j\pi(\omega - \omega_1)}{\omega_1 b_c^2}. \tag{7.87}$$

So far we have ignored losses, but for a high-Q cavity loss can be included by replacing the resonant frequency ω_1 with the complex resonant frequency given by $\omega_1(1 + j/2Q)$, which follows from (7.10). Applying this procedure to (7.87) gives the input

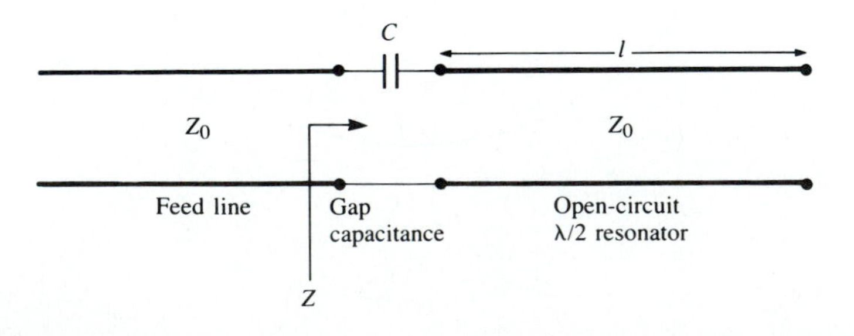

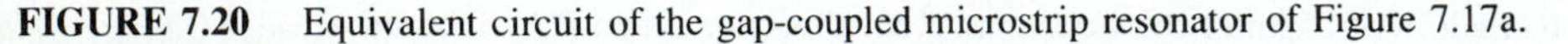

FIGURE 7.20 Equivalent circuit of the gap-coupled microstrip resonator of Figure 7.17a.

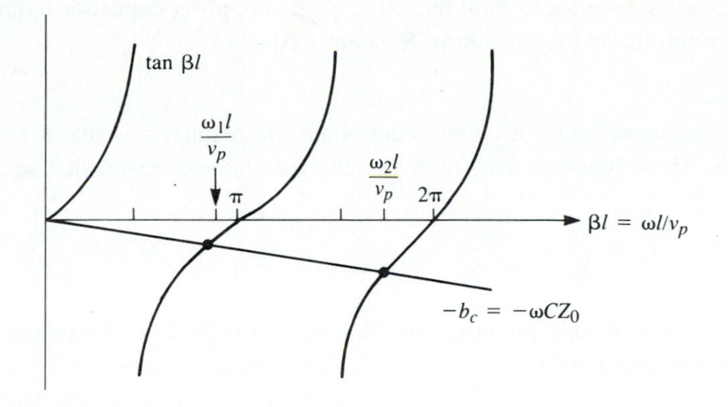

FIGURE 7.21 Solutions to (7.85) for the resonant frequencies of the gap-coupled microstrip resonator.

impedance of gap-coupled lossy resonator as

$$z(\omega) = \frac{\pi}{2Qb_c^2} + j\frac{\pi(\omega - \omega_1)}{\omega_1 b_c^2}. \tag{7.88}$$

Note that an uncoupled $\lambda/2$ open-circuited transmission line resonator looks like a parallel RLC circuit near resonance, but the present case of a capacitive coupled $\lambda/2$ resonator looks like a series RLC circuit near resonance. This is because the series coupling capacitor has the effect of inverting the driving point impedance of the resonator (see the discussion of impedance inverters in Section 9.5).

At resonance, then, the input resistance is $R = Z_0\pi/2Qb_c^2$. For critical coupling we must have $R = Z_0$, or

$$b_c = \sqrt{\frac{\pi}{2Q}}. \tag{7.89}$$

The coupling coefficient of (7.83) is

$$g = \frac{R}{Z_0} = \frac{\pi}{2Qb_c^2}. \tag{7.90}$$

If $b_c > \sqrt{\pi/2Q}$, then $g < 1$ and the resonator is undercoupled; if $b_c < \sqrt{\pi/2Q}$, then $g > 1$ and the resonator is overcoupled.

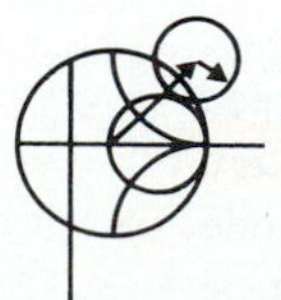

EXAMPLE 7.7

A resonator is made from an open-circuited 50 Ω microstrip line, and is gap-coupled to a 50 Ω feed line, as in Figure 7.17a. The resonator has a length of 2.175 cm, an effective dielectric constant of 1.9, and an attenuation of 0.01

dB/cm near its resonance. Find the value of the coupling capacitor required for critical coupling, and the resulting resonant frequency.

Solution

The first resonant frequency will occur when the resonator is about $\ell = \lambda_g/2$ in length. Thus, ignoring fringing fields, the approximate resonant frequency is

$$f_0 = \frac{v_p}{\lambda_g} = \frac{c}{2\ell\sqrt{\epsilon_e}} = \frac{3 \times 10^8}{2(0.02175)\sqrt{1.9}} = 5.00 \text{ GHz},$$

which does not include the effect of the coupling capacitor. Then from (7.35) the Q of this resonator is

$$Q = \frac{\beta}{2\alpha} = \frac{\pi}{\lambda_g \alpha} = \frac{\pi}{2\ell\alpha} = \frac{\pi(8.7 \text{ dB/Np})}{2(0.02175 \text{ m})(1\text{dB/m})} = 628.$$

From (7.89) the normalized coupling capacitor susceptance is

$$b_c = \sqrt{\frac{\pi}{2Q}} = \sqrt{\frac{\pi}{2(628)}} = 0.05,$$

so the coupling capacitor has a value of

$$C = \frac{b_c}{\omega Z_0} = \frac{0.05}{2\pi(5 \times 10^9)(50)} = 0.032 \text{ pF},$$

which should result in the critical coupling of the resonator to the 50 Ω feed line.

Now that C is determined, the resonant frequency can be found by solving the transcendental equation of (7.85). Since we know from the graphical solution of Figure 7.21 that the actual resonant frequency is slightly lower than the unloaded resonant frequency of 5.0 GHz, it is an easy matter to calculate (7.85) for several frequencies in this vicinity, which leads to a value of about 4.918 GHz. This is about 1.6% lower than the unloaded resonant frequency. Figure 7.22 shows a Smith chart plot of the input impedance of the gap-coupled resonator for coupling capacitor values that lead to under, critical, and overcoupled resonators. ○

An Aperture Coupled Cavity

As a final example of resonator excitation, we will consider the aperture coupled waveguide cavity shown in Figure 7.23. As discussed in Section 5.9, a small aperture in the transverse wall acts as a shunt inductance. If we consider the first resonant mode of the cavity, which occurs when the cavity length $\ell = \lambda_g/2$, then the cavity can be considered as a transmission line resonator shorted at one end. The aperture coupled

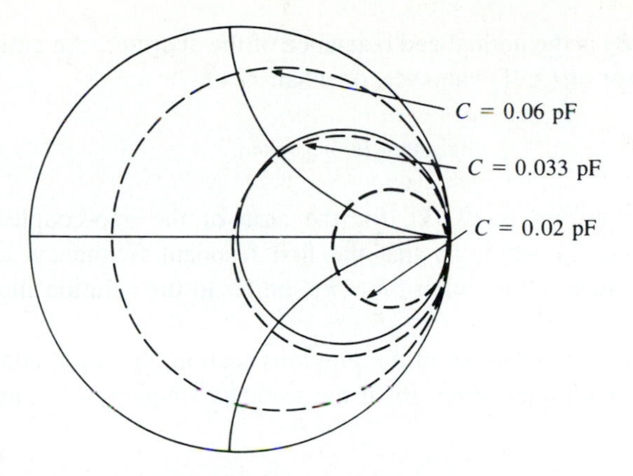

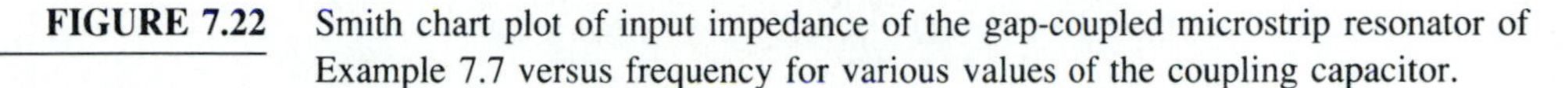

FIGURE 7.22 Smith chart plot of input impedance of the gap-coupled microstrip resonator of Example 7.7 versus frequency for various values of the coupling capacitor.

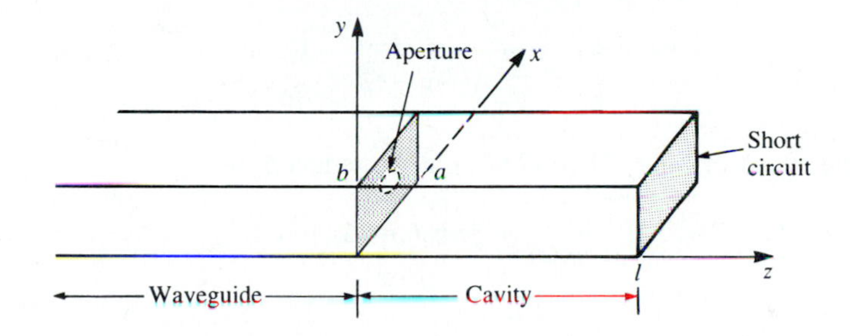

FIGURE 7.23 A rectangular waveguide aperture coupled to a rectangular cavity.

cavity can then be modeled by the equivalent circuit shown in Figure 7.24. This circuit is basically the dual of the equivalent circuit of Figure 7.20, for the gap-coupled microstrip resonator, so we will approach the solution in the same manner.

The normalized input admittance seen by the feed line is

$$y = Z_0 Y = -jZ_0 \left(\frac{1}{X_L} + \cos \beta \ell \right) = -j \left(\frac{\tan\ \beta \ell + x_L}{x_L \tan\ \beta \ell} \right), \qquad 7.91$$

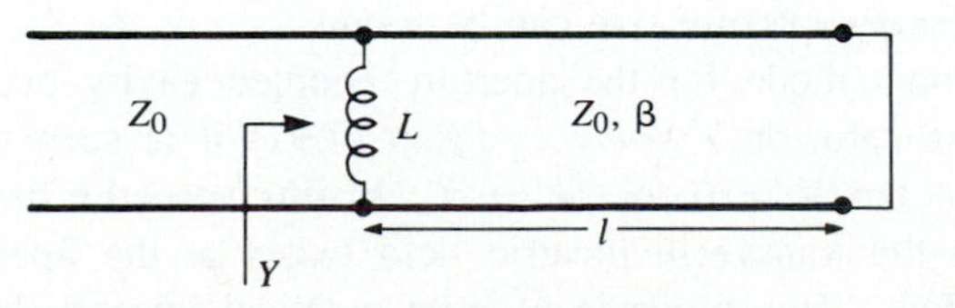

FIGURE 7.24 Equivalent circuit of the aperture coupled cavity.

where $x_L = \omega L/Z_0$ is the normalized reactance of the aperture. An antiresonance occurs when the numerator of (7.91) vanishes, or when

$$\tan \beta\ell + x_L = 0, \tag{7.92}$$

which is similar in form to (7.85), for the case of the gap-coupled microstrip resonator. In practice, $x_L \ll 1$, so that the first resonant frequency, ω_1, will be close to the resonant frequency for which $\beta\ell = \pi$, similar to the solution illustrated in Figure 7.21.

Using the same procedure as in the previous section, the input admittance of (7.91) can be expanded in a Taylor series about the resonant frequency, ω_1, assuming $x_L \ll 1$, to obtain

$$y(\omega) = y(\omega_1) + (\omega - \omega_1)\left.\frac{dy(\omega)}{d\omega}\right|_{\omega_1} + \cdots \simeq \frac{j\ell}{x_L^2}(\omega - \omega_1)\left.\frac{d\beta}{d\omega}\right|_{\omega 1}, \tag{7.93}$$

since $y(\omega_1) = 0$. For a rectangular waveguide,

$$\frac{d\beta}{d\omega} = \frac{d}{d\omega}\sqrt{k_0^2 - k_c^2} = \frac{k_0}{\beta c},$$

where c is the speed of light. Then (7.93) can be reduced to

$$y(\omega) = \frac{j\pi k_0(\omega - \omega_1)}{\beta^2 c x_L^2}. \tag{7.94}$$

In (7.94), k_0, β, and x_L should be evaluated at the resonant frequency ω_1.

Loss can now be included by assuming a high-Q cavity and replacing ω_1 in the numerator of (7.94) with $\omega_1(1 + j/2Q)$, to obtain

$$y(\omega) \simeq \frac{\pi k_0 \omega_1}{2Q\beta^2 c x_L^2} + j\frac{\pi k_0(\omega - \omega_1)}{\beta^2 c x_L^2}. \tag{7.95}$$

At resonance, the input resistance is $R = 2Q\beta^2 c x_L^2 Z_0/\pi k_0 \omega_1$. To obtain critical coupling we must have $R = Z_0$, which yields the required aperture reactance as

$$X_L = Z_0\sqrt{\frac{\pi k_0 \omega_1}{2Q\beta^2 c}}. \tag{7.96}$$

From X_L, the necessary aperture size can be found.

The next resonant mode for the aperture coupled cavity occurs when the input impedance becomes zero, or $Y \rightarrow \infty$. From (7.91) it is seen that this occurs at a frequency such that $\tan \beta\ell = 0$, or $\beta\ell = \pi$. In this case the cavity is exactly $\lambda_g/2$ long, so a null in the transverse electric field exists at the aperture plane, and the aperture has no effect. This mode is of little practical interest, because of this loose coupling.

The excitation of a cavity resonator by an electric current probe or loop can be analyzed by the method of modal analysis, similar to that discussed in Sections 5.8 and 5.9. The procedure is complicated, however, by the fact that a complete modal expansion requires fields having irrotational (zero curl) components. The interested reader is referred to references [1] and [6].

7.8 CAVITY PERTURBATIONS

In practical applications cavity resonators are often modified by making small changes in their shape, or by the introduction of small pieces of dielectric or metallic materials. For example, the resonant frequency of a cavity can be easily tuned with a small screw (dielectric or metallic) that enters the cavity volume, or by changing the size of the cavity with a movable wall. Another application involves the determination of dielectric constant by measuring the shift in resonant frequency when a small dielectric sample is introduced into the cavity.

In some cases, the effect of such perturbations on the cavity performance can be calculated exactly, but often approximations must be made. One useful technique for doing this is the perturbational method, which assumes that the actual fields of a cavity with a small shape or material perturbation are not greatly different from those of the unperturbed cavity. Thus, this technique is similar in concept to the perturbational method introduced in Section 3.9 for treating loss in good conductors, where it was assumed that there was not a significant difference between the fields of a component with good conductors and one with perfect conductors.

In this section we will derive expressions for the approximate change in resonant frequency when a cavity is perturbed by small changes in the material filling the cavity, or by small changes in its shape.

Material Perturbations

Figure 7.25 shows a cavity perturbed by a change in the permittivity ($\Delta\epsilon$), or permeability ($\Delta\mu$), of all or part of the material filling the cavity. If $\bar{E}_0, \bar{H}_0$ are the fields of the original cavity, and $\bar{E}, \bar{H}$ are the fields of the perturbed cavity, then Maxwell's curl equations can be written for the two cases as

$$\nabla \times \bar{E}_0 = -j\omega_0\mu\bar{H}_0, \tag{7.97a}$$

$$\nabla \times \bar{H}_0 = j\omega_0\epsilon\bar{E}_0, \tag{7.97b}$$

$$\nabla \times \bar{E} = -j\omega(\mu + \Delta\mu)\bar{H}, \tag{7.98a}$$

$$\nabla \times \bar{H} = j\omega(\epsilon + \Delta\epsilon)\bar{E}, \tag{7.98b}$$

where ω_0 is the resonant frequency of the original cavity and ω is the resonant frequency of the perturbed cavity.

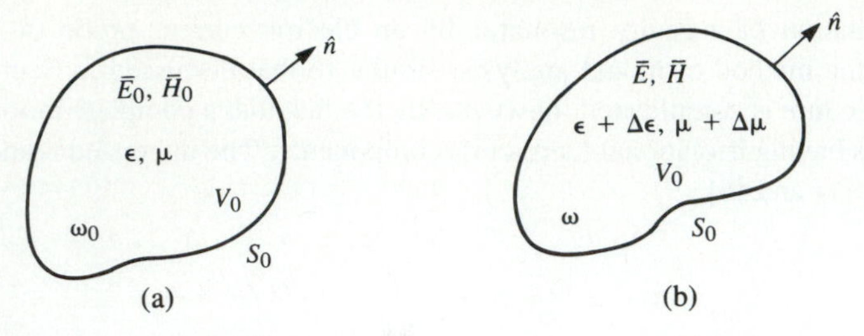

FIGURE 7.25 A resonant cavity perturbed by a change in the permittivity or permeability of the material in the cavity. (a) Original cavity. (b) Perturbed cavity.

Now multiply the conjugate of (7.97a) by $\bar{H}$ and multiply (7.98b) by $\bar{E}_0^*$ to get

$$\bar{H} \cdot \nabla \times \bar{E}_0^* = j\omega_0 \mu \bar{H} \cdot \bar{H}_0^*,$$
$$\bar{E}_0^* \cdot \nabla \times \bar{H} = j\omega(\epsilon + \Delta\epsilon)\bar{E}_0^* \cdot \bar{E}.$$

Subtracting these two equations and using the vector identity (B.8) that $\nabla \cdot (\bar{A} \times \bar{B}) = \bar{B} \cdot \nabla \times \bar{A} - \bar{A} \cdot \nabla \times \bar{B}$ gives

$$\nabla \cdot (\bar{E}_0^* \times \bar{H}) = j\omega_0 \mu \bar{H} \cdot \bar{H}_0^* - j\omega(\epsilon + \Delta\epsilon)\bar{E}_0^* \cdot \bar{E}. \qquad 7.99a$$

Similarly, we multiply the conjugate of (7.97b) by $\bar{E}$ and multiply (7.98a) by $\bar{H}_0^*$ to get

$$\bar{E} \cdot \nabla \times \bar{H}_0^* = -j\omega_0 \epsilon \bar{E}_0^* \cdot \bar{E},$$
$$\bar{H}_0^* \cdot \nabla \times \bar{E} = -j\omega(\mu + \Delta\mu)\bar{H}_0^* \cdot \bar{H}.$$

Subtracting these two equations and using vector identity (B.8) gives

$$\nabla \cdot (\bar{E} \times \bar{H}_0^*) = -j\omega(\mu + \Delta\mu)\bar{H}_0^* \cdot \bar{H} + j\omega_0 \epsilon \bar{E}_0^* \cdot \bar{E}. \qquad 7.99b$$

Now add (7.99a) and (7.99b), integrate over the volume V_0, and use the divergence theorem to obtain

$$\int_{V_0} \nabla \cdot (\bar{E}_0^* \times \bar{H} + \bar{E} \times \bar{H}_0^*) dv = \oint_{S_0} (\bar{E}_0^* \times \bar{H} + \bar{E} \times \bar{H}_0^*) \cdot d\bar{s} = 0$$
$$= j \int_{V_0} \{[\omega_0 \epsilon - \omega(\epsilon + \Delta\epsilon)]\bar{E}_0^* \cdot \bar{E} + [\omega_0 \mu - \omega(\mu + \Delta\mu)]\bar{H}_0^* \cdot \bar{H}\} dv, \qquad 7.100$$

where the surface integral is zero because $\hat{n} \times \bar{E} = 0$ on S_0. Rewriting gives

$$\frac{\omega - \omega_0}{\omega} = \frac{-\int_{V_0} (\Delta\epsilon \bar{E} \cdot \bar{E}_0^* + \Delta\mu \bar{H} \cdot \bar{H}_0^*) dv}{\int_{V_0} (\epsilon \bar{E} \cdot \bar{E}_0^* + \mu \bar{H} \cdot \bar{H}_0^*) dv}. \qquad 7.101$$

This is an exact equation for the change in resonant frequency due to material perturbations, but is not in a very usable form since we generally do not know $\bar{E}$ and $\bar{H}$, the exact fields in the perturbed cavity. But, if we assume that $\Delta\epsilon$ and $\Delta\mu$ are small, then we can approximate the perturbed fields $\bar{E}, \bar{H}$ by the original fields $\bar{E}_0, \bar{H}_0$, and ω in the denominator of (7.101) by ω_0, to give the fractional change in resonant frequency as

$$\frac{\omega - \omega_0}{\omega_0} \simeq \frac{-\int_{V_0} \left(\Delta\epsilon|\bar{E}_0|^2 + \Delta\mu|\bar{H}_0|^2\right) dv}{\int_{V_0} \left(\epsilon|\bar{E}_0|^2 + \mu|\bar{H}_0|^2\right) dv}. \tag{7.102}$$

This result shows that any increase in ϵ or μ at any point in the cavity will decrease the resonant frequency. The reader may also observe that the terms in (7.102) can be related to the stored electric and magnetic energies in the original and perturbed cavities, so that the decrease in resonant frequency can be related to the increase in stored energy of the perturbed cavity.

EXAMPLE 7.8

A rectangular cavity operating in the TE_{101} mode is perturbed by the insertion of a thin dielectric slab into the bottom of the cavity, as shown in Figure 7.26. Use the perturbational result of (7.102) to derive an expression for the change in resonant frequency.

Solution

From (7.42a-c), the fields for the unperturbed TE_{101} cavity mode can be written as

$$E_y = A \sin\frac{\pi x}{a} \sin\frac{\pi z}{d},$$

$$H_x = \frac{-jA}{Z_{TE}} \sin\frac{\pi x}{a} \cos\frac{\pi z}{d},$$

$$H_z = \frac{j\pi A}{k\eta a} \cos\frac{\pi x}{a} \sin\frac{\pi z}{d}.$$

In the numerator of (7.102), $\Delta\epsilon = (\epsilon_r - 1)\epsilon_0$ for $0 \le y \le t$, and zero elsewhere.

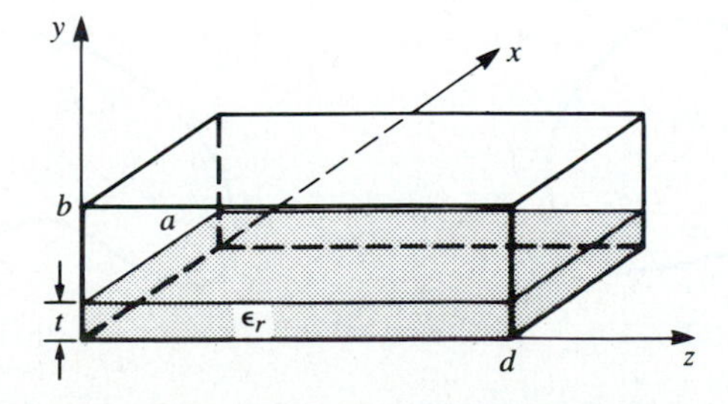

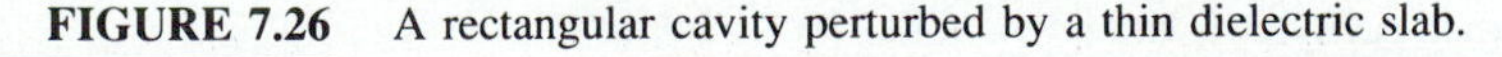

FIGURE 7.26 A rectangular cavity perturbed by a thin dielectric slab.

The integral can then be evaluated as

$$\int_V (\Delta\epsilon|\bar{E}_0|^2 + \Delta\mu|\bar{H}_0|^2)dv = (\epsilon_r - 1)\epsilon_0 \int_{x=0}^{a}\int_{y=0}^{t}\int_{z=0}^{d} |E_y|^2 dz\, dy\, dx$$

$$= \frac{(\epsilon_r - 1)\epsilon_0 A^2 atd}{4}.$$

The denominator of (7.102) is proportional to the total energy in the unperturbed cavity, which was evaluated in (7.43), thus,

$$\int_V (\epsilon|\bar{E}_0|^2 + \mu|\bar{H}_0|^2)dv = \frac{abd\epsilon_0}{2}A^2.$$

Then (7.102) gives the fractional change (decrease) in resonant frequency as

$$\frac{\omega - \omega_0}{\omega_0} = \frac{-(\epsilon_r - 1)t}{2b}.$$

○

Shape Perturbations

Changing the size of a cavity or inserting a tuning screw can be considered as a change in the shape of the cavity and, for small changes, can also be treated by the perturbation technique. Figure 7.27 shows an arbitrary cavity with a perturbation in its shape; we will derive an expression for the change in resonant frequency.

As in the case of material perturbations, let $\bar{E}_0, \bar{H}_0, \omega_0$ be the fields and resonant frequency of the original cavity and let $\bar{E}, \bar{H}, \omega$ be the fields and resonant frequency of the perturbed cavity. Then Maxwell's curl equations can be written for the two cases as

$$\nabla \times \bar{E}_0 = -j\omega_0\mu\bar{H}_0, \qquad 7.103a$$

$$\nabla \times \bar{H}_0 = j\omega_0\epsilon\bar{E}_0, \qquad 7.103b$$

$$\nabla \times \bar{E} = -j\omega\mu\bar{H}, \qquad 7.104a$$

$$\nabla \times \bar{H} = j\omega\epsilon\bar{E}. \qquad 7.104b$$

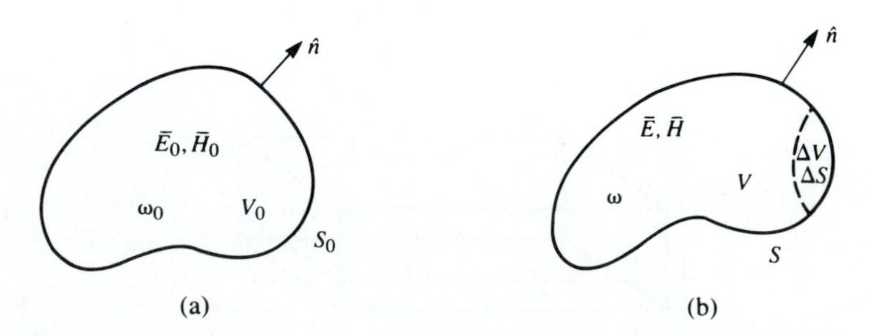

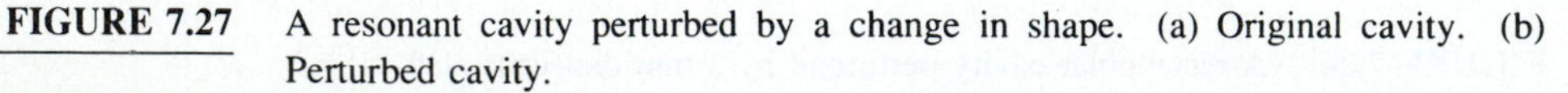

FIGURE 7.27 A resonant cavity perturbed by a change in shape. (a) Original cavity. (b) Perturbed cavity.

Now multiply the conjugate of (7.103a) by $\bar{H}$ and multiply (7.104b) by $\bar{E}_0^*$ to get

$$\bar{H} \cdot \nabla \times \bar{E}_0^* = j\omega_0 \mu \bar{H} \cdot \bar{H}_0^*,$$
$$\bar{E}_0^* \cdot \nabla \times \bar{H} = j\omega\epsilon \bar{E}_0^* \cdot \bar{E}.$$

Subtracting these two equations and using vector identity (B.8) then gives

$$\nabla \cdot (\bar{E}_0^* \times \bar{H}) = j\omega_0 \mu \bar{H} \cdot \bar{H}_0^* - j\omega\epsilon \bar{E}_0^* \cdot \bar{E}. \qquad 7.105a$$

Similarly, we multiply the conjugate of (7.103b) by $\bar{E}$ and (7.104a) by $\bar{H}_0^*$ to get

$$\bar{E} \cdot \nabla \times \bar{H}_0^* = -j\omega_0 \epsilon \bar{E} \cdot \bar{E}_0^*,$$
$$\bar{H}_0^* \cdot \nabla \times \bar{E} = -j\omega\mu \bar{H}_0^* \cdot \bar{H}.$$

Subtracting and applying vector identity (B.8) gives

$$\nabla \cdot (\bar{E} \times \bar{H}_0^*) = -j\omega\mu \bar{H}_0^* \cdot \bar{H} + j\omega_0 \epsilon \bar{E} \cdot \bar{E}_0^* \qquad 7.105b$$

Now add (7.105a) and (7.105b), integrate over the volume V, and use the divergence theorem to obtain

$$\int_V \nabla \cdot (\bar{E} \times \bar{H}_0^* + \bar{E}_0^* \times \bar{H}) dv = \oint_S (\bar{E} \times \bar{H}_0^* + \bar{E}_0^* \times \bar{H}) \cdot d\bar{s}$$
$$= \oint_S \bar{E}_0^* \times \bar{H} \cdot d\bar{s} = -j(\omega - \omega_0) \int_V (\epsilon \bar{E} \cdot \bar{E}_0^* + \mu \bar{H} \cdot \bar{H}_0^*) dv, \qquad 7.106$$

since $\hat{n} \times \bar{E} = 0$ on S.

Since the perturbed surface $S = S_0 - \Delta S$, we can write

$$\oint_S \bar{E}_0^* \times \bar{H} \cdot d\bar{s} = \oint_{S_0} \bar{E}_0^* \times \bar{H} \cdot d\bar{s} - \oint_{\Delta S} \bar{E}_0^* \times \bar{H} \cdot d\bar{s} = -\oint_{\Delta S} \bar{E}_0^* \times \bar{H} \cdot ds,$$

because $\hat{n} \times \bar{E}_0 = 0$ on S_0. Using this result in (7.106) gives

$$\omega - \omega_0 = \frac{-j \oint_{\Delta S} \bar{E}_0^* \times \bar{H} \cdot d\bar{s}}{\int_V (\epsilon \bar{E} \cdot \bar{E}_0^* + \mu \bar{H} \cdot \bar{H}_0^*) dv}, \qquad 7.107$$

which is an exact expression for the new resonant frequency, but not a very usable one since we generally do not initially know $\bar{E}, \bar{H}$, or ω. If we assume ΔS is small, and approximate $\bar{E}, \bar{H}$ by the unperturbed values of $\bar{E}_0, \bar{H}_0$, then the numerator of (7.107) can be reduced as follows:

$$\oint_{\Delta S} \bar{E}_0^* \times \bar{H} \cdot d\bar{s} \simeq \oint_{\Delta S} \bar{E}_0^* \times \bar{H}_0 \cdot d\bar{s} = -j\omega_0 \int_{\Delta V} (\epsilon |\bar{E}_0|^2 - \mu |\bar{H}_0|^2) dv, \qquad 7.108$$

where the last identity follows from conservation of power, as derived from the conjugate of (2.87) with σ, $\bar{J}_s$, and $\bar{M}_s$ set to zero. Using this result in (7.107) gives an expression for the fractional change in resonant frequency as

$$\frac{\omega - \omega_0}{\omega_0} \simeq \frac{\int_{V_0} (\mu |\bar{H}_0|^2 - \epsilon |\bar{E}_0|^2) dv}{\int_{V_0} (\mu |\bar{H}_0|^2 + \epsilon |\bar{E}_0|^2) dv}, \qquad 7.109$$

where we have also assumed that the denominator of (7.107), which represents the total energy stored in the perturbed cavity, is approximately the same as that for the unperturbed cavity.

Equation (7.109) can be written in terms of stored energies as follows:

$$\frac{\omega - \omega_0}{\omega_0} = \frac{\Delta W_m - \Delta W_e}{W_m + W_e}, \qquad 7.110$$

where ΔW_m and ΔW_e are the changes in the stored magnetic energy and electric energy, respectively, after the shape perturbation, and $W_m + W_e$ is the total stored energy in the cavity. These results show that the resonant frequency may either increase or decrease, depending on where the perturbation is located and whether it increases or decreases the cavity volume.

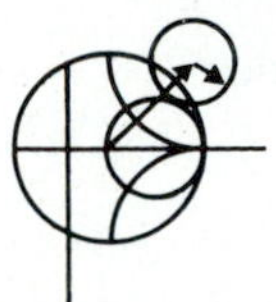

EXAMPLE 7.9

A thin screw of radius r_0 extends a distance ℓ through the center of the top wall of a rectangular cavity operating in the TE_{101} mode, as shown in Figure 7.28. If the cavity is air-filled, use (7.109) to derive an expression for the change in resonant frequency from the unperturbed cavity.

Solution

From (7.42a-b), the fields for the unperturbed TE_{101} cavity can be written as

$$E_y = A \sin \frac{\pi x}{a} \sin \frac{\pi z}{d},$$

$$H_x = \frac{-jA}{Z_{TE}} \sin \frac{\pi x}{a} \cos \frac{\pi z}{d},$$

$$H_z = \frac{j\pi A}{k\eta a} \cos \frac{\pi x}{a} \sin \frac{\pi z}{d}.$$

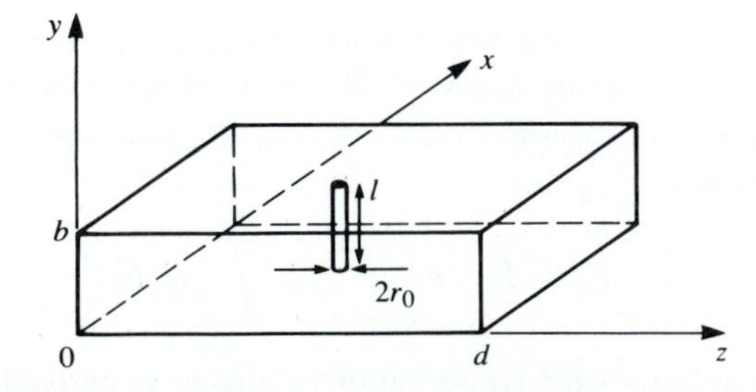

FIGURE 7.28 A rectangular cavity perturbed by a tuning post in the center of the top wall.

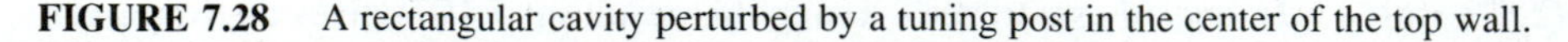

Now if the screw is thin, we can assume that the fields are constant over the cross-section of the screw and can be represented by the fields at $x = a/2, z = d/2$:

$$E_y\left(x = \frac{a}{2}, y, z = \frac{d}{2}\right) = A,$$

$$H_x\left(x = \frac{a}{2}, y, z = \frac{d}{2}\right) = 0,$$

$$H_z\left(x = \frac{a}{2}, y, z = \frac{d}{2}\right) = 0.$$

Then the numerator of (7.109) can be evaluated as

$$\int_{\Delta V} (\mu|\bar{H}_0|^2 - \epsilon|\bar{E}_0|^2)dv = -\epsilon_0 \int_{\Delta V} A^2 dv = -\epsilon_0 A^2 \Delta V,$$

where $\Delta V = \pi \ell r_0^2$ is the volume of the screw. The denominator of (7.109) is, from (7.43),

$$\int_{V_0} (\mu|\bar{H}_0|^2 + \epsilon|\bar{E}_0|^2)dv = \frac{abd\epsilon_0 A^2}{2} = \frac{V_0 \epsilon_0 A^2}{2},$$

where $V_0 = abd$ is the volume of the unperturbed cavity. Then (7.109) gives

$$\frac{\omega - \omega_0}{\omega_0} = \frac{-2\ell\pi r_0}{abd} = \frac{-2\Delta V}{V_0},$$

which indicates a lowering of the resonant frequency. ○

REFERENCES

[1] R. E. Collin, *Foundations for Microwave Engineering*, McGraw-Hill, N. Y., 1966.

[2] S. B. Cohn, "Microwave Bandpass Filters Containing High-Q Dielectric Resonators," *IEEE Trans. Microwave Theory and Techniques,* vol. MTT-16, pp. 218–227, April 1968.

[3] M. W. Pospieszalski, "Cylindrical Dielectric Resonators and Their Applications in TEM Line Microwave Circuits," *IEEE Trans. Microwave Theory and Techniques*, vol. MTT-27, pp. 233–238, March 1979.

[4] J. E. Degenford and P. D. Coleman, "A Quasi-Optics Perturbation Technique for Measuring Dielectric Constants," *Proc. IEEE*, vol. 54, pp. 520–522, April 1966.

[5] S. Ramo, J. R. Whinnery, and T. Van Duzer, *Fields and Waves in Communication Electronics*, John Wiley & Sons, N. Y., 1965.

[6] R. E. Collin, *Field Theory of Guided Waves*, McGraw-Hill, N. Y., 1960.

PROBLEMS

7.1 Consider the loaded parallel resonant RLC circuit shown below. Compute the resonant frequency, unloaded Q, and loaded Q.

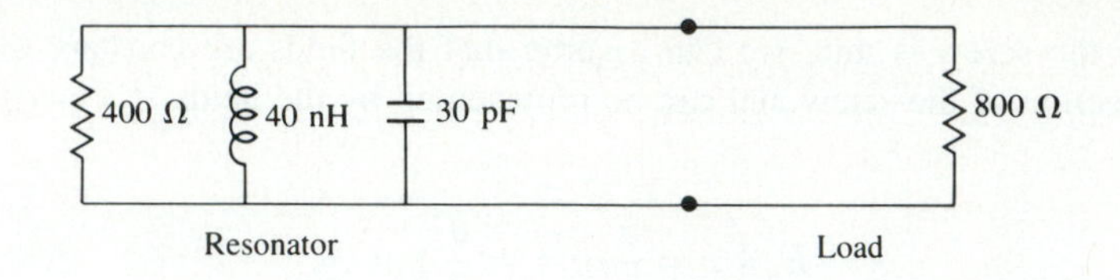

7.2 Calculate the Q of a transmission line resonator consisting of a short-circuited transmission line 1λ long.

7.3 A transmission line resonator is fabricated from a $\lambda/4$ length of open-circuited line. Find the Q of this resonator if the complex propagation constant of the line is $\alpha + j\beta$.

7.4 Consider the resonator shown below, consisting of a $\lambda/2$ length of lossless transmission line shorted at both ends. At an arbitrary point z on the line, compute the impedances Z_L and Z_R seen looking to the left and to the right, and show that $Z_L = Z_R^*$. (This condition holds true for any lossless resonator and is the basis for the transverse resonance technique discussed in Section 4.9).

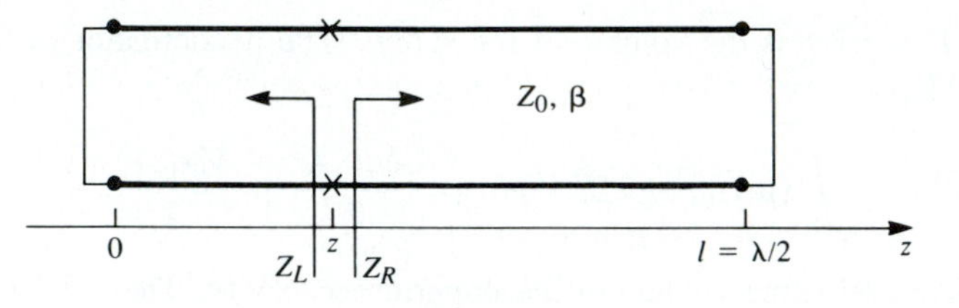

7.5 A resonator is constructed from a 3.0 cm length of 100 Ω air-filled coaxial line, shorted at one end and terminated with a capacitor at the other end, as shown, (a) Determine the capacitor value to achieve the lowest-order resonance at 6.0 GHz. (b) Now assume that loss is introduced by placing a 10,000 Ω resistor in parallel with the capacitor. Calculate the Q.

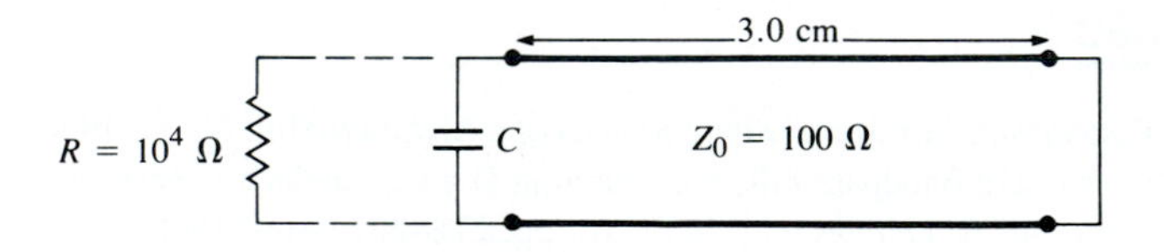

7.6 A transmission line resonator is made from a length ℓ of lossless transmission line of characteristic impedance $Z_0 = 100$ Ω. If the line is terminated at both ends as shown, find ℓ/λ for the first resonance, and the Q of this resonator.

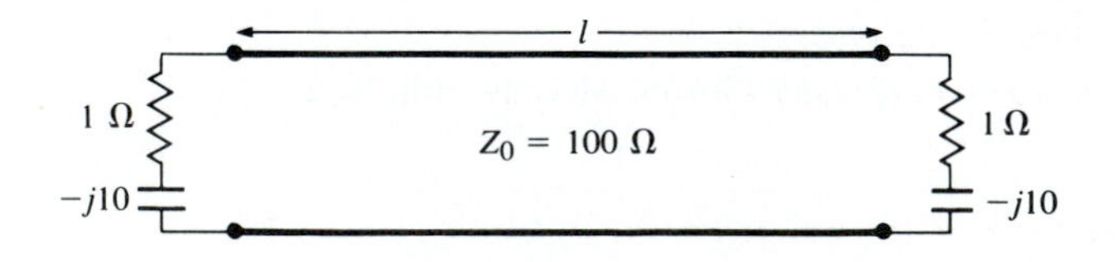

7.7 Write the expressions for the $\bar{E}$ and $\bar{H}$ fields for a short-circuited $\lambda/2$ coaxial line resonator, and show that the time–average stored electric and magnetic energies are equal.

7.8 A series RLC resonant circuit is connected to a length of transmission line that is $\lambda/4$ long at its resonant frequency. Show that, in the vicinity of resonance, the input impedance behaves like that of a parallel RLC circuit.

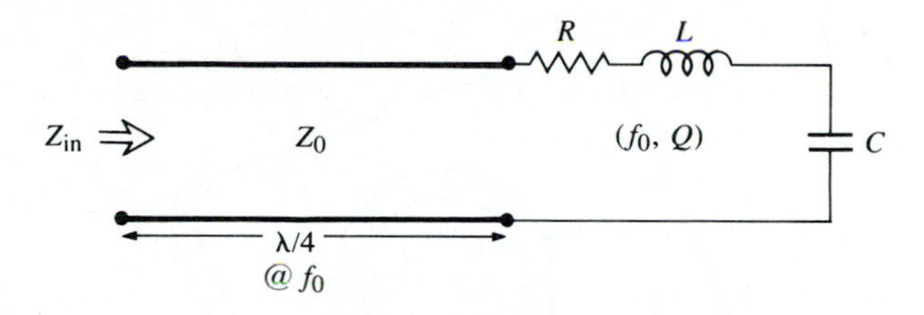

7.9 An air-filled, silver-plated rectangular waveguide cavity has dimensions $a = b = d = 5$ cm. Find the resonant frequency and Q of the TE_{101} and TE_{102} modes.

7.10 Derive the Q for the TM_{111} mode of a rectangular cavity, assuming lossy conducting walls and lossless dielectric.

7.11 Consider the rectangular cavity resonator shown below, partially filled with dielectric. Derive a transcendental equation for the resonant frequency of the dominant mode by writing the fields in the air- and dielectric-filled regions in terms of TE_{10} waveguide modes, and enforcing boundary conditions at $z = 0,\ d - t$, and d.

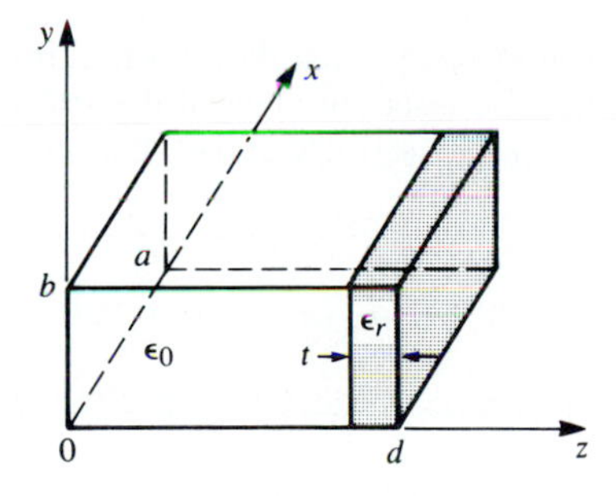

7.12 Determine the resonant frequencies of the rectangular cavity by carrying out a full separation of variables solution to the wave equation for E_z (for TM modes) and H_z (for TE modes), subject to the appropriate boundary conditions of the cavity. (Assume a solution of the form $X(x)Y(y)Z(z)$.)

7.13 Find the Q for the TM_{nm0} resonant mode of a circular cavity. Consider both conductor and dielectric losses.

7.14 Design an air-filled circular cavity to operate in the TE_{111} mode with the maximum Q at $f = 7$ GHz. If the cavity is silver-plated, calculate the resulting Q.

7.15 An air-filled rectangular cavity resonator has its first three resonant modes at the frequencies 5.2 GHz, 6.5 GHz, and 7.2 GHz. Find the dimensions of the cavity.

7.16 Consider the microstrip ring resonator shown below. If the effective dielectric constant of the microstrip line is ϵ_e, find an equation for the frequency of the first resonance. Suggest some methods of coupling to this resonator.

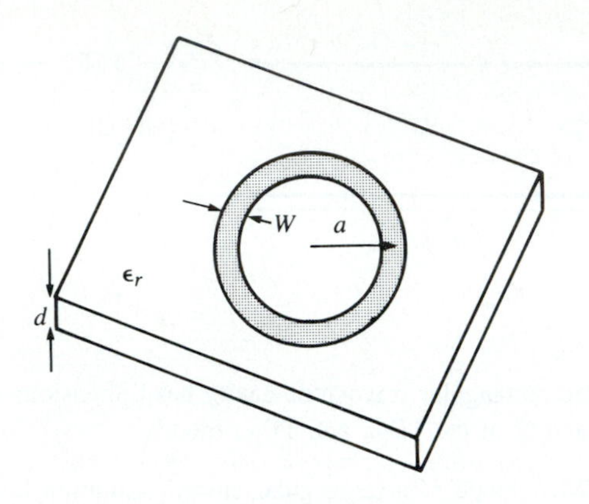

7.17 Compute the resonant frequency of a cylindrical dielectric resonator with $\epsilon_r = 36.2$, $2a = 7.99$ mm, and $L = 2.14$ mm.

7.18 Extend the analysis of Section 7.5 to derive a transcendental equation for the resonant frequency of the next resonant mode of the cylindrical dielectric resonator. (H_z odd in z.)

7.19 In the cylindrical dielectric resonator configuration shown below, parallel metal plates are in close proximity to the ends of the dielectric resonator. Extend the analysis of Section 7.5 to derive a transcendental equation for the resonant frequency of the $TE_{10\delta}$ mode for this geometry.

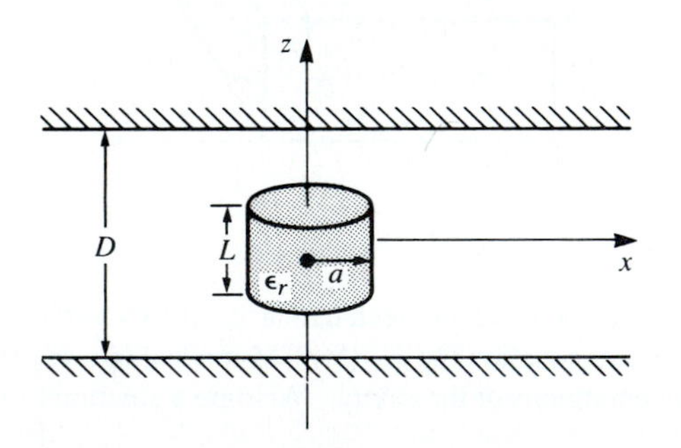

7.20 Consider the rectangular dielectric resonator shown below. Assume a magnetic wall boundary condition around the edges of the cavity, and allow for evanescent fields in the $\pm z$ directions away from the dielectric, similar to the analysis of Section 7.5. Derive a transcendental equation for the resonant frequency.

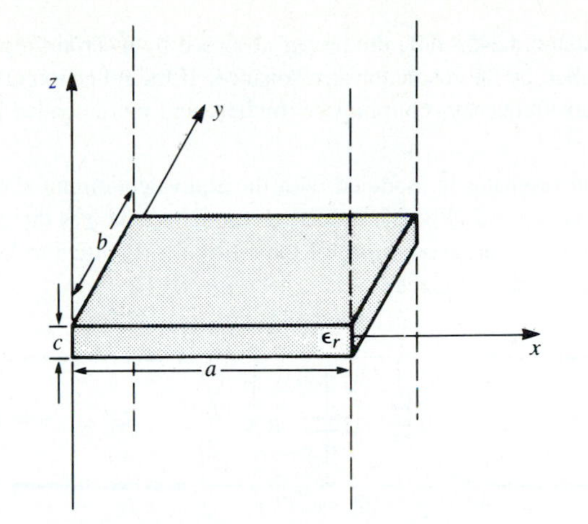

7.21 Consider an open resonator consisting of a plane mirror and a spherical mirror of radius 0.5 m, as shown below. Find the spacing d such that this resonator operates in as stable a mode as possible.

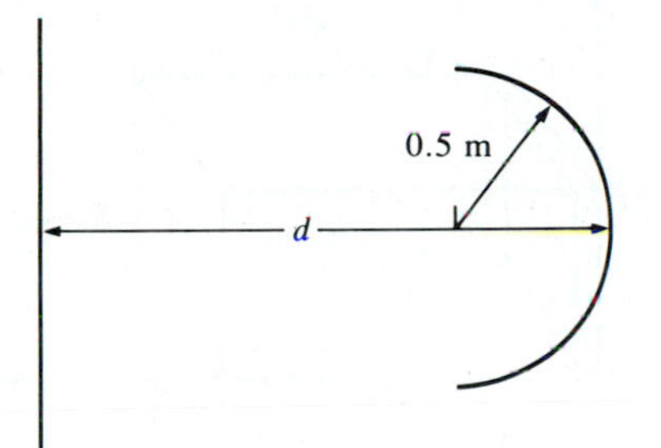

7.22 Consider a parallel RLC circuit connected to a feed line of characteristic impedance Z_0. Sketch input admittance loci on a Smith chart corresponding to the conditions of under, critical, and overcoupling.

7.23 A parallel RLC circuit, with $R = 500\ \Omega$, $L = 1.26$ nH, $C = 0.804$ pF, is coupled with a series capacitor, C_0, to a 50 Ω transmission line, as shown below. Determine C_0 for critical coupling to the line. What is the resonant frequency?

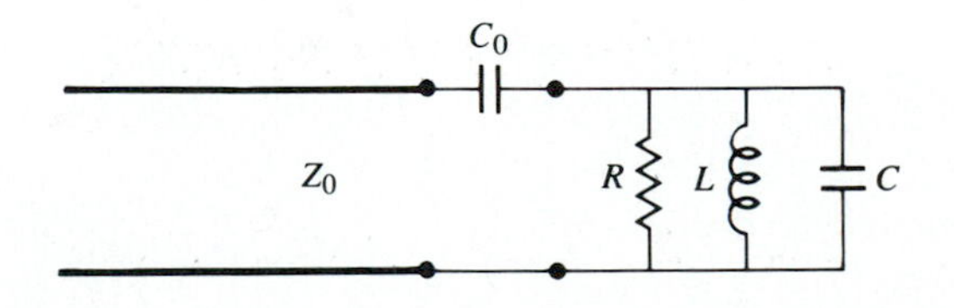

7.24 An aperture coupled rectangular waveguide cavity has a resonant frequency of 9.0 GHz, and a Q of 11,000. If the waveguide dimensions are $a = 2.5$ cm, $b = 1.25$ cm, find the normalized aperture reactance required for critical coupling.

7.25 At frequencies of 8.220 and 8.245 GHz, the power absorbed by a certain resonator is exactly one-half of the power absorbed by the resonator at resonance. If the reflection coefficient at resonance is 0.33, find the resonant frequency, coupling coefficient, and the unloaded and loaded Qs of the resonator.

7.26 A two-port transmission resonator is modeled with the equivalent circuit shown below. If ω_0 and Q are the resonant frequency and Q of the unloaded resonator, and g is the coupling coefficient to either transmission line, derive an expression for the ratio of transmitted to incident power, P_t/P_i, and sketch P_t/P_i versus g, at resonance.

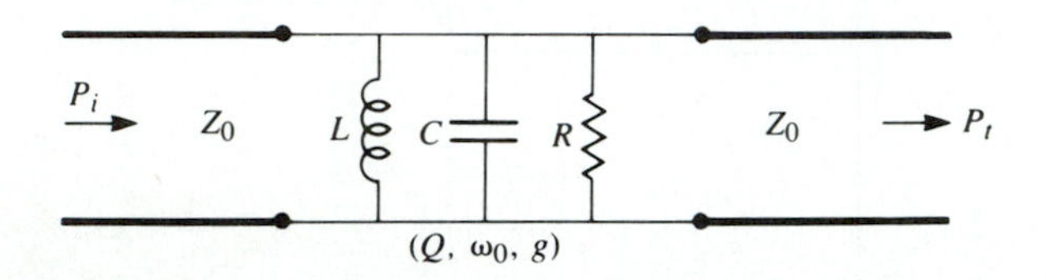

7.27 A thin slab of magnetic material is inserted next to the $z = 0$ wall of the rectangular cavity shown below. If the cavity is operating in the TE_{101} mode, derive a perturbational expression for the change in resonant frequency caused by the magnetic material.

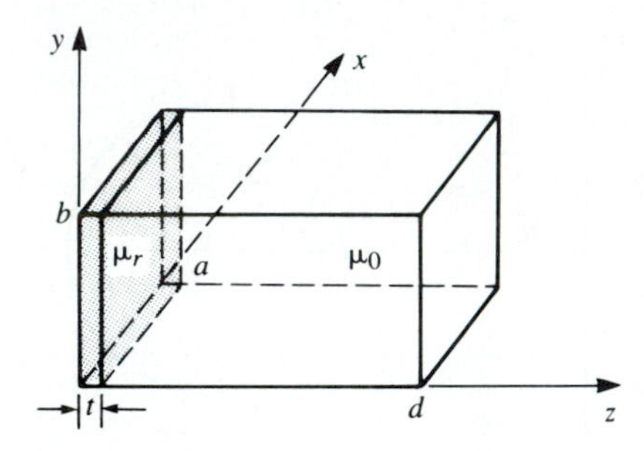

7.28 Derive an expression for the change in resonant frequency for the screw-tuned rectangular cavity of Example 7.9 if the screw is located at $x = a/2, z = 0$, where H_x is maximum and E_y is minimum.

CHAPTER 8

Power Dividers, Directional Couplers, and Hybrids

Power dividers, directional couplers, and hybrid junctions are passive microwave components used for power division or power combining, as illustrated in Figure 8.1. In power division, an input signal is divided by the coupler into two (or more) signals of lesser power. The coupler may be a three-port component as shown, with or without loss, or may be a four-port component. Three-port networks take the form of T-junctions and other power dividers, while four-port networks take the form of directional couplers and hybrids. Power dividers often are of the equal-division (3 dB) type, but unequal power division ratios are also possible. Directional couplers can be designed for arbitrary power division, while hybrid junctions usually have equal power division. Hybrid junctions have either a 90° (quadrature) or a 180° (magic-T) phase shift between the outport ports.

We will first discuss some of the general properties of three- and four-port networks, and then treat the analysis and design of several of the most common types of dividers, couplers, and hybrids.

8.1 BASIC PROPERTIES OF DIVIDERS, COUPLERS, AND HYBRIDS

In this section we will use the scattering matrix theory of Section 5.4 to derive some basic properties of three- and four-port networks. We will also define the terms isolation, coupling, and directivity, which are useful quantities for the characterization of couplers and hybrids.

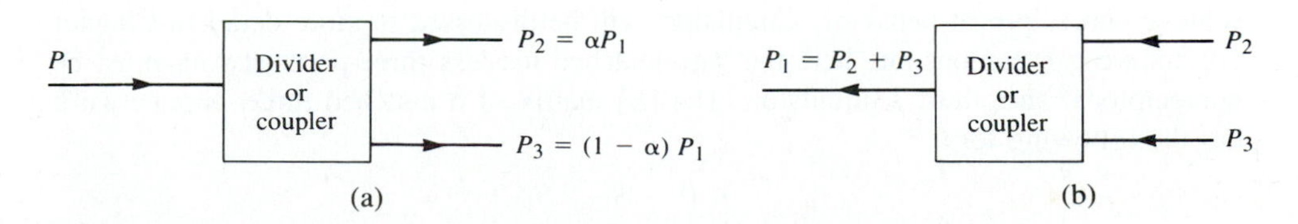

FIGURE 8.1 Power division and combining. (a) Power division. (b) Power combining.

Three-Port Networks (T-Junctions)

The simplest type of power divider is a T-junction, which is a three-port network with two inputs and one output. The scattering matrix of an arbitrary three-port network has nine independent elements:

$$[S] = \begin{bmatrix} S_{11} & S_{12} & S_{13} \\ S_{21} & S_{22} & S_{23} \\ S_{31} & S_{32} & S_{33} \end{bmatrix}. \quad 8.1$$

If the component is passive and contains no anisotropic materials, then it must be reciprocal and its $[S]$ matrix must be symmetric ($S_{ij} = S_{ji}$). Usually, to avoid power loss, we would like to have a junction which is lossless and matched at all ports. We can easily show, however, that it is impossible to construct such a three-port lossless reciprocal network that is matched at all ports.

If all ports are matched, then $S_{ii} = 0$, and if the network is reciprocal the scattering matrix of (8.1) reduces to

$$[S] = \begin{bmatrix} 0 & S_{12} & S_{13} \\ S_{12} & 0 & S_{23} \\ S_{13} & S_{23} & 0 \end{bmatrix}. \quad 8.2$$

Now if the network is also lossless, then energy conservation (5.60a) requires that the scattering matrix be unitary, which leads to the following conditions [1], [2]:

$$|S_{12}|^2 + |S_{13}|^2 = 1, \quad 8.3a$$

$$|S_{12}|^2 + |S_{23}|^2 = 1, \quad 8.3b$$

$$|S_{13}|^2 + |S_{23}|^2 = 1, \quad 8.3c$$

$$S_{13}^* S_{23} = 0, \quad 8.3d$$

$$S_{23}^* S_{12} = 0, \quad 8.3e$$

$$S_{12}^* S_{13} = 0. \quad 8.3f$$

Equations (8.3d–f) shows that at least two of the three parameters (S_{12}, S_{13}, S_{23}) must be zero. But this condition will always be inconsistent with one of equations (8.3a–c), implying that a three-port network cannot be lossless, reciprocal, and matched at all ports. If any one of these three conditions are relaxed, then a physically realizable device is possible.

If the three-port network is nonreciprocal, then $S_{ij} \neq S_{ji}$, and the conditions of input matching at all ports and energy conservation can be satisfied. Such a device is known as a circulator [1], and generally relies on an anisotropic material, such as ferrite, to achieve nonreciprocal behavior. Circulators will be discussed in more detail in Chapter 10, but we can demonstrate here that any matched lossless three-port network must be nonreciprocal and, thus, a circulator. The $[S]$ matrix of a matched three–port network has the following form:

$$[S] = \begin{bmatrix} 0 & S_{12} & S_{13} \\ S_{21} & 0 & S_{23} \\ S_{31} & S_{32} & 0 \end{bmatrix}. \quad 8.4$$

Then if the network is lossless, $[S]$ must be unitary, which implies the following:

$$S_{31}^* S_{32} = 0, \tag{8.5a}$$

$$S_{21}^* S_{23} = 0, \tag{8.5b}$$

$$S_{12}^* S_{13} = 0, \tag{8.5c}$$

$$|S_{12}|^2 + |S_{13}|^2 = 1, \tag{8.5d}$$

$$|S_{21}|^2 + |S_{23}|^2 = 1, \tag{8.5e}$$

$$|S_{31}|^2 + |S_{32}|^2 = 1. \tag{8.5f}$$

These equations can be satisfied in one of two ways. Either

$$S_{12} = S_{23} = S_{31} = 0, \qquad |S_{21}| = |S_{32}| = |S_{13}| = 1, \tag{8.6a}$$

or

$$S_{21} = S_{32} = S_{13} = 0, \qquad |S_{12}| = |S_{23}| = |S_{31}| = 1. \tag{8.6b}$$

This result shows that $S_{ij} \neq S_{ji}$ for $i \neq j$, which implies that the device must be nonreciprocal. The $[S]$ matrices for the two solutions of (8.6) are shown in Figure 8.2, together with the symbols for the two possible types of circulators. The only difference is in the direction of power flow between the ports. Thus, solution (8.6a) corresponds to a circulator that allows power flow only from port 1 to 2, or port 2 to 3, or port 3 to 1, while solution (8.6b) corresponds to a circulator with the opposite direction of power flow.

Alternatively, a lossless and reciprocal three-port network can be physically realized if only two of its ports are matched [1]. If ports 1 and 2 are these matched ports, then the $[S]$ matrix can be written as

$$[S] = \begin{bmatrix} 0 & S_{12} & S_{13} \\ S_{12} & 0 & S_{23} \\ S_{13} & S_{23} & S_{33} \end{bmatrix}. \tag{8.7}$$

To be lossless, the following unitarity conditions must be satisfied:

$$S_{13}^* S_{23} = 0, \tag{8.8a}$$

$$S_{12}^* S_{13} + S_{23}^* S_{33} = 0, \tag{8.8b}$$

$$S_{23}^* S_{12} + S_{33}^* S_{13} = 0, \tag{8.8c}$$

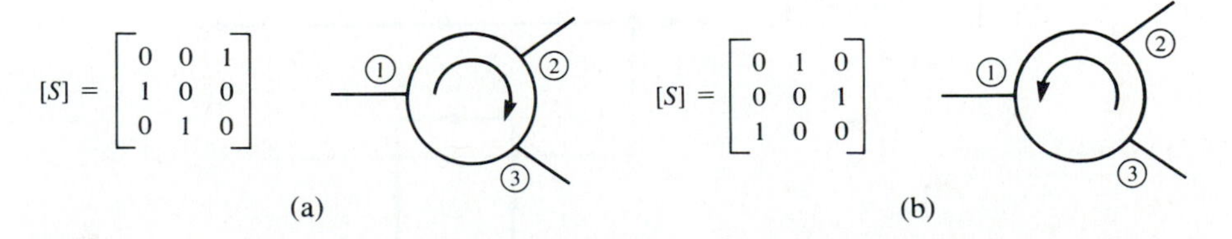

FIGURE 8.2 The two types of circulators and their $[S]$ matrices. (The phase references for the ports are arbitrary.) (a) Clockwise circulation. (b) Counter-clockwise circulation.

$$|S_{12}|^2 + |S_{13}|^2 = 1, \tag{8.8d}$$

$$|S_{12}|^2 + |S_{23}|^2 = 1, \tag{8.8e}$$

$$|S_{13}|^2 + |S_{23}|^2 + |S_{33}|^2 = 1. \tag{8.8f}$$

Equations (8.8d–e) show that $|S_{13}| = |S_{23}|$, so (8.8a) leads to the result that $S_{13} = S_{23} = 0$. Then, $|S_{12}| = |S_{33}| = 1$. The scattering matrix and corresponding signal flow graph for this network are shown in Figure 8.3, where it is seen that the network actually consists of two separate components, one a matched two-port line and the other a totally mismatched one-port.

Finally, if the three-port network is allowed to be lossy, then it can be reciprocal and matched at all ports; this is the case of the resistive divider, which will be discussed in Section 8.2. In addition, a lossy three-port can be made to have isolation between its output ports (for example $S_{23} = S_{32} = 0$).

Four-Port Networks (Directional Couplers and Hybrids)

The $[S]$ matrix of a reciprocal four-port network matched at all ports has the following form:

$$[S] = \begin{bmatrix} 0 & S_{12} & S_{13} & S_{14} \\ S_{12} & 0 & S_{23} & S_{24} \\ S_{13} & S_{23} & 0 & S_{34} \\ S_{14} & S_{24} & S_{34} & 0 \end{bmatrix}. \tag{8.9}$$

If the network is lossless, 10 equations result from the unitarity, or energy conservation, condition [1], [2]. Let us consider the multiplication of row 1 and row 2, and the multiplication of row 4 and row 3:

$$S_{13}^* S_{23} + S_{14}^* S_{24} = 0, \tag{8.10a}$$

$$S_{14}^* S_{13} + S_{24}^* S_{23} = 0. \tag{8.10b}$$

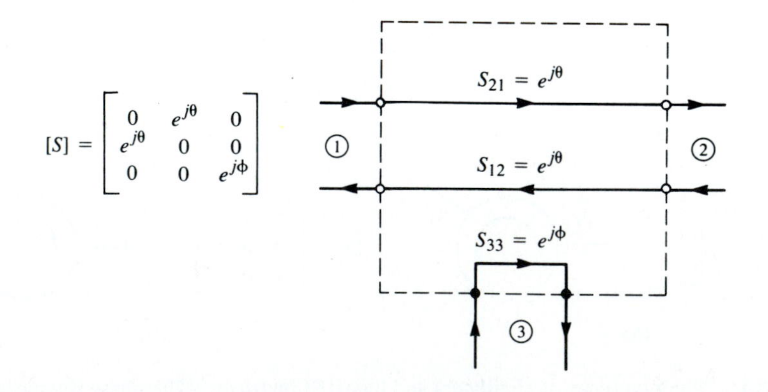

FIGURE 8.3 A reciprocal, lossless three-port network matched at ports 1 and 2.

Now multiply (8.10a) by S_{24}^* and (8.10b) by S_{13}^*, and subtract to obtain

$$S_{14}^*(|S_{13}|^2 - |S_{24}|^2) = 0. \tag{8.11}$$

Similarly, the multiplication of row 1 and row 3, and the multiplication of row 4 and row 2, gives

$$S_{12}^* S_{23} + S_{14}^* S_{34} = 0, \tag{8.12a}$$

$$S_{14}^* S_{12} + S_{34} S_{23} = 0. \tag{8.12b}$$

Now multiply (8.12a) by S_{12} and (8.12b) by S_{34}, and subtract to obtain

$$S_{23}(|S_{12}|^2 - |S_{34}|^2) = 0. \tag{8.13}$$

One way for (8.11) and (8.13) to be satisfied is if $S_{14} = S_{23} = 0$, which results in a directional coupler. Then the self-products of the rows of the unitary $[S]$ matrix of (8.9) yield the following equations:

$$|S_{12}|^2 + |S_{13}|^2 = 1, \tag{8.14a}$$

$$|S_{12}|^2 + |S_{24}|^2 = 1, \tag{8.14b}$$

$$|S_{13}|^2 + |S_{34}|^2 = 1, \tag{8.14c}$$

$$|S_{24}|^2 + |S_{34}|^2 = 1, \tag{8.14d}$$

which imply that $|S_{13}| = |S_{24}|$ (using 8.14a and 8.14b), and that $|S_{12}| = |S_{34}|$ (using 8.14b and 8.14d).

Further simplification can be made by choosing the phase references on three of the four ports. Thus, we choose $S_{12} = S_{34} = \alpha$, $S_{13} = \beta e^{j\theta}$, and $S_{24} = \beta e^{j\phi}$, where α and β are real, and θ and ϕ are phase constants to be determined (one of which we are still free to choose). The dot product of rows 2 and 3 gives

$$S_{12}^* S_{13} + S_{24}^* S_{34} = 0, \tag{8.15}$$

which yields a relation between the remaining phase constants as

$$\theta + \phi = \pi \pm 2n\pi. \tag{8.16}$$

If we ignore integer multiples of 2π, there are two particular choices that commonly occur in practice:

1. The Symmetrical Coupler: $\theta = \phi = \pi/2$. The phases of the terms having amplitude β are chosen equal. Then the scattering matrix has the following form:

$$[S] = \begin{bmatrix} 0 & \alpha & j\beta & 0 \\ \alpha & 0 & 0 & j\beta \\ j\beta & 0 & 0 & \alpha \\ 0 & j\beta & \alpha & 0 \end{bmatrix}. \tag{8.17}$$

2. The Antisymmetrical Coupler: $\theta = 0, \phi = \pi$. The phases of the terms having amplitude β are chosen to be 180° apart. Then the scattering matrix has the

following form:

$$[S] = \begin{bmatrix} 0 & \alpha & \beta & 0 \\ \alpha & 0 & 0 & -\beta \\ \beta & 0 & 0 & \alpha \\ 0 & -\beta & \alpha & 0 \end{bmatrix}. \quad 8.18$$

Note that the two couplers differ only in the choice of reference planes. Also, the amplitudes α and β are not independent, as (8.14a) requires that

$$\alpha^2 + \beta^2 = 1. \quad 8.19$$

Thus, apart from phase references, an ideal directional coupler has only one degree of freedom.

Another way for (8.11) and (8.13) to be satisfied is if $|S_{13}| = |S_{24}|$ and $|S_{12}| = |S_{34}|$. If we choose phase references, however, such that $S_{13} = S_{24} = \alpha$, and $S_{12} = S_{34} = j\beta$ (which satisfies (8.16)), then (8.10a) yields $\alpha(S_{23} + S_{14}^*) = 0$, and (8.12a) yields $\beta(S_{14}^* - S_{23}) = 0$. These two equations have two possible solutions. First, $S_{14} = S_{23} = 0$, which is the same as the above solution for the directional coupler. The other solution occurs for $\alpha = \beta = 0$, which implies that $S_{12} = S_{13} = S_{24} = S_{34} = 0$. This is the case of two decoupled two-port networks (between ports 1 and 4, and ports 2 and 3), which is of trivial interest and will not be considered further. We are thus left with the conclusion that any reciprocal, lossless, matched four-port network is a directional coupler.

The basic operation of a directional coupler can then be illustrated with the aid of Figure 8.4, which shows two commonly used symbols for a directional coupler and the port definitions. Power supplied to port 1 is coupled to port 3 (the coupled port) with the coupling factor $|S_{13}|^2 = \beta^2$, while the remainder of the input power is delivered to port 2 (the through port) with the coefficient $|S_{12}|^2 = \alpha^2 = 1 - \beta^2$. In an ideal directional coupler, no power is delivered to port 4 (the isolated port).

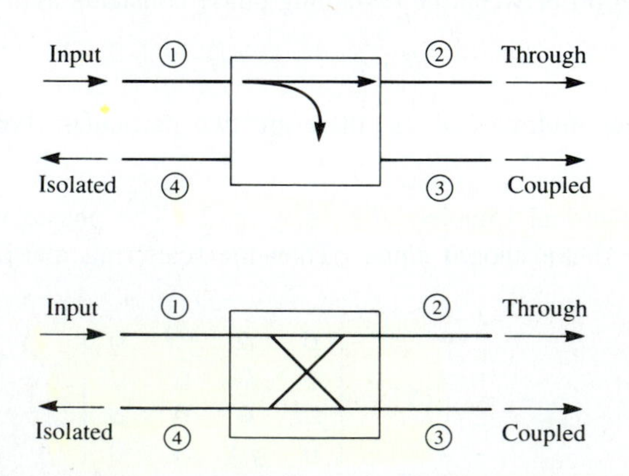

FIGURE 8.4 Two commonly used symbols for directional couplers, and power flow conventions.

The following three quantities are generally used to characterize a directional coupler:

$$\text{Coupling} = C = 10 \log \frac{P_1}{P_3} = -20 \log \beta \text{ dB}, \quad 8.20a$$

$$\text{Directivity} = D = 10 \log \frac{P_3}{P_4} = 20 \log \frac{\beta}{|S_{14}|} \text{ dB}, \quad 8.20b$$

$$\text{Isolation} = I = 10 \log \frac{P_1}{P_4} = -20 \log |S_{14}| \text{ dB}. \quad 8.20c$$

The coupling factor indicates the fraction of the input power which is coupled to the output port. The directivity is a measure of the coupler's ability to isolate forward and backward waves, as is the isolation. These quantities are then related as

$$I = D + C \text{ dB}. \quad 8.21$$

The ideal coupler would have infinite directivity and isolation ($S_{14} = 0$). Then both α and β could be determined from the coupling factor, C.

Hybrid couplers are special cases of directional couplers, where the coupling factor is 3 dB, which implies that $\alpha = \beta = 1/\sqrt{2}$. There are two types of hybrids. The quadrature hybrid has a 90° phase shift between ports 2 and 3 ($\theta = \phi = \pi/2$) when fed at port 1, and is an example of a symmetrical coupler. Its $[S]$ matrix has the following form:

$$[S] = \frac{1}{\sqrt{2}} \begin{bmatrix} 0 & 1 & j & 0 \\ 1 & 0 & 0 & j \\ j & 0 & 0 & 1 \\ 0 & j & 1 & 0 \end{bmatrix}. \quad 8.22$$

The magic-T hybrid or rat-race hybrid has a 180° phase difference between ports 2 and 3 when fed at port 4, and is an example of an antisymmetrical coupler. Its $[S]$ matrix has the following form:

$$[S] = \frac{1}{\sqrt{2}} \begin{bmatrix} 0 & 1 & 1 & 0 \\ 1 & 0 & 0 & -1 \\ 1 & 0 & 0 & 1 \\ 0 & -1 & 1 & 0 \end{bmatrix}. \quad 8.23$$

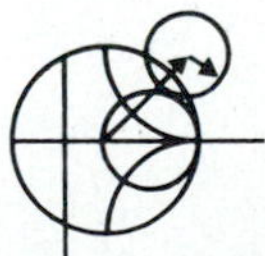

POINT OF INTEREST: Measuring Coupler Directivity

The directivity of a directional coupler is a measure of the coupler's ability to separate forward and reverse wave components, so applications of directional couplers often require high (35 dB or greater) directivity. Poor directivity will limit the accuracy of a reflectometer, and can cause variations in the coupled power level from a coupler when there is even a small mismatch on the through line.

The directivity of a coupler generally cannot be measured directly because it involves a low-level signal that can be masked by coupled power from a reflected wave on the through arm. For example, if a coupler has $C = 20$ dB and $D = 35$ dB, with a load having $RL = 30$ dB, the signal level through the directivity path will be $D+C = 55$ dB below the input power, but the reflected power through the coupled arm will only be $RL + C = 50$ dB below the input power.

One way of measuring coupler directivity uses a sliding matched load, as follows. First, the coupler is connected to a source and matched load, as shown in the left-hand figure below, and the coupled output power is measured. If we assume an input power P_i, this power will be $P_c = C^2 P_i$, where $C = 10^{(-C\text{ dB})/20}$ is the numerical voltage coupling factor of the coupler. Now reverse the position of the coupler as shown in the right-hand figure below and terminate the through line with a sliding load.

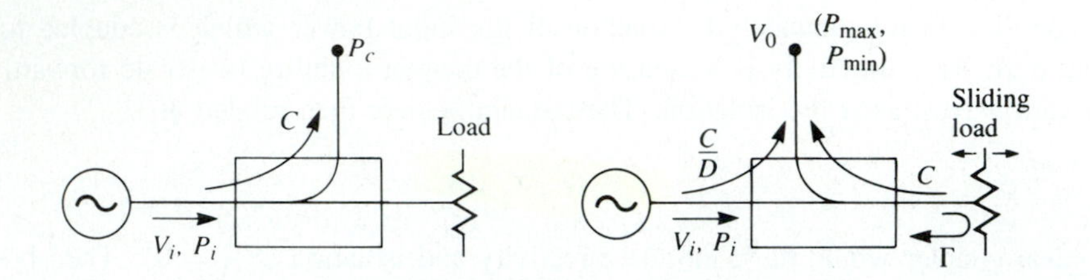

Changing the position of the sliding load introduces a variable phase shift in the signal reflected from the load and coupled to the output port. Thus the voltage at the output port can be written as

$$V_0 = V_i\left(\frac{C}{D} + C|\Gamma|e^{-j\theta}\right),$$

where V_i is the input voltage, $D = 10^{(D\text{ dB})/20} \geq 1$ is the numerical value of the directivity, $|\Gamma|$ is the reflection coefficient magnitude of the load, and θ is the path length difference between the directivity and reflected signals. Moving the sliding load changes θ, so the two signals will combine to trace out a circular locus, as shown in the following figure.

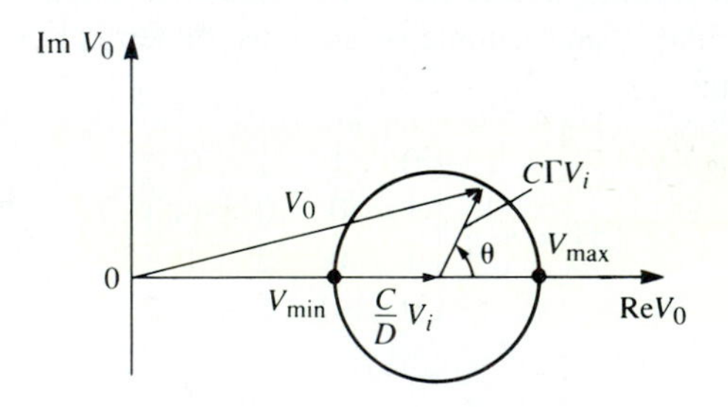

The minimum and maximum output powers are given by

$$P_{min} = P_i\left(\frac{C}{D} - C|\Gamma|\right)^2, \qquad P_{max} = P_i\left(\frac{C}{D} + C|\Gamma|\right)^2.$$

Now let M and m be defined in terms of these powers as follows:

$$M = \frac{P_c}{P_{max}} = \left(\frac{D}{1+|\Gamma|D}\right)^2, \qquad m = \frac{P_{max}}{P_{min}} = \left(\frac{1+|\Gamma|D}{1-|\Gamma|D}\right)^2.$$

These ratios can be accurately measured directly by using a variable attenuator between the source and coupler. The directivity (numerical) can then be found as

$$D = M\left(\frac{2m}{m+1}\right).$$

This method requires that $|\Gamma| < 1/D$ or, in dB, $RL > D$.

Reference: M. Sucher and J. Fox, editors, *Handbook of Microwave Measurements*, third edition, volume II, Polytechnic Press, New York, 1963.

8.2 THE T-JUNCTION POWER DIVIDER

The T-junction power divider is a simple three-port network that can be used for power division or power combining, and can be implemented in virtually any type of transmission line media. Figure 8.5 shows some commonly used T-junctions in waveguide and microstrip or stripline form. The junctions shown here are, in the absence of transmission line loss, lossless junctions. Thus, as discussed in the preceding section, such junctions cannot be matched simultaneously at all ports. We will treat such junctions below, followed by a discussion of the resistive divider, which can be matched at all ports but is not lossless.

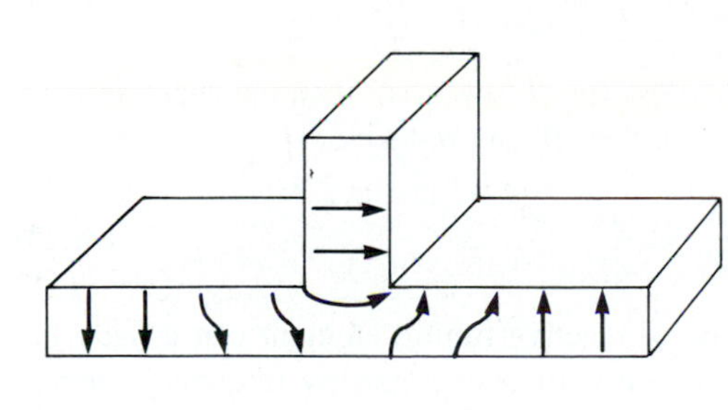

(a)

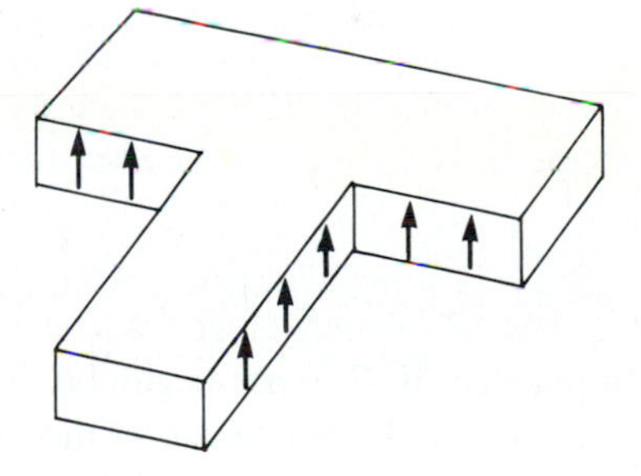

(b)

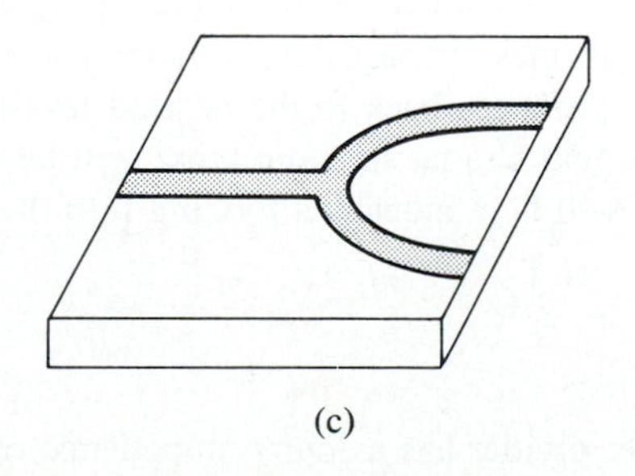

(c)

FIGURE 8.5 Various T-junction power dividers. (a) E plane waveguide T. (b) H plane waveguide T. (c) Microstrip T-junction.

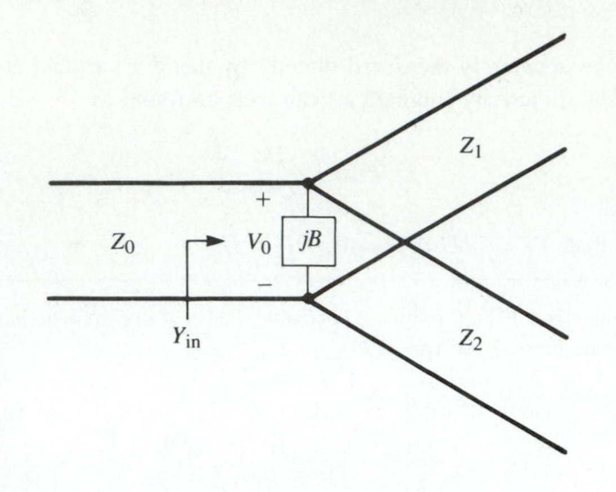

FIGURE 8.6 Transmission line model of a lossless T-junction.

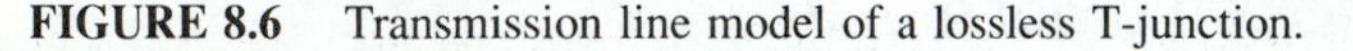

Lossless Divider

The lossless T-junctions of Figure 8.5 can all be modelled as a junction of three transmission lines, as shown in Figure 8.6 [3]. In general, there are fringing fields and higher-order modes associated with the discontinuity at such a junction, leading to stored energy which can be accounted for by a lumped susceptance, B. In order for the divider to be matched to the input line of characteristic impedance Z_0, we must have

$$Y_{in} = jB + \frac{1}{Z_1} + \frac{1}{Z_2} = \frac{1}{Z_0}. \tag{8.24}$$

If the transmission lines are assumed to be lossless (or of low loss), then the characteristic impedances are real. If we also assume $B = 0$, then (8.24) reduces to

$$\frac{1}{Z_1} + \frac{1}{Z_2} = \frac{1}{Z_0}. \tag{8.25}$$

In practice, if B is not negligible, some type of reactive tuning element can usually be added to the divider to cancel this susceptance, at least over a narrow frequency range.

The output line impedances Z_1 and Z_2 can then be selected to provide various power division ratios. Thus, for a 50 Ω input line, a 3 dB (equal split) power divider can be made by using two 100 Ω output lines. If necessary, quarter-wave transformers can be used to bring the output line impedances back to the desired levels. If the output lines are matched, then the input line will be matched, but there will be no isolation between the two output ports, and there will be a mismatch looking into the output ports.

EXAMPLE 8.1

A lossless T-junction power divider has a source impedance of 50 Ω. Find the output characteristic impedances so that the input power is divided in a 2:1 ratio. Compute the reflection coefficients seen looking into the output ports.

Solution

If the voltage at the junction is V_0, as shown in Figure 8.6, the input power to the matched divider is

$$P_{in} = \frac{1}{2}\frac{V_0^2}{Z_0},$$

while the output powers are

$$P_1 = \frac{1}{2}\frac{V_0^2}{Z_1} = \frac{1}{3}P_{in},$$

$$P_2 = \frac{1}{2}\frac{V_0^2}{Z_2} = \frac{2}{3}P_{in}.$$

These results yield the characteristic impedances as

$$Z_1 = 3Z_0 = 150\,\Omega,$$

$$Z_2 = \frac{3Z_0}{2} = 75\,\Omega,$$

Then the input impedance to the junction is

$$Z_{in} = 75||150 = 50\,\Omega,$$

so that the input is matched to the $50\,\Omega$ source.

Looking into the $150\,\Omega$ output line, we see an impedance of $50||75 = 30\,\Omega$, while at the $75\,\Omega$ output line we see an impedance of $50||105 = 37.5\,\Omega$. Thus, the reflection coefficients seen looking into these ports are

$$\Gamma_1 = \frac{30 - 150}{30 + 150} = -0.666,$$

$$\Gamma_2 = \frac{37.5 - 75}{37.5 + 75} = -0.333.$$

○

Resistive Divider

If a three-port divider contains lossy components it can be made to be matched at all ports, although the two output ports may not be isolated [3]. The circuit for such a divider is illustrated in Figure 8.7, using lumped-element resistors. An equal-split (−3 dB) divider is shown, but unequal power division ratios are also possible.

The resistive divider of Figure 8.7 can easily be analyzed using circuit theory. Assuming that all ports are terminated in the characteristic impedance Z_0, the impedance Z, seen looking into the $Z_0/3$ resistor followed by the output line, is

$$Z = \frac{Z_0}{3} + Z_0 = \frac{4Z_0}{3}. \tag{8.26}$$

Then the input impedance of the divider is

$$Z_{in} = \frac{Z_0}{3} + \frac{2Z_0}{3} = Z_0, \tag{8.27}$$

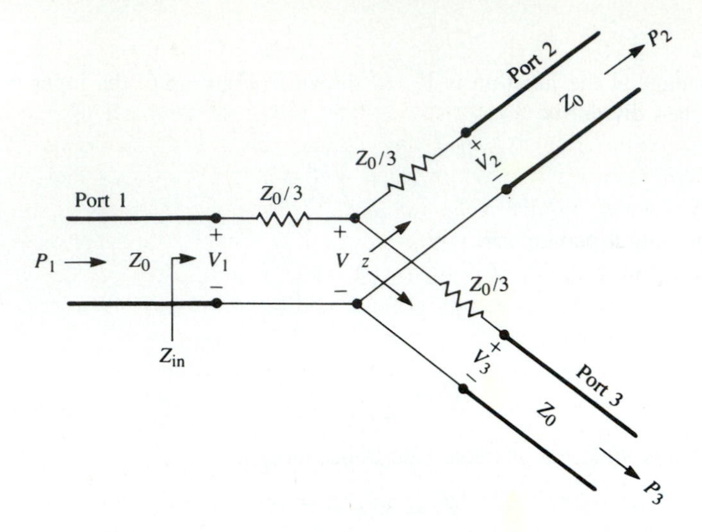

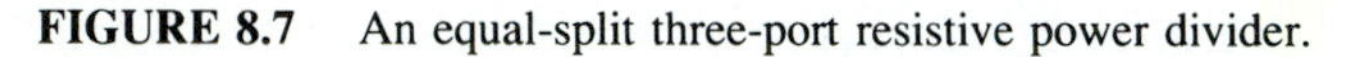

FIGURE 8.7 An equal-split three-port resistive power divider.

which shows that the input is matched to the feed line. Since the network is symmetric from all three ports, the output ports are also matched. Thus, $S_{11} = S_{22} = S_{33} = 0$.

If the voltage at port 1 is V_1, then by voltage division the voltage V at the center of the junction is

$$V = V_1 \frac{2Z_0/3}{Z_0/3 + 2Z_0/3} = \frac{2}{3} V_1, \tag{8.28}$$

and the output voltages are, again by voltage division

$$V_2 = V_3 = V \frac{Z_0}{Z_0 + Z_0/3} = \frac{3}{4} V = \frac{1}{2} V_1. \tag{8.29}$$

Thus, $S_{21} = S_{31} = S_{23} = 1/2$, which is -6 dB below the input power level. The network is reciprocal, so the scattering matrix is symmetric, and can be written as

$$[S] = \frac{1}{2} \begin{bmatrix} 0 & 1 & 1 \\ 1 & 0 & 1 \\ 1 & 1 & 0 \end{bmatrix}. \tag{8.30}$$

The reader may verify that this is not a unitary matrix.

The power delivered to the input of the divider is

$$P_{in} = \frac{1}{2} \frac{V_1^2}{Z_0}, \tag{8.31}$$

while the output powers are

$$P_2 = P_3 = \frac{1}{2} \frac{(1/2 V_1)^2}{Z_0} = \frac{1}{8} \frac{V_1^2}{Z_0} = \frac{1}{4} P_{in}, \tag{8.32}$$

which shows that half of the supplied power is dissipated in the resistors.

8.3 THE WILKINSON POWER DIVIDER

→ lossy

The lossless T-junction divider suffers from the problem of not being matched at all ports and, in addition, does not have any isolation between output ports. The resistive divider can be matched at all ports, but even though it is not lossless, isolation is still not achieved. From the discussion in Section 8.1, however, we know that a lossy three-port network can be made having all ports matched with isolation between the output ports. The Wilkinson power divider [4] is such a network and is the subject of the present section. It has the useful property of being lossless when the output ports are matched; that is, only reflected power is dissipated.

The Wilkinson power divider can be made to give arbitrary power division, but we will first consider the equal-split (3 dB) case. This divider is often made in microstrip or stripline form, as depicted in Figure 8.8a; the corresponding transmission line circuit is given in Figure 8.8b. We will analyze this circuit by reducing it to two simpler circuits driven by symmetric and antisymmetric sources at the output ports. This "even-odd" mode analysis technique [5] will also be useful for other networks which we will analyze in later sections.

Even-Odd Mode Analysis

For simplicity, we can normalize all impedances to the characteristic impedance Z_0, and redraw the circuit of Figure 8.8b with voltage generators at the output ports as shown in Figure 8.9. This network has been drawn in a form that is symmetric across the midplane; the two source resistors of normalized value 2 combine in parallel to give a resistor of normalized value 1, representing the impedance of a matched source. The quarter-wave lines have a normalized characteristic impedance Z, and the shunt resistor has a normalized value of r; we shall show that, for the equal-split power divider, these values should be $Z = \sqrt{2}$ and $r = 2$, as given in Figure 8.8.

Now we define two separate modes of excitation for the circuit of Figure 8.9: the even mode, where $V_{g2} = V_{g3} = 2$ V, and the odd mode, where $V_{g2} = -V_{g3} = 2$ V.

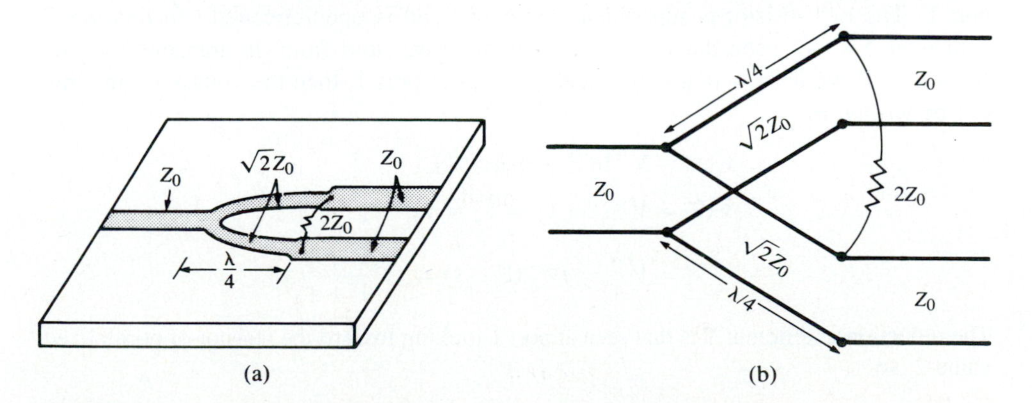

FIGURE 8.8 The Wilkinson power divider. (a) An equal-split Wilkinson power divider in microstrip form. (b) Equivalent transmission line circuit.

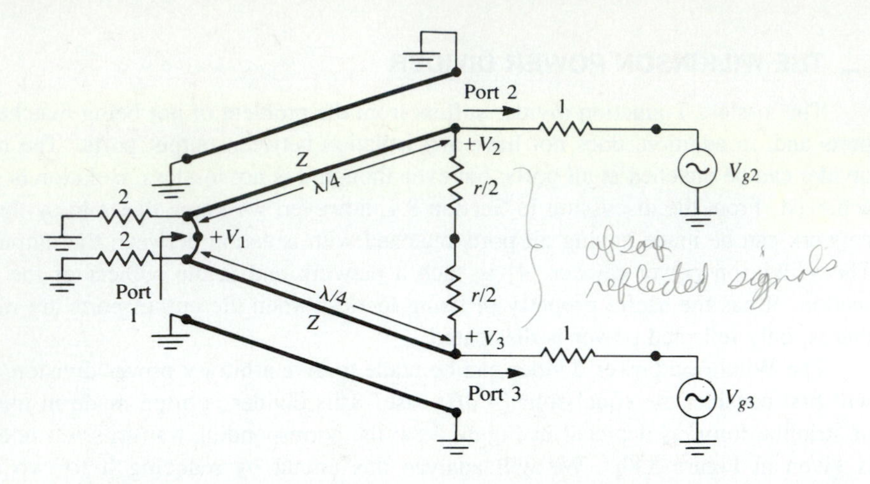

FIGURE 8.9 The Wilkinson power divider circuit in normalized and symmetric form.

Then by superposition of these two modes, we effectively have an excitation of $V_{g2} = 4$ V, $V_{g3} = 0$, from which we can find the S parameters of the network. We now treat these two modes separately.

Even mode. For the even-mode excitation, $V_{g2} = V_{g3} = 2$ V, and so $V_2 = V_3$ and there is no current flow through the $r/2$ resistors or the short circuit between the inputs of the two transmission lines at port 1. Thus we can bisect the network of Figure 8.9 with open-circuits at these points, to obtain the network of Figure 8.10a (the grounded side of the $\lambda/4$ line is not shown). Then, looking into port 2, we see an impedance

$$Z_{\text{in}}^{e} = \frac{Z^2}{2}, \tag{8.33}$$

since the transmission line looks like a quarter-wave transformer. Thus, if $Z = \sqrt{2}$, port 2 will be matched ($S_{22} = 0$), and all power will be delivered to the load connected at port 1. The $r/2$ resistor is superfluous, since one end is open-circuited. To find the S parameter S_{12}, we need the voltage V_1, which we can find from the transmission line equations. If we let $x = 0$ at port 2, and $x = \lambda/4$ at port 1, then the voltage on the line can be written as

$$V(x) = V^{+}(e^{-j\beta x} + \Gamma e^{j\beta x}),$$

$$V(0) = V^{+}(1 + \Gamma) = V_2 = V,$$

$$V_1 = V\left(\frac{\lambda}{4}\right) = jV^{+}(\Gamma - 1) = jV\frac{\Gamma - 1}{\Gamma + 1}.$$

The reflection coefficient Γ is that seen at port 1 looking toward the resistor of normalized value 2, so

$$\Gamma = \frac{2 - \sqrt{2}}{2 + \sqrt{2}},$$

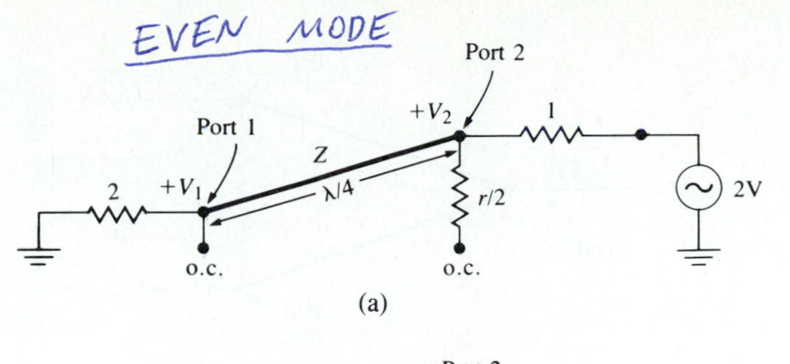

(a)

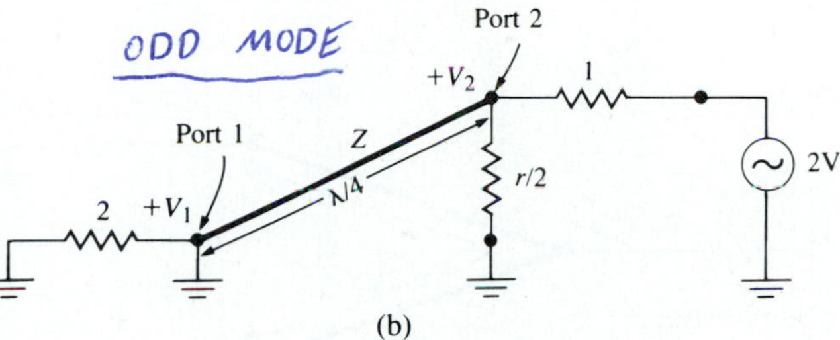

(b)

FIGURE 8.10 Bisection of the circuit of Figure 8.9. (a) For even-mode excitation. (b) For odd-mode excitation.

and
$$V_1 = jV\frac{-1}{\sqrt{2}}. \qquad 8.34$$

Thus,
$$S_{12} = \frac{V_1}{V_2} = \frac{-j}{\sqrt{2}} = -j0.707. \qquad 8.35$$

From symmetry, we also have that $S_{33} = 0$ and $S_{13} = -j0.707$.

Odd mode. For the odd-mode excitation, $V_{g2} = -V_{g3} = 2$ V, and so $V_2 = -V_3$, and there is a voltage null along the middle of the circuit in Figure 8.9. Thus, we can bisect this circuit by grounding it at two points on its midplane, to give the network of Figure 8.10b. Looking into port 2, we see an impedance of $r/2$, since the parallel-connected transmission line is $\lambda/4$ long and shorted at port 1, and so looks like an open circuit at port 2. Thus, port 2 will be matched ($S_{22} = 0$) if we select $r = 2$. In this mode of excitation, all power is delivered to the $r/2$ resistors, with none going to port 1.

In summary, then, we have deduced the following S parameters:

$S_{22} = S_{33} = 0$ (since ports 2 and 3 were matched for both modes of excitation),
$S_{12} = S_{21} = -j0.707$ (symmetry since reciprocal network),
$S_{13} = S_{31} = -j0.707$ (symmetry since reciprocal network),
$S_{23} = S_{32} = 0$ (because of short or open at bisections).

It is this last result that implies isolation between ports 2 and 3.

Finally, we must derive S_{11}, by determining the input impedance at port 1 of the Wilkinson divider when ports 2 and 3 are terminated in matched loads. The resulting circuit is shown in Figure 8.11a, where it is seen that this is similar to the even mode of excitation, as $V_2 = V_3$. Thus, no current flows through the resistor of normalized value

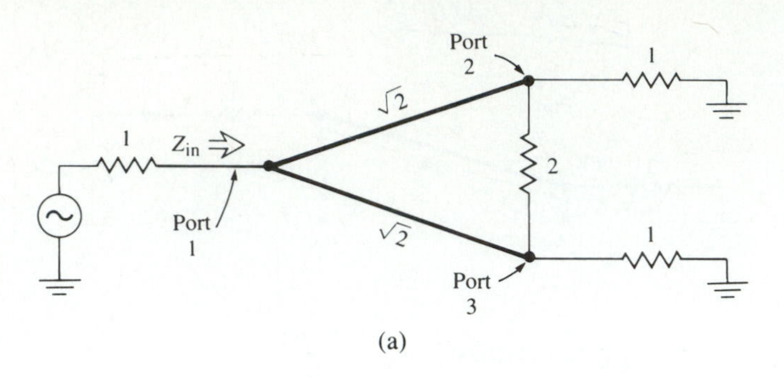

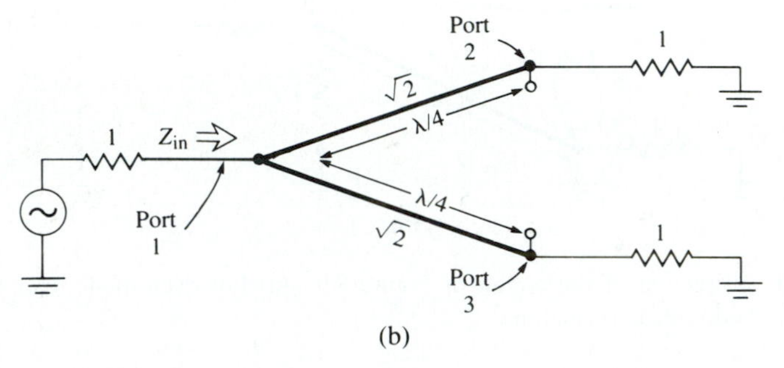

FIGURE 8.11 Analysis of the Wilkinson divider to find S_{11}. (a) The terminated Wilkinson divider. (b) Bisection of the circuit in (a).

2, so it can be removed, leaving the circuit of Figure 8.11b. We now have the parallel connection of two quarter-wave transformers terminated in loads of unity (normalized). The input impedance is then

$$Z_{in} = \frac{1}{2}\frac{(\sqrt{2})^2}{1} = 1, \tag{8.36}$$

and so $S_{11} = 0$. Note that when the divider is driven at port 1 and the output ports are matched, no power is dissipated in the resistor. Thus the divider is lossless when the outputs are matched; only reflected power from ports 2 or 3 is dissipated in the resistor.

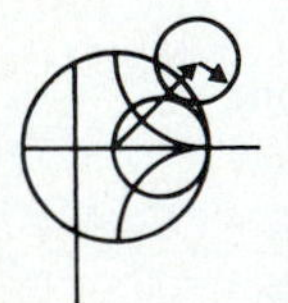

EXAMPLE 8.2

Design an equal-split Wilkinson power divider for a 50 Ω system impedance at frequency f_0, and plot the return loss (S_{11}), insertion loss ($S_{21} = S_{31}$), and isolation ($S_{23} = S_{32}$) versus frequency from $0.5f_0$ to $1.5f_0$.

Solution

From Figure 8.8 and the above derivation, we have that the quarter-wave transmission lines in the divider should have a characteristic impedance of

$$Z = \sqrt{2}Z_0 = 70.7\,\Omega,$$

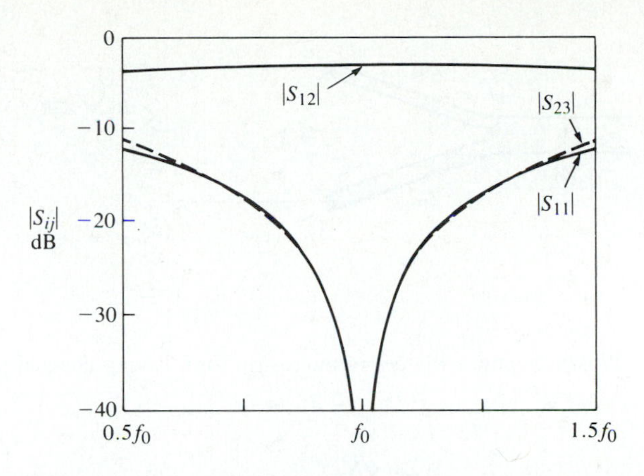

FIGURE 8.12 Frequency response of an equal-split Wilkinson power divider. Port 1 is the input port; ports 2 and 3 are the output ports.

and the shunt resistor a value of

$$R = 2Z_0 = 100\,\Omega.$$

The transmission lines are $\lambda/4$ long at the frequency f_0. Using a computer-aided design program for the analysis of microwave circuits, the S parameter magnitudes were calculated and plotted in Figure 8.12. ○

Unequal Power Division and N-Way Wilkinson Dividers

Wilkinson-type power dividers can also be made with unequal power splits; a microstrip version is shown in Figure 8.13. If the power ratio between ports 2 and 3 is $K^2 = P_3/P_2$, then the following design equations apply:

$$Z_{03} = Z_0\sqrt{\frac{1+K^2}{K^3}}, \tag{8.37a}$$

$$Z_{02} = K^2 Z_{03} = Z_0\sqrt{K(1+K^2)}, \tag{8.37b}$$

$$R = Z_0\left(K + \frac{1}{K}\right). \tag{8.37c}$$

Note that the above results reduce to the equal split case for $K = 1$. Also observe that the output lines are matched to the impedances $R_2 = Z_0 K$ and $R_3 = Z_0/K$, as opposed to the impedance Z_0; matching transformers can be used to transform these output impedances.

The Wilkinson divider can also be generalized to an N-way divider or combiner [4], as shown in Figure 8.14. This circuit can be matched at all ports, with isolation between all ports. A disadvantage, however, is the fact that the divider requires crossovers for the resistors for $N \geq 3$. This makes fabrication difficult in planar form. The Wilkinson

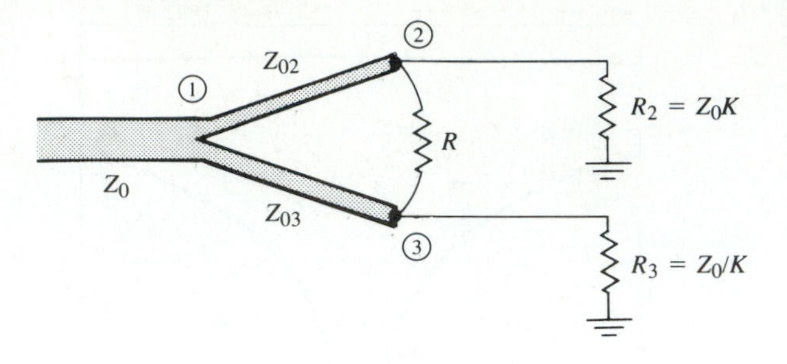

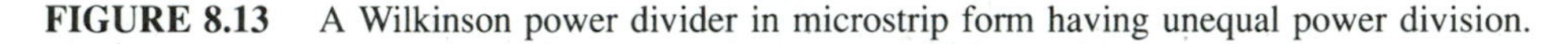
FIGURE 8.13 A Wilkinson power divider in microstrip form having unequal power division.

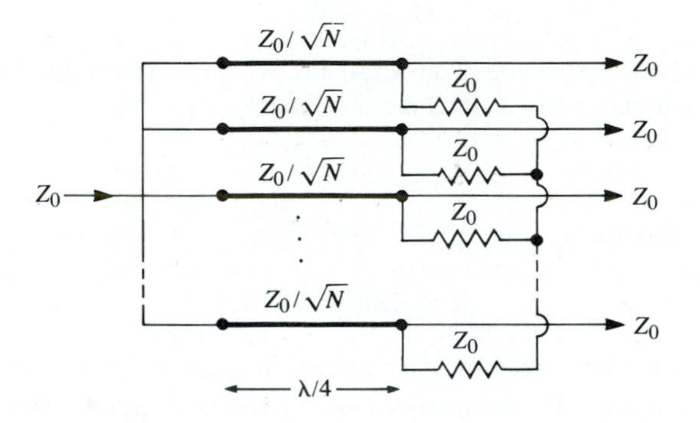

FIGURE 8.14 An N-way, equal-split Wilkinson power divider.

divider can also be made with stepped multiple sections, for increased bandwidth. A photograph of a four-section Wilkinson divider is shown in Figure 8.15.

8.4 WAVEGUIDE DIRECTIONAL COUPLERS

We now turn our attention to directional couplers, which are four-port devices with the characteristics discussed in Section 8.1. To review the basic operation, consider the directional coupler schematic symbols shown in Figure 8.4. Power incident at port 1 will couple to port 2 (the through port) and to port 3 (the coupled port), but not to port 4 (the isolated port). Similarly, power incident in port 2 will couple to ports 1 and 4, but not 3. Thus, ports 1 and 4 are decoupled, as are ports 2 and 3. The fraction of power coupled from port 1 to port 3 is given by C, the coupling, as defined in (8.20a), and the leakage of power from port 1 to port 4 is given by I, the isolation, as defined in (8.20c). Another quantity that can be used to characterize a coupler is the directivity, $D = I - C$ (dB), which is the ratio of the power delivered to the coupled port and the isolated port. The ideal coupler is characterized solely by the coupling factor, as the isolation and directivity are infinite. The ideal coupler is also lossless, and matched at all ports.

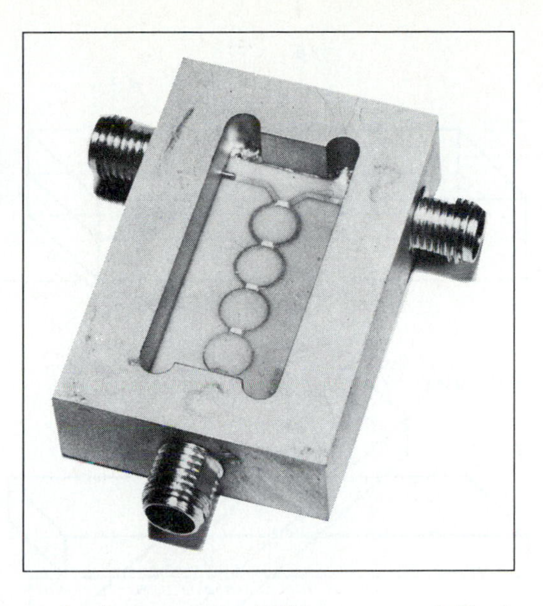

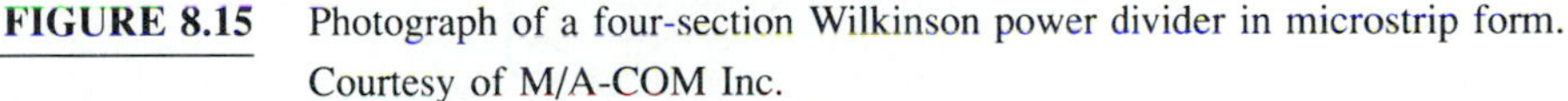

FIGURE 8.15 Photograph of a four-section Wilkinson power divider in microstrip form. Courtesy of M/A-COM Inc.

Directional couplers can be made in many different forms. We will first discuss waveguide couplers, followed by hybrid junctions. A hybrid junction is a special case of a directional coupler, where the coupling factor is 3 dB (equal split), and the phase relation between the output ports is either 90° (*quadrature hybrid*), or 180° (*magic-T* or *rat-race hybrid*). Then we will discuss the implementation of directional couplers in coupled transmission line form.

Bethe-Hole Coupler

The directional property of all directional couplers is produced through the use of two separate waves or wave components, which add in phase at the coupled port and are cancelled at the isolated port. One of the simplest ways of doing this is to couple one waveguide to another through a single small hole in the common broad wall between the two guides. Such a coupler is known as a Bethe-hole coupler, two versions of which are shown in Figure 8.16. From the small-aperture coupling theory of Section 5.9, we know that an aperture can be replaced with equivalent sources consisting of electric and magnetic dipole moments [6]. The normal electric dipole moment and the axial magnetic dipole moment radiate with even symmetry in the coupled guide, while the transverse magnetic dipole moment radiates with odd symmetry. Thus, by adjusting the relative amplitudes of these two equivalent sources, we can cancel the radiation in the direction of the isolated port, while enhancing the radiation in the direction of the coupled port. Figure 8.16 shows two ways in which these wave amplitudes can be controlled; in the coupler shown in Figure 8.16a, the two guides are parallel and the coupling is controlled

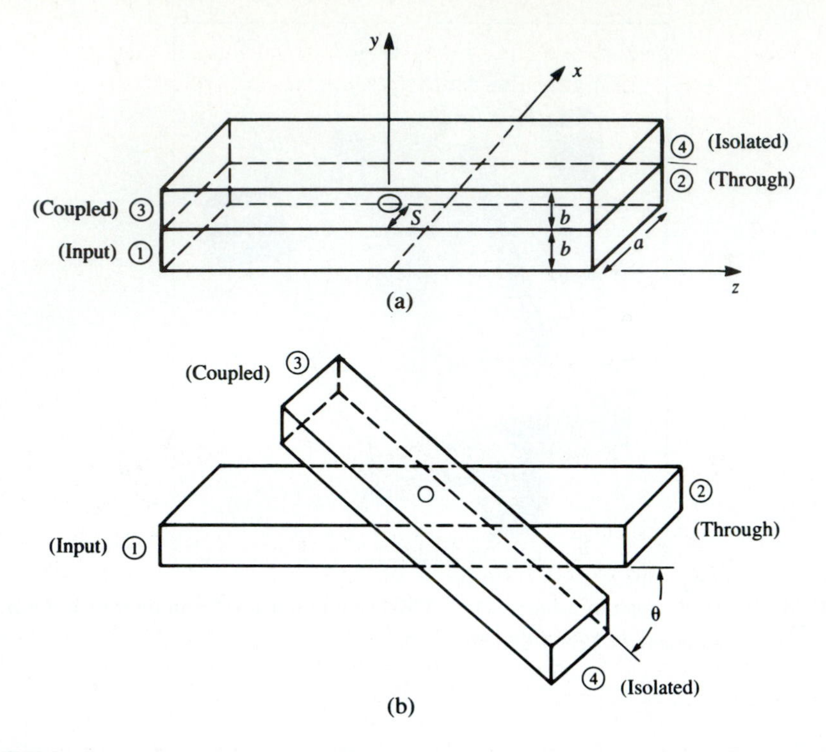

FIGURE 8.16 Two versions of the Bethe-hole directional coupler. (a) Parallel guides. (b) Skewed guides.

by S, the aperture offset from the sidewall of the guide. For the coupler of Figure 8.16b, the wave amplitudes are controlled by the angle, θ, between the two guides.

First consider the configuration of Figure 8.16a, with an incident TE_{10} mode into port 1. These fields can be written as

$$E_y = A \sin \frac{\pi x}{a} e^{-j\beta z}, \tag{8.38a}$$

$$H_x = \frac{-A}{Z_{10}} \sin \frac{\pi x}{a} e^{-j\beta z}, \tag{8.38b}$$

$$H_z = \frac{j\pi A}{\beta a Z_{10}} \cos \frac{\pi x}{a} e^{-j\beta z}, \tag{8.38c}$$

where $Z_{10} = k_0\eta_0/\beta$ is the wave impedance of the TE_{10} mode. Then, from (5.137) and (5.138), this incident wave generates the following equivalent polarization currents at the aperture at $x = s, y = b, z = 0$:

$$\bar{P}_e = \epsilon_0 \alpha_e \hat{y} A \sin \frac{\pi s}{a} \delta(x - s)\delta(y - b)\delta(z), \tag{8.39a}$$

$$\bar{P}_m = -\alpha_m A \left[\frac{-\hat{x}}{Z_{10}} \sin \frac{\pi s}{a} + \hat{z} \frac{j\pi}{\beta a Z_{10}} \cos \frac{\pi s}{a} \right] \delta(x - s)\delta(y - b)\delta(z). \tag{8.39b}$$

Using (5.141a,b) to relate $\bar{P}_e$ and $\bar{P}_m$ to the currents $\bar{J}$ and $\bar{M}$, and then using (5.131), (5.133), (5.135), and (5.136) gives the amplitudes of the forward and reverse traveling waves in the bottom guide as

$$A_{10}^{+} = \frac{-1}{P_{10}} \int_v \bar{E}_{10}^{-} \cdot \bar{J}\, dv + \frac{1}{P_{10}} \int_v \bar{H}_{10}^{-} \cdot \bar{M}\, dv$$
$$= \frac{-j\omega A}{P_{10}} \left[\epsilon_0 \alpha_e \sin^2 \frac{\pi s}{a} - \frac{\mu_0 \alpha_m}{Z_{10}^2} \left(\sin^2 \frac{\pi s}{a} + \frac{\pi^2}{\beta^2 a^2} \cos^2 \frac{\pi s}{a} \right) \right], \tag{8.40a}$$

$$A_{10}^{-} = \frac{-1}{P_{10}} \int_v \bar{E}_{10}^{+} \cdot \bar{J}\, dv + \frac{1}{P_{10}} \int_v \bar{H}_{10}^{+} \cdot \bar{M}\, dv$$
$$= \frac{-j\omega A}{P_{10}} \left[\epsilon_0 \alpha_e \sin^2 \frac{\pi s}{a} + \frac{\mu_0 \alpha_m}{Z_{10}^2} \left(\sin^2 \frac{\pi s}{a} - \frac{\pi^2}{\beta^2 a^2} \cos^2 \frac{\pi s}{a} \right) \right], \tag{8.40b}$$

where $P_{10} = ab/Z_{10}$ is the power normalization constant. Note from (8.40a,b) that the amplitude of the wave excited toward port 4 (A_{10}^{+}) is generally different from that excited toward port 3 (A_{10}^{-}) (because $H_x^{+} = -H_x^{-}$) so we can cancel the power delivered to port 4 by setting $A_{10}^{+} = 0$. If we assume that the aperture is round, then Table 5.3 gives the polarizabilities as $\alpha_e = 2r_0^3/3$ and $\alpha_m = 4r_0^3/3$, where r_0 is the radius of the aperture. Then from (8.40a) we obtain the following condition:

$$\left(2\epsilon_0 - \frac{4\mu_0}{Z_{10}^2} \right) \sin^2 \frac{\pi s}{a} - \frac{4\pi^2 \mu_0}{\beta^2 a^2 Z_{10}^2} \cos^2 \frac{\pi s}{a} = 0,$$

$$(k_0^2 - 2\beta^2) \sin^2 \frac{\pi s}{a} = \frac{2\pi^2}{a^2} \cos^2 \frac{\pi s}{a},$$

$$\left(\frac{4\pi^2}{a^2} - k_0^2 \right) \sin^2 \frac{\pi s}{a} = \frac{2\pi^2}{a^2},$$

or

$$\sin \frac{\pi s}{a} = \pi \sqrt{\frac{2}{4\pi^2 - k_0^2 a^2}} = \frac{\lambda_0}{\sqrt{2(\lambda_0^2 - a^2)}}. \tag{8.41}$$

The coupling factor is then given by

$$C = 20 \log \left| \frac{A}{A_{10}^{-}} \right| \text{ dB}, \tag{8.42a}$$

and the directivity by

$$D = 20 \log \left| \frac{A_{10}^{-}}{A_{10}^{+}} \right| \text{ dB}. \tag{8.42b}$$

Thus, a Bethe-hole coupler of the type shown in Figure 8.16a can be designed by first using (8.41) to find s, the position of the aperture, and then using (8.42a) to determine the aperture size, r_0, to give the required coupling factor.

For the skewed geometry of Figure 8.16b, the aperture may be centered at $s = a/2$, and the skew angle θ adjusted for cancellation at port 4. In this case, the normal electric field does not change with θ, but the transverse magnetic field components are reduced by $\cos\theta$. We can thus account for the skew by replacing α_m in the previous derivation by $\alpha_m \cos\theta$. The wave amplitudes of (8.40a,b) then become, for $s = a/2$,

$$A_{10}^{+} = \frac{-j\omega A}{P_{10}}\left(\epsilon_0\alpha_e - \frac{\mu_0\alpha_m}{Z_{10}^2}\cos\theta\right), \quad 8.43a$$

$$A_{10}^{-} = \frac{-j\omega A}{P_{10}}\left(\epsilon_0\alpha_e + \frac{\mu_0\alpha_m}{Z_{10}^2}\cos\theta\right). \quad 8.43b$$

Setting $A_{10}^{+} = 0$ results in the following condition for the angle θ:

$$2\epsilon_0 - \frac{4\mu_0}{Z_{10}^2}\cos\theta = 0,$$

or

$$\cos\theta = \frac{k_0^2}{2\beta^2}. \quad 8.44$$

The coupling factor then simplifies to

$$C = 20\log\left|\frac{A}{A_{10}^{-}}\right| = -20\log\frac{4k_0^2 r_0^3}{3ab\beta}\ \text{dB}. \quad 8.45$$

The geometry of the skewed Bethe-hole coupler is often a disadvantage in terms of fabrication and application. Also, both coupler designs only operate properly at the design frequency; deviation from this frequency will alter the coupling level and the directivity.

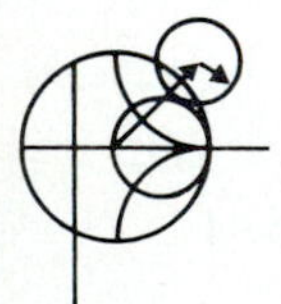

EXAMPLE 8.3

Design a Bethe-hole coupler of the type shown in Figure 8.16a for X-band waveguide operating at 9 GHz, with a coupling of 20 dB. Calculate and plot the coupling and directivity from 7 to 11 GHz. Assume a round aperture.

Solution

For X-band waveguide at 9 GHz, we have the following constants:

$$a = 0.02286\,\text{m},$$

$$b = 0.01016\,\text{m},$$

$$\lambda_0 = 0.0333\,\text{m},$$

$$k_0 = 188.5\,\text{m}^{-1},$$

$$\beta = 129.0\,\text{m}^{-1},$$

$$Z_{10} = 550.9\,\Omega,$$

$$P_{10} = 4.22 \times 10^{-7}\,\text{m}^2/\Omega.$$

Then (8.41) can be used to find the aperture position s:

$$\sin\frac{\pi s}{a} = \frac{\lambda_0}{\sqrt{2(\lambda_0^2 - a^2)}} = 0.972,$$

$$s = \frac{a}{\pi}\sin^{-1} 0.972 = 0.424a = 9.69\,\text{mm}.$$

The coupling is 20 dB, so

$$C = 20 \text{ dB} = 20 \log \left|\frac{A}{A_{10}^-}\right|,$$

or

$$\left|\frac{A}{A_{10}^-}\right| = 10^{20/20} = 10,$$

thus, $|A_{10}^-/A| = 1/10$. We now use (8.40b) to find r_0:

$$\left|\frac{A_{10}^-}{A}\right| = \frac{1}{10} = \frac{\omega}{P_{10}}\left[\left(\epsilon_0\alpha_e + \frac{\mu_0\alpha_m}{Z_{10}^2}\right)(0.944) - \frac{\pi^2\mu_0\alpha_m}{\beta^2 a^2 Z_{10}^2}(0.056)\right].$$

Since $\alpha_e = 2r_0^3/3$ and $\alpha_m = 4r_0^3/3$, we obtain,

$$0.1 = 1.44 \times 10^6 r_0^3,$$

or

$$r_0 = 4.15 \text{ mm}.$$

This completes the design of the Bethe-hole coupler. To evaluate the coupling and directivity versus frequency, we evaluate (8.42a) and (8.42b), using the expressions for A_{10}^- and A_{10}^+ given in (8.40a) and (8.40b). In these expressions, the aperture position and size are fixed at $s = 9.69$ mm and $r_0 = 4.15$ mm, and the frequency is varied. A short computer program was used to calculate the data shown in Figure 8.17. Observe that the coupling varies by less than 1 dB over the band. The directivity is very large (>60 dB) at the design

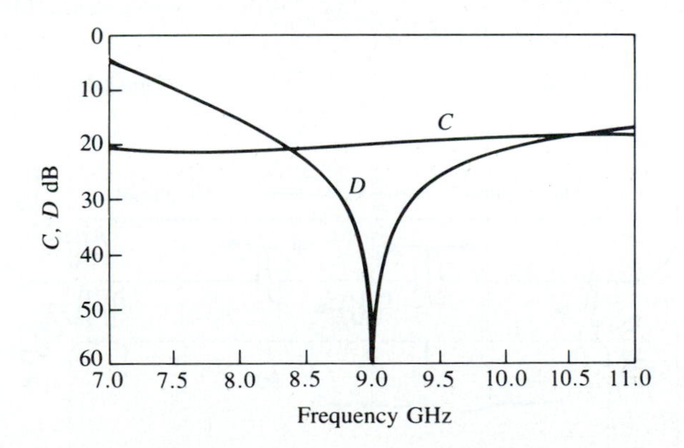

FIGURE 8.17 Coupling and directivity versus frequency for the Bethe-hole coupler of Example 8.3.

frequency, but decreases to 15–20 dB at the band edges. The directivity is a more sensitive function of frequency because it depends on the cancellation of two wave components. ○

Design of Multihole Couplers

As seen from Example 8.3, a single-hole coupler has a relatively narrow bandwidth, at least in terms of its directivity. But if the coupler is designed with a series of coupling holes, the extra degrees of freedom can be used to increase this bandwidth. The principle of operation and design of such a multihole waveguide coupler is very similar to that of the multisection matching transformer.

First let us consider the operation of the two-hole coupler shown in Figure 8.18. Two parallel waveguides sharing a common broad wall are shown, although the same type of structure could be made in microstrip or stripline form. Two small apertures are spaced $\lambda_g/4$ apart, and couple the two guides. A wave entering at port 1 is mostly transmitted through to port 2, but some power is coupled through the two apertures. If a phase reference is taken at the first aperture, then the phase of the wave incident at the second aperture will be $-90°$. Each aperture will radiate a forward wave component and a backward wave component into the upper guide; in general, the forward and backward amplitudes are different. In the direction of port 3, both components are in phase, since both have traveled $\lambda_g/4$ to the second aperture. But we obtain a cancellation in the direction of port 4, since the wave coming through the second aperture travels $\lambda_g/2$ further than the wave component coming through the first aperture. Clearly, this cancellation is frequency sensitive, making the directivity a sensitive function of frequency. The coupling is less frequency dependent, since the path lengths from port 1 to port 3 are always the same. Thus, in the multihole coupler design, we synthesize the directivity response, as opposed to the coupling response, as a function of frequency.

We now consider the general case of the multihole coupler shown in Figure 8.19, where $N+1$ equally spaced apertures couple two parallel waveguides. The amplitude of the incident wave in the lower left guide is A and, for small coupling, is essentially the same as the amplitude of the through wave. For instance, a 20 dB coupler has a power

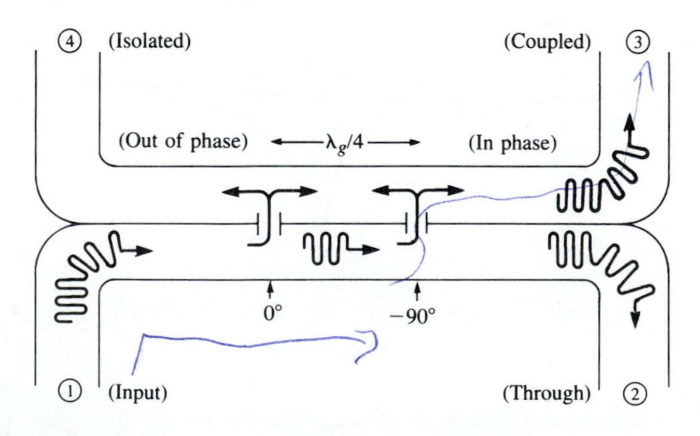

FIGURE 8.18 Basic operation of a two-hole directional coupler.

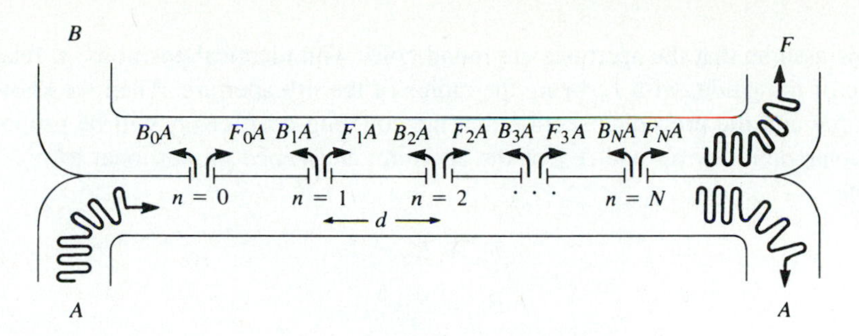

FIGURE 8.19 Geometry of an $N + 1$ hole waveguide directional coupler.

coupling factor of $10^{-20/10} = 0.01$, so the power transmitted through waveguide A is $1 - 0.01 = 0.99$ of the incident power (1% coupled to the upper guide). The voltage (or field) drop in waveguide A is $\sqrt{0.99} = 0.995$, or 0.5%. Thus, the assumption that the amplitude of the incident field is identical at each aperture is a good one. Of course, the phase will change from one aperture to the next.

As we saw in the previous section for the Bethe-hole coupler, an aperture generally excites forward and backward traveling waves with different amplitudes. Thus, let

F_n denote the coupling coefficient of the nth aperture in the forward direction.

B_n denote the coupling coefficient of the nth aperture in the backward direction.

Then the amplitude of the forward wave can be written as

$$F = Ae^{-j\beta N d} \sum_{n=0}^{N} F_n, \tag{8.46}$$

since all components travel the same path length. The amplitude of the backward wave is

$$B = A \sum_{n=0}^{N} B_n e^{-2j\beta n d}, \tag{8.47}$$

since the path length for the nth component is $2\beta nd$, where d is the spacing between the apertures. In (8.46) and (8.47) the phase reference is taken at the $n = 0$ aperture.

From the definitions in (8.20a) and (8.20b) the coupling and directivity can be computed as

$$C = -20 \log \left| \frac{F}{A} \right| = -20 \log \left| \sum_{n=0}^{N} F_n \right| \text{ dB}, \tag{8.48}$$

$$\begin{aligned} D &= -20 \log \left| \frac{B}{F} \right| = -20 \log \left| \frac{\sum_{n=0}^{N} B_n e^{-2j\beta n d}}{\sum_{n=0}^{N} F_n} \right| \\ &= -C - 20 \log \left| \sum_{n=0}^{N} B_n e^{-2j\beta n d} \right| \text{ dB}. \end{aligned} \tag{8.49}$$

Now assume that the apertures are round holes with identical positions, s, relative to the edge of the guide, with r_n being the radius of the nth aperture. Then we know from Section 5.9 and the preceding section that the coupling coefficients will be proportional to the polarizabilities α_e and α_m of the aperture, and hence proportional to r_n^3. So we can write

$$F_n = K_f r_n^3, \tag{8.50a}$$

$$B_n = K_b r_n^3, \tag{8.50b}$$

where K_f and K_b are constants for the forward and backward coupling coefficients that are the same for all apertures, but are functions of frequency. Then (8.48) and (8.49) reduce to

$$C = -20 \log |K_f| - 20 \log \sum_{n=0}^{N} r_n^3 \text{ dB}, \tag{8.51}$$

$$D = -C - 20 \log |K_b| - 20 \log \left| \sum_{n=0}^{N} r_n^3 e^{-2j\beta nd} \right|$$

$$= -C - 20 \log |K_b| - 20 \log S \text{ dB}. \tag{8.52}$$

In (8.51), the second term is constant with frequency. The first term is not affected by the choice of r_ns, but is a relatively slowly varying function of frequency. Similarly, in (8.52) the first two terms are slowly varying functions of frequency, representing the directivity of a single aperture, but the last term (S) is a sensitive function of frequency due to phase cancellation in the summation. Thus we can choose the r_ns to synthesize a desired frequency response for the directivity, while the coupling should be relatively constant with frequency.

Observe that the last term in (8.52),

$$S = \left| \sum_{n=0}^{N} r_n^3 e^{-2j\beta nd} \right|, \tag{8.53}$$

is very similar in form to the expression obtained in Section 6.5 for multisection quarter-wave matching transformers. As in that case, we will develop coupler designs that yield either a binomial (maximally flat) or a Chebyshev (equal ripple) response for the directivity. Another interpretation of (8.53) may be recognizable to the student familiar with basic antenna theory, as this expression is identical to the array pattern factor of an $N + 1$ element array with element weights r_n^3. In that case, too, the pattern may be synthesized in terms of binomial or Chebyshev polynomials.

Binomial response. As in the case of the multisection quarter-wave matching transformers, we can obtain a binomial, or maximally flat, response for the directivity of the multihole coupler by making the coupling coefficients proportional to the binomial coefficients. Thus,

$$r_n^3 = kC_n^N, \tag{8.54}$$

where k is a constant to be determined, and C_n^N is a binomial coefficient given in (6.51). To find k, we evaluate the coupling using (8.51) to give

$$C = -20 \log |K_f| - 20 \log k - 20 \log \sum_{n=0}^{N} C_n^N \text{ dB}. \quad 8.55$$

Since we know K_f, N, and C, we can solve for k, and then find the required aperture radii from (8.54). The spacing d should be $\lambda_g/4$ at the center frequency.

Chebyshev response. First assume that N is even (an odd number of holes), and that the coupler is symmetric, so that $r_0 = r_N, r_1 = r_{N-1}$, etc. Then from (8.53) we can write S as

$$S = \left| \sum_{n=0}^{N} r_n^3 e^{-2jn\theta} \right| = 2 \sum_{n=0}^{N/2} r_n^3 \cos(N-2n)\theta,$$

where $\theta = \beta d$. To achieve a Chebyshev response we equate this to the Chebyshev polynomial of degree N:

$$S = 2 \sum_{n=0}^{N/2} r_n^3 \cos(N-2n)\theta = k|T_N(\sec\theta_m \cos\theta)|, \quad 8.56$$

where k and θ_m are constants to be determined. From (8.53) and (8.56), we see that for $\theta = 0$, $S = \sum_{n=0}^{N} r_n^3 = k|T_N(\sec\theta_m)|$. Using this result in (8.51) gives the coupling as

$$\begin{aligned} C &= -20 \log |K_f| - 20 \log S \Big|_{\theta=0} \\ &= -20 \log |K_f| - 20 \log k - 20 \log |T_N(\sec\theta_m)| \text{ dB}. \end{aligned} \quad 8.57$$

From (8.52) the directivity is

$$\begin{aligned} D &= -C - 20 \log |K_b| - 20 \log S \\ &= 20 \log \frac{K_f}{K_b} + 20 \log \frac{T_N(\sec\theta_m)}{T_N(\sec\theta_m \cos\theta)} \text{ dB}. \end{aligned} \quad 8.58$$

The term $\log K_f/K_b$ is a function of frequency, so D will not have an exact Chebyshev response. This error is usually small, however. Thus, we can assume that the smallest value of D will occur when $T_N(\sec\theta_m \cos\theta) = 1$, since $|T_N(\sec\theta_m)| \geq |T_N(\sec\theta_m \cos\theta)|$. So if $D_{\min}$ is the specified minimum value of directivity in the passband, then θ_m can be found from the relation

$$D_{\min} = 20 \log T_N(\sec\theta_m) \text{ dB}. \quad 8.59$$

Alternatively, we could specify the bandwidth, which then dictates θ_m and $D_{\min}$. In either case, (8.57) can then be used to find k, and then (8.56) solved for the radii, r_m.

If N is odd (an even number of holes), the results for C, D, and $D_{\min}$ in (8.57), (8.58), and (8.59) still apply, but instead of (8.56), the following relation is used to find the aperture radii:

$$S = 2 \sum_{n=0}^{(N-1)/2} r_n^3 \cos(N-2n)\theta = k|T_N(\sec\theta_m \cos\theta)|. \qquad 8.60$$

EXAMPLE 8.4

Design a four-hole Chebyshev coupler in X-band waveguide using round apertures located at $s = a/4$. The center frequency is 9 GHz, the coupling is 20 dB, and the minimum directivity is 40 dB. Plot the directivity response from 7 to 11 GHz.

Solution

For X-band waveguide at 9 GHz, we have the following constants:

$$\begin{aligned} a &= 0.02286 \text{ m}, \\ b &= 0.01016 \text{ m}, \\ \lambda_0 &= 0.0333 \text{ m}, \\ k_0 &= 188.5 \text{ m}^{-1}, \\ \beta &= 129.0 \text{ m}^{-1}, \\ Z_{10} &= 550.9\ \Omega, \\ P_{10} &= 4.22 \times 10^{-7} \text{ m}^2/\Omega. \end{aligned}$$

From (8.40a) and (8.40b), we obtain for an aperture at $s = a/4$:

$$|K_f| = \frac{2k_0}{3\eta_0 P_{10}}\left[\sin^2\frac{\pi s}{a} - \frac{2\beta^2}{k_0^2}\left(\sin^2\frac{\pi s}{a} + \frac{\pi^2}{\beta^2 a^2}\cos^2\frac{\pi s}{a}\right)\right] = 3.953 \times 10^5,$$

$$|K_b| = \frac{2k_0}{3\eta_0 P_{10}}\left[\sin^2\frac{\pi s}{a} + \frac{2\beta^2}{k_0^2}\left(\sin^2\frac{\pi s}{a} - \frac{\pi^2}{\beta^2 a^2}\cos^2\frac{\pi s}{a}\right)\right] = 3.454 \times 10^5.$$

For a four-hole coupler, $N = 3$, so (8.59) gives

$$\begin{aligned} 40 &= 20\log T_3(\sec\theta_m) \text{ dB}, \\ 100 &= T_3(\sec\theta_m) = \cosh(3\cosh^{-1}(\sec\theta_m)), \\ \sec\theta_m &= 3.01, \end{aligned}$$

where (6.58b) was used. Thus $\theta_m = 70.6°$ and $109.4°$ at the band edges. Then from (8.57) we can solve for k:

$$C = 20 = -20\log(3.953 \times 10^5) - 20\log k - 40 \text{ dB},$$

$$\begin{aligned} 20\log k &= -171.94, \\ k &= 2.53 \times 10^{-9}. \end{aligned}$$

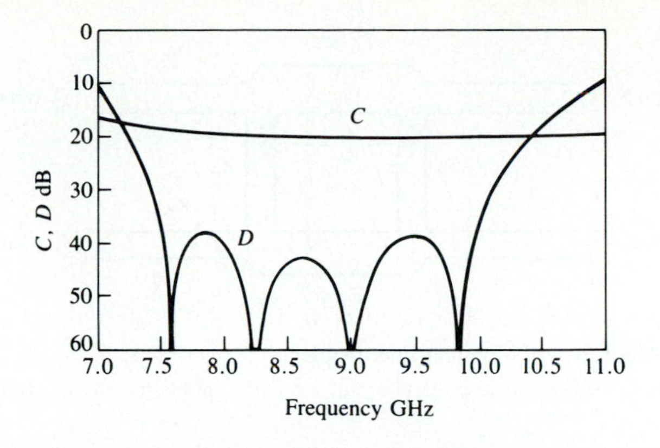

FIGURE 8.20 Coupling and directivity versus frequency for the four-hole coupler of Example 8.4.

Finally, (8.60) and the expansion from (6.60c) for T_3 allows us to solve for the radii as follows:

$$S = 2[r_0^3 \cos 3\theta + r_1^3 \cos\theta] = k[\sec^3\theta_m(\cos 3\theta + 3\cos\theta) - 3\sec\theta_m \cos\theta],$$

$$2r_0^3 = k\sec^3\theta_m \quad \Rightarrow \quad r_0 = r_3 = 3.26 \text{ mm},$$

$$2r_1^3 = 3k(\sec^3\theta_m - \sec\theta_m) \quad \Rightarrow \quad r_1 = r_2 = 4.51 \text{ mm}.$$

The resulting coupling and directivity are plotted in Figure 8.20; note the increased directivity bandwidth compared to that of the Bethe-hole coupler of Example 8.3. ○

8.5 THE QUADRATURE (90°) HYBRID

Quadrature hybrids are 3 dB directional couplers with a 90° phase difference in the outputs of the through and coupled arms. This type of hybrid is often made in microstrip or stripline form as shown in Figure 8.21, and is also known as a branch-line hybrid. Other 3 dB couplers, such as coupled line couplers or Lange couplers, can also be used as quadrature couplers; these components will be discussed in later sections. Here we will analyze the operation of the quadrature hybrid using an even-odd mode decomposition technique similar to that used for the Wilkinson power divider.

With reference to Figure 8.21 the basic operation of the branch-line coupler is as follows. With all ports matched, power entering port 1 is evenly divided between ports 2 and 3, with a 90° phase shift between these outputs. No power is coupled to port 4 (the isolated port). Thus, the $[S]$ matrix will have the following form:

$$[S] = \frac{-1}{\sqrt{2}} \begin{bmatrix} 0 & j & 1 & 0 \\ j & 0 & 0 & 1 \\ 1 & 0 & 0 & j \\ 0 & 1 & j & 0 \end{bmatrix}. \tag{8.61}$$

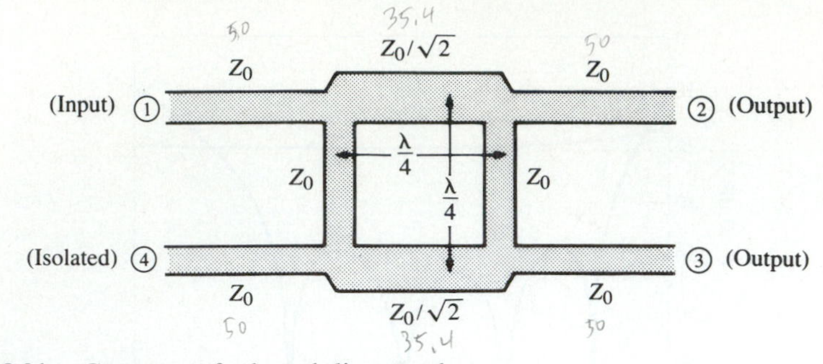

FIGURE 8.21 Geometry of a branch-line coupler.

Observe that the branch-line hybrid has a high degree of symmetry, as any port can be used as the input port. The output ports will always be on the opposite side of the junction from the input port, and the isolated port will be the remaining port on the same side as the input port. This symmetry is reflected in the scattering matrix, as each row can be obtained as a transposition of the first row.

Even-Odd Mode Analysis

We first draw the schematic circuit of the branch-line coupler in normalized form, as in Figure 8.22, where it is understood that each line represents a transmission line with indicated characteristic impedance normalized to Z_0. The common ground return for each transmission line is not shown. We assume that a wave of unit amplitude $A_1 = 1$ is incident at port 1.

Now the circuit of Figure 8.22 can be decomposed into the superposition of an even-mode excitation and an odd-mode excitation [5], as shown in Figure 8.23. Note that adding the two sets of excitations produces the original excitation of Figure 8.22, and since the circuit is linear, the actual response (the scattered waves) can be obtained from the sum of the responses to the even and odd excitations.

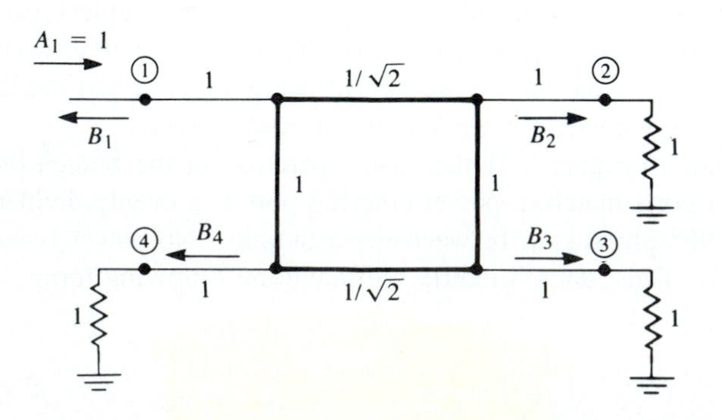

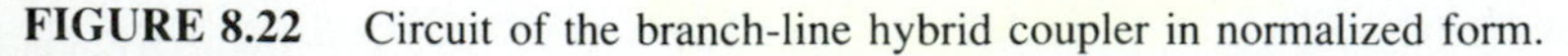
FIGURE 8.22 Circuit of the branch-line hybrid coupler in normalized form.

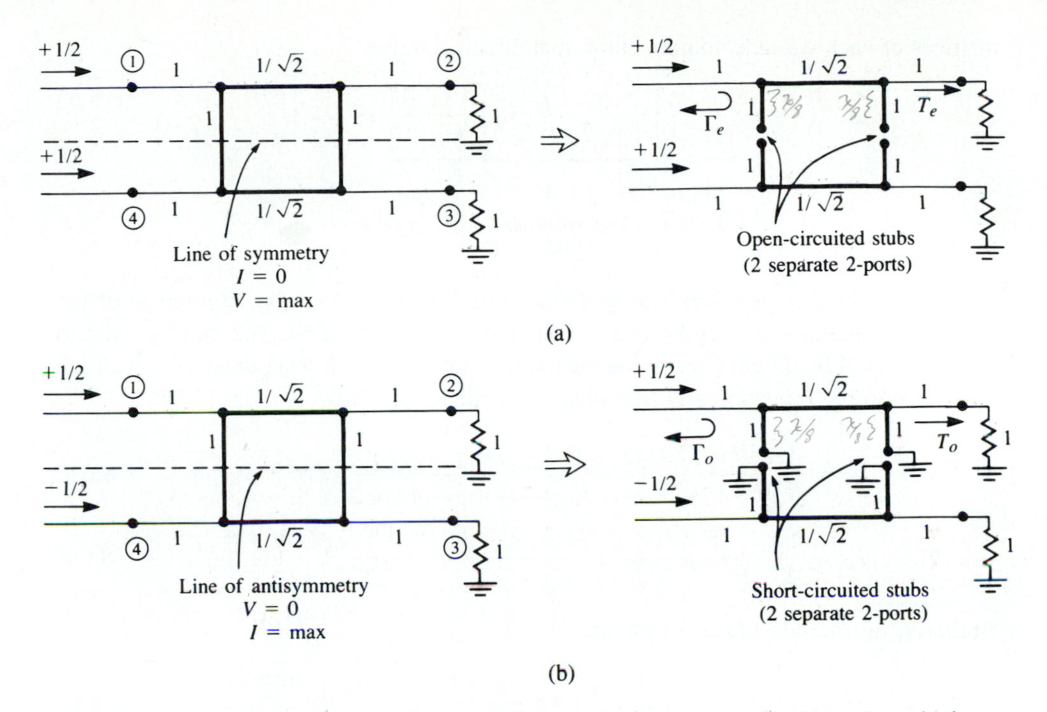

FIGURE 8.23 Decomposition of the branch-line coupler into even- and odd-mode excitations. (a) Even mode (e). (b) Odd mode (o).

Because of the symmetry or antisymmetry of the excitation, the four-port network can be decomposed into a set of two decoupled two-port networks, as shown in Figure 8.23. Since the amplitudes of the incident waves for these two-ports are $\pm 1/2$, the amplitudes of the emerging wave at each port of the branch-line hybrid can be expressed as

$$B_1 = \frac{1}{2}\Gamma_e + \frac{1}{2}\Gamma_o, \tag{8.62a}$$

$$B_2 = \frac{1}{2}T_e + \frac{1}{2}T_o, \tag{8.62b}$$

$$B_3 = \frac{1}{2}T_e - \frac{1}{2}T_o, \tag{8.62c}$$

$$B_4 = \frac{1}{2}\Gamma_e - \frac{1}{2}\Gamma_o, \tag{8.62d}$$

where $\Gamma_{e,o}$ and $T_{e,o}$ are the even- and odd-mode reflection and transmission coefficients for the two-port networks of Figure 8.23. First consider the calculation of Γ_e and T_e, for the even-mode two-port circuit. This can best be done by multiplying the $ABCD$

matrices of each cascade component in that circuit, to give

$$\begin{bmatrix} A & B \\ C & D \end{bmatrix}_e = \underbrace{\begin{bmatrix} 1 & 0 \\ j & 1 \end{bmatrix}}_{\substack{\text{Shunt} \\ Y=j}} \underbrace{\begin{bmatrix} 0 & j/\sqrt{2} \\ j\sqrt{2} & 0 \end{bmatrix}}_{\substack{\lambda/4 \\ \text{Transmission} \\ \text{line}}} \underbrace{\begin{bmatrix} 1 & 0 \\ j & 1 \end{bmatrix}}_{\substack{\text{Shunt} \\ Y=j}} = \frac{1}{\sqrt{2}} \begin{bmatrix} -1 & j \\ j & -1 \end{bmatrix}, \quad 8.63$$

where the individual matrices can be found from Table 5.1, and the admittance of the shunt open-circuited $\lambda/8$ stubs is $Y = j\tan\beta\ell = j$. Then Table 5.2 can be used to convert from $ABCD$ parameters (defined here with $Z_o = 1$) to S parameters, which are equivalent to the reflection and transmission coefficients. Thus,

$$\Gamma_e = \frac{A+B-C-D}{A+B+C+D} = \frac{(-1+j-j+1)/\sqrt{2}}{(-1+j+j-1)/\sqrt{2}} = 0, \quad 8.64a$$

$$T_e = \frac{2}{A+B+C+D} = \frac{2}{(-1+j+j-1)/\sqrt{2}} = \frac{-1}{\sqrt{2}}(1+j). \quad 8.64b$$

Similarly, for the odd mode we obtain

$$\begin{bmatrix} A & B \\ C & D \end{bmatrix}_o = \frac{1}{\sqrt{2}} \begin{bmatrix} 1 & j \\ j & 1 \end{bmatrix}, \quad 8.65$$

which gives the reflection and transmission coefficients as

$$\Gamma_o = 0, \quad 8.66a$$

$$T_o = \frac{1}{\sqrt{2}}(1-j). \quad 8.66b$$

Then using (8.64) and (8.66) in (8.62) gives the following results:

$$B_1 = 0 \quad \text{(port 1 is matched)}, \quad 8.67a$$

$$B_2 = -\frac{j}{\sqrt{2}} \quad \text{(half-power, } -90^\circ \text{ phase shift from port 1 to 2)}, \quad 8.67b$$

$$B_3 = -\frac{1}{\sqrt{2}} \quad \text{(half-power, } -180^\circ \text{ phase shift from port 1 to 3)}, \quad 8.67c$$

$$B_4 = 0 \quad \text{(no power to port 4)}. \quad 8.67d$$

These results agree with the first row and column of the $[S]$ matrix given in (8.61); the remaining elements can be easily found by transposition.

In practice, due to the quarter-wave length requirement, the bandwidth of a branch-line hybrid is limited to 10–20%. But as with multisection matching transformers and multihole directional couplers, the bandwidth of a branch-line hybrid can be increased to a decade or more by using multiple sections in cascade. Figure 8.24 shows a photograph of a three-section quadrature hybrid. In addition, the basic design can be modified for unequal power division and/or different characteristic impedances at the output ports.

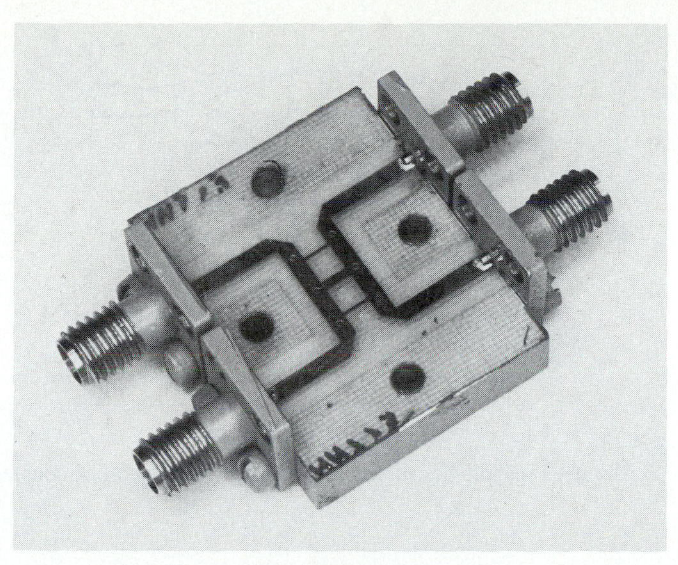

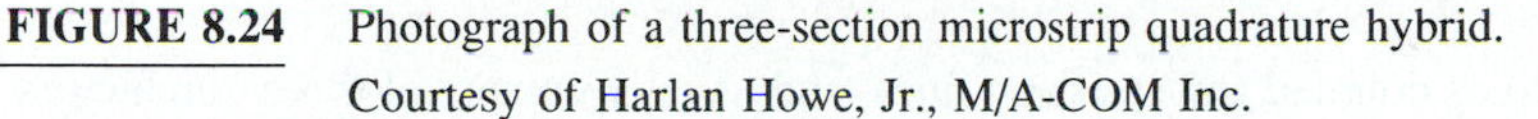

FIGURE 8.24 Photograph of a three-section microstrip quadrature hybrid. Courtesy of Harlan Howe, Jr., M/A-COM Inc.

Another practical point to be aware of is the fact that discontinuity effects at the junctions of the branch-line coupler may require that the shunt arms be lengthened by 10°–20°.

EXAMPLE 8.5

Design a 50 Ω branch-line quadrature hybrid junction, and plot the S parameter magnitudes from $0.5f_0$ to $1.5f_0$, where f_0 is the design frequency.

Solution

After the preceding analysis, the design of a quadrature hybrid is trivial. The lines are $\lambda/4$ at the design frequency f_0, and the branch-line impedances are

$$\frac{Z_0}{\sqrt{2}} = \frac{50}{\sqrt{2}} = 35.4\,\Omega.$$

The calculated frequency response is plotted in Figure 8.25. Note that we obtain perfect 3 dB power division in ports 2 and 3, and perfect isolation and return loss at ports 4 and 1, respectively, at the design frequency f_0. All of these quantities, however, degrade quickly as the frequency departs from f_0. ○

8.6 COUPLED LINE DIRECTIONAL COUPLERS

When two unshielded transmission lines are close together, power can be coupled between the lines due to the interaction of the electromagnetic fields of each line. Such

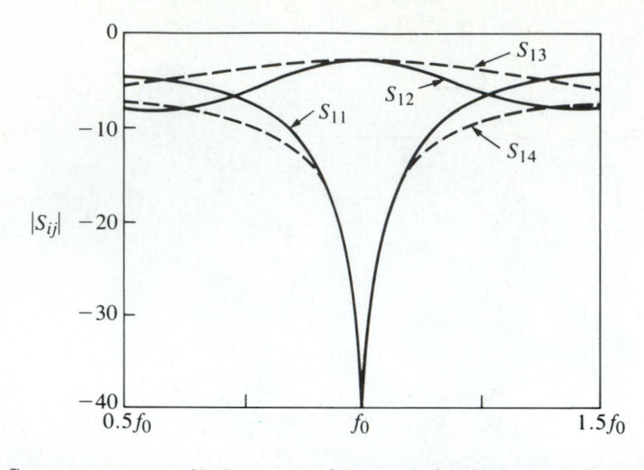

FIGURE 8.25 S parameter magnitudes versus frequency for the branch-line coupler of Example 8.5.

lines are referred to as coupled transmission lines, and usually consist of three conductors in close proximity, although more conductors can be used. Figure 8.26 shows several examples of coupled transmission lines. Coupled transmission lines are usually assumed to operate in the TEM mode, which is rigorously valid for stripline structures and approximately valid for microstrip structures. In general, a three-wire line, like those of Figure 8.26, can support two distinct propagating modes. This feature can be used to implement directional couplers, hybrids, and filters.

We will first discuss the theory of coupled lines, and present some design data for coupled stripline and coupled microstrip. Then we will analyze the operation of a single-section directional coupler, and extend these results to multisection coupler design.

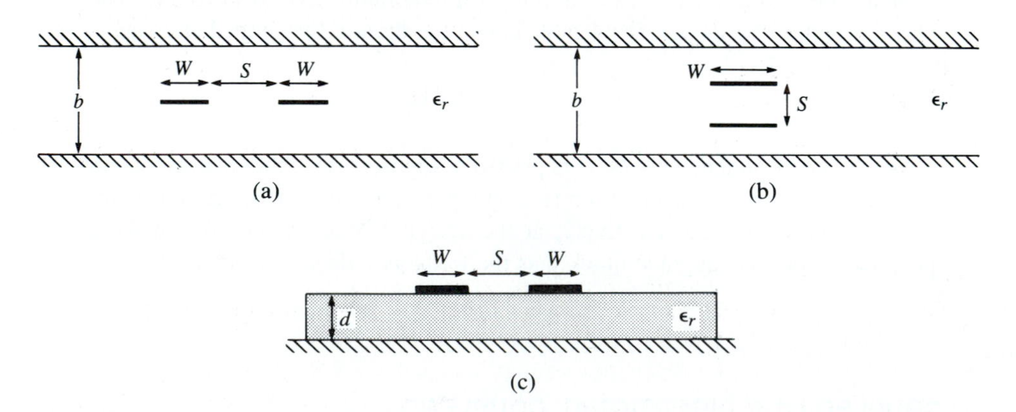

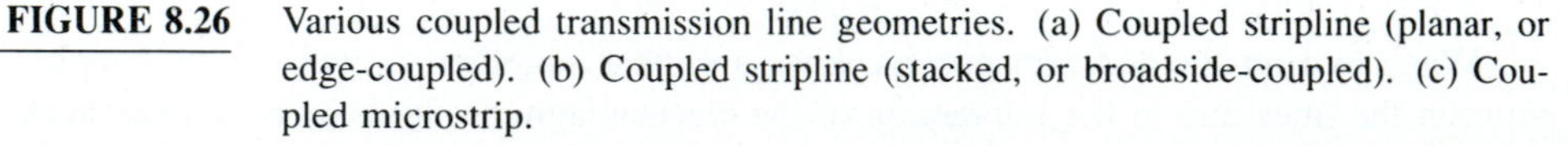

FIGURE 8.26 Various coupled transmission line geometries. (a) Coupled stripline (planar, or edge-coupled). (b) Coupled stripline (stacked, or broadside-coupled). (c) Coupled microstrip.

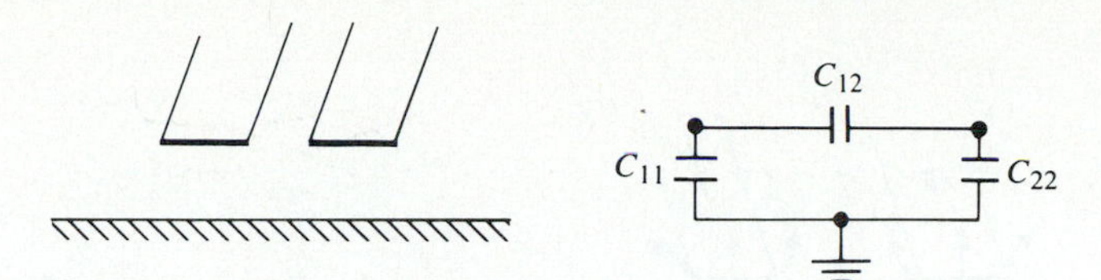

FIGURE 8.27 A three-wire coupled transmission line and its equivalent capacitance network.

Coupled Line Theory

The coupled lines of Figure 8.26, or any other three-wire line, can be represented by the structure shown in Figure 8.27. If we assume a TEM-type of propagation, then the electrical characteristics of the coupled lines can be completely determined from the effective capacitances between the lines and the velocity of propagation on the line. As depicted in Figure 8.27, C_{12} represents the capacitance between the two strip conductors in the absence of the ground conductor, while C_{11} and C_{22} represent the capacitance between one strip conductor and ground, in the absence of the other strip conductor. If the strip conductors are identical in size and location relative to the ground conductor, then $C_{11} = C_{22}$. Note that the designation of "ground" for the third conductor has no special relevance beyond the fact that it is convenient, since in many applications this conductor is the ground plane of a stripline or microstrip circuit.

Now consider two special types of excitations for the coupled line: the even mode, where the currents in the strip conductors are equal in amplitude and in the same direction, and the odd mode, where the currents in the strip conductors are equal in amplitude but in opposite directions. The electric field lines for these two cases are sketched in Figure 8.28.

For the even mode, the electric field has even symmetry about the center line, and no current flows between the two strip conductors. This leads to the equivalent circuit shown, where C_{12} is effectively open-circuited. Then the resulting capacitance of either line to ground for the even mode is

$$C_e = C_{11} = C_{22}, \tag{8.68}$$

assuming that the two strip conductors are identical in size and location. Then the characteristic impedance for the even mode is

$$Z_{0e} = \sqrt{\frac{L}{C_e}} = \frac{\sqrt{LC_e}}{C_e} = \frac{1}{vC_e}, \tag{8.69}$$

where v is the velocity of propagation on the line.

For the odd mode, the electric field lines have an odd symmetry about the center line, and a voltage null exists between the two strip conductors. We can imagine this as a ground plane through the middle of C_{12}, which leads to the equivalent circuit as shown. In this case, the effective capacitance between either strip conductor and ground is

$$C_o = C_{11} + 2C_{12} = C_{22} + 2C_{12}, \tag{8.70}$$

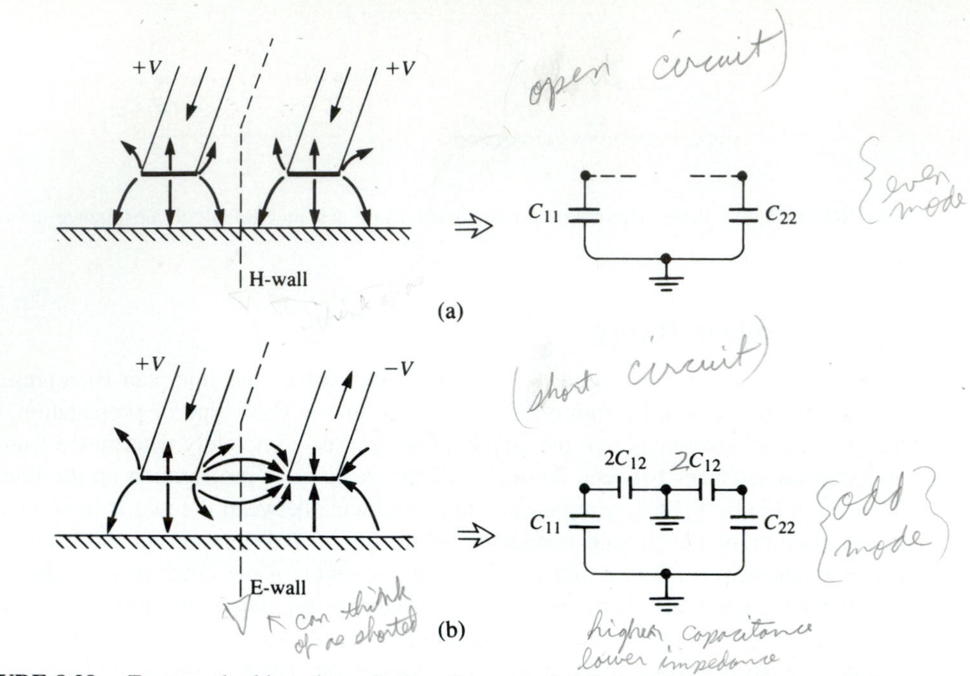

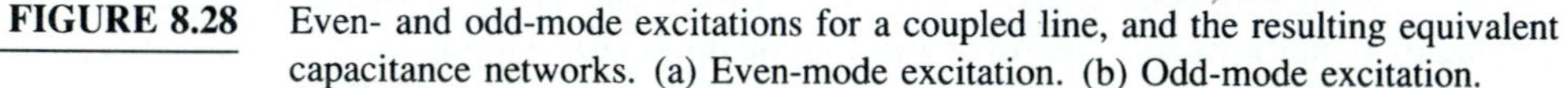

FIGURE 8.28 Even- and odd-mode excitations for a coupled line, and the resulting equivalent capacitance networks. (a) Even-mode excitation. (b) Odd-mode excitation.

and the characteristic impedance for the odd mode is

$$Z_{0o} = \frac{1}{vC_o}. \tag{8.71}$$

In words, $Z_{0e}(Z_{0o})$ is the characteristic impedance of one of the strip conductors relative to ground when the coupled line is operated in the even (odd) mode. An arbitrary excitation of a coupled line can always be treated as a superposition of appropriate amplitudes of even and odd modes.

If the coupled line is purely TEM, such as coaxial, parallel-plate, or stripline, analytical techniques such as conformal mapping [7] can be used to evaluate the capacitances per unit length of line, and the even- and odd-mode characteristic impedances can then be determined. For quasi-TEM lines, such as microstrip, these results can be obtained numerically or by approximate quasi-static techniques [8]. In either case, such calculations are generally too involved for our consideration, so we will only present two examples of design data for coupled lines.

For a symmetric coupled stripline of the type shown in Figure 8.26a, the design graph in Figure 8.29 can be used to determine the necessary strip widths and spacing for a given set of characteristic impedances, Z_{0e} and Z_{0o}, and the dielectric constant. This graph should cover ranges of parameters for most practical applications, and can be used for any dielectric constant, since stripline supports a purely TEM mode.

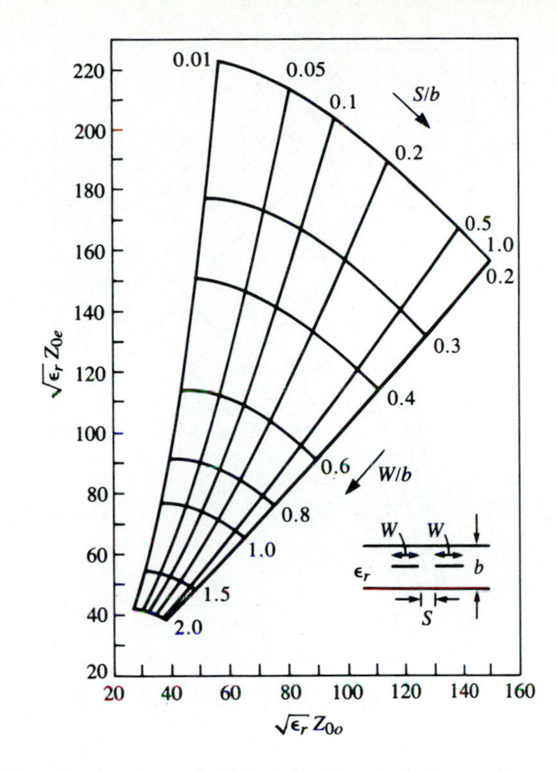

FIGURE 8.29 Normalized even- and odd-mode characteristic impedance design data for edge-coupled striplines.

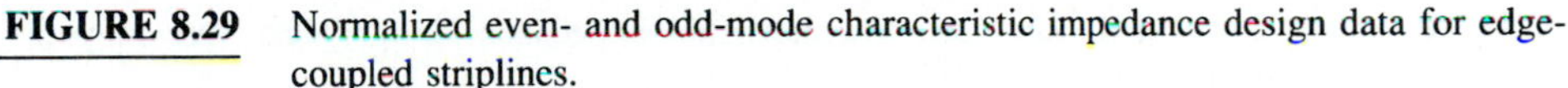

For microstrip, the results do not scale with dielectric constant, so design graphs must be made for specific values of dielectric constant. Figure 8.30 shows such a design graph for coupled microstrip lines on a substrate with $\epsilon_r = 10$. Another difficulty with microstrip coupled lines is the fact that the phase velocity is usually different for the two modes of propagation, since the two modes operate with different field configurations in the vicinity of the air-dielectric interface. This can have a degrading effect on coupler directivity.

EXAMPLE 8.6

For the coupled stripline geometry of Figure 8.26b, let $W >> S$ and $W >> b$, so that fringing fields can be ignored, and determine the even- and odd-mode characteristic impedances.

Solution

We first find the equivalent network capacitances, C_{11} and C_{12} (since the line is symmetric, $C_{22} = C_{11}$). C_{11} is the capacitance of one of the strip conductors to the ground planes, in the absence of the other strip conductor.

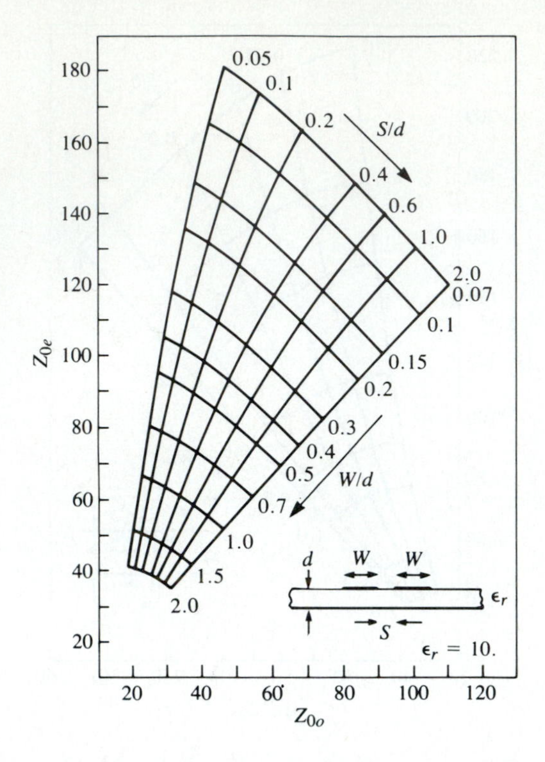

FIGURE 8.30 Even- and odd-mode characteristic impedance design data for coupled microstrip lines.

The capacitance of a parallel plate capacitor with plate area, A, and plate separation, d, is

$$C = \frac{\epsilon A}{d},$$

with ϵ being the permittivity of the material between the plates. This formula ignores fringing fields.

Now C_{11} is formed by the parallel combination of two capacitors from one strip to the two ground planes. Thus, the capacitance per unit length is

$$\bar{C}_{11} = \frac{\epsilon_r \epsilon_0 W}{(b-S)/2} + \frac{\epsilon_r \epsilon_0 W}{(b+S)/2} = \frac{4b\epsilon_r \epsilon_0 W}{b^2 - S^2} \text{ Fd/m.}$$

The capacitance between the strips is, per unit length,

$$\bar{C}_{12} = \frac{\epsilon_r \epsilon_0 W}{S} \text{ Fd/m.}$$

Then from (8.68) and (8.70), the even- and odd-mode capacitances are

$$\bar{C}_e = \bar{C}_{11} = \frac{4b\epsilon_r\epsilon_0 W}{b^2 - S^2} \quad \text{Fd/m},$$

$$\bar{C}_o = \bar{C}_{11} + 2\bar{C}_{12} = 2\epsilon_r\epsilon_0 W\left(\frac{2b}{b^2 - S^2} + \frac{1}{S}\right) \quad \text{Fd/m}.$$

The phase velocity on the line is

$$v = 1/\sqrt{\epsilon_r\epsilon_0\mu_0},$$

so the characteristic impedances are

$$Z_{0e} = \frac{1}{v\bar{C}_e} = Z_0\frac{b^2 - S^2}{4bW\sqrt{\epsilon_r}},$$

$$Z_{0o} = \frac{1}{v\bar{C}_o} = Z_0\frac{1}{2W\sqrt{\epsilon_r}[2b/(b^2 - S^2) + 1/S]}.$$

Design of Coupled Line Couplers

With the preceding definitions of the even- and odd-mode characteristic impedances, we can apply an even-odd mode analysis to a length of coupled line to arrive at the design equations for a single-section coupled line coupler. Such a line is shown in Figure 8.31. This four-port network is terminated in the impedance Z_0 at three of its ports, and driven with a voltage generator of 2 V and internal impedance Z_0 at port 1. We will show that a coupler can be designed with arbitrary coupling such that the input (port 1) is matched, while port 4 is isolated. Port 2 is the through port, and port 3 is the coupled port. In Figure 8.31, a ground conductor is understood to be common to both strip conductors.

For this problem we will apply the even-odd mode analysis technique in conjunction with the input impedances of the line, as opposed to the reflection and transmission coefficients of the line. So by superposition, the excitation at port 1 in Figure 8.31 can be treated as the sum of the even- and odd-mode excitations shown in Figure 8.32. From symmetry, we can see that $I_{1e} = I_{3e}, I_{4e} = I_{2e}, V_{1e} = V_{3e}$, and $V_{4e} = V_{2e}$ for the even modes, while $I_{1o} = -I_{3o}, I_{4o} = -I_{2o}, V_{1o} = -V_{3o}$, and $V_{4o} = -V_{2o}$ for the odd mode.

The input impedance at port 1 of the coupler of Figure 8.31 can thus be expressed as

$$Z_{\text{in}} = \frac{V_1}{I_1} = \frac{V_{1e} + V_{1o}}{I_{1e} + I_{1o}}. \tag{8.72}$$

Now if we let Z_{in}^e be the input impedance at port 1 for the even mode, and Z_{in}^o be the input impedance for the odd mode, then we have

$$Z_{\text{in}}^e = Z_{0e}\frac{Z_0 + jZ_{0e}\tan\theta}{Z_{0e} + jZ_0\tan\theta}, \tag{8.73a}$$

$$Z_{\text{in}}^o = Z_{0o}\frac{Z_0 + jZ_{0o}\tan\theta}{Z_{0o} + jZ_0\tan\theta}, \tag{8.73b}$$

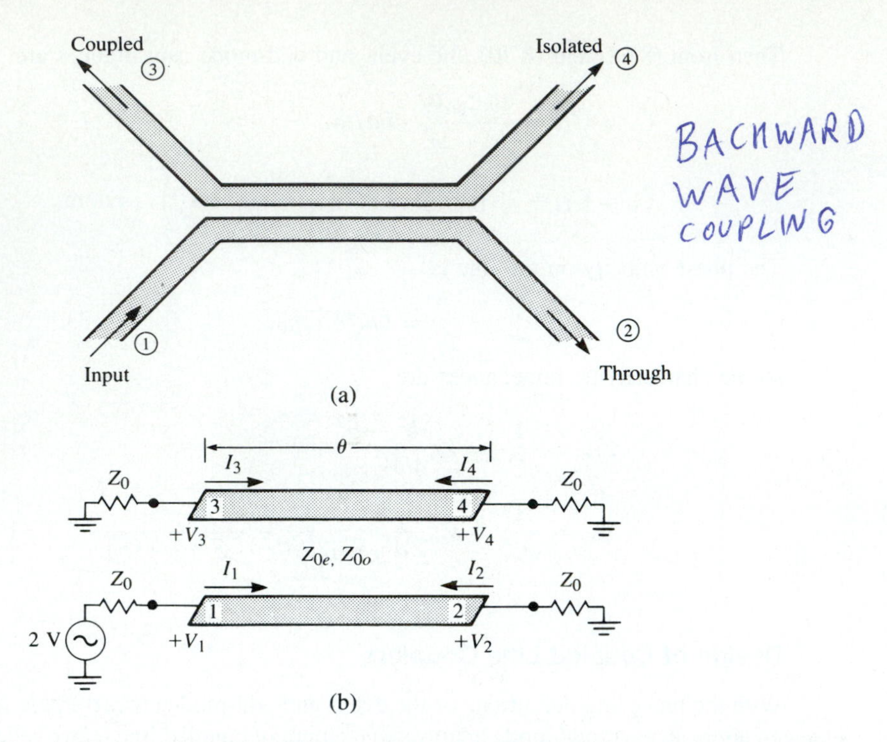

FIGURE 8.31 A single-section coupled line coupler. (a) Geometry and port designations. (b) The schematic circuit.

since, for each mode, the line looks like a transmission line of characteristic impedance Z_{0e} or Z_{0o}, terminated in a load impedance, Z_0. Then by voltage division

$$V_{1o} = V \frac{Z_{\text{in}}^o}{Z_{\text{in}}^o + Z_0}, \tag{8.74a}$$

$$V_{1e} = V \frac{Z_{\text{in}}^e}{Z_{\text{in}}^e + Z_0}, \tag{8.74b}$$

$$I_{1o} = \frac{V}{Z_{\text{in}}^o + Z_0}, \tag{8.75a}$$

$$I_{1e} = \frac{V}{Z_{\text{in}}^e + Z_0}. \tag{8.75b}$$

Using these results in (8.72) yields

$$Z_{\text{in}} = \frac{Z_{\text{in}}^o(Z_{\text{in}}^e + Z_0) + Z_{\text{in}}^e(Z_{\text{in}}^o + Z_0)}{Z_{\text{in}}^e + Z_{\text{in}}^o + 2Z_0} = Z_0 + \frac{2(Z_{\text{in}}^o Z_{\text{in}}^e - Z_o^2)}{Z_{\text{in}}^e + Z_{\text{in}}^o + 2Z_0}. \tag{8.76}$$

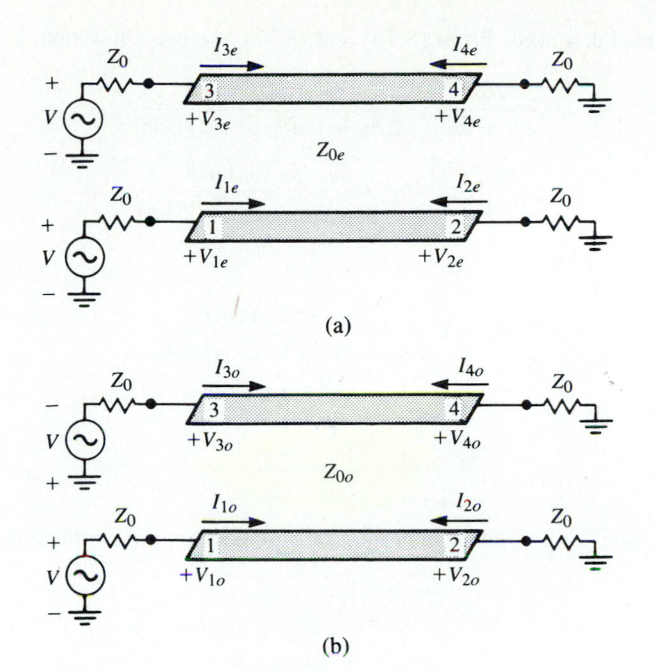

FIGURE 8.32 Decomposition of the coupled line coupler circuit of Figure 8.31 into even- and odd-mode excitations. (a) Even mode. (b) Odd mode.

Now if we let

$$Z_0 = \sqrt{Z_{0e} Z_{0o}}, \tag{8.77}$$

then (8.73a,b) reduce to

$$Z_{in}^e = Z_{0e} \frac{\sqrt{Z_{0o}} + j\sqrt{Z_{0e}} \tan\theta}{\sqrt{Z_{0e}} + j\sqrt{Z_{0o}} \tan\theta},$$

$$Z_{in}^o = Z_{0o} \frac{\sqrt{Z_{0e}} + j\sqrt{Z_{0o}} \tan\theta}{\sqrt{Z_{0o}} + j\sqrt{Z_{0e}} \tan\theta},$$

so that $Z_{in}^e Z_{in}^o = Z_{0e} Z_{0o} = Z_0^2$, and (8.76) reduces to

$$Z_{in} = Z_0. \tag{8.78}$$

Thus, as long as (8.77) is satisfied, port 1 (and, by symmetry, all other ports) will be matched.

Now if (8.77) is satisfied, so that $Z_{in} = Z_0$, we have that $V_1 = V$, by voltage division. The voltage at port 3 is

$$V_3 = V_{3e} + V_{3o} = V_{1e} - V_{1o} = V \left[\frac{Z_{in}^e}{Z_{in}^e + Z_0} - \frac{Z_{in}^o}{Z_{in}^o + Z_0} \right], \tag{8.79}$$

where (8.74) has been used. From (8.73) and (8.77), we can show that

$$\frac{Z_{\text{in}}^e}{Z_{\text{in}}^e + Z_0} = \frac{Z_0 + jZ_{0e}\tan\theta}{2Z_0 + j(Z_{0e} + Z_{0o})\tan\theta},$$

$$\frac{Z_{\text{in}}^o}{Z_{\text{in}}^o + Z_0} = \frac{Z_0 + jZ_{0o}\tan\theta}{2Z_0 + j(Z_{0e} + Z_{0o})\tan\theta},$$

so that (8.79) reduces to

$$V_3 = V\frac{j(Z_{0e} - Z_{0o})\tan\theta}{2Z_0 + j(Z_{0e} + Z_{0o})\tan\theta}. \tag{8.80}$$

Now define C as

$$C = \frac{Z_{0e} - Z_{0o}}{Z_{0e} + Z_{0o}}, \tag{8.81}$$

which we will soon see is actually the midband voltage coupling coefficient, V_3/V. Then,

$$\sqrt{1 - C^2} = \frac{2Z_0}{Z_{0e} + Z_{0o}},$$

so that

$$V_3 = V\frac{jC\tan\theta}{\sqrt{1 - C^2} + j\tan\theta}. \tag{8.82}$$

Similarly, we can show that

$$V_4 = V_{4e} + V_{4o} = V_{2e} - V_{2o} = 0, \tag{8.83}$$

and

$$V_2 = V_{2e} + V_{2o} = V\frac{\sqrt{1 - C^2}}{\sqrt{1 - C^2}\cos\theta + j\sin\theta}. \tag{8.84}$$

Equations (8.82) and (8.84) can be used to plot the coupled and through port voltages versus frequency, as shown in Figure 8.33. At very low frequencies ($\theta << \pi/2$), virtually all power is transmitted through port 2, with none being coupled to port 3. For $\theta = \pi/2$, the coupling to port 3 is at its first maximum; this is where the coupler is generally operated, for small size and minimum line loss. Otherwise, the response is periodic, with maxima in V_3 for $\theta = \pi/2, 3\pi/2, \cdots$.

For $\theta = \pi/2$, the coupler is $\lambda/4$ long, and (8.82) and (8.84) reduce to

$$\frac{V_3}{V} = C, \tag{8.85}$$

$$\frac{V_2}{V} = -j\sqrt{1 - C^2}, \tag{8.86}$$

which shows that $C < 1$ is the voltage coupling factor at the design frequency, $\theta = \pi/2$. Note that these results satisfy power conservation, since $P_{\text{in}} = (1/2)|V|^2/Z_0$, while the output powers are, $P_2 = (1/2)|V_2|^2/Z_0 = (1/2)(1 - C^2)|V|^2/Z_0$, $P_3 = (1/2)|C|^2|V|^2/Z_0$, $P_4 = 0$, so that $P_{\text{in}} = P_2 + P_3 + P_4$. Also observe that there is a 90° phase shift between the two output port voltages; thus this coupler can be used as

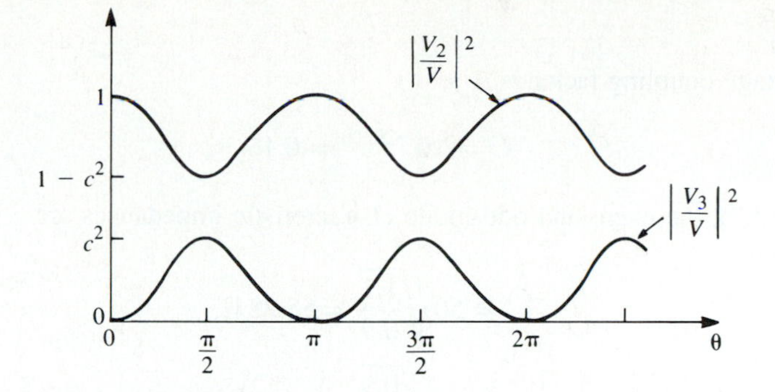

FIGURE 8.33 Coupled and through port voltages (squared) versus frequency for the coupled line coupler of Figure 8.31.

a quadrature hybrid. And, as long as (8.77) is satisfied, the coupler will be matched at the input and have perfect isolation, at any frequency.

Finally, if the characteristic impedance, Z_0, and the voltage coupling coefficient, C, are specified, then the following design equations for the required even- and odd-mode characteristic impedances can be easily derived from (8.77) and (8.81):

$$Z_{0e} = Z_0\sqrt{\frac{1+C}{1-C}}, \qquad 8.87a$$

$$Z_{0o} = Z_0\sqrt{\frac{1-C}{1+C}}. \qquad 8.87b$$

In the above analysis, it was assumed that the even and odd modes of the coupled line structure have the same velocities of propagation, so that the line has the same electrical length for both modes. For a coupled microstrip, or other non-TEM, line this condition will generally not be satisfied, and the coupler will have poor directivity. The fact that coupled microstrip lines have unequal even- and odd-mode phase velocities can be intuitively explained by considering the field line plots of Figure 8.28, which show that the even mode has less fringing field in the air region than the odd mode. Thus its effective dielectric constant should be higher, indicating a smaller phase velocity for the even mode. Techniques for compensating coupled microstrip lines to achieve equal even- and odd-mode phase velocities include the use of dielectric overlays and anistropic substrates.

This type of coupler is best suited for weak couplings, as tight coupling requires lines that are too close together to be practical, or a combination of even- and odd-mode characteristic impedances that is nonrealizable.

EXAMPLE 8.7

Design a 20 dB single-section coupled line coupler in stripline with a 0.158 cm ground plane spacing, dielectric constant of 2.56, a characteristic impedance of 50 Ω, and a center frequency of 3 GHz. Plot the coupling and directivity from 1 to 5 GHz.

Solution

The voltage coupling factor is

$$C = 10^{-20/20} = 0.1.$$

From (8.87), the even- and odd-mode characteristic impedances are

$$Z_{0e} = 50\sqrt{\frac{1.1}{0.9}} = 55.28\,\Omega,$$

$$Z_{0o} = 50\sqrt{\frac{0.9}{1.1}} = 45.23\,\Omega.$$

To use Figure 8.29, we have that

$$\sqrt{\epsilon_r}Z_{0e} = 88.4,$$

$$\sqrt{\epsilon_r}Z_{0o} = 72.4,$$

and so, $W/b = 0.72, S/b = 0.34$. This gives a conductor width of

$$W = 0.72b = 0.114\,\text{cm},$$

and a conductor separation of

$$S = 0.34b = 0.054\,\text{cm}.$$

Note that these lines are quite close together, which may make fabrication difficult.

The coupling and directivity are plotted in Figure 8.34. ○

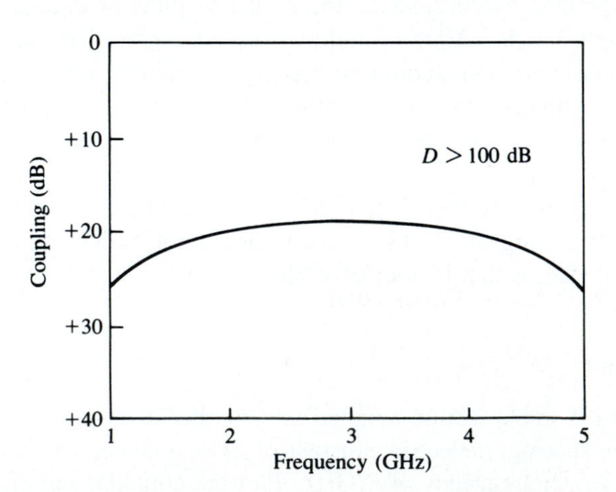

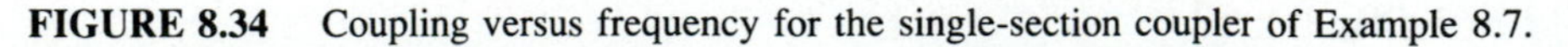

FIGURE 8.34 Coupling versus frequency for the single-section coupler of Example 8.7.

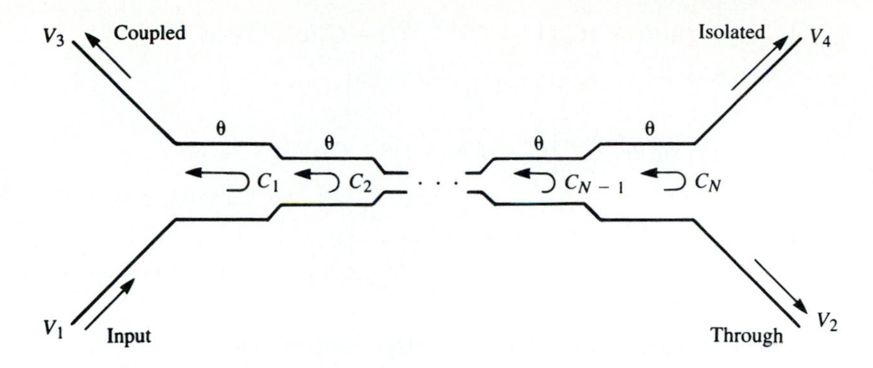

FIGURE 8.35 An N-section coupled line coupler.

Design of Multisection Coupled Line Couplers

As Figure 8.33 shows, the coupling of a single-section coupled line coupler is limited in bandwidth due to the $\lambda/4$ length requirement. As in the case of matching transformers and waveguide couplers, bandwidth can be increased by using multiple sections. As a matter of fact, there is a very close relation between multisection coupled line couplers and multisection quarter-wave transformers [9].

Because the phase characteristics are usually better, multisection coupled line couplers are generally made with an odd number of sections, as shown in Figure 8.35. Thus, we will assume that N is odd. We will also assume that the coupling is weak ($C \leq 10$ dB), and that each section is $\lambda/4$ long ($\theta = \pi/2$) at the center frequency.

Now for a single coupled line section, with $C << 1$, (8.82) and (8.84) simplify to

$$\frac{V_3}{V_1} = \frac{jC\tan\theta}{\sqrt{1-C^2}+j\tan\theta} \simeq \frac{jC\tan\theta}{1+j\tan\theta} = jC\sin\theta e^{-j\theta}, \tag{8.88a}$$

$$\frac{V_2}{V_1} = \frac{\sqrt{1-C^2}}{\sqrt{1-C^2}\cos\theta + j\sin\theta} \simeq e^{-j\theta}. \tag{8.88b}$$

Then for $\theta = \pi/2$, we have that $V_3/V_1 = C$ and $V_2/V_1 = -j$. This approximation is equivalent to assuming that no power is lost on the through path from one section to the next, and is similar to the multisection waveguide coupler analysis. It is a good assumption for small C, even though power conservation is violated.

Using these results, the total voltage at the coupled port (port 3) of the cascaded coupler in Figure 8.35 can be expressed as

$$\begin{aligned} V_3 &= (jC_1\sin\theta e^{-j\theta})V_1 + (jC_2\sin\theta e^{-j\theta})V_1 e^{-2j\theta} \\ &\quad + \cdots + (jC_N\sin\theta e^{-j\theta})V_1 e^{-2j(N-1)\theta}, \end{aligned} \tag{8.89}$$

where C_n is the voltage coupling coefficient of the nth section. If we assume that the coupler is symmetric, so that $C_1 = C_N, C_2 = C_{N-1}$, etc., (8.89) can be simplified to

$$\begin{aligned} V_3 &= jV_1 \sin\theta e^{-j\theta}[C_1(1+e^{-2j(N-1)\theta}) + C_2(e^{-2j\theta} + e^{-2j(N-2)\theta}) \\ &\qquad + \cdots + C_M e^{-j(N-1)\theta}] \\ &= 2jV_1 \sin\theta e^{-jN\theta}\left[C_1 \cos(N-1)\theta + C_2 \cos(N-3)\theta \right. \\ &\qquad \left. + \cdots + \frac{1}{2}C_M\right], \end{aligned} \tag{8.90}$$

where $M = (N+1)/2$.

At the center frequency, we define the voltage coupling factor C_0:

$$C_0 = \left|\frac{V_3}{V_1}\right|_{\theta=\pi/2}. \tag{8.91}$$

Equation (8.90) is in the form of a Fourier series for the coupling, as a function of frequency. Thus, we can synthesize a desired coupling response by choosing the coupling coefficients, C_n. Note that in this case, we synthesize the coupling response, while in the case of the multihole waveguide coupler we synthesized the directivity response. This is because the path for the uncoupled arm of the multisection coupled line coupler is in the forward direction, and so is less dependent on frequency than the coupled arm path, which is in the reverse direction; this is the opposite situation from the multihole waveguide coupler.

Multisection couplers of this form can achieve decade bandwidths, but coupling levels must be low. Because of the longer electrical length, it is more critical to have equal even- and odd-mode phase velocities than it is for the single-section coupler. This usually means that stripline is the preferred medium for such couplers. Mismatched phase velocities will degrade the coupler directivity, as will junction discontinuities, load mismatches, and fabrication tolerances. A photograph of a three-section coupled line coupler is shown in Figure 8.36.

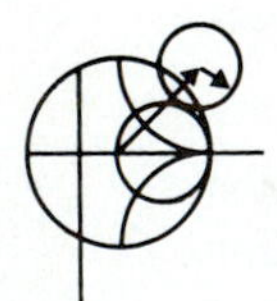

EXAMPLE 8.8

Design a three-section 20 dB coupler with a binomial (maximally flat) response, a system impedance of 50 Ω, and a center frequency of 3 GHz. Plot the coupling and directivity from 1 to 5 GHz.

Solution

For a maximally flat response for a three-section ($N = 3$) coupler, we require that

$$\left.\frac{d^n}{d\theta^n}C(\theta)\right|_{\theta=\pi/2} = 0, \qquad \text{for } n = 1, 2.$$

From (8.90),

FIGURE 8.36 Photograph of a three-section coupled line coupler in microstrip form. Courtesy of Harlan Howe, Jr., M/A-COM Inc.

$$C = \left|\frac{V_3}{V_1}\right| = 2\sin\theta\left[C_1\cos 2\theta + \frac{1}{2}C_2\right]$$
$$= C_1(\sin 3\theta - \sin\theta) + C_2\sin\theta$$
$$= C_1\sin 3\theta + (C_2 - C_1)\sin\theta,$$

so
$$\frac{dC}{d\theta} = [3C_1\cos 3\theta + (C_2 - C_1)\cos\theta]\Big|_{\pi/2} = 0,$$
$$\frac{d^2C}{d\theta^2} = [-9C_1\sin 3\theta - (C_2 - C_1)\sin\theta]\Big|_{\pi/2} = 10C_1 - C_2 = 0.$$

Now at midband, $\theta = \pi/2$ and $C_0 = 20$ db. Thus, $C = 10^{-20/20} = 0.1 = C_2 - 2C_1$. Solving these two equations for C_1 and C_2 gives

$$C_1 = C_3 = 0.0125,$$
$$C_2 = 0.125.$$

Then from (8.87) the even- and odd-mode characteristic impedances for each section are

$$Z_{0e}^1 = Z_{0e}^3 = 50\sqrt{\frac{1.0125}{0.9875}} = 50.63\,\Omega,$$
$$Z_{0o}^1 = Z_{0o}^3 = 50\sqrt{\frac{0.9875}{1.0125}} = 49.38\,\Omega,$$

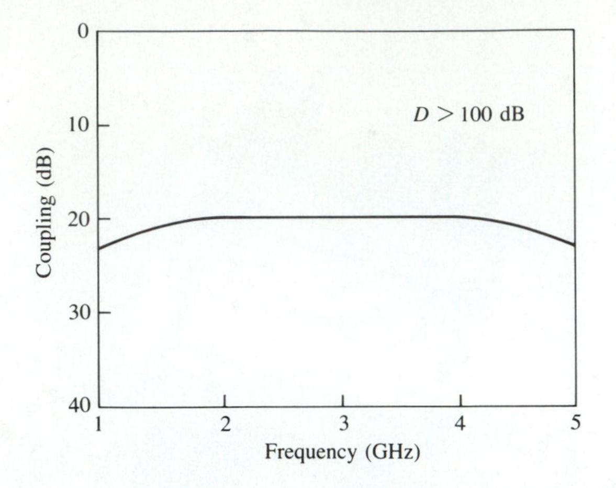

FIGURE 8.37 Coupling versus frequency for the three-section binomial coupler of Example 8.8.

$$Z_{0e}^2 = 50\sqrt{\frac{1.125}{0.875}} = 56.69\,\Omega,$$

$$Z_{0o}^2 = 50\sqrt{\frac{0.875}{1.125}} = 44.10\,\Omega.$$

The coupling and directivity for this coupler are plotted in Figure 8.37. ○

8.7 THE LANGE COUPLER

Generally, the coupling in a coupled line coupler is too loose to accommodate coupling factors of 3 dB or 6 dB. One way to increase the coupling between edge-coupled lines is to use several lines parallel to each other, so that the fringing fields at both edges of a line contribute to the coupling. Probably the most practical implementation of this idea is the Lange coupler [10], shown in Figure 8.38a. Here, four coupled lines are used with interconnections to provide tight coupling. This coupler can easily achieve 3 dB coupling ratios, with an octave or more bandwidth. The design tends to compensate for unequal even- and odd-mode phase velocities, which also improves the bandwidth. There is a 90° phase difference between the output lines (ports 2 and 3), so the Lange coupler is a type of quadrature coupler. The main disadvantage of the Lange coupler is probably practical, as the lines are very narrow, close together, and it is difficult to fabricate the necessary bonding wires across the lines. This type of coupled line geometry is also referred to as interdigitated; such structures can also be used for filter circuits.

The *unfolded* Lange coupler [11], shown in Figure 8.38b, operates essentially the same as the original Lange coupler, but is easier to model with an equivalent circuit. Such an equivalent circuit consists of a four-wire coupled line structure, as shown in Figure 8.39a. All the lines have the same width and spacing. If we make the reasonable

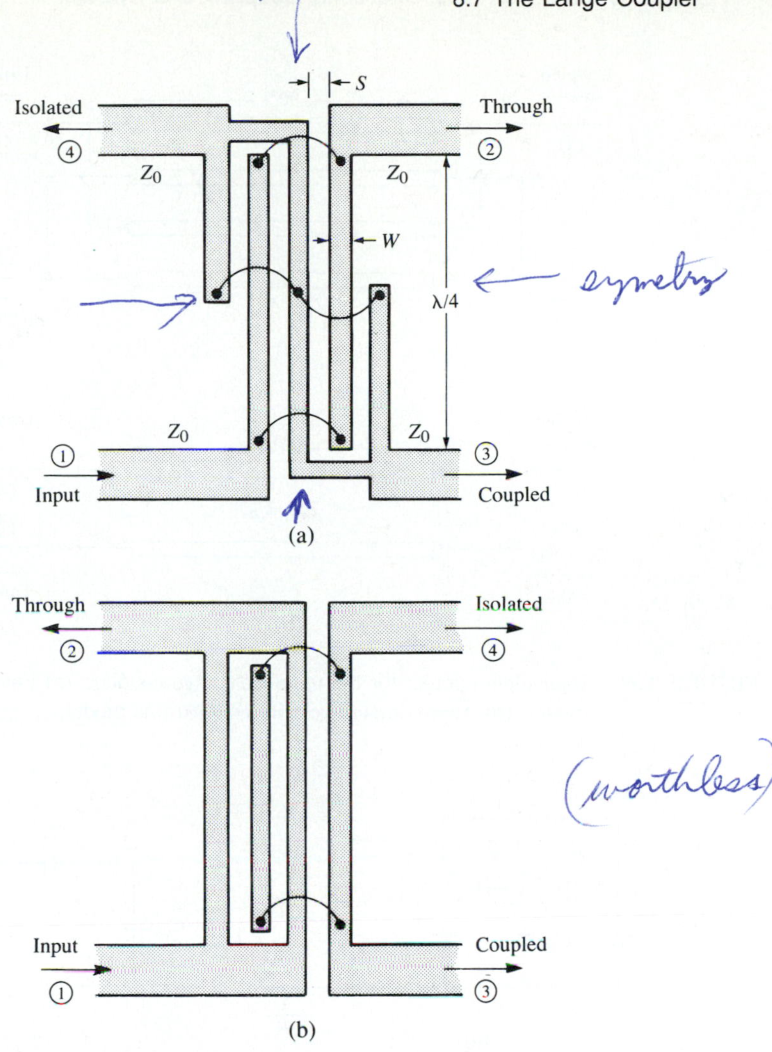

FIGURE 8.38 The Lange coupler. (a) Layout in microstrip form. (b) The unfolded Lange coupler.

assumption that each line couples only to its nearest neighbor, and ignore more distant couplings, then we effectively have a two-wire coupled line circuit, as shown in Figure 8.39b. Then, if we can derive the even- and odd-mode characteristic impedances, Z_{e4} and Z_{o4}, of the four-wire circuit of Figure 8.39a in terms of Z_{0e} and Z_{0o}, the even- and odd-mode characteristic impedances of any adjacent pair of lines, we can apply the coupled line coupler results of Section 8.6 to analyze the Lange coupler.

Figure 8.40a shows the effective capacitances between the conductors of the four-wire coupled line of Figure 8.39a. Unlike the two-line case of Section 8.6, the capacitances of the four lines to ground are different depending on whether the line is on the outside (1 and 4), or on the inside (2 and 3). An approximate relation between these

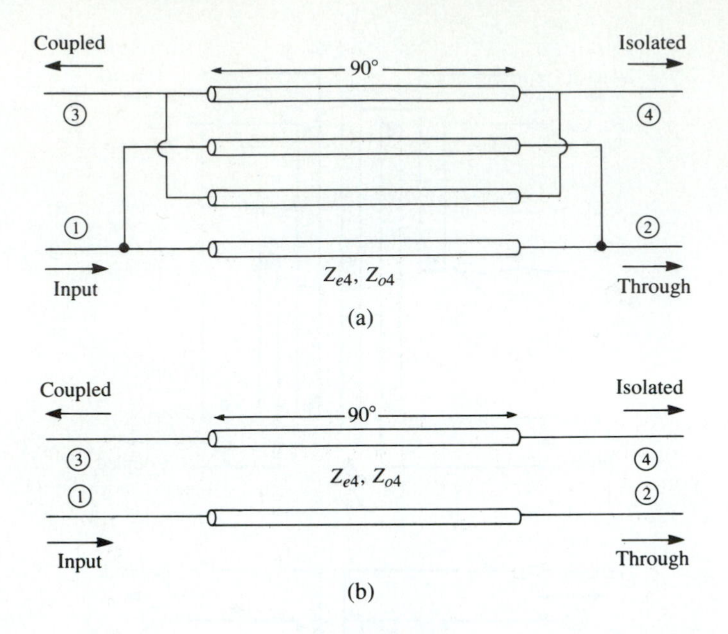

FIGURE 8.39 Equivalent circuits for the unfolded Lange coupler. (a) Four-wire coupled line model. (b) Approximate two-wire coupled line model.

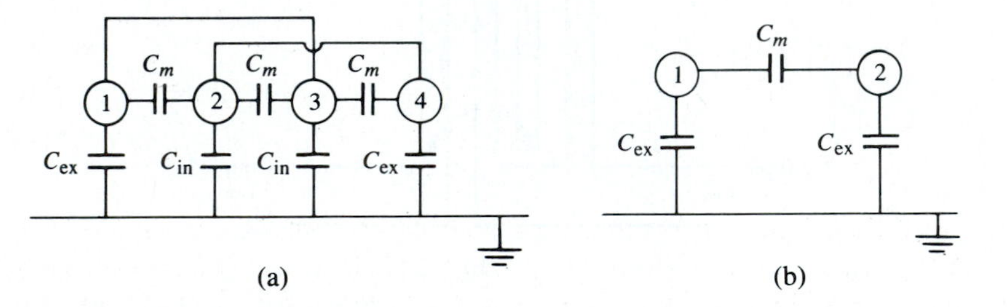

(a) (b)

FIGURE 8.40 Effective capacitance networks for the unfolded Lange coupler equivalent circuits of Figure 8.39. (a) Effective capacitance for the four-wire model. (b) Effective capacitance for the two-wire model.

capacitances is given as [12]

$$C_{\text{in}} = C_{\text{ex}} - \frac{C_{\text{ex}} C_m}{C_{\text{ex}} + C_m}. \qquad 8.92$$

For an even-mode excitation, all four conductors in Figure 8.40a are at the same potential, so C_m has no effect and the total capacitance of any line to ground is

$$C_{e4} = C_{\text{ex}} + C_{\text{in}}. \qquad 8.93a$$

For an odd-mode excitation, electric walls effectively exist through the middle of each C_m, so the capacitance of any line to ground is

$$C_{o4} = C_{\text{ex}} + C_{\text{in}} + 6C_m. \tag{8.93b}$$

The even- and odd-mode characteristic impedances are then

$$Z_{e4} = \frac{1}{vC_{e4}}, \tag{8.94a}$$

$$Z_{o4} = \frac{1}{vC_{o4}}, \tag{8.94b}$$

where v is the velocity of propagation on the line.

Now consider any isolated pair of adjacent conductors in the four-line model; the effective capacitances are as shown in Figure 8.40b. The even- and odd-mode capacitances are

$$C_e = C_{\text{ex}}, \tag{8.95a}$$

$$C_o = C_{\text{ex}} + 2C_m. \tag{8.95b}$$

Solving (8.95) for C_{ex} and C_m, and substituting into (8.93) with the aid of (8.92) gives the even-odd mode capacitances of the four-wire line in terms of a two-wire coupled line:

$$C_{e4} = \frac{C_e(3C_e + C_o)}{C_e + C_o}, \tag{8.96a}$$

$$C_{o4} = \frac{C_o(3C_o + C_e)}{C_e + C_o}. \tag{8.96b}$$

Since characteristic impedances are related to capacitance as $Z_0 = 1/vC$, we can rewrite (8.96) to give the even/odd mode characteristic impedances of the Lange coupler in terms of the characteristic impedances of a two-conductor line which is identical to any pair of adjacent lines in the coupler:

$$Z_{e4} = \frac{Z_{0o} + Z_{0e}}{3Z_{0o} + Z_{0e}} Z_{0e}, \tag{8.97a}$$

$$Z_{o4} = \frac{Z_{0o} + Z_{0e}}{3Z_{0e} + Z_{0o}} Z_{0o}, \tag{8.97b}$$

where Z_{0e}, Z_{0o} are the even- and odd-mode characteristic impedances of the two-conductor pair.

Now we can apply the results of Section 8.6 to the coupler of Figure 8.39b. From (8.77) the characteristic impedance is

$$Z_0 = \sqrt{Z_{e4}Z_{o4}} = \sqrt{\frac{Z_{0e}Z_{0o}(Z_{0o} + Z_{0e})^2}{(3Z_{0o} + Z_{0e})(3Z_{0e} + Z_{0o})}}, \tag{8.98}$$

while the voltage coupling coefficient is, from (8.81),

$$C = \frac{Z_{e4} - Z_{o4}}{Z_{e4} + Z_{o4}} = \frac{3(Z_{0e}^2 - Z_{0o}^2)}{3(Z_{0e}^2 + Z_{0o}^2) + 2Z_{0e}Z_{0o}}, \qquad 8.99$$

where (8.97) was used. For design purposes, it is useful to invert these results to give the necessary even- and odd-mode impedances for a desired characteristic impedance and coupling coefficient:

$$Z_{0e} = \frac{4C - 3 + \sqrt{9 - 8C^2}}{2C\sqrt{(1-C)/(1+C)}} Z_0, \qquad 8.100a$$

$$Z_{0o} = \frac{4C + 3 - \sqrt{9 - 8C^2}}{2C\sqrt{(1+C)/(1-C)}} Z_0. \qquad 8.100b$$

These results are approximate because of the simplifications involved with the application of two-line characteristic impedances to the four-line circuit, and because of the assumption of equal even- and odd-mode phase velocities. In practice, however, these results generally give sufficient acuracy. If necessary, a more complete analysis can be made to directly determine Z_{e4} and Z_{o4} for the four-line circuit, as in reference [13].

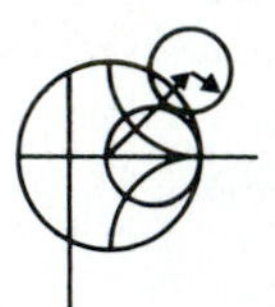

EXAMPLE 8.9

Design a 3 dB 50 Ω Lange coupler for operation at 5 GHz. If the coupler is to be fabricated in microstrip on an alumina substrate with $\epsilon_r = 10$ and $d = 1.0$ mm, compute Z_{0e} and Z_{0o} for two adjacent lines, and find the necessary spacing and widths of the lines.

Solution
For a 3 dB coupler the voltage coupling coefficient is

$$C = 10^{-3/20} = 0.707,$$

so from (8.100) the even- and odd-mode characteristic impedances of a pair of adjacent coupled lines is

$$Z_{0e} = \frac{4C - 3 + \sqrt{9 - 8C^2}}{2C\sqrt{(1-C)/(1+C)}} Z_0 = 176.2\,\Omega,$$

$$Z_{0o} = \frac{4C + 3 - \sqrt{9 - 8C^2}}{2C\sqrt{(1+C)/(1-C)}} Z_0 = 52.6\,\Omega.$$

From Figure 8.30, the line width and spacings are $W = 0.07$ mm and $S = 0.075$ mm. The length of the coupled line sections is $\lambda/4$ at 5 GHz, or about 5.0 mm. ○

8.8 THE 180° HYBRID

The 180° hybrid junction is a four-port network with a 180° phase shift between the two output ports. It can also be operated so that the outputs are in phase. With reference to the 180° hybrid symbol shown in Figure 8.41, a signal applied to port 1 will be evenly split into two in-phase components at ports 2 and 3, and port 4 will be isolated. If the input is applied to port 4, it will be equally split into two components with a 180° phase difference at ports 2 and 3, and port 1 will be isolated. When operated as a combiner, with input signals applied at ports 2 and 3, the sum of the inputs will be formed at port 1, while the difference will be formed at port 4. Hence, ports 1 and 4 are referred to as the sum and difference ports, respectively. The scattering matrix for the ideal 3 dB 180° hybrid thus has the following form:

$$[S] = \frac{-j}{\sqrt{2}} \begin{bmatrix} 0 & 1 & 1 & 0 \\ 1 & 0 & 0 & -1 \\ 1 & 0 & 0 & 1 \\ 0 & -1 & 1 & 0 \end{bmatrix}. \tag{8.101}$$

The reader may verify that this matrix is unitary and symmetric.

The 180° hybrid can be fabricated in several forms. The ring hybrid, or rat-race, shown in Figures 8.42 and 8.43a, can easily be constructed in planar (microstrip or stripline) form, although waveguide versions are also possible. Another type of planar 180° hybrid uses tapered matching lines and coupled lines, as shown in Figure 8.43b. Yet another type of hybrid is the hybrid waveguide junction, or magic-T, shown in Figure 8.43c. We will first analyze the ring hybrid, using an even-odd mode analysis similar to that used for the branch-line hybrid, and use a similar technique for the analysis of the tapered line hyrid. Then we will qualitatively discuss the operation of the waveguide magic-T.

Even-Odd Mode Analysis of the Ring Hybrid

First consider a unit amplitude wave incident at port 1 (the sum port) of the ring hybrid of Figure 8.43a. At the ring junction this wave will divide into two components, which both arrive in phase at ports 2 and 3, and 180° out of phase at port 4. Using the even-odd mode analysis technique [5], we can decompose this case into a superposition of the two simpler circuits and excitations shown in Figure 8.44. Then the amplitudes

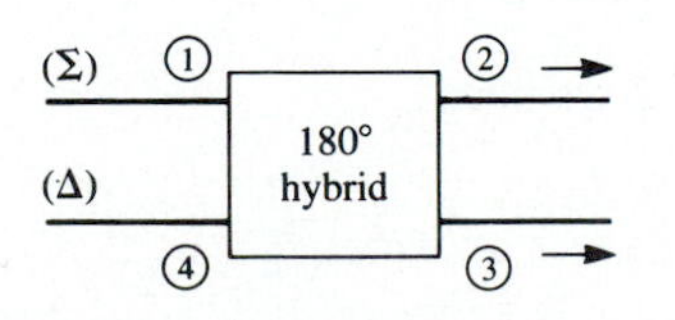

FIGURE 8.41 Symbol for a 180° hybrid junction.

FIGURE 8.42 Photograph of a microstrip ring hybrid.
Courtesy of M. D. Abouzahra, MIT Lincoln Laboratory, Lexington, Mass.

of the scattered waves from the ring hybrid will be

$$B_1 = \frac{1}{2}\Gamma_e + \frac{1}{2}\Gamma_o, \tag{8.102a}$$

$$B_2 = \frac{1}{2}T_e + \frac{1}{2}T_o, \tag{8.102b}$$

$$B_3 = \frac{1}{2}\Gamma_e - \frac{1}{2}\Gamma_o, \tag{8.102c}$$

$$B_4 = \frac{1}{2}T_e - \frac{1}{2}T_o. \tag{8.102d}$$

We can evaluate the required reflection and transmission coefficients defined in Figure 8.44 using the $ABCD$ matrix for the even- and odd-mode two-port circuits in Figure 8.44. The results are

$$\begin{bmatrix} A & B \\ C & D \end{bmatrix}_e = \begin{bmatrix} 1 & j\sqrt{2} \\ j\sqrt{2} & -1 \end{bmatrix}, \tag{8.103a}$$

$$\begin{bmatrix} A & B \\ C & D \end{bmatrix}_o = \begin{bmatrix} -1 & j\sqrt{2} \\ j\sqrt{2} & 1 \end{bmatrix}. \tag{8.103b}$$

Then with the aid of Table 5.2 we have

$$\Gamma_e = \frac{-j}{\sqrt{2}}, \tag{8.104a}$$

$$T_e = \frac{-j}{\sqrt{2}}, \tag{8.104b}$$

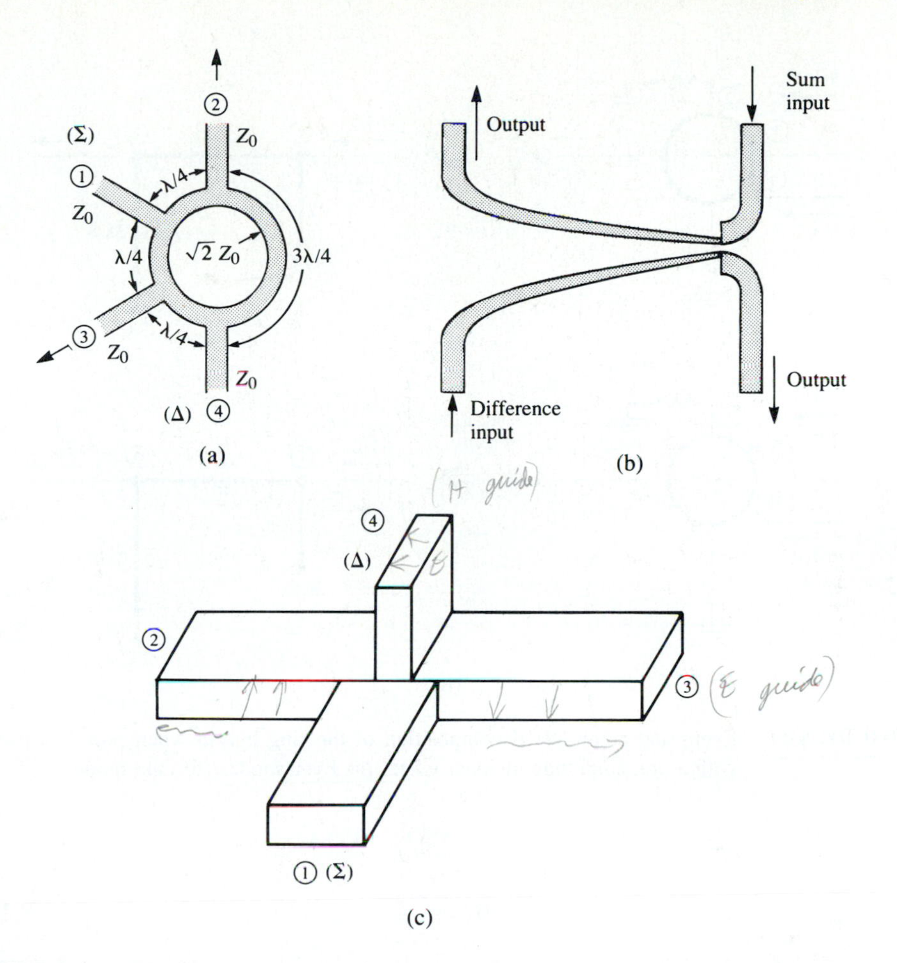

FIGURE 8.43 Hybrid junctions. (a) A ring hybrid, or *rat-race*, in microstrip or stripline form. (b) A tapered coupled line hybrid. (c) A waveguide hybrid junction, or *magic-T*.

$$\Gamma_o = \frac{j}{\sqrt{2}}, \qquad 8.104c$$

$$T_o = \frac{-j}{\sqrt{2}}. \qquad 8.104d$$

Using these results in (8.102) gives

$$B_1 = 0, \qquad 8.105a$$

$$B_2 = \frac{-j}{\sqrt{2}}, \qquad 8.105b$$

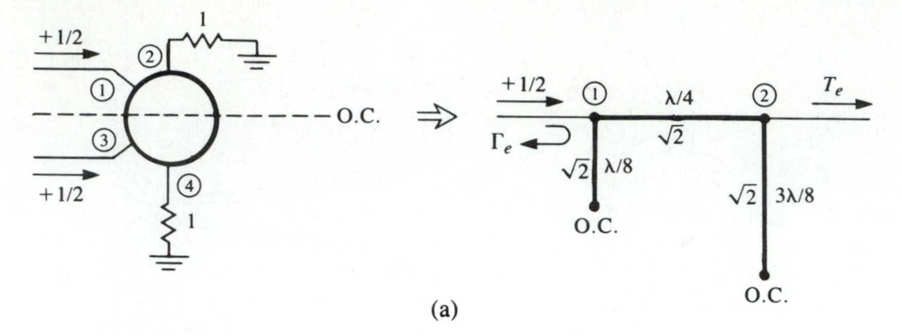

(a)

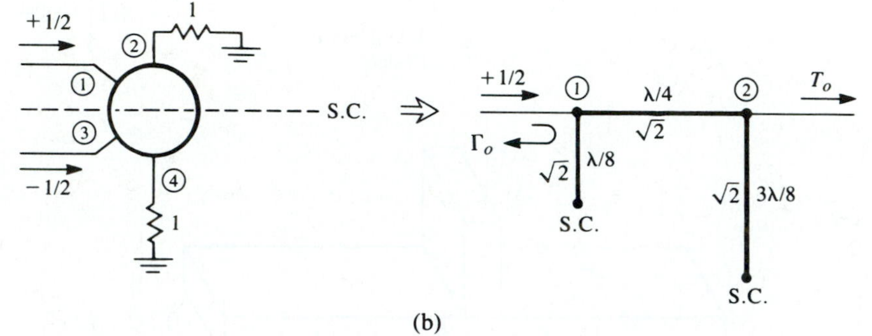

(b)

FIGURE 8.44 Even- and odd-mode decomposition of the ring hybrid when port 1 is excited with a unit amplitude incident wave. (a) Even mode. (b) Odd mode.

$$B_3 = \frac{-j}{\sqrt{2}}, \qquad 8.105c$$

$$B_4 = 0, \qquad 8.105d$$

which shows that the input port is matched, port 4 is isolated, and the input power is evenly divided in phase between ports 2 and 3. These results form the first row and column of the scattering matrix in (8.101).

Now consider a unit amplitude wave incident at port 4 (the difference port) of the ring hybrid of Figure 8.43a. The two wave components on the ring will arrive in phase at ports 2 and 3, with a net phase difference of 180° between these ports. The two wave components will be 180° out of phase at port 1. This case can be decomposed into a superposition of the two simpler circuits and excitations shown in Figure 8.45. Then the amplitudes of the scattered waves will be

$$B_1 = \frac{1}{2}T_e - \frac{1}{2}T_o, \qquad 8.106a$$

$$B_2 = \frac{1}{2}\Gamma_e - \frac{1}{2}\Gamma_o, \qquad 8.106b$$

$$B_3 = \frac{1}{2}T_e + \frac{1}{2}T_o, \qquad 8.106c$$

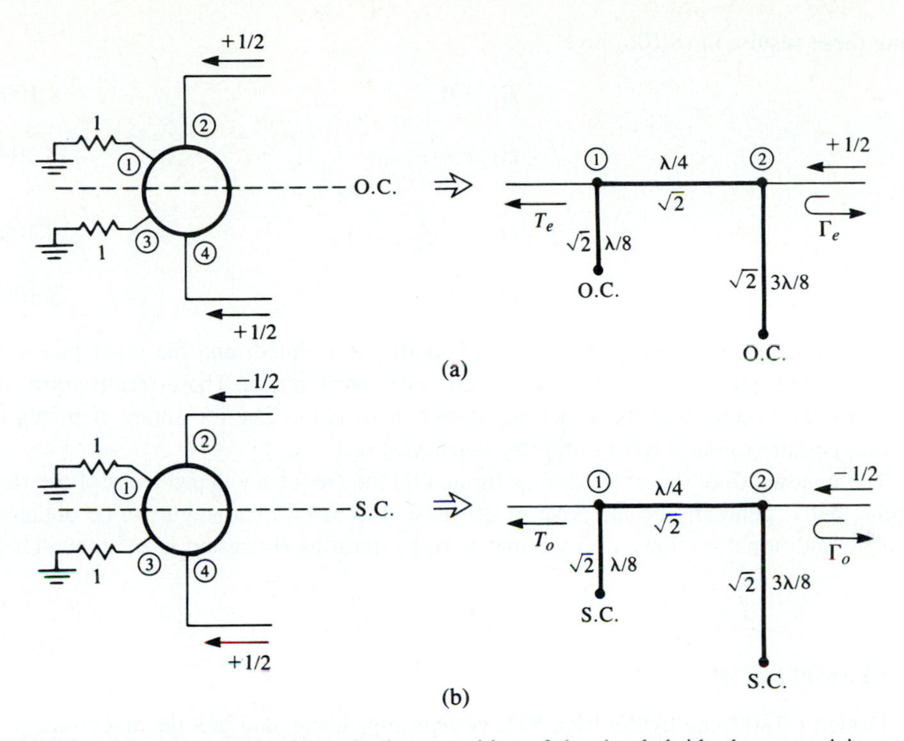

FIGURE 8.45 Even- and odd-mode decomposition of the ring hybrid when port 4 is excited with a unit amplitude incident wave. (a) Even mode. (b) Odd mode.

$$B_4 = \frac{1}{2}\Gamma_e + \frac{1}{2}\Gamma_o. \tag{8.106d}$$

The $ABCD$ matrices for the even- and odd-mode circuits of Figure 8.45 are

$$\begin{bmatrix} A & B \\ C & D \end{bmatrix}_e = \begin{bmatrix} -1 & j\sqrt{2} \\ j\sqrt{2} & 1 \end{bmatrix}, \tag{8.107a}$$

$$\begin{bmatrix} A & B \\ C & D \end{bmatrix}_o = \begin{bmatrix} 1 & j\sqrt{2} \\ j\sqrt{2} & 1 \end{bmatrix}. \tag{8.107b}$$

Then from Table 5.2, the necessary reflection and transmission coefficients are

$$\Gamma_e = \frac{j}{\sqrt{2}}, \tag{8.108a}$$

$$T_e = \frac{-j}{\sqrt{2}}, \tag{8.108b}$$

$$\Gamma_o = \frac{-j}{\sqrt{2}}, \tag{8.108c}$$

$$T_o = \frac{-j}{\sqrt{2}}. \tag{8.108d}$$

Using these results in (8.106) gives

$$B_1 = 0, \tag{8.109a}$$

$$B_2 = \frac{j}{\sqrt{2}}, \tag{8.109b}$$

$$B_3 = \frac{-j}{\sqrt{2}}, \tag{8.109c}$$

$$B_4 = 0, \tag{8.109d}$$

which shows that the input port is matched, port 1 is isolated, and the input power is evenly divided into ports 2 and 3 with a 180° phase difference. These results form the fourth row and column of the scattering matrix of (8.101). The remaining elements in this matrix can be found from symmetry considerations.

The bandwidth of the ring hybrid is limited by the frequency dependence of the ring lengths, but is generally on the order of 20–30%. Increased bandwidth can be obtained by using additional sections, or a symmetric ring circuit as suggested in reference [14].

EXAMPLE 8.10

Design a 180° ring hybrid for a 50 Ω system impedance, and plot the magnitude of the S parameters (S_{1j}) from 0.5 f_0 to 1.5 f_0, where f_0 is the design frequency.

Solution

With reference to Figure 8.43a, the characteristic impedance of the ring transmission line is

$$\sqrt{2}Z_0 = 70.7\,\Omega,$$

while the feed line impedances are 50 Ω. The S parameter magnitudes are plotted versus frequency in Figure 8.46. ○

Even-Odd Mode Analysis of the Tapered Coupled Line Hybrid

The tapered coupled line 180° hybrid [15], shown in Figure 8.43b, can provide any power division ratio with a bandwidth of a decade or more. This hybrid is also referred to as an asymmetric tapered coupled line coupler.

The schematic circuit of this coupler is shown in Figure 8.47; the ports have been numbered to correspond functionally to the ports of the 180° hybrids in Figures 8.41 and 8.43. The coupler consists of two coupled lines with tapering characteristic impedances over the length $0 < z < L$. At $z = 0$ the lines are very weakly coupled so that $Z_{0e}(0) = Z_{0o}(0) = Z_0$, while at $z = L$ the coupling is such that $Z_{0e}(L) = Z_0/k$ and $Z_{0o}(L) = kZ_0$, where $0 \leq k \leq 1$ is a coupling factor which we will relate to the voltage coupling factor. The even mode of the coupled line thus matches a load impedance of Z_0/k (at $z = L$) to Z_0, while the odd mode matches a load of kZ_0 to Z_0; note that $Z_{0e}(z)Z_{0o}(z) = Z_0^2$ for all z. The Klopfenstein taper is generally used for

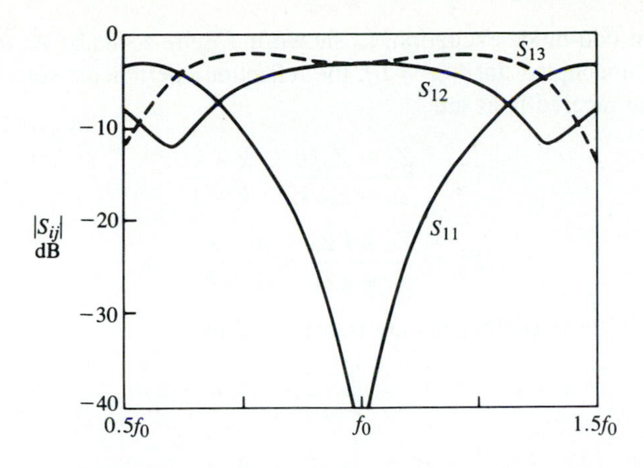

FIGURE 8.46 S parameter magnitudes versus frequency for the ring hybrid of Example 8.10.

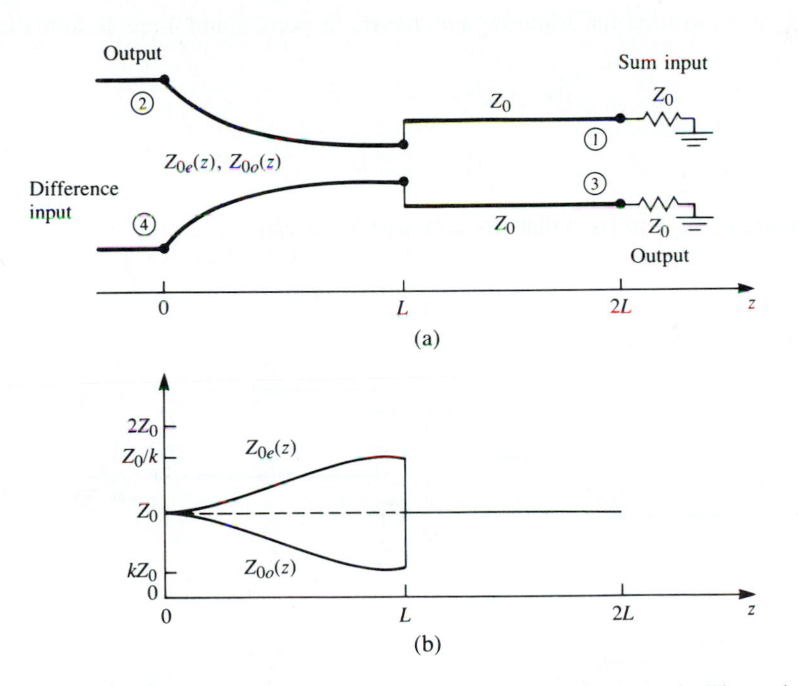

FIGURE 8.47 (a) Schematic diagram of the tapered coupled line hybrid. (b) The variation of characteristic impedances.

these tapered matching lines. For $L < z < 2L$, the lines are uncoupled, and both have a characteristic impedance Z_0; these lines are required for phase compensation of the coupled line section. The length of each section, $\theta = \beta L$, must be the same, and should be electrically long to provide a good impedance match over the desired bandwidth.

First consider an incident voltage wave of amplitude V applied to port 4, the difference input. This excitation can be reduced to the superposition of an even-mode excitation and an odd-mode excitation, as shown in Figure 8.48a,b. At the junctions of the coupled and uncoupled lines ($z = L$), the reflection coefficients seen by the even or odd modes of the tapered lines are

$$\Gamma'_e = \frac{Z_0 - Z_0/k}{Z_0 + Z_0/k} = \frac{k-1}{k+1}, \tag{8.110a}$$

$$\Gamma'_o = \frac{Z_0 - kZ_0}{Z_0 + kZ_0} = \frac{1-k}{1+k}. \tag{8.110b}$$

Then at $z = 0$ these coefficients are transformed to

$$\Gamma_e = \frac{k-1}{k+1} e^{-2j\theta}, \tag{8.111a}$$

$$\Gamma_o = \frac{1-k}{1+k} e^{-2j\theta}. \tag{8.111b}$$

Then by superposition the scattering parameters of ports 2 and 4 are as follows:

$$S_{44} = \frac{1}{2}(\Gamma_e + \Gamma_o) = 0, \tag{8.112a}$$

$$S_{24} = \frac{1}{2}(\Gamma_e - \Gamma_o) = \frac{k-1}{k+1} e^{-2j\theta}. \tag{8.112b}$$

By symmetry, we also have that $S_{22} = 0$, and $S_{42} = S_{24}$.

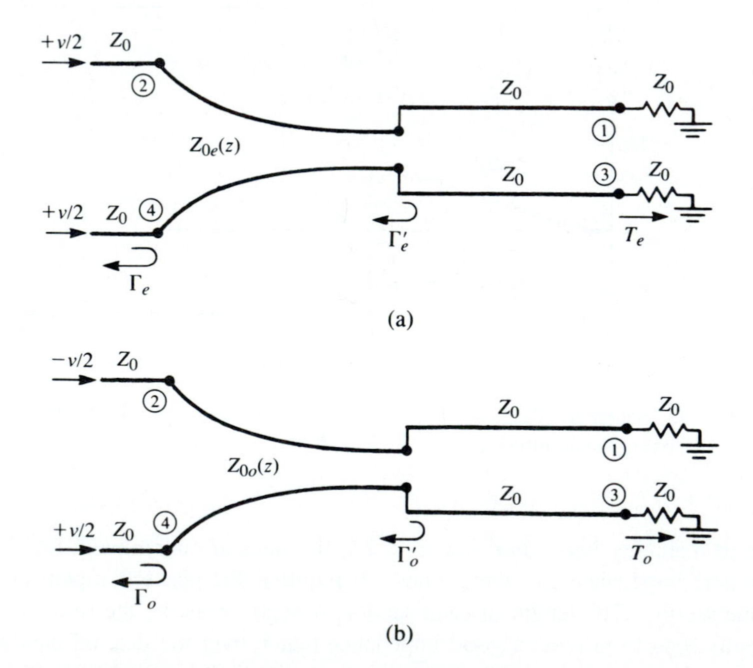

FIGURE 8.48 Excitation of the tapered coupled line hybrid. (a) Even-mode excitation. (b) Odd-mode excitation.

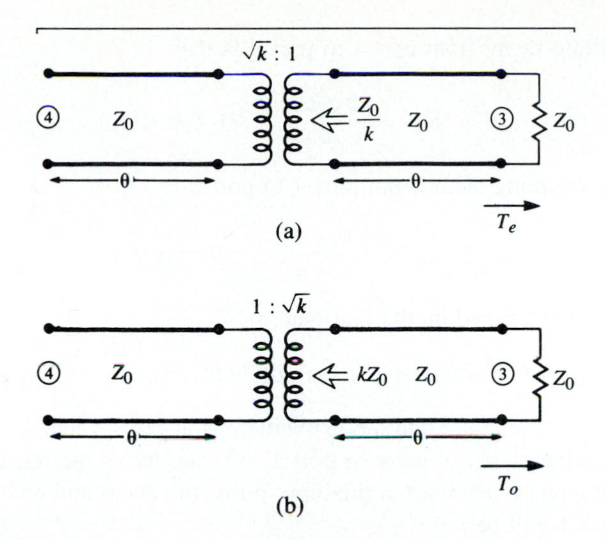

FIGURE 8.49 Equivalent circuits for the tapered coupled line hybrid, for transmission from port 4 to port 3. (a) Even-mode case. (b) Odd-mode case.

To evaluate the transmission coefficients into ports 1 and 3, we will use the $ABCD$ parameters for the equivalent circuits shown in Figure 8.49, where the tapered matching sections have been assumed to be ideal, and replaced with transformers. The $ABCD$ matrix of the transmission line-transformer-transmission line cascade can be found by multiplying the three individual $ABCD$ matrices for these components, but it is easier to use the fact that the transmission line sections only affect the phase of the transmission coefficients. The $ABCD$ matrix of the transformer is

$$\begin{bmatrix} \sqrt{k} & 0 \\ 0 & 1/\sqrt{k} \end{bmatrix},$$

for the even mode, and

$$\begin{bmatrix} 1/\sqrt{k} & 0 \\ 0 & \sqrt{k} \end{bmatrix},$$

for the odd mode. Then the even- and odd-mode transmission coefficients are

$$T_e = T_o = \frac{2\sqrt{k}}{k+1} e^{-2j\theta}, \qquad 8.113$$

since $T = 2/(A + B/Z_0 + CZ_0 + D) = 2\sqrt{k}/(k+1)$ for both modes; the $e^{-2j\theta}$ factor accounts for the phase delay of the two transmission line sections. We can then evaluate the following S parameters:

$$S_{34} = \frac{1}{2}(T_e + T_o) = \frac{2\sqrt{k}}{k+1} e^{-2j\theta}, \qquad 8.114a$$

$$S_{14} = \frac{1}{2}(T_e - T_o) = 0. \qquad 8.114b$$

The voltage coupling factor from port 4 to port 3 is then

$$\beta = |S_{34}| = \frac{2\sqrt{k}}{k+1}, \qquad 0 < \beta < 1, \tag{8.115a}$$

while the voltage coupling factor from port 4 to port 2 is

$$\alpha = |S_{24}| = -\frac{k-1}{k+1}, \qquad 0 < \alpha < 1. \tag{8.115b}$$

Power conservation is verified by the fact that

$$|S_{24}|^2 + |S_{34}|^2 = \alpha^2 + \beta^2 = 1.$$

If we now apply even- and odd-mode excitations at ports 1 and 3, so that superposition yields an incident voltage wave at port 1, we can derive the remaining scattering parameters. With a phase reference at the input ports, the even- and odd-mode reflection coefficients at port 1 will be

$$\Gamma_e = \frac{1-k}{1+k} e^{-2j\theta}, \tag{8.116a}$$

$$\Gamma_o = \frac{k-1}{k+1} e^{-2j\theta}. \tag{8.116b}$$

Then we can calculate the following S parameters:

$$S_{11} = \frac{1}{2}(\Gamma_e + \Gamma_o) = 0, \tag{8.117a}$$

$$S_{31} = \frac{1}{2}(\Gamma_e - \Gamma_o) = \frac{1-k}{1+k} e^{-2j\theta} = \alpha e^{-2j\theta}. \tag{8.117b}$$

From symmetry, we also have that $S_{33} = 0, S_{13} = S_{31}$, and that $S_{14} = S_{32}, S_{12} = S_{34}$. The tapered coupled line 180° hybrid thus has the following scattering matrix:

$$[S] = \begin{bmatrix} 0 & \beta & \alpha & 0 \\ \beta & 0 & 0 & -\alpha \\ \alpha & 0 & 0 & \beta \\ 0 & -\alpha & \beta & 0 \end{bmatrix} e^{-2j\theta}. \tag{8.118}$$

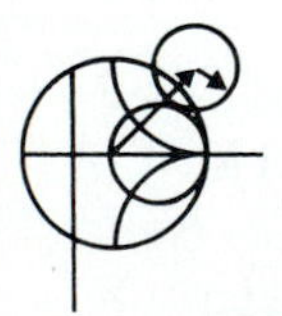

EXAMPLE 8.11

Calculate the even- and odd-mode characteristic impedances for a tapered coupled line 180° hybrid for a 3 dB coupling ratio and a $50\,\Omega$ characteristic impedance.

Solution

For a 3 dB coupler, $\alpha = \beta = 0.707$.

From (8.115b), we have that

$$k = \frac{1-\alpha}{1+\alpha} = 0.1716.$$

So at $z = L$ (the point of strongest coupling), the characteristic impedances of the coupled line must be

$$Z_{0e}(L) = \frac{Z_0}{k} = 291.4\,\Omega, \qquad Z_{0o}(L) = kZ_0 = 8.6\,\Omega.$$

At $z = 0$ (no coupling), the characteristic impedances are $Z_{0e}(0) = Z_{0o}(0) = Z_0 = 50\,\Omega$. So the taper of the coupled line section should be designed to match $291.4\,\Omega$ to $50\,\Omega$ with $Z_{0e}(z)$, and to match $8.6\,\Omega$ to $50\,\Omega$ with $Z_{0o}(z)$. In practice the required characteristic impedances can be calculated at several points along the line and joined with a smooth curve. Knowing $Z_{0e}(z)$ and $Z_{0o}(z)$ then determines the width and separation of the coupled lines versus z. ○

Waveguide Magic-T

The waveguide magic-T hybrid junction in Figure 8.43c has terminal properties similar to the ring hybrid, and a scattering matrix similar in form to (8.101). A rigorous analysis of this junction is too complicated to present here, but we can explain its operation in a qualitative sense by considering the field lines for excitations at the sum and difference ports.

First consider a TE_{10} mode incident at port 1. The resulting E_y field lines are illustrated in Figure 8.50a, where it is seen that there is an odd symmetry about guide 4. Since the field lines of a TE_{10} mode in guide 4 would have even symmetry, there is no coupling between ports 1 and 4. There is identical coupling to ports 2 and 3, however, resulting in an in-phase, equal-split power division.

For a TE_{10} mode incident at port 4, the field lines are as shown in Figure 8.50b. Again ports 1 and 4 are decoupled, due to symmetry (or reciprocity). Ports 2 and 3 are excited equally by the incident wave, but with a 180° phase difference.

In practice, tuning posts or irises are often used for matching; such components must be placed symmetrically to maintain proper operation of the hybrid.

8.9 OTHER COUPLERS

While we have discussed the general properties of couplers, and have analyzed and derived design data for several of the most frequently used couplers, there are many other types of couplers which we have not treated in detail. In this section, we will briefly describe some of these.

Moreno crossed-guide coupler. This is a waveguide directional coupler, consisting of two waveguides at right angles, with coupling provided by two apertures in the common broad wall of the guides. See Figure 8.51. By proper design [16], the two wave components excited by these apertures can be made to cancel in the back direction. The apertures usually consist of crossed slots, in order to couple tightly to the fields of both guides.

Schwinger reversed-phase coupler. This waveguide coupler is designed so that the path lengths for the two coupling apertures are the same for the uncoupled port, so that

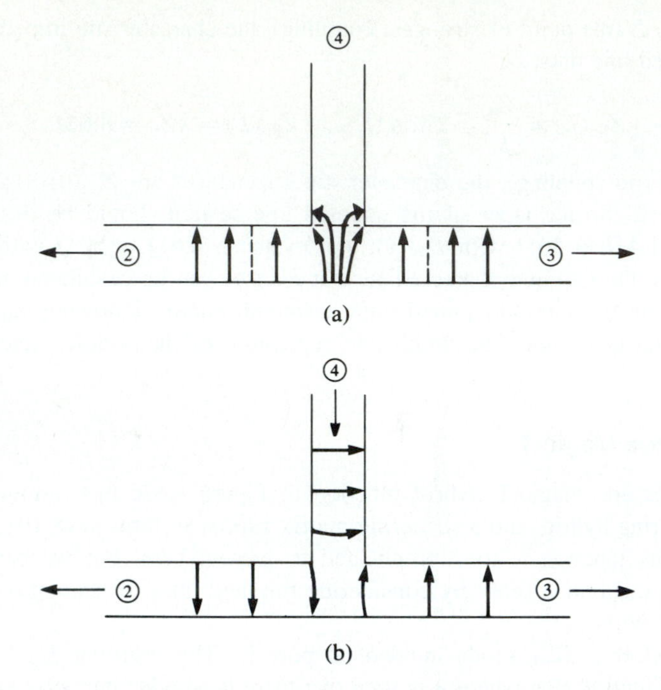

FIGURE 8.50 Electric field lines for a waveguide hybrid junction. (a) Incident wave at port 1. (b) Incident wave at port 4.

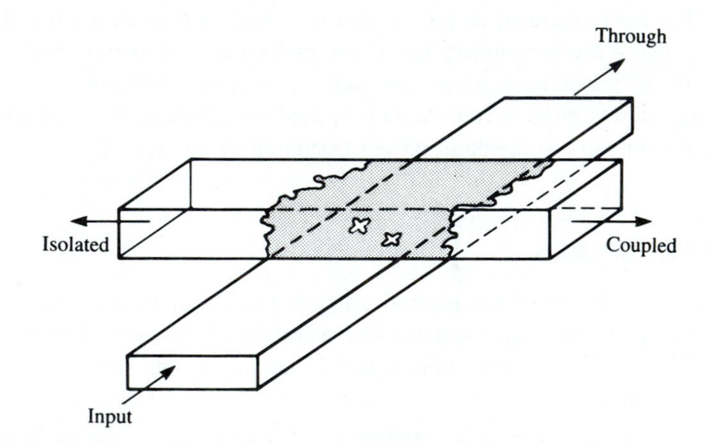

FIGURE 8.51 The Moreno crossed-guide coupler.

the directivity is essentially independent of frequency. Cancellation in the isolated port is accomplished by placing the slots on opposite sides of the centerline of the waveguide walls, as shown in Figure 8.52, which couple to magnetic dipoles with a 180° phase difference. Then, the $\lambda g/4$ slot spacing leads to in–phase combining at the coupled

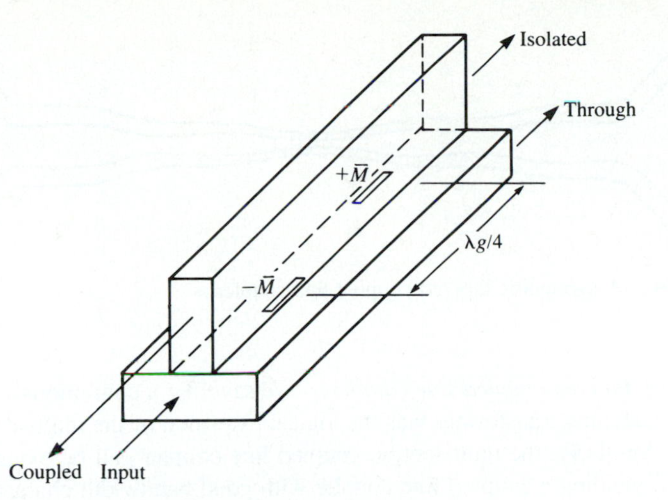

FIGURE 8.52 The Schwinger reversed-phase coupler.

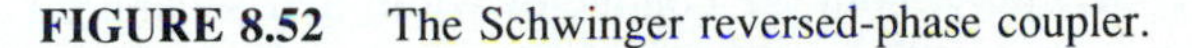

(backward) port, but this coupling is very frequency sensitive. This is the opposite situation from that of the multihole waveguide coupler discussed in Section 8.4.

Riblet short-slot coupler. Figure 8.53 shows a Riblet short-slot coupler, consisting of two waveguides with a common sidewall. Coupling takes place in the region where part of the common wall has been removed. In this region, both the TE_{10} (even) and the TE_{20} (odd) mode are excited, and by proper design can be made to cause cancellation at the isolated port and addition at the coupled port. The width of the interaction region must generally be reduced to prevent propagation of the undesired TE_{30} mode. This coupler can usually be made smaller than other waveguide couplers.

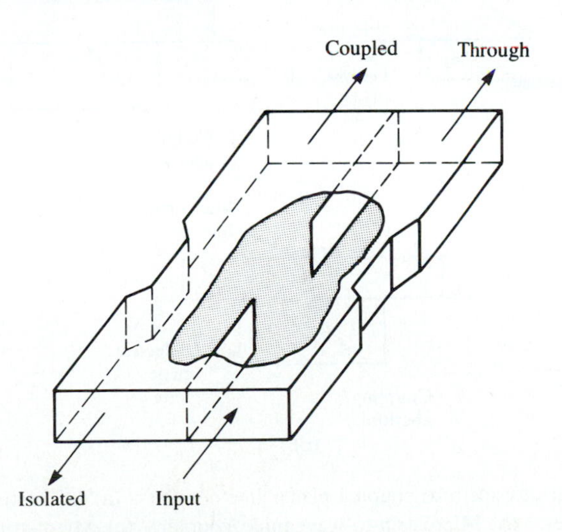

FIGURE 8.53 The Riblet short-slot coupler.

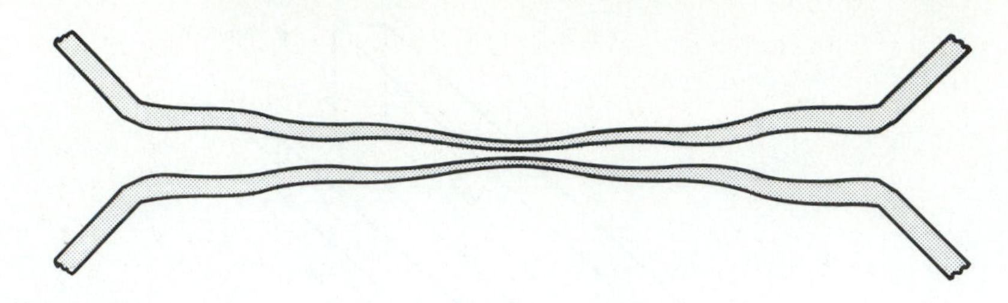

FIGURE 8.54 A symmetric tapered coupled line coupler.

Symmetric tapered coupled line coupler. We saw that a continuously tapered transmission line matching transformer was the logical extension of the multisection matching transformer. Similarly, the multisection coupled line coupler can be extended to a continuous taper, yielding a coupled line coupler with good bandwidth characteristics. Such a coupler is shown in Figure 8.54. Generally, both the conductor width and separation can be adjusted to provide a synthesized coupling or directivity response. One way to do this involves the computer optimization of a stepped-section approximation to the continuous taper [17]. This coupler provides a 90° phase shift between the outputs.

Couplers with apertures in planar lines. Many of the above-mentioned waveguide couplers can also be fabricated with planar lines such as microstrip, stripline, dielectric image lines, or various combinations of these. Some possibilities are illustrated in Figure 8.55. In principle, the design of such couplers can be carried out using the small-hole coupling theory and analysis techniques used in this chapter. The evaluation of the

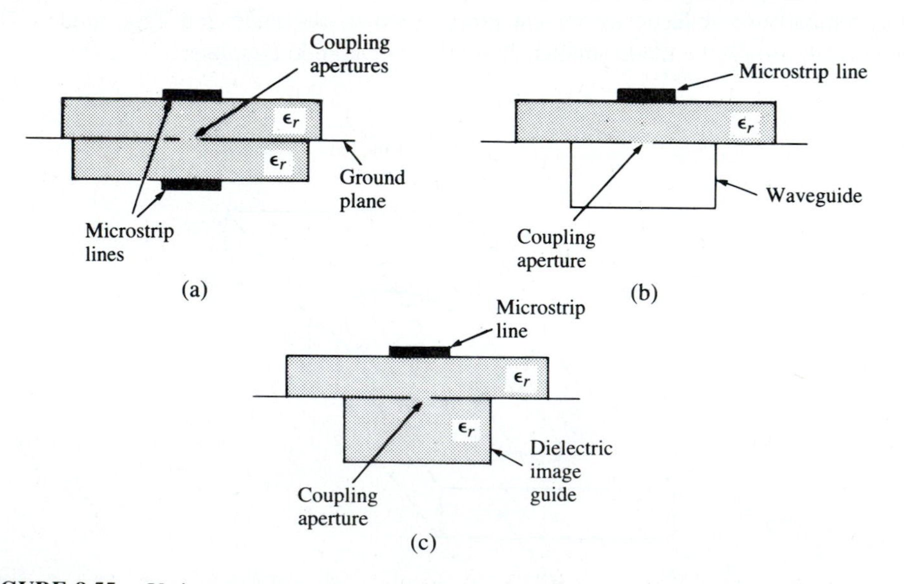

FIGURE 8.55 Various aperture coupled planar line couplers. (a) Microstrip-to-microstrip coupler. (b) Microstrip-to-waveguide coupler. (c) Microstrip-to-dielectric image line coupler.

fields of planar lines, however, is usually much more complicated than for rectangular waveguides.

POINT OF INTEREST: The Reflectometer

A reflectometer is a circuit that uses a directional coupler to isolate and sample the incident and reflected powers from a mismatched load. It forms the heart of a scalar or vector network analyzer, as it can be used to measure the reflection coefficient of a one-port network and, in a more general configuration, the S parameters of a two-port network. It can also be used as an SWR meter, or as a power monitor in systems applications.

The basic reflectometer circuit shown below can be used to measure the reflection coefficient magnitude of an unknown load. If we assume a reasonably matched coupler with loose coupling ($C << 1$), so that $\sqrt{1 - C^2} \simeq 1$, then the circuit can be represented by the signal flow graph shown below. In operation, the directional coupler provides a sample, V_i, of the incident wave, and a sample, V_r, of the reflected wave. A ratio meter with an appropriately calibrated scale can then measure these voltages and provide a reading in terms of reflection coefficient magnitude, or SWR.

Realistic directional couplers, however, have finite directivity, which means that both the incident and reflected powers will contribute to both V_i and V_r, leading to an error. If we assume a unit incident wave from the source, inspection of the signal flow graph leads to the following expressions for V_i and V_r:

$$V_i = C + \frac{C}{D}\Gamma e^{j\theta},$$

$$V_r = \frac{C}{D} + C\Gamma e^{j\phi},$$

where Γ is the reflection coefficient of the load, $D = 10^{(D\,\text{dB}/20)}$ is the numerical directivity of the coupler, and θ, ϕ are unknown phase delay differences through the circuit. Then the maximum and minimum values of the magnitude of V_r/V_i can be written as

$$\left|\frac{V_r}{V_i}\right|_{\substack{\max \\ \min}} = \frac{|\Gamma| \pm \dfrac{1}{D}}{1 \mp \dfrac{|\Gamma|}{D}}.$$

For a coupler with infinite directivity this reduces to the desired result of $|\Gamma|$. Otherwise a measurement uncertainty of approximately $\pm(1 + |\Gamma|)/D$ is introduced. Good accuracy thus requires a coupler with high directivity, preferably greater than 40 dB.

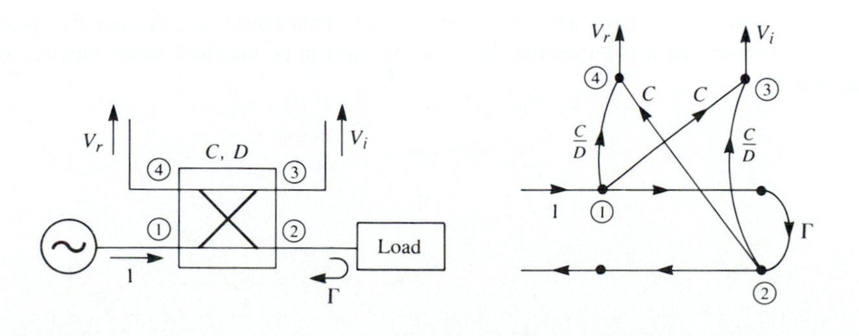

REFERENCES

[1] A. E. Bailey, Ed., *Microwave Measurement*, Peter Peregrinus, London, 1985.

[2] R. E. Collin, *Foundations for Microwave Engineering*, McGraw-Hill, N. Y., 1966.

[3] F. E. Gardiol, *Introduction to Microwaves*, Artech House, Dedham, Mass., 1984.

[4] E. Wilkinson, "An N-Way Hybrid Power Divider," *IRE Trans. on Microwave Theory and Techniques*, vol. MTT-8, pp. 116–118, January 1960.

[5] J. Reed and G. J. Wheeler, "A Method of Analysis of Symmetrical Four-Port Networks," *IRE Trans. on Microwave Theory and Techniques*, vol. MTT-4, pp. 246–252, October 1956.

[6] C. G. Montgomery, R. H. Dicke, and E. M. Purcell, *Principles of Microwave Circuits*, MIT Radiation Laboratory Series, vol. 8, McGraw-Hill, N. Y., 1948.

[7] H. Howe, *Stripline Circuit Design*, Artech House, Dedham, Mass., 1974.

[8] K. C. Gupta, R. Garg, and I. J. Bahl, *Microstrip Lines and Slot Lines*, Artech House, Dedham, Mass., 1979.

[9] L. Young, "The Analytical Equivalence of the TEM-Mode Directional Couplers and Transmission-Line Stepped Impedance Filters," *Proc. IEEE*, vol. 110, pp. 275–281, February 1963.

[10] J. Lange, "Interdigitated Stripline Quadrature Hybrid," *IEEE Trans. Microwave Theory and Techniques*, vol. MTT-17, pp. 1150–1151, December 1969.

[11] R. Waugh and D. LaCombe, "Unfolding the Lange Coupler," *IEEE Trans. Microwave Theory and Techniques*, vol. MTT-20, pp. 777–779, November 1972.

[12] W. P. Ou, "Design Equations for an Interdigitated Directional Coupler," *IEEE Trans. Microwave Theory and Techniques*, vol. MTT-23, pp. 253–255, February 1973.

[13] D. Paolino, "Design More Accurate Interdigitated Couplers," *Microwaves*, vol. 15, pp. 34–38, May 1976.

[14] J. Hughes and K. Wilson, "High Power Multiple IMPATT Amplifiers," *Proc. European Microwave Conference*, pp. 118–122, 1974.

[15] R. H. DuHamel and M. E. Armstrong, "The Tapered-Line Magic-T," *Abstracts of 15th Annual Symposium of the USAF Antenna Research and Development Program*, Monticello, Ill., October 12–14, 1965.

[16] T. N. Anderson, "Directional Coupler Design Nomograms," *Microwave Journal*, vol. 2, pp. 34–38, May 1959.

[17] D. W. Kammler, "The Design of Discrete N-Section and Continuously Tapered Symmetrical Microwave TEM Directional Couplers," *IEEE Trans. on Microwave Theory and Techniques*, vol. MTT-17, pp. 577–590, August 1969.

PROBLEMS

8.1 Consider the T-junction of three lines with characteristic impedances Z_1, Z_2, and Z_3, as shown below. Demonstrate that it is impossible for all three lines to be matched, when looking toward the junction.

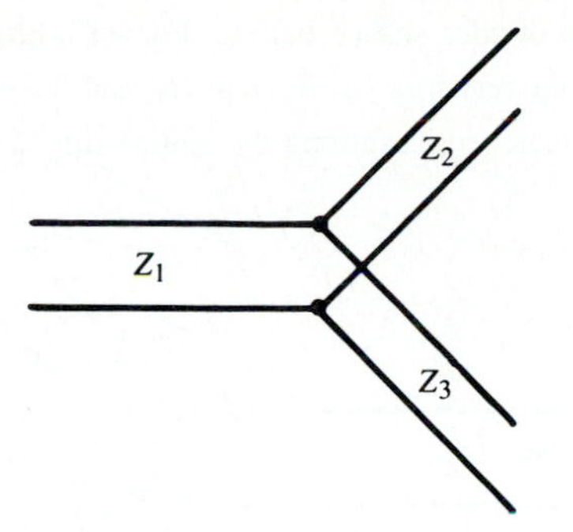

8.2 Show that a lossless, reciprocal and symmetrical three-port network can be matched at ports 1 and 2 by appropriately positioning a short circuit at port 3.

8.3 A directional coupler has the scattering matrix given below. Find the directivity, coupling, isolation, and return loss at the input port when the other ports are terminated in matched loads.

$$[S] = \begin{bmatrix} 0.05\angle 30 & 0.96\angle 0 & 0.1\angle 90 & 0.05\angle 90 \\ 0.96\angle 0 & 0.05\angle 30 & 0.05\angle 90 & 0.1\angle 90 \\ 0.1\angle 90 & 0.05\angle 90 & 0.04\angle 30 & 0.96\angle 0 \\ 0.05\angle 90 & 0.1\angle 90 & 0.96\angle 0 & 0.05\angle 30 \end{bmatrix}.$$

8.4 Two identical 90° couplers with $C = 8.34$ dB are connected as shown below. Find the resulting phase and amplitudes at ports $2'$ and $3'$, relative to port 1.

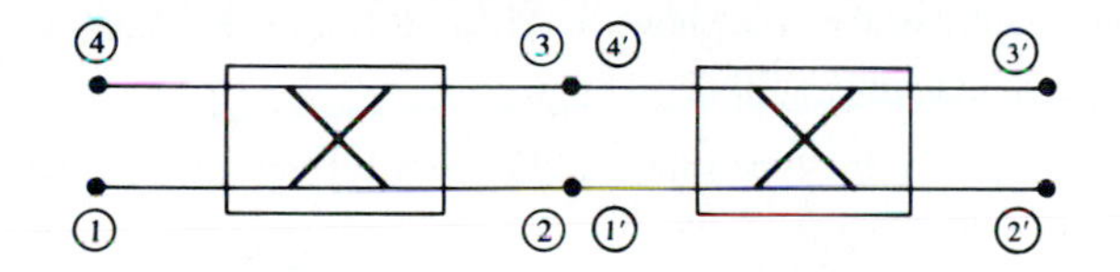

8.5 A 90 W power source is connected to the input of a directional coupler with $C = 20$ dB, $D = 35$ dB, and an insertion loss of 0.5 dB. Find the output powers at the through, coupled, and isolated ports. Assume all ports to be matched.

8.6 Design a lossless T-junction divider with a 30 Ω source impedance to give a 3:1 power split. Design quarter-wave matching transformers to convert the impedances of the output lines to 30 Ω. Determine the magnitude of the S parameters for this circuit, using a 30 Ω characteristic impedance.

8.7 Design a three-port resistive divider for an equal power split and a 100 Ω system impedance. If one of the output ports is connected to a load with a mismatch of $\Gamma = 0.3$, calculate what fraction of the output power will be reflected and coupled through to the adjacent output port.

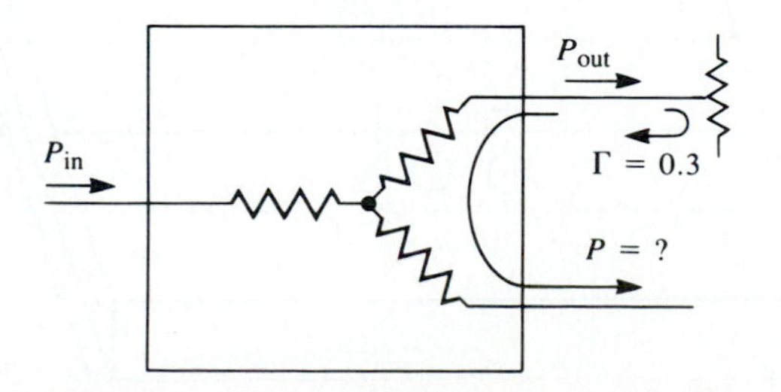

8.8 Consider the general resistive divider shown below. For an arbitrary power division ratio, $\alpha = P_2/P_3$, derive expressions for the resistors R_1, R_2, and R_3, and the output characteristic impedances Z_{o2}, Z_{o3} so that all ports are matched, assuming the source impedance is Z_0.

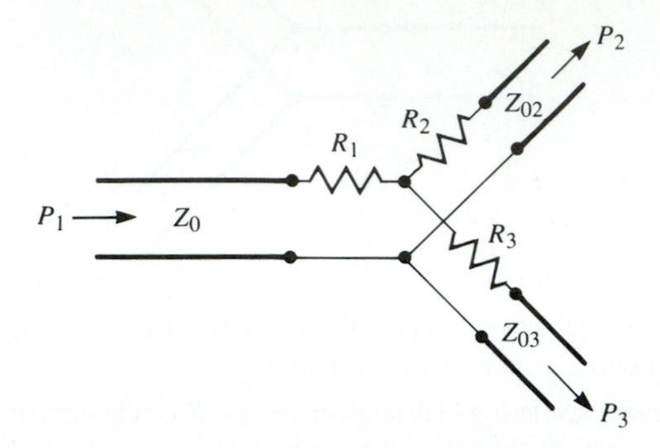

8.9 Design a Wilkinson power divider with a power division ratio of $P_3/P_2 = 1/3$, and a source impedance of 50 Ω.

8.10 Derive the design equations in (8.37a,b,c) for the unequal-split Wilkinson divider.

8.11 For the Bethe-hole coupler of the type shown in Figure 8.16a, derive a design for s so that port 3 is the isolated port.

8.12 Design a Bethe-hole coupler of the type shown in Figure 8.16a for Ku-band waveguide operating at 11 GHz. The required coupling is 20 dB.

8.13 Design a Bethe-hole coupler of the type shown in Figure 8.16b for Ku-band waveguide operating at 17 GHz. The required coupling is 30 dB.

8.14 Design a five-hole directional coupler in Ku-band waveguide with a binomial directivity response. The center frequency is 17.5 GHz, and the required coupling is 20 dB. Use round apertures centered across the broad wall of the waveguides.

8.15 Repeat Problem 8.14 for a design with a Chebyshev response, having a minimum directivity of 30 dB.

8.16 Develop the necessary equations required to design a two-hole directional coupler using two waveguides with apertures in a common sidewall, as shown below.

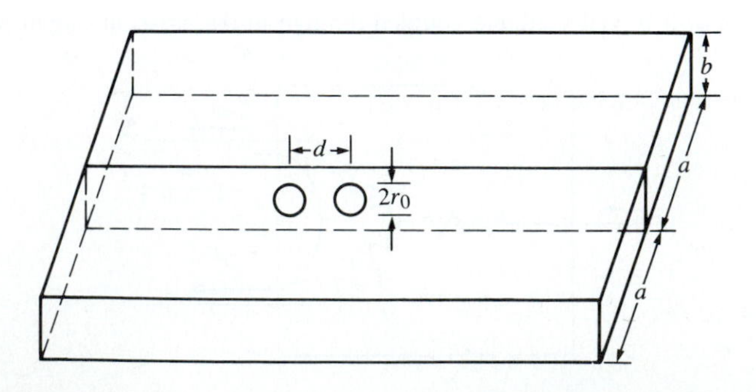

8.17 Consider the general branch-line coupler shown below, having shunt arm characteristic impedances Z_a, and series arm characteristic impedances Z_b. Using an even-odd mode analysis, derive design equations for a quadrature hybrid coupler with an arbitrary power division ratio of $\alpha = P_2/P_3$, and with the input port (port 1) matched. Assume all arms are $\lambda/4$ long. Is port 4 isolated, in general?

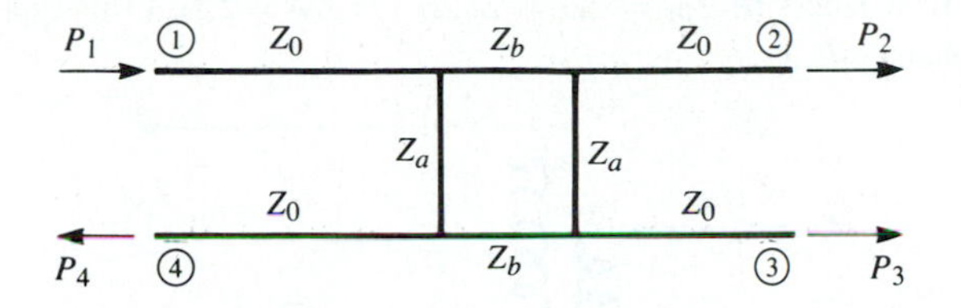

8.18 An edge-coupled stripline with a ground plane spacing of 0.32 cm and a dielectric constant of 2.2 is required to have even- and odd-mode characteristic impedances of $Z_{0e} = 70\,\Omega$ and $Z_{0o} = 40\,\Omega$. Find the necessary strip widths and spacing.

8.19 A coupled microstrip line on a substrate with $\epsilon_r = 10$ and $d = 0.16$ cm has strip widths of 0.16 cm and a strip spacing of 0.064 cm. Find the even- and odd-mode characteristic impedances.

8.20 Derive expressions for the even- and odd-mode characteristic impedances of the three-wire line shown below. (Use the formula given in Table 3.1 for the capacitance of a two-wire line.) Use these results to find the even- and odd-mode characteristic impedance of a three-wire line in air with $a = 1$ mm and $s = 3$ mm.

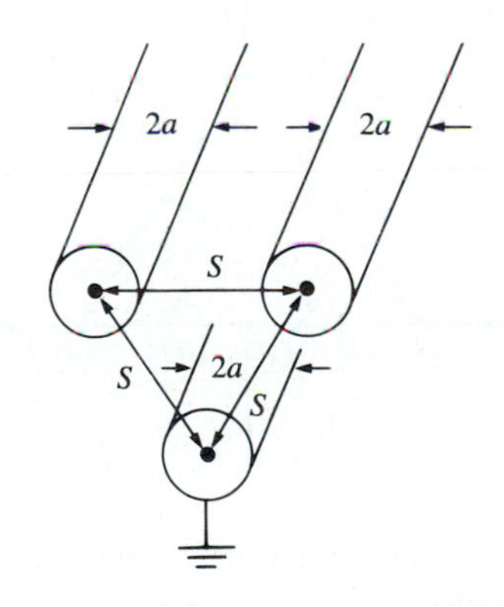

8.21 Repeat the derivation in Section 8.6 for the design equations of a single-section coupled line coupler using reflection and transmission coefficients, instead of voltages and currents.

8.22 Design a single-section coupled line coupler with a coupling of 19.1 dB, a system impedance of $60\,\Omega$, and a center frequency of 8 GHz. If the coupler is to be made in stripline (edge-coupled), with $\epsilon_r = 2.2$ and $b = 0.32$ cm, find the necessary strip widths and separation.

8.23 Repeat Problem 8.22 for a coupling factor of 6 dB. Is this a practical design?

8.24 Derive equations (8.83) and (8.84).

8.25 Design a 25 dB three-section coupled line coupler with a maximally flat coupling response. Assume $Z_0 = 50\,\Omega$, and find Z_{0e}, Z_{0o} for each section.

8.26 Repeat Problem 8.25 for a coupler with an equal-ripple coupling response, where the ripple in the coupling is 1 dB over the passband.

8.27 For the Lange coupler, derive the design equations (8.100) for Z_{0e} and Z_{0o} from (8.98) and (8.99).

8.28 Consider the four-port hybrid transformer shown below. Determine the scattering matrix for this device, and show that it is similar in form to the scattering matrix for the 180° hybrid. Let the port characteristic impedances be $Z_{01} = Z_{04} = Z_0$; $Z_{02} = Z_{03} = 2Z_0$. (This type of transformer is often used in telephone circuits.)

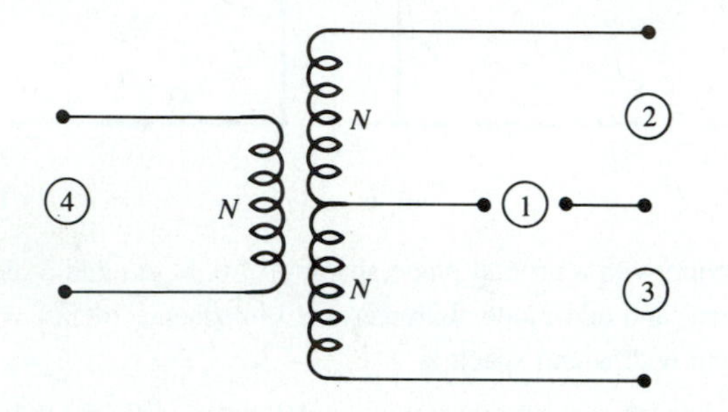

8.29 An input signal $V_1 = 3\angle 80°$ is applied to the sum port of a 180° hybrid, and another signal $V_4 = 2\angle 150°$ is applied to the difference port. What are the output signals?

8.30 Find the S parameters for the four-port Bagley polygon power divider shown below.

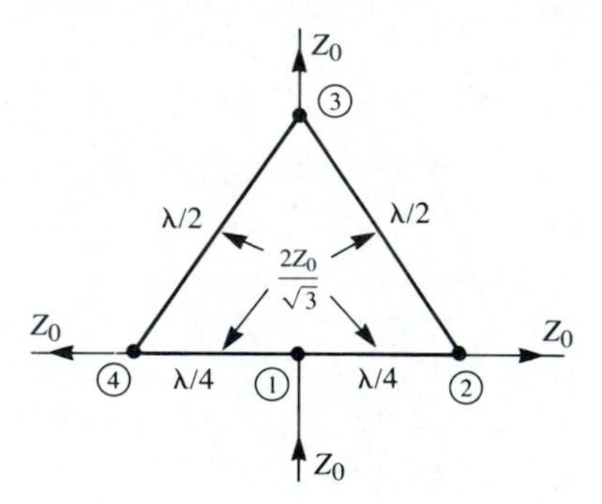

8.31 For the symmetric hybrid shown below, calculate the output voltages if port 1 is fed with an incident wave of $1\angle 0$ V. Assume the outputs are matched.

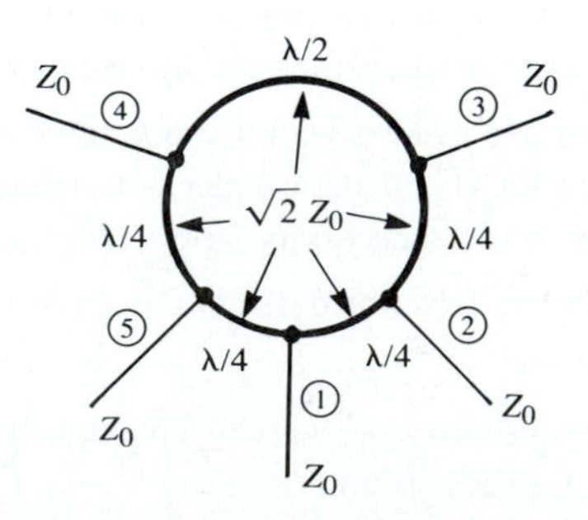

CHAPTER 9

Microwave Filters

A microwave filter is a two-port network used to control the frequency response at a certain point in a microwave system by providing transmission at frequencies within the passband of the filter, and attenuation in the stopband of the filter. Typical frequency responses include low-pass, high-pass, bandpass, and band-reject characteristics. Applications can be found in virtually any type of microwave communication, radar, or test and measurement system.

We begin our discussion of filter theory and design with the frequency characteristics of periodic structures, which consist of a transmission line or waveguide periodically loaded with reactive elements. These structures are of interest in themselves, because of the application to slow-wave components and traveling-wave amplifier design, and also because they exhibit basic passband-stopband responses which lead to the image parameter method of filter design.

Filters designed using the *image parameter method* consist of a cascade of simpler two-port filter sections to provide the desired cutoff frequencies and attenuation characteristics, but do not allow the specification of a frequency response over the complete operating range. Thus, although the procedure is relatively simple, the design of filters by the image parameter method often must be iterated many times to achieve the desired results.

A more modern procedure, called the *insertion loss method*, uses network synthesis techniques to design filters with a completely specified frequency response. The design is simplified by beginning with low-pass filter prototypes which are normalized in terms of impedance and frequency. Transformations are then applied to convert the prototype designs to the desired frequency range and impedance level.

Both the image parameter and insertion loss method of filter design provide lumped-element circuits. For microwave applications such designs usually must be modified to use distributed elements consisting of transmission line sections. The Richard's transformation and the Kuroda identities provide this step. We also will discuss transmission line filters using stepped impedances and coupled lines; filters using coupled resonators will also be briefly described.

The subject of microwave filters is quite extensive, due to the importance of these components in practical systems, and the wide variety of possible implementations. We give here a treatment of only the basic principles and some of the more common filter designs, and refer the reader to references such as [1], [2], [3], and [4] for further discussion.

9.1 PERIODIC STRUCTURES

An infinite transmission line or waveguide periodically loaded with reactive elements is referred to as a periodic structure. As shown in Figure 9.1, periodic structures can take various forms, depending on the transmission line media being used. Often the loading elements are formed as discontinuities in the line, but in any case they can be modelled as lumped reactances across a transmission line as shown in Figure 9.2. Periodic structures support slow-wave (slower than the phase velocity of the unloaded line), and have passband and stopband characteristics similar to filters; they find application in traveling-wave tubes, masers, phase shifters, and antennas.

Analysis of Infinite Periodic Structures

We begin by studying the propagation characteristics of the infinite loaded line shown in Figure 9.2. Each unit cell of this line consists of a length d of transmission line

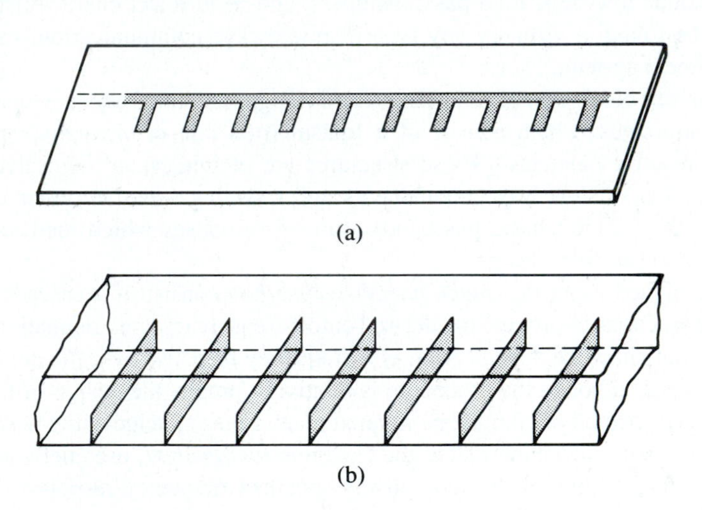

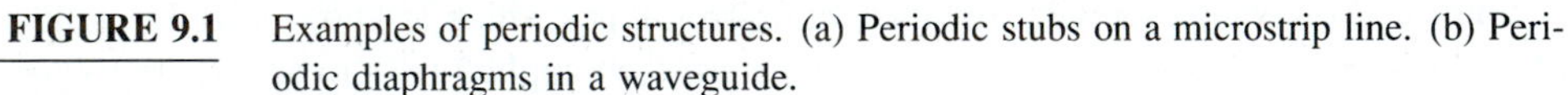

FIGURE 9.1 Examples of periodic structures. (a) Periodic stubs on a microstrip line. (b) Periodic diaphragms in a waveguide.

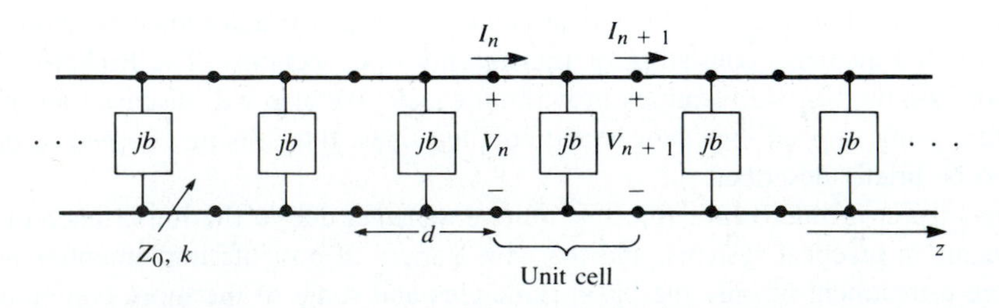

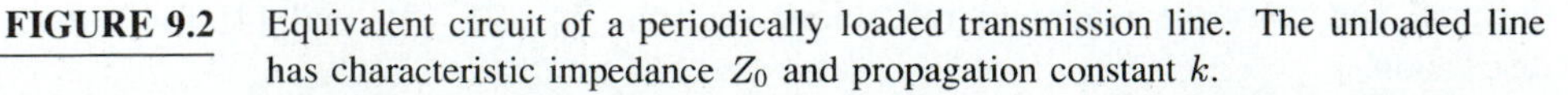

FIGURE 9.2 Equivalent circuit of a periodically loaded transmission line. The unloaded line has characteristic impedance Z_0 and propagation constant k.

with a shunt susceptance across the midpoint of the line; the susceptance b is normalized to the characteristic impedance, Z_0. If we consider the infinite line as being composed of a cascade of identical two-port networks, we can relate the voltages and currents on either side of the nth unit cell using the $ABCD$ matrix:

$$\begin{bmatrix} V_n \\ I_n \end{bmatrix} = \begin{bmatrix} A & B \\ C & D \end{bmatrix} \begin{bmatrix} V_{n+1} \\ I_{n+1} \end{bmatrix}, \tag{9.1}$$

where $A, B, C,$ and D are the matrix parameters for a cascade of a transmission line section of length $d/2$, a shunt susceptance b, and another transmission line section of length $d/2$. From Table 5.1 we then have, in normalized form,

$$\begin{aligned} \begin{bmatrix} A & B \\ C & D \end{bmatrix} &= \begin{bmatrix} \cos\frac{\theta}{2} & j\sin\frac{\theta}{2} \\ j\sin\frac{\theta}{2} & \cos\frac{\theta}{2} \end{bmatrix} \begin{bmatrix} 1 & 0 \\ jb & 1 \end{bmatrix} \begin{bmatrix} \cos\frac{\theta}{2} & j\sin\frac{\theta}{2} \\ j\sin\frac{\theta}{2} & \cos\frac{\theta}{2} \end{bmatrix} \\ &= \begin{bmatrix} \left(\cos\theta - \frac{b}{2}\sin\theta\right) & j\left(\sin\theta + \frac{b}{2}\cos\theta - \frac{b}{2}\right) \\ j\left(\sin\theta + \frac{b}{2}\cos\theta + \frac{b}{2}\right) & \left(\cos\theta - \frac{b}{2}\sin\theta\right) \end{bmatrix}, \end{aligned} \tag{9.2}$$

where $\theta = kd$, and k is the propagation constant of the unloaded line. The reader can verify that $AD - BC = 1$, as required for reciprocal networks.

Now for any wave propagating in the $+z$ direction, we must have

$$V(z) = V(0)e^{-\gamma z}, \tag{9.3a}$$

$$I(z) = I(0)e^{-\gamma z}, \tag{9.3b}$$

for a phase reference at $z = 0$. Since the structure is infinitely long, the voltage and current at the nth terminals can differ from the voltage and current at the $n+1$ terminals only by the propagation factor, $e^{-\gamma d}$. Thus,

$$V_{n+1} = V_n e^{-\gamma d}, \tag{9.4a}$$

$$I_{n+1} = I_n e^{-\gamma d}. \tag{9.4b}$$

Using this result in (9.1) gives the following:

$$\begin{bmatrix} V_n \\ I_n \end{bmatrix} = \begin{bmatrix} A & B \\ C & D \end{bmatrix} \begin{bmatrix} V_{n+1} \\ I_{n+1} \end{bmatrix} = \begin{bmatrix} V_{n+1}e^{\gamma d} \\ I_{n+1}e^{\gamma d} \end{bmatrix},$$

or

$$\begin{bmatrix} A - e^{\gamma d} & B \\ C & D - e^{\gamma d} \end{bmatrix} \begin{bmatrix} V_{n+1} \\ I_{n+1} \end{bmatrix} = 0. \tag{9.5}$$

For a nontrivial solution, the determinant of the above matrix must vanish:

$$AD + e^{2\gamma d} - (A + D)e^{\gamma d} - BC = 0, \tag{9.6}$$

or, since $AD - BC = 1$,

$$1 + e^{2\gamma d} - (A + D)e^{\gamma d} = 0,$$

$$e^{-\gamma d} + e^{\gamma d} = A + D,$$

$$\cosh \gamma d = \frac{A + D}{2} = \cos\theta - \frac{b}{2}\sin\theta, \tag{9.7}$$

where (9.2) was used for the values of A and D. Now if $\gamma = \alpha + j\beta$, we have that

$$\cosh \gamma d = \cosh \alpha d \cos \beta d + j \sinh \alpha d \sin \beta d = \cos\theta - \frac{b}{2}\sin\theta. \tag{9.8}$$

Since the right-hand side of (9.8) is purely real, we must have either $\alpha = 0$ or $\beta = 0$.

Case 1: $\alpha = 0, \beta \neq 0$. This case corresponds to a nonattenuating, propagating wave on the periodic structure, and defines the passband of the structure. Then (9.8) reduces to

$$\cos \beta d = \cos\theta - \frac{b}{2}\sin\theta, \tag{9.9a}$$

which can be solved for β if the magnitude of the right-hand side is less than or equal to unity. Note that there are an infinite number of values of β that can satisfy (9.9a).

Case 2: $\alpha \neq 0, \beta = 0, \pi$. In this case the wave does not propagate, but is attenuated along the line; this defines the stopband of the structure. Because the line is lossless, power is not dissipated, but is reflected back to the input of the line. The magnitude of (9.8) reduces to

$$\cosh \alpha d = |\cos\theta - \frac{b}{2}\sin\theta| \geq 1, \tag{9.9b}$$

which has only one solution ($\alpha > 0$) for positively traveling waves; $\alpha < 0$ applies for negatively traveling waves. If $\cos\theta - (b/2)\sin\theta \leq -1$, (9.9b) is obtained from (9.8) by letting $\beta = \pi$; then all the lumped loads on the line are $\lambda/2$ apart, yielding an input impedance the same as if $\beta = 0$.

Thus, depending on the frequency and normalized susceptance values, the periodically loaded line will exhibit either passbands or stopbands, and so can be considered as a type of filter. It is important to note that the voltage and current waves defined in (9.3) and (9.4) are only meaningful when measured at the terminals of the unit cells, and do not apply to voltages and currents that may exist at points within a unit cell. These waves are sometimes referred to as *Bloch waves* because of their similarity to elastic waves that propagate through periodic crystal lattices.

Besides the propagation constant of the waves on the periodically loaded line, we will also be interested in the characteristic impedance for these waves. We can define a characteristic impedance at the unit cell terminals as

$$Z_B = Z_0 \frac{V_{n+1}}{I_{n+1}}, \tag{9.10}$$

since V_{n+1} and I_{n+1} in the above derivation were normalized quantities. This impedance is also referred to as the Bloch impedance. From (9.5) we have that

$$(A - e^{\gamma d})V_{n+1} + BI_{n+1} = 0,$$

so (9.10) yields

$$Z_B = \frac{-BZ_0}{A - e^{\gamma d}}.$$

From (9.6) we can solve for $e^{\gamma d}$ in terms of A and D as follows:

$$e^{\gamma d} = \frac{(A+D) \pm \sqrt{(A+D)^2 - 4}}{2}.$$

Then the Bloch impedance has two solutions given by

$$Z_B^{\pm} = \frac{-2BZ_0}{A - D \mp \sqrt{(A+D)^2 - 4}}. \qquad 9.11$$

For symmetrical unit cells (as assumed in Figure (9.2)) we will always have $A = D$. In this case (9.11) reduces to

$$Z_B^{\pm} = \frac{\pm BZ_0}{\sqrt{A^2 - 1}}. \qquad 9.12$$

The $\pm$ solutions correspond to the characteristic impedance for positively and negatively traveling waves, respectively. For symmetrical networks these impedances are the same except for the sign; the characteristic impedance for a negatively traveling wave turns out to be negative because we have defined I_n in Figure 9.2 as always being in the positive direction.

From (9.2) we see that B is always purely imaginary. If $\alpha = 0, \beta \neq 0$ (passband), then (9.7) shows that $\cosh \gamma d = A \leq 1$ (for symmetrical networks) and (9.12) shows that Z_B will be real. If $\alpha \neq 0, \beta = 0$ (stopband), then (9.7) shows that $\cosh \gamma d = A \geq 1$, and (9.12) shows that Z_B is imaginary. This situation is similar to that for the wave impedance of a waveguide, which is real for propagating modes and imaginary for cutoff, or evanescent, modes.

Terminated Periodic Structures

Next consider a truncated periodic structure, terminated in a load impedance Z_L, as shown in Figure 9.3. At the terminals of an arbitrary unit cell, the incident and reflected voltages and currents can be written as (assuming operation in the passband)

$$V_n = V_0^+ e^{-j\beta nd} + V_0^- e^{j\beta nd}, \qquad 9.13a$$

$$I_n = I_0^+ e^{-j\beta nd} + I_0^- e^{j\beta nd} = \frac{V_0^+}{Z_B^+} e^{-j\beta nd} + \frac{V_0^-}{Z_B^-} e^{j\beta nd}, \qquad 9.13b$$

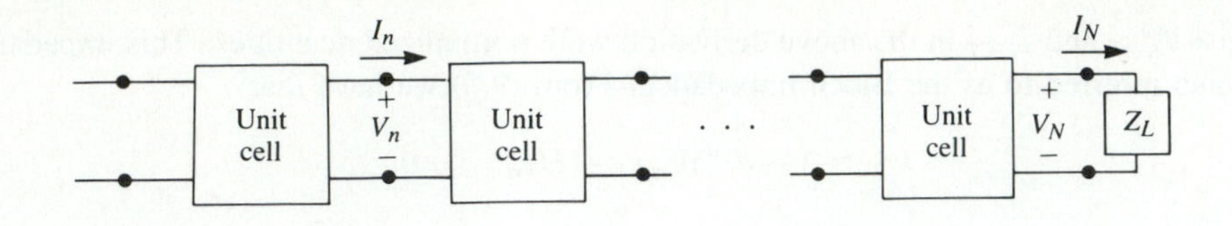

FIGURE 9.3 A periodic structure terminated in a normalized load impedance Z_L.

where we have replaced γz in (9.3) with $j\beta nd$, since we are only interested in terminal quantities.

Now define the following incident and reflected voltages at the nth unit cell:

$$V_n^+ = V_0^+ e^{-j\beta nd}, \tag{9.14a}$$

$$V_n^- = V_0^- e^{j\beta nd}. \tag{9.14b}$$

Then (9.13) can be written as

$$V_n = V_n^+ + V_n^-, \tag{9.15a}$$

$$I_n = \frac{V_n^+}{Z_B^+} + \frac{V_n^-}{Z_B^-}. \tag{9.15b}$$

At the load, where $n = N$, we have

$$V_N = V_N^+ + V_N^- = Z_L I_N = Z_L \left(\frac{V_N^+}{Z_B^+} + \frac{V_N^-}{Z_B^-} \right), \tag{9.16}$$

so the reflection coefficient at the load can be found as

$$\Gamma = \frac{V_N^-}{V_N^+} = -\frac{Z_L/Z_B^+ - 1}{Z_L/Z_B^- - 1}. \tag{9.17}$$

If the unit cell network is symmetrical ($A = D$), then $Z_B^+ = -Z_B^- = Z_B$, which reduces (9.17) to the familiar result that

$$\Gamma = \frac{Z_L - Z_B}{Z_L + Z_B}. \tag{9.18}$$

So to avoid reflections on the terminated periodic structure, we must have $Z_L = Z_B$, which is real for a lossless structure operating in a passband. If necessary, a quarter-wave transformer can be used between the periodically loaded line and the load.

k-β Diagrams and Wave Velocities

When studying the passband and stopband characteristics of a periodic structure, it is useful to plot the propagation constant, β, versus the propagation constant of the unloaded line, k (or ω). Such a graph is called a k-β *diagram*, or Brillouin diagram (after L. Brillouin, a physicist who studied wave propagation in periodic crystal structures).

The k-β diagram can be plotted from (9.9a), which is the dispersion relation for a general periodic structure. In fact, a k-β diagram can be used to study the dispersion characteristics of many types of microwave components and transmission lines. For instance, consider the dispersion relation for a waveguide mode:

$$\beta = \sqrt{k^2 - k_c^2},$$

or

$$k = \sqrt{\beta^2 + k_c^2}, \quad 9.19$$

where k_c is the cutoff wavenumber of the mode, k is the free-space wavenumber, and β is the propagation constant of the mode. Relation (9.19) is plotted in the k-β diagram of Figure 9.4. For values of $k < k_c$, there is no real solution for β, so the mode is nonpropagating. For $k > k_c$, the mode propagates, and k approaches β for large values of β (TEM propagation).

The k-β diagram is also useful for interpreting the various wave velocities associated with a dispersive structure. The phase velocity is

$$v_p = \frac{\omega}{\beta} = c\frac{k}{\beta}, \quad 9.20$$

which is seen to be equal to c (speed of light) times the slope of the line from the origin to the operating point on the k-β diagram. The group velocity is

$$v_g = \frac{d\omega}{d\beta} = c\frac{dk}{d\beta}, \quad 9.21$$

which is the slope of the k-β curve at the operating point. Thus, referring to Figure 9.4, we see that the phase velocity for a propagating waveguide mode is infinite at cutoff and approaches c (from above) as k increases. The group velocity, however, is zero at cutoff and approaches c (from below) as k increases. We finish our discussion of periodic structures with a practical example of a capacitively loaded line.

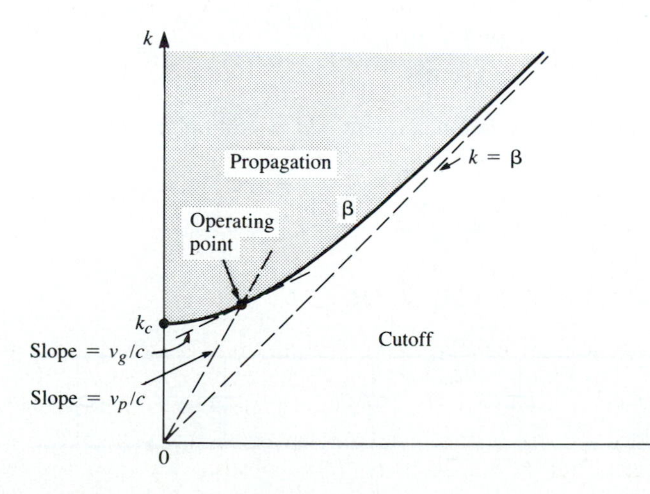

FIGURE 9.4 k-β diagram for a waveguide mode.

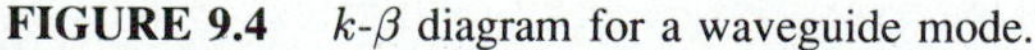

EXAMPLE 9.1

Consider a periodic capacitively loaded line, as shown in Figure 9.5 (such a line may be implemented as in Figure 9.1 with short capacitive stubs). If $Z_0 = 50\ \Omega, d = 1.0$ cm, and $C_0 = 2.666$ pF, sketch the k-β diagram and compute the propagation constant, phase velocity, and Bloch impedance at $f = 3.0$ GHz. Assume $k = k_0$.

Solution
We can rewrite the dispersion relation of (9.9a) as

$$\cos \beta d = \cos k_0 d - \left(\frac{C_0 Z_0 c}{2d}\right) k_0 d \sin k_0 d.$$

Then

$$\frac{C_0 Z_0 c}{2d} = \frac{(2.666 \times 10^{-12})(50)(3 \times 10^8)}{2(0.01)} = 2.0,$$

so we have

$$\cos \beta d = \cos k_0 d - 2 k_0 d \sin k_0 d.$$

The most straightforward way to proceed at this point is to numerically evaluate the right-hand side of the above equation for a set of values of $k_0 d$ starting at zero. When the magnitude of the right-hand side is unity or less, we have a passband and can solve for βd. Otherwise we have a stopband. Calculation shows that the first passband exists for $0 \leq k_0 d \leq 0.96$. The second passband does not begin until the $\sin k_0 d$ term changes sign at $k_0 d = \pi$. As $k_0 d$ increases, an infinite number of passbands are possible, but they become narrower. Figure 9.6 shows the k-β diagram for the first two passbands.

At 3.0 GHz, we have

$$k_0 d = \frac{2\pi(3 \times 10^9)}{3 \times 10^8}(0.01) = 0.6283 = 36°,$$

so $\beta d = 1.5$ and the propagation constant is $\beta = 150$ rad/m. The phase velocity is

$$v_p = \frac{k_0 c}{\beta} = \frac{0.6283}{1.5} c = 0.42c,$$

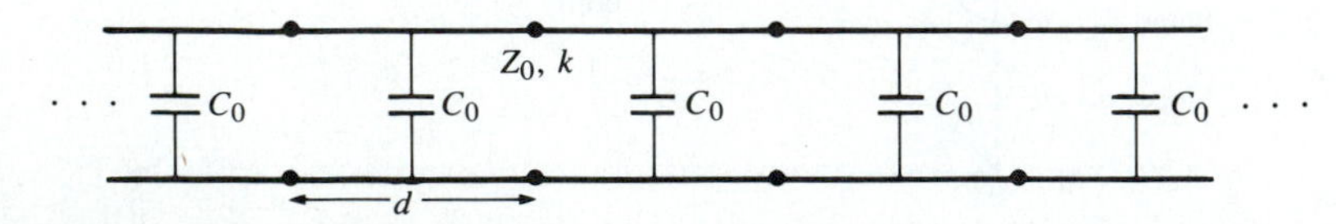

FIGURE 9.5 A capacitively loaded line.

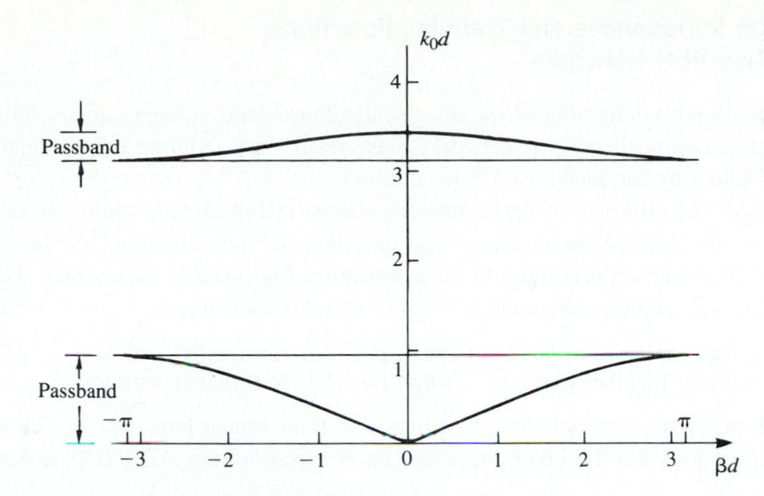

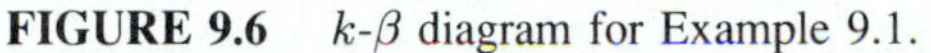

FIGURE 9.6 k-β diagram for Example 9.1.

which is much less than the speed of light, indicating that this is a slow-wave structure. To evaluate the Bloch impedance, we use (9.2) and (9.12):

$$\frac{b}{2} = \frac{\omega C_0 Z_0}{2} = 1.256,$$

$$\theta = k_0 d = 36^\circ,$$

$$A = \cos\theta - \frac{b}{2}\sin\theta = 0.0707,$$

$$B = j\left(\sin\theta + \frac{b}{2}\cos\theta - \frac{b}{2}\right) = j0.3479.$$

Then,

$$Z_B = \frac{BZ_0}{\sqrt{A^2 - 1}} = \frac{(j0.3479)(50)}{j\sqrt{1 - (0.0707)^2}} = 17.4\ \Omega.$$

○

9.2 FILTER DESIGN BY THE IMAGE PARAMETER METHOD

The image parameter method of filter design involves the specification of passband and stopband characteristics for a cascade of two-port networks, and so is similar in concept to the periodic structures which were studied in Section 9.1. The method is relatively simple, but has the disadvantage that an arbitrary frequency response cannot be incorporated into the design. This is in contrast to the insertion loss method, which is the subject of the following section. Nevertheless, the image parameter method is useful for simple filters, and provides a link between infinite periodic structures and practical filter design. The image parameter method also finds application in solid-state traveling wave amplifier design.

Image Impedances and Transfer Functions for Two-Port Networks

We begin with definitions of the image impedances and voltage transfer function for an arbitrary reciprocal two-port network; these results are required for the analysis and design of filters by the image parameter method.

Consider the arbitrary two-port network shown in Figure 9.7, where the network is specified by its $ABCD$ parameters. Note that the reference direction for the current at port 2 has been chosen according to the convention for $ABCD$ parameters. The image impedances, Z_{i1} and Z_{i2}, are defined for this network as follows:

Z_{i1} = input impedance at port 1 when port 2 is terminated with Z_{i2},
Z_{i2} = input impedance at port 2 when port 1 is terminated with Z_{i1}.

Thus both ports are matched when terminated in their image impedances. We now will derive expressions for the image impedances in terms of the $ABCD$ parameters of a network.

The port voltages and currents are related as

$$V_1 = AV_2 + BI_2, \quad 9.22a$$

$$I_1 = CV_2 + DI_2. \quad 9.22b$$

The input impedance at port 1, with port 2 terminated in Z_{i2}, is

$$Z_{\text{in1}} = \frac{V_1}{I_1} = \frac{AV_2 + BI_2}{CV_2 + DI_2} = \frac{AZ_{i2} + B}{CZ_{i2} + D}, \quad 9.23$$

since $V_2 = Z_{i2}I_2$.

Now solve (9.22) for V_2, I_2 by inverting the $ABCD$ matrix. Since $AD - BC = 1$ for a reciprocal network, we obtain

$$V_2 = DV_1 - BI_1, \quad 9.24a$$

$$I_2 = -CV_1 + AI_1. \quad 9.24b$$

Then the input impedance at port 2, with port 1 terminated in Z_{i1}, can be found as

$$Z_{\text{in2}} = \frac{-V_2}{I_2} = -\frac{DV_1 - BI_1}{-CV_1 + AI_1} = \frac{DZ_{i1} + B}{CZ_{i1} + A}, \quad 9.25$$

since $V_1 = -Z_{i1}I_1$ (circuit of Figure 9.7).

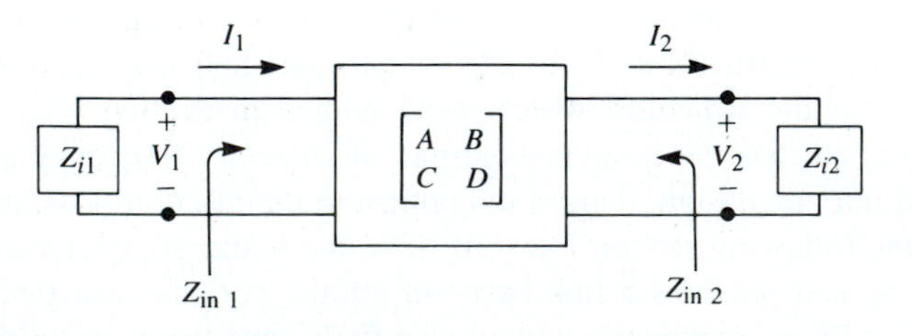

FIGURE 9.7 A two-port network terminated in its image impedances.

We desire that $Z_{\text{in}1} = Z_{i1}$ and $Z_{\text{in}2} = Z_{i2}$, so (9.23) and (9.25) give two equations for the image impedances:

$$Z_{i1}(CZ_{i2} + D) = AZ_{i2} + B, \tag{9.26a}$$

$$Z_{i1}D - B = Z_{i2}(A - CZ_{i1}). \tag{9.26b}$$

Solving for Z_{i1} and Z_{i2} gives

$$Z_{i1} = \sqrt{\frac{AB}{CD}}, \tag{9.27a}$$

$$Z_{i2} = \sqrt{\frac{BD}{AC}}, \tag{9.27b}$$

with $Z_{i2} = DZ_{i1}/A$. If the network is symmetric, then $A = D$ and $Z_{i1} = Z_{i2}$ as expected.

Now consider the voltage transfer function for a two-port network terminated in its image impedances. With reference to Figure 9.8 and (9.24a), the output voltage at port 2 can be expressed as

$$V_2 = DV_1 - BI_1 = \left(D - \frac{B}{Z_{i1}}\right)V_1 \tag{9.28}$$

(since we now have $V_1 = I_1 Z_{i1}$) so the voltage ratio is

$$\frac{V_2}{V_1} = D - \frac{B}{Z_{i1}} = D - B\sqrt{\frac{CD}{AB}} = \sqrt{\frac{D}{A}}(\sqrt{AD} - \sqrt{BC}). \tag{9.29a}$$

Similarly, the current ratio is

$$\frac{I_2}{I_1} = -C\frac{V_1}{I_1} + A = -CZ_{i1} + A = \sqrt{\frac{A}{D}}(\sqrt{AD} - \sqrt{BC}). \tag{9.29b}$$

The factor $\sqrt{D/A}$ occurs in reciprocal positions in (9.29a) and (9.29b), and so can be interpreted as a transformer turns ratio. Apart from this factor we can define a propagation factor for the network as

$$e^{-\gamma} = \sqrt{AD} - \sqrt{BC}, \tag{9.30}$$

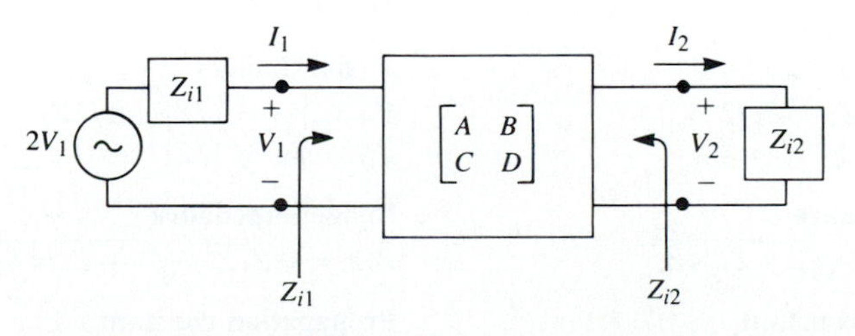

FIGURE 9.8 A two-port network terminated in its image impedances and driven with a voltage generator.

with $\gamma = \alpha + j\beta$ as usual. Since $e^{\gamma} = 1/(\sqrt{AD} - \sqrt{BC}) = (AD - BC)/(\sqrt{AD} - \sqrt{BC}) = \sqrt{AD} + \sqrt{BC}$, and $\cosh\gamma = (e^{\gamma} + e^{-\gamma})/2$, we also have that

$$\cosh\gamma = \sqrt{AD}. \tag{9.31}$$

Two important types of two-port networks are the T and π circuits, which can be made in symmetric form. Table 9.1 lists the image impedances and propagation factors, along with other useful parameters, for these two networks.

Constant-k Filter Sections

Now we are ready to develop low-pass and high-pass filter sections. First consider the T network shown in Figure 9.9; intuitively, we can see that this is a low-pass filter network because the series inductors and shunt capacitor tend to block high-frequency signals while passing low-frequency signals. Comparing with the results given in Table 9.1, we have $Z_1 = j\omega L$ and $Z_2 = 1/j\omega C$, so the image impedance is

$$Z_{iT} = \sqrt{\frac{L}{C}}\sqrt{1 - \frac{\omega^2 LC}{4}}. \tag{9.32}$$

If we define a cutoff frequency, ω_c, as

$$\omega_c = \frac{2}{\sqrt{LC}}, \tag{9.33}$$

TABLE 9.1 Image Parameters for T and π Networks

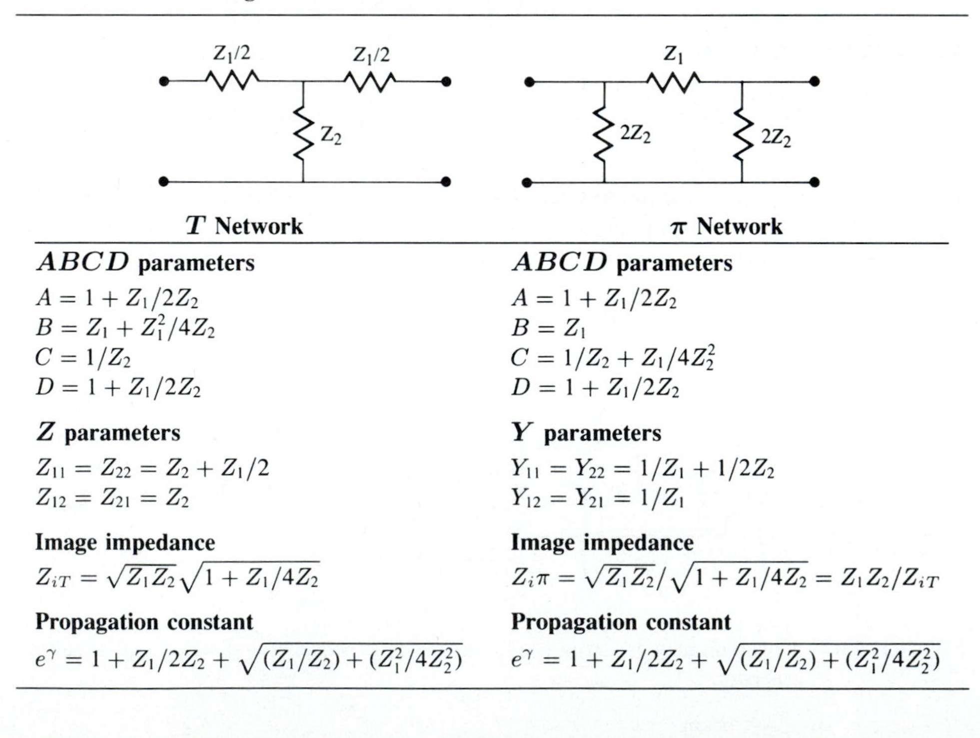

T Network	**π Network**
***ABCD* parameters**	***ABCD* parameters**
$A = 1 + Z_1/2Z_2$	$A = 1 + Z_1/2Z_2$
$B = Z_1 + Z_1^2/4Z_2$	$B = Z_1$
$C = 1/Z_2$	$C = 1/Z_2 + Z_1/4Z_2^2$
$D = 1 + Z_1/2Z_2$	$D = 1 + Z_1/2Z_2$
Z parameters	**Y parameters**
$Z_{11} = Z_{22} = Z_2 + Z_1/2$	$Y_{11} = Y_{22} = 1/Z_1 + 1/2Z_2$
$Z_{12} = Z_{21} = Z_2$	$Y_{12} = Y_{21} = 1/Z_1$
Image impedance	**Image impedance**
$Z_{iT} = \sqrt{Z_1 Z_2}\sqrt{1 + Z_1/4Z_2}$	$Z_{i\pi} = \sqrt{Z_1 Z_2}/\sqrt{1 + Z_1/4Z_2} = Z_1 Z_2/Z_{iT}$
Propagation constant	**Propagation constant**
$e^{\gamma} = 1 + Z_1/2Z_2 + \sqrt{(Z_1/Z_2) + (Z_1^2/4Z_2^2)}$	$e^{\gamma} = 1 + Z_1/2Z_2 + \sqrt{(Z_1/Z_2) + (Z_1^2/4Z_2^2)}$

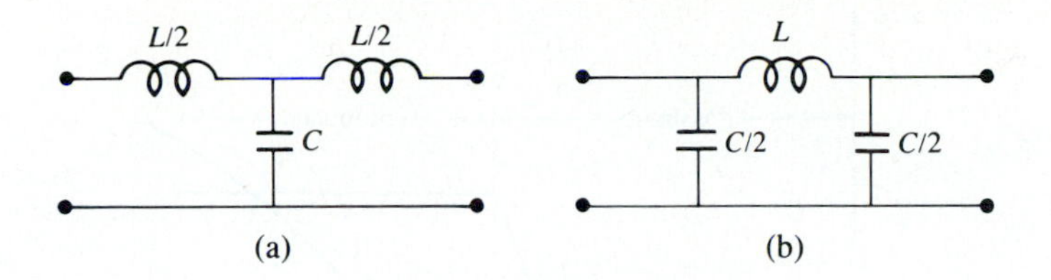

FIGURE 9.9 Low-pass constant-k filter sections in T and π form. (a) T-section. (b) π-section.

and a nominal characteristic impedance, R_0, as

$$R_0 = \sqrt{\frac{L}{C}} = k, \tag{9.34}$$

where k is a constant, then (9.32) can be rewritten as

$$Z_{iT} = R_0\sqrt{1 - \frac{\omega^2}{\omega_c^2}}. \tag{9.35}$$

Then $Z_{iT} = R_0$ for $\omega = 0$.

The propagation factor, also from Table 9.1, is

$$e^\gamma = 1 - \frac{2\omega^2}{\omega_c^2} + \frac{2\omega}{\omega_c}\sqrt{\frac{\omega^2}{\omega_c^2} - 1}. \tag{9.36}$$

Now consider two frequency regions:

1. **For $\omega < \omega_c$:** This is the passband of the filter section. Equation (9.35) shows that Z_{iT} is real, and (9.36) shows that γ is imaginary, since $\omega^2/\omega_c^2 - 1$ is negative and $|e^\gamma| = 1$:

$$|e^\gamma|^2 = \left(1 - \frac{2\omega^2}{\omega_c^2}\right)^2 + \frac{4\omega^2}{\omega_c^2}\left(1 - \frac{\omega^2}{\omega_c^2}\right) = 1.$$

2. **For $\omega > \omega_c$:** This is the stopband of the filter section. Equation (9.35) shows that Z_{iT} is imaginary, and (9.36) shows that γ is real, since e^γ is real and $-1 < e^\gamma < 0$ (as seen from the limits as $\omega \rightarrow \omega_c$ and $\omega \rightarrow \infty$). The attenuation rate for $\omega >> \omega_c$ is 40 dB/decade.

Typical phase and attenuation constants are sketched in Figure 9.10. Observe that the attenuation, α, is zero or relatively small near the cutoff frequency, although $\alpha \rightarrow \infty$ as $\omega \rightarrow \infty$. This type of filter is known as a constant-k low-pass prototype. There are only two parameters to choose (L and C), which are determined by ω_c, the cutoff frequency, and R_0, the image impedance at zero frequency.

The above results are only valid when the filter section is terminated in its image impedance at both ports. This is a major weakness of the design, because the image impedance is a function of frequency which is not likely to match a given source or load impedance. This disadvantage, as well as the fact that the attenuation is rather small near cutoff, can be remedied with the modified m-derived sections to be discussed shortly.

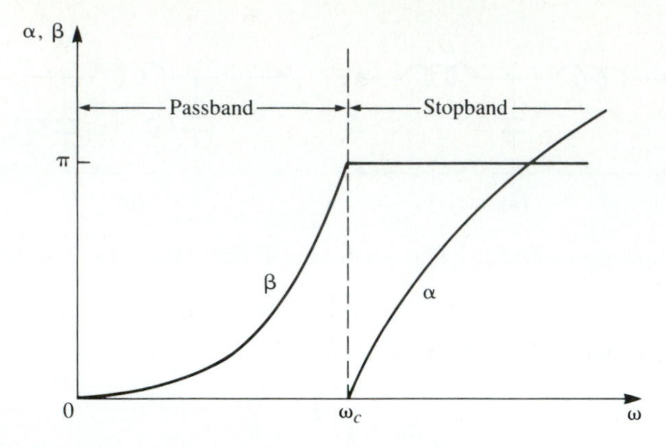

FIGURE 9.10 Typical passband and stopband characteristics of the low-pass constant-k sections of Figure 9.9.

For the low-pass π network of Figure 9.9, we have that $Z_1 = j\omega L$ and $Z_2 = 1/j\omega C$, so the propagation factor is the same as that for the low-pass T network. The cutoff frequency, ω_c, and nominal characteristic impedance, R_0, are the same as the corresponding quantities for the T network as given in (9.33) and (9.34). At $\omega = 0$ we have that $Z_{iT} = Z_{i\pi} = R_0$, where $Z_{i\pi}$ is the image impedance of the low-pass π network, but Z_{iT} and $Z_{i\pi}$ are generally not equal at other frequencies.

High-pass constant-k sections are shown in Figure 9.11; we see that the positions of the inductors and capacitors are reversed from those in the low-pass prototype. The design equations are easily shown to be

$$R_0 = \sqrt{\frac{L}{C}}, \tag{9.37}$$

$$\omega_c = \frac{1}{2\sqrt{LC}}. \tag{9.38}$$

m-Derived Filter Sections

We have seen that the constant-k filter section suffers from the disadvantages of a relatively slow attenuation rate past cutoff, and a nonconstant image impedance. The

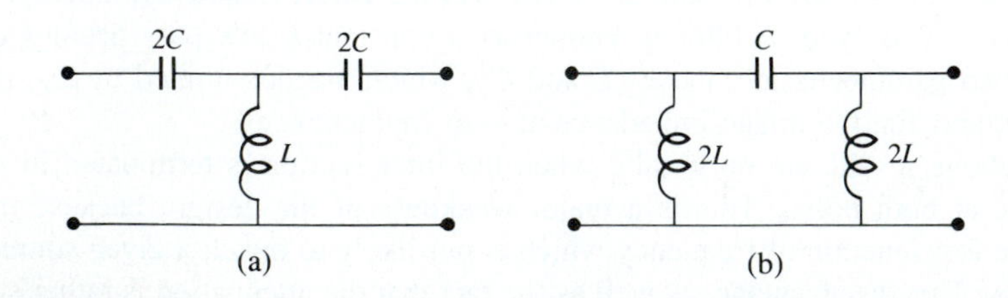

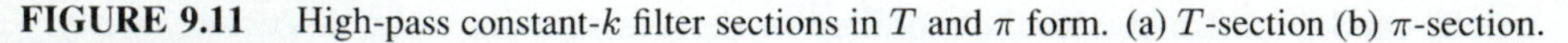

FIGURE 9.11 High-pass constant-k filter sections in T and π form. (a) T-section (b) π-section.

m-derived filter section is a modification of the constant-k section designed to overcome these problems. As shown in Figure 9.12a,b the impedances Z_1 and Z_2 in a constant-k T-section are replaced with Z_1' and Z_2', and we let

$$Z_1' = mZ_1. \tag{9.39}$$

Then we choose Z_2' to obtain the same value of Z_{iT} as for the constant-k section. Thus, from Table 9.1,

$$Z_{iT} = \sqrt{Z_1 Z_2 + \frac{Z_1^2}{4}} = \sqrt{Z_1' Z_2' + \frac{Z_1'^2}{4}} = \sqrt{mZ_1 Z_2' + \frac{m^2 Z_1^2}{4}}. \tag{9.40}$$

Solving for Z_2' gives

$$Z_2' = \frac{Z_2}{m} + \frac{Z_1}{4m} - \frac{mZ_1}{4} = \frac{Z_2}{m} + \frac{(1-m^2)}{4m} Z_1. \tag{9.41}$$

Because the impedances Z_1 and Z_2 represent reactive elements, Z_2' represents two elements in series, as indicated in Figure 9.12c. Note that $m = 1$ reduces to the original constant-k section.

For a low-pass filter, we have $Z_1 = j\omega L$ and $Z_2 = 1/j\omega C$. Then (9.39) and (9.41) give the m-derived components as

$$Z_1' = j\omega L m, \tag{9.42a}$$

$$Z_2' = \frac{1}{j\omega C m} + \frac{(1-m^2)}{4m} j\omega L, \tag{9.42b}$$

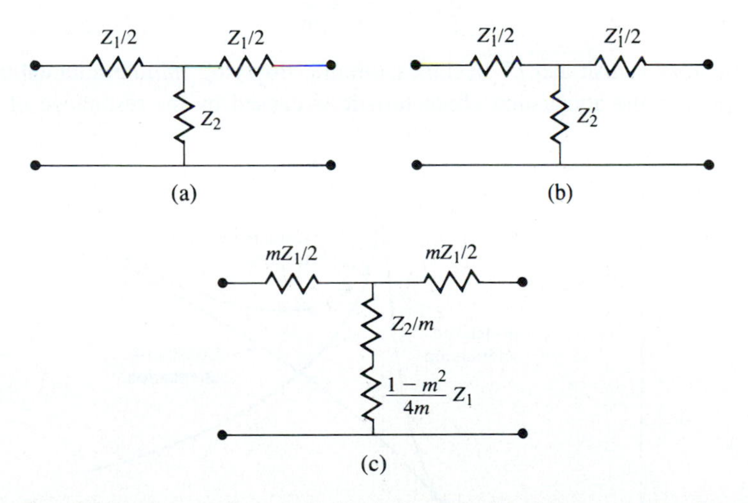

FIGURE 9.12 Development of an m-derived filter section from a constant-k section. (a) Constant-k section. (b) General m-derived section. (c) Final m-derived section.

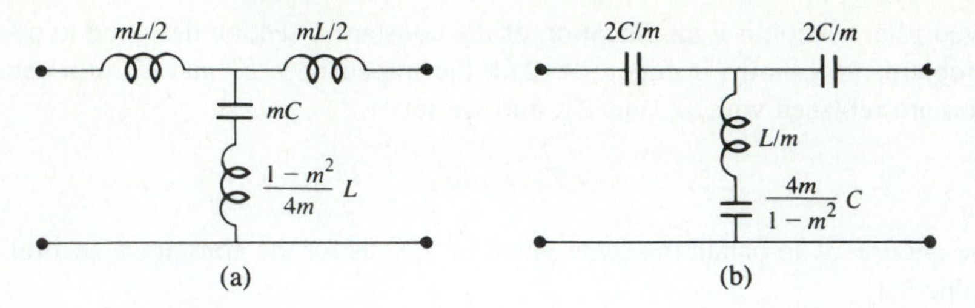

FIGURE 9.13 m-derived filter sections. (a) Low-pass T-section. (b) High-pass T-section.

which results in the circuit of Figure 9.13. Now consider the propagation factor for the m-derived section. From Table 9.1,

$$e^{\gamma} = 1 + \frac{Z_1'}{2Z_2'} + \sqrt{\frac{Z_1'}{Z_2'}\left(1 + \frac{Z_1'}{4Z_2'}\right)}. \tag{9.43}$$

For the low-pass m-derived filter,

$$\frac{Z_1'}{Z_2'} = \frac{j\omega Lm}{(1/j\omega Cm) + j\omega L(1-m^2)/4m} = \frac{-(2\omega m/\omega_c)^2}{1-(1-m^2)(\omega/\omega_c)^2},$$

where $\omega_c = 2/\sqrt{LC}$ as before. Then,

$$1 + \frac{Z_1'}{4Z_2'} = \frac{1-(\omega/\omega_c)^2}{1-(1-m^2)(\omega/\omega_c)^2}.$$

If we restrict $0 < m < 1$, then these results show that e^{γ} is real and $|e^{\gamma}| > 1$ for $\omega > \omega_c$. Thus the stopband begins at $\omega = \omega_c$, as for the constant-k section. However, when $\omega = \omega_{\infty}$, where

$$\omega_{\infty} = \frac{\omega_c}{\sqrt{1-m^2}}, \tag{9.44}$$

the denominators vanish and e^{γ} becomes infinite, implying infinite attenuation. Physically, this pole in the attenuation characteristic is caused by the resonance of the series

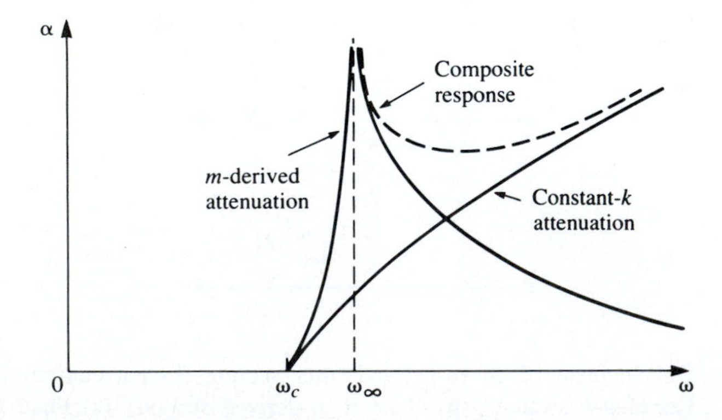

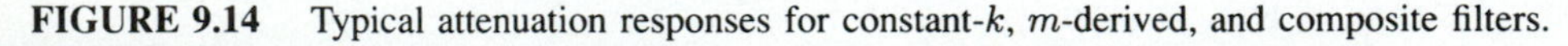
FIGURE 9.14 Typical attenuation responses for constant-k, m-derived, and composite filters.

LC resonator in the shunt arm of the T; this is easily verified by showing that the resonant frequency of this LC resonator is ω_∞. Note that (9.44) indicates that $\omega_\infty > \omega_c$, so infinite attenuation occurs after the cutoff frequency, ω_c, as illustrated in Figure 9.14. The position of the pole at ω_∞ can be controlled with the value of m.

We now have a very sharp cutoff response, but one problem with the m-derived section is that its attenuation decreases for $\omega > \omega_\infty$. Since it is often desirable to have infinite attenuation as $\omega \to \infty$, the m-derived section can be cascaded with a constant-k section to give the composite attenuation response shown in Figure 9.14.

The m-derived T-section was designed so that its image impedance was identical to that of the constant-k section (independent of m), so we still have the problem of a non-constant image impedance. But the image impedance of the π-equivalent will depend on m, and this extra degree of freedom can be used to design an optimum matching section.

The easiest way to obtain the corresponding π-section is to consider it as a piece of an infinite cascade of m-derived T-sections, as shown in Figure 9.15a,b. Then the image impedance of this network is, using the results of Table 9.1 and (9.35),

$$Z_{i\pi} = \frac{Z_1' Z_2'}{Z_{iT}} = \frac{Z_1 Z_2 + Z_1^2(1-m^2)/4}{R_0\sqrt{1-(\omega/\omega_c)^2}}. \qquad 9.45$$

Now $Z_1 Z_2 = L/C = R_0^2$ and $Z_1^2 = -\omega^2 L^2 = -4R_0^2(\omega/\omega_c)^2$, so (9.45) reduces to

$$Z_{i\pi} = \frac{1-(1-m^2)(\omega/\omega_c)^2}{\sqrt{1-(\omega/\omega_c)^2}} R_0. \qquad 9.46$$

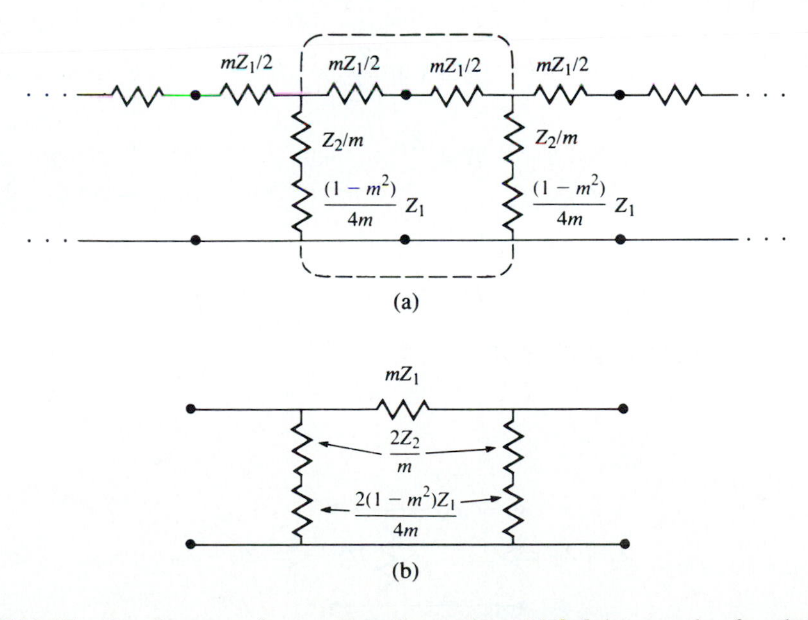

FIGURE 9.15 Development of an m-derived π-section. (a) Infinite cascade of m-derived T-sections. (b) A de-embedded π-equivalent.

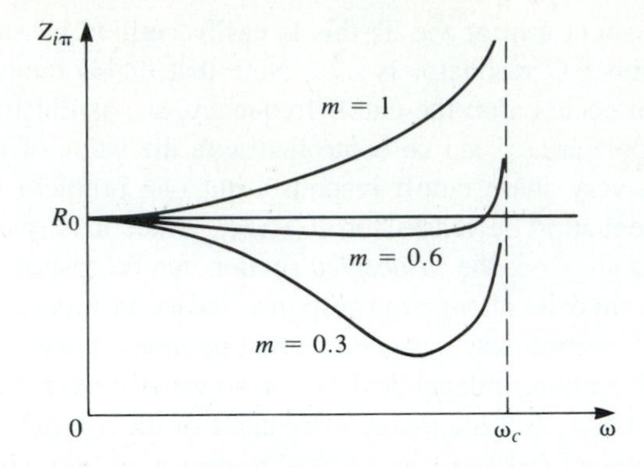

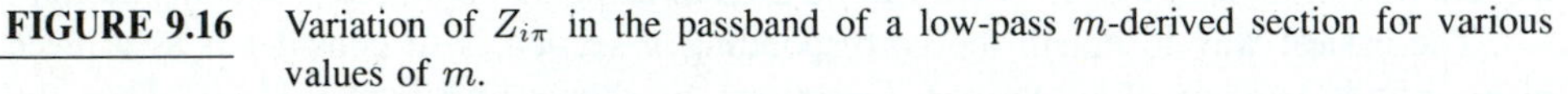

FIGURE 9.16 Variation of $Z_{i\pi}$ in the passband of a low-pass m-derived section for various values of m.

Since this impedance is a function of m, we can choose m to minimize the variation of $Z_{i\pi}$ over the passband of the filter. Figure 9.16 shows this variation with frequency for several values of m; a value of $m = 0.6$ generally gives the best results.

This type of m-derived section can then be used at the input and output of the filter to provide a nearly constant impedance match to and from R_0. But the image impedance of the constant-k and m-derived T-sections, Z_{iT}, does not match $Z_{i\pi}$; this problem can be surmounted by bisecting the π-sections, as shown in Figure 9.17. The image impedances of this circuit are $Z_{i1} = Z_{iT}$ and $Z_{i2} = Z_{i\pi}$, which can be shown by finding its $ABCD$ parameters:

$$A = 1 + \frac{Z_1'}{4Z_2'}, \quad 9.47a$$

$$B = \frac{Z_1'}{2}, \quad 9.47b$$

$$C = \frac{1}{2Z_2'}, \quad 9.47c$$

$$D = 1, \quad 9.47d$$

and then using (9.27) for Z_{i1} and Z_{i2}:

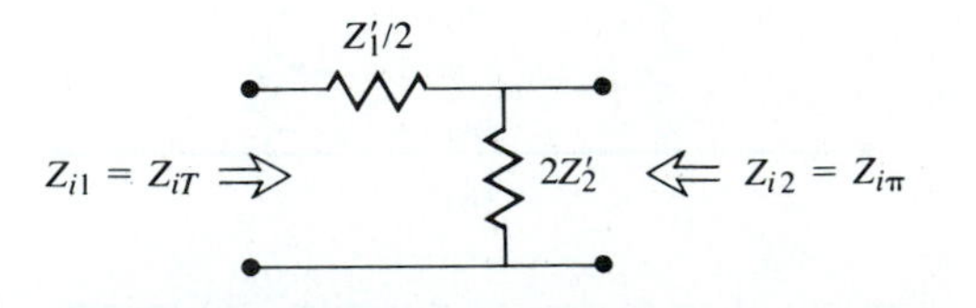

FIGURE 9.17 A bisected π-section used to match $Z_{i\pi}$ to Z_{iT}.

$$Z_{i1} = \sqrt{Z_1' Z_2' + \frac{Z_1'^2}{4}} = Z_{iT}, \tag{9.48a}$$

$$Z_{i2} = \sqrt{\frac{Z_1' Z_2'}{1 + Z_1'/4Z_2'}} = \frac{Z_1' Z_2'}{Z_{iT}} = Z_{i\pi}, \tag{9.48b}$$

where (9.40) has been used for Z_{iT}.

Composite Filters

By combining in cascade the constant-k, m-derived sharp cutoff, and the m-derived matching sections we can realize a filter with the desired attenuation and matching properties. This type of design is called a composite filter, and is shown in Figure 9.18. The sharp-cutoff section, with $m < 0.6$, places an attenuation pole near the cutoff frequency to provide a sharp attenuation response; the constant-k section provides high attenuation further into the stopband. The bisected-π sections at the ends of the filter match the nominal source and load impedance, R_0, to the internal image impedances, Z_{iT}, of the constant-k and m-derived sections. Table 9.2 summarizes the design equations for low- and high-pass composite filters; notice that once the cutoff frequency and impedance are specified, there is only one degree of freedom (the value of m for the sharp-cutoff section) left to control the filter response. The following example illustrates the design procedure.

EXAMPLE 9.2

Design a low-pass composite filter with a cutoff frequency of 2 MHz and impedance of 75 Ω. Place the infinite attenuation pole at 2.05 MHz, and plot the frequency response from 0 to 4 MHz.

Solution

All the component values can be found from Table 9.2. For the constant-k section:

$$L = \frac{2R_0}{\omega_c} = 11.94\ \mu\text{H}, \qquad C = \frac{2}{R_0\omega_c} = 2.122\ \text{nF.}$$

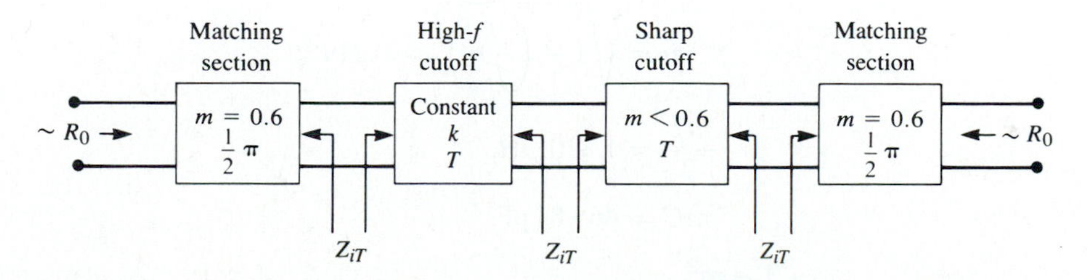

FIGURE 9.18 The final four-stage composite filter.

TABLE 9.2 Summary of Composite Filter Design

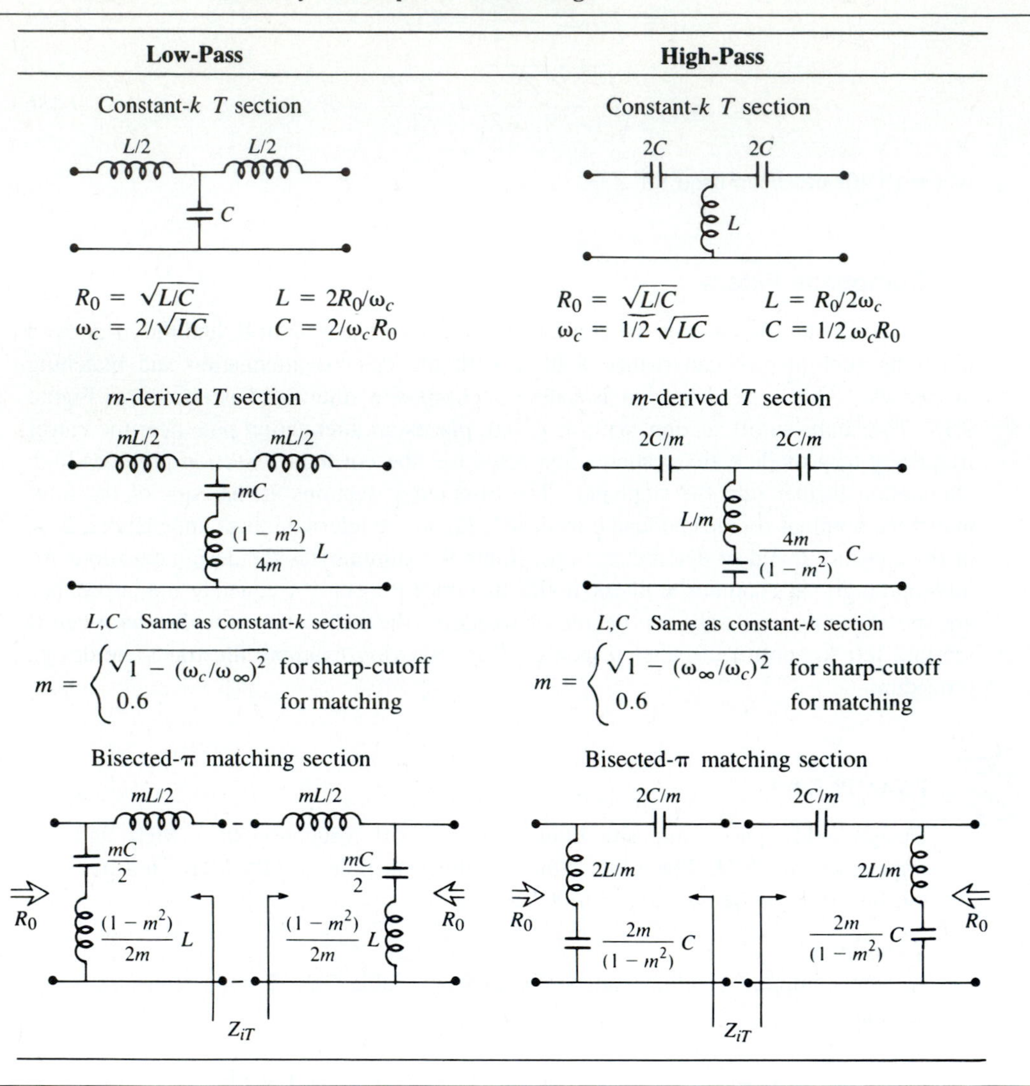

For the m-derived sharp-cutoff section:

$$m = \sqrt{1 - \left(\frac{f_c}{f_\infty}\right)^2} = 0.2195,$$

$$\frac{mL}{2} = 1.310\ \mu\text{H},$$

$$mC = 465.8\ \text{pF},$$

$$\frac{1-m^2}{4m}L = 12.94\ \mu\text{H}.$$

For the $m = 0.6$ matching sections:

$$\frac{mL}{2} = 3.582\ \mu\text{H},$$

$$\frac{mC}{2} = 636.5\ \text{pF},$$

$$\frac{1 - m^2}{2m} L = 6.368\ \mu\text{H}.$$

The completed filter circuit is shown in Figure 9.19; the series pairs of inductors between the sections can be combined. Figure 9.20 shows the resulting frequency response for $|S_{12}|$. Note the sharp dip at $f = 2.05$ MHz due to the $m = 0.2195$ section, and the pole at 2.56 MHz, which is due to the $m = 0.6$ matching sections. ○

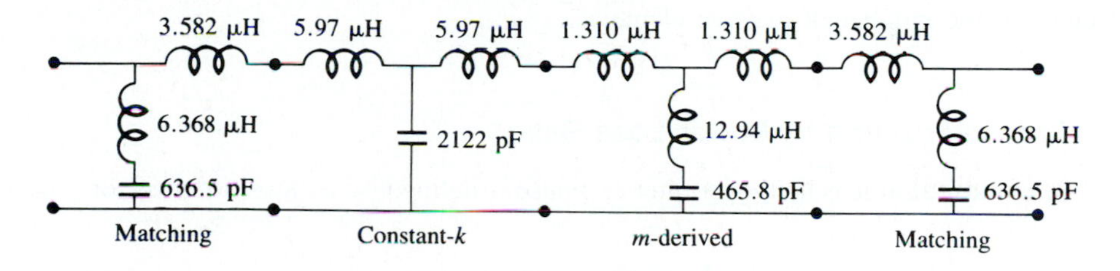

FIGURE 9.19 Low-pass composite filter for Example 9.2.

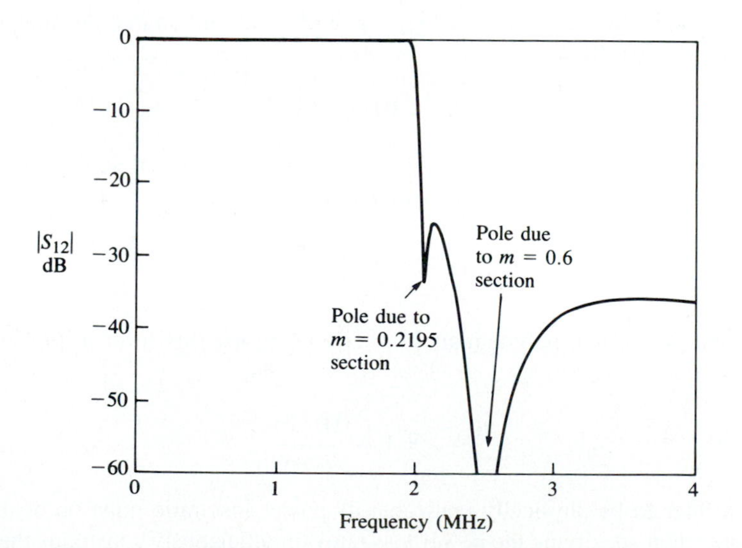

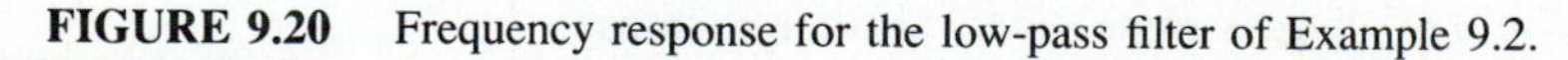

FIGURE 9.20 Frequency response for the low-pass filter of Example 9.2.

9.3 FILTER DESIGN BY THE INSERTION LOSS METHOD

The perfect filter would have zero insertion loss in the passband, infinite attenuation in the stopband, and a linear phase response (to avoid signal distortion) in the passband. Of course, such filters do not exist in practice, so compromises must be made; herein lies the art of filter design.

The image parameter method of the previous section may yield a usable filter response, but if not there is no clear-cut way to improve the design. The insertion loss method, however, allows a high degree of control over the passband and stopband amplitude and phase characteristics, with a systematic way to synthesize a desired response. The necessary design trade-offs can be evaluated to best meet the application requirements. If, for example, a minimum insertion loss is most important, a binomial response could be used; a Chebyshev response would satisfy a requirement for the sharpest cutoff. If it is possible to sacrifice the attenuation rate, a better phase response can be obtained by using a linear phase filter design. And in all cases, the insertion loss method allows filter performance to be improved in a straightforward manner, at the expense of a higher-order filter. For the filter prototypes to be discussed below, the order of the filter is equal to the number of reactive elements.

Characterization by Power Loss Ratio

In the insertion loss method a filter response is defined by its insertion loss, or *power loss ratio*, P_{LR}:

$$P_{\mathrm{LR}} = \frac{\text{Power available from source}}{\text{Power delivered to load}} = \frac{P_{\mathrm{inc}}}{P_{\mathrm{load}}} = \frac{1}{1 - |\Gamma(\omega)|^2}. \tag{9.49}$$

(Observe that this quantity is the reciprocal of the transducer power gain defined in Section 5.6, and is the reciprocal of $|S_{12}|^2$ if both load and source are matched.) The insertion loss (IL) in dB is

$$\mathrm{IL} = 10 \log P_{\mathrm{LR}}. \tag{9.50}$$

From Section 5.2 we know that $|\Gamma(\omega)|^2$ is an even function of ω, therefore it can be expressed as a polynomial in ω^2. Thus we can write

$$|\Gamma(\omega)|^2 = \frac{M(\omega^2)}{M(\omega^2) + N(\omega^2)}, \tag{9.51}$$

where M and N are real polynomials in ω^2. Substituting this form in (9.49) gives the following:

$$P_{\mathrm{LR}} = 1 + \frac{M(\omega^2)}{N(\omega^2)}. \tag{9.52}$$

Thus, for a filter to be physically realizable its power loss ratio must be of the form in (9.52). Notice that specifying the power loss ratio simultaneously constrains the reflection coefficient, $\Gamma(\omega)$. We now discuss some practical filter responses.

Maximally flat. This characteristic is also called the binomial or Butterworth response, and is optimum in the sense that it provides the flattest possible passband response for a given filter complexity, or order. For a low-pass filter, it is specified by

$$P_{\text{LR}} = 1 + k^2 \left(\frac{\omega}{\omega_c}\right)^{2N}, \tag{9.53}$$

where N is the order of the filter, and ω_c is the cutoff frequency. The passband extends from $\omega = 0$ to $\omega = \omega_c$; at the band edge the power loss ratio is $1 + k^2$. If we choose this as the -3 dB point, as is common, we have $k = 1$, which we will assume from now on. For $\omega > \omega_c$, the attenuation increases monotonically with frequency, as shown in Figure 9.21. For $\omega >> \omega_c$, $P_{\text{LR}} \simeq k^2(\omega/\omega_c)^{2N}$, which shows that the insertion loss increases at the rate of $20N$ dB/decade. Like the binomial response for multisection quarter-wave matching transformers, the first $(2N - 1)$ derivatives of (9.53) are zero at $\omega = 0$.

Equal ripple. If a Chebyshev polynomial is used to specify the insertion loss of an N-order low-pass filter as

$$P_{\text{LR}} = 1 + k^2 T_N^2 \left(\frac{\omega}{\omega_c}\right), \tag{9.54}$$

then a sharper cutoff will result, although the passband response will have ripples of amplitude $1+k^2$, as shown in Figure 9.21, since $T_N(x)$ oscillates between ± 1 for $|x| \leq 1$. Thus, k^2 determines the passband ripple level. For large x, $T_N(x) \simeq 1/2(2x)^N$, so for $\omega >> \omega_c$ the insertion loss becomes

$$P_{\text{LR}} \simeq \frac{k^2}{4} \left(\frac{2\omega}{\omega_c}\right)^{2N},$$

which also increases at the rate of $20N$ dB/decade. But the insertion loss for the Chebyshev case is $(2^{2N})/4$ greater than the binomial response, at any given frequency where $\omega >> \omega_c$.

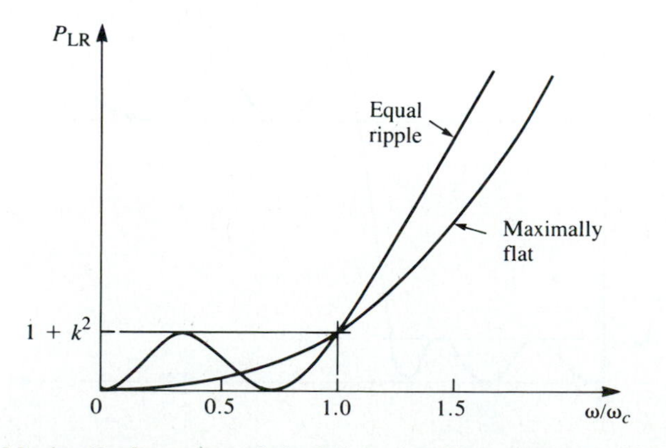

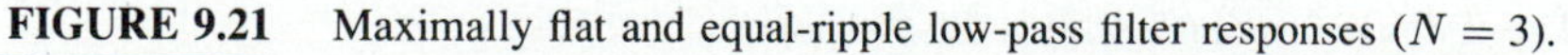

FIGURE 9.21 Maximally flat and equal-ripple low-pass filter responses ($N = 3$).

Elliptic function. The maximally flat and equal-ripple responses both have monotonically increasing attenuation in the stopband. In many applications it is adequate to specify a minimum stopband attenuation, in which case a better cutoff rate can be obtained. Such filters are called elliptic function filters [3], and have equal-ripple responses in the passband as well as the stopband, as shown in Figure 9.22. The maximum attenuation in the passband, A_{max}, can be specified, as well as the minimum attenuation in the stopband, A_{min}. Elliptic function filters are difficult to synthesize, so we will not consider them further; the interested reader is referred to reference [3].

Linear phase. The above filters specify the amplitude response, but in some applications (such as multiplexing filters for communication systems) it is important to have a linear phase response in the passband to avoid signal distortion. It turns out that a sharp-cutoff response is generally incompatible with a good phase response, so the phase response of the filter must be deliberately synthesized, usually resulting in an inferior amplitude cutoff characteristic. A linear phase characteristic can be achieved with the following phase response:

$$\phi(\omega) = A\omega \left[1 + p\left(\frac{\omega}{\omega_c}\right)^{2N}\right], \tag{9.55}$$

where $\phi(\omega)$ is the phase of the voltage transfer function of the filter, and p is a constant. A related quantity is the group delay, defined as

$$\tau_d = \frac{d\phi}{d\omega} = A\left[1 + p(2N+1)\left(\frac{\omega}{\omega_c}\right)^{2N}\right], \tag{9.56}$$

which shows that the group delay for a linear phase filter is a maximally flat function.

More general filter specifications can be obtained, but the above cases are the most common. We will next discuss the design of low-pass filter prototypes which are normalized in terms of impedance and frequency; this normalization simplifies the design of filters for arbitrary frequency, impedance, and type (low-pass, high-pass, bandpass,

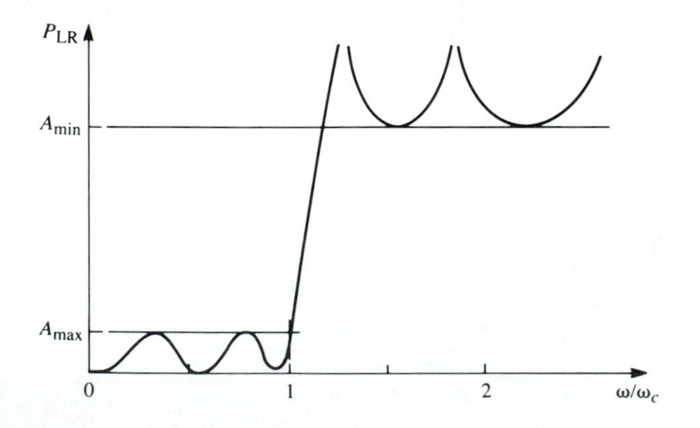

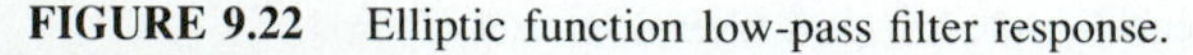
FIGURE 9.22 Elliptic function low-pass filter response.

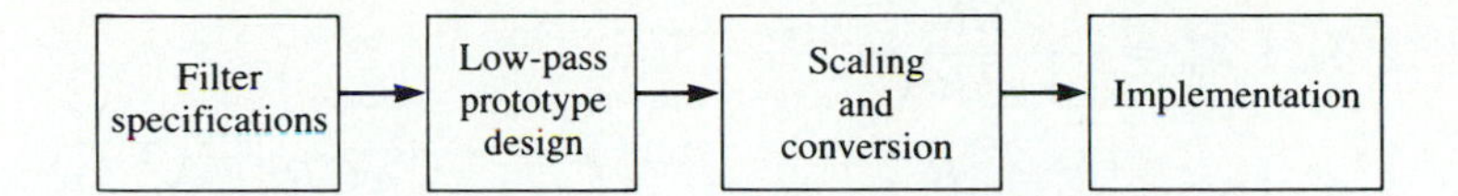

FIGURE 9.23 The process of filter design by the insertion loss method.

or bandstop). The low-pass prototypes are then scaled to the desired frequency and impedance; and the lumped-element components replaced with distributed circuit elements for implementation at microwave frequencies. This design process is illustrated in Figure 9.23.

Maximally Flat Low-Pass Filter Prototype

Consider the two-element low-pass filter prototype shown in Figure 9.24; we will derive the normalized element values, L and C, for a maximally flat response. We assume a source impedance of 1 Ω, and a cutoff frequency $\omega_c = 1$. From (9.53), the desired power loss ratio will be, for $N = 2$,

$$P_{\text{LR}} = 1 + \omega^4. \tag{9.57}$$

The input impedance of this filter is

$$Z_{\text{in}} = j\omega L + \frac{R(1 - j\omega RC)}{1 + \omega^2 R^2 C^2}. \tag{9.58}$$

Since

$$\Gamma = \frac{Z_{\text{in}} - 1}{Z_{\text{in}} + 1},$$

the power loss ratio can be written as

$$P_{\text{LR}} = \frac{1}{1 - |\Gamma|^2} = \frac{1}{1 - [(Z_{\text{in}} - 1)/(Z_{\text{in}} + 1)][(Z_{\text{in}}^* - 1)/(Z_{\text{in}}^* + 1)]} = \frac{|Z_{\text{in}} + 1|^2}{2(Z_{\text{in}} + Z_{\text{in}}^*)}. \tag{9.59}$$

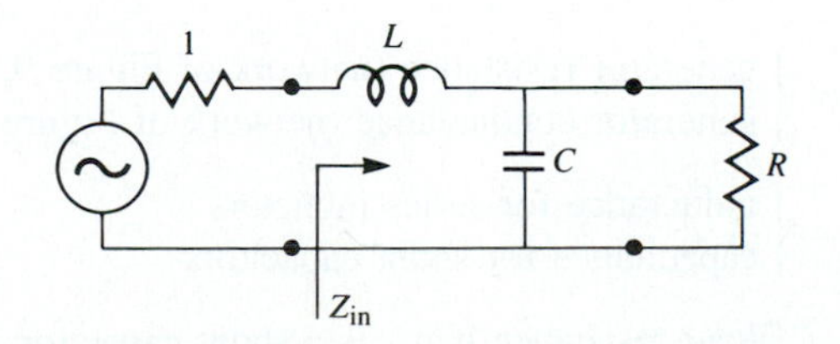

FIGURE 9.24 Low-pass filter prototype, $N = 2$.

Now, $$Z_{\text{in}} + Z_{\text{in}}^* = \frac{2R}{1+\omega^2R^2C^2},$$

and $$|Z_{\text{in}}+1|^2 = \left(\frac{R}{1+\omega^2R^2C^2}+1\right)^2 + \left(\omega L - \frac{\omega C R^2}{1+\omega^2R^2C^2}\right)^2,$$

so (9.59) becomes

$$\begin{aligned} P_{\text{LR}} &= \frac{1+\omega^2R^2C^2}{4R}\left[\left(\frac{R}{1+\omega^2R^2C^2}+1\right)^2 + \left(\omega L - \frac{\omega C R^2}{1+\omega^2R^2C^2}\right)^2\right] \\ &= \frac{1}{4R}(R^2+2R+1+R^2\omega^2C^2+\omega^2L^2+\omega^4L^2C^2R^2-2\omega^2LCR^2) \\ &= 1+\frac{1}{4R}\left[(1-R)^2+(R^2C^2+L^2-2LCR^2)\omega^2+L^2C^2R^2\omega^4\right]. \end{aligned} \tag{9.60}$$

Notice that this expression is a polynomial in ω^2. Comparing to the desired response of (9.57) shows that $R = 1$, since $P_{\text{LR}} = 1$ for $\omega = 0$. In addition, the coefficient of ω^2 must vanish, so

$$C^2 + L^2 - 2LC = (C-L)^2 = 0,$$

or $L = C$. Then for the coefficient of ω^4 to be unity we must have

$$\frac{1}{4}L^2C^2 = \frac{1}{4}L^4 = 1,$$

or $$L = C = \sqrt{2}.$$

In principle, this procedure can be extended to find the element values for filters with an arbitrary number of elements, N, but clearly this is not practical for large N. For a normalized low-pass design where the source impedance is 1 Ω and the cutoff frequency is $\omega_c = 1$, however, the element values for the ladder-type circuits of Figure 9.25 can be tabulated [2]. Table 9.3 gives such element values for maximally flat low-pass filter prototypes for $N = 1$ to 10. (Notice that the values for $N = 2$ agree with the above analytical solution.) This data is used with either of the ladder circuits of Figure 9.25 in the following way. The element values are numbered from g_0 at the generator impedance to g_{N+1} at the load impedance, for a filter having N reactive elements. The elements alternate between series and shunt connections, and g_k has the following definition:

$$g_0 = \begin{cases} \text{generator resistance (network of Figure 9.25a)} \\ \text{generator conductance (network of Figure 9.25b)} \end{cases}$$

$$\underset{(k=1 \text{ to } N)}{g_k} = \begin{cases} \text{inductance for series inductors} \\ \text{capacitance for shunt capacitors} \end{cases}$$

$$g_{N+1} = \begin{cases} \text{load resistance if } g_N \text{ is a shunt capacitor} \\ \text{load conductance if } g_N \text{ is a series inductor} \end{cases}$$

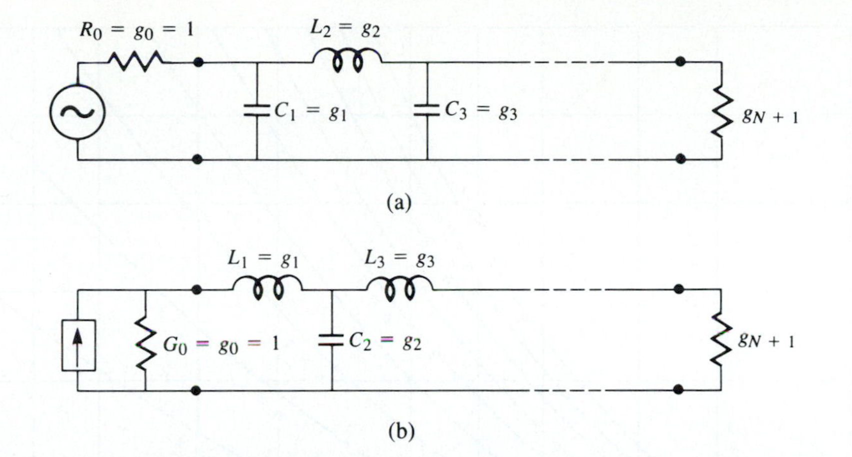

FIGURE 9.25 Ladder circuits for low-pass filter prototypes and their element definitions. (a) Prototype beginning with a shunt element. (b) Prototype beginning with a series element.

TABLE 9.3 Element Values for Maximally Flat Low-Pass Filter Prototypes ($g_0 = 1$, $\omega_c = 1$, $N = 1$ to 10)

N	g_1	g_2	g_3	g_4	g_5	g_6	g_7	g_8	g_9	g_{10}	g_{11}
1	2.0000	1.0000									
2	1.4142	1.4142	1.0000								
3	1.0000	2.0000	1.0000	1.0000							
4	0.7654	1.8478	1.8478	0.7654	1.0000						
5	0.6180	1.6180	2.0000	1.6180	0.6180	1.0000					
6	0.5176	1.4142	1.9318	1.9318	1.4142	0.5176	1.0000				
7	0.4450	1.2470	1.8019	2.0000	1.8019	1.2470	0.4450	1.0000			
8	0.3902	1.1111	1.6629	1.9615	1.9615	1.6629	1.1111	0.3902	1.0000		
9	0.3473	1.0000	1.5321	1.8794	2.0000	1.8794	1.5321	1.0000	0.3473	1.0000	
10	0.3129	0.9080	1.4142	1.7820	1.9754	1.9754	1.7820	1.4142	0.9080	0.3129	1.0000

Source: Reprinted from G. L. Matthaei, L. Young, and E. M. T. Jones, *Microwave Filters, Impedance-Matching Networks, and Coupling Structures* (Dedham, Mass.: Artech House, 1980) with permission.

Then the circuits of Figure 9.25 can be considered as the dual of each other, and both will give the same response.

Finally, as a matter of practical design procedure, it will be necessary to determine the size, or order, of the filter. This is usually dictated by a specification on the insertion loss at some frequency in the stopband of the filter. Figure 9.26 shows the attenuation characteristics for various N, versus normalized frequency. If a filter with $N > 10$ is required, a good result can usually be obtained by cascading two designs of lower order.

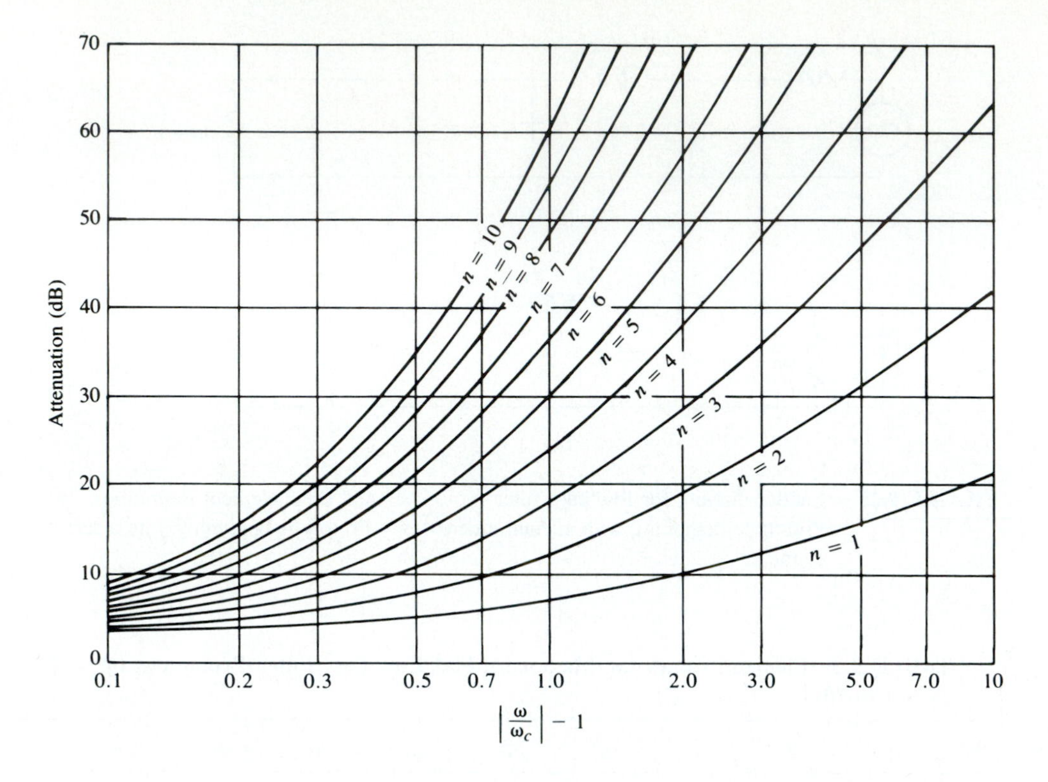

FIGURE 9.26 Attenuation versus normalized frequency for maximally flat filter prototypes. Adapted from G. L. Matthaei, L. Young, and E. M. T. Jones, *Microwave Filters, Impedance-Matching Networks, and Coupling Structures* (Dedham, Mass.: Artech House, 1980) with permission.

EXAMPLE 9.3

A maximally flat low-pass filter is to be designed with a cutoff frequency of 8 GHz and a minimum attenuation of 20 dB at 11 GHz. How many filter elements are required?

Solution

We have $\omega/2\pi = 11$ GHz and $\omega_c/2\pi = 8$ GHz, so

$$\left|\frac{\omega}{\omega_c}\right| - 1 = \frac{11}{8} - 1 = 0.375.$$

Then from Figure 9.26 we see that an attenuation of 20 dB at this frequency requires that $N \geq 8$. Further design details will be discussed in Section 9.4. ○

Equal-Ripple Low-Pass Filter Prototype

For an equal-ripple low-pass filter with a cutoff frequency $\omega_c = 1$, the power loss ratio from (9.54) is

$$P_{\rm LR} = 1 + k^2 T_N^2(\omega), \qquad 9.61$$

where $1 + k^2$ is the ripple level in the passband. Since the Chebyshev polynomials have the property that

$$T_N(0) = \begin{cases} 0 & \text{for } N \text{ odd,} \\ 1 & \text{for } N \text{ even,} \end{cases}$$

equation (9.61) shows that the filter will have a unity power loss ratio at $\omega = 0$ for N odd, but a power loss ratio of $1 + k^2$ at $\omega = 0$ for N even. Thus, there are two cases to consider, depending on N.

For the two-element filter of Figure 9.24, the power loss ratio is given in terms of the component values in (9.60). From (6.56b), we see that $T_2(x) = 2x^2 - 1$, so equating (9.61) to (9.60) gives

$$1+k^2(4\omega^4-4\omega^2+1) = 1+\frac{1}{4R}[(1-R)^2+(R^2C^2+L^2-2LCR^2)\omega^2+L^2C^2R^2\omega^4], \qquad 9.62$$

which can be solved for R, L, and C if the ripple level (as determined by k^2) is known. Thus, at $\omega = 0$ we have that

$$k^2 = \frac{(1-R)^2}{4R},$$

or

$$R = 1 + 2k^2 - 2k\sqrt{1+k^2} \qquad \text{(for } N \text{ even).} \qquad 9.63$$

Equating coefficients of ω^2 and ω^4 yields the additional relations,

$$4k^2 = \frac{1}{4R}L^2C^2R^2,$$

$$-4k^2 = \frac{1}{4R}(R^2C^2 + L^2 - 2LCR^2),$$

which can be used to find L and C. Note that (9.63) gives a value for R that is not unity, so there will be an impedance mismatch if the load actually has a unity (normalized) impedance; this can be corrected with a quarter-wave transformer, or by using an additional filter element to make N odd. For odd N, it can be shown that $R = 1$. (This is because there is a unity power loss ratio at $\omega = 0$ for N odd).

Tables exist for designing equal-ripple low-pass filters with a normalized source impedance and cutoff frequency ($\omega_c' = 1$) [2], and can be applied to either of the ladder circuits of Figure 9.25. This design data depends on the specified passband ripple level; Table 9.4 lists element values for normalized low-pass filter prototypes having 0.5 dB

TABLE 9.4 Element Values for Equal-Ripple Low-Pass Filter Prototypes (g_0 = 1, ω_c = 1, N = 1 to 10, 0.5 dB and 3.0 dB ripple)

0.5 dB Ripple

N	g_1	g_2	g_3	g_4	g_5	g_6	g_7	g_8	g_9	g_{10}	g_{11}
1	0.6986	1.0000									
2	1.4029	0.7071	1.9841								
3	1.5963	1.0967	1.5963	1.0000							
4	1.6703	1.1926	2.3661	0.8419	1.9841						
5	1.7058	1.2296	2.5408	1.2296	1.7058	1.0000					
6	1.7254	1.2479	2.6064	1.3137	2.4758	0.8696	1.9841				
7	1.7372	1.2583	2.6381	1.3444	2.6381	1.2583	1.7372	1.000			
8	1.7451	1.2647	2.6564	1.3590	2.6964	1.3389	2.5093	0.8796	1.9841		
9	1.7504	1.2690	2.6678	1.3673	2.7239	1.3673	2.6678	1.2690	1.7504	1.0000	
10	1.7543	1.2721	2.6754	1.3725	2.7392	1.3806	2.7231	1.3485	2.5239	0.8842	1.9841

3.0 dB Ripple

N	g_1	g_2	g_3	g_4	g_5	g_6	g_7	g_8	g_9	g_{10}	g_{11}
1	1.9953	1.0000									
2	3.1013	0.5339	5.8095								
3	3.3487	0.7117	3.3487	1.0000							
4	3.4389	0.7483	4.3471	0.5920	5.8095						
5	3.4817	0.7618	4.5381	0.7618	3.4817	1.0000					
6	3.5045	0.7685	4.6061	0.7929	4.4641	0.6033	5.8095				
7	3.5182	0.7723	4.6386	0.8039	4.6386	0.7723	3.5182	1.0000			
8	3.5277	0.7745	4.6575	0.8089	4.6990	0.8018	4.4990	0.6073	5.8095		
9	3.5340	0.7760	4.6692	0.8118	4.7272	0.8118	4.6692	0.7760	3.5340	1.0000	
10	3.5384	0.7771	4.6768	0.8136	4.7425	0.8164	4.7260	0.8051	4.5142	0.6091	5.8095

Source: Reprinted from G. L. Matthaei, L. Young, and E. M. T. Jones, *Microwave Filters, Impedance-Matching Networks, and Coupling Structures* (Dedham, Mass.: Artech House, 1980) with permission.

or 3.0 dB ripple, for $N = 1$ to 10. Notice that the load impedance $g_{N+1} \neq 1$ for even N. If the stopband attenuation is specified, the curves in Figures 9.27a,b can be used to determine the necessary value of N for these ripple values.

Linear Phase Low-Pass Filter Prototypes

Filters having a maximally flat time delay, or a linear phase response, can be designed in the same way, but things are somewhat more complicated because the phase of the voltage transfer function is not as simply expressed as is its amplitude. Design values have been derived for such filters [2], however, again for the ladder circuits of Figure 9.25, and are given in Table 9.5 for a normalized source impedance and cutoff frequency ($\omega_c' = 1$). The resulting group delay in the passband will be $\tau_d = 1/\omega_c' = 1$.

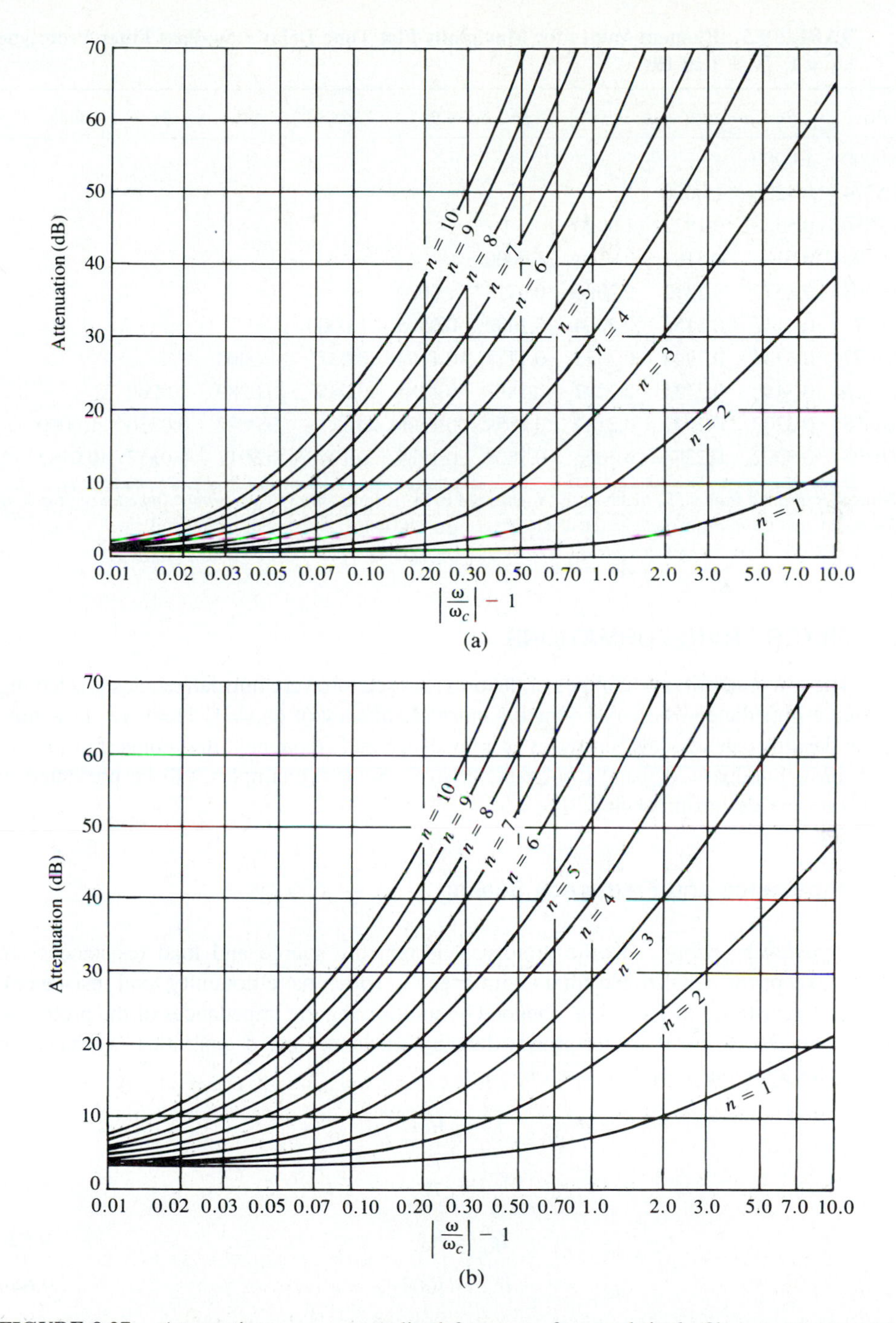

FIGURE 9.27 Attenuation versus normalized frequency for equal-ripple filter prototypes. (a) 0.5 dB ripple level. (b) 3.0 dB ripple level.

Adapted from G. L. Matthaei, L. Young, and E. M. T. Jones, *Microwave Filters, Impedance-Matching Networks, and Coupling Structures* (Dedham, Mass.: Artech House, 1980) with permission.

TABLE 9.5 Element Values for Maximally Flat Time Delay Low-Pass Filter Prototypes (g_0 = 1, ω_c = 1, N = 1 to 10)

N	g_1	g_2	g_3	g_4	g_5	g_6	g_7	g_8	g_9	g_{10}	g_{11}
1	2.0000	1.0000									
2	1.5774	0.4226	1.0000								
3	1.2550	0.5528	0.1922	1.0000							
4	1.0598	0.5116	0.3181	0.1104	1.0000						
5	0.9303	0.4577	0.3312	0.2090	0.0718	1.0000					
6	0.8377	0.4116	0.3158	0.2364	0.1480	0.0505	1.0000				
7	0.7677	0.3744	0.2944	0.2378	0.1778	0.1104	0.0375	1.0000			
8	0.7125	0.3446	0.2735	0.2297	0.1867	0.1387	0.0855	0.0289	1.0000		
9	0.6678	0.3203	0.2547	0.2184	0.1859	0.1506	0.1111	0.0682	0.0230	1.0000	
10	0.6305	0.3002	0.2384	0.2066	0.1808	0.1539	0.1240	0.0911	0.0557	0.0187	1.0000

Source: Reprinted from G. L. Matthaei, L. Young, and E. M. T. Jones, *Microwave Filters, Impedance-Matching Networks, and Coupling Structures* (Dedham, Mass.: Artech House, 1980) with permission.

9.4 FILTER TRANSFORMATIONS

The low-pass filter prototypes of the previous section were normalized designs having a source impedance of $R_s = 1\ \Omega$ and a cutoff frequency of $\omega_c = 1$. Here we show how these designs can be scaled in terms of impedance and frequency, and converted to give high-pass, bandpass, or bandstop characteristics. Several examples will be presented to illustrate the design procedure.

Impedance and Frequency Scaling

Impedance scaling. In the prototype design, the source and load resistances are unity (except for equal-ripple filters with even N, which have nonunity load resistance). A source resistance of R_0 can be obtained by multiplying the impedances of the prototype design by R_0. Then, if we let primes denote impedance scaled quantities, we have the new filter component values given by

$$L' = R_0 L, \qquad 9.64a$$

$$C' = \frac{C}{R_0}, \qquad 9.64b$$

$$R'_s = R_0, \qquad 9.64c$$

$$R'_L = R_0 R_L, \qquad 9.64d$$

where L, C, and R_L are the component values for the original prototype.

Frequency scaling for low-pass filters. To change the cutoff frequency of a low-pass prototype from unity to ω_c requires that we scale the frequency dependence of the filter by the factor $1/\omega_c$, which is accomplished by replacing ω by ω/ω_c:

$$\omega \leftarrow \frac{\omega}{\omega_c}. \tag{9.65}$$

Then the new power loss ratio will be

$$P'_{\mathrm{LR}}(\omega) = P_{\mathrm{LR}}\left(\frac{\omega}{\omega_c}\right),$$

where ω_c is the new cutoff frequency; cutoff occurs when $\omega/\omega_c = 1$, or $\omega = \omega_c$. This transformation can be viewed as a stretching, or expansion, of the original passband, as illustrated in Figure 9.28.

The new element values are determined by applying the substitution of (9.65) to the series reactances, $j\omega L_k$, and shunt susceptances, $j\omega C_k$, of the prototype filter. Thus,

$$jX_k = j\frac{\omega}{\omega_c}L_k = j\omega L'_k,$$

$$jB_k = j\frac{\omega}{\omega_c}C_k = j\omega C'_k,$$

which shows that the new element values are given by

$$L'_k = \frac{L_k}{\omega_c}, \tag{9.66a}$$

$$C'_k = \frac{C_k}{\omega_c}. \tag{9.66b}$$

When both impedance and frequency scaling are required, the results of (9.64) can be combined with (9.66) to give

$$L'_k = \frac{R_0 L}{\omega_c}, \tag{9.67a}$$

$$C'_k = \frac{C_k}{R_0\omega_c}. \tag{9.67b}$$

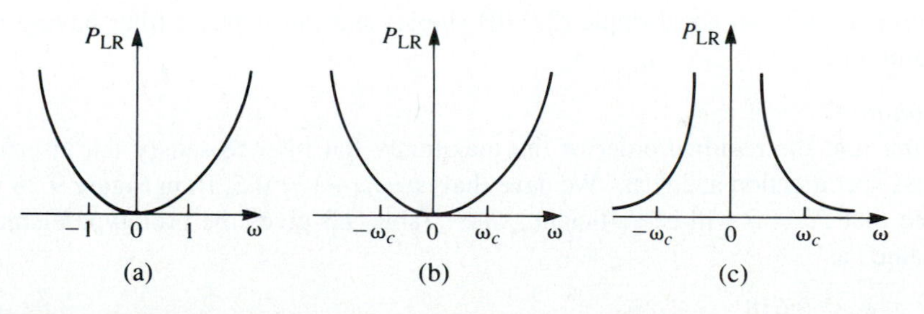

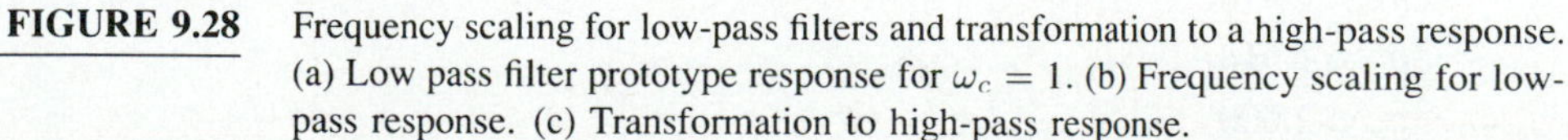

FIGURE 9.28 Frequency scaling for low-pass filters and transformation to a high-pass response. (a) Low pass filter prototype response for $\omega_c = 1$. (b) Frequency scaling for low-pass response. (c) Transformation to high-pass response.

Low-pass to high-pass transformation. The frequency substitution where,

$$\omega \leftarrow -\frac{\omega_c}{\omega}, \tag{9.68}$$

can be used to convert a low-pass response to a high-pass response, as shown in Figure 9.28. This substitution maps $\omega = 0$ to $\omega = \pm\infty$, and vice versa; cutoff occurs when $\omega = \pm\omega_c$. The negative sign is needed to convert inductors (and capacitors) to realizable capacitors (and inductors). Applying (9.68) to the series reactances, $j\omega L_k$, and the shunt susceptances, $j\omega C_k$, of the prototype filter gives

$$jX_k = -j\frac{\omega_c}{\omega}L_k = \frac{1}{j\omega C'_k},$$

$$jB_k = -j\frac{\omega_c}{\omega}C_k = \frac{1}{j\omega L'_k},$$

which shows that series inductors L_k must be replaced with capacitors C'_k, and shunt capacitors C_k must be replaced with inductors L'_k. The new component values are given by

$$C'_k = \frac{1}{\omega_c L_k}, \tag{9.69a}$$

$$L'_k = \frac{1}{\omega_c C_k}. \tag{9.69b}$$

Impedance scaling can be included by using (9.64) to give

$$C'_k = \frac{1}{R_0 \omega_c L_k}, \tag{9.70a}$$

$$L'_k = \frac{R_0}{\omega_c C_k}. \tag{9.70b}$$

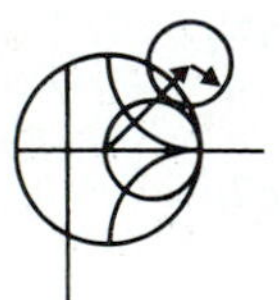

EXAMPLE 9.4

Design a maximally flat low-pass filter with a cutoff frequency of 2 GHz, impedance of 50 Ω, and at least 15 dB insertion loss at 3 GHz. Compute and plot the amplitude response and group delay for $f = 0$ to 4 GHz, and compare with an equal-ripple (3.0 dB ripple) and linear phase filter having the same order.

Solution

First find the required order of the maximally flat filter to satisfy the insertion loss specification at 3 GHz. We have that $|\omega/\omega_c| - 1 = 0.5$; from Figure 9.26 we see that $N = 5$ will be sufficient. Then Table 9.3 gives the prototype element values as

$$g_1 = 0.618,$$
$$g_2 = 1.618,$$
$$g_3 = 2.000,$$

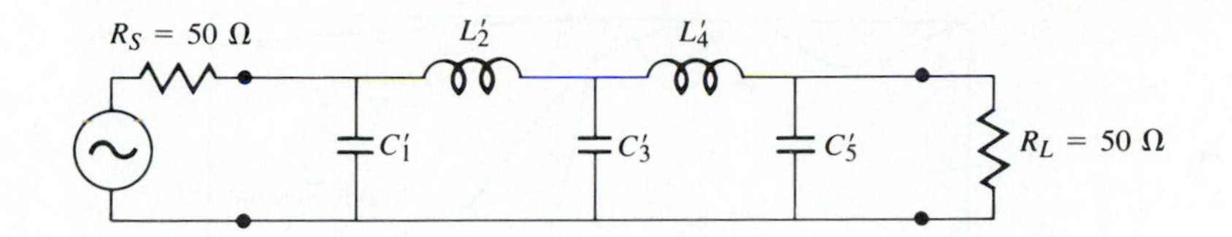

FIGURE 9.29 Low-pass maximally flat filter circuit for Example 9.4.

$$g_4 = 1.618,$$
$$g_5 = 0.618.$$

Then (9.67) can be used to obtain the scaled element values:

$$C_1' = 0.984 \text{ pF},$$
$$L_2' = 6.438 \text{ nH},$$
$$C_3' = 3.183 \text{ pF},$$
$$L_4' = 6.438 \text{ nH},$$
$$C_5' = 0.984 \text{ pF}.$$

The final filter circuit is shown in Figure 9.29; the ladder circuit of Figure 9.25a was used, but that of Figure 9.25b could have been used just as well.

The component values for the equal-ripple filter and the linear phase filter, for $N = 5$, can be determined from Tables 9.4 and 9.5. The amplitude and group delay results for these three filters are shown in Figure 9.30. These results clearly show the trade-offs involved with the three types of filters. The equal-ripple response has the sharpest cutoff, but the worst group delay characteristics. The maximally flat response has a flatter attenuation characteristic in the passband, but a slightly lower cutoff rate. The linear phase filter has the worst cutoff rate, but a very good group delay characteristic. ○

Bandpass and Bandstop Transformation

Low-pass prototype filter designs can also be transformed to have the bandpass or bandstop responses illustrated in Figure 9.31. If ω_1 and ω_2 denote the edges of the passband, then a bandpass response can be obtained using the following frequency substitution:

$$\omega \longleftarrow \frac{\omega_0}{\omega_2 - \omega_1}\left(\frac{\omega}{\omega_0} - \frac{\omega_0}{\omega}\right) = \frac{1}{\Delta}\left(\frac{\omega}{\omega_0} - \frac{\omega_0}{\omega}\right), \tag{9.71}$$

where

$$\Delta = \frac{\omega_2 - \omega_1}{\omega_0} \tag{9.72}$$

is the fractional bandwidth of the passband. The center frequency, ω_0, could be chosen as the arithmetic mean of ω_1 and ω_2, but the equations are simpler if it is chosen as the

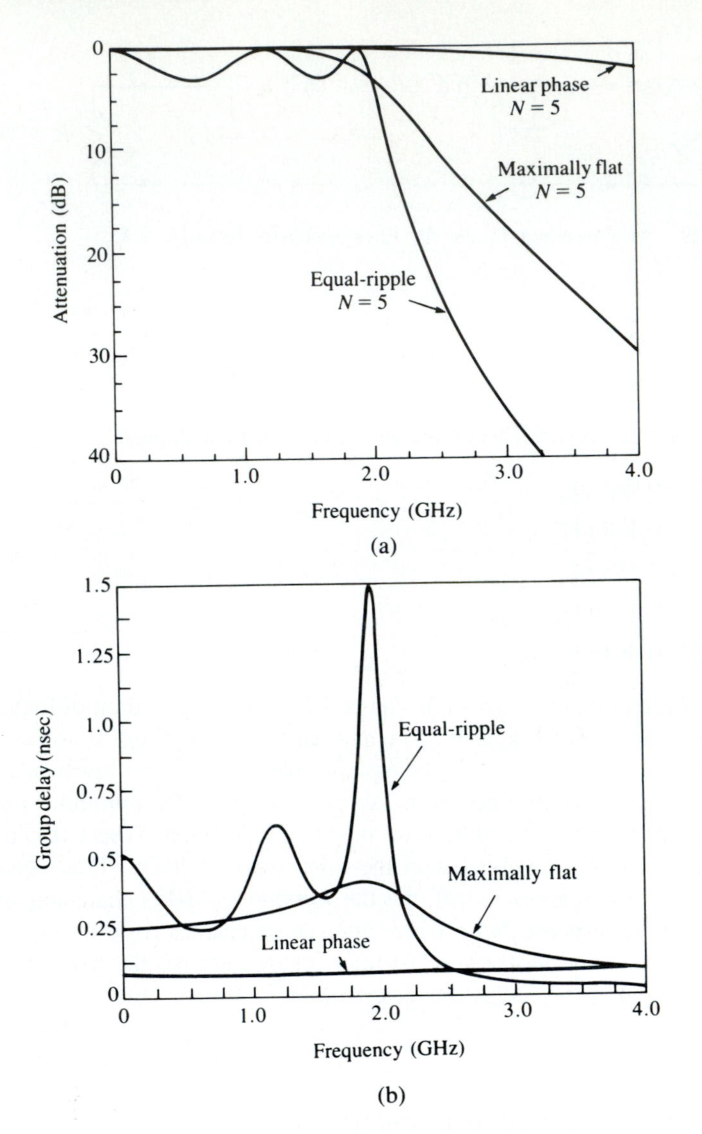

FIGURE 9.30 Frequency response of the filter design of Example 9.4. (a) Amplitude response. (b) Group delay response.

geometric mean:

$$\omega_0 = \sqrt{\omega_1 \omega_2}. \tag{9.73}$$

Then the transformation of (9.71) maps the bandpass characteristics of Figure 9.31b to the low-pass response of Figure 9.31a as follows:

When $\omega = \omega_0$, $\qquad \dfrac{1}{\Delta}\left(\dfrac{\omega}{\omega_0} - \dfrac{\omega_0}{\omega}\right) = 0;$

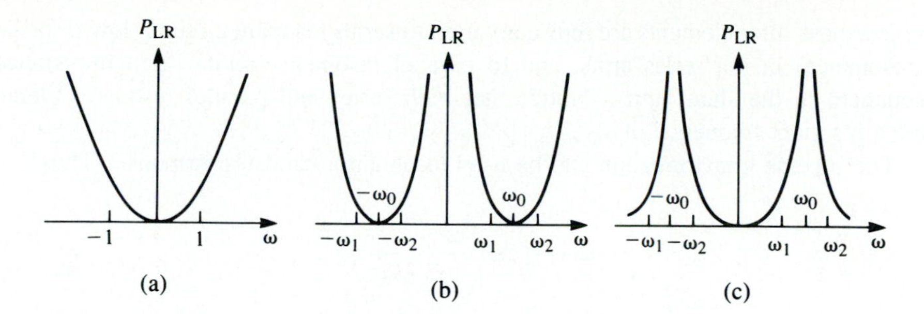

FIGURE 9.31 Bandpass and bandstop frequency transformations. (a) Low-pass filter prototype response for $\omega_c = 1$. (b) Transformation to bandpass response. (c) Transformation to bandstop response.

When $\omega = \omega_1$,
$$\frac{1}{\Delta}\left(\frac{\omega}{\omega_0} - \frac{\omega_0}{\omega}\right) = \frac{1}{\Delta}\left(\frac{\omega_1^2 - \omega_0^2}{\omega_0\omega_1}\right) = -1;$$

When $\omega = \omega_2$,
$$\frac{1}{\Delta}\left(\frac{\omega}{\omega_0} - \frac{\omega_0}{\omega}\right) = \frac{1}{\Delta}\left(\frac{\omega_2^2 - \omega_0^2}{\omega_0\omega_1}\right) = 1.$$

The new filter elements are determined by using (9.71) in the expressions for the series reactance and shunt susceptances. Thus,

$$jX_k = \frac{j}{\Delta}\left(\frac{\omega}{\omega_0} - \frac{\omega_0}{\omega}\right)L_k = j\frac{\omega L_k}{\Delta\omega_0} - j\frac{\omega_0 L_k}{\Delta\omega} = j\omega L_k' - j\frac{1}{\omega C_k'},$$

which shows that a series inductor, L_k, is transformed to a series LC circuit with element values,

$$L_k' = \frac{L_k}{\Delta\omega_0}, \qquad 9.74a$$

$$C_k' = \frac{\Delta}{\omega_0 L_k}. \qquad 9.74b$$

Similarly,

$$jB_k = \frac{j}{\Delta}\left(\frac{\omega}{\omega_0} - \frac{\omega_0}{\omega}\right)C_k = j\frac{\omega C_k}{\Delta\omega_0} - j\frac{\omega_0 C_k}{\Delta\omega} = j\omega C_k' - j\frac{1}{\omega L_k'},$$

which shows that a shunt capacitor, C_k, is transformed to a shunt LC circuit with element values,

$$L_k' = \frac{\Delta}{\omega_0 C_k}, \qquad 9.74c$$

$$C_k' = \frac{C_k}{\Delta\omega_0}. \qquad 9.74d$$

The low-pass filter elements are thus converted to series resonant circuits (low impedance at resonance) in the series arms, and to parallel resonant circuits (high impedance at resonance) in the shunt arms. Notice that both series and parallel resonator elements have a resonant frequency of ω_0.

The inverse transformation can be used to obtain a bandstop response. Thus,

$$\omega \longleftarrow \Delta \left(\frac{\omega}{\omega_0} - \frac{\omega_0}{\omega} \right)^{-1}, \tag{9.75}$$

where Δ and ω_0 have the same definitions as in (9.72) and (9.73). Then series inductors of the low-pass prototype is converted to parallel LC circuits having element values given by

$$L'_k = \frac{\Delta L_k}{\omega_0}, \tag{9.76a}$$

$$C'_k = \frac{1}{\omega_0 \Delta L_k}. \tag{9.76b}$$

The shunt capacitor of the low-pass prototype is converted to series LC circuits having element values given by

$$L'_k = \frac{1}{\omega_0 \Delta C_k}, \tag{9.76c}$$

$$C'_k = \frac{\Delta C_k}{\omega_0}. \tag{9.76d}$$

The element transformations from a low-pass prototype to a highpass, bandpass, or bandstop filter are summarized in Table 9.6. These results do not include impedance scaling, which can be made using (9.64).

EXAMPLE 9.5

Design a bandpass filter having a 0.5 dB equal-ripple response, with $N = 3$. The center frequency is 1 GHz, the bandwidth is 10%, and the impedance is 50 Ω.

Solution

From Table 9.4 the element values for the low-pass prototype circuit of Figure 9.25b are given as

$$g_1 = 1.5963 = L_1,$$

$$g_2 = 1.0967 = C_2,$$

TABLE 9.6 Summary of Prototype Filter Transformations

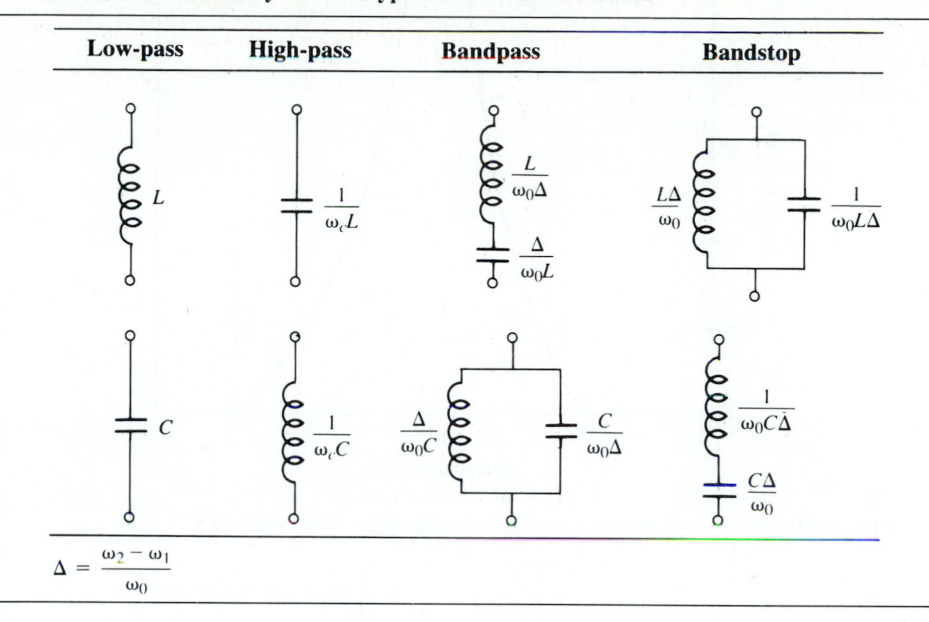

$$g_3 = 1.5963 = L_3,$$
$$g_4 = 1.000 = R_L.$$

Then (9.64) and (9.74) give the impedance-scaled and frequency-transformed element values for the circuit of Figure 9.32 as

$$L_1' = \frac{L_1 Z_o}{\omega_o \Delta} = 127.0 \text{ nH},$$

$$C_1' = \frac{\Delta}{\omega_o L_1 Z_o} = 0.199 \text{ pF},$$

$$L_2' = \frac{\Delta Z_o}{\omega_o C_2} = 0.726 \text{ nH},$$

$$C_2' = \frac{C_2}{\omega_o \Delta Z_o} = 34.91 \text{ pF},$$

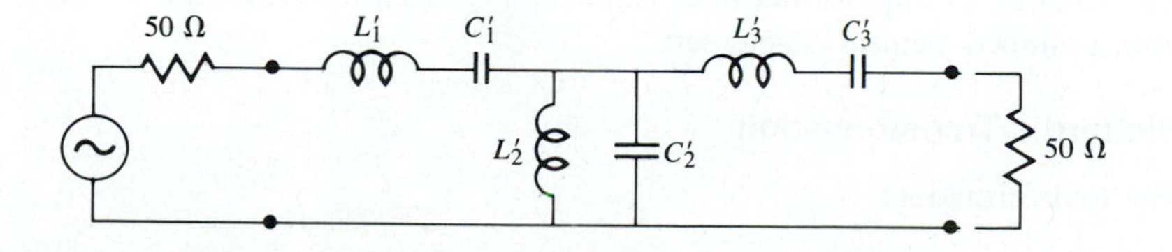

FIGURE 9.32 Bandpass filter circuit for Example 9.5.

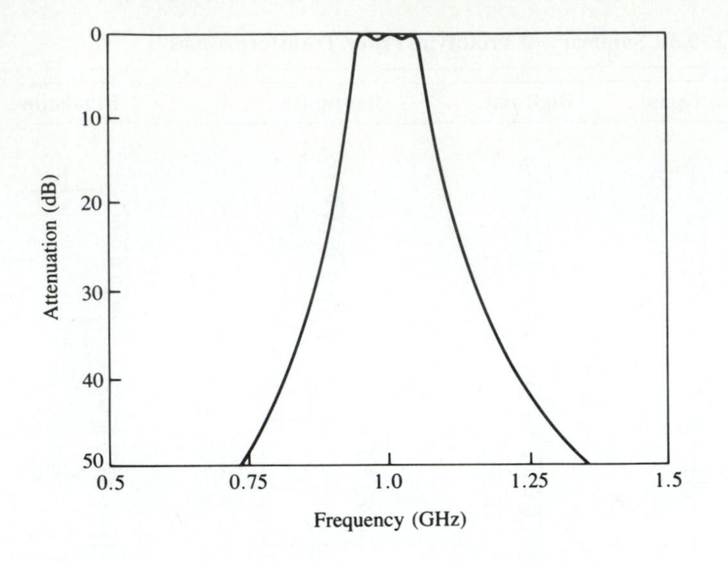

FIGURE 9.33 Amplitude response for the bandpass filter of Example 9.5.

$$L_3' = \frac{L_3 Z_o}{\omega_o \Delta} = 127.0 \text{ nH},$$

$$C_3' = \frac{\Delta}{\omega_o L_3 Z_o} = 0.199 \text{ pF}.$$

The resulting amplitude response is shown in Figure 9.33. ○

9.5 FILTER IMPLEMENTATION

The lumped-element filter design discussed in the previous sections generally works well at low frequencies, but two problems arise at microwave frequencies. First, lumped elements such as inductors and capacitors are generally available only for a limited range of values and are difficult to implement at microwave frequencies, but must be approximated with distributed components. In addition, at microwave frequencies the distances between filter components is not negligible. Richard's transformation is used to convert lumped elements to transmission line sections, while Kuroda's identities can be used to separate filter elements by using transmission line sections. Because such additional transmission line sections do not affect the filter response, this type of design is called *redundant* filter synthesis. It is possible to design microwave filters that take advantage of these sections to improve the filter response [4]; such *nonredundant* synthesis does not have a lumped-element counterpart.

Richard's Transformation

The transformation,

$$\Omega = \tan \beta\ell = \tan\left(\frac{\omega\ell}{v_p}\right), \tag{9.77}$$

maps the ω plane to the Ω plane, which repeats with a period of $\omega\ell/v_p = 2\pi$. This transformation was introduced by P. Richard [6] to synthesize an LC network using open- and short-circuited transmission lines. Thus, if we replace the frequency variable ω with Ω, the reactance of an inductor can be written as

$$jX_L = j\Omega L = jL \tan \beta\ell, \tag{9.78a}$$

and the susceptance of a capacitor can be written as

$$jB_C = j\Omega C = jC \tan \beta\ell. \tag{9.78b}$$

These results indicate that an inductor can be replaced with a short-circuited stub of length $\beta\ell$ and characteristic impedance L, while a capacitor can be replaced with an open-circuited stub of length $\beta\ell$ and characteristic impedance $1/C$. A unity filter impedance is assumed.

Cutoff occurs at unity frequency for a low-pass filter prototype; to obtain the same cutoff frequency for the Richard's-transformed filter, (9.77) shows that

$$\Omega = 1 = \tan \beta\ell,$$

which gives a stub length of $\ell = \lambda/8$, where λ is the wavelength of the line at the cutoff frequency, ω_c. At the frequency $\omega_0 = 2\omega_c$, the lines will be $\lambda/4$ long, and an attenuation pole will occur. At frequencies away from ω_c, the impedances of the stubs will no longer match the original lumped-element impedances, and the filter response will differ from the desired prototype response. Also, the response will be periodic in frequency, repeating every $4\omega_c$.

In principle, then, the inductors and capacitors of a lumped-element filter design can be replaced with short-circuited and open-circuited stubs, as illustrated in Figure 9.34. Since the lengths of all the stubs are the same ($\lambda/8$ at ω_c), these lines are called *commensurate lines*.

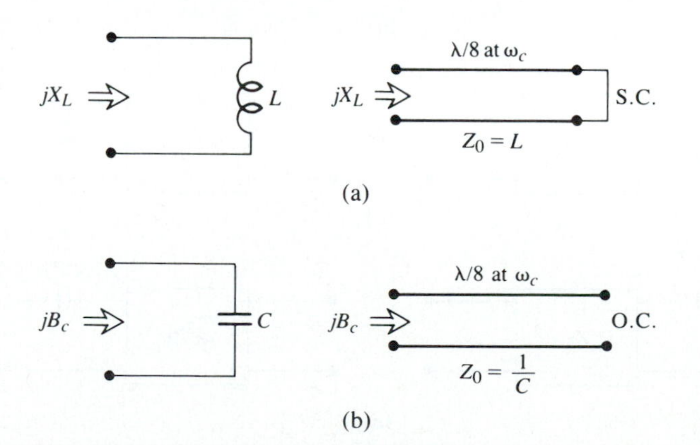

FIGURE 9.34 Richard's transformation. (a) For an inductor to a short-circuited stub. (b) For a capacitor to an open-circuited stub.

Kuroda's Identities

The four Kuroda identities use redundant transmission line sections to achieve a more practical microwave filter implementation by performing any of the following operations:

- Physically separate transmission line stubs
- Transform series stubs into shunt stubs, or vice versa
- Change impractical characteristic impedances into more realizable ones

The additional transmission line sections are called *unit elements* and are $\lambda/8$ long at ω_c; the unit elements are thus commensurate with the stubs used to implement the inductors and capacitors of the prototype design.

The four identities are illustrated in Table 9.7, where each box represents a unit element, or transmission line, of the indicated characteristic impedance and length ($\lambda/8$ at ω_c). The inductors and capacitors represent short-circuit and open-circuit stubs, respectively. We will prove the equivalence of the first case, and then show how to use these identities in Example 9.6.

The two circuits of identity (a) in Table 9.7 can be redrawn as shown in Figure 9.35; we will show that these two networks are equivalent by showing that their $ABCD$

TABLE 9.7 The Four Kuroda Identities

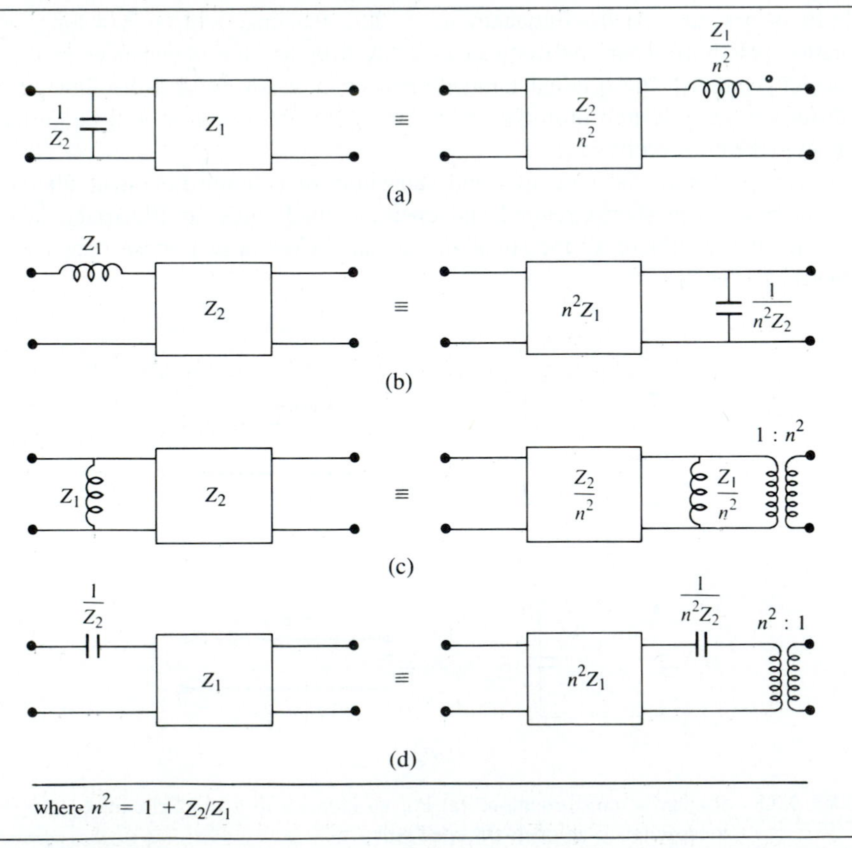

where $n^2 = 1 + Z_2/Z_1$

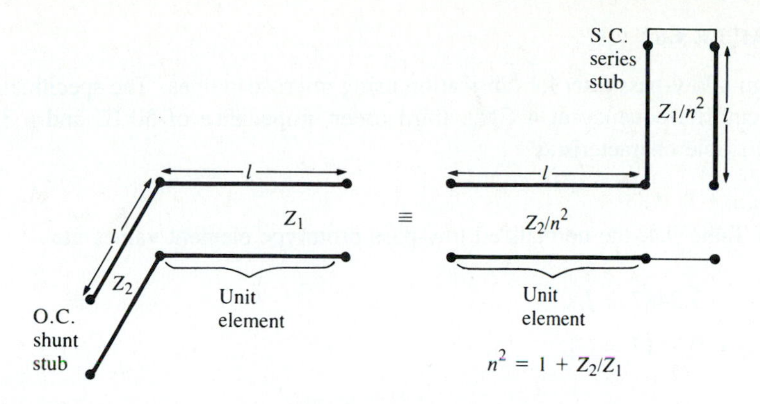

FIGURE 9.35 Equivalent circuits illustrating Kuroda identity (a) in Table 9.7.

matrices are identical. From Table 5.1, the $ABCD$ matrix of a length ℓ of transmission line with characteristic impedance Z_0 is

$$\begin{bmatrix} A & B \\ C & D \end{bmatrix} = \begin{bmatrix} \cos\beta\ell & jZ_1 \sin\beta\ell \\ \dfrac{j}{Z_1}\sin\beta\ell & \cos\beta\ell \end{bmatrix} = \frac{1}{\sqrt{1+\Omega^2}} \begin{bmatrix} 1 & j\Omega Z_1 \\ \dfrac{j\Omega}{Z_1} & 1 \end{bmatrix}, \tag{9.79}$$

where $\Omega = \tan\beta\ell$. Now the open-circuited shunt stub in the first circuit in Figure 9.35 has an impedance of $-jZ_2 \cot\beta\ell = -jZ_2/\Omega$, so the $ABCD$ matrix of the entire circuit is

$$\begin{aligned} \begin{bmatrix} A & B \\ C & D \end{bmatrix}_L &= \begin{bmatrix} 1 & 0 \\ \dfrac{j\Omega}{Z_2} & 1 \end{bmatrix} \begin{bmatrix} 1 & j\Omega Z_1 \\ \dfrac{j\Omega}{Z_1} & 1 \end{bmatrix} \frac{1}{\sqrt{1+\Omega^2}} \\ &= \frac{1}{\sqrt{1+\Omega^2}} \begin{bmatrix} 1 & j\Omega Z_1 \\ j\Omega\left(\dfrac{1}{Z_1}+\dfrac{1}{Z_2}\right) & 1-\Omega^2\dfrac{Z_1}{Z_2} \end{bmatrix}. \end{aligned} \tag{9.80a}$$

The short-circuited series stub in the second circuit in Figure 9.35 has an impedance of $j(Z_1/n^2)\tan\beta\ell = j(\Omega Z_1/n^2)$, so the $ABCD$ matrix of the entire circuit is

$$\begin{aligned} \begin{bmatrix} A & B \\ C & D \end{bmatrix}_R &= \begin{bmatrix} 1 & j\dfrac{\Omega Z_2}{n^2} \\ \dfrac{j\Omega n^2}{Z_2} & 1 \end{bmatrix} \begin{bmatrix} 1 & \dfrac{j\Omega Z_1}{n^2} \\ 0 & 1 \end{bmatrix} \frac{1}{\sqrt{1+\Omega^2}} \\ &= \frac{1}{\sqrt{1+\Omega^2}} \begin{bmatrix} 1 & \dfrac{j\Omega}{n^2}(Z_1+Z_2) \\ \dfrac{j\Omega n^2}{Z_2} & 1-\Omega^2\dfrac{Z_1}{Z_2} \end{bmatrix}. \end{aligned} \tag{9.80b}$$

The results in (9.80a) and (9.80b) are identical if we choose $n^2 = 1 + Z_2/Z_1$. The other identities in Table 9.7 can be proven in the same way.

EXAMPLE 9.6

Design a low-pass filter for fabrication using microstrip lines. The specifications are: cutoff frequency of 4 GHz, third order, impedance of 50 Ω, and a 3 dB equal-ripple characteristic.

Solution

From Table 9.4, the normalized low-pass prototype element values are

$$g_1 = 3.3487 = L_1,$$

$$g_2 = 0.7117 = C_2,$$

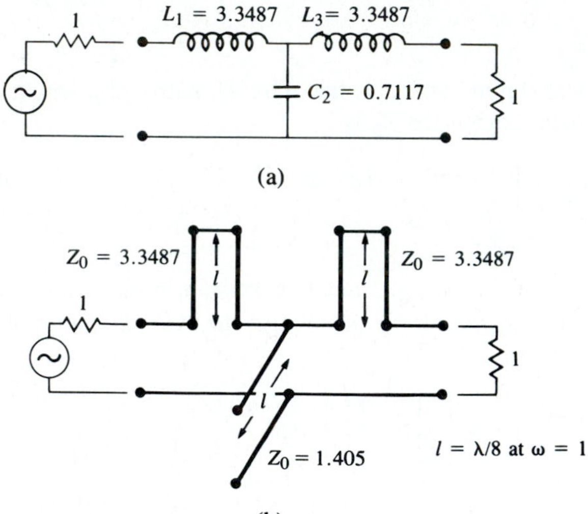

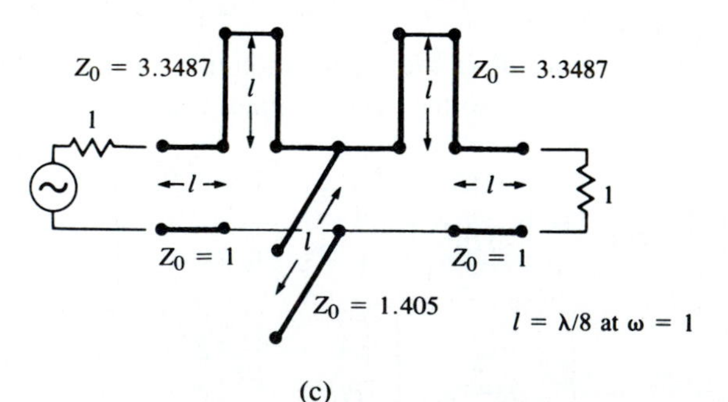

(continues)

FIGURE 9.36 Filter design procedure for Example 9.6. (a) Lumped-element low-pass filter prototype. (b) Using Richard's transformations to convert inductors and capacitors to series and shunt stubs. (c) Adding unit elements at ends of filter.

$$g_3 = 3.3487 = L_3,$$
$$g_4 = 1.0000 = R_L,$$

with the lumped-element circuit shown in Figure 9.36a.

The next step is to use Richard's transformations to convert series inductors to series stubs, and shunt capacitors to shunt stubs, as shown in Figure 9.36b. According to (9.78), the characteristic impedance of a series stub (inductor) is L, and the characteristic impedance of a shunt stub (capacitor) is $1/C$. For commensurate line synthesis, all stubs are $\lambda/8$ long at $\omega = \omega_c$. (It is usually most convenient to work with normalized quantities until the last step in the design.)

The series stubs of Figure 9.36b would be very difficult to implement in microstrip form, so we will use one of the Kuroda identities to convert these to shunt stubs. First, we must add unit elements at either end of the filter, as shown in Figure 9.36c. These redundant elements do not affect filter performance since

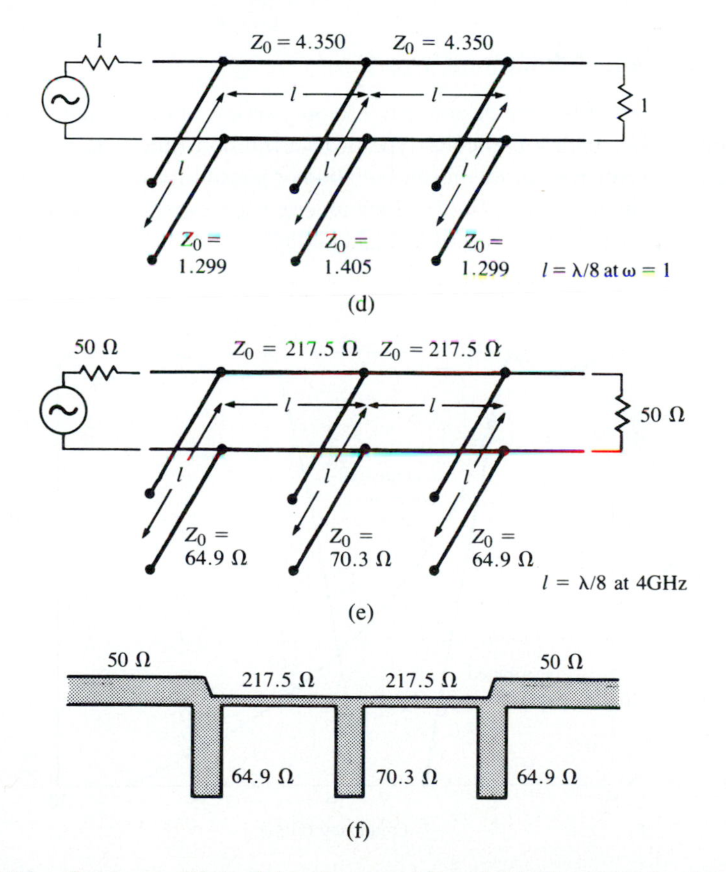

FIGURE 9.36 Continued. (d) Applying the second Kuroda identity. (e) After impedance and frequency scaling. (f) Microstrip fabrication of final filter.

they are matched to the source and load ($Z_0 = 1$). Then we can apply Kuroda identity (b) from Table 9.7 to both ends of the filter. In both cases we have that

$$n^2 = 1 + \frac{Z_2}{Z_1} = 1 + \frac{1}{3.3487} = 1.299.$$

The result is shown in Figure 9.36d.

Finally, we impedance and frequency scale the circuit, which simply involves multiplying the normalized characteristic impedances by 50 Ω and choosing the line and stub lengths to be $\lambda/8$ at 4 GHz. The final circuit is shown in Figure 9.36e, with a microstrip layout in Figure 9.36f.

The calculated amplitude response of this design is plotted in Figure 9.37, along with the response of the lumped-element version. Note that the passband characteristics are very similar up to 4 GHz, but the distributed-element filter has a sharper cutoff. Also notice that the distributed-element filter has a response which repeats every 16 GHz, as a result of the periodic nature of Richard's transformation. ○

Impedance and Admittance Inverters

As we have seen, it is often desirable to use only series, or only shunt, elements when implementing a filter with a particular type of transmission line. The Kuroda identities can be used for conversions of this form, but another possibility is to use impedance (K) or admittance (J) inverters [2], [4], [7]. Such inverters are especially useful for bandpass or bandstop filters with narrow ($< 10\%$) bandwidths.

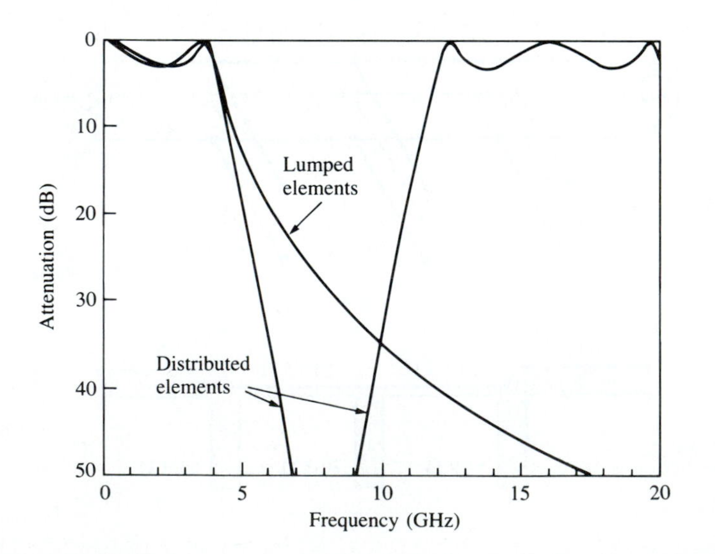

FIGURE 9.37 Amplitude responses of lumped-element and distributed-element low-pass filter of Example 9.6.

The conceptual operation of impedance and admittance is illustrated in Figure 9.38; since these inverters essentially form the inverse of the load impedance or admittance, they can be used to transform series-connected elements to shunt-connected elements, or vice versa. This procedure will be illustrated in later sections for bandpass and bandstop filters.

In its simplest form, a J or K inverter can be constructed using a quarter-wave transformer of the appropriate characteristic impedance, as shown in Figure 9.38b. This implementation also allows the $ABCD$ matrix of the inverter to be easily found from the $ABCD$ parameters for a length of transmission line given in Table 5.1. Many other types of circuits can also be used as J or K inverters, with one such alternative being shown in Figure 9.38c. Inverters of this form turn out to be useful for modelling the coupled resonator filters of Section 9.8. The lengths, $\theta/2$, of the transmission line sections are generally required to be negative for this type of inverter, but this poses no problem if these lines can be absorbed into connecting transmission lines on either side.

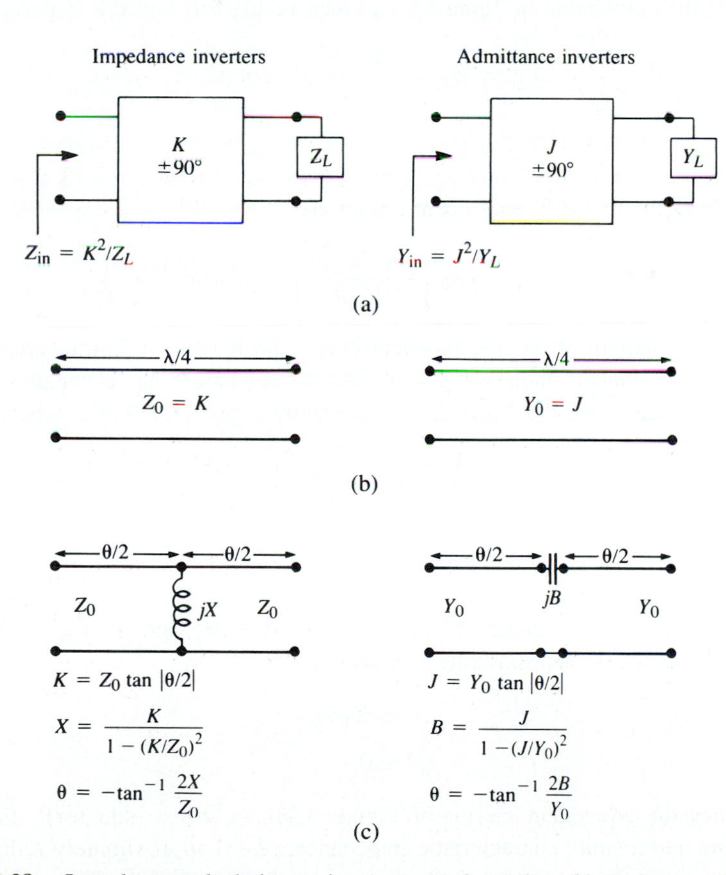

FIGURE 9.38 Impedance and admittance inverters. (a) Operation of impedance and admittance inverters. (b) Implementation as quarter-wave transformers. (c) An alternative implementation.

9.6 STEPPED-IMPEDANCE LOW-PASS FILTERS

A relatively easy way to implement low-pass filters in microstrip or stripline is to use alternating sections of very high and very low characteristic impedance lines. Such filters are usually referred to as *stepped-impedance*, or hi-Z, low-Z filters, and are popular because they are easier to design and take up less space than a similar low-pass filter using stubs. Because of the approximations involved, however, their electrical performance is not as good, so the use of such filters is usually limited to applications where a sharp cutoff is not required (for instance, in rejecting out-of-band mixer products).

Approximate Equivalent Circuits for Short Transmission Line Sections

We begin by finding the approximate equivalent circuits for a short length of transmission line having either a very large or very small characteristic impedance. The $ABCD$ parameters of a length, ℓ, of line having characteristic impedance Z_0 are given in Table 5.1; the conversion in Table 5.2 can then be used to find the Z-parameters as

$$Z_{11} = Z_{22} = \frac{A}{C} = -jZ_0 \cot \beta\ell, \tag{9.81a}$$

$$Z_{12} = Z_{21} = \frac{1}{C} = -jZ_0 \csc \beta\ell. \tag{9.81b}$$

The series elements of the T-equivalent circuit are

$$Z_{11} - Z_{12} = -jZ_0 \left[\frac{\cos \beta\ell - 1}{\sin \beta\ell}\right] = jZ_0 \tan\left(\frac{\beta\ell}{2}\right), \tag{9.82}$$

while the shunt element of the T-equivalent is Z_{12}. So if $\beta\ell < \pi/2$, the series elements have a positive reactance (inductors), while the shunt element has a negative reactance (capacitor). We thus have the equivalent circuit shown in Figure 9.39a, where

$$\frac{X}{2} = Z_0 \tan\left(\frac{\beta\ell}{2}\right), \tag{9.83a}$$

$$B = \frac{1}{Z_0} \sin \beta\ell. \tag{9.83b}$$

Now assume a short length of line (say $\beta\ell < \pi/4$) and a large characteristic impedance. Then (9.83) approximately reduces to

$$X \simeq Z_0\beta\ell, \tag{9.84a}$$

$$B \simeq 0, \tag{9.84b}$$

which implies the equivalent circuit of Figure 9.39b (a series inductor). For a short length of line and a small characteristic impedance, (9.83) approximately reduces to

$$X \simeq 0, \tag{9.85a}$$

$$B \simeq Y_0\beta\ell, \tag{9.85b}$$

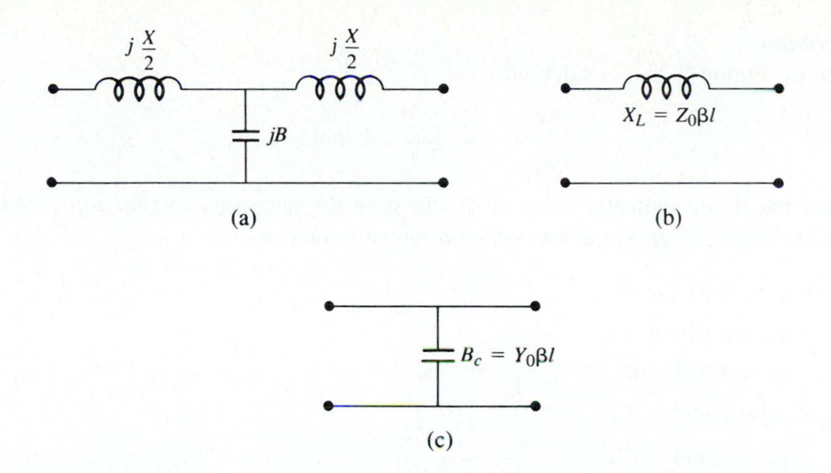

FIGURE 9.39 Approximate equivalent circuits for short sections of transmission lines. (a) T equivalent circuit for a transmission line section having $\beta\ell << \pi/2$. (b) Equivalent circuit for small $\beta\ell$ and large Z_0. (c) Equivalent circuit for small $\beta\ell$ and small Z_0.

which implies the equivalent circuit of Figure 9.39c (a shunt capacitor). So the series inductors of a low-pass prototype can be replaced with high-impedance line sections ($Z_0 = Z_h$), and the shunt capacitors can be replaced with low-impedance line sections ($Z_0 = Z_\ell$). The ratio Z_h/Z_ℓ should be as high as possible, so the actual values of Z_h and Z_ℓ are usually set to the highest and lowest characteristic impedance that can be practically fabricated. The lengths of the lines can then be determined from (9.84) and (9.85); to get the best response near cutoff, these lengths should be evaluated at $\omega = \omega_c$. Combining the results of (9.84) and (9.85) with the scaling equations of (9.67) allows the electrical lengths of the inductor sections to be calculated as

$$\beta\ell = \frac{LR_0}{Z_h} \quad \text{(inductor)}, \tag{9.86a}$$

and the electrical length of the capacitor sections as

$$\beta\ell = \frac{CZ_\ell}{R_0} \quad \text{(capacitor)}, \tag{9.86b}$$

where R_0 is the filter impedance and L and C are the normalized element values (the g_ks) of the low-pass prototype.

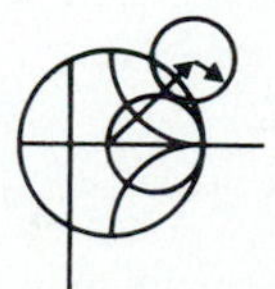

EXAMPLE 9.7

Design a stepped-impedance low-pass filter having a maximally flat response and a cutoff frequency of 2.5 GHz. It is necessary to have more than 20 dB insertion loss at 4.0 GHz. The filter impedance is $50\,\Omega$; the highest practical line impedance is $150\,\Omega$, and the lowest is $10\,\Omega$.

Solution

To use Figure 9.26, we calculate

$$\frac{\omega}{\omega_c} - 1 = \frac{4.0}{2.5} - 1 = 0.6,$$

then the figure indicates $N = 6$ should give the necessary attenuation at 4.0 GHz. Table 9.3 gives the low-pass prototype values as

$$g_1 = 0.517 = C_1,$$
$$g_2 = 1.414 = L_2,$$
$$g_3 = 1.932 = C_3,$$
$$g_4 = 1.932 = L_4,$$
$$g_5 = 1.414 = C_5,$$
$$g_6 = 0.517 = L_6.$$

The low-pass prototype circuit is shown in Figure 9.40a.

Next, we use (9.86) to find the electrical lengths of the hi-Z, low-Z transmission line sections to replace the series inductors and shunt capacitors:

$$\beta\ell_1 = g_1 \frac{Z_\ell}{R_0} = 5.9°,$$

$$\beta\ell_2 = g_2 \frac{R_0}{Z_h} = 27.0°,$$

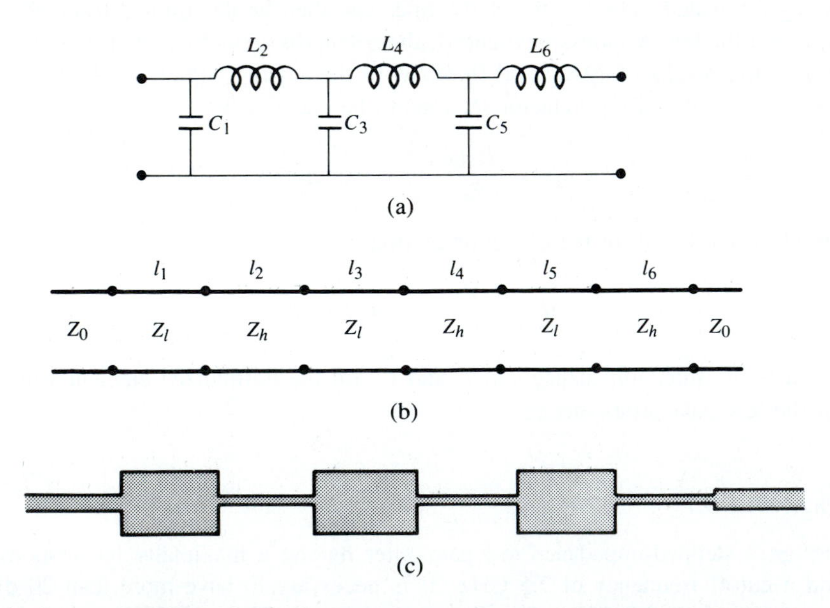

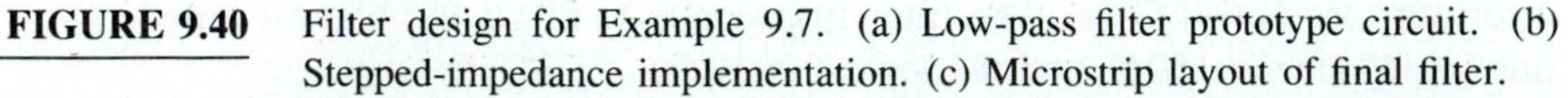

FIGURE 9.40 Filter design for Example 9.7. (a) Low-pass filter prototype circuit. (b) Stepped-impedance implementation. (c) Microstrip layout of final filter.

$$\beta\ell_3 = g_3 \frac{Z_\ell}{R_0} = 22.1^\circ,$$

$$\beta\ell_4 = g_4 \frac{R_0}{Z_h} = 36.9^\circ,$$

$$\beta\ell_5 = g_5 \frac{Z_\ell}{R_0} = 16.2^\circ,$$

$$\beta\ell_6 = g_6 \frac{R_0}{Z_h} = 9.9^\circ.$$

The final filter circuit is shown in Figure 9.40b, where $Z_\ell = 10\,\Omega$ and $Z_h = 150\,\Omega$. Note that $\beta\ell < \pi/4$ in all cases. A layout in microstrip is shown in Figure 9.40c.

Figure 9.41 shows the calculated amplitude response, compared with the response of the corresponding lumped-element filter. The passband characteristics are very similar, but the lumped-element circuit gives more attenuation at higher frequencies. This is because the stepped-impedance filter elements depart significantly from the lumped-element values at the higher frequencies. The stepped-impedance filter may have other passbands at higher frequencies, but the response will not be perfectly periodic because the lines are not commensurate. ○

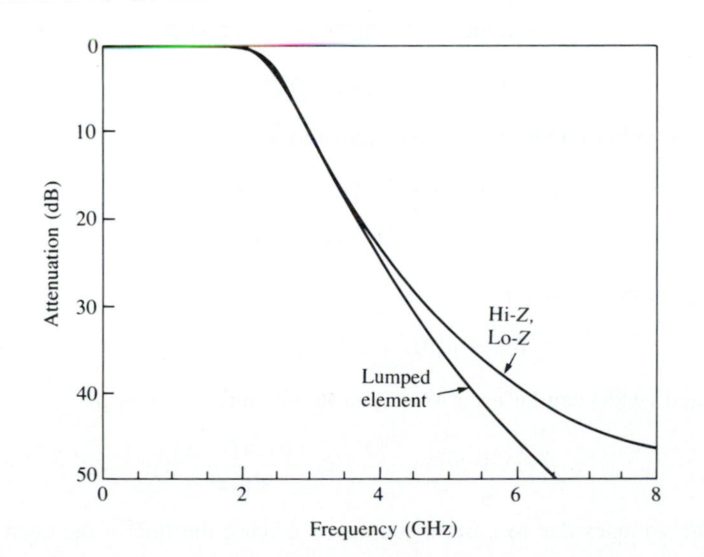

FIGURE 9.41 Amplitude response of the stepped-impedance low-pass filter of Example 9.7, compared with the corresponding lumped-element design.

9.7 COUPLED LINE FILTERS

The parallel coupled transmission lines discussed in Section 8.6 (for directional couplers) can also be used to construct many types of filters. Fabrication of multisection bandpass or bandstop coupled line filters is particularly easy in microstrip or stripline form, for bandwidths less than about 20%. Wider bandwidth filters generally require very tightly coupled lines, which are difficult to fabricate. We will first study the filter characteristics of a single quarter-wave coupled line section, and then show how these sections can be used to design a bandpass filter [7]. Other filter designs using coupled lines can be found in reference [2].

Filter Properties of a Coupled Line Section

A parallel coupled line section is shown in Figure 9.42a, with port voltage and current definitions. We will derive the open-circuit impedance matrix for this four-port network by considering the superposition of even- and odd-mode excitations [8], which are shown in Figure 9.42b. Thus, the current sources i_1 or i_3 drive the line in the even mode, while i_2 or i_4 drive the line in the odd mode. By superposition, we see that the total port currents, I_i, can be expressed in terms of the even- and odd-mode currents as

$$I_1 = i_1 + i_2, \tag{9.87a}$$

$$I_2 = i_1 - i_2, \tag{9.87b}$$

$$I_3 = i_3 - i_4, \tag{9.87c}$$

$$I_4 = i_3 + i_4. \tag{9.87d}$$

First consider the line as being driven in the even mode by the i_1 current sources. If the other ports are open-circuited, the impedance seen at port 1 or 2 is

$$Z_{\text{in}}^{e} = -jZ_{0e}\cot\beta\ell. \tag{9.88}$$

The voltage on either conductor can be expressed as

$$\begin{aligned} v_a^1(z) = v_b^1(z) &= V_e^+[e^{-j\beta(z-\ell)} + e^{j\beta(z-\ell)}] \\ &= 2V_e^+\cos\beta(\ell - z), \end{aligned} \tag{9.89}$$

so the voltage at port 1 or 2 is

$$v_a^1(0) = v_b^1(0) = 2V_e^+\cos\beta\ell = i_1 Z_{\text{in}}^e.$$

This result and (9.88) can be used to rewrite (9.89) in terms of i_1 as

$$v_a^1(z) = v_b^1(z) = -jZ_{0e}\frac{\cos\beta(\ell - z)}{\sin\beta\ell}i_1. \tag{9.90}$$

Similarly, the voltages due to current sources i_3 driving the line in the even mode are

$$v_a^3(z) = v_b^3(z) = -jZ_{0e}\frac{\cos\beta z}{\sin\beta\ell}i_3. \tag{9.91}$$

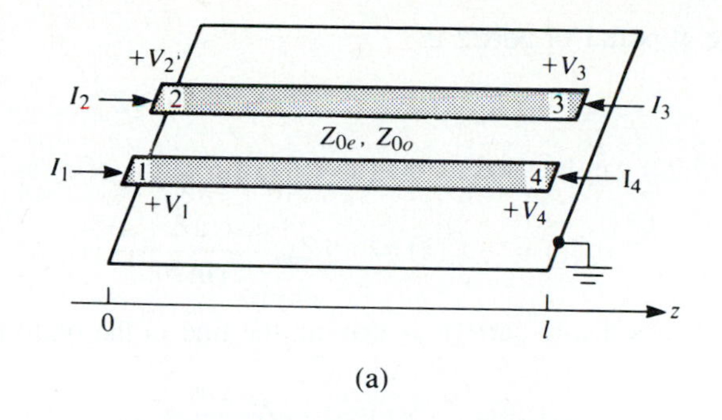

(a)

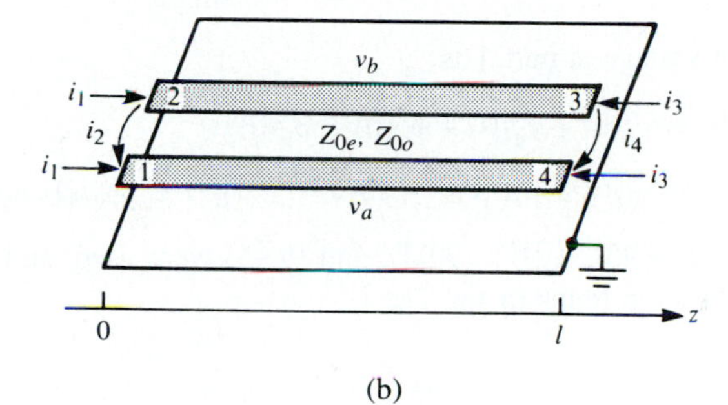

(b)

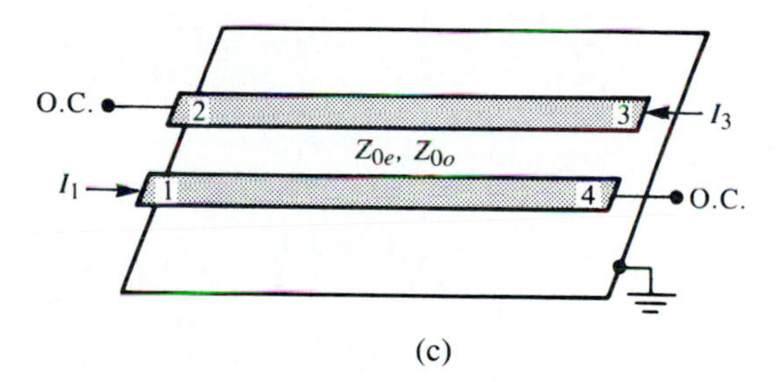

(c)

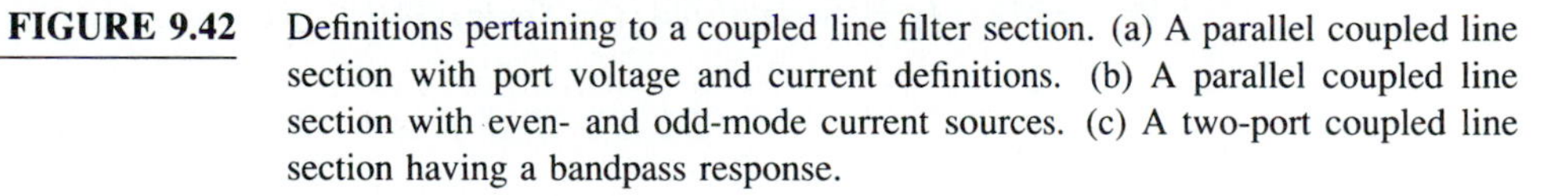

FIGURE 9.42 Definitions pertaining to a coupled line filter section. (a) A parallel coupled line section with port voltage and current definitions. (b) A parallel coupled line section with even- and odd-mode current sources. (c) A two-port coupled line section having a bandpass response.

Now consider the line as being driven in the odd mode by current i_2. If the other ports are open-circuited, the impedance seen at port 1 or 2 is

$$Z_{\text{in}}^{o} = -jZ_{0o}\cot\beta\ell. \tag{9.92}$$

The voltage on either conductor can be expressed as

$$v_a^2(z) = -v_b^2(z) = V_0^+[e^{-j\beta(z-\ell)} + e^{j\beta(z-\ell)}] = 2V_0^+\cos\beta(\ell - z). \tag{9.93}$$

Then the voltage at port 1 or port 2 is

$$v_a^2(0) = -v_b^2(0) = 2V_0^+ \cos\beta\ell = i_2 Z_{in}^o.$$

This result and (9.92) can be used to rewrite (9.93) in terms of i_2 as

$$v_a^2(z) = -v_b^2(z) = -jZ_{0o}\frac{\cos\beta(\ell - z)}{\sin\beta\ell} i_2. \tag{9.94}$$

Similarly, the voltages due to current i_4 driving the line in the odd mode are

$$v_a^4(z) = -v_b^4(z) = -jZ_{0o}\frac{\cos\beta z}{\sin\beta\ell} i_4. \tag{9.95}$$

Now the total voltage at port 1 is

$$\begin{aligned} V_1 &= v_a^1(0) + v_a^2(0) + v_a^3(0) + v_a^4(0) \\ &= -j(Z_{0e}i_1 + Z_{0o}i_2)\cot\theta - j(Z_{0e}i_3 + Z_{0o}i_4)\csc\theta, \end{aligned} \tag{9.96}$$

where the results of (9.90), (9.91), (9.94), and (9.95) were used, and $\theta = \beta\ell$. Next, we solve (9.87) for the i_j in terms of the Is:

$$i_1 = \frac{1}{2}(I_1 + I_2), \tag{9.97a}$$

$$i_2 = \frac{1}{2}(I_1 - I_2), \tag{9.97b}$$

$$i_3 = \frac{1}{2}(I_3 + I_4), \tag{9.97c}$$

$$i_4 = \frac{1}{2}(I_4 - I_3), \tag{9.97d}$$

and use these results in (9.96):

$$\begin{aligned} V_1 =& \frac{-j}{2}(Z_{0e}I_1 + Z_{0e}I_2 + Z_{0o}I_1 - Z_{0o}I_2)\cot\theta \\ & \frac{-j}{2}(Z_{0e}I_3 + Z_{0e}I_4 + Z_{0o}I_4 - Z_{0o}I_3)\csc\theta. \end{aligned} \tag{9.98}$$

This result yields the top row of the open-circuit impedance matrix $[Z]$ that describes the coupled line section. From symmetry, all other matrix elements can be found once the first row is known. The matrix elements are then

$$Z_{11} = Z_{22} = Z_{33} = Z_{44} = \frac{-j}{2}(Z_{0e} + Z_{0o})\cot\theta, \tag{9.99a}$$

$$Z_{12} = Z_{21} = Z_{34} = Z_{43} = \frac{-j}{2}(Z_{0e} - Z_{0o})\cot\theta, \tag{9.99b}$$

$$Z_{13} = Z_{31} = Z_{24} = Z_{42} = \frac{-j}{2}(Z_{0e} - Z_{0o})\csc\theta, \tag{9.99c}$$

$$Z_{14} = Z_{41} = Z_{23} = Z_{32} = \frac{-j}{2}(Z_{0e} + Z_{0o})\csc\theta. \tag{9.99d}$$

A two-port network can be formed from the coupled line section by terminating two of the four ports in either open or short circuits; there are ten possible combinations, as illustrated in Table 9.8. As indicated in this table, the various circuits have different frequency responses, including low-pass, bandpass, all pass, and all stop. For bandpass filters, we are most interested in the case shown in Figure 9.42c, as open circuits are easier to fabricate than are short circuits. In this case, $I_2 = I_4 = 0$, so the four-port impedance matrix equations reduce to

$$V_1 = Z_{11}I_1 + Z_{13}I_3, \tag{9.100a}$$

$$V_3 = Z_{31}I_1 + Z_{33}I_3, \tag{9.100b}$$

where Z_{ij} is given in (9.99).

We can analyze the filter characteristics of this circuit by calculating the image impedance (which is the same at ports 1 and 3), and the propagation constant. From Table 9.1, the image impedance in terms of the Z-parameters is

$$\begin{aligned} Z_i &= \sqrt{Z_{11}^2 - \frac{Z_{11}Z_{13}^2}{Z_{33}}} \\ &= \frac{1}{2}\sqrt{(Z_{0e} - Z_{0o})^2 \csc^2\theta - (Z_{0e} + Z_{0o})^2 \cot^2\theta}. \end{aligned} \tag{9.101}$$

When the coupled line section is $\lambda/4$ long ($\theta = \pi/2$), the image impedance reduces to

$$Z_i = \frac{1}{2}(Z_{0e} - Z_{0o}), \tag{9.102}$$

which is real and positive, since $Z_{0e} > Z_{0o}$. But when $\theta \to 0$ or $\pi, Z_i \to \pm j\infty$, indicating a stopband. The real part of the image impedance is sketched in Figure 9.43, where the cutoff frequencies can be found from (9.101) as

$$\cos\theta_1 = -\cos\theta_2 = \frac{Z_{0e} - Z_{0o}}{Z_{0e} + Z_{0o}}.$$

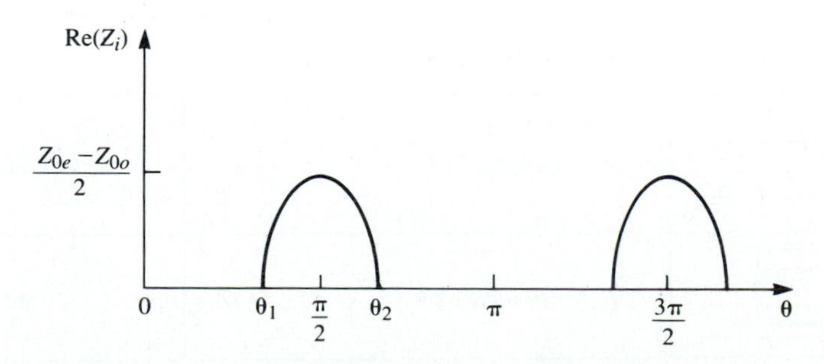

FIGURE 9.43 The real part of the image impedance of the bandpass network of Figure 9.42c.

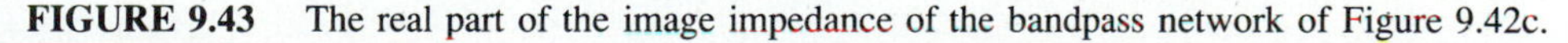

TABLE 9.8 Ten Canonical Coupled Line Circuits

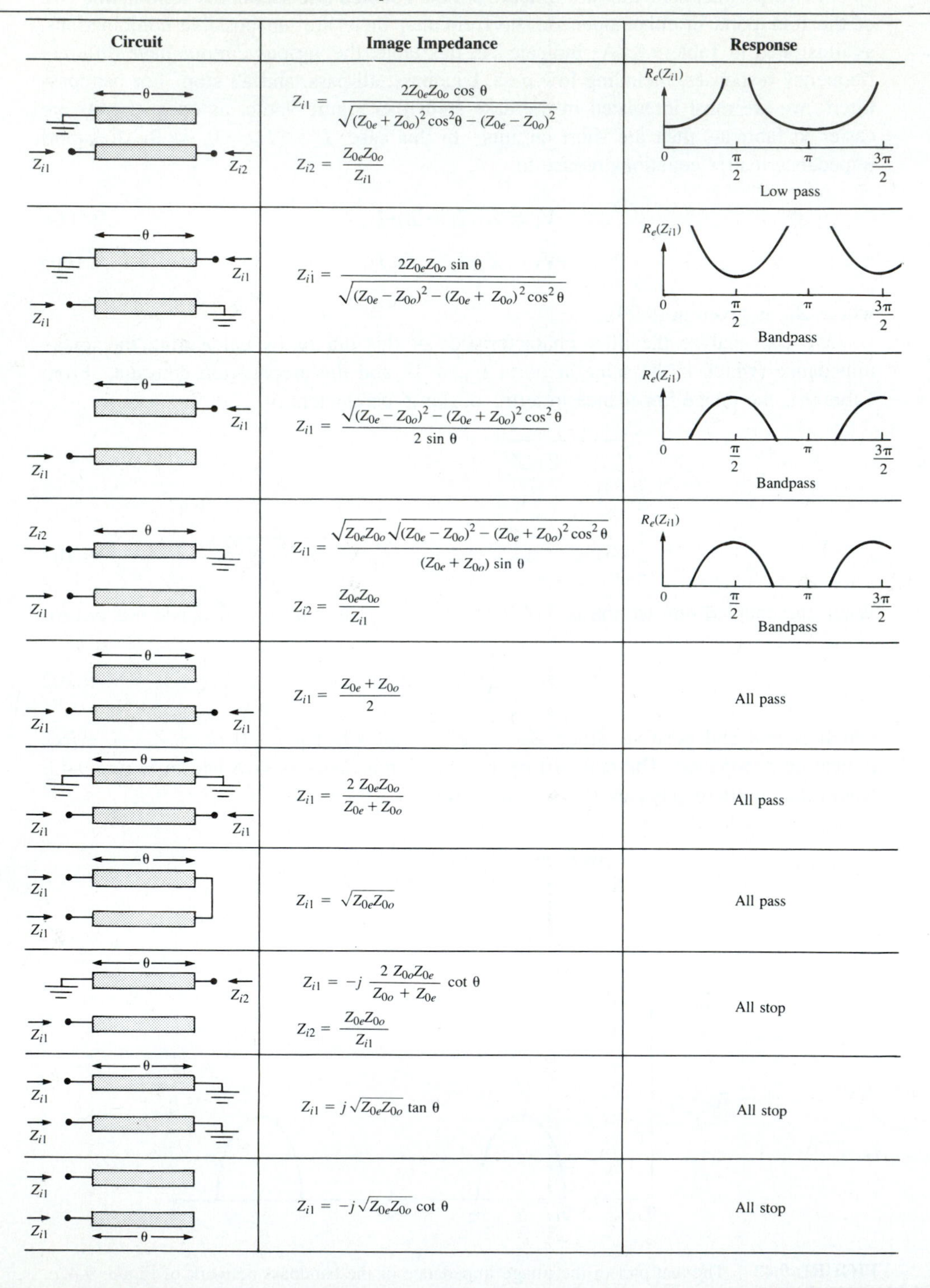

Circuit	Image Impedance	Response
θ, Z_{i1}, Z_{i2}	$Z_{i1} = \dfrac{2Z_{0e}Z_{0o}\cos\theta}{\sqrt{(Z_{0e}+Z_{0o})^2\cos^2\theta-(Z_{0e}-Z_{0o})^2}}$ $Z_{i2} = \dfrac{Z_{0e}Z_{0o}}{Z_{i1}}$	$R_e(Z_{i1})$; 0, $\frac{\pi}{2}$, π, $\frac{3\pi}{2}$; Low pass
θ, Z_{i1}, Z_{i1}	$Z_{i1} = \dfrac{2Z_{0e}Z_{0o}\sin\theta}{\sqrt{(Z_{0e}-Z_{0o})^2-(Z_{0e}+Z_{0o})^2\cos^2\theta}}$	$R_e(Z_{i1})$; 0, $\frac{\pi}{2}$, π, $\frac{3\pi}{2}$; Bandpass
θ, Z_{i1}, Z_{i1}	$Z_{i1} = \dfrac{\sqrt{(Z_{0e}-Z_{0o})^2-(Z_{0e}+Z_{0o})^2\cos^2\theta}}{2\sin\theta}$	$R_e(Z_{i1})$; 0, $\frac{\pi}{2}$, π, $\frac{3\pi}{2}$; Bandpass
Z_{i2}, θ, Z_{i1}	$Z_{i1} = \dfrac{\sqrt{Z_{0e}Z_{0o}}\sqrt{(Z_{0e}-Z_{0o})^2-(Z_{0e}+Z_{0o})^2\cos^2\theta}}{(Z_{0e}+Z_{0o})\sin\theta}$ $Z_{i2} = \dfrac{Z_{0e}Z_{0o}}{Z_{i1}}$	$R_e(Z_{i1})$; 0, $\frac{\pi}{2}$, π, $\frac{3\pi}{2}$; Bandpass
θ, Z_{i1}, Z_{i1}	$Z_{i1} = \dfrac{Z_{0e}+Z_{0o}}{2}$	All pass
θ, Z_{i1}, Z_{i1}	$Z_{i1} = \dfrac{2\,Z_{0e}Z_{0o}}{Z_{0e}+Z_{0o}}$	All pass
θ, Z_{i1}, Z_{i1}	$Z_{i1} = \sqrt{Z_{0e}Z_{0o}}$	All pass
θ, Z_{i2}, Z_{i1}	$Z_{i1} = -j\,\dfrac{2\,Z_{0o}Z_{0e}}{Z_{0o}+Z_{0e}}\cot\theta$ $Z_{i2} = \dfrac{Z_{0e}Z_{0o}}{Z_{i1}}$	All stop
θ, Z_{i1}, Z_{i1}	$Z_{i1} = j\sqrt{Z_{0e}Z_{0o}}\tan\theta$	All stop
Z_{i1}, Z_{i1}, θ	$Z_{i1} = -j\sqrt{Z_{0e}Z_{0o}}\cot\theta$	All stop

The propagation constant can also be calculated from the results of Table 9.1 as

$$\cos\beta = \sqrt{\frac{Z_{11}Z_{33}}{Z_{13}^2}} = \frac{Z_{11}}{Z_{13}} = \frac{Z_{0e}+Z_{0o}}{Z_{0e}-Z_{0o}}\cos\theta, \tag{9.103}$$

which shows β is real for $\theta_1 < \theta < \theta_2 = \pi - \theta_1$, where $\cos\theta_1 = (Z_{0e}-Z_{0o})/(Z_{0e}+Z_{0o})$.

Design of Coupled Line Bandpass Filters

Narrowband bandpass filters can be made with cascaded coupled line sections of the form shown in Figure 9.42c. To derive the design equations for filters of this type, we first show that a single coupled line section can be approximately modeled by the equivalent circuit shown in Figure 9.44. We will do this by calculating the image impedance and propagation constant of the equivalent circuit and showing that they are approximately equal to those of the coupled line section for $\theta = \pi/2$, which will correspond to the center frequency of the bandpass response.

The $ABCD$ parameters of the equivalent circuit can be computed using the $ABCD$ matrices for transmission lines from Table 5.1:

$$\begin{aligned}\begin{bmatrix} A & B \\ C & D \end{bmatrix} &= \begin{bmatrix} \cos\theta & jZ_0\sin\theta \\ \dfrac{j\sin\theta}{Z_0} & \cos\theta \end{bmatrix}\begin{bmatrix} 0 & -j/J \\ -jJ & 0 \end{bmatrix}\begin{bmatrix} \cos\theta & jZ_0\sin\theta \\ \dfrac{j\sin\theta}{Z_0} & \cos\theta \end{bmatrix} \\ &= \begin{bmatrix} \left(JZ_0 + \dfrac{1}{JZ_0}\right)\sin\theta\cos\theta & j\left(JZ_0^2\sin^2\theta - \dfrac{\cos^2\theta}{J}\right) \\ j\left(\dfrac{1}{JX_0^2}\sin^2\theta - J\cos^2\theta\right) & \left(JZ_0 + \dfrac{1}{JZ_0}\right)\sin\theta\cos\theta \end{bmatrix}. \end{aligned} \tag{9.104}$$

The $ABCD$ parameters of the admittance inverter were obtained by considering it as a quarter-wave length of transmission of characteristic impedance, $1/J$. From (9.27) the image impedance of the equivalent circuit is

$$Z_i = \sqrt{\frac{B}{C}} = \sqrt{\frac{JZ_0^2\sin^2\theta - 1/J\cos^2\theta}{(1/JZ_0^2)\sin^2\theta - J\cos^2\theta}}, \tag{9.105}$$

which reduces to the following value at the center frequency, $\theta = \pi/2$:

$$Z_i = JZ_0^2. \tag{9.106}$$

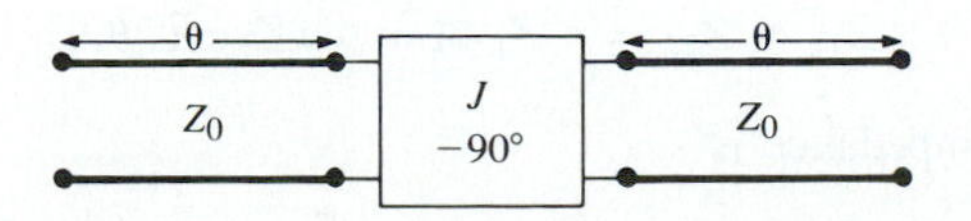

FIGURE 9.44 Equivalent circuit of the coupled line section of Figure 9.42c.

From (9.31) the propagation constant is

$$\cos\beta = A = \left(JZ_0 + \frac{1}{JZ_0}\right)\sin\theta\cos\theta. \tag{9.107}$$

Equating the image impedances in (9.102) and (9.106), and the propagation constants of (9.103) and (9.107) yields the following equations:

$$\frac{1}{2}(Z_{0e} - Z_{0o}) = JZ_0^2,$$

$$\frac{Z_{0e} + Z_{0o}}{Z_{0e} - Z_{0o}} = JZ_0 + \frac{1}{JZ_0},$$

where we have assumed $\sin\theta \simeq 1$ for θ near $\pi/2$. These equations can be solved for the even- and odd-mode line impedances to give

$$Z_{0e} = Z_0[1 + JZ_0 + (JZ_0)^2], \tag{9.108a}$$

$$Z_{0o} = Z_0[1 - JZ_0 + (JZ_0)^2]. \tag{9.108b}$$

Now consider a bandpass filter composed of a cascade of $N + 1$ coupled line sections, as shown in Figure 9.45a. The sections are numbered from left to right, with the load on the right, but the filter can be reversed without affecting the response. Since each coupled line section has an equivalent circuit of the form shown in Figure 9.44, the equivalent circuit of the cascade is as shown in Figure 9.45b. Between any two consecutive inverters we have a transmission line section that is effectively 2θ in length. This line is approximately $\lambda/2$ long in the vicinity of the bandpass region of the filter, and has an approximate equivalent circuit that consists of a shunt parallel LC resonator, as in Figure 9.45c.

The first step in establishing this equivalence is to find the parameters for the T-equivalent and ideal transformer circuit of Figure 9.45c (an exact equivalent). The $ABCD$ matrix for this circuit can be calculated using the results in Table 5.1 for a T-circuit and an ideal transformer:

$$\begin{bmatrix} A & B \\ C & D \end{bmatrix} = \begin{bmatrix} \dfrac{Z_{11}}{Z_{12}} & \dfrac{Z_{11}^2 - Z_{12}^2}{Z_{12}} \\ \dfrac{1}{Z_{12}} & \dfrac{Z_{11}}{Z_{12}} \end{bmatrix} \begin{bmatrix} -1 & 0 \\ 0 & -1 \end{bmatrix} = \begin{bmatrix} \dfrac{-Z_{11}}{Z_{12}} & \dfrac{Z_{12}^2 - Z_{12}^2}{Z_{12}} \\ \dfrac{-1}{Z_{12}} & \dfrac{-Z_{11}}{Z_{12}} \end{bmatrix}. \tag{9.109}$$

Equating this result to the $ABCD$ parameters for a transmission line of length 2θ and characteristic impedance Z_0 gives the parameters of the equivalent circuit as

$$Z_{12} = \frac{-1}{C} = \frac{jZ_0}{\sin 2\theta}, \tag{9.110a}$$

$$Z_{11} = Z_{22} = -Z_{12}A = -jZ_0 \cot 2\theta. \tag{9.110b}$$

Then the series arm impedance is

$$Z_{11} - Z_{12} = -jZ_0\frac{\cos 2\theta + 1}{\sin 2\theta} = -jZ_0 \cot\theta. \tag{9.111}$$

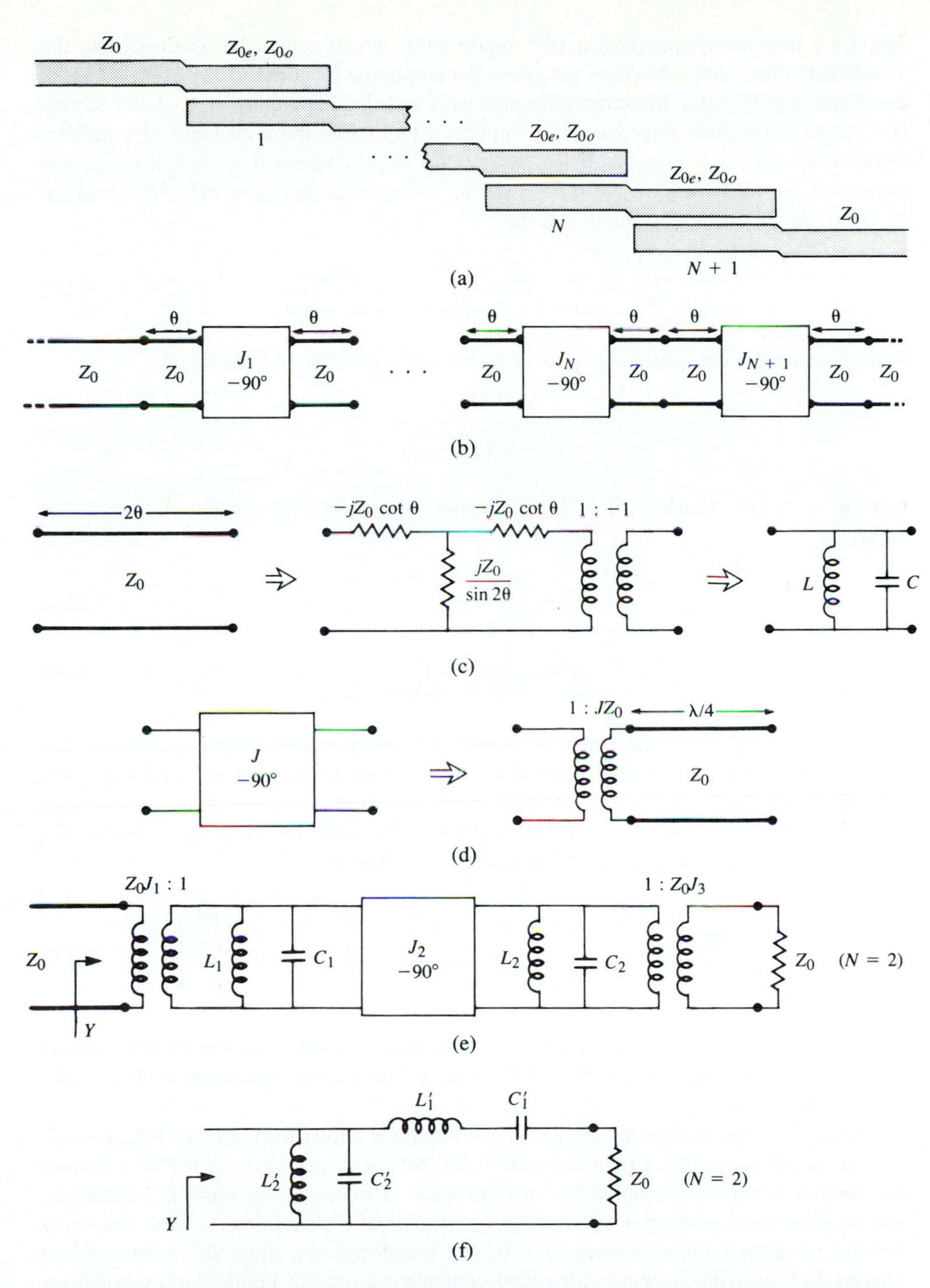

FIGURE 9.45 Development of an equivalent circuit for derivation of design equations for a coupled-line bandpass filter. (a) Layout of an $N+1$ section coupled-line bandpass filter. (b) Using equivalent circuit of Figure 9.44 for each coupled-line section. (c) Equivalent circuit for transmission lines of length 2θ. (d) Equivalent circuit of the admittance inverters. (e) Using results of (c) and (d) for the $N = 2$ case. (f) Lumped-element circuit for a bandpass filter for $N = 2$.

The $1{:}-1$ transformer provides a 180° phase shift, which cannot be obtained with the T-network alone; since this does not affect the amplitude response of the filter, it can be discarded. For $\theta \sim \pi/2$ the series arm impedances of (9.111) are near zero, and can also be ignored. The shunt impedance Z_{12}, however, looks like the impedance of a parallel resonant circuit for $\theta \sim \pi/2$. If we let $\omega = \omega_0 + \Delta\omega$, where $\theta = \pi/2$ at the center frequency ω_0, then we have $2\theta = \beta\ell = \omega\ell/v_p = (\omega_0 + \Delta\omega)\pi/\omega_0 = \pi(1 + \Delta\omega/\omega_0)$, so (9.110a) can be written for small $\Delta\omega$ as

$$Z_{12} = \frac{jZ_0}{\sin \pi(1 + \Delta\omega/\omega_0)} \simeq \frac{-jZ_0\omega_0}{\pi(\omega - \omega_0)}. \qquad 9.112$$

From Section 7.1 the impedance near resonance of a parallel LC circuit is

$$Z = \frac{-jL\omega_0^2}{2(\omega - \omega_0)}, \qquad 9.113$$

with $\omega_0^2 = 1/LC$. Equating this to (9.112) gives the equivalent inductor and capacitor values as

$$L = \frac{2Z_0}{\pi\omega_0}, \qquad 9.114a$$

$$C = \frac{1}{\omega_0^2 L} = \frac{\pi}{2Z_0\omega_0}. \qquad 9.114b$$

The end sections of the circuit of Figure 9.45b require a different treatment. The lines of length θ on either end of the filter are matched to Z_0, and so can be ignored. The end inverters, J_1 and J_{N+1}, can each be represented as a transformer followed by a $\lambda/4$ section of line, as shown in Figure 9.45d. The $ABCD$ matrix of a transformer with a turns ratio N in cascade with a quarter-wave line is,

$$\begin{bmatrix} A & B \\ C & D \end{bmatrix} = \begin{bmatrix} \frac{1}{N} & 0 \\ 0 & N \end{bmatrix} \begin{bmatrix} 0 & -jZ_0 \\ \frac{-j}{Z_0} & 0 \end{bmatrix} = \begin{bmatrix} 0 & \frac{-jZ_0}{N} \\ \frac{-jN}{Z_0} & 0 \end{bmatrix}. \qquad 9.115$$

Comparing this to the $ABCD$ matrix of an admittance inverter (part of (9.104)) shows that the necessary turns ratio is $N = JZ_0$. The $\lambda/4$ line merely produces a phase shift, and so can be ignored.

Using these results for the interior and end sections allows the circuit of Figure 9.45b to be transformed into the circuit of Figure 9.45e, which is specialized to the $N = 2$ case. We see that each pair of coupled line sections leads to an equivalent shunt LC resonator, and an admittance inverter occurs between each pair of LC resonators. Next, we show that the admittance inverters have the effect of transforming a shunt LC resonator into a series LC resonator, leading to the final equivalent circuit of Figure 9.45f (shown for $N = 2$). This will then allow the admittance inverter constants, J_n, to be determined from the element values of a low-pass prototype. We will demonstrate this for the $N = 2$ case.

With reference to Figure 9.45e, the admittance just to the right of the J_2 inverter is

$$j\omega C_2 + \frac{1}{j\omega L_1} + Z_0 J_3^2 = j\sqrt{\frac{C_2}{L_2}}\left(\frac{\omega}{\omega_0} - \frac{\omega_0}{\omega}\right) + Z_0 J_3^2,$$

since the transformer scales the load admittance by the square of the turns ratio. Then the admittance seen at the input of the filter is

$$Y = \frac{1}{J_1^2 Z_0^2}\left\{j\omega C_1 + \frac{1}{j\omega L_1} + \frac{J_2^2}{j\sqrt{C_2/L_2}\left[(\omega/\omega_0) - (\omega_0/\omega)\right] + Z_0 J_3^2}\right\}$$

$$= \frac{1}{J_1^2 Z_0^2}\left\{j\sqrt{\frac{C_1}{L_1}}\left(\frac{\omega}{\omega_0} - \frac{\omega_0}{\omega}\right) + \frac{J_2^2}{j\sqrt{C_2/L_2}\left[(\omega/\omega_0) - (\omega_0/\omega)\right] + Z_0 J_3^2}\right\}. \quad 9.116$$

These results also use the fact, from (9.114), that $L_n C_n = 1/\omega_0^2$ for all LC resonators. Now the admittance seen looking into the circuit of Figure 9.45f is

$$Y = j\omega C_1' + \frac{1}{j\omega L_1'} + \frac{1}{j\omega L_2' + (1/j\omega C_2') + Z_0}$$

$$= j\sqrt{\frac{C_1'}{L_1'}}\left(\frac{\omega}{\omega_0} - \frac{\omega_0}{\omega}\right) + \frac{1}{j\sqrt{L_2'/C_2'}\left[(\omega/\omega_0) - (\omega_0/\omega)\right] + Z_0}, \quad 9.117$$

which is identical in form to (9.116). Thus, the two circuits will be equivalent if the following conditions are met:

$$\frac{1}{J_1^2 Z_0^2}\sqrt{\frac{C_1}{L_1}} = \sqrt{\frac{C_1'}{L_1'}}, \quad 9.118a$$

$$\frac{J_1^2 Z_0^2}{J_2^2}\sqrt{\frac{C_2}{L_2}} = \sqrt{\frac{L_2'}{C_2'}}, \quad 9.118b$$

$$\frac{J_1^2 Z_0^3 J_3^2}{J_2^2} = Z_0. \quad 9.118c$$

We know L_n and C_n from (9.114); L_n' and C_n' are determined from the element values of a lumped-element low-pass prototype which has been impedance scaled and frequency transformed to a bandpass filter. Using the results in Table 9.6 and the impedance scaling formulas of (9.64) allows the L_n' and C_n' values to be written as

$$L_1' = \frac{\Delta Z_0}{\omega_0 g_1}, \quad 9.119a$$

$$C_1' = \frac{g_1}{\Delta \omega_0 Z_0}, \quad 9.119b$$

$$L_2' = \frac{g_2 Z_0}{\Delta \omega_0}, \quad 9.119c$$

$$C_2' = \frac{\Delta}{\omega_0 g_2 Z_0}, \quad 9.119d$$

where $\Delta = (\omega_2 - \omega_1)/\omega_0$ is the fractional bandwidth of the filter. Then (9.118) can be solved for the inverter constants with the following results (for $N = 2$):

$$J_1 Z_0 = \left(\frac{C_1 L_1'}{L_1 C_1'}\right)^{1/4} = \sqrt{\frac{\pi\Delta}{2g_1}}, \tag{9.120a}$$

$$J_2 Z_0 = J_1 Z_0^2 \left(\frac{C_2 C_2'}{L_2 L_2'}\right)^{1/4} = \frac{\pi\Delta}{2\sqrt{g_1 g_2}}, \tag{9.120b}$$

$$J_3 Z_0 = \frac{J_2}{J_1} = \sqrt{\frac{\pi\Delta}{2g_2}}. \tag{9.120c}$$

After the J_ns are found, Z_{0e} and Z_{0o} for each coupled line section can be calculated from (9.108).

The above results were derived for the special case of $N = 2$ (three coupled line sections), but more general results can be derived for any number of sections, and for the case where $Z_L \neq Z_0$ (or $g_{N+1} \neq 1$, as in the case of an equal-ripple response with N even). Thus, the design equations for a bandpass filter with $N + 1$ coupled line sections are

$$Z_0 J_1 = \sqrt{\frac{\pi\Delta}{2g_1}}, \tag{9.121a}$$

$$Z_0 J_n = \frac{\pi\Delta}{2\sqrt{g_{n-1} g_n}}, \qquad \text{for } n = 2, 3, \ldots, N, \tag{9.121b}$$

$$Z_0 J_{N+1} = \sqrt{\frac{\pi\Delta}{2g_N g_{N+1}}}. \tag{9.121c}$$

The even and odd mode characteristic impedances for each section are then found from (9.108).

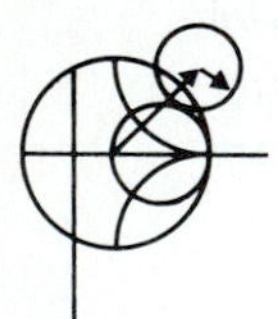

EXAMPLE 9.8

Design a coupled line bandpass filter with $N = 3$ and a 0.5 dB equal-ripple response. The center frequency is 2.0 GHz, the bandwidth is 10%, and $Z_0 = 50\,\Omega$. What is the attenuation at 1.8 GHz?

Solution

The fractional bandwidth is $\Delta = 0.1$. We can use Figure 9.27a to obtain the attenuation at 1.8 GHz, but first we must use (9.71) to convert this frequency to the normalized low-pass form ($\omega_c = 1$):

$$\omega \leftarrow \frac{1}{\Delta}\left(\frac{\omega}{\omega_0} - \frac{\omega_0}{\omega}\right) = \frac{1}{0.1}\left(\frac{1.8}{2.0} - \frac{2.0}{1.8}\right) = -2.11.$$

Then the value on the horizontal scale of Figure 9.27a is

$$\left|\frac{\omega}{\omega_c}\right| - 1 = |-2.11| - 1 = 1.11,$$

which indicates an attenuation of about 20 dB for $N = 3$.

The low-pass prototype values, g_n, are given in Table 9.4; then (9.121) can be used to calculate the admittance inverter constants, J_n. Finally, the even- and odd-mode characteristic impedances can be found from (9.108). These results are summarized in the following table:

n	g_n	$Z_0 J_n$	$Z_{0e}(\Omega)$	$Z_{0o}(\Omega)$
1	1.5963	0.3137	70.61	39.24
2	1.0967	0.1187	56.64	44.77
3	1.5963	0.1187	56.64	44.77
4	1.0000	0.3137	70.61	39.24

Note that the filter sections are symmetric about the midpoint. The calculated response of this filter is shown in Figure 9.46; passbands also occur at 6 GHz, 10 GHz, etc.

Many other types of filters can be constructed using coupled line sections; most of these are of the bandpass or bandstop variety. One particularly compact design is the interdigitated filter, which can be obtained from a coupled line filter by folding the lines at their midpoints; see [2] and [3] for details. ○

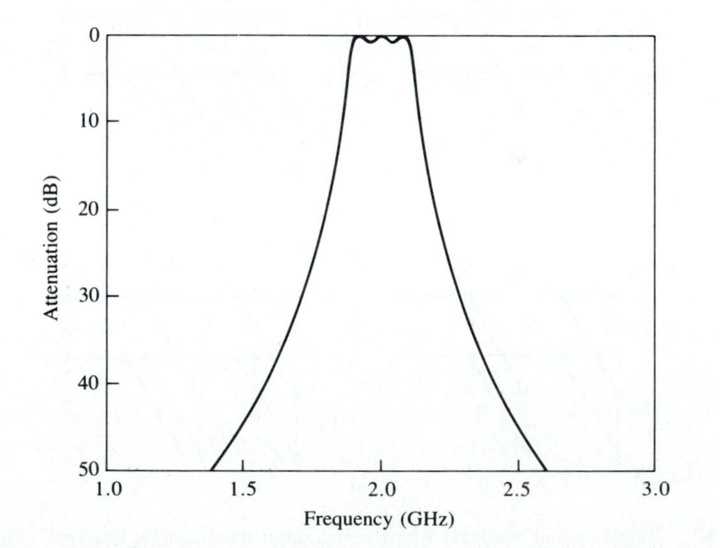

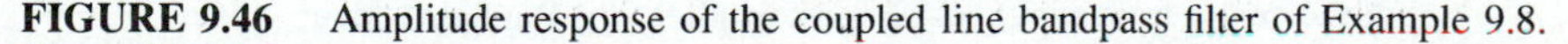
FIGURE 9.46 Amplitude response of the coupled line bandpass filter of Example 9.8.

9.8 FILTERS USING COUPLED RESONATORS

We have seen that bandpass and bandstop filters require elements which behave as series or parallel resonant circuits; the coupled line bandpass filters of the previous section were of this type. Here we will consider several other types of microwave filters that use transmission line or cavity resonators.

Bandstop and Bandpass Filters Using Quarter-Wave Resonators

From Chapter 7 we know that quarter-wave open-circuited or short-circuited transmission line stubs look like series or parallel resonant circuits, respectively. Thus we can use such stubs in shunt along a transmission line to implement bandpass or bandstop filters, as shown in Figure 9.47. Quarter-wavelength sections of line between the stubs act as admittance inverters to effectively convert alternate shunt resonators to series resonators. The stubs and the transmission line sections are $\lambda/4$ long at the center frequency, ω_0.

For narrow bandwidths the response of such a filter using N stubs is essentially the same as that of a coupled line filter using $N + 1$ sections. The internal impedance of the stub filter is Z_0, while in the case of the coupled line filter end sections are required to transform the impedance level. This makes the stub filter more compact and easier to design. A disadvantage, however, is that a filter using stub resonators often requires characteristic impedances that are difficult to realize in practice.

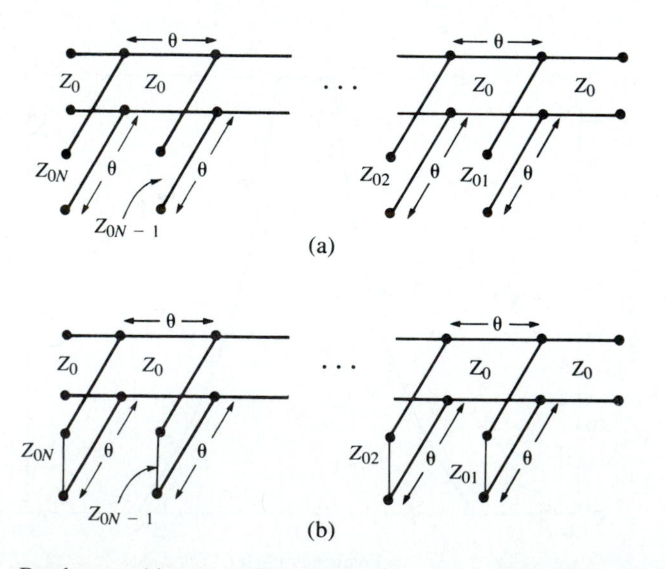

FIGURE 9.47 Bandstop and bandpass filters using shunt transmission line resonators ($\theta = \pi/2$ at the center frequency). (a) Bandstop filter. (b) Bandpass filter.

We first consider a bandstop filter using N open-circuited stubs, as shown in Figure 9.47a. The design equations for the required stub characteristic impedances, Z_{0n}, will be derived in terms of the element values of a low-pass prototype through the use of an equivalent circuit. The analysis of the bandpass version, using short-circuited stubs, follows the same procedure so the design equations for this case are presented without detailed derivation.

As indicated in Figure 9.48a, an open-circuited stub can be approximated as a series LC resonator when its length is near 90°. The input impedance of an open-circuited transmission line of characteristic impedance Z_{0n} is

$$Z = -jZ_{0n} \cot\theta,$$

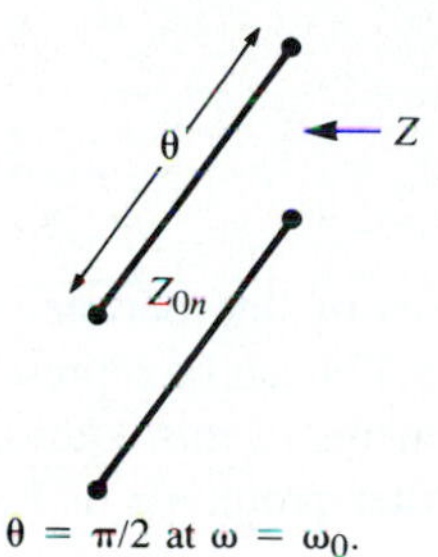

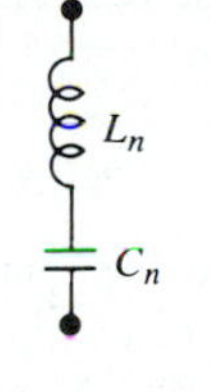

(a)

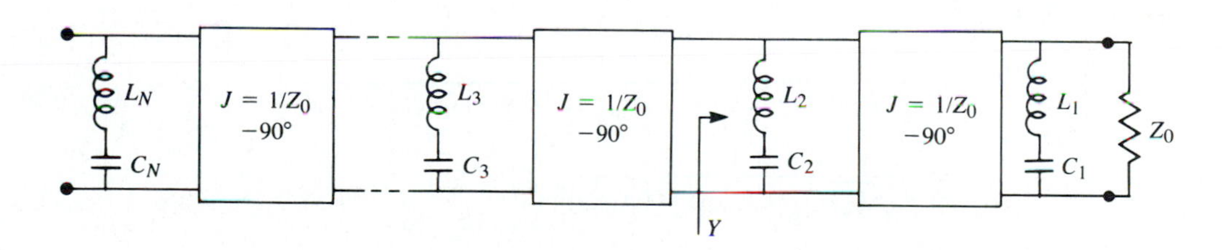

(b)

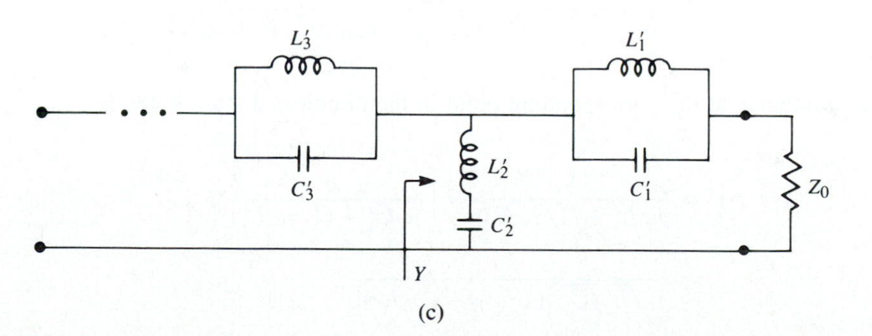

(c)

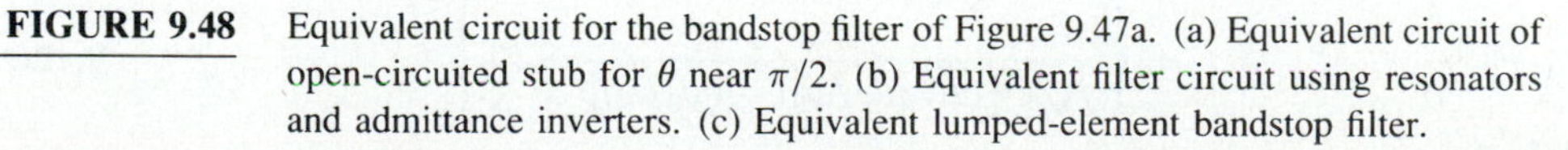

FIGURE 9.48 Equivalent circuit for the bandstop filter of Figure 9.47a. (a) Equivalent circuit of open-circuited stub for θ near $\pi/2$. (b) Equivalent filter circuit using resonators and admittance inverters. (c) Equivalent lumped-element bandstop filter.

where $\theta = \pi/2$ for $\omega = \omega_0$. If we let $\omega = \omega_0 + \Delta\omega$, where $\Delta\omega << \omega_0$, then $\theta = \pi/2(1 + \Delta\omega/\omega_0)$, and this impedance can be approximated as

$$Z = jZ_{0n} \tan \frac{\pi \Delta\omega}{2\omega_0} \simeq \frac{jZ_{0n}\pi(\omega - \omega_0)}{2\omega_0}, \tag{9.122}$$

for frequencies in the vicinity of the center frequency, ω_0. The impedance of a series LC circuit is

$$Z = j\omega L_n + \frac{1}{j\omega C_n} = j\sqrt{\frac{L_n}{C_n}}\left(\frac{\omega}{\omega_0} - \frac{\omega_0}{\omega}\right) \simeq 2j\sqrt{\frac{L_n}{C_n}}\frac{\omega - \omega_0}{\omega_0} \simeq 2jL_n(\omega - \omega_0), \tag{9.123}$$

where $L_nC_n = 1/\omega_0^2$. Equating (9.122) and (9.123) gives the characteristic impedance of the stub in terms of the resonator parameters:

$$Z_{0n} = \frac{4\omega_0 L_n}{\pi}. \tag{9.124}$$

Then, if we consider the quarter-wave sections of line between the stubs as ideal admittance inverters, the bandstop filter of Figure 9.47a can be represented by the equivalent circuit of Figure 9.48b. Next, the circuit elements of this equivalent circuit can be related to those of the lumped-element bandstop filter prototype of Figure 9.48c.

With reference to Figure 9.48b, the admittance, Y, seen looking toward the L_2C_2 resonator is

$$\begin{aligned} Y &= \frac{1}{j\omega L_2 + (1/j\omega C_2)} + \frac{1}{Z_0^2}\left[\frac{1}{j\omega L_1 + (1/j\omega C_1)} + \frac{1}{Z_0}\right]^{-1} \\ &= \frac{1}{j\sqrt{L_2/C_2}\left[(\omega/\omega_0) - (\omega_0/\omega)\right]} \\ &\quad + \frac{1}{Z_0^2}\left\{\frac{1}{j\sqrt{L_1/C_1}\left[(\omega/\omega_0) - (\omega_0/\omega)\right] + (1/Z_0)}\right\}^{-1}. \end{aligned} \tag{9.125}$$

The admittance at the corresponding point in the circuit of Figure 9.48c is

$$\begin{aligned} Y &= \frac{1}{j\omega L_2' + (1/j\omega C_2)} + \left[\frac{1}{j\omega C_1' + (1/j\omega L_1')} + Z_0\right]^{-1} \\ &= \frac{1}{j\sqrt{L_2'/C_2'}\left[(\omega/\omega_0) - (\omega_0/\omega)\right]} \\ &\quad + \left\{\frac{1}{j\sqrt{C_1'/L_1'}\left[(\omega/\omega_0) - (\omega_0/\omega)\right]} + Z_0\right\}^{-1}. \end{aligned} \tag{9.126}$$

These two results will be equivalent if the following conditions are satisfied:

$$\frac{1}{Z_0^2}\sqrt{\frac{L_1}{C_1}} = \sqrt{\frac{C_1'}{L_1'}}, \tag{9.127a}$$

$$\sqrt{\frac{L_2}{C_2}} = \sqrt{\frac{L_2'}{C_2'}}. \tag{9.127b}$$

Since $L_nC_n = L_n'C_n' = 1/\omega_0^2$, these results can be solved for L_n:

$$L_1 = \frac{Z_0^2}{\omega_0^2 L_1'}, \tag{9.128a}$$

$$L_2 = L_2'. \tag{9.128b}$$

Then using (9.124) and the impedance-scaled bandstop filter elements from Table 9.6 gives the stub characteristic impedances as

$$Z_{01} = \frac{4Z_0^2}{\pi\omega_0 L_1'} = \frac{4Z_0}{\pi g_1 \Delta}, \tag{9.129a}$$

$$Z_{02} = \frac{4\omega_0 L_2'}{\pi} = \frac{4Z_0}{\pi g_2 \Delta}, \tag{9.129b}$$

where $\Delta = (\omega_2 - \omega_1)/\omega_0$ is the fractional bandwidth of the filter. It is easy to show that the general result for the characteristic impedances of a bandstop filter is

$$Z_{0n} = \frac{4Z_0}{\pi g_n \Delta}. \tag{9.130}$$

For a bandpass filter using short-circuited stub resonators the corresponding result is,

$$Z_{0n} = \frac{\pi Z_0 \Delta}{4g_n}. \tag{9.131}$$

These results only apply to filters having input and output impedances of Z_0, and so cannot be used for equal-ripple designs with N even.

EXAMPLE 9.9

Design a bandstop filter using three quarter-wave open-circuit stubs. The center frequency is 2.0 GHz, the bandwidth is 15%, and the impedance is 50 Ω. Use an equal-ripple response, with a 0.5 dB ripple level.

Solution

The fractional bandwidth is $\Delta = 0.15$. Table 9.4 gives the low-pass prototype values, g_n, for $N = 3$. Then the characteristic impedances of the stubs can be found from (9.130). The results are listed in the following table:

n	g_n	Z_{0n}
1	1.5963	265.9 Ω
2	1.0967	387.0 Ω
3	1.5963	265.9 Ω

The filter circuit is shown in Figure 9.47a, with all stubs and transmission line sections $\lambda/4$ long at 2.0 GHz. The calculated attenuation for this filter is shown in Figure 9.49; the ripple in the passbands is somewhat greater than 0.5 dB, as a result of the approximations involved in the development of the design equations. ○

The performance of quarter-wave resonator filters can be improved by allowing the characteristic impedances of the interconnecting lines to be variable; then an exact correspondence with coupled line bandpass or bandstop filters can be demonstrated. Design details for this case can be found in reference [2].

Bandpass Filters Using Capacitively-Coupled Resonators

Another type of bandpass filter that can be conveniently fabricated in microstrip or stripline form is the capacitive-gap coupled resonator filter shown in Figure 9.50. An Nth order filter of this form will use N resonant sections of transmission line with $N + 1$ capacitive gaps between them. These gaps can be approximated as series capacitors; design data relating the capacitance to the gap size and transmission line parameters

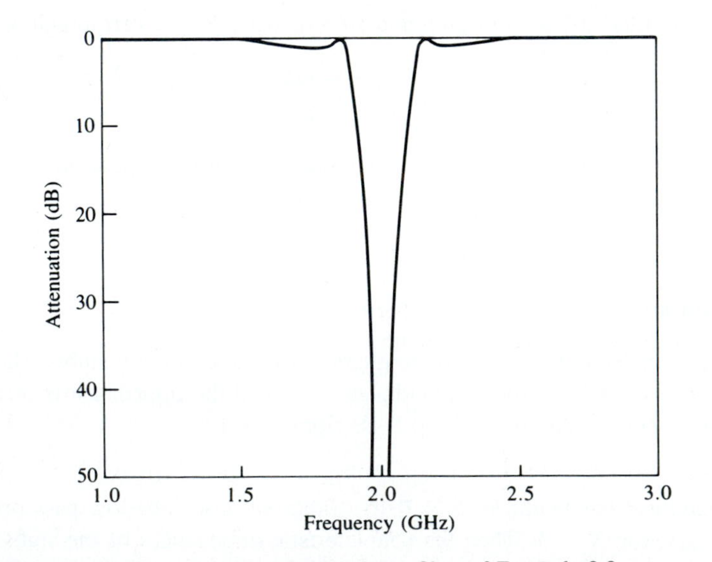

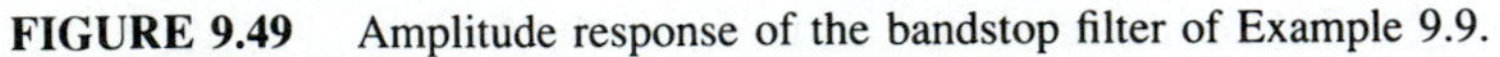

FIGURE 9.49 Amplitude response of the bandstop filter of Example 9.9.

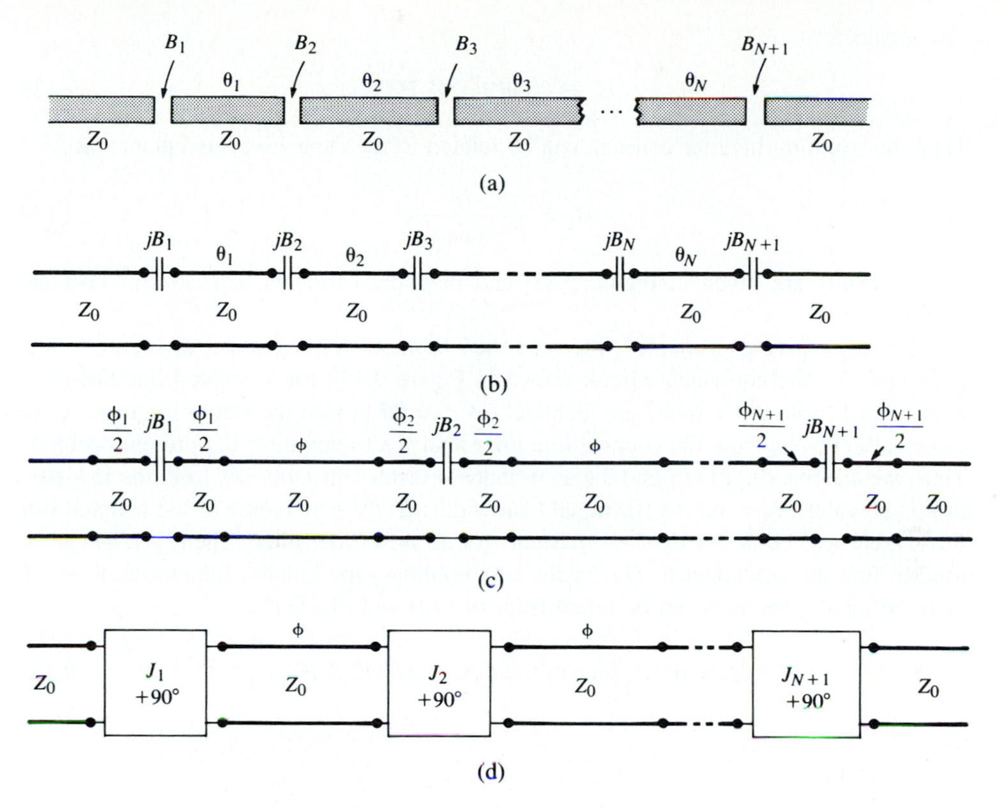

FIGURE 9.50 Development of the equivalence of a capacitive-gap coupled resonator bandpass filter to the coupled line bandpass filter of Figure 9.45. (a) The capacitive-gap coupled resonator bandpass filter. (b) Transmission line model. (c) Transmission line model with negative-length sections forming admittance inverters ($\phi_i/2 < 0$). (d) Equivalent circuit using inverters and $\lambda/2$ resonators ($\phi = \pi$ at ω_0). This circuit is now identical in form with the coupled line bandpass filter equivalent circuit in Figure 9.45b.

is given in graphical form in reference [2]. The filter can then be modelled as shown in Figure 9.50b. The resonators are approximately $\lambda/2$ long at the center frequency, ω_0.

Next, we redraw the equivalent circuit of Figure 9.50b with negative-length transmission line sections on either side of the series capacitors. The lines of length ϕ will be $\lambda/2$ long at ω_0, so the electrical length, θ_i, of the ith section in Figures 9.50a,b is

$$\theta_i = \pi + \frac{1}{2}\phi_i + \frac{1}{2}\phi_{i+1}, \qquad \text{for } i = 1, 2, \ldots, N, \tag{9.132}$$

with $\phi_i < 0$. The reason for doing this is that the combination of series capacitor and negative-length transmission lines forms the equivalent circuit of an admittance inverter, as seen from Figure 9.38c. In order for this equivalence to be valid, the following relationship must hold between the electrical length of the lines and the capacitive susceptance:

$$\phi_i = -\tan^{-1}(2Z_0B_i). \qquad 9.133$$

Then the resulting inverter constant can be related to the capacitive susceptance as

$$B_i = \frac{J_i}{1-(Z_0J_i)^2}. \qquad 9.134$$

(These results are given in Figure 9.38, and their derivation is requested in Problem 9.15.)

The capacitive-gap coupled filter can then be modelled as shown in Figure 9.50d. Now consider the equivalent circuit shown in Figure 9.45b for a coupled line bandpass filter. Since these two circuits are identical (as $\phi = 2\theta = \pi$ at the center frequency), we can use the results from the coupled line filter analysis to complete the present problem. Thus, we can use (9.121) to find the admittance inverter constants, J_i, from the low-pass prototype values (g_i) and the fractional bandwidth, Δ. As in the case of the coupled line filter, there will be $N+1$ inverter constants for an Nth order filter. Then (9.134) can be used to find the susceptance, B_i, for the ith coupling gap. Finally, the electrical length of the resonator sections can be found from (9.132) and (9.133):

$$\theta_i = \pi - \frac{1}{2}[\tan^{-1}(2Z_0B_i) + \tan^{-1}(2Z_0B_{i+1})]. \qquad 9.135$$

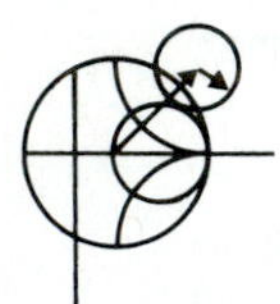

EXAMPLE 9.10

Design a bandpass filter using capacitive coupled resonators, with a 0.5 dB equal-ripple passband characteristic. The center frequency is 2.0 GHz, the bandwidth is 10%, and the impedance is 50 Ω. At least 20 dB of attenuation is required at 2.2 GHz.

Solution

We first determine the order of the filter to satisfy the attenuation specification at 2.2 GHz. Using (9.71) to convert to normalized frequency gives

$$\omega \leftarrow \frac{1}{\Delta}\left(\frac{\omega}{\omega_0} - \frac{\omega_0}{\omega}\right) = \frac{1}{0.1}\left(\frac{2.2}{2.0} - \frac{2.0}{2.2}\right) = 1.91.$$

Then,

$$\left|\frac{\omega}{\omega_c}\right| - 1 = 1.91 - 1.0 = 0.91.$$

From Figure 9.27a, we see that $N = 3$ should satisfy the attenuation specification at 2.2 GHz. The low-pass prototype values are given in Table 9.4, from which the inverter constants can be calculated using (9.121). Then the coupling susceptances can be found from (9.134), and the coupling capacitor values as

$$C_n = \frac{B_n}{\omega_0}.$$

Finally, the resonator lengths can be calculated from (9.135). The following table summarizes these results.

n	g_n	Z_0J_n	B_n	C_n	θ_n
1	1.5963	0.3137	6.96×10^{-3}	0.554 pF	155.8°
2	1.0967	0.1187	2.41×10^{-3}	0.192 pF	166.5°
3	1.5963	0.1187	2.41×10^{-3}	0.192 pF	155.8°
4	1.0000	0.3137	6.96×10^{-3}	0.554 pF	----

The calculated amplitude response is plotted in Figure 9.51. The specifications of this filter are the same as the coupled line bandpass filter of Example 9.8, and comparison of the results in Figures 9.51 and 9.46 shows that the responses are identical near the passband region. ○

Direct-Coupled Waveguide Cavity Filters

Another type of bandpass filter that is commonly made in waveguide form is the direct-coupled cavity filter shown in Figure 9.52a. Inductive irises spaced along the waveguide form resonators that are approximately $\lambda_g/2$ in length. The equivalent circuit is shown in Figure 9.52b. Observe that this circuit is the dual of the capacitive-gap coupled resonator filter circuit of Figure 9.50b; therefore, the design and operation of these two filters will be very similar. One difference is that the waveguide filter must

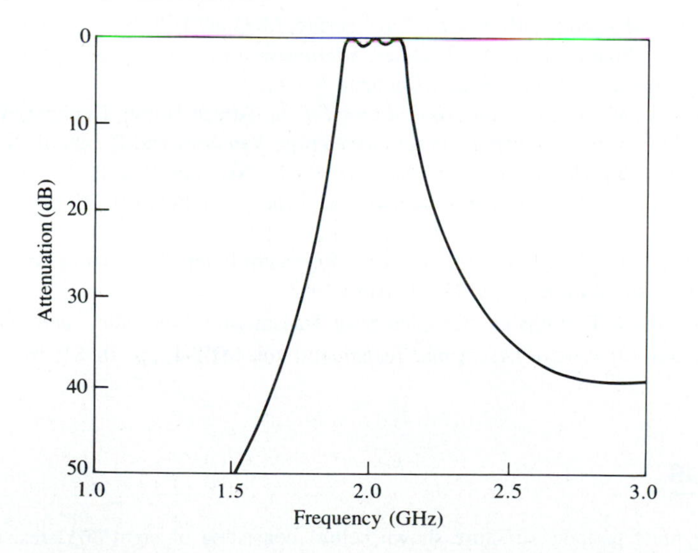

FIGURE 9.51 Amplitude response for the capacitive-gap coupled resonator bandpass filter of Example 9.10.

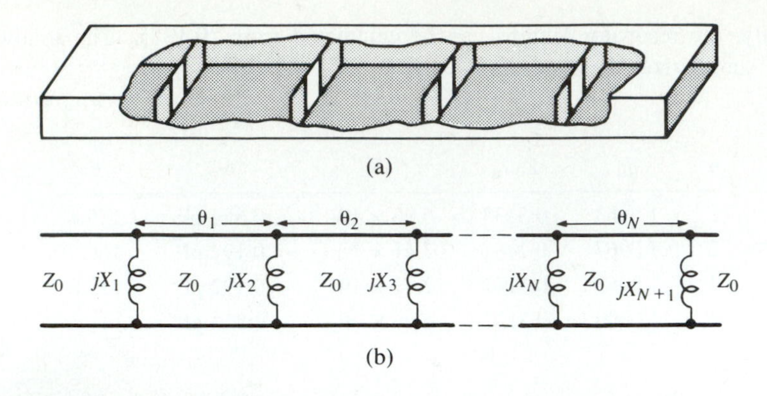

FIGURE 9.52 (a) A direct-coupled waveguide cavity filter. (b) Its equivalent circuit.

be designed in terms of $(\beta/k_0)\omega = \beta c$, instead of ω, because the electrical lengths of the transmission line sections are proportional to β, the propagation constant of the waveguide. In addition, the reactance of the inductive diaphragms is proportional to β, rather than ω. Design equations and other details for waveguide cavity filters can be found in the literature [2].

REFERENCES

[1] R. E. Collin, *Foundations for Microwave Engineering*, McGraw-Hill, N. Y., 1966.

[2] G. L. Matthaei, L. Young, and E. M. T. Jones, *Microwave Filters, Impedance-Matching Networks, and Coupling Structures*, Artech House, Dedham, Mass., 1980.

[3] J. A. G. Malherbe, *Microwave Transmission Line Filters*, Artech House, Dedham, Mass., 1979.

[4] W. A. Davis, *Microwave Semiconductor Circuit Design*, Van Nostrand Reinhold, N. Y., 1984.

[5] R. F. Harrington, *Time-Harmonic Electromagnetic Fields*, McGraw-Hill, N. Y., 1961.

[6] P. I. Richard, "Resistor-Transmission Line Circuits," *Proc. of the IRE*, vol. 36, pp. 217–220, February 1948.

[7] S. B. Cohn, "Parallel-Coupled Transmission-Line-Resonator Filters," *IRE Trans. Microwave Theory and Techniques*, vol. MTT-6, pp. 223–231, April 1958.

[8] E. M. T. Jones and J. T. Bolljahn, "Coupled-Strip-Transmission Line Filters and Directional Couplers," *IRE Trans. Microwave Theory and Techniques*, vol. MTT-4, pp. 78–81, April 1956.

PROBLEMS

9.1 Consider the finite periodic structure shown below, consisting of eight $80\,\Omega$ resistors spaced at intervals of $\lambda/2$ along a transmission line with $Z_0 = 50\ \Omega$. Find the voltage $V(z)$ along the line, and plot $|V(z)|$ versus z.

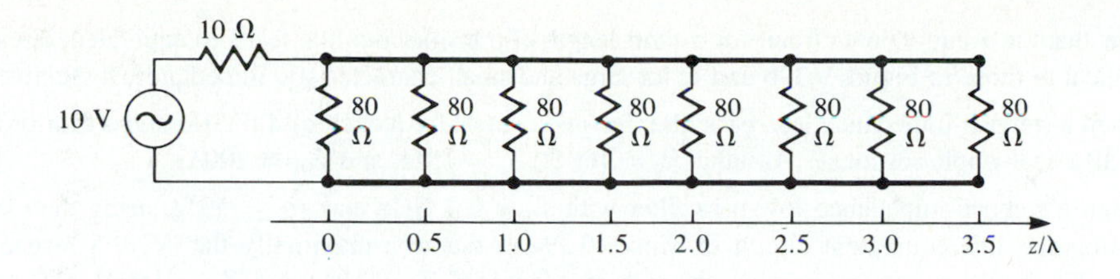

9.2 Sketch the k-β diagram for the infinite periodic structure shown below. Assume $Z_0 = 100\,\Omega$, $d = 1.0$ cm, $k = k_0$, and $L_0 = 3.0$ nH.

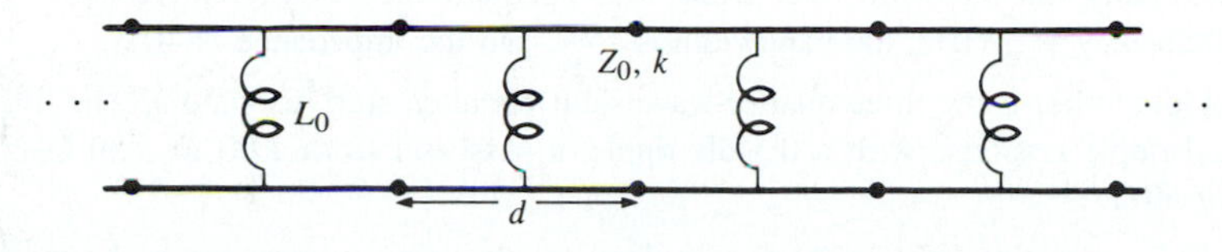

9.3 Verify the expression for the image impedance of a π-network given in Table 9.1.

9.4 Compute the image impedances and propagation factor for the network shown below.

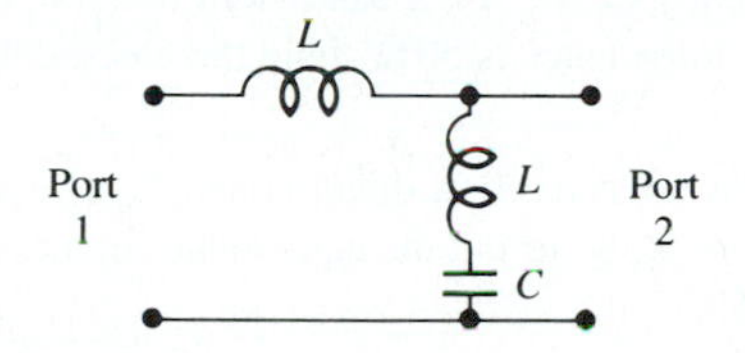

9.5 Design a composite low-pass filter by the image parameter method with the following specifications: $R_0 = 50\,\Omega$, $f_c = 50$ MHz, and $f_\infty = 52$ MHz.

9.6 Design a composite high-pass filter by the image parameter method with the following specifications: $R_0 = 75\,\Omega$, $f_c = 50$ MHz, and $f_\infty = 48$ MHz.

9.7 Solve the design equations in Section 9.3 for the elements of an $N = 2$ equal-ripple filter if the ripple specification is 1.0 dB.

9.8 Design a low-pass maximally flat filter having a passband of 0 to 3 GHz, and an attenuation of 20 dB at 5 GHz. The characteristic impedance is $75\,\Omega$.

9.9 Design a five-section high-pass filter with a 3 dB equal-ripple response, a cutoff frequency of 1 GHz, and an impedance of $50\,\Omega$. What is the resulting attenuation at 0.6 GHz?

9.10 Design a four-section bandpass filter having a maximally flat group delay response. The bandwidth should be 5% with a center frequency of 2 GHz. The impedance is $50\,\Omega$.

9.11 Design a three-section bandstop filter with a 0.5 dB equal-ripple response, a bandwidth of 10% centered at 3 GHz, and an impedance of $75\,\Omega$. What is the resulting attenuation at 3.1 GHz?

9.12 Verify the second Kuroda identity in Table 9.7 by calculating the $ABCD$ matrices for both circuits.

9.13 Design a low-pass third-order maximally flat filter using only series stubs. The cutoff frequency is 6 GHz and the impedance is $50\,\Omega$.

9.14 Design a low-pass fourth-order maximally flat filter using only shunt stubs. The cutoff frequency is 8 GHz and the impedance is $50\,\Omega$.

9.15 Verify the operation of the admittance inverter of Figure 9.38c by calculating its $ABCD$ matrix and comparing it to the $ABCD$ matrix of the admittance inverter made from a quarter-wave line.

9.16 Show that the π equivalent circuit for a short length of transmission line leads to equivalent circuits identical to those in Figure 9.39b and c, for large and small characteristic impedance, respectively.

9.17 Design a stepped-impedance low-pass filter having a cutoff frequency of 4.0 GHz and a fifth-order, 0.5 dB equal-ripple response. Assume $R_0 = 100\,\Omega$, $Z_\ell = 15\,\Omega$, and $Z_h = 200\,\Omega$.

9.18 Design a stepped-impedance low-pass filter with $f_c = 2.0$ GHz and $R_0 = 50\,\Omega$, using the exact transmission line equivalent circuit of Figure 9.39a. Assume a maximally flat $N = 5$ response, and solve for the necessary line lengths and impedances if $Z_\ell = 10\,\Omega$ and $Z_h = 150\,\Omega$.

9.19 Design a four-section coupled line bandpass filter with a maximally flat response. The passband is 3.00 to 3.50 GHz, and the impedance is $50\,\Omega$. What is the attenuation at 2.9 GHz?

9.20 Design a maximally flat bandstop filter using four open-circuited quarter-wave stub resonators. The center frequency is 3 GHz, the bandwidth is 15%, and the impedance is $40\,\Omega$.

9.21 Design a bandpass filter using three quarter-wave short-circuited stub resonators. The filter should have an equal-ripple response with a 0.5 dB ripple, a passband from 3.00 to 3.50 GHz, and an impedance of $50\,\Omega$.

9.22 Derive the design equation of (9.131) for bandpass filters using quarter-wave shorted stub resonators.

9.23 Design a bandpass filter using capacitive-gap coupled resonators. The response should be maximally flat, with a center frequency of 4 GHz, a bandwidth of 12%, and at least 12 dB attenuation at 3.6 GHz. The characteristic impedance is $50\,\Omega$. Find the electrical line lengths and the coupling capacitor values.

9.24 Derive the design equations for an N-section direct-coupled waveguide cavity filter. (*Hint*: Use the impedance inverter of Figure 9.38c to put the equivalent circuit in the form of the coupled line filter equivalent in Figure 9.45b.)

CHAPTER 10

Theory and Design of Ferrimagnetic Components

The components and networks discussed up to this point have all been reciprocal. That is, the response between any two ports, i and j, of a component did not depend on the direction of signal flow (thus, $S_{ij} = S_{ji}$). This will always be the case when the component consists of passive and isotropic material, but if anisotropic (different properties in different directions) materials are used, nonreciprocal behavior can be obtained. This allows the implementation of a wide variety of devices having directional properties.

In Chapter 2 we discussed materials with electric anisotropy (tensor permittivity), and magnetic anisotropy (tensor permeability). The most practical anisotropic materials for microwave applications are ferromagnetic compounds such as YIG (yttrium iron garnet), and ferrites composed of iron oxides and various other elements such as aluminum, cobalt, manganese, and nickel. In contrast to ferromagnetic materials (e.g., iron, steel), ferrimagnetic compounds have high resistivity and a significant amount of anisotropy at microwave frequencies. As we will see, the magnetic anisotropy of a ferrimagnetic material is actually induced by applying a DC magnetic bias field. This field aligns the magnetic dipoles in the ferrite material to produce a net (nonzero) magnetic dipole moment, and causes the magnetic dipoles to precess at a frequency controlled by the strength of the bias field. A microwave signal circularly polarized in the same direction as this precession will interact strongly with the dipole moments, while an oppositely polarized field will interact less strongly. Since, for a given direction of rotation, the sense of polarization changes with the direction of propagation, a microwave signal will propagate through a ferrite differently in different directions. This effect can be utilized to fabricate directional devices such as isolators, circulators, and gyrators. Another useful characteristic of ferrimagnetic materials is that the interaction with an applied microwave signal can be controlled by adjusting the strength of the bias field. This effect leads to a variety of control devices such as phase shifters, switches, and tunable resonators and filters.

We will begin by considering the microscopic behavior of a ferrimagnetic material and its interaction with a microwave signal to derive the permeability tensor. This macroscopic description of the material can then be used with Maxwell's equations to analyze wave propagation in an infinite ferrite medium, and in a ferrite-loaded waveguide. These canonical problems will illustrate the nonreciprocal propagation properties of ferrimagnetic materials, including Faraday rotation and birefringence effects, and will be used in later sections when discussing the operation and design of waveguide phase shifters and isolators.

10.1 BASIC PROPERTIES OF FERRIMAGNETIC MATERIALS

In this section we will show how the permeability tensor for a ferrimagnetic material can be deduced from a relatively simple microscopic view of the atom. We will also discuss how loss affects the permeability tensor, and the demagnetization field inside a finite-sized piece of ferrite.

The Permeability Tensor

The magnetic properties of a material are due to the existence of magnetic dipole moments, which arise primarily from electron spin. From quantum mechanical considerations [1], the magnetic dipole moment of an electron due to its spin is given by

$$m = \frac{q\hbar}{2m_e} = 9.27 \times 10^{-24} \text{ A-m}^2, \tag{10.1}$$

where $\hbar$ is Planck's constant divided by 2π, q is the electron charge, and m_e is the mass of the electron. An electron in orbit around a nucleus gives rise to an effective current loop, and thus an additional magnetic moment, but this effect is generally insignificant compared to the magnetic moment due to spin. The Landé g factor is a measure of the relative contributions of the orbital moment and the spin moment to the total magnetic moment; $g = 1$ when the moment is due only to orbital motion, and $g = 2$ when the moment is due only to spin. For most microwave ferrite materials, g is in the range of 1.98 to 2.01, so $g = 2$ is a good approximation.

In most solids, electron spins occur in pairs with opposite signs so the overall magnetic moment is negligible. In a magnetic material, however, a large fraction of the electron spins are unpaired (more left-hand spins than right-hand spins, or vice versa), but are generally oriented in random directions so that the net magnetic moment is still small. An external magnetic field, however, can cause the dipole moments to align in the same direction to produce a large overall magnetic moment. The existence of exchange forces can keep adjacent electron spins aligned after the external field is removed; the material is then said to be permanently magnetized.

A spinning electron also has a spin angular momentum given in terms of Planck's constant as [1], [2]

$$s = \frac{\hbar}{2}. \tag{10.2}$$

The vector direction of this momentum is opposite the direction of the spin magnetic dipole moment, as indicated in Figure 10.1. The ratio of the spin magnetic moment to the spin angular momentum is a constant called the *gyromagnetic ratio*:

$$\gamma = \frac{m}{s} = \frac{q}{m_e} = 1.759 \times 10^{11} \text{ C/Kg}, \tag{10.3}$$

where (10.1) and (10.2) have been used. Then we can write the following vector relation between the magnetic moment and the angular momentum:

$$\bar{m} = -\gamma \bar{s}, \tag{10.4}$$

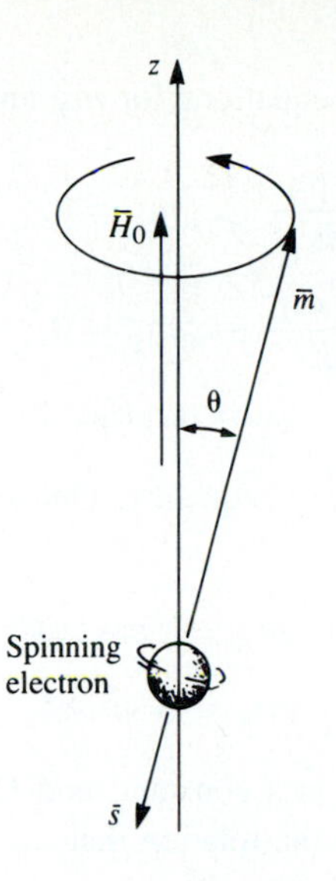

FIGURE 10.1 Spin magnetic dipole moment and angular momentum vectors for a spinning electron.

where the negative sign is due to the fact that these vectors are oppositely directed.

When a magnetic bias field $\bar{H}_0 = \hat{z}H_0$ is present, a torque will be exerted on the magnetic dipole:

$$\bar{T} = \bar{m} \times \bar{B}_0 = \mu_0 \bar{m} \times \bar{H}_0 = -\mu_0 \gamma \bar{s} \times \bar{H}_0. \qquad 10.5$$

Since torque is equal to the time rate of change of angular momentum, we have

$$\frac{d\bar{s}}{dt} = \frac{-1}{\gamma}\frac{d\bar{m}}{dt} = \bar{T} = \mu_0 \bar{m} \times \bar{H}_0,$$

or

$$\frac{d\bar{m}}{dt} = -\mu_0 \gamma \bar{m} \times \bar{H}_0. \qquad 10.6$$

This is the equation of motion for the magnetic dipole moment, $\bar{m}$. We will solve this equation to show that the magnetic dipole precesses around the H_0-field vector, like a spinning top precesses around a vertical axis.

Writing (10.6) in terms of its three vector components gives

$$\frac{dm_x}{dt} = -\mu_0 \gamma m_y H_0, \qquad 10.7a$$

$$\frac{dm_y}{dt} = \mu_0 \gamma m_x H_0, \qquad 10.7b$$

$$\frac{dm_z}{dt} = 0. \qquad 10.7c$$

Now use (10.7a,b) to obtain two equations for m_x and m_y:

$$\frac{d^2 m_x}{dt^2} + \omega_0^2 m_x = 0, \qquad 10.8a$$

$$\frac{d^2 m_y}{dt^2} + \omega_0^2 m_y = 0, \qquad 10.8b$$

where

$$\omega_0 = \mu_0 \gamma H_0, \qquad 10.9$$

is called the *Larmor*, or *precession*, frequency. One solution to (10.8) that is compatible with (10.7a,b) is given by

$$m_x = A \cos \omega_0 t, \qquad 10.10a$$

$$m_y = A \sin \omega_0 t. \qquad 10.10b$$

Equation (10.7c) shows that m_z is a constant, and (10.1) shows that the magnitude of $\bar{m}$ is also a constant, so we have the relation that

$$|\bar{m}|^2 = \left(\frac{g\hbar}{2m_e}\right)^2 = m_x^2 + m_y^2 + m_z^2 = A^2 + m_z^2. \qquad 10.11$$

Thus the precession angle, θ, between $\bar{m}$ and $\bar{H}_0$ (the z-axis) is given by

$$\sin\theta = \frac{\sqrt{m_x^2 + m_y^2}}{|\bar{m}|} = \frac{A}{|\bar{m}|}. \qquad 10.12$$

The projection of $\bar{m}$ on the xy plane is given by (10.10), which shows that $\bar{m}$ traces a circular path in this plane. The position of this projection at time t is given by $\phi = \omega_0 t$, so the angular rate of rotation is $d\phi/dt = \omega_0$, the precession frequency. In the absence of any damping forces, the actual precession angle will be determined by the initial position of the magnetic dipole, and the dipole will precess about $\bar{H}_0$ at this angle indefinitely (free precession). In reality, however, the existence of damping forces will cause the magnetic dipole moment to spiral in from its initial angle until $\bar{m}$ is aligned with $\bar{H}_0$ ($\theta = 0$).

Now assume that there are N unbalanced electron spins (magnetic dipoles) per unit volume, so that the total magnetization is

$$\bar{M} = N\bar{m}, \qquad 10.13$$

and the equation of motion in (10.6) becomes,

$$\frac{d\bar{M}}{dt} = -\mu_0 \gamma \bar{M} \times \bar{H}, \qquad 10.14$$

where $\bar{H}$ is the internal applied field. (Note: In Chapter 2 we used $\bar{P}_m$ for magnetization and $\bar{M}$ for magnetic currents; here we use $\bar{M}$ for magnetization, as this is common practice in ferrimagnetics work. Since we will not be using magnetic currents in this chapter, there should be no confusion.) As the strength of the bias field H_0 is increased, more magnetic dipole moments will align with H_0 until all are aligned, and $\bar{M}$ reaches

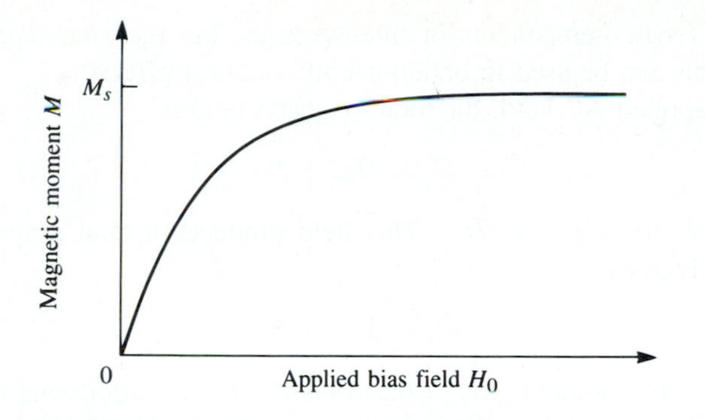

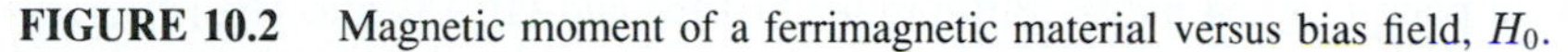

FIGURE 10.2 Magnetic moment of a ferrimagnetic material versus bias field, H_0.

an upper limit. See Figure 10.2. The material is then said to be magnetically saturated, and M_s is denoted as the *saturation magnetization.* M_s is thus a physical property of the ferrite material, and typically ranges from $4\pi M_s = 300$ to 5000 Gauss. (Appendix H lists the saturation magnetization and other physical properties of several types of microwave ferrite materials.) Below saturation, ferrite materials can be very lossy at microwave frequencies, and the RF interaction is reduced. Thus ferrites are usually operated in the saturated state, and this assumption is made for the remainder of this chapter.

The saturation magnetization of a material is a strong function of temperature, decreasing as temperature increases, as illustrated in Figure 10.3. This effect can be understood by noting that the vibrational energy of an atom increases with temperature, making it more difficult to align all the magnetic dipoles. At a high enough temperature the thermal energy is greater than the energy supplied by the internal magnetic field, and a zero net magnetization results. This temperature is called the Curie temperature, T_C.

We now consider the interaction of a small AC (microwave) magnetic field with a magnetically saturated ferrite material. Such a field will cause a forced precession of the dipole moments around the $\bar{H}_0$ $(\hat{z})$ axis at the frequency of the applied AC field, much like the operation of an AC synchronous motor. The small-signal approximation will

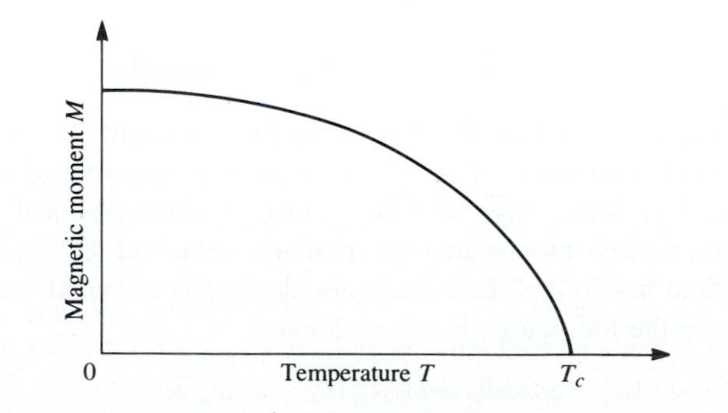

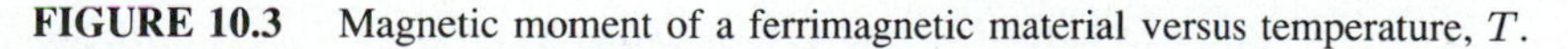

FIGURE 10.3 Magnetic moment of a ferrimagnetic material versus temperature, T.

apply to all the ferrite components of interest to us, but there are applications where high-power signals can be used to obtain useful nonlinear effects.

If $\bar{H}$ is the applied AC field, the total magnetic field is

$$\bar{H}_t = H_0\hat{z} + \bar{H}, \tag{10.15}$$

where we assume that $|\bar{H}| << H_0$. This field produces a total magnetization in the ferrite material given by

$$\bar{M}_t = M_s\hat{z} + \bar{M}, \tag{10.16}$$

where M_s is the (DC) saturation magnetization and $\bar{M}$ is the additional (AC) magnetization (in the xy plane) caused by $\bar{H}$. Substituting (10.16) and (10.15) into (10.14) gives the following component equations of motion:

$$\frac{dM_x}{dt} = -\mu_0\gamma M_y(H_0 + H_z) + \mu_0\gamma(M_s + M_z)H_y, \tag{10.17a}$$

$$\frac{dM_y}{dt} = \mu_0\gamma M_x(H_0 + H_z) - \mu_0\gamma(M_s + M_z)H_x, \tag{10.17b}$$

$$\frac{dM_z}{dt} = -\mu_0\gamma M_x H_y + \mu_0\gamma M_y H_x, \tag{10.17c}$$

since $dM_s/dt = 0$. Since $|\bar{H}| << H_0$, we have $|\bar{M}||\bar{H}| << |\bar{M}|H_0$ and $|\bar{M}||\bar{H}| << M_s|\bar{H}|$, so we can ignore MH products. Then (10.17) reduces to

$$\frac{dM_x}{dt} = -\omega_0 M_y + \omega_m H_y, \tag{10.18a}$$

$$\frac{dM_y}{dt} = \omega_0 M_x - \omega_m H_x, \tag{10.18b}$$

$$\frac{dM_z}{dt} = 0, \tag{10.18c}$$

where $\omega_0 = \mu_0\gamma H_0$ and $\omega_m = \mu_0\gamma M_s$. Solving (10.18a,b) for M_x and M_y gives the following equations:

$$\frac{d^2M_x}{dt^2} + \omega_0^2 M_x = \omega_m\frac{dH_y}{dt} + \omega_0\omega_m H_x, \tag{10.19a}$$

$$\frac{d^2M_y}{dt^2} + \omega_0^2 M_y = -\omega_m\frac{dH_x}{dt} + \omega_0\omega_m H_y. \tag{10.19b}$$

These are the equations of motion for the forced precession of the magnetic dipoles, assuming small-signal conditions. It is now an easy step to arrive at the permeability tensor for ferrites; after doing this, we will try to gain some physical insight into the magnetic interaction process by considering circularly polarized AC fields.

If the AC $\bar{H}$ field has an $e^{j\omega t}$ time-harmonic dependence, the AC steady-state form of (10.19) reduces to the following phasor equations:

$$(\omega_0^2 - \omega^2)M_x = \omega_0\omega_m H_x + j\omega\omega_m H_y, \tag{10.20a}$$

$$(\omega_0^2 - \omega^2)M_y = -j\omega\omega_m H_x + \omega_0\omega_m H_y, \tag{10.20b}$$

which shows the linear relationship between $\bar{H}$ and $\bar{M}$. As in (2.24), (10.20) can be written with a tensor susceptibility, $[\chi]$, to relate $\bar{H}$ and $\bar{M}$:

$$\bar{M} = [\chi]\bar{H} = \begin{bmatrix} \chi_{xx} & \chi_{xy} & 0 \\ \chi_{yx} & \chi_{yy} & 0 \\ 0 & 0 & 0 \end{bmatrix} \bar{H}, \tag{10.21}$$

where the elements of $[\chi]$ are given by

$$\chi_{xx} = \chi_{yy} = \frac{\omega_0 \omega_m}{\omega_0^2 - \omega^2}, \tag{10.22a}$$

$$\chi_{xy} = -\chi_{yx} = \frac{j\omega\omega_m}{\omega_0^2 - \omega^2}. \tag{10.22b}$$

The $\hat{z}$ component of $\bar{H}$ does not affect the magnetic moment of the material, under the above assumptions.

To relate $\bar{B}$ and $\bar{H}$, we have from (2.23) that

$$\bar{B} = \mu_0(\bar{M} + \bar{H}) = [\mu]\bar{H}, \tag{10.23}$$

where the tensor permeability $[\mu]$ is given by

$$[\mu] = \mu_0([U] + [\chi]) = \begin{bmatrix} \mu & j\kappa & 0 \\ -j\kappa & \mu & 0 \\ 0 & 0 & \mu_0 \end{bmatrix} \quad (\hat{z} \text{ bias}). \tag{10.24}$$

The elements of the permeability tensor are then

$$\mu = \mu_0(1 + \chi_{xx}) = \mu_0(1 + \chi_{yy}) = \mu_0\left(1 + \frac{\omega_0\omega_m}{\omega_0^2 - \omega^2}\right), \tag{10.25a}$$

$$\kappa = -j\mu_0\chi_{xy} = j\mu_0\chi_{yx} = \mu_0\frac{\omega\omega_m}{\omega_0^2 - \omega^2}. \tag{10.25b}$$

A material having a permeability tensor of this form is called gyrotropic; note that an $\hat{x}$ (or $\hat{y}$) component of $\bar{H}$ gives rise to both $\hat{x}$ and $\hat{y}$ components of $\bar{B}$, with a 90° phase shift between them.

If the direction of bias is reversed, both H_0 and M_s will change signs, so ω_0 and ω_m will change signs. Equation (10.25) then shows that μ will be unchanged, but κ will change sign. If the bias field is suddenly removed ($H_0 = 0$), the ferrite will generally remain magnetized ($0 < |M| < M_s$); only by demagnetizing the ferrite (with a decreasing AC bias field, for example) can $M = 0$ be obtained. Since the results of (10.22) and (10.25) assume a saturated ferrite sample, both M_s and H_0 should be set to zero for the unbiased, demagnetized case. Then $\omega_0 = \omega_m = 0$, and (10.25) show that $\mu = \mu_0$ and $\kappa = 0$, as expected for a nonmagnetic material.

The tensor results of (10.24) assume bias in the $\hat{z}$ direction. If the ferrite is biased in a different direction the permeability tensor will be transformed according to the change in coordinates. Thus, if $\bar{H}_0 = \hat{x}H_0$, the permeability tensor will be

$$[\mu] = \begin{bmatrix} \mu_0 & 0 & 0 \\ 0 & \mu & j\kappa \\ 0 & -j\kappa & \mu \end{bmatrix} \quad (\hat{x} \text{ bias}), \tag{10.26}$$

while if $\bar{H}_0 = \hat{y}H_0$ the permeability tensor will be

$$[\mu] = \begin{bmatrix} \mu & 0 & -j\kappa \\ 0 & \mu_0 & 0 \\ j\kappa & 0 & \mu \end{bmatrix} \quad (\hat{y} \text{ bias}). \tag{10.27}$$

A comment must be made about units. By tradition most practical work in magnetics is done with CGS units, with magnetization measured in Gauss (1 Gauss = 10^{-4} Weber/m^2), and field strength measured in Oersteds ($4\pi \times 10^{-3}$ Oersted = 1 A/m). Thus, $\mu_0 = 1$ Gauss/Oersted in CGS units, implying that B and H have the same numerical values in a nonmagnetic material. Saturation magnetization is usually expressed as $4\pi M_s$ Gauss; the corresponding MKS value is then $\mu_0 M_s$ Weber/m^2 $= 10^{-4}$ ($4\pi M_s$ Gauss). In CGS units, the Larmor frequency can be expressed as $f_0 = \omega_0/2\pi = \mu_0\gamma H_0/2\pi =$ (2.8 MHz/Oersted) (H_0 Oersted), and $f_m = \omega_m/2\pi = \mu_0\gamma M_s/2\pi =$ (2.8 MHz/Oersted)$\cdot$($4\pi M_s$ Gauss). In practice, these units are convenient and easy to use.

Circularly Polarized Fields

To get a better physical understanding of the interaction of an AC signal with a saturated ferrimagnetic material we will consider circularly polarized fields. As discussed in Section 2.5, a right-hand circularly polarized field can be expressed in phasor form as

$$\bar{H}^+ = H^+(\hat{x} - j\hat{y}), \tag{10.28a}$$

and in time-domain form as

$$\bar{\mathcal{H}}^+ = \text{Re}\{\bar{H}^+ e^{j\omega t}\} = H^+(\hat{x}\cos\omega t + \hat{y}\sin\omega t), \tag{10.28b}$$

where we have assumed the amplitude H^+ as real. This latter form shows that $\bar{\mathcal{H}}^+$ is a vector which rotates with time, such that at time t it is oriented at the angle ωt from the x-axis; thus its angular velocity is ω. (Also note that $|\bar{\mathcal{H}}^+| = H^+ \neq |\bar{H}^+|$.) Applying the RHCP field of (10.28a) to (10.20) gives the magnetization components as

$$M_x^+ = \frac{\omega_m}{\omega_0 - \omega} H^+,$$

$$M_y^+ = \frac{-j\omega_m}{\omega_0 - \omega} H^+,$$

so the magnetization vector resulting from $\bar{H}^+$ can be written as

$$\bar{M}^+ = M_x^+\hat{x} + M_y^+\hat{y} = \frac{\omega_m}{\omega_0 - \omega} H^+(\hat{x} - j\hat{y}), \tag{10.29}$$

which shows that the magnetization is also RHCP, and so rotates with angular velocity ω in synchronism with the driving field, $\bar{H}^+$. Since $\bar{M}^+$ and $\bar{H}^+$ are vectors in the same direction, we can write $\bar{B}^+ = \mu_0(\bar{M}^+ + \bar{H}^+) = \mu^+\bar{H}^+$, where μ^+ is the effective permeability for a RHCP wave given by

$$\mu^+ = \mu_0\left(1 + \frac{\omega_m}{\omega_0 - \omega}\right). \tag{10.30}$$

The angle, θ_M, between $\bar{M}^+$ and the z-axis is given by

$$\tan\theta_M = \frac{|\bar{\mathcal{M}}^+|}{M_s} = \frac{\omega_m H^+}{(\omega_0 - \omega)M_s} = \frac{\omega_0 H^+}{(\omega_0 - \omega)H_0}, \qquad 10.31$$

while the angle, θ_H, between $\bar{H}^+$ and the z-axis is given by

$$\tan\theta_H = \frac{|\bar{\mathcal{H}}^+|}{H_0} = \frac{H^+}{H_0}. \qquad 10.32$$

For frequencies such that $\omega < 2\omega_0$, (10.31) and (10.32) show that $\theta_M > \theta_H$, as illustrated in Figure 10.4a. In this case the magnetic dipole is precessing in the same direction as it would freely precess in the absence of $\bar{H}^+$.

Now consider a left-hand circularly polarized field, expressed in phasor form as

$$\bar{H}^- = H^-(\hat{x} + j\hat{y}), \qquad 10.33a$$

and in time-domain form as

$$\bar{\mathcal{H}}^- = \text{Re}\{\bar{H}^- e^{j\omega t}\} = H^-(\hat{x}\cos\omega t - \hat{y}\sin\omega t). \qquad 10.33b$$

Equation (10.33b) shows that $\bar{\mathcal{H}}^-$ is a vector rotating in the $-\omega$ (left-hand) direction. Applying the LHCP field of (10.33a) to (10.20) gives the magnetization components as

$$M_x^- = \frac{\omega_m}{\omega_0 + \omega} H^-,$$

$$M_y^- = \frac{j\omega_m}{\omega_0 + \omega} H^-,$$

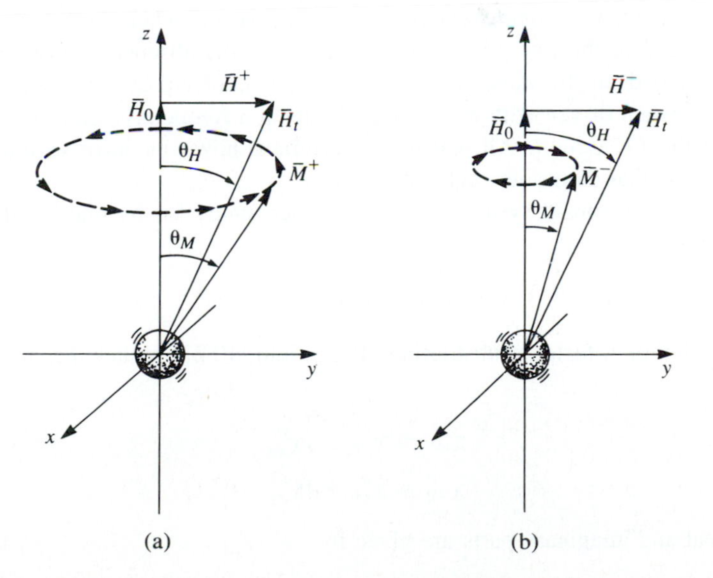

FIGURE 10.4 Forced precession of a magnetic dipole with circularly polarized fields. (a) RHCP, $\theta_M > \theta_H$. (b) LHCP, $\theta_M < \theta_H$.

so the vector magnetization can be written as

$$\bar{M}^- = M_x^- \hat{x} + M_y^- \hat{y} = \frac{\omega_m}{\omega_0 + \omega} H^-(\hat{x} + j\hat{y}), \qquad 10.34$$

which shows that the magnetization is LHCP, rotating in synchronism with $\bar{H}^-$. Writing $\bar{B}^- = \mu_0(\bar{M}^- + \bar{H}^-) = \mu^- \bar{H}^-$, gives the effective permeability for a LHCP wave as

$$\mu^- = \mu_0 \left(1 + \frac{\omega_m}{\omega_0 + \omega}\right). \qquad 10.35$$

The angle, θ_M, between $\bar{M}^-$ and the z-axis is given by

$$\tan\theta_M = \frac{|\bar{M}^-|}{M_s} = \frac{\omega_m H^-}{(\omega_0 + \omega) M_s} = \frac{\omega_0 H^-}{(\omega_0 + \omega) H_0}, \qquad 10.36$$

which is seen to be less than θ_H of (10.32), as shown in Figure 10.4b. In this case the magnetic dipole is precessing in the opposite direction to its free precession.

Thus we see that the interaction of a circularly polarized wave with a biased ferrite depends on the sense of the polarization (RHCP or LHCP). This is because the bias field sets up a preferential precession direction coinciding with the direction of forced precession for a RHCP wave but opposite to that of a LHCP wave. As we will see in Section 10.2, this effect leads to nonreciprocal propagation characteristics.

Effect of Loss

Equations (10.22) and (10.25) show that the elements of the susceptibility or permeability tensors become infinite when the frequency, ω, equals the Larmor frequency, ω_0. This effect is known as *gyromagnetic resonance*, and occurs when the forced precession frequency is equal to the free precession frequency. In the absence of loss the response may be unbounded, in the same way that the response of an LC resonant circuit will be unbounded when driven with an AC signal having a frequency equal to the resonant frequency of the LC circuit. All real ferrite materials, however, have various magnetic loss mechanisms that damp out such singularities.

As with other resonant systems, loss can be accounted for by making the resonant frequency complex:

$$\omega_0 \longleftarrow \omega_0 + j\alpha\omega, \qquad 10.37$$

where α is a damping factor. Substituting (10.37) into (10.22) makes the susceptibilities complex:

$$\chi_{xx} = \chi'_{xx} - j\chi''_{xx} \qquad 10.38a$$

$$\chi_{xy} = \chi''_{xy} + j\chi'_{xy} \qquad 10.38b$$

where the real and imaginary parts are given by

$$\chi'_{xx} = \frac{\omega_0\omega_m(\omega_0^2 - \omega^2) + \omega_0\omega_m\omega^2\alpha^2}{[\omega_0^2 - \omega^2(1+\alpha^2)]^2 + 4\omega_0^2\omega^2\alpha^2}, \qquad 10.39a$$

$$\chi''_{xx} = \frac{\alpha\omega\omega_m[\omega_0^2 + \omega^2(1+\alpha^2)]}{[\omega_0^2 - \omega^2(1+\alpha^2)]^2 + 4\omega_0^2\omega^2\alpha^2}, \quad 10.39b$$

$$\chi'_{xy} = \frac{\omega\omega_m[\omega_0^2 - \omega^2(1+\alpha^2)]}{[\omega_0^2 - \omega^2(1+\alpha^2)]^2 + 4\omega_0^2\omega^2\alpha^2}, \quad 10.39c$$

$$\chi''_{xy} = \frac{2\omega_0\omega_m\omega^2\alpha}{[\omega_0^2 - \omega^2(1+\alpha^2)]^2 + 4\omega_0^2\omega^2\alpha^2}. \quad 10.39d$$

Equation (10.37) can also be applied to (10.25) to give a complex $\mu = \mu' - j\mu''$, and $\kappa = \kappa' - j\kappa''$; this is why (10.38b) appears to define χ'_{xy} and χ''_{xy} backwards, as $\chi_{xy} = j\kappa/\mu_0$. For most ferrite materials the loss is small, so $\alpha << 1$, and the $(1+\alpha^2)$ terms in (10.39) can be approximated as unity. The real and imaginary parts of the susceptibilities of (10.39) are sketched in Figure 10.5 for a typical ferrite.

The damping factor, α, is related to the *linewidth*, ΔH, of the susceptibility curve near resonance. Consider the plot of χ''_{xx} versus bias field, H_0, shown in Figure 10.6. For a fixed frequency, ω, resonance occurs when $H_0 = H_r$, such that $\omega_0 = \mu_0\gamma H_r$. The linewidth, ΔH, is defined as the width of the curve of χ''_{xx} versus H_0 where χ''_{xx} has decreased to half its peak value. If we assume $(1+\alpha^2) \simeq 1$, (10.39b) shows that the maximum value of χ''_{xx} is $\omega_m/2\alpha\omega$, and occurs when $\omega = \omega_0$. Now let ω_{02} be the Larmor frequency for which $H_0 = H_2$, where χ''_{xx} has decreased to half its maximum

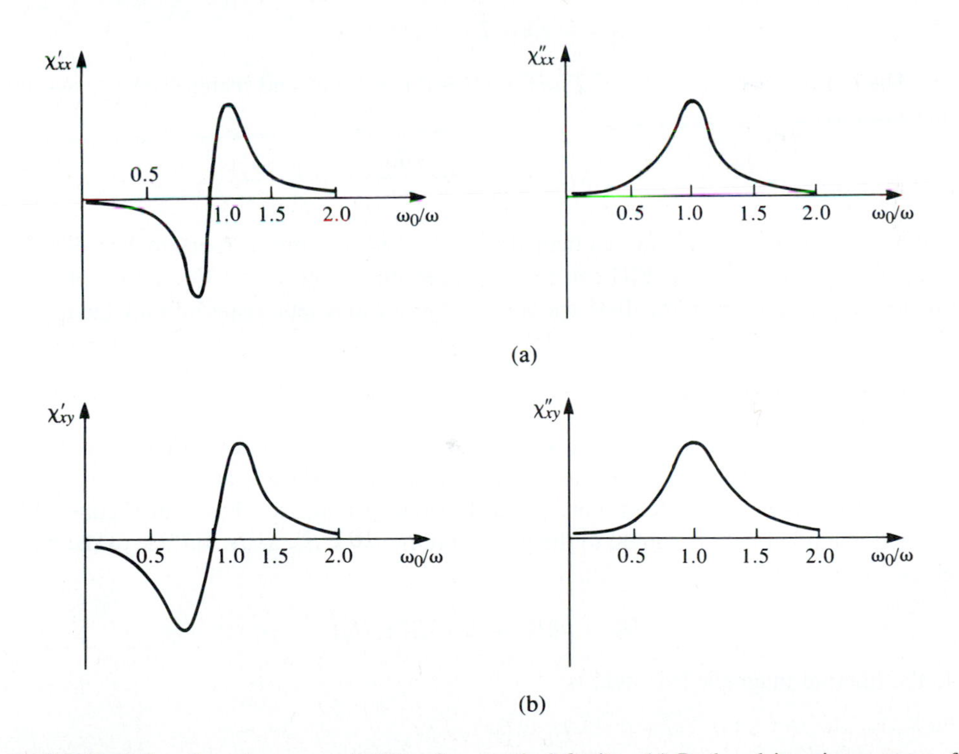

FIGURE 10.5 Complex susceptibilities for a typical ferrite. (a) Real and imaginary parts of χ_{xx}. (b) Real and imaginary parts of χ_{xy}.

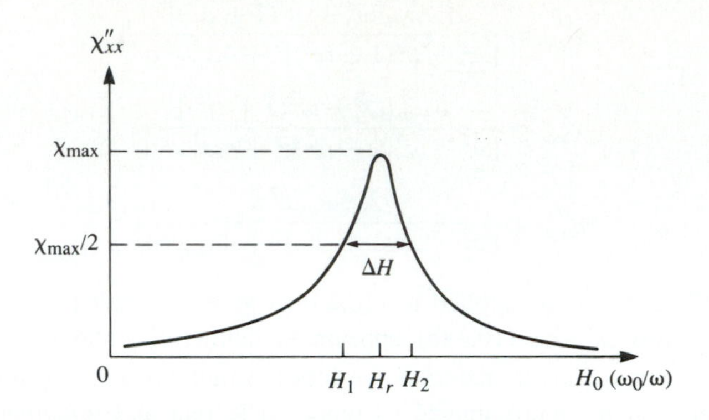

FIGURE 10.6 Definition of the linewidth, ΔH, of the gyromagnetic resonance.

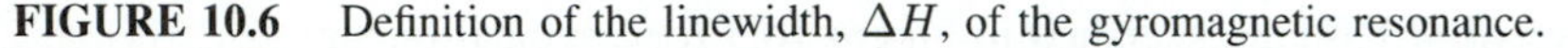

value. Then we can solve (10.39b) for α in terms of ω_{02}:

$$\frac{\alpha\omega\omega_m(\omega_{02}^2+\omega^2)}{(\omega_{02}^2-\omega^2)^2+4\omega_{02}^2\omega^2\alpha^2}=\frac{\omega_m}{4\alpha\omega},$$

$$4\alpha^2\omega^4=(\omega_{02}^2-\omega^2)^2,$$

$$\omega_{02}=\omega\sqrt{1+2\alpha}\simeq\omega(1+\alpha).$$

Then $\Delta\omega_0=2(\omega_{02}-\omega_0)\simeq 2[\omega(1+\alpha)-\omega]=2\alpha\omega$, and using (10.9) gives the linewidth as

$$\Delta H=\frac{\Delta\omega_0}{\mu_0\gamma}=\frac{2\alpha\omega}{\mu_0\gamma}. \qquad 10.40$$

Typical linewidths range from less than 100 Oe (for yttrium iron garnet), to 100–500 Oe (for ferrites); single-crystal YIG can have a linewidth as low as 0.3 Oe. Also note that this loss is separate from the dielectric loss that a ferrimagnetic material may have.

Demagnetization Factors

The DC bias field, H_0, internal to a ferrite sample is generally different from the externally applied field, H_a, because of the boundary conditions at the surface of the ferrite. To illustrate this effect, consider a thin ferrite plate, as shown in Figure 10.7. When the applied field is normal to the plate, continuity of B_n at the surface of the plate gives

$$B_n=\mu_0H_a=\mu_0(M_s+H_0),$$

so the internal magnetic bias field is

$$H_0=H_a-M_s.$$

This shows that the internal field is less than the applied field by an amount equal to the saturation magnetization. When the applied field is parallel to the ferrite plate, continuity

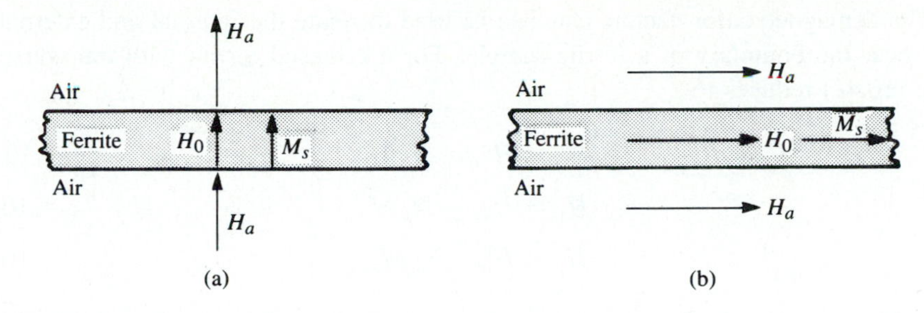

FIGURE 10.7 Internal and external fields for a thin ferrite plate. (a) Normal bias. (b) Tangential bias.

of H_t at the surfaces of the plate gives

$$H_t = H_a = H_0.$$

In this case the internal field is not reduced. In general, the internal field (AC or DC), $\bar{H}$, is affected by the shape of the ferrite sample and its orientation with respect to the external field, $\bar{H}_e$, and can be expressed as

$$\bar{H} = \bar{H}_e - N\bar{M}, \tag{10.41}$$

where $N = N_x$, N_y, or N_z is called the *demagnetization factor* for that direction of the external field. Different shapes have different demagnetization factors, which depend on the direction of the applied field. Table 10.1 lists the demagnetization factors for a few simple shapes. The demagnetization factors are defined such that $N_x + N_y + N_z = 1$.

TABLE 10.1 Demagnetization Factors for Some Simple Shapes

Shape	N_x	N_y	N_z
Thin disk or plate	0	0	1
Thin rod	$\frac{1}{2}$	$\frac{1}{2}$	0
Sphere	$\frac{1}{3}$	$\frac{1}{3}$	$\frac{1}{3}$

The demagnetization factors can also be used to relate the internal and external RF fields near the boundary of a ferrite sample. For a z-biased ferrite with transverse RF fields, (10.41) reduces to

$$H_x = H_{xe} - N_x M_x, \tag{10.42a}$$

$$H_y = H_{ye} - N_y M_y, \tag{10.42b}$$

$$H_z = H_a - N_z M_s, \tag{10.42c}$$

where H_{xe}, H_{ye} are the RF fields external to the ferrite, and H_a is the externally applied bias field. Equation (10.21) relates the internal transverse RF fields and magnetization as

$$M_x = \chi_{xx} H_x + \chi_{xy} H_y,$$

$$M_y = \chi_{yx} H_x + \chi_{yy} H_y.$$

Using (10.42a,b) to eliminate H_x and H_y gives

$$M_x = \chi_{xx} H_{xe} + \chi_{xy} H_{ye} - \chi_{xx} N_x M_x - \chi_{xy} N_y M_y,$$

$$M_y = \chi_{yx} H_{xe} + \chi_{yy} H_{ye} - \chi_{yx} N_x M_x - \chi_{yy} N_y M_y.$$

These equations can be solved for M_x, M_y to give

$$M_x = \frac{\chi_{xx}(1 + \chi_{yy} N_y) - \chi_{xy}\chi_{yx} N_y}{D} H_{xe} + \frac{\chi_{xy}}{D} H_{ye}, \tag{10.43a}$$

$$M_y = \frac{\chi_{yx}}{D} H_{xe} + \frac{\chi_{yy}(1 + \chi_{xx} N_x) - \chi_{yx}\chi_{xy} N_x}{D} H_{ye}, \tag{10.43b}$$

where

$$D = (1 + \chi_{xx} N_x)(1 + \chi_{yy} N_y) - \chi_{yx}\chi_{xy} N_x N_y. \tag{10.44}$$

This result is of the form $\bar{M} = [\chi_e]\bar{H}$, where the coefficients of H_{xe} and H_{ye} in (10.43) can be defined as "external" susceptibilities since they relate magnetization to the external RF fields.

For an infinite ferrite medium gyromagnetic resonance occurs when the denominator of the susceptibilities of (10.22) vanishes, at the frequency $\omega_r = \omega = \omega_0$. But for a finite-sized ferrite sample the gyromagnetic resonance frequency is altered by the demagnetization factors, and given by the condition that $D = 0$ in (10.43). Using the expressions in (10.22) for the susceptibilities in (10.44), and setting the result equal to zero gives

$$\left(1 + \frac{\omega_0 \omega_m N_x}{\omega_0^2 - \omega^2}\right)\left(1 + \frac{\omega_0 \omega_m N_y}{\omega_0^2 - \omega^2}\right) - \frac{\omega^2 \omega_m^2}{(\omega_0^2 - \omega^2)^2} N_x N_y = 0.$$

After some algebraic manipulations this result can be reduced to give the resonance frequency, ω_r, as

$$\omega_r = \omega = \sqrt{(\omega_0 + \omega_m N_x)(\omega_0 + \omega_m N_y)} \tag{10.45}$$

Since $\omega_0 = \mu_0\gamma H_0 = \mu_0\gamma(H_a - N_z M_s)$, and $\omega_m = \mu_0\gamma M_s$, (10.45) can be rewritten in terms of the applied bias field strength and saturation magnetization as

$$\omega_r = \mu_0\gamma\sqrt{[H_a + (N_x - N_z)M_s][H_a + (N_y - N_z)M_s]}. \tag{10.46}$$

This result is known as Kittel's equation [4].

POINT OF INTEREST: Permanent Magnets

Since ferrite components such as isolators, gyrators, and circulators generally use permanent magnets to supply the required DC bias field, it may be useful to mention some of the important characteristics of permanent magnets.

A permanent magnet is made by placing the magnetic material in a strong magnetic field, and then removing the field, to leave the material magnetized in a remanent state. Unless the magnet shape forms a closed path (like a toroid), the demagnetization factors at the magnet ends will cause a slightly negative H field to be induced in the magnet. Thus the "operating point" of a permanent magnet will be in the second quadrant of the B-H hysteresis curve for the magnet material. This portion of the curve is called the demagnetization curve. A typical example is shown below.

The residual magnetization, for $H = 0$, is called the remanence, B_r of the material. This quantity characterizes the strength of the magnet, so generally a magnet material is chosen to have a large remanence. Another important parameter is the coercivity, H_c, which is the value of the negative H field required to reduce the magnetization to zero. A good permanent magnet should have a high coercivity to reduce the effects of vibration, temperature changes, and external fields, which can lead to a loss of magnetization. An overall figure of merit for a permanent magnet is sometimes given as the maximum value of the BH product, $(BH)_{max}$, on the demagnetization curve. This quantity is essentially the maximum magnetic energy density that can be stored by the magnet, and can be useful in electromechanical applications. The following table lists the remanence, coercivity, and $(BH)_{max}$ for some of the most common permanent magnet materials.

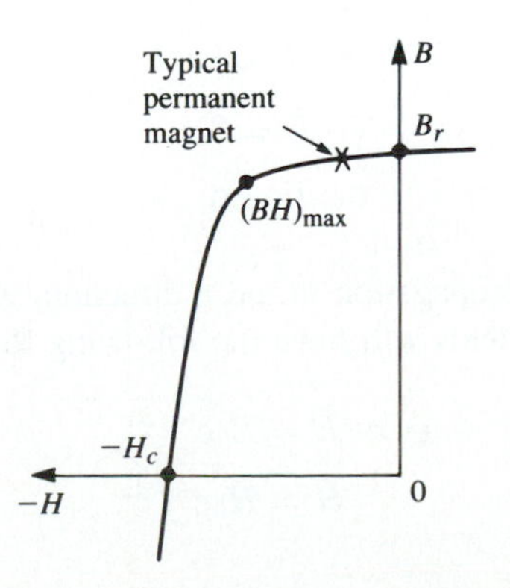

Material	Composition	B_r (Oe)	H_c (G)	$(BH)_{\max}$ (G-Oe)$\times 10^6$
ALNICO 5	Al, Ni, Co, Cu	12,000	720	5.0
ALNICO 8	Al, Ni, Co, Cu, T_i	7,100	2,000	5.5
ALNICO 9	Al, Ni, Co, Cu, T_i	10,400	1,600	8.5
Remalloy	Mo, Co, Fe	10,500	250	1.1
Platinum Cobalt	Pt, Co	6,450	4,300	9.5
Ceramic	$BaO_6Fe_2O_3$	3,950	2,400	3.5
Cobalt Samarium	Co, Sm	8,400	7,000	16.0

10.2 PLANE WAVE PROPAGATION IN A FERRITE MEDIUM

The previous section gives an explanation of the microscopic phenomena that occur inside a biased ferrite material to produce a tensor permeability of the form given in (10.24) (or in (10.26) or (10.27), depending on the bias direction). Once we have this macroscopic description of the ferrite material, we can solve Maxwell's equations for wave propagation in various geometries involving ferrite materials. We begin with plane wave propagation in an infinite ferrite medium, for propagation either in the direction of bias, or propagation transverse to the bias field. These problems will illustrate the important effects of Faraday rotation and birefringence.

Propagation in Direction of Bias (Faraday Rotation)

Consider an infinite ferrite-filled region with a DC bias field given by $\bar{H}_0 = \hat{z}H_0$, and a tensor permittivity $[\mu]$ given by (10.24). Maxwell's equations can be written as

$$\nabla \times \bar{E} = -j\omega[\mu]\bar{H}, \tag{10.47a}$$

$$\nabla \times \bar{H} = j\omega\epsilon\bar{E}, \tag{10.47b}$$

$$\nabla \cdot \bar{D} = 0, \tag{10.47c}$$

$$\nabla \cdot \bar{B} = 0. \tag{10.47d}$$

Now assume plane wave propagation in the z direction, with $\partial/\partial x = \partial/\partial y = 0$. Then the electric and magnetic fields will have the following form:

$$\bar{E} = \bar{E}_0 e^{-j\beta z}, \tag{10.48a}$$

$$\bar{H} = \bar{H}_0 e^{-j\beta z}. \tag{10.48b}$$

The two curl equations of (10.47a,b) then reduce to the following, after using (10.24):

$$j\beta E_y = -j\omega(\mu H_x + j\kappa H_y), \tag{10.49a}$$

$$-j\beta E_x = -j\omega(-j\kappa H_x + \mu H_y), \tag{10.49b}$$

$$0 = -j\omega\mu_0 H_z, \tag{10.49c}$$

$$j\beta H_y = j\omega\epsilon E_x, \tag{10.49d}$$

$$-j\beta H_x = j\omega\epsilon E_y, \tag{10.49e}$$

$$0 = j\omega\epsilon E_z. \tag{10.49f}$$

Equations (10.49c) and (10.49f) show that $E_z = H_z = 0$, as expected for TEM planes. Then we also have $\nabla \cdot \bar{D} = \nabla \cdot \bar{B} = 0$, since $\partial/\partial x = \partial/\partial y = 0$. Equations (10.49d,e) give relations between the transverse field components as

$$Y = \frac{H_y}{E_x} = \frac{-H_x}{E_y} = \frac{\omega\epsilon}{\beta}, \tag{10.50}$$

where Y is the wave admittance. Using (10.50) in (10.49a) and (10.49b) to eliminate H_x and H_y gives the following results:

$$j\omega^2\epsilon\kappa E_x + (\beta^2 - \omega^2\mu\epsilon)E_y = 0, \tag{10.51a}$$

$$(\beta^2 - \omega^2\mu\epsilon)E_x - j\omega^2\epsilon\kappa E_y = 0. \tag{10.51b}$$

For a nontrivial solution for E_x and E_y the determinant of this set of equations must vanish:

$$\omega^4\epsilon^2\kappa^2 - (\beta^2 - \omega^2\mu\epsilon)^2 = 0,$$

or

$$\beta_\pm = \omega\sqrt{\epsilon(\mu \pm \kappa)}. \tag{10.52}$$

So there are two possible propagation constants, β_+ and β_-.

First consider the fields associated with β_+, which can be found by substituting β_+ into (10.51a), or (10.51b):

$$j\omega^2\epsilon\kappa E_x + \omega^2\epsilon\kappa E_y = 0,$$

or

$$E_y = -jE_x.$$

Then the electric field of (10.48a) must have the following form:

$$\bar{E}_+ = E_0(\hat{x} - j\hat{y})e^{-j\beta_+ z}, \tag{10.53a}$$

which is seen to be a right-hand circularly polarized plane wave. Using (10.50) gives the associated magnetic field as

$$\bar{H}_+ = E_0 Y_+(j\hat{x} + \hat{y})e^{-j\beta_+ z}, \tag{10.53b}$$

where Y_+ is the wave admittance for this wave:

$$Y_+ = \frac{\omega\epsilon}{\beta_+} = \sqrt{\frac{\epsilon}{\mu + \kappa}}. \tag{10.53c}$$

Similarly, the fields associated with β_- are left-hand circularly polarized:

$$\bar{E}_- = E_0(\hat{x} + j\hat{y})e^{-j\beta_- z}, \tag{10.54a}$$

$$\bar{H}_- = E_0 Y_-(-j\hat{x} + \hat{y})e^{-j\beta_- z}, \tag{10.54b}$$

where Y_- is the wave admittance for this wave:

$$Y_- = \frac{\omega\epsilon}{\beta_-} = \sqrt{\frac{\epsilon}{\mu - \kappa}}. \tag{10.54c}$$

Thus we see that RHCP and LHCP plane waves are the source-free modes of the $\hat{z}$-biased ferrite medium, and these waves propagate through the ferrite medium with different propagation constants. As discussed in the previous section, the physical explanation for this effect is that the magnetic bias field creates a preferred direction for magnetic dipole precession, and one sense of circular polarization causes precession in this preferred direction while the other sense of polarization causes precession in the opposite direction. Also note that for a RHCP wave, the ferrite material can be represented with an effective permeability of $\mu + \kappa$, while for a LHCP wave the effective permeability is $\mu - \kappa$. In mathematical terms, we can state that $(\mu + \kappa)$ and $(\mu - \kappa)$, or β_+ and β_-, are the *eigenvalues* of the system of equations in (10.51), and that $\bar{E}_+$ and $\bar{E}_-$ are the associated *eigenvectors*. When losses are present, the attenuation constants for RHCP and LHCP waves will also be different.

Now consider a linearly polarized electric field at $z = 0$, represented as the sum of a RHCP and a LHCP wave:

$$\bar{E}|_{z=0} = \hat{x}E_0 = \frac{E_0}{2}(\hat{x} - j\hat{y}) + \frac{E_0}{2}(\hat{x} + j\hat{y}). \tag{10.55}$$

The RHCP component will propagate in the z direction as $e^{-j\beta_+ z}$, and the LHCP component will propagate as $e^{-j\beta_- z}$, so the total field of (10.55) will propagate as

$$\begin{aligned}\bar{E} &= \frac{E_0}{2}(\hat{x} - j\hat{y})e^{-j\beta_+ z} + \frac{E_0}{2}(\hat{x} + j\hat{y})e^{-j\beta_- z}\\ &= \frac{E_0}{2}\hat{x}(e^{-j\beta_+ z} + e^{-j\beta_- z}) - j\frac{E_0}{2}\hat{y}(e^{-j\beta_+ z} - e^{-j\beta_- z})\\ &= E_0\left[\hat{x}\cos\left(\frac{\beta_+ - \beta_-}{2}\right)z - \hat{y}\sin\left(\frac{\beta_+ - \beta_-}{2}\right)z\right]e^{-j(\beta_+ + \beta_-)z/2}.\end{aligned} \tag{10.56}$$

This is still a linearly polarized wave, but one whose polarization rotates as the wave propagates along the z-axis. At a given point along the z-axis the polarization direction measured from the x-axis is given by

$$\phi = \tan^{-1}\frac{E_y}{E_x} = \tan^{-1}\left[-\tan\left(\frac{\beta_+ - \beta_-}{2}\right)z\right] = -\left(\frac{\beta_+ - \beta_-}{2}\right)z. \tag{10.57}$$

This effect is called *Faraday rotation*, after Michael Faraday, who first observed this phenomenon during his study of the propagation of light through liquids which had magnetic properties. Note that for a fixed position on the z-axis, the polarization angle is fixed, unlike the case for a circularly polarized wave where the polarization would rotate with time.

For $\omega < \omega_0$, μ and κ are positive and $\mu > \kappa$. Then $\beta_+ > \beta_-$, and (10.57) shows that ϕ becomes more negative as z increases, meaning that the polarization (direction of $\bar{E}$) rotates counterclockwise as we look in the $+z$ direction. Reversing the bias direction (sign of H_0 and M_s) changes the sign of κ, which changes the direction of rotation to clockwise. Similarly, for $+z$ bias, a wave traveling in the $-z$ direction will rotate its polarization clockwise as we look in the direction of propagation $(-z)$; if we were looking in the $+z$ direction, however, the direction of rotation would be counterclockwise (same as a wave propagating in the $+z$ direction). Thus, a wave that travels from $z = 0$ to $z = L$ and back again to $z = 0$ undergoes a total polarization rotation of 2ϕ, where ϕ is given in (10.57) with $z = L$. So, unlike the situation of a screw being driven into a block of wood and then backed out, the polarization does not "unwind" when the direction of propagation is reversed. Faraday rotation is thus seen to be a nonreciprocal effect.

EXAMPLE 10.1

Consider an infinite ferrite medium with $4\pi M_s = 1800$ Gauss, $\Delta H = 75$ Oersted, $\epsilon_r = 14$, and $\tan\delta = 0.001$. If the bias field strength is $H_0 = 3570$ Oersted, calculate and plot the phase and attenuation constants for RHCP and LHCP plane waves versus frequency, for $f = 0$ to 20 GHz.

Solution

The Larmor precession frequency is

$$f_0 = \frac{\omega_0}{2\pi} = (2.8 \text{ MHz/Oersted})(3570 \text{ Oersted}) = 10.0 \text{ GHz},$$

and $$f_m = \frac{\omega_m}{2\pi} = (2.8 \text{ MHz/Oersted})\ (1800 \text{ Gauss}) = 5.04 \text{ GHz}.$$

At each frequency we can compute the complex propagation constant as

$$\gamma_\pm = \alpha_\pm + j\beta_\pm = j\omega\sqrt{\epsilon(\mu \pm \kappa)},$$

where $\epsilon = \epsilon_0\epsilon_r(1 - j\tan\delta)$ is the complex permittivity, and μ, κ are given by (10.25). The following substitution for ω_0 is used to account for ferrimagnetic loss:

$$\omega_0 \longleftarrow \omega_0 + j\frac{\mu_0\gamma\Delta H}{2},$$

or $$f_0 \longleftarrow f_0 + j\frac{(2.8 \text{ MHz/Oe})(75 \text{ Oe})}{2} = (10. + j0.105) \text{ GHz},$$

which is derived from (10.37) and (10.40). The quantities $(\mu \pm \kappa)$ can be simplified to the following, by using (10.25):

$$\mu + \kappa = \mu_0 \left(1 + \frac{\omega_m}{\omega_0 - \omega}\right),$$

$$\mu - \kappa = \mu_0 \left(1 + \frac{\omega_m}{\omega_0 + \omega}\right).$$

The phase and attenuation constants are plotted in Figure 10.8, normalized to the free-space wavenumber, k_0.

Observe that β_+ and α_+ (for a RHCP wave) show a resonance near $f = f_0 = 10$ GHz; β_- and α_- (for a LHCP wave) do not, however, because the singularities in μ and κ cancel in the $(\mu - \kappa)$ term contained in γ_-. Also note from Figure 10.8 that a stopband (β_+ near zero, large α_+) exists for RHCP waves for frequencies between f_0 and $f_0 + f_m$ (between ω_0 and $\omega_0 + \omega_m$). For frequencies in this range, the above expression for $(\mu + \kappa)$ shows that this quantity is negative, and $\beta_+ = 0$ (in the absence of loss), so a RHCP wave incident on such a ferrite medium would be totally reflected. ○

Propagation Transverse to Bias (Birefringence)

Now consider the case where an infinite ferrite region is biased in the $\hat{x}$ direction, transverse to the direction of propagation; the permeability tensor is given in (10.26).

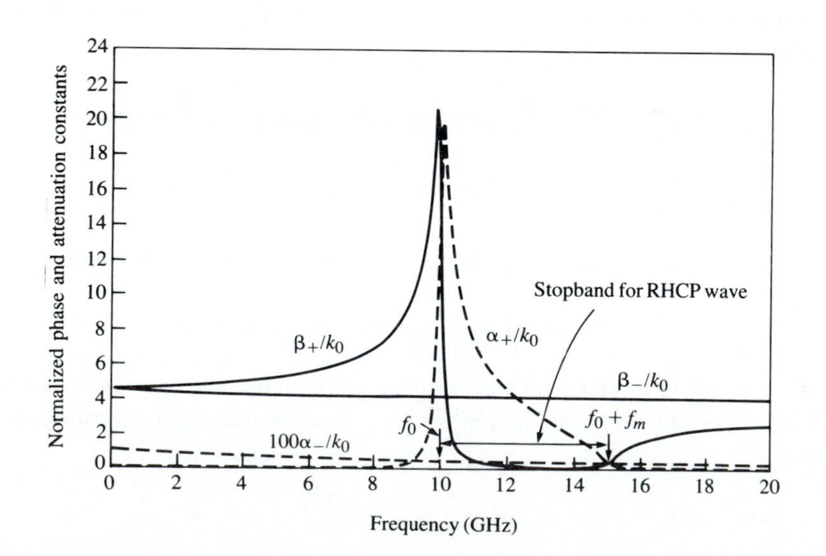

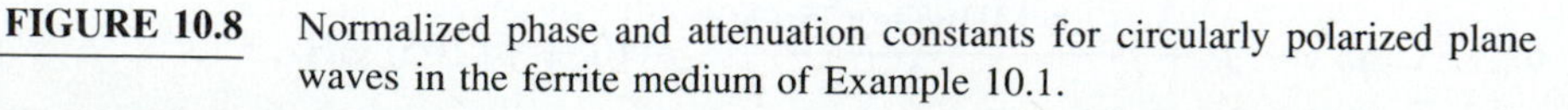

FIGURE 10.8 Normalized phase and attenuation constants for circularly polarized plane waves in the ferrite medium of Example 10.1.

For plane wave fields of the form in (10.48), Maxwell's curl equations reduce to

$$j\beta E_y = -j\omega\mu_0 H_x, \tag{10.58a}$$

$$-j\beta E_x = -j\omega(\mu H_y + j\kappa H_z), \tag{10.58b}$$

$$0 = -j\omega(-j\kappa H_y + \mu H_z), \tag{10.58c}$$

$$j\beta H_y = j\omega\epsilon E_x, \tag{10.58d}$$

$$-j\beta H_x = j\omega\epsilon E_y, \tag{10.58e}$$

$$0 = j\omega\epsilon E_z. \tag{10.58f}$$

Then $E_z = 0$, and $\nabla \cdot \bar{D} = 0$ since $\partial/\partial x = \partial/\partial y = 0$. Equations (10.58d,e) give an admittance relation between the transverse field components:

$$Y = \frac{H_y}{E_x} = \frac{-H_x}{E_y} = \frac{\omega\epsilon}{\beta}, \tag{10.59}$$

Using (10.59) in (10.58a,b) to eliminate H_x and H_y, and using (10.58c) in (10.58b) to eliminate H_z gives the following results:

$$\beta^2 E_y = \omega^2 \mu_0 \epsilon E_y, \tag{10.60a}$$

$$\mu(\beta^2 - \omega^2\mu\epsilon)E_x = -\omega^2\epsilon\kappa^2 E_x. \tag{10.60b}$$

One solution to (10.60) occurs for

$$\beta_o = \omega\sqrt{\mu_0\epsilon}, \tag{10.61}$$

with $E_x = 0$. Then the complete fields are

$$\bar{E}_o = \hat{y}E_0 e^{-j\beta_o z}, \tag{10.62a}$$

$$\bar{H}_o = -\hat{x}E_0 Y_o e^{-j\beta_o z}, \tag{10.62b}$$

since (10.59) shows that $H_y = 0$ when $E_x = 0$, and (10.58c) shows that $H_z = 0$ when $H_y = 0$. The admittance is

$$Y_o = \frac{\omega\epsilon}{\beta_o} = \sqrt{\frac{\epsilon}{\mu_0}}. \tag{10.63}$$

This wave is called the *ordinary wave*, because it is unaffected by the magnetization of the ferrite. This happens whenever the magnetic field components transverse to the bias direction are zero ($H_y = H_z = 0$). The wave propagates in either the $+z$ or $-z$ direction with the same propagation constant, which is independent of H_0.

Another solution to (10.60) occurs for

$$\beta_e = \omega\sqrt{\mu_e\epsilon}, \tag{10.64}$$

with $E_y = 0$, where μ_e is an effective permeability given by

$$\mu_e = \frac{\mu^2 - \kappa^2}{\mu}. \tag{10.65}$$

This wave is called the *extraordinary wave*, and is affected by the ferrite magnetization. Note that the effective permeability may be negative for certain values of ω, ω_0. The electric field is

$$\bar{E}_e = \hat{x} E_0 e^{-j\beta_e z}. \tag{10.66a}$$

Since $E_y = 0$, (10.58e) shows that $H_x = 0$. H_y can be found from (10.58d), and H_z from (10.58c), giving the complete magnetic field as

$$\bar{H}_e = E_0 Y_e \left(\hat{y} + \hat{z} \frac{j\kappa}{\mu} \right) e^{-j\beta_e z}, \tag{10.66b}$$

where

$$Y_e = \frac{\omega\epsilon}{\beta_e} = \sqrt{\frac{\epsilon}{\mu_e}}. \tag{10.67}$$

These fields constitute a linearly polarized wave, but note that the magnetic field has a component in the direction of propagation. Except for the existence of H_z, the extraordinary wave has electric and magnetic fields that are perpendicular to the corresponding fields of the ordinary wave. Thus, a wave polarized in the y direction will have a propagation constant β_o (ordinary wave), but a wave polarized in the x direction will have a propagation constant β_e (extraordinary wave). This effect, where the propagation constant depends on the polarization direction, is called *birefringence* [2]. Birefringence often occurs in optics work, where the index of refraction can have different values depending on the polarization. The double image seen through a calcite crystal is an example of this effect.

From (10.65) we can see that μ_e, the effective permeability for the extraordinary wave, can be negative if $\kappa^2 > \mu^2$. This condition depends on the values of ω, ω_0, and ω_m, or f, H_0, and M_s, but for a fixed frequency and saturation magnetization there will always be some range of bias field for which $\mu_e < 0$ (ignoring loss). When this occurs β_e will become imaginary, as seen from (10.64), which implies that the wave will be cutoff, or evanescent. An $\hat{x}$ polarized plane wave incident at the interface of such a ferrite region would be totally reflected. The effective permeability is plotted versus bias field strength in Figure 10.9, for several values of frequency and saturation magnetization.

10.3 PROPAGATION IN A FERRITE-LOADED RECTANGULAR WAVEGUIDE

In the previous section we introduced the effects of a ferrite material on electromagnetic waves by considering the propagation of plane waves in an infinite ferrite medium. In practice, however, most ferrite components use waveguide or other types of transmission lines loaded with ferrite material. Most of these geometries are very difficult to analyze. Nevertheless, it is worth the effort to treat some of the easier cases, involving ferrite-loaded rectangular waveguides, in order to quantitatively demonstrate the operation and design of several types of practical ferrite components.

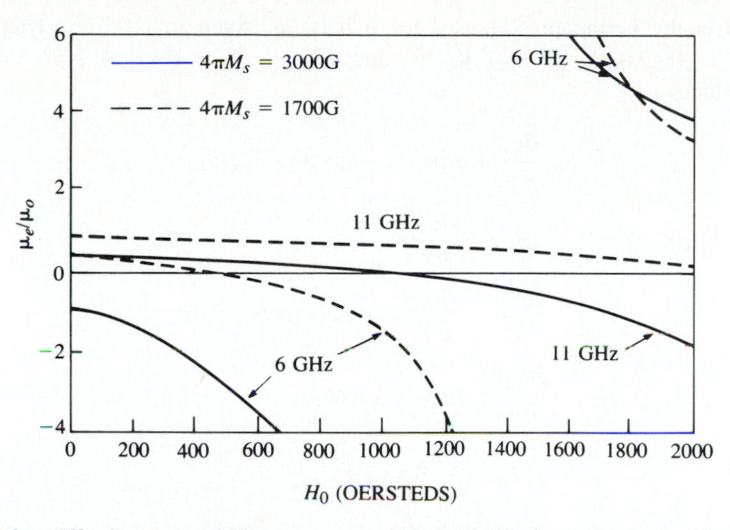

FIGURE 10.9 Effective permeability, μ_e, versus bias field, H_0, for various saturation magnetizations and frequencies.

TE$_{mo}$ Modes of Waveguide with a Single Ferrite Slab

We first consider the geometry shown in Figure 10.10, where a rectangular waveguide is loaded with a vertical slab of ferrite material, biased in the $\hat{y}$ direction. This geometry and its analysis will be used in later sections to treat the operation and design of resonance isolators, field-displacement isolators, and remanent (nonreciprocal) phase shifters.

In the ferrite slab, Maxwell's equations can be written as

$$\nabla \times \bar{E} = -j\omega[\mu]\bar{H}, \tag{10.68a}$$

$$\nabla \times \bar{H} = j\omega\epsilon\bar{E}, \tag{10.68b}$$

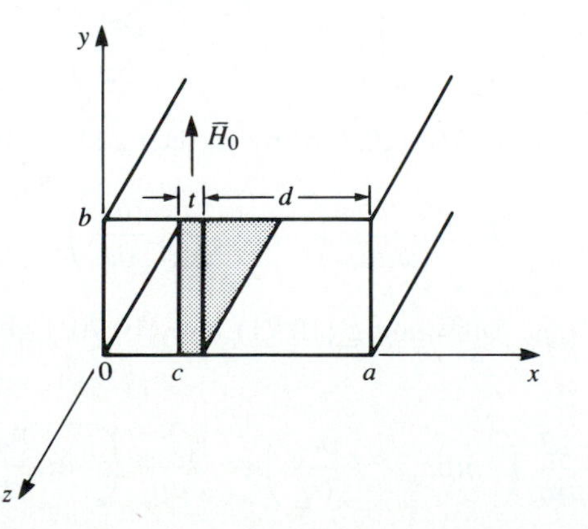

FIGURE 10.10 Geometry of a rectangular waveguide loaded with a transversely biased ferrite slab.

where $[\mu]$ is the permeability tensor for $\hat{y}$ bias, as given in (10.27). Then if we let $\bar{E}(x,y,z) = [\bar{e}(x,y) + \hat{z}e_z(x,y)]e^{-j\beta z}$ and $\bar{H}(x,y,z) = [\bar{h}(x,y) + \hat{z}h_z(x,y)]e^{-j\beta z}$, (10.68) reduces to

$$\frac{\partial e_z}{\partial y} + j\beta e_y = -j\omega(\mu h_x + j\kappa h_z), \quad 10.69a$$

$$-j\beta e_x - \frac{\partial e_z}{\partial x} = -j\omega\mu_0 h_y, \quad 10.69b$$

$$\frac{\partial e_y}{\partial x} - \frac{\partial e_x}{\partial y} = -j\omega(-j\kappa h_x + \mu h_z), \quad 10.69c$$

$$\frac{\partial h_z}{\partial y} + j\beta h_y = j\omega\epsilon e_x, \quad 10.69d$$

$$-j\beta h_x - \frac{\partial h_z}{\partial x} = j\omega\epsilon e_y, \quad 10.69e$$

$$\frac{\partial h_y}{\partial x} - \frac{\partial h_x}{\partial y} = j\omega\epsilon e_z. \quad 10.69f$$

For TE_{mo} modes, we know that $E_z = 0$ and $\partial/\partial y = 0$. Then (10.69b) and (10.69d) imply that $e_x = h_y = 0$ (since $\beta \neq \omega^2\mu_0\epsilon$ for a waveguide mode) and so (10.69) reduces to three equations:

$$j\beta e_y = -j\omega(\mu h_x + j\kappa h_z), \quad 10.70a$$

$$\frac{\partial e_y}{\partial x} = -j\omega(-j\kappa h_x + \mu h_z), \quad 10.70b$$

$$j\omega\epsilon e_y = -j\beta h_x - \frac{\partial h_z}{\partial x}. \quad 10.70c$$

We can solve (10.70a,b) for h_x and h_z as follows. Multiply (10.70a) by μ and (10.70b) by $-j\kappa$, then add to obtain

$$h_x = \frac{1}{\omega\mu\mu_e}\left(-\mu\beta e_y + \kappa\frac{\partial e_y}{\partial x}\right). \quad 10.71a$$

Now multiply (10.70a) by $j\kappa$ and (10.71b) by μ, then add to obtain

$$h_z = \frac{j}{\omega\mu\mu_e}\left(-\kappa\beta e_y + \mu\frac{\partial e_y}{\partial x}\right), \quad 10.71b$$

where $\mu_e = (\mu^2 - \kappa^2)/\mu$. Substituting (10.71) into (10.70c) gives a wave equation for e_y:

$$j\omega\epsilon e_y = \frac{-j\beta}{\omega\mu\mu_e}\left(-\mu\beta e_y + \kappa\frac{\partial e_y}{\partial x}\right) - \frac{j}{\omega\mu\mu_e}\left(-\kappa\beta\frac{\partial e_y}{\partial x} + \mu\frac{\partial^2 e_y}{\partial x^2}\right),$$

or

$$\left(\frac{\partial^2}{\partial x^2} + k_f^2\right)e_y = 0, \quad 10.72$$

where k_f is defined as a cutoff wavenumber for the ferrite:

$$k_f^2 = \omega^2 \mu_e \epsilon - \beta^2. \qquad 10.73$$

We can obtain the corresponding results for the air regions by letting $\mu = \mu_0, \kappa = 0$, and $\epsilon_r = 1$, to obtain

$$\left(\frac{\partial^2}{\partial x^2} + k_a^2 \right) e_y = 0, \qquad 10.74$$

where k_a is the cutoff wavenumber for the air regions:

$$k_a^2 = k_0^2 - \beta^2 \qquad 10.75$$

The magnetic field in the air region is given by

$$h_x = \frac{-\beta}{\omega \mu_0} e_y = \frac{-1}{Z_w} e_y, \qquad 10.76a$$

$$h_z = \frac{j}{\omega \mu_0} \frac{\partial e_y}{\partial x}. \qquad 10.76b$$

The solutions for e_y in the air-ferrite-air regions of the waveguide are then

$$e_y = \begin{cases} A \sin k_a x & \text{for } 0 < x < c, \\ B \sin k_f(x - c) + C \sin k_f(c + t - x) & \text{for } c < x < c + t, \\ D \sin k_a(a - x) & \text{for } c + t < x < a, \end{cases} \qquad 10.77a$$

which have been constructed to facilitate the enforcement of boundary conditions at $x = 0, c, c + t$, and a [3]. We will also need h_z, which can be found from (10.77a), (10.71b), and (10.76b):

$$h_z = \begin{cases} (jk_a A/\omega\mu_o) \cos k_a x & \text{for } 0 < x < c, \\ (j/\omega\mu\mu_e)\{-\kappa\beta[B \sin k_f(x - c) + C \sin k_f(c + t - x)] \\ \qquad +\mu k_f[B \cos k_f(x - c) - C \cos k_f(c + t - x)]\} & \text{for } c < x < c + t, \\ (-jk_a D/\omega\mu_o) \cos k_a(a - x) & \text{for } c + t < x < a. \end{cases} \qquad 10.77b$$

Matching e_y and h_z at $x = c$ and $x = c+t = a-d$ gives four equations for the constants A, B, C, D:

$$A \sin k_a c = C \sin k_f t, \qquad 10.78a$$

$$B \sin k_f t = D \sin k_a d, \qquad 10.78b$$

$$A \frac{k_a}{\mu_o} \cos k_a c = B \frac{k_f}{\mu_e} - C \frac{1}{\mu\mu_e} (\kappa\beta \sin k_f t + \mu k_f \cos k_f t), \qquad 10.78c$$

$$B \frac{1}{\mu\mu_e} (-\kappa\beta \sin k_f t + \mu k_f \cos k_f t) - C \frac{k_f}{\mu_e} = -D \frac{k_a}{\mu_o} \cos k_a d. \qquad 10.78d$$

Solving (10.78a) and (10.78b) for C and D, substituting into (10.78c) and (10.78d), and then eliminating A or B gives the following transcendental equation for the propagation constant, β:

$$\left(\frac{k_f}{\mu_e}\right)^2 + \left(\frac{\kappa\beta}{\mu\mu_e}\right)^2 - k_a \cot k_a c \left(\frac{k_f}{\mu_o\mu_e} \cot k_f t - \frac{\kappa\beta}{\mu_o\mu\mu_e}\right) - \left(\frac{k_a}{\mu_o}\right)^2$$
$$\times \cot k_a c \cot k_a d - k_a \cot k_a d \left(\frac{k_f}{\mu_o\mu_e} \cot k_f t + \frac{\kappa\beta}{\mu_o\mu\mu_e}\right) = 0. \qquad 10.79$$

After using (10.73) and (10.75) to express the cutoff wavenumbers k_f and k_a in terms of β, (10.79) can be solved numerically. The fact that (10.79) contains terms which are odd in $\kappa\beta$ indicates that the resulting wave propagation will be nonreciprocal, since changing the direction of the bias field (which is equivalent to changing the direction of propagation) changes the sign of κ, which leads to a different solution for β. We will identify these two solutions as β_+ and β_-, for positive bias and propagation in the $+z$ direction (positive κ), or in the $-z$ direction (negative κ), respectively. The effects of magnetic loss can easily be included by allowing ω_o to be complex, as in (10.37).

In later sections we will also need to evaluate the electric field in the guide, as given in (10.77a). If we choose the arbitrary amplitude constant as A, then B, C, and D can be found in terms of A by using (10.78a), (10.78b), and (10.78c). Note from (10.75) that if $\beta > k_o$, then k_a will be imaginary. In this case, the $\sin k_a x$ function of (10.77a) becomes $j \sinh |k_a| x$, indicating an almost exponential variation in the field distribution.

A useful approximate result can be obtained for the differential phase shift, $\beta_+ - \beta_-$, by expanding β in (10.79) in a Taylor series about $t = 0$. This can be accomplished with implicit differentiation after using (10.73) and (10.75) to express k_f and k_a in terms of β [4]. The result is

$$\beta_+ - \beta_- = \frac{-2k_c t\kappa}{a\mu} \sin 2k_c c = -2k_c \frac{\kappa}{\mu}\frac{\Delta S}{S} \sin 2k_c c, \qquad 10.80$$

where $k_c = \pi/a$ is the cutoff frequency of the empty guide, and $\Delta S/S = t/a$ is the *filling factor*, or ratio of slab cross-sectional area to waveguide cross-sectional area. Thus, this formula can be applied to other geometries such as waveguides loaded with small ferrite strips or rods, although the appropriate demagnetization factors may be required for some ferrite shapes. The result in (10.80) is only accurate, however, for very small ferrite cross-sections, typically for $\Delta S/S < 0.01$.

This same technique can be used to obtain an approximate expression for the forward and reverse attenuation constants, in terms of the imaginary parts of the susceptibilities defined in (10.39):

$$\alpha_\pm = \frac{\Delta S}{S\beta_o}(\beta_o^2 \chi''_{xx} \sin^2 k_c x + k_c^2 \chi''_{zz} \cos^2 k_c x \mp \chi''_{xy} k_c \beta_o \sin 2k_c x), \qquad 10.81$$

where $\beta_o = \sqrt{k_o^2 - k_c^2}$ is the propagation constant of the empty guide. This result will be useful in the design of resonance isolators. Both (10.80) and (10.81) can also be derived using a perturbation method with the empty waveguide fields [4], and so are usually referred to as the perturbation theory results.

TE$_{mo}$ Modes of Waveguide with Two Symmetrical Ferrite Slabs

A related geometry is the rectangular waveguide loaded with two symmetrically placed ferrite slabs, as shown in Figure 10.11. With equal but opposite $\hat{y}$-directed bias fields on the ferrite slabs, this configuration provides a useful model for the nonreciprocal remanent phase shifter which will be discussed in Section 10.5. Its analysis is very similar to that of the single-slab geometry.

Since the h_y and h_z fields (including the bias fields) are antisymmetric about the midplane of the waveguide at $x = a/2$, a magnetic wall can be placed at this point. Then we only need to consider the region for $0 < x < a/2$. The electric field in this region can be written as

$$e_y = \begin{cases} A \sin k_a x & 0 < x < c, \\ B \sin k_f(x-c) + C \sin k_f(c+t-x) & c < x < c+t, \\ D \cos k_a(a/2 - x) & c+t < x < a/2, \end{cases} \qquad 10.82a$$

which is similar in form to (10.77a), except that the expression for $c+t < x < a/2$ was constructed to have a maximum at $x = a/2$ (since h_z must be zero at $x = a/2$). The cutoff wavenumbers k_f and k_a are defined in (10.73) and (10.75).

Using (10.71) and (10.76) gives the h_z field as,

$$h_z = \begin{cases} (jk_a A/\omega\mu_o)\cos k_a x & 0 < x < c, \\ (j/\omega\mu\mu_e)\{-\kappa\beta[B \sin k_f(x-c) + C \sin k_f(c+t-x)] \\ \qquad +\mu k_f[B \cos k_f(x-c) - C \cos k_f(c+t-x)]\} & c < x < c+t, \\ (jk_a D/\omega\mu_o)\sin k_a(a/2 - x) & c+t < x < a/2. \end{cases} \qquad 10.82b$$

Matching e_y and h_z at $x = c$ and $x = c + t = a/2 - d$ gives four equations for the constants A, B, C, D:

$$A \sin k_a c = C \sin k_f t, \qquad 10.83a$$

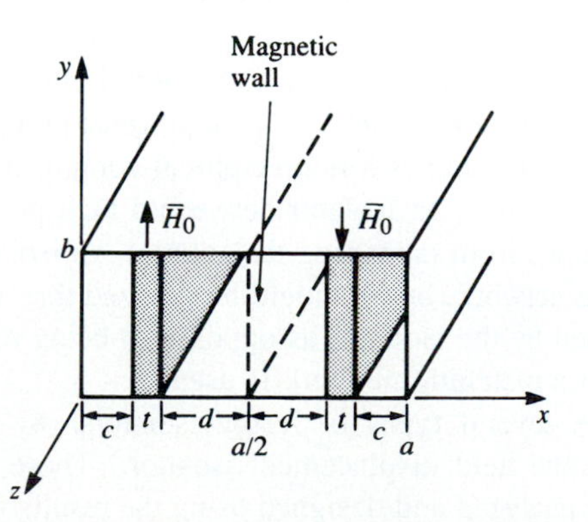

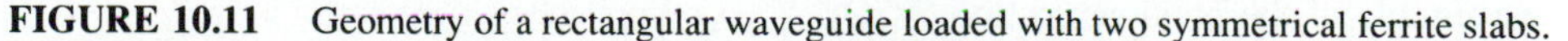

FIGURE 10.11 Geometry of a rectangular waveguide loaded with two symmetrical ferrite slabs.

$$B \sin k_f t = D \cos k_a d, \tag{10.83b}$$

$$A \frac{k_a}{\mu_o} \cos k_a c = B \frac{k_f}{\mu_e} - C \frac{1}{\mu \mu_e} (\kappa \beta \sin k_f t + \mu k_f \cos k_f t), \tag{10.83c}$$

$$\frac{B}{\mu \mu_e} (-\kappa \beta \sin k_f t + \mu k_f \cos k_f t) - C \frac{k_f}{\mu_e} = D \frac{k_a}{\mu_o} \sin k_a d. \tag{10.83d}$$

Reducing these results gives a transcendental equation for the propagation constant, β:

$$\left(\frac{k_f}{\mu_e}\right)^2 + \left(\frac{\kappa\beta}{\mu\mu_e}\right)^2 - k_a \cot k_a c \left(\frac{k_f}{\mu_o \mu_e} \cot k_f t - \frac{\kappa \beta}{\mu_o \mu \mu_e}\right) + \left(\frac{k_a}{\mu_o}\right)^2$$

$$\times \cot k_a c \tan k_a d + k_a \tan k_a d \left(\frac{k_f}{\mu_o \mu_e} \cot k_f t + \frac{\kappa\beta}{\mu_o \mu \mu_e}\right) = 0 \tag{10.84}$$

This equation can be solved numerically for β. Like (10.79) for the single-slab case, κ and β appear in (10.84) only as $\kappa\beta, \kappa^2$, or β^2, which implies nonreciprocal propagation since changing the sign of κ (or bias fields) necessitates a change in sign for β (propagation direction) for the same root. At first glance it may seem that, for the same waveguide and slab dimensions and parameters, two slabs would give twice the phase shift as one slab, but this is generally untrue because the fields are highly concentrated in the ferrite regions.

10.4 FERRITE ISOLATORS

One of the most useful microwave ferrite components is the *isolator*, which is a two-port device having unidirectional transmission characteristics. The S matrix for an ideal isolator has the form

$$[S] = \begin{bmatrix} 0 & 0 \\ 1 & 0 \end{bmatrix}, \tag{10.85}$$

indicating that both ports are matched, but transmission occurs only in the direction from port 1 to port 2. Since $[S]$ is not unitary, the isolator must be lossy. And, of course, $[S]$ is not symmetric, since an isolator is a nonreciprocal component.

A common application uses an isolator between a high-power source and a load to prevent possible reflections from damaging the source. An isolator can be used in place of a matching or tuning network, but it should be realized that any power reflected from the load will be absorbed by the isolator, as opposed to being reflected back to the load, which is the case when a matching network is used.

Although there are several types of ferrite isolators, we will concentrate on the resonance isolator and the field displacement isolator. These devices are of practical importance, and can be analyzed and designed using the results for the ferrite slab-loaded waveguide of the previous section.

Resonance Isolators

We have seen that a circularly polarized plane wave rotating in the same direction as the precessing magnetic dipoles of a ferrite medium will have a strong interaction with the material, while a circularly polarized wave rotating in the opposite direction will have a weaker interaction. Such a result was illustrated in Example 10.1, where the attenuation of a circularly polarized wave was very large near the gyromagnetic resonance of the ferrite, while the attenuation of a wave propagating in the opposite direction was very small. This effect can be used to construct an isolator; such isolators must operate near gyromagnetic resonance, and so are called *resonance isolators*. Resonance isolators usually consist of a ferrite slab or strip mounted at a certain point in a waveguide. We will discuss the two isolator geometries shown in Figure 10.12.

Ideally, the RF fields inside the ferrite material should be circularly polarized. In an empty rectangular waveguide the magnetic fields of the TE_{10} mode can be written as

$$H_x = \frac{j\beta_o}{k_c} A \sin k_c x e^{-j\beta_o z},$$

$$H_z = A \cos k_c x e^{-j\beta_o z},$$

where $k_c = \pi/a$ is the cutoff wavenumber and $\beta_o = \sqrt{k_o^2 - k_c^2}$ is the propagation constant of the empty guide. Since a circularly polarized wave must satisfy the condition that $H_x/H_z = \pm j$, the location, x, of the CP point of the empty guide is given by

$$\tan k_c x = \pm \frac{k_c}{\beta_o}. \tag{10.86}$$

Ferrite loading, however, may perturb the fields so that (10.86) may not give the actual optimum position, or it may prevent the internal fields from being circularly polarized for any position.

First consider the full-height E-plane slab geometry of Figure 10.12a; we can analyze this case using the exact results from the previous section. Alternatively, we could use

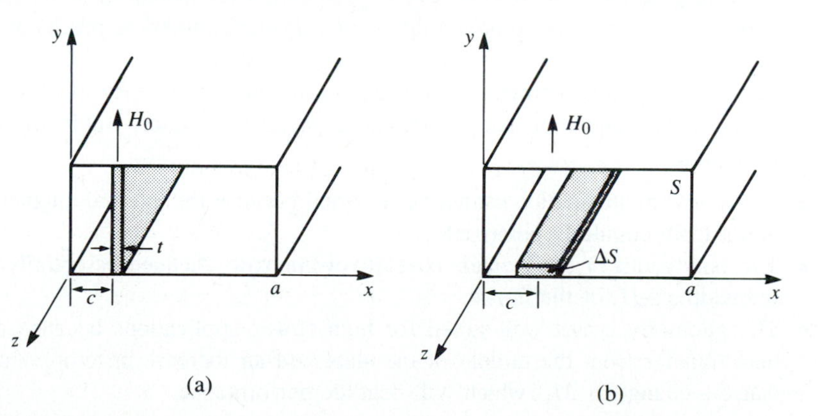

FIGURE 10.12 Two resonance isolator geometries. (a) E-plane, full-height slab. (b) H-plane slab.

the perturbation result of (10.81), but this would require the use of a demagnetization factor for h_x, and would be less accurate than the exact results. Thus, for a given set of parameters, (10.79) can be solved numerically for the complex propagation constants of the forward and reverse waves of the ferrite-loaded guide. It is necessary to include the effect of magnetic loss, which can be done by using (10.37) for the complex resonant frequency, ω_0, in the expressions for μ and κ. The imaginary part of ω_0 can be related to the linewidth, ΔH, of the ferrite through (10.40). Usually the waveguide width, a, frequency, ω, and ferrite parameters $4\pi M_s$, and ϵ_r will be fixed, and the bias field and slab position and thickness will be determined to give the optimum design.

Ideally, the forward attenuation constant (α_+) would be zero, with a nonzero attenuation constant (α_-) in the reverse direction. But for the E-plane ferrite slab there is no position $x = c$ where the fields are perfectly CP in the ferrite (this is because the demagnetization factor $N_x \simeq 1$, [4]). Hence the forward and reverse waves both contain a RHCP component and a LHCP component, so ideal attenuation characteristics cannot be obtained. The optimum design, then, generally minimizes the forward attenuation, which determines the slab position. Alternatively, it may be desired to maximize the ratio of the reverse to forward attenuations. Since the maximum reverse attenuation generally does not occur at the same slab position as the minimum forward attenuation, such a design will involve a tradeoff of the forward loss.

For a long, thin slab, the demagnetization factors are approximately those of a thin disk: $N_x \simeq 1, N_y = N_z = 0$. It can then be shown via the Kittel equation of (10.45) that the gyromagnetic resonance frequency of the slab is given by

$$\omega = \sqrt{\omega_0(\omega_0 + \omega_m)}, \tag{10.87}$$

which determines H_0, given the operating frequency and saturation magnetization. This is an approximate result; the transcendental equation of (10.79) accounts for demagnetization exactly, so the actual internal bias field, H_0, can be found by numerically solving (10.79) for the attenuation constants for values of H_0 near the approximate value given by (10.87).

Once the slab position, c, and bias field, H_o, have been found the slab length, L, can be chosen to give the desired total reverse attenuation (or isolation) as $(\alpha_-)L$. The slab thickness can also be used to adjust this value. Typical numerical results are given in Example 10.2.

One advantage of this geometry is that the full-height slab is easy to bias with an external C-shaped permanent magnet, with no demagnetization factor. But it suffers from several disadvantages:

- Zero forward attenuation cannot be obtained because the internal magnetic field is not truly circularly polarized.
- The bandwidth of the isolator is relatively narrow, dictated essentially by the linewidth, ΔH, of the ferrite.
- The geometry is not well-suited for high-power applications because of poor heat transfer from the middle of the slab, and an increase in temperature will cause a change in M_s, which will degrade performance.

The first two problems noted above can be remedied to a significant degree by adding a dielectric loading slab; see F. E. Gardiol and A. S. Vander Vorst [5] for details.

EXAMPLE 10.2

Design an E-plane resonance isolator in X-band waveguide to operate at 10 GHz with a minimum forward insertion loss and 30 dB reverse attenuation. Use a 0.5 mm thick ferrite slab with $4\pi M_s = 1700\,\text{G}, \Delta H = 200$ Oe, and $\epsilon_r = 13$. Determine the bandwidth for which the reverse attenuation is at least 27 dB.

Solution

The complex roots of (10.79) were found numerically using an interval-halving routine followed by a Newton-Rhapson iteration. The approximate bias field, H_0, given by (10.87) is 2820 Oe, but numerical results indicate the actual field to be closer to 2840 Oe for resonance at 10 GHz. Figure 10.13a shows the calculated forward (α_+) and reverse (α_-) attenuation constants at 10 GHz versus slab position, and it can be seen that the minimum forward attenuation occurs for $c/a = 0.125$; the reverse attenuation at this point is $\alpha_- = 12.4$ dB/cm. Figure 10.13b shows the attenuation constants versus frequency, for this slab position. For a total reverse attenuation of 20 dB, the length of the slab must be

$$L = \frac{30\text{ dB}}{12.4\text{ dB/cm}} = 2.4\text{ cm}.$$

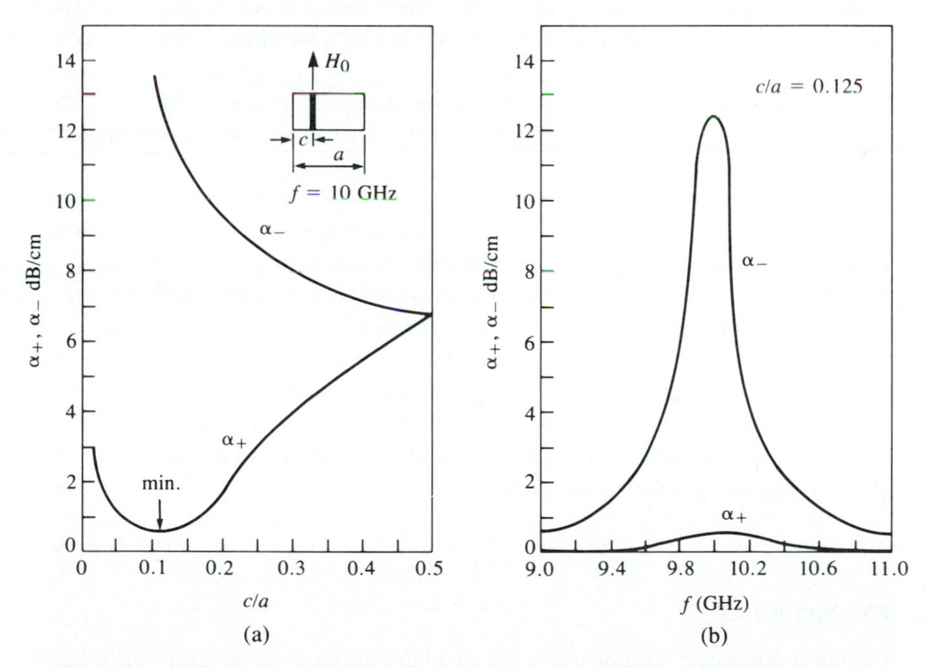

$4\pi M_s = 1700\text{G}$ $H_0 = 2840$ Oe $\Delta H = 200$ Oe $a = 2.286$ cm $t = 0.05$ cm $\epsilon_r = 13$

FIGURE 10.13 Forward and reverse attenuation constants for the resonance isolator of Example 10.2. (a) Versus slab position. (b) Versus frequency.

For the total reverse attenuation to be at least 27 dB, we must have

$$\alpha_- > \frac{27\ \text{dB}}{2.4\ \text{cm}} = 11.3\ \text{dB/cm}.$$

So the bandwidth according to the above definition is, from the data of Figure 10.13b, less than 2%. This figure could be improved by using a ferrite with a larger linewidth, at the expense of a longer or thicker slab and a higher forward attenuation. ○

Next we consider a resonance isolator using the H-plane slab geometry of Figure 10.12b. If the slab is much thinner than it is wide, the demagnetization factors will approximately be $N_x = N_z = 0$, $N_y = 1$. This means that a stronger applied bias field will be required to produce the internal field, H_0, in the y direction. But the RF magnetic field components, h_x and h_z, will not be affected by the air-ferrite boundary since $N_x = N_z = 0$, and perfect circular polarized fields will exist in the ferrite when it is positioned at the CP point of the empty guide, as given by (10.86). Another advantage of this geometry is that it has better thermal properties than the E-plane version, since the ferrite slab has a large surface area in contact with a waveguide wall for heat dissipation.

Unlike the full-height E-plane slab case, the H-plane geometry of Figure 10.12b cannot be analyzed exactly. But if the slab occupies only a very small fraction of the total guide cross-section ($\Delta S/S << 1$, where ΔS and S are the cross-sectional areas of the slab and waveguide, respectively), the perturbational result for α_+ in (10.81) can be used with reasonable results. This expression is given in terms of the susceptibilities $\chi_{xx} = \chi'_{xx} - j\chi''_{xx}$, $\chi_{zz} = \chi'_{zz} - j\chi''_{zz}$, and $\chi_{xy} = \chi''_{xy} + j\chi'_{xy}$, as defined for a $\hat{y}$-biased ferrite in a manner similar to (10.22). For ferrite shapes other than a thin H-plane slab, these susceptibilities would have to be modified with the appropriate demagnetization factors, as in (10.43) [4].

As seen from the susceptibility expressions of (10.22), gyromagnetic resonance for this geometry will occur when $\omega = \omega_0$, which determines the internal bias field, H_0. The center of the slab is positioned at the circular polarization point of the empty guide, as given by (10.86). This should result in a near-zero forward attenuation constant. The total reverse attenuation, or isolation, can be controlled with either the length, L, of the ferrite slab or its cross-section ΔS, since (10.81) shows $\alpha_\pm$ is proportional to $\Delta S/S$. If $\Delta S/S$ is too large, however, the purity of circular polarization over the slab cross-section will be degraded, and forward loss will increase. One practical alternative is to use a second identical ferrite slab on the top wall of the guide, to double $\Delta S/S$ without significantly degrading polarization purity.

EXAMPLE 10.3

Design a resonance isolator using the H-plane ferrite slab geometry in X-band waveguide. The isolator should have minimum forward insertion loss and a reverse attenuation of 30 dB at 10 GHz. Use a ferrite slab having $\Delta S/S = 0.01$, $4\pi M_s = 1700\,\text{G}$, and $\Delta H = 200$ Oe.

Solution

We first determine the necessary internal bias field as

$$H_0 = \frac{10,000 \text{ MHz}}{2.8 \text{ MHz/Oe}} = 3570 \text{ Oe}.$$

For X-band waveguide, $a = 2.286$ cm, so the cutoff wavenumber and propagation constant are

$$k_c = \frac{\pi}{a} = 137.4 \text{ m}^{-1},$$

$$\beta_0 = \sqrt{k_0^2 - k_c^2} = 158.1 \text{ m}^{-1}.$$

Then (10.86) gives the optimum position of the slab as

$$\frac{c}{a} = \frac{1}{\pi} \tan^{-1} \frac{k_c}{\beta_0} = 0.23.$$

Next, we evaluate (10.81) for the forward and reverse attenuation constants, $\alpha_{\pm}$, using the complex susceptibilities from (10.22). As in the previous example, loss is accounted for by making ω_0 complex. Note that (10.81) does not depend on the dielectric constant of the ferrite, since $\Delta S/S$ is assumed to be small. Figure 10.14a shows α_+ and α_- versus slab position, for $f = 10$ GHz and $H_0 = 3570$ Oe. As expected, $\alpha_+ \simeq 0$ for $c/a = 0.23$; at this point the reverse attenuation is $\alpha_- \simeq 2.0$ dB/cm, but the peak value of α_- occurs for $c/a \simeq 0.25$.

The slab length required for 30 dB total reverse attenuation, for $c/a = 0.23$, is

$$L = \frac{30 \text{ dB}}{2.0 \text{ dB/cm}} = 15 \text{ cm}.$$

This length could be halved if a second H-plane slab were used on the top waveguide wall.

The frequency dependence is shown in Figure 10.14b, for $c/a = 0.23$. The bandwidth for which the isolation is reduced by 3 dB is about 2.5%, which is about the same as the E-plane isolator of Example 10.2. Figure 10.14b also shows the attenuation for a ferrite having a linewidth $\Delta H = 400$ Oe (more loss). It is seen that the Q of the resonance is lowered; this will increase the bandwidth of the isolator, although a longer slab will be needed to obtain the same isolation. ○

The Field Displacement Isolator

Another type of isolator uses the fact that the electric field distributions of the forward and reverse waves in a ferrite slab-loaded waveguide can be quite different. As illustrated in Figure 10.15, the electric field for the forward wave can be made to vanish at the side of the ferrite slab at $x = c + t$, while the electric field of the reverse wave can be quite large at this same point. Then if a thin resistive sheet is placed in this position, the

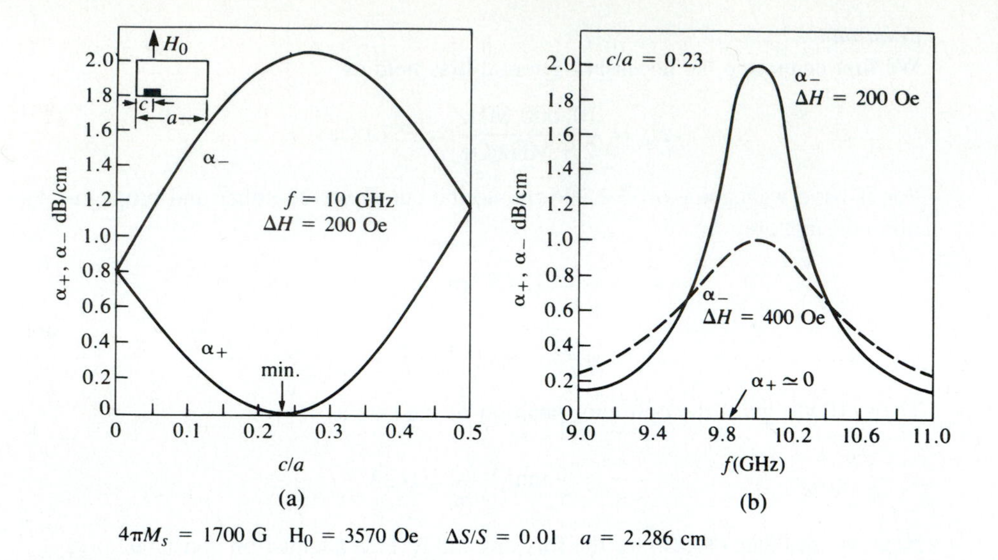

FIGURE 10.14 Forward and reverse attenuation constants for the resonance isolator of Example 10.3. (a) Versus slab position. (b) Versus frequency.

forward wave will be essentially unaffected while the reverse wave will be attenuated. Such an isolator is called a *field displacement isolator*; high values of isolation with a relatively compact device can be obtained with bandwidths on the order of 10%. Another advantage of the field displacement isolator over the resonance isolator is that a much smaller bias field is required, since it operates well below resonance.

The main problem in designing a field displacement isolator is to determine the design parameters that produce field distributions like those shown in Figure 10.15. The

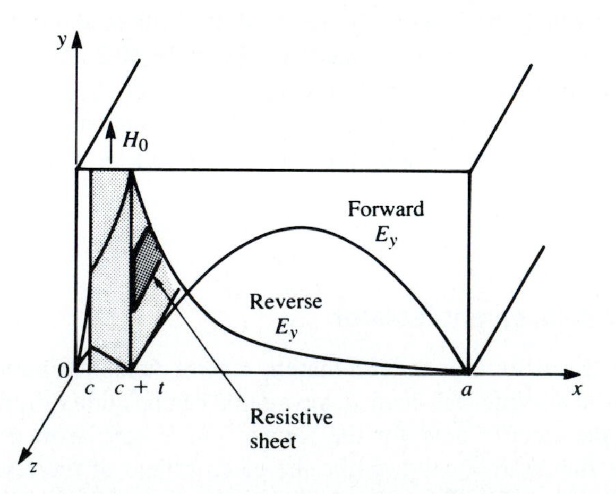

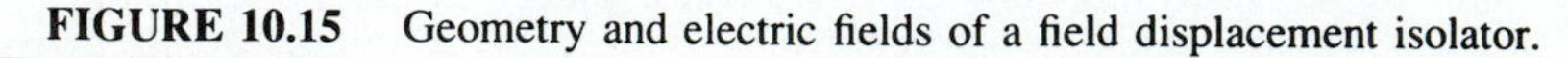
FIGURE 10.15 Geometry and electric fields of a field displacement isolator.

general form of the electric field is given in (10.77a), from the analysis of the ferrite slab-loaded waveguide. This shows that for the electric field of the forward wave to have a sinusoidal dependence for $c + t < x < a$, and to vanish at $x = c + t$, the cutoff wavenumber k_a^+ must be real and satisfy the condition that

$$k_a^+ = \frac{\pi}{d}, \qquad 10.88$$

where $d = a - c - t$. In addition, the electric field of the reverse wave should have a hyperbolic dependence for $c + t < x < a$, which implies that k_a^- must be imaginary. Since from (10.75), $k_a^2 = k_0^2 - \beta^2$, the above conditions imply that $\beta^+ < k_0$ and $\beta^- > k_0$, where $k_0 = \omega\sqrt{\mu_0\epsilon_0}$. These conditions on $\beta_\pm$ depend critically on the slab position, which must be determined by numerically solving (10.79) for the propagation constants. The slab thickness also affects this result, but less critically; a typical value is $t = a/10$.

It also turns out that in order to satisfy (10.88), to force $E_y = 0$ at $x = c + t$, $\mu_e = (\mu^2 - \kappa^2)/\mu$ must be negative. This requirement can be intuitively understood by thinking of the waveguide mode for $c + t < x < a$ as a superposition of two obliquely traveling plane waves. The magnetic field components H_x and H_z of these waves are both perpendicular to the bias field, a situation which is similar to the extraordinary plane waves discussed in Section 10.2, where it was seen that propagation would not occur for $\mu_e < 0$. Applying this cutoff condition to the ferrite-loaded waveguide will allow a null in E_y for the forward wave to be formed at $x = c + t$.

The condition that μ_e be negative depends on the frequency, saturation magnetization, and bias field. Figure 10.9 shows the dependence of μ_e versus bias field for several frequencies and saturation magnetization. This type of data can be used to select the saturation magnetization and bias field to give $\mu_e < 0$ at the design frequency. Observe that higher frequencies will require a ferrite with higher saturation magnetization, and a higher bias field, but $\mu_e < 0$ always occurs before the resonance in μ_e at $\sqrt{\omega_0(\omega_0 + \omega_m)}$. Further design details will be given in the following example.

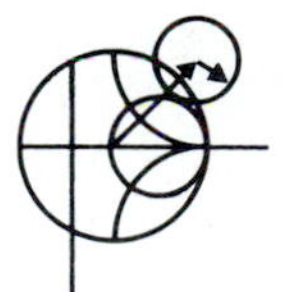

EXAMPLE 10.4

Design a field displacement isolator in X-band waveguide to operate at 11 GHz. The ferrite has $4\pi M_s = 3000$ G, and $\epsilon_r = 13$. Ferrite loss can be ignored.

Solution

We first determine the internal bias field, H_0, such that $\mu_e < 0$. This can be found from Figure 10.9, which shows μ_e/μ_o versus H_0 for $4\pi M_s = 3000$ G at 11 GHz. We see that $H_0 = 1200$ Oe should be sufficient. Also note from this figure that a ferrite with a smaller saturation magnetization would require a much larger bias field.

Next we determine the slab position, c/a, by numerically solving (10.79) for the propagation constants, $\beta_\pm$, as a function of c/a. The slab thickness was set to $t = 0.25$ cm, which is approximately $a/10$. Figure 10.16a shows the resulting propagation constants, as well as the locus of points where β_+ and c/a satisfy the condition of (10.88). The intersection of β_+ with this locus will insure that $E_y = 0$ at $x = c + t$ for the forward wave; this intersection occurs for a slab position of $c/a = 0.028$. The resulting propagation constants are $\beta_+ = 0.724k_0 < k_0$ and $\beta_- = 1.607k_0 > k_0$.

The electric fields are plotted in Figure 10.16b. Note that the forward wave has a null at the face of the ferrite slab, while the reverse wave has a peak (the relative amplitudes of these fields are arbitrary). Then a resistive sheet can

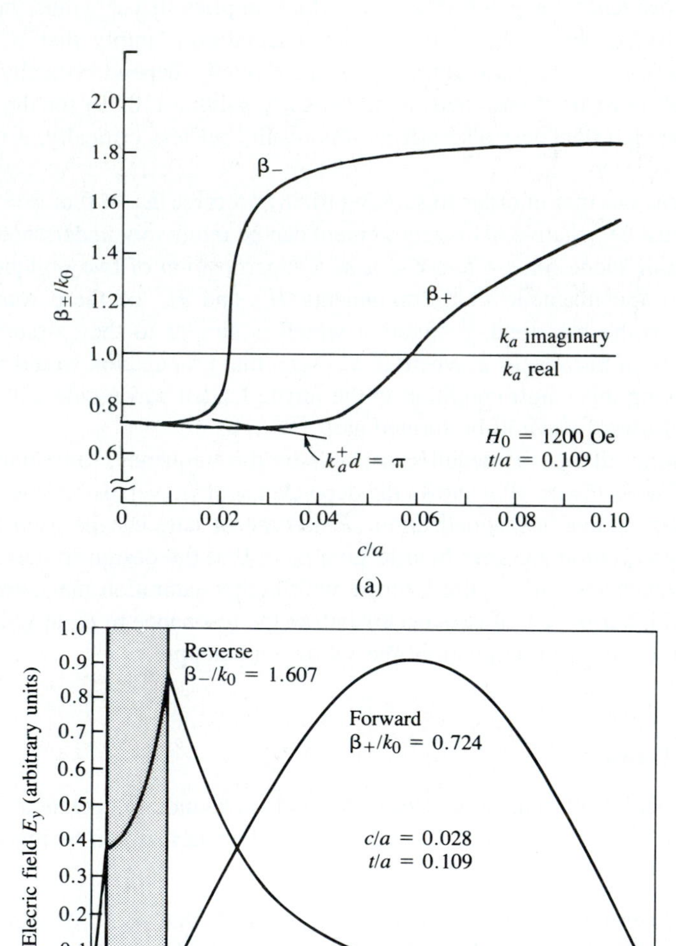

FIGURE 10.16 Propagation constants and electric field distribution for the field displacement isolator of Example 10.4. (a) Forward and reverse propagation constants versus slab position. (b) Electric field amplitudes for the forward and reverse waves.

be placed at this point to attenuate the reverse wave. The actual isolation will depend on the resistivity of this sheet; a value of 75 Ω per square is typical. ○

10.5 FERRITE PHASE SHIFTERS

Another important application of ferrite materials is in *phase shifters*, which are two-port components that provide variable phase shift by changing the bias field of the ferrite. (Microwave diodes and FETs can also be used to implement phase shifters; see Section 11.5.) Phase shifters find application in test and measurements systems, but the most significant use is in phased array antennas where the antenna beam can be steered in space by electronically controlled phase shifters. Because of this demand, many different types of phase shifters have been developed, both reciprocal (same phase shift in either direction) and nonreciprocal [2], [6]. One of the most useful designs is the latching (or *remanent*) nonreciprocal phase shifter using a ferrite toroid in a rectangular waveguide; we can analyze this geometry with a reasonable degree of approximation using the double ferrite slab geometry discussed in Section 10.3. Then we will qualitatively discuss the operation of a few other types of phase shifters.

Nonreciprocal Latching Phase Shifter

The geometry of a latching phase shifter is shown in Figure 10.17; it consists of a toroidal ferrite core symmetrically located in the waveguide with a bias wire passing through its center. When the ferrite is magnetized, the magnetization of the side walls of the toroid will be oppositely directed, and perpendicular to the plane of circular polarization of the RF fields. Since the sense of circular polarization is also opposite on opposite sides of the waveguide, a strong interaction between the RF fields and the ferrite can be obtained. Of course, the presence of the ferrite perturbs the waveguide fields (the fields tend to concentrate in the ferrite), so the circular polarization point does not occur at $\tan k_c x = k_c/\beta_0$, as it does for an empty guide.

In principle, such a geometry can be used to provide a continuously variable (analog) phase shift by varying the bias current. But a more useful technique employs the magnetic

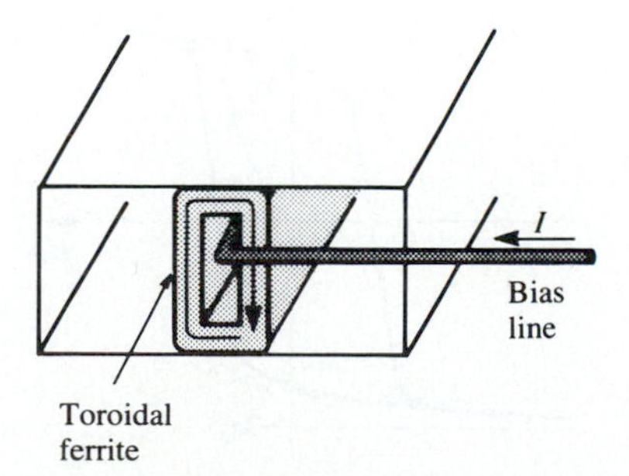

FIGURE 10.17 Geometry of a nonreciprocal latching phase shifter using a ferrite toroid.

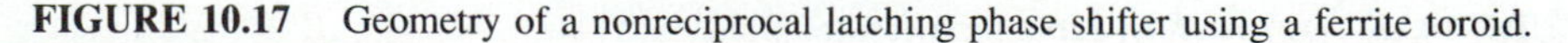

hysteresis of the ferrite to provide a phase shift that can be switched between two values (digital). A typical hysteresis curve is shown in Figure 10.18, showing the variation in magnetization, M, with bias field, H_0. When the ferrite is initially demagnetized and the bias field is off, both M and H_0 are zero. As the bias field is increased, the magnetization increases along the dotted line path until the ferrite is magnetically saturated, and $M = M_s$. If the bias field is now reduced to zero, the magnetization will decrease to a remanent condition (like a permanent magnet), where $M = M_r$. A bias field in the opposite direction will saturate the ferrite with $M = -M_s$, whereupon the removal of the bias field will leave the ferrite in a remanent state with $M = -M_r$. Thus we can "latch" the ferrite magnetization in one of two states, where $M = \pm M_r$, giving a digital phase shift. The amount of differential phase shift between these two states is controlled by the length of the ferrite toroid. In practice, several sections having individual bias lines and decreasing lengths are used in series to give binary differential phase shifts of 180°, 90°, 45°, etc. to as fine a resolution as desired (or can be afforded). An important advantage of the latching mode of operation is that the bias current does not have to be continuously applied, but only pulsed with one polarity or the other to change the polarity of the remanent magnetization; switching speeds can be on the order of a few microseconds. The bias wire can be oriented perpendicular to the electric field in the guide, with a negligible perturbing effect. The top and bottom walls of the ferrite toroid have very little magnetic interaction with the RF fields because the magnetization is not perpendicular to the plane of circular polarization, and the top and bottom magnetizations are oppositely directed. So these walls provide mainly a dielectric loading effect, and the essential operating features of the remanent phase shifters can be obtained by considering the simpler dual ferrite slab geometry of Section 10.3.

For a given operating frequency and waveguide size, the design of a remanent dual slab phase shifter mainly involves the determination of the slab thickness, t, the spacing between the slabs, $s = 2d = a - 2c - 2t$ (see Figure 10.11), and the length of the slabs for the desired phase shift. This requires the propagation constants, $\beta_\pm$, for the dual slab geometry, which can be numerically evaluated from the transcendental equation of (10.84). This equation requires values for μ and κ, which can be determined from (10.25)

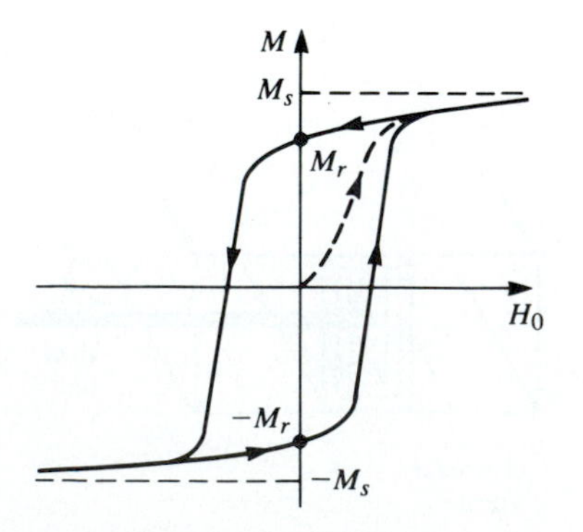

FIGURE 10.18 A hysteresis curve for a ferrite toroid.

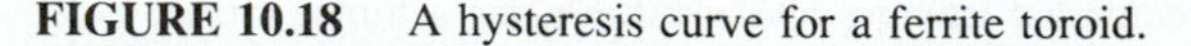

for the remanent state by setting $H_0 = 0$ ($\omega_0 = 0$) and $M_s = M_r$ ($\omega_m = \mu_0\gamma M_r$):

$$\mu = \mu_0, \tag{10.89a}$$

$$\kappa = -\mu_0 \frac{\omega_m}{\omega}. \tag{10.89b}$$

The differential phase shift, $\beta_+ - \beta_-$, is linearly proportional to κ, for κ/μ_0 up to about 0.5. Then, since κ is proportional to M_r, as seen by (10.89b), it follows that a shorter ferrite can be used to provide a given phase shift if a ferrite with a higher remanent magnetization is selected. The insertion loss of the phase shifter decreases with length, but is a function of the ferrite linewidth, ΔH. A figure of merit commonly used to characterize phase shifters is the ratio of phase shift to insertion loss, measured in degrees/dB.

EXAMPLE 10.5

Design a two-slab remanent phase shifter at 10 GHz using X-band waveguide with ferrite having $4\pi M_r = 1786\,\text{G}$ and $\epsilon_r = 13$. Assume that the ferrite slabs are spaced 1 mm apart. Determine the slab thicknesses for maximum differential phase shift, and the lengths of the slabs for 180° and 90° phase shifter sections.

Solution
From (10.89) we have that

$$\frac{\mu}{\mu_0} = 1,$$

$$\frac{\kappa}{\mu_0} = \pm\frac{\omega_m}{\omega} = \pm\frac{(2.8\ \text{MHz/Oe})(1786\,\text{G})}{10,000\ \text{MHz}} = \pm 0.5$$

Using a numerical root-finding technique, such as interval-halving, we can solve (10.84) for the propagation constants β_+ and β_- by using positive and negative values of κ. Figure 10.19 shows the resulting differential phase shift, $(\beta_+ - \beta_-)/k_0$, versus slab thickness, t, for several slab spacings. Observe that the phase shift increases as the spacing, s, between the slabs decreases, and as the slab thickness increases, for t/a up to about 0.12.

From the curve in Figure 10.19 for $s = 1$ mm, we see that the optimum slab thickness for maximum phase shift is $t/a = 0.12$, or $t = 2.74$ mm, since $a = 2.286$ cm for X-band guide. The corresponding normalized differential phase shift is 0.40, so,

$$\beta_+ - \beta_- = 0.4k_0 = 0.4\left(\frac{2.09\ \text{rad}}{\text{cm}}\right)$$

$$= 0.836\ \text{rad/cm} = 48^\circ/\text{cm}$$

The ferrite length required for the 180° phase shift section is then

$$L = \frac{180^\circ}{48^\circ/\text{cm}} = 3.75\,\text{cm},$$

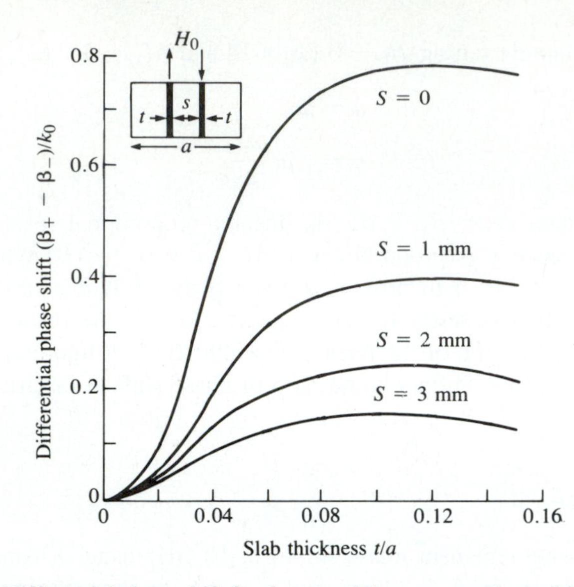

FIGURE 10.19 Differential phase shift for the two-slab remanent phase shifter of Example 10.5.

while the length required for a 90° section is

$$L = \frac{90^\circ}{48^\circ/\text{cm}} = 1.88 \text{ cm}.$$

○

Other Types of Ferrite Phase Shifters

Many other types of ferrite phase shifters have been developed, with various combinations of rectangular or circular waveguide, transverse or longitudinal biasing, latching or continuous phase variation, and reciprocal or nonreciprocal operation. Phase shifters using printed transmission lines have also been proposed. Even though PIN diode and FET circuits offer a less bulky and more integratable alternative to ferrite components, ferrite phase shifters often have advantages in terms of cost, power handling capacity, and power requirements. But there is still a great need for a low-cost, compact phase shifter.

Several waveguide phase shifter designs are derived from the nonreciprocal Faraday rotation phase shifter shown in Figure 10.20. In operation, a rectangular waveguide TE_{10} mode entering at the left is converted to a TE_{11} circular waveguide mode with a short transition section. Then a quarter-wave dielectric plate, oriented 45° from the electric field vector, converts the wave to a RHCP wave by providing a 90° phase difference between the field components that are parallel and perpendicular to the plate. In the ferrite-loaded region the phase delay is $\beta_+ z$, which can be controlled with the bias field strength. The second quarter-wave plate converts the wave back to a linearly polarized field. The operation is similar for a wave entering at the right, except now the phase delay is $\beta_- z$; the phase shift is thus nonreciprocal. The ferrite rod is biased longitudinally, in

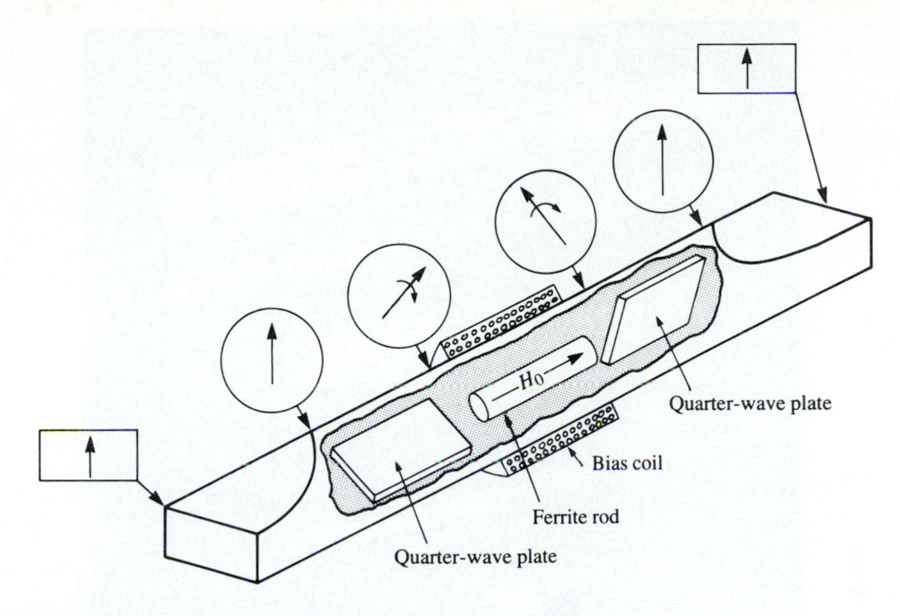

FIGURE 10.20 Nonreciprocal Faraday rotation phase shifter.

the direction of propagation, with a solenoid coil. This type of phase shifter can be made reciprocal by using nonreciprocal quarter–wave plates to convert a linearly polarized wave to the same sense of circular polarization for either propagation direction.

The Reggia-Spencer phase shifter, shown in Figure 10.21, is a popular reciprocal phase shifter. In either rectangular or circular waveguide form, a longitudinally biased ferrite rod is centered in the guide. When the diameter of the rod is greater than a certain critical size, the fields become tightly bound to the ferrite, and are circularly polarized. A large reciprocal phase shift can be obtained over relatively short lengths, although the phase shift is rather frequency sensitive. Figure 10.22 shows a photograph of several different types of ferrite phase shifters.

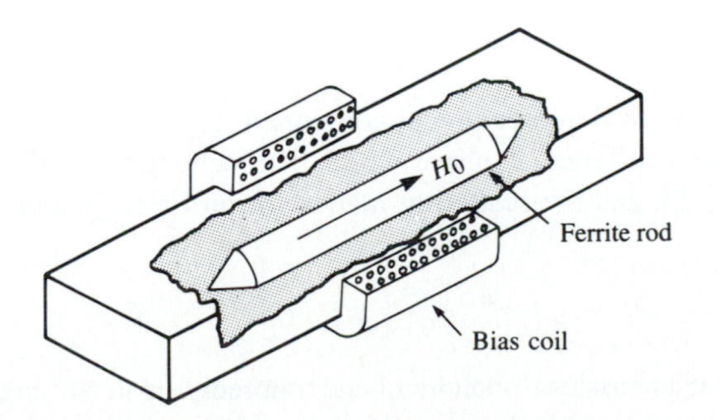

FIGURE 10.21 Reggia-Spencer reciprocal phase shifter.

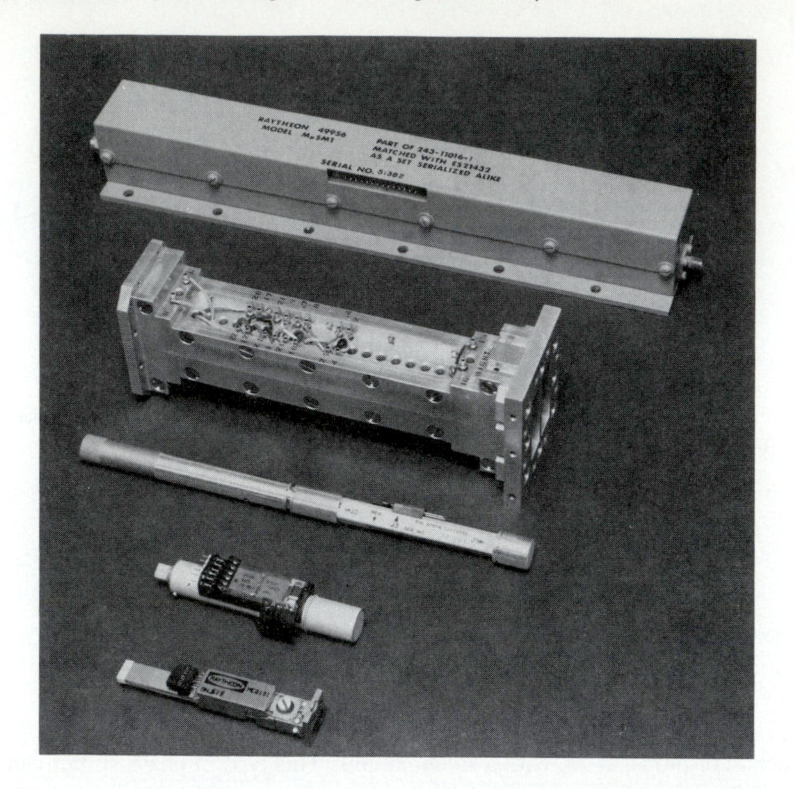

FIGURE 10.22 A photograph of several types of ferrite phase shifters. The top unit is an S-band nonreciprocal latching phase shifter. Second from the top is a dual C-band waveguide nonreciprocal latching phase shifter. The middle unit is an X-band nonreciprocal latching phase shifter combined with a circular waveguide antenna element. Second from the bottom is an X-band reciprocal Faraday rotation phase shifter, with a circular waveguide antenna. The bottom unit is a nonreciprocal latching phase shifter with a rectangular waveguide antenna element.
Courtesy of John Mather, Raytheon Corporation, Microwave and Power Tube Division.

The Gyrator

An important canonical nonreciprocal component is the *gyrator*, which is a two-port device having a 180° differential phase shift. The schematic symbol for a gyrator is shown in Figure 10.23, and the scattering matrix for an ideal gyrator is

$$[S] = \begin{bmatrix} 0 & 1 \\ -1 & 0 \end{bmatrix}, \qquad 10.90$$

which shows that it is lossless, matched, and nonreciprocal. Using the gyrator as a basic nonreciprocal building block in combination with reciprocal dividers and couplers can lead to useful equivalent circuits for nonreciprocal components such as isolators

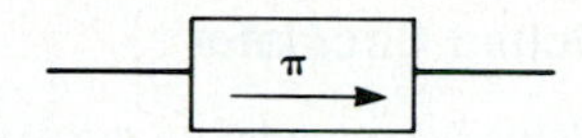

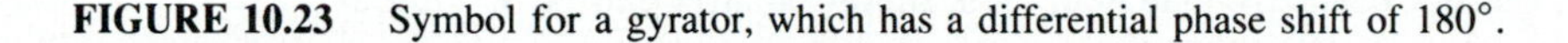

FIGURE 10.23 Symbol for a gyrator, which has a differential phase shift of 180°.

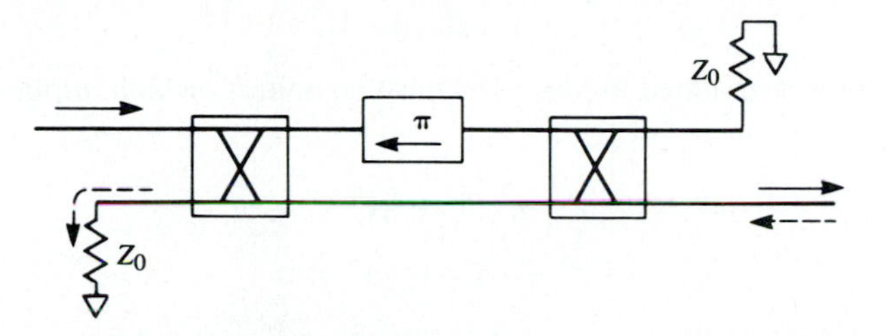

FIGURE 10.24 An isolator constructed with a gyrator and two quadrature hybrids. The forward wave (→) is passed, while the reverse wave (⇠) is absorbed in the matched load of the first hybrid.

and circulators. Figure 10.24, for example, shows an equivalent circuit for an isolator using a gyrator and two quadrature hybrids.

The gyrator can be implemented as a phase shifter with a 180° differential phase shift; bias can be provided with a permanent magnet, making the gyrator a passive device.

10.6 FERRITE CIRCULATORS

As we discussed in Section 8.1, a *circulator* is a three-port device that can be lossless and matched at all ports; by using the unitary properties of the scattering matrix we were able to show how such a device must be nonreciprocal. The scattering matrix for an ideal circulator thus has the following form:

$$[S] = \begin{bmatrix} 0 & 0 & 1 \\ 1 & 0 & 0 \\ 0 & 1 & 0 \end{bmatrix}, \qquad 10.91$$

which shows that power flow can occur from ports 1 to 2, 2 to 3, and 3 to 1, but not in the reverse direction. By transposing the port indices, the opposite circularity can be obtained. In practice, this result can be produced by changing the polarity of the ferrite bias field. Most circulators use permanent magnets for the bias field, but if an electromagnet is used the circulator can operate in a latching (remanent) mode as a single-pole double-throw (SPDT) switch. A circulator can also be used as an isolator by terminating one of the ports with a matched load.

We will first discuss the properties of an imperfectly matched circulator in terms of its scattering matrix. Then we will analyze the operation of the stripline junction circulator. The operation of waveguide circulators is similar in principle.

Properties of a Mismatched Circulator

If we assume that a circulator has circular symmetry around its three ports and is lossless, but not perfectly matched, its scattering matrix can be written as

$$[S] = \begin{bmatrix} \Gamma & \beta & \alpha \\ \alpha & \Gamma & \beta \\ \beta & \alpha & \Gamma \end{bmatrix}. \tag{10.92}$$

Since the circulator is assumed lossless, $[S]$ must be unitary, which implies the following two conditions:

$$|\Gamma|^2 + |\beta|^2 + |\alpha|^2 = 1, \tag{10.93a}$$

$$\Gamma\beta^* + \alpha\Gamma^* + \beta\alpha^* = 0. \tag{10.93b}$$

If the circulator were matched ($\Gamma = 0$), then (10.93) shows that either $\alpha = 0$ and $|\beta| = 1$, or $\beta = 0$ and $|\alpha| = 1$; this describes the ideal circulator with its two possible circularity states. Observe that this condition only depends on a lossless and matched device.

Now assume small imperfections, such that $|\Gamma| << 1$. To be specific, consider the circularity state where power flows primarily in the 1-2-3 direction, so that $|\alpha|$ is close to unity and $|\beta|$ is small. Then $\beta\Gamma \sim 0$, and (10.93b) shows that $\alpha\Gamma^* + \beta\alpha^* \simeq 0$, so $|\Gamma| \simeq |\beta|$. Then (10.93a) shows that $|\alpha|^2 \simeq 1 - 2|\beta|^2 \simeq 1 - 2|\Gamma|^2$, or $|\alpha| \simeq 1 - |\Gamma|^2$. Then the scattering matrix of (10.92) can be written as

$$[S] = \begin{bmatrix} \Gamma & \Gamma & 1-\Gamma^2 \\ 1-\Gamma^2 & \Gamma & \Gamma \\ \Gamma & 1-\Gamma^2 & \Gamma \end{bmatrix}, \tag{10.94}$$

ignoring phase factors. This result shows that circulator isolation, $\beta \simeq \Gamma$, and transmission, $\alpha \simeq 1 - \Gamma^2$, both deteriorate as the input ports become mismatched.

Junction Circulator

The stripline junction circulator geometry is shown in Figure 10.25. Two ferrite disks fill the spaces between the center metallic disk and the ground planes of the stripline. Three stripline conductors are attached to the periphery of the center disk at 120° intervals, forming the three ports of the circulator. The DC bias field is applied normal to the ground planes.

In operation, the ferrite disks form a resonant cavity; in the absence of a bias field, this cavity has a single lowest-order resonant mode with a $\cos\phi$ (or $\sin\phi$) dependence. When the ferrite is biased this mode breaks into two resonant modes with slightly different resonant frequencies. The operating frequency of the circulator can then be chosen so that the superposition of these two modes add at the output port and cancel at the isolated port.

We can analyze the junction circulator by treating it as a thin cavity resonator with electric walls on the top and bottom, and an approximate magnetic wall on the side. Then $E_\rho = E_\phi \simeq 0$, and $\partial/\partial z = 0$, so we have TM modes. Since E_z on either side of the center conducting disk is antisymmetric, we need only consider the solution for one of the ferrite disks [7].

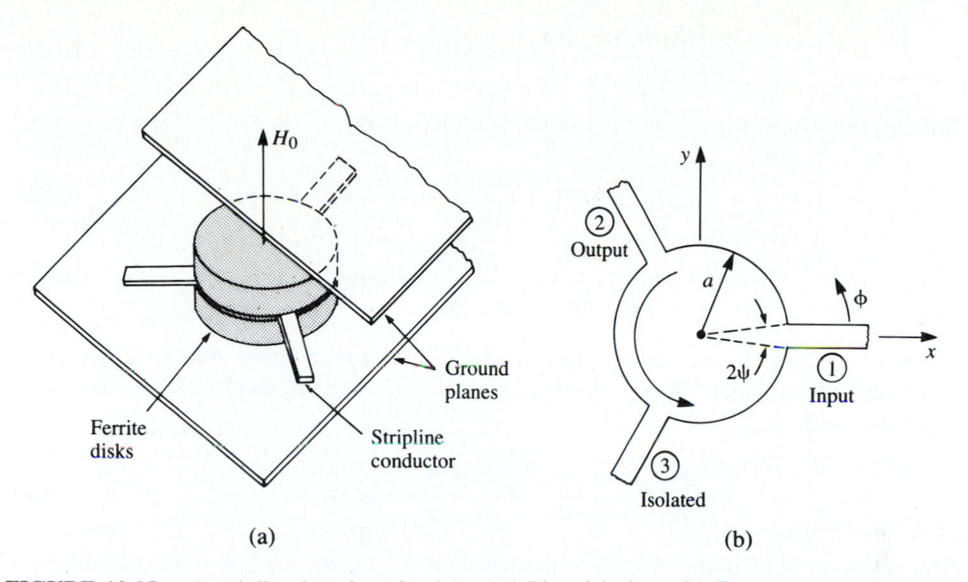

FIGURE 10.25 A stripline junction circulator. (a) Pictorial view. (b) Geometry.

We begin by transforming (10.23), $\bar{B} = [\mu]\bar{H}$, from rectangular to cylindrical coordinates:

$$\begin{aligned} B_\rho &= B_x \cos\phi + B_y \sin\phi \\ &= (\mu H_x + j\kappa H_y)\cos\phi + (-j\kappa H_x + \mu H_y)\sin\phi \\ &= \mu H_\rho + j\kappa H_\phi, \end{aligned} \qquad 10.95a$$

$$\begin{aligned} B_\phi &= -B_x \sin\phi + B_y \cos\phi \\ &= -(\mu H_x + j\kappa H_y)\sin\phi + (-j\kappa H_x + \mu H_y)\cos\phi \\ &= -j\kappa H_\rho + \mu H_\phi. \end{aligned} \qquad 10.95b$$

So we have that

$$\begin{bmatrix} B_\rho \\ B_\phi \\ B_z \end{bmatrix} = [\mu] \begin{bmatrix} H_\rho \\ H_\phi \\ H_z \end{bmatrix}, \qquad 10.96$$

where $[\mu]$ is the same matrix as for rectangular coordinates, as given in (10.24).

In cylindrical coordinates, with $\partial/\partial z = 0$, Maxwell's curl equations reduce to the following:

$$\frac{1}{\rho}\frac{\partial E_z}{\partial \phi} = -j\omega(\mu H_\rho + j\kappa H_\phi), \qquad 10.97a$$

$$-\frac{\partial E_z}{\partial \rho} = -j\omega(-j\kappa H_\rho + \mu H_\phi), \qquad 10.97b$$

$$\frac{1}{\rho}\left[\frac{\partial(\rho H_\phi)}{\partial\rho} - \frac{\partial H_\rho}{\partial\phi}\right] = j\omega\epsilon E_z. \tag{10.97c}$$

Solving (10.97a,b) for H_ρ and H_ϕ in terms of E_z gives

$$H_\rho = \frac{jY}{k\mu}\left(\frac{\mu}{\rho}\frac{\partial E_z}{\partial\phi} + j\kappa\frac{\partial E_z}{\partial\rho}\right), \tag{10.98a}$$

$$H_\phi = \frac{-jY}{k\mu}\left(\frac{-j\kappa}{\rho}\frac{\partial E_z}{\partial\phi} + \mu\frac{\partial E_z}{\partial\rho}\right), \tag{10.98b}$$

where $k^2 = \omega^2\epsilon(\mu^2 - \kappa^2)/\mu = \omega^2\epsilon\mu_e$ is an effective wavenumber, and $Y = \sqrt{\epsilon/\mu_e}$ is an effective admittance. Using (10.98) to eliminate H_ρ and H_ϕ in (10.97c) gives a wave equation for E_z:

$$\frac{\partial^2 E_z}{\partial\rho^2} + \frac{1}{\rho}\frac{\partial E_z}{\partial\rho} + \frac{1}{\rho^2}\frac{\partial^2 E_z}{\partial\phi^2} + k^2 E_z = 0. \tag{10.99}$$

This equation is identical in form to the equation for E_z for the TM mode of a circular waveguide, so the general solution can be written as

$$E_{zn} = \left[A_{+n}e^{jn\phi} + A_{-n}e^{-jn\phi}\right] J_n(k\rho), \tag{10.100a}$$

where we have excluded the solution with $Y_n(k\rho)$ because E_z must be finite at $\rho = 0$. We will also need $H_{\phi n}$, which can be found using (10.98b):

$$H_{\phi n} = -jY\left\{A_{+n}e^{jn\phi}\left[J_n'(k\rho) + \frac{n\kappa}{k\rho\mu}J_n(k\rho)\right] + A_{-n}e^{-jn\phi}\left[J_n'(k\rho) - \frac{n\kappa}{k\rho\mu}J_n(k\rho)\right]\right\}. \tag{10.100b}$$

The resonant modes can now be found by enforcing the boundary condition that $H_\phi = 0$ at $\rho = a$.

If the ferrite is not magnetized, then $H_0 = M_s = 0$ and $\omega_0 = \omega_m = 0$ so that $\kappa = 0$ and $\mu = \mu_e = \mu_o$, and resonance occurs when

$$J_n'(ka) = 0,$$

or $ka = x_0 = p_{11}' = 1.841$. Define this frequency as ω_0 (not to be confused with $\omega_0 = \gamma\mu_0 H_0$):

$$\omega_0 = \frac{x_0}{a\sqrt{\epsilon\mu_e}} = \frac{1.841}{a\sqrt{\epsilon\mu_o}}. \tag{10.101}$$

When the ferrite is magnetized there are two possible resonant modes for each value of n, as associated with either a $e^{jn\phi}$ variation or $e^{-jn\phi}$ variation. The resonance condition for the two $n = 1$ modes is

$$\frac{\kappa}{\mu x}J_1(x) \pm J_1'(x) = 0, \tag{10.102}$$

where $x = ka$. This result shows the nonreciprocal property of the circulator, since changing the sign of κ (the polarity of the bias field) in (10.102) leads to the other root and propagation in the opposite direction in ϕ.

If we let x_+ and x_- be the two roots of (10.102), then the resonant frequencies for these two $n = 1$ modes can be expressed as

$$\omega_\pm = \frac{x_\pm}{a\sqrt{\epsilon\mu_e}} \tag{10.103}$$

We can develop an approximate result for $\omega_\pm$ if we assume that κ/μ is small, so that $\omega_\pm$ will be close to ω_0 of (10.101). Using a Taylor series about x_0 for the two terms in (10.102) gives the following results, since $J_1'(x_0) = 0$:

$$\begin{aligned} J_1(x) &\simeq J_1(x_0) + (x - x_0)J_1'(x_0) = J_1(x_0), \\ J_1'(x) &\simeq J_1'(x_0) + (x - x_0)J_1''(x_0) \\ &= (x - x_0)\left(1 - \frac{1}{x_0^2}\right) J_1(x_0). \end{aligned}$$

Then (10.102) becomes

$$\frac{\kappa}{\mu x_0} \pm (x_\pm - x_0)\left(1 - \frac{1}{x_0^2}\right) = 0,$$

or

$$x_\pm = x_0\left(1 \mp 0.418\frac{\kappa}{\mu}\right), \tag{10.104}$$

since $x_0 = 1.841$. This result gives the resonant frequencies as

$$\omega_\pm = \omega_0\left(1 \mp 0.418\frac{\kappa}{\mu}\right). \tag{10.105}$$

Note that $\omega_\pm$ approaches ω_0 as $\kappa \to 0$, and that

$$\omega_+ \leq \omega_0 \leq \omega_-.$$

Now we can use these two modes to design a circulator. The amplitudes of these modes provide two degrees of freedom that can be used to provide coupling from the input to the output port, and to provide cancellation at the isolated port. It will turn out that ω_0 will be the operating frequency, between the resonances of the $\omega_\pm$ modes. Thus, $H_\phi \neq 0$ over the periphery of the ferrite disks, since $\omega \neq \omega_\pm$. If we select port 1 as the input, port 2 as the output, and port 3 as the isolated port, as in Figure 10.25, we can assume the following E_z field at the ports at $\rho = a$:

$$E_z(\rho = a, \phi) = \begin{cases} E_0, & \text{for } \phi = 0 \quad \text{(Port 1)}, \\ -E_0, & \text{for } \phi = 120° \text{ (Port 2)}, \\ 0, & \text{for } \phi = 240° \text{ (Port 3)}. \end{cases} \tag{10.106a}$$

If the feedlines are narrow, the E_z field will be relatively constant across their width.

The corresponding H_ϕ field should be

$$H_\phi(\rho = a, \phi) = \begin{cases} H_0 & \text{for } -\psi < \phi < \psi, \\ H_0 & \text{for } 120° - \psi < \phi < 120° + \psi, \\ 0 & \text{elsewhere.} \end{cases} \quad 10.106b$$

Equating (10.106a) to E_z of (10.100a) gives the mode amplitude constants as

$$A_{+1} = \frac{E_0(1 + j/\sqrt{3})}{2J_1(ka)}, \quad 10.107a$$

$$A_{-1} = \frac{E_0(1 - j/\sqrt{3})}{2J_1(ka)}. \quad 10.107b$$

Then (10.100a,b) can be reduced to give the electric and magnetic fields as

$$E_{z1} = \frac{E_0 J_1(k\rho)}{2J_1(ka)} \left[\left(1 + \frac{j}{\sqrt{3}}\right) e^{j\phi} + \left(1 - \frac{j}{\sqrt{3}}\right) e^{-j\phi} \right]$$
$$= \frac{E_0 J_1(k\rho)}{J_1(ka)} \left(\cos\phi - \frac{\sin\phi}{\sqrt{3}} \right), \quad 10.108a$$

$$H_{\phi 1} = \frac{-jYE_0}{2J_1(ka)} \left\{ \left(1 + \frac{j}{\sqrt{3}}\right) \left[J_1'(k\rho) + \frac{\kappa}{k\rho\mu} J_1(k\rho) \right] e^{j\phi} \right.$$
$$\left. + \left(1 - \frac{j}{\sqrt{3}}\right) \left[J_1'(k\rho) - \frac{\kappa}{k\rho\mu} J_1(k\rho) \right] e^{-j\phi} \right\}. \quad 10.108b$$

To approximately equate $H_{\phi 1}$ to H_ϕ in (10.106b) requires that H_ϕ be expanded in a Fourier series:

$$H_\phi(\rho = a, \phi) = \sum_{n=-\infty}^{\infty} C_n e^{jn\phi} = \frac{2H_0\psi}{\pi}$$
$$+ \frac{H_0}{\pi} \sum_{n=1}^{\infty} \left[(1 + e^{-j2\pi n/3}) e^{jn\phi} + (1 + e^{j2\pi n/3}) e^{-jn\phi} \right]$$
$$\times \frac{\sin n\psi}{n}. \quad 10.109$$

The $n = 1$ term of this result is

$$H_{\phi 1}(\rho = a, \phi) = \frac{-j\sqrt{3} H_0 \sin\psi}{2\pi} \left[\left(1 + \frac{j}{\sqrt{3}}\right) e^{j\phi} - \left(1 - \frac{j}{\sqrt{3}}\right) e^{-j\phi} \right],$$

which can now be equated to (10.108b) for $\rho = a$. Equivalence can be obtained if two conditions are met:

$$J_1'(ka) = 0,$$

and

$$\frac{YE_0\kappa}{ka\mu} = \frac{\sqrt{3} H_0 \sin\psi}{\pi}.$$

The first condition is identical to the condition for resonance in the absence of bias, which implies that the operating frequency is ω_0, as given by (10.101). For a given operating frequency, (10.101) can then be used to find the disk radius, a. The second condition can be related to the wave impedance at port 1 or 2:

$$Z_w = \frac{E_0}{H_0} = \frac{\sqrt{3}ka\mu \sin\psi}{\pi Y \kappa} \simeq \frac{\mu \sin\psi}{\kappa Y}, \qquad 10.110$$

since $\sqrt{3}ka/\pi = \sqrt{3}(1.841)/\pi \simeq 1.0$. Thus, Z_w can be controlled for impedance matching by adjusting κ/μ via the bias field.

We can compute the power flows at the three ports as follows:

$$P_{\text{in}} = P_1 = -\hat{\rho} \cdot \bar{E} \times \bar{H}^* = E_z H_\phi \Big|_{\phi=0} = \frac{E_0 H_0 \sin\psi}{\pi} = \frac{E_0^2 \kappa Y}{\pi\mu}, \qquad 10.111a$$

$$P_{\text{out}} = P_2 = \hat{\rho} \cdot \bar{E} \times \bar{H}^* = -E_z H_\phi \Big|_{\phi=120^\circ} = \frac{E_0 H_0 \sin\psi}{\pi} = \frac{E_0^2 \kappa Y}{\pi\mu}, \qquad 10.111b$$

$$P_{\text{iso}} = P_3 = \hat{\rho} \cdot \bar{E} \times \bar{H}^* = -E_z H_\phi \Big|_{\phi=240^\circ} = 0. \qquad 10.111c$$

This shows that power flow occurs from port 1 to 2, but not from 1 to 3. By the azimuthal symmetry of the circulator, this also implies that power can be coupled from port 2 to 3, or from port 3 to 1, but not in the reverse directions.

The electric field of (10.108a) is sketched in Figure 10.26 along the periphery of the circulator, showing that the amplitudes and phases of the $e^{\pm j\phi}$modes are such that their superposition gives a null at the isolated port, with equal voltages at the input and output ports. This result ignores the loading effect of the input and output lines, which will distort the field from that shown in Figure 10.26. This design is narrowband, but bandwidth can be improved using dielectric loading; the analysis then requires consideration of higher-order modes.

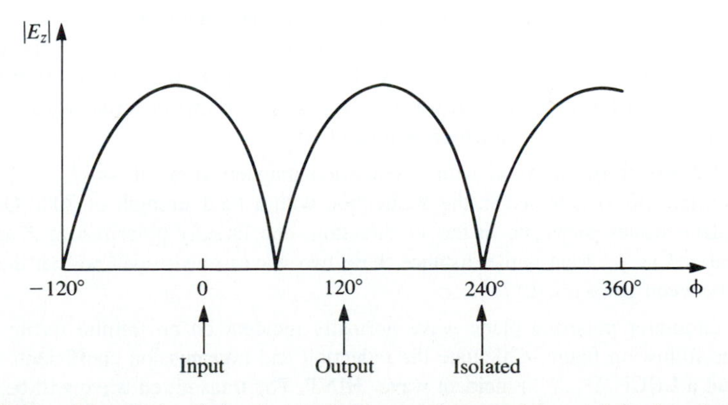

FIGURE 10.26 Magnitude of the electric field around the periphery of the junction circulator.

REFERENCES

[1] R. F. Soohoo, *Microwave Magnetics*, Harper and Row, N. Y., 1985.

[2] A. J. Baden Fuller, *Ferrites at Microwave Frequencies*, Peter Peregrinus, London, 1987.

[3] R. E. Collin, *Field Theory of Guided Waves*, McGraw-Hill, N. Y., 1960.

[4] B. Lax and K. J. Button, *Microwave Ferrites and Ferrimagnetics*, McGraw-Hill, N. Y., 1962.

[5] F. E. Gardiol and A. S. Vander Vorst, "Computer Analysis of E-plane Resonance Isolators," *IEEE Trans. Microwave Theory and Techniques*, vol. MTT-19, pp. 315–322, March 1971.

[6] G. P. Rodrigue, "A Generation of Microwave Ferrite Devices," *Proc. IEEE*, vol. 76, pp. 121–137, February 1988.

[7] C. E. Fay and R. L. Comstock, "Operation of the Ferrite Junction Circulator," *IEEE Trans. Microwave Theory and Techniques*, vol. MTT-13, pp. 15–27, January 1965.

PROBLEMS

10.1 A certain ferrite material has a saturation magnetization of $4\pi M_s = 1780\,\text{G}$. Ignoring loss, calculate the elements of the permeability tensor at $f = 10$ GHz for two cases: (1) no bias field and ferrite demagnetized ($M_s = H_0 = 0$) and (2) a z-directed bias field of 1000 Oersted.

10.2 Consider the following field transformations from rectangular to circular polarized components:

$$\begin{aligned} B^+ &= (B_x + jB_y)/2, \qquad & H^+ &= (H_x + jH_y)/2, \\ B^- &= (B_x - jB_y)/2, \qquad & H^- &= (H_x - jH_y)/2, \\ B_z &= B_z, \qquad & H_z &= H_z. \end{aligned}$$

For a z-biased ferrite medium, show that the relation between $\bar{B}$ and $\bar{H}$ can be expressed in terms of a diagonal tensor permeability as follows:

$$\begin{bmatrix} B^+ \\ B^- \\ B_z \end{bmatrix} = \begin{bmatrix} (\mu+\kappa) & 0 & 0 \\ 0 & (\mu-\kappa) & 0 \\ 0 & 0 & \mu_o \end{bmatrix} \begin{bmatrix} H^+ \\ H^- \\ H_z \end{bmatrix}.$$

10.3 A YIG sphere with $4\pi M_s = 1780\,\text{G}$ lies in a uniform magnetic field having a strength of 1200 Oe. What is the magnetic field strength inside the YIG sphere?

10.4 A thin rod is biased along its axis with an external applied field of $H_a = 1000$ Oe. If $4\pi M_s = 600\,\text{G}$, calculate the gyromagnetic resonance frequency for the rod.

10.5 An infinite lossless ferrite medium with a saturation magnetization of $4\pi M_s = 1200\,\text{G}$ and a dielectric constant of 10 is biased to a field strength of 500 Oersted. At 8 GHz, calculate the differential phase shift per meter between a RHCP and a LHCP plane wave propagating in the direction of bias. If a linearly polarized wave is propagating in this material, what is the distance it must travel in order that its polarization is rotated 90°?

10.6 An infinite lossless ferrite medium with a saturation magnetization of $4\pi M_s = 1200\,\text{G}$ and a dielectric constant of 10 is biased in the $\hat{x}$ direction with a field strength of 2000 Oersted. At 4 GHz, two plane waves propagate in the +z direction, one linearly polarized in x and the other linearly polarized in y. What is the distance these two waves must travel so that the differential phase shift between them is 270°?

10.7 Consider a circularly polarized plane wave normally incident on an infinite ferrite medium, as shown in the following figure. Calculate the reflection and transmission coefficients for a RHCP (Γ^+, T^+) and a LHCP (Γ^-, T^-) incident wave. HINT: The transmitted wave will be polarized in the same sense as the incident wave, but the reflected wave will be oppositely polarized.

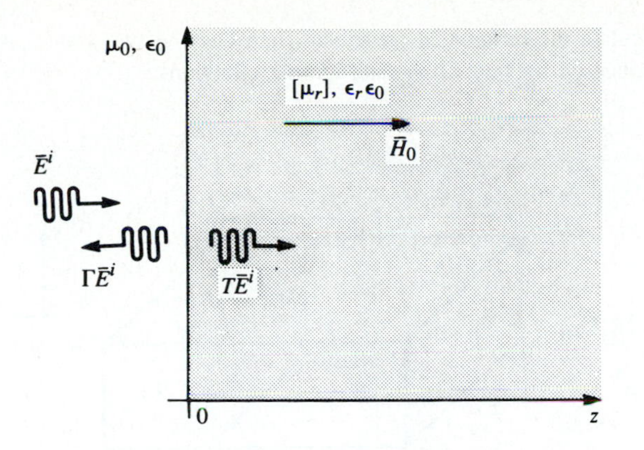

10.8 An infinite lossless ferrite material with $4\pi M_s = 1200\,\text{G}$ is biased in the $\hat{x}$ direction with $\bar{H}_0 = H_0\hat{x}$. Determine the range of H_0, in Oersteds, where an extraordinary wave (polarized in $\hat{x}$, propagating in $\hat{z}$) will be cutoff. The frequency is 4 GHz.

10.9 Find the forward and reverse propagation constants for a waveguide half-filled with a transversely biased ferrite. (The geometry of Figure 10.10 with $c = 0$ and $t = a/2$.) Assume $a = 1.0$ cm, $f = 10$ GHz, $4\pi M_s = 1700\,\text{G}$, and $\epsilon_r = 13$. Plot versus $H_0 = 0$ to 1500 Oe. Ignore loss, and the fact that the ferrite may not be saturated for small H_0.

10.10 Find the forward and reverse propagation constants for a waveguide filled with two pieces of oppositely biased ferrite. (The geometry of Figure 10.11 with $c = 0$ and $t = a/2$.) Assume $a = 1.0$ cm, $f = 10$ GHz, $4\pi M_s = 1700$ G, and $\epsilon_r = 13$. Plot versus $H_0 = 0$ to 1500 Oe. Ignore loss, and the fact that the ferrite may not be saturated for small H_0.

10.11 Consider a wide, thin ferrite slab in a rectangular X-band waveguide, as shown below. If $f = 10$ GHz, $4\pi M_s = 1700\,\text{G}$, $c = a/4$, and $\Delta S = 2\,\text{mm}^2$, use the perturbation formula of (10.80) to plot the differential phase shift, $(\beta_+ - \beta_-)/k_0$, versus the bias field for $H_0 = 0$ to 1200 Oe. Ignore loss.

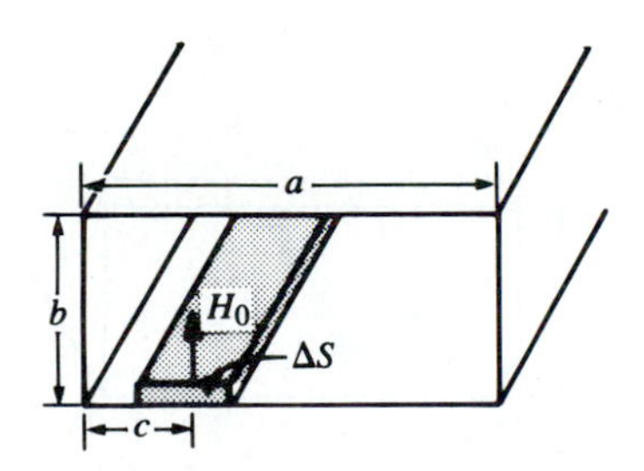

10.12 An E-plane resonance isolator with the geometry of Figure 10.12a is to be designed to operate at 8 GHz, with a ferrite having a saturation magnetization of $4\pi M_s = 1500\,\text{G}$. What is the approximate bias field, H_0, required for resonance? What is the required bias field if the H-plane geometry of Figure 10.12b is used?

10.13 Calculate and plot the two normalized positions, x/a, where the magnetic fields of the TE_{10} mode of an empty rectangular waveguide are circularly polarized, for $k_0 = k_c$ to $2k_c$.

10.14 The latching ferrite phase shifter shown in the figure on page 580 uses the birefringence effect. In state 1, the ferrite is magnetized so that $H_0 = 0$ and $\bar{M} = M_r\hat{x}$. In state 2, the ferrite is magnetized so that $H_0 = 0$ and $\bar{M} = M_r\hat{y}$. If $f = 6$ GHz, $\epsilon_r = 10$, $4\pi M_r = 1500$ G, and

$L = 2.78$ cm, calculate the differential phase shift between the two states. Assume the incident plane wave is $\hat{x}$ polarized for both states, and ignore reflections.

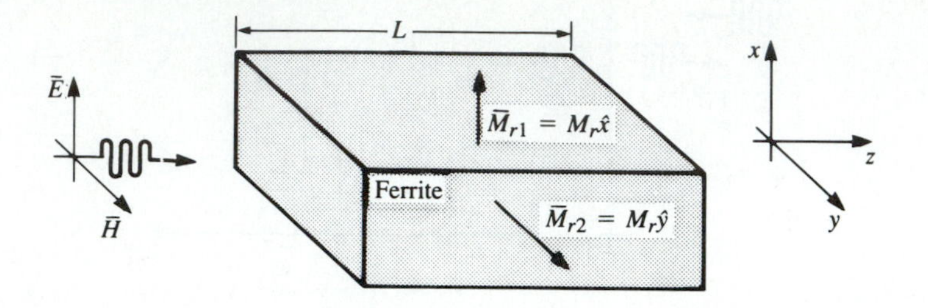

10.15 Rework Example 10.5 with a slab spacing of $s = 2$ mm, and a remanent magnetization of 1000 G. (Assume all other parameters as unchanged, and that the differential phase shift is linearly proportional to κ.)

10.16 Consider a latching phase shifter constructed with a wide, thin H-plane ferrite slab in an X-band waveguide, as shown. If $f = 9$ GHz, $4\pi M_r = 1200$ G, $c = a/4$, and $\Delta S = 2\,\text{mm}^2$, use the perturbation formula of (10.80) to calculate the required length for a differential phase shift of 22.5°.

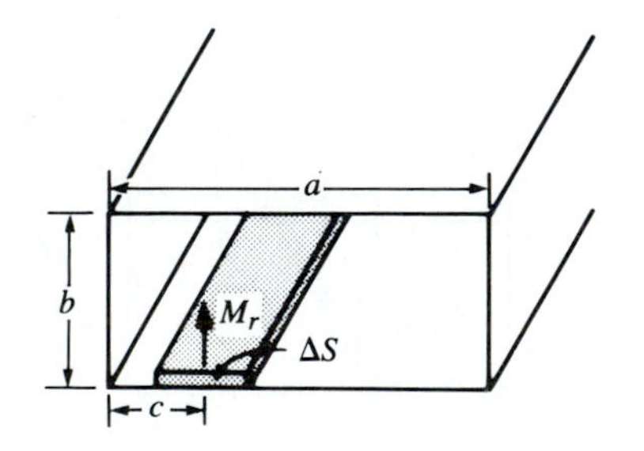

10.17 Design a gyrator using the twin H-plane ferrite slab geometry shown below. The frequency is 9.0 GHz, and the saturation magnetization is $4\pi M_s = 1700$ G. The cross-sectional area of each slab is 3.0 mm^2, and the guide is X-band waveguide. The permanent magnet has a field strength of $H_a = 2000$ Oe. Determine the internal field in the ferrite, H_0, and use the perturbation formula of (10.80) to determine the optimum location of the slabs and the length, L, to give the necessary 180° differential phase shift.

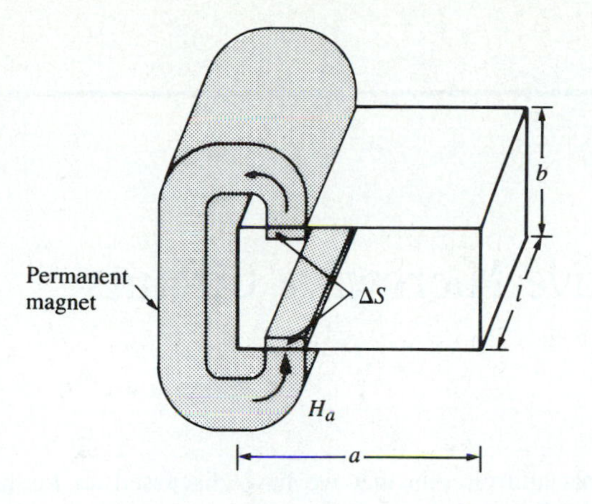

10.18 Draw an equivalent circuit for a circulator using a gyrator and two couplers.

10.19 A certain lossless circulator has a return loss of 10 dB. What is the isolation? What is the isolation if the return loss is 20 dB?

CHAPTER 11

Active Microwave Circuits

The components and circuits that we have discussed so far have been linear and passive, but any useful microwave system will require some nonlinear and active components. Such devices include diodes, transistors, and tubes, which can be used for detection, mixing, amplification, frequency multiplication, switching, and as sources. Active circuit design is a broad and rapidly evolving field, so we can only present some of the basic concepts and principles here, and refer the reader to the references for more detail. We will also avoid any discussion of the physics of diodes, transistors, or tubes, since for our purposes it will be adequate to characterize these devices in terms of their terminal properties.

The electrical performance of a microwave system can be affected by many factors, but the effect of noise is probably one of the most fundamental. Thus we begin the chapter with a discussion of the sources of noise, and the characterization of components in terms of noise temperature and noise figure. Next we discuss the small-signal characteristics of detector diodes, and their application to the frequency conversion functions of rectification, detection, and mixing. Then we present an introduction to transistor amplifier design, including stability criteria, conjugate matching, constant gain circles, and constant noise figure circles. Several examples in this section are used to illustrate various amplifier design procedures. In Section 11.4 we discuss some of the basic requirements for one- and two-port oscillator design. Then we show how PIN diodes can be used for a variety of control circuits, including switches and phase shifters. Finally, we give brief overviews of microwave integrated circuits, and microwave solid-state and tube sources.

11.1 NOISE IN MICROWAVE SYSTEMS

Noise power is a result of random processes such as the flow of charges or holes in an electron tube or solid state device, propagation through the ionosphere or other ionized gas or, most basic of all, the thermal vibrations in any component at a temperature above absolute zero. Noise can be passed into a microwave system from external sources, or generated within the system itself. In either case the noise level of a system sets the lower limit on the strength of a signal that can be detected in the presence of the noise. Thus, it is generally desired to minimize the residual noise level of a radar or communications receiver, to achieve the best performance. In some cases, such as radiometers or radio astronomy systems, the desired signal is actually the noise power received by an antenna, and it is necessary to distinguish between the received noise power and the undesired noise generated by the receiver system itself.

Dynamic Range and Sources of Noise

In previous chapters we have implicitly assumed that all components were *linear*, meaning that the output is directly proportional to the input, and *deterministic*, meaning that the output is predictable from the input. In reality no component can perform in this way over an unlimited range of input/output signal levels. In practice, however, there is a range of signal levels over which such assumptions are valid; this range is called the dynamic range of the component.

As an example, consider a realistic microwave transistor amplifier having a gain of 10 dB, as shown in Figure 11.1. If the amplifier were ideal, the output power would be related to the input power as

$$P_{\text{out}} = 10P_{\text{in}},$$

and this relation would hold true for any value of P_{in}. Thus if $P_{\text{in}} = 0$, we would have $P_{\text{out}} = 0$, and if $P_{\text{in}} = 10^6$ W, we would have $P_{\text{out}} = 10^7$ W. Obviously neither of

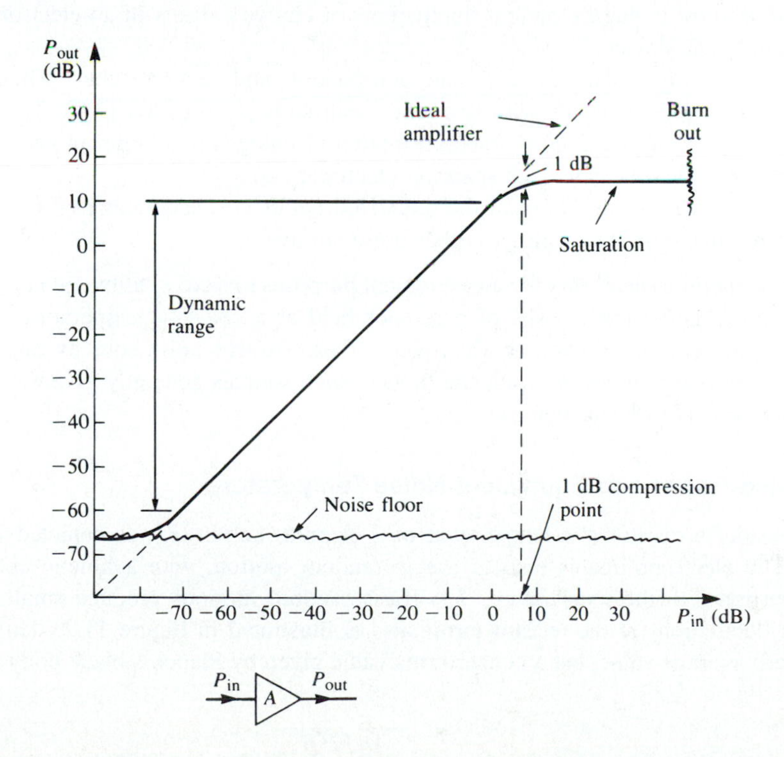

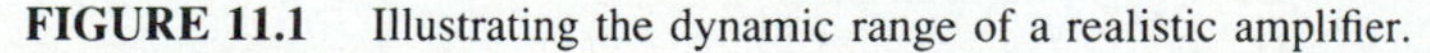

FIGURE 11.1 Illustrating the dynamic range of a realistic amplifier.

these conditions will be true in practice. Because of noise generated by the amplifier itself, as well as external noise that might be fed into the amplifier, a certain nonzero noise power will be delivered by the amplifier even when the input power is zero. For very high input powers, the amplifier will be destroyed. Thus, the actual relation between the input and output power will be as shown in Figure 11.1. At very low input power levels, the output will be dominated by the noise of the amplifier. This level is often called the noise floor of the component or system; typical values may range from -60 dBm to -100 dBm over the bandwidth of the system, with lower values being obtainable with cooled components. Above the noise floor, the amplifier has a range of input powers for which $P_{\text{out}} = 10P_{\text{in}}$ is closely approximated. This is the usable dynamic range of the component. At the upper end of the dynamic range, the output begins to saturate, meaning that the output power no longer increases linearly as the input power increases. A quantitative measure of the onset of saturation is given by the 1 dB *compression point*, which is defined as the input power for which the output is 1 dB below that of the ideal amplifier (the corresponding output power level can also be used to specify this point). If the input power is excessive, the amplifier can be destroyed.

Noise is usually generated by the random motions of charges or charge carriers in devices and materials. Such motions can be caused by any of several mechanisms, leading to various sources of noise:

- *Thermal noise* is the most basic type of noise, being caused by thermal vibration of bound charges. Also known as Johnson or Nyquist noise.
- *Shot noise* is due to random fluctuations of charge carriers in an electron tube or solid state device.
- *Flicker noise* occurs in solid-state components and vacuum tubes. Flicker noise power varies inversely with frequency, and so is often called $1/f$-noise.
- *Plasma noise* is caused by random motion of charges in an ionized gas, such as a plasma, the ionosphere, or sparking electrical contacts.
- *Quantum noise* results from the quantized nature of charge carriers and photons; often insignificant relative to other noise sources.

It is sometimes necessary for measurement purposes to have a calibrated noise source. Passive noise generators consist of a resistor held at a constant temperature, either in a temperature-controlled oven or a cryogenic flask. Active noise sources can be made using gas-discharge tubes or avalanche diodes; such sources generally give much higher noise power than passive sources.

Noise Power and Equivalent Noise Temperature

Consider a resistor at a temperature of T degrees kelvin (K), as depicted in Figure 11.2. The electrons in this resistor are in random motion, with a kinetic energy that is proportional to the temperature, T. These random motions produce small, random voltage fluctuations at the resistor terminals, as illustrated in Figure 11.2. This voltage has a zero average value, but a nonzero rms value given by Planck's black body radiation law,

$$v_n = \sqrt{\frac{4hfBR}{e^{hf/kT} - 1}}, \tag{11.1}$$

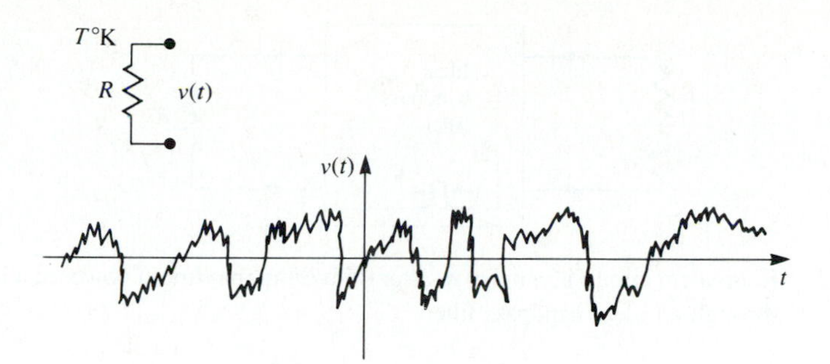

FIGURE 11.2 A random voltage generated by a noisy resistor.

where

$h = 6.546 \times 10^{-34}$ J-sec is Planck's constant.

$k = 1.380 \times 10^{-23}$J/°K is Boltzmann's constant.

T is the temperature in degrees kelvin (K).

B is the bandwidth of the system in Hz.

f is the center frequency of the bandwidth in Hz.

R is the resistance, in Ω.

This result comes from quantum mechanical considerations, and is valid for any frequency, f. At microwave frequencies the above result can be simplified by making use of the fact that $hf << kT$. (As a worst-case example, let $f = 100$ GHz and $T = 100$ K. Then $hf = 6.5 \times 10^{-23} << kT = 1.4 \times 10^{-21}$.) Using the first two terms of a Taylor series expansion for the exponential gives

$$e^{hf/kt} - 1 \simeq \frac{hf}{kT},$$

so that (11.1) reduces to

$$v_n = \sqrt{4kTBR}. \qquad 11.2$$

This is the *Rayleigh-Jeans approximation*, and is the form most commonly used in microwave work [1]. For very high frequencies or very low temperatures, however, this approximation may be invalid, in which case (11.1) should be used. Note that this noise power is independent of frequency; such a noise source has a power spectral density that is constant with frequency, and is referred to as a *white noise source*. The noise power is directly proportional to the bandwidth, which in practice is usually limited by the passband of the microwave system. Since independent white noise sources can be treated as Gaussian distributed random variables, the noise powers (variances) are additive.

The noisy resistor of Figure 11.2 can be replaced with a Thévenin equivalent circuit consisting of a noiseless resistor and a generator with a voltage given by (11.2), as shown in Figure 11.3. Connecting a load resistor R results in maximum power transfer from

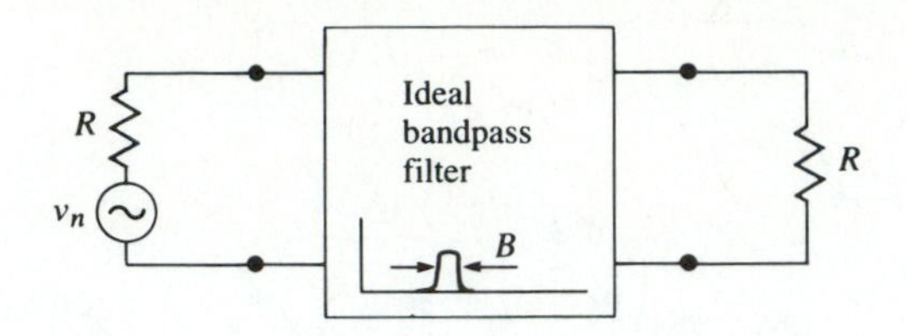

FIGURE 11.3 Equivalent circuit of a noisy resistor delivering maximum power to a load resistor through an ideal bandpass filter.

the noisy resistor, with the result that power delivered to the load in a bandwidth B, is

$$P_n = \left(\frac{v_n}{2R}\right)^2 R = \frac{v_n^2}{4R} = kTB, \tag{11.3}$$

since v_n is an rms voltage. This important result gives the maximum available noise power from the noisy resistor at temperature T. Observe the following trends:

- As $B \to 0, P_n \to 0$. This means that systems with smaller bandwidths collect less noise power.
- As $T \to 0, P_n \to 0$. This means that cooler devices and components generate less noise power.
- As $B \to \infty, P_n \to \infty$. This is the so-called *ultraviolet catastrophe*, which does not occur in reality because (11.2–11.3) are not valid as f (or B) $\to \infty$; (11.1) must be used in this case.

If an arbitrary source of noise (thermal or nonthermal) is "white," so that the noise power is not a strong function of frequency, it can be modelled as an equivalent thermal noise source, and characterized with an *equivalent noise temperature*. Thus, consider the arbitrary white noise source of Figure 11.4, which has a driving-point impedance of R and delivers a noise power P_s to a load resistor R. This noise source can be replaced by a noisy resistor of value R, at temperature T_e, where T_e is an equivalent temperature selected so that the same noise power is delivered to the load. That is,

$$T_e = \frac{P_s}{kB}. \tag{11.4}$$

Components and systems can then be characterized by saying that they have an equivalent noise temperature of T_e; this implies some fixed bandwidth, B, which is generally the bandwidth of the component or system.

For example, consider a noisy amplifier with a bandwidth B and gain G. Let the amplifier be matched to noiseless source and load resistors, as shown in Figure 11.5. If the source resistor is at a (hypothetical) temperature of $T_s = 0\,K$, then the input power to the amplifier will be $P_i = 0$, and the output noise power P_o will be due only to the noise generated by the amplifier itself. We can obtain the same load noise power by driving an ideal noiseless amplifier with a resistor at a temperature,

$$T_e = \frac{P_o}{GkB}, \tag{11.5}$$

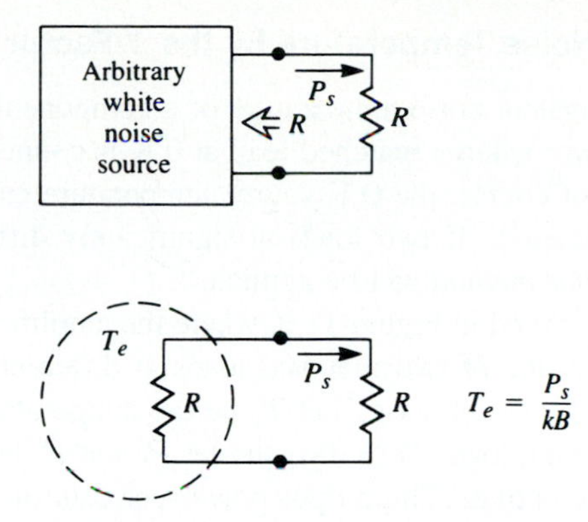

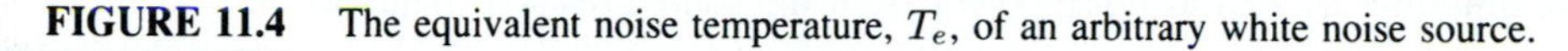

FIGURE 11.4 The equivalent noise temperature, T_e, of an arbitrary white noise source.

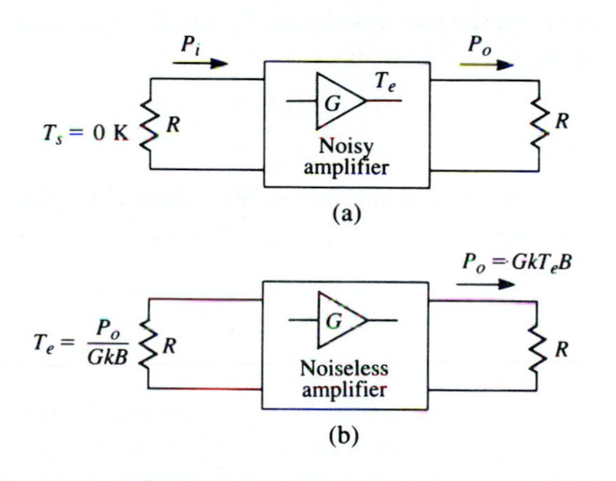

FIGURE 11.5 Defining the equivalent noise temperature of a noisy amplifier. (a) Noisy amplifier. (b) Noiseless amplifier.

so that the output power in both cases is $P_o = GkT_eB$. Then T_e is the equivalent noise temperature of the amplifier.

Active noise sources use a diode or tube to provide a calibrated noise power output, and are useful for test and measurement applications. Active noise generators can be characterized by an equivalent noise temperature, but a more common measure of noise power for such components is the *excess noise ratio* (ENR), defined as

$$\text{ENR(dB)} = 10 \log \frac{P_N - P_o}{P_o} = 10 \log \frac{T_N - T_0}{T_0}, \qquad 11.6$$

where P_N and T_N are the noise power and equivalent temperature of the generator, and P_o and T_0 are the noise power and temperature associated with a room-temperature passive source (a matched load). Solid state noise generators typically have ENRs ranging from 20 to 40 dB.

Measurement of Noise Temperature by the *Y*-Factor Method

In principle, the equivalent noise temperature of a component can be determined by measuring the output power when a matched load at 0 K is connected at the input of the component. In practice, of course, the 0 K source temperature cannot be achieved, so a different method must be used. If two loads at significantly different temperatures are available, then the *Y-factor method* can be applied.

This technique is illustrated in Figure 11.6, where the amplifier (or other component) under test is connected to one of two matched loads at different temperatures, and the output power is measured for each case. Let T_1 be the temperature of the hot load, and T_2 the temperature of the cold load ($T_1 > T_2$), and let P_1 and P_2 be the respective powers measured at the amplifier output. The output power consists of noise power generated by the amplifier as well as noise power from the source resistor. Thus we have

$$P_1 = GkT_1B + GkT_eB, \tag{11.7a}$$

$$P_2 = GkT_2B + GkT_eB, \tag{11.7b}$$

which are two equations for the two unknowns, T_e and GB (the gain-bandwidth product of the amplifier). Define the Y-factor as

$$Y = \frac{P_1}{P_2} = \frac{T_1 + T_e}{T_2 + T_e} > 1, \tag{11.8}$$

which is determined via the power measurements. Then (11.7) can be solved for the equivalent noise temperature,

$$T_e = \frac{T_1 - YT_2}{Y - 1}, \tag{11.9}$$

in terms of the load temperatures and the Y-factor.

Observe that to obtain accurate results from this method, the two source temperatures must not be too close together. If they are, P_1 will be close to P_2, Y will be close to unity, and the evaluation of (11.9) will involve the subtractions of numbers close to each other, resulting in a loss of accuracy. In practice, one noise source is usually a load resistor at room temperature (T_0), while the other noise source is either "hotter" or "colder," depending if T_e is greater or lesser than T_0. An active noise generator can be used as a "hotter" source, while a "colder" source can be obtained by immersing a load resistor in liquid nitrogen ($T = 77\,\mathrm{K}$), or liquid helium ($T = 4\,\mathrm{K}$).

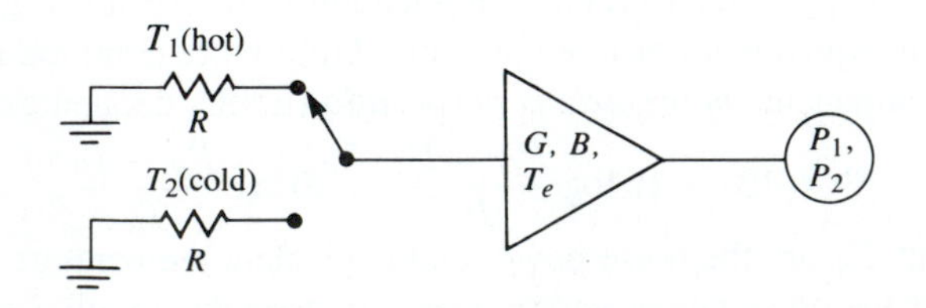

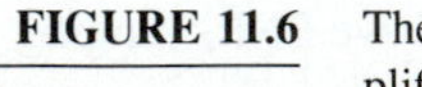

FIGURE 11.6 The Y-factor method for measuring the equivalent noise temperature of an amplifier.

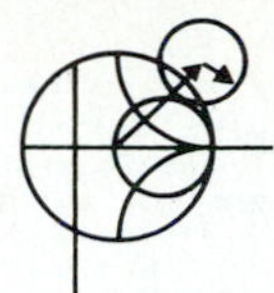

EXAMPLE 11.1

An X-band amplifier has a gain of 20 dB and a 1 GHz bandwidth. Its equivalent noise temperature is to be measured via the Y-factor method. The following data is obtained:

$$\text{for } T_1 = 290\,\text{K}, \quad P_1 = -62.0\text{ dBm}.$$
$$\text{for } T_2 = 77\,\text{K}, \quad P_2 = -64.7\text{ dBm}.$$

Determine the equivalent noise temperature of the amplifier. If the amplifier is used with a source having an equivalent noise temperature of $T_s = 450\,\text{K}$, what is the output noise power in dBm?

Solution
From (11.8), the Y-factor in dB is

$$Y = (P_1 - P_2)\text{ dB } = (-62.0) - (-64.7) = 2.7\text{ dB},$$

which is a numeric value of $Y = 1.86$. Then using (11.9) gives the equivalent noise temperature as

$$T_e = \frac{T_1 - YT_2}{Y - 1} = \frac{290 - (1.86)(77)}{1.86 - 1} = 170\text{ K}.$$

If a source with an equivalent noise temperature of $T_s = 450\,\text{K}$ drives the amplifier, the noise power into the amplifier will be kT_sB. The total noise power out of the amplifier will be

$$P_o = GkT_sB + GkT_eB = 100(1.38 \times 10^{-23})(10^9)(450 + 170)$$
$$= 8.56 \times 10^{-10}\,\text{W} = -60.7\,\text{dBm}.$$

○

Noise Figure

We have seen that a noisy microwave component can be characterized by an equivalent noise temperature. An alternative characterization is the *noise figure* of the component, which is a measure of the degradation in the signal-to-noise ratio between the input and output of the component. The signal-to-noise ratio is the ratio of desired signal power to undesired noise power, and so is dependent on the signal power. When noise and a desired signal are applied to the input of a noiseless network, both noise and signal will be attenuated or amplified by the same factor, so that the signal-to-noise ratio will be unchanged. But if the network is noisy, the output noise power will be increased more than the output signal power, so that the output signal-to-noise ratio will be reduced. The noise figure, F, is a measure of this reduction in signal-to-noise ratio, and is defined as

$$F = \frac{S_i/N_i}{S_o/N_o} \geq 1, \tag{11.10}$$

where S_i, N_i are the input signal and noise powers, and S_o, N_o are the output signal and noise powers. By definition, the input noise power is assumed to be the noise power resulting from a matched resistor at $T_0 = 290\,\text{K}$; that is, $N_i = kT_0B$.

Consider Figure 11.7, which shows noise power N_i and signal power S_i being fed into a noisy two-port network. The network is characterized by a gain G, a bandwidth B, and an equivalent noise temperature, T_e. The input noise power is $N_i = kT_0B$, and the output noise power is a sum of the amplified input noise and the internally generated noise: $N_o = kGB(T_0 + T_e)$. The output signal power is $S_o = GS_i$. Using these results in (11.10) gives the noise figure as

$$F = \frac{S_i}{kT_0B}\frac{kGB(T_0 + T_e)}{GS_i} = 1 + \frac{T_e}{T_0} \geq 1. \qquad 11.11$$

In dB, $F = 10\log(1 + T_e/T_0)$ dB ≥ 0. If the network were noiseless, T_e would be zero, giving $F = 1$, or 0 dB. Solving (11.11) for T_e gives

$$T_e = (F - 1)T_0. \qquad 11.12$$

It is important to keep in mind two things concerning the definition of noise figure: noise figure is defined for a matched input source, and for a noise source that consists of a resistor at temperature $T_0 = 290$ K. Noise figure and equivalent noise temperatures are interchangeable characterizations of the noise properties of a component.

An important special case occurs in practice when the two-port network is a passive, lossy component, such as an attenuator or lossy transmission line, held at a temperature, T. Consider such a network with a matched source resistor, which is also at temperature T, as shown in Figure 11.8. The gain, G, of a lossy network is less than unity; the loss factor, L, can be defined as $L = 1/G > 1$. Because the entire system is in thermal equilibrium at the temperature T, and has a driving point impedance of R, the output noise power must be $P_o = kTB$. But we can also think of this power as coming from the source resistor (through the lossy line), and from the noise generated by the line itself. Thus we also have that

$$P_o = kTB = GkTB + GN_{\text{added}}, \qquad 11.13$$

where N_{added} is the noise generated by the line, as if it appeared at the input terminals of the line. Solving (11.13) for this power gives

$$N_{\text{added}} = \frac{1 - G}{G}kTB = (L - 1)kTB. \qquad 11.14$$

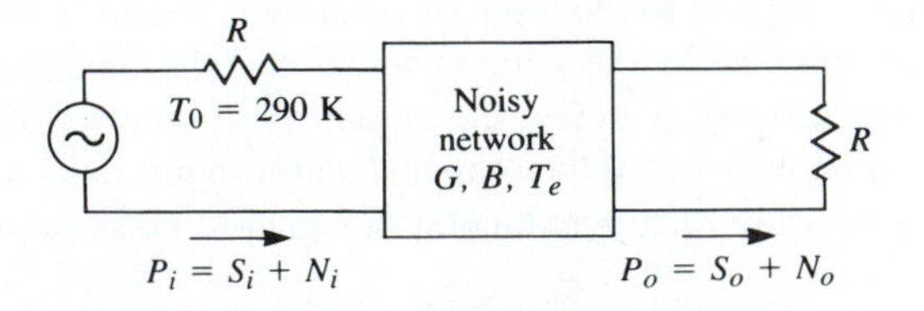

FIGURE 11.7 Determining the noise figure of a noisy network.

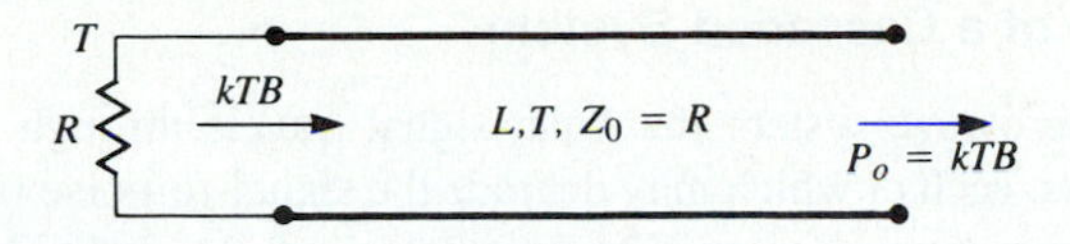

FIGURE 11.8 Determining the noise figure of a lossy line or attenuator with loss L and temperature T.

Then (11.4) shows that the lossy line has an equivalent noise temperature (as referred to the input) given by

$$T_e = \frac{1-G}{G}T = (L-1)T. \tag{11.15}$$

Then from (11.11) the noise figure is

$$F = 1 + (L-1)\frac{T}{T_0}. \tag{11.16}$$

If the line is at temperature T_0, then $F = L$. For instance, a 6 dB attenuator at room temperature has a noise figure of $F = 6$ dB.

EXAMPLE 11.2

A 10–12 GHz amplifier has a gain of 20 dB, a noise figure of 3.5 dB, and an output power of 10 dBm at its 1 dB compression point. What is the dynamic range of this amplifier?

Solution

The upper end of the dynamic range is set by the 1 dB compression point, which corresponds to an output power of 10 dBm. The lower end is set by the output noise power, N_o, due to the amplifier itself. From the definition of noise figure in (11.10), we have

$$F = \frac{S_i/N_i}{S_o/N_o} = \frac{S_i/kT_0B}{GS_i/N_o} = \frac{N_o}{GkT_0B}.$$

So the output noise power is

$$N_o = GFkT_0B.$$

In dBm,

$$N_o = 20 + 3.5 + 10\log\frac{(1.38\times10^{-23})(290)(2\times10^9)\text{ W}}{10^{-3}\text{ W}}$$

$$= -57.5 \text{ dBm}.$$

So the dynamic range is 10 dBm $-(-57.5$ dBm$) = 67.5$ dB. ○

Noise Figure of a Cascaded System

In a typical microwave system the input signal travels through a cascade of many different components, each of which may degrade the signal-to-noise ratio to some degree. If we know the noise figure (or noise temperature) of the individual stages, we can determine the noise figure (or noise temperature) of the cascade connection of stages. We will see that the noise performance of the first stage is usually the most critical, an interesting result that is very important in practice.

Consider the cascade of two components, having gains G_1, G_2, noise figures F_1, F_2, and noise temperature T_{e1}, T_{e2}, as shown in Figure 11.9. We wish to find the overall noise figure and noise temperature of the cascade, as if it were a single component. The overall gain of the cascade is $G_1 G_2$.

Using noise temperatures, the noise power at the output of the first stage is

$$N_1 = G_1 k T_0 B + G_1 k T_{e1} B, \tag{11.17}$$

since $N_i = kT_0B$ for noise figure calculations. The noise power at the output of the second stage is

$$\begin{aligned} N_o &= G_2 N_1 + G_2 k T_{e2} B \\ &= G_1 G_2 k B\left(T_0 + T_{e1} + \frac{1}{G_1} T_{e2}\right). \end{aligned} \tag{11.18}$$

For the equivalent system we have

$$N_o = G_1 G_2 k B(T_{cas} + T_0), \tag{11.19}$$

so comparison with (11.18) gives the noise temperature of the cascade system as

$$T_{cas} = T_{e1} + \frac{1}{G_1} T_{e2}. \tag{11.20}$$

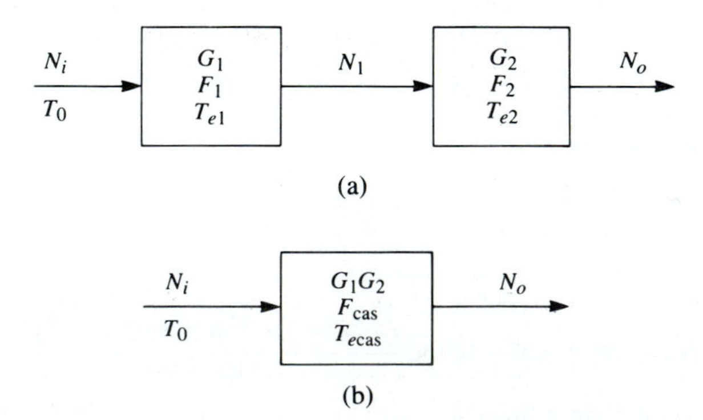

FIGURE 11.9 Noise figure and equivalent noise temperature of a cascaded system. (a) Two cascaded networks. (b) Equivalent network.

Using (11.12) to convert the temperatures in (11.20) to noise figures yields the noise figure of the cascade system as

$$F_{\text{cas}} = F_1 + \frac{1}{G_1}(F_2 - 1). \tag{11.21}$$

Equations (11.20) and (11.21) show that the noise characteristics of a cascaded system are dominated by the characteristics of the first stage, since the effect of the second stage is reduced by the gain of the first. Thus, for the best overall system noise performance, the first stage should have a low noise figure and at least moderate gain. Expense and effort should be devoted primarily to the first stage, as opposed to later stages, since later stages have a diminished impact on the overall noise performance.

Equations (11.20) and (11.21) can be generalized to an arbitrary number of stages, as follows:

$$T_{\text{cas}} = T_{e1} + \frac{T_{e2}}{G_1} + \frac{T_{e3}}{G_1 G_2} + \cdots, \tag{11.22}$$

$$F_{\text{cas}} = F_1 + \frac{F_2 - 1}{G_1} + \frac{F_3 - 1}{G_1 G_2} + \cdots. \tag{11.23}$$

EXAMPLE 11.3

An antenna is connected to a low-noise amplifier with a piece of coaxial transmission line. The amplifier has a gain of 15 dB, a bandwidth of 100 MHz, and a noise temperature of 150 K. The coaxial line has an attenuation of 2 dB. Find the noise figure of the transmission line-amplifier cascade. What would be the noise figure if the amplifier were placed at the antenna, eliminating the transmission line? Assume all components are at an ambient temperature of $T = 300$ K.

Solution

The loss factor of the coaxial line is $L = 10^{2/10} = 1.58$, so from (11.16) the noise figure of the line is

$$F_\ell = 1 + (L - 1)\frac{T}{T_0} = 1 + (1.58 - 1)\frac{300}{290} = 1.60 = 2.04 \text{ dB}.$$

From (11.11), the noise figure of the amplifier is

$$F_a = 1 + \frac{T_e}{T_0} = 1 + \frac{150}{290} = 1.52 = 1.81 \text{ dB}.$$

Then (11.21) gives the noise figure of the cascade as

$$F_{\text{cas}} = F_\ell + \frac{1}{G_\ell}(F_a - 1) = 1.60 + 1.58(1.52 - 1) = 2.42 = 3.84 \text{ dB},$$

since $1/G_\ell = L = 1.58$ for the coaxial line. Without the transmission line, the noise figure would be that of the amplifier itself, or 1.81 dB. So we see that the effect of the lossy feed line reduces the noise figure of the system by about 2 dB—a substantial amount. Sometimes such a line cannot be avoided in the front end of a receiver. Its effect, however, will be deleterious, since not only does the line itself add noise but, since its gain is less than unity, it increases the effect of the noise of the next stage. ○

11.2 DETECTORS AND MIXERS

Detectors and mixers use a nonlinear device to achieve *frequency conversion* of an input signal. Microwave diodes are most commonly used as the nonlinear element, but transistors can also be used. Figure 11.10 illustrates the three basic frequency conversion functions of rectification, detection, and mixing. We will first discuss the nonlinear voltage-current characteristics of a diode, and then use a small-signal analysis to describe the operation of various circuits that perform these functions.

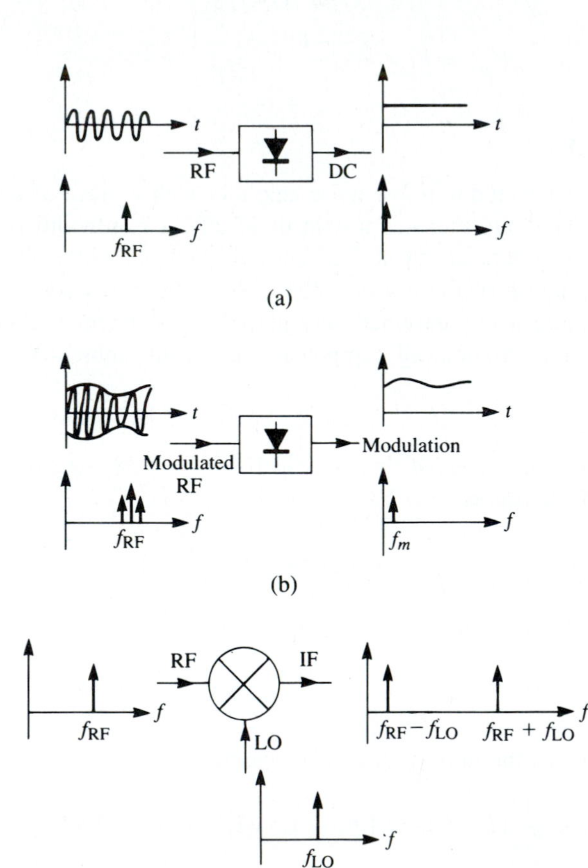

FIGURE 11.10 Basic operations of rectification, detection, and mixing. (a) Diode rectifier. (b) Diode detector. (c) Mixer.

Diode Rectifiers and Detectors

A diode is basically a nonlinear resistor, with a DC V-I characteristic that can be expressed as

$$I(V) = I_s(e^{\alpha V} - 1), \tag{11.24}$$

where $\alpha = q/nkT$, and q is the charge of an electron, k is Boltzmann's constant, T is temperature, n is the idealty factor, and I_s is the saturation current [2], [3]. Typically, I_s is between 10^{-6} and 10^{-15} A, and $\alpha = q/nkT$ is approximately 1/(25 mV) for $T = 290$ K. The idealty factor, n, depends on the structure of the diode itself, and can vary from 1.2 for Schottky barrier diodes to about 2.0 for point-contact silicon diodes. Figure 11.11 shows a typical diode V-I charactersitic. Now let the diode voltage be

$$V = V_0 + v, \tag{11.25}$$

where V_0 is a DC bias voltage and v is a small AC signal voltage. Then (11.24) can be expanded in a Taylor series about V_0 as follows:

$$I(V) = I_0 + v\left.\frac{dI}{dV}\right|_{V_0} + \frac{1}{2}v^2\left.\frac{d^2I}{dV^2}\right|_{V_0} + \cdots, \tag{11.26}$$

where $I_0 = I(V_0)$ is the DC bias current. The first derivative can be evaluated as

$$\left.\frac{dI}{dV}\right|_{V_0} = \alpha I_s e^{\alpha V_0} = \alpha(I_0 + I_s) = G_d = \frac{1}{R_j}, \tag{11.27}$$

which defines R_j, the junction resistance of the diode, and $G_d = 1/R_j$, which is called the dynamic conductance of the diode. The second derivative is

$$\left.\frac{d^2I}{dV^2}\right|_{V_0} = \left.\frac{dG_d}{dV}\right|_{V_0} = \alpha^2 I_s e^{\alpha V_0} = \alpha^2(I_0 + I_s) = \alpha G_d = G'_d. \tag{11.28}$$

Then (11.26) can be rewritten as the sum of the DC bias current, I_0, and an AC current, i:

$$I(V) = I_0 + i = I_0 + vG_d + \frac{v^2}{2}G'_d + \cdots. \tag{11.29}$$

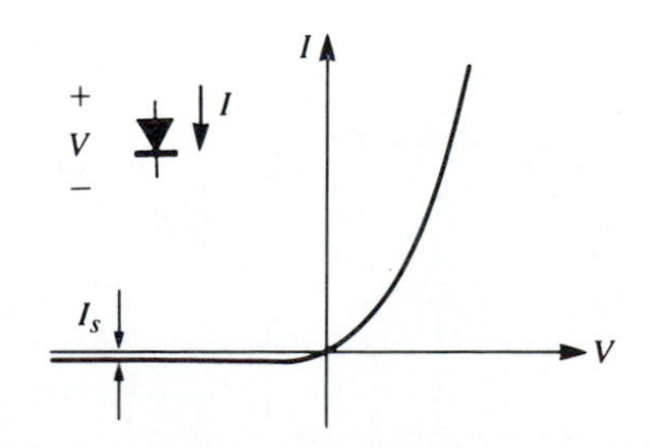

FIGURE 11.11 V-I characteristics of a diode.

The three-term approximation for the diode current in (11.29) is called the *small-signal approximation*, and will be adequate for most of our purposes.

The small-signal approximation is based on the DC voltage-current relationship of (11.24), and shows that the equivalent circuit of a diode will involve a nonlinear resistance. In practice, however, the AC characteristics of a diode also involve reactive effects due to the structure and packaging of the diode. A typical equivalent circuit for a diode is shown in Figure 11.12. The leads and contacts of the diode package lead to a series inductance, L_p, and shunt capacitance, C_p. The series resistor, R_s, accounts for contact and current-spreading resistance. C_j and R_j are the junction capacitance and resistance, and are bias-dependent.

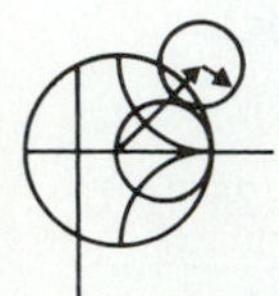

EXAMPLE 11.4

A diode in an axial-lead package has the following equivalent circuit parameters: $C_p = 0.10$ pF, $L_p = 2.0$ nH, $C_j = 0.15$ pF, $R_s = 10\,\Omega$, and $I_s = 0.1\,\mu$A. Calculate and plot the impedance of this diode from 4 to 14 GHz, for a bias current $I_0 = 0$ and $I_0 = 60\,\mu$A. Ignore the change in C_j with bias, and assume $\alpha = 1/25$ mV.

Solution

From (11.27) the junction resistance for the two bias states is

$$\text{for } I_0 = 0, \quad R_j = \frac{1}{\alpha(I_0 + I_s)} = \frac{25\,\text{mV}}{0.1\,\mu\text{A}} = 2.5 \times 10^5\,\Omega,$$

$$\text{for } I_0 = 60\,\mu\text{A}, \quad R_j = \frac{1}{\alpha(I_0 + I_s)} = \frac{25\,\text{mV}}{(60 + 0.1)\,\mu\text{A}} = 417\,\Omega.$$

Then the input impedance can be calculated from the equivalent circuit of Figure 11.12; the result is plotted versus frequency on a $50\,\Omega$ Smith chart in Figure 11.13. ○

In a rectifier application, a diode is used to convert a fraction of an RF input signal to DC power. Rectification is a very common function, and is used for power monitors, automatic gain control circuits, and signal strength indicators. If the diode voltage consists of a DC bias voltage and a small-signal RF voltage,

$$V = V_0 + v_0 \cos \omega_0 t, \tag{11.30}$$

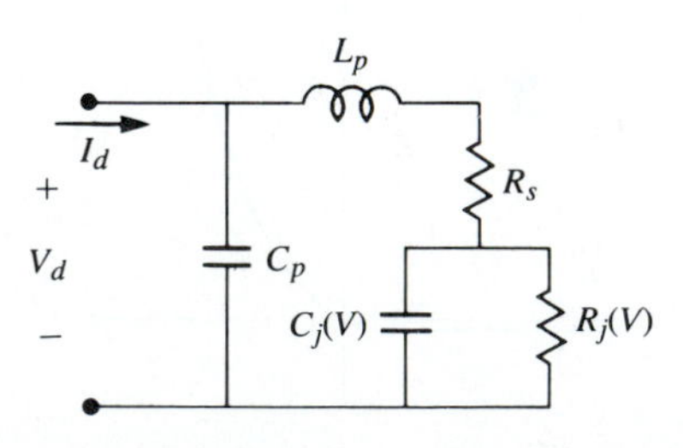

FIGURE 11.12 Equivalent AC circuit model for a diode.

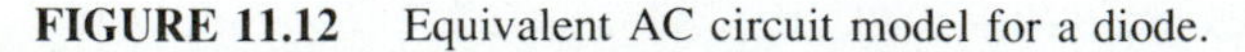

then (11.29) shows that the diode current will be

$$\begin{aligned} I &= I_0 + v_0 G_d \cos \omega_0 t + \frac{v_0^2}{2} G_d' \cos^2 \omega_0 t \\ &= I_0 + \frac{v_0^2}{4} G_d' + v_0 G_d \cos \omega_0 t + \frac{v_0^2}{4} G_d' \cos 2\omega_0 t. \end{aligned} \tag{11.31}$$

I_0 is the bias current and $v_0^2 G_d'/4$ is the DC rectified current. The output also contains AC signals of frequency ω_0, and $2\omega_0$ (and higher-order harmonics), which are usually filtered out with a simple low-pass filter. A current sensitivity, β_i, can be defined as a measure of the change in DC output current for a given input RF power. From (11.29) the RF input power is $v_0^2 G_d/2$ (using only the first term), while (11.31) shows the change in DC current is $v_0^2 G_d'/4$. The current sensitivity is then

$$\beta_i = \frac{\Delta I_{\rm dc}}{P_{\rm in}} = \frac{G_d'}{2G_d} \text{ A/W}. \tag{11.32}$$

An open-circuit voltage sensitivity, β_v, can be defined in terms of the voltage drop across the junction resistance when the diode is open-circuited. Thus,

$$\beta_v = \beta_i R_j. \tag{11.33}$$

Typical values for the voltage sensitivity of a diode range from 400 to 1500 mV/mW.

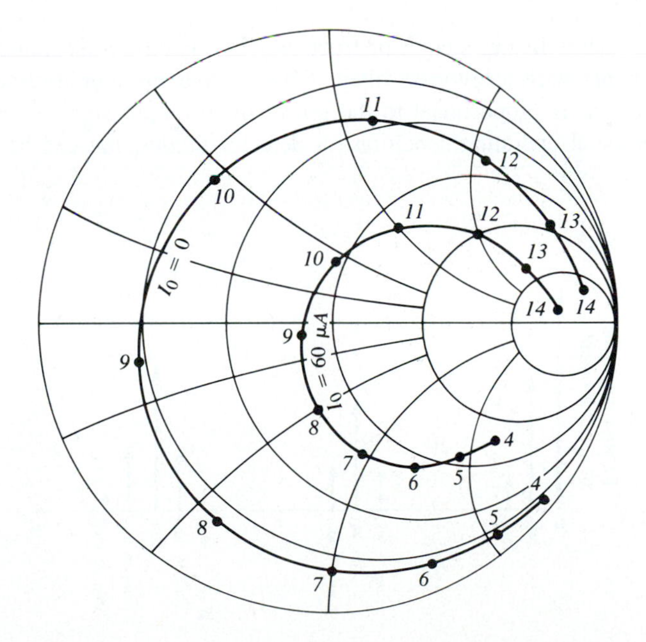

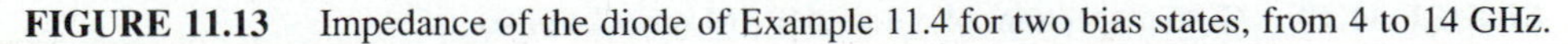

FIGURE 11.13 Impedance of the diode of Example 11.4 for two bias states, from 4 to 14 GHz.

In a detector application the nonlinearity of a diode is used to demodulate an amplitude modulated RF carrier. For this case, the diode voltage can be expressed as

$$v(t) = v_0(1 + m \cos \omega_m t) \cos \omega_0 t, \tag{11.34}$$

where ω_m is the modulation frequency, ω_0 is the RF carrier frequency ($\omega_0 >> \omega_m$), and m is defined as the modulation index ($0 \leq m \leq 1$). Using (11.34) in (11.29) gives the diode current:

$$\begin{aligned} i(t) &= v_0 G_d (1 + m \cos \omega_m t) \cos \omega_0 t + \frac{v_0^2}{2} G'_d (1 + m \cos \omega_m t)^2 \cos^2 \omega_0 t \\ &= v_0 G_d \left[\cos \omega_0 t + \frac{m}{2} \sin(\omega_0 + \omega_m)t + \frac{m}{2} \sin(\omega_0 - \omega_m)t \right] \\ &\quad + \frac{v_0^2}{4} G'_d \left[1 + \frac{m^2}{2} + 2m \cos \omega_m t + \frac{m^2}{2} \cos 2\omega_m t + \cos 2\omega_0 t \right. \\ &\quad + m \sin(2\omega_0 + \omega_m)t + m \sin(2\omega_0 - \omega_m)t + \frac{m^2}{2} \cos 2\omega_0 t \\ &\quad \left. + \frac{m^2}{4} \sin 2(\omega_0 + \omega_m)t + \frac{m^2}{4} \sin 2(\omega_0 - \omega_m)t \right]. \end{aligned} \tag{11.35}$$

The frequency spectrum of this output is shown in Figure 11.14. The output current terms which are linear in the diode voltage (terms multiplying $v_0 G_d$) have frequencies of ω_0 and $\omega_0 \pm \omega_m$, while the terms that are proportional to the square of the diode voltage (terms multiplying $v_0^2 G'_d/2$) include the frequencies and relative amplitudes listed in Table 11.1.

The desired demodulated output of frequency ω_m is easily separated from the undesired components with a low-pass filter. Observe that the amplitude of this current is $m v_0^2 G'_d/2$, which is proportional to the power of the input signal. This *square-law* behavior is the usual operating condition for detector diodes, but can be obtained only

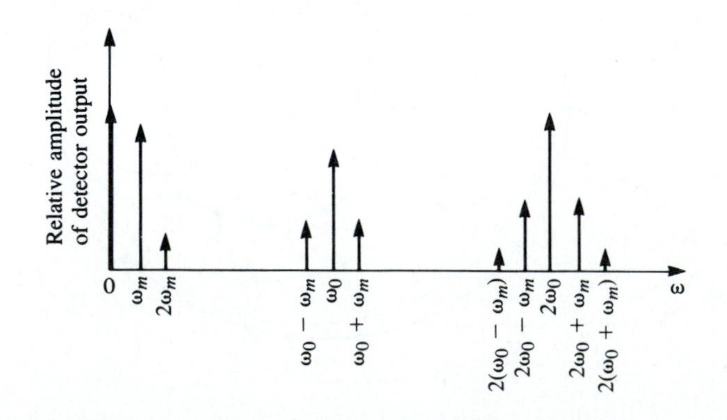

FIGURE 11.14 Output spectrum of a detected AM modulated signal.

TABLE 11.1 Frequencies and Relative Amplitudes of the Square-Law Output of a Detected AM Signal

Frequency	Relative Amplitude
0	$1 + m^2/2$
ω_m	$2m$
$2\omega_m$	$m^2/2$
$2\omega_0$	$1 + m^2/2$
$2\omega_0 \pm \omega_m$	m
$2(\omega_0 \pm \omega_m)$	$m^2/4$

over a restricted range of input powers. If the input power is too large, small-signal conditions will not apply, and the output will become saturated and approach a linear, and then a constant, i versus P characteristic. At very low signal levels the input signal will be lost in the noise floor of the device. Figure 11.15 shows the typical v_{out} versus P_{in} characteristic, where the output voltage can be considered as the voltage drop across a resistor in series with the diode. Square-law operation is particularly important for applications where power levels are inferred from detector voltage, as in SWR indicators and signal level indicators. Detectors may be DC biased to an operating point that provides the best sensitivity.

Single-Ended Mixer

A mixer uses the nonlinearity of a diode to generate an output spectrum consisting of the sum and difference frequencies of two input signals. In a receiver application, a low-level RF signal and an RF local oscillator (LO) signal are mixed together to produce

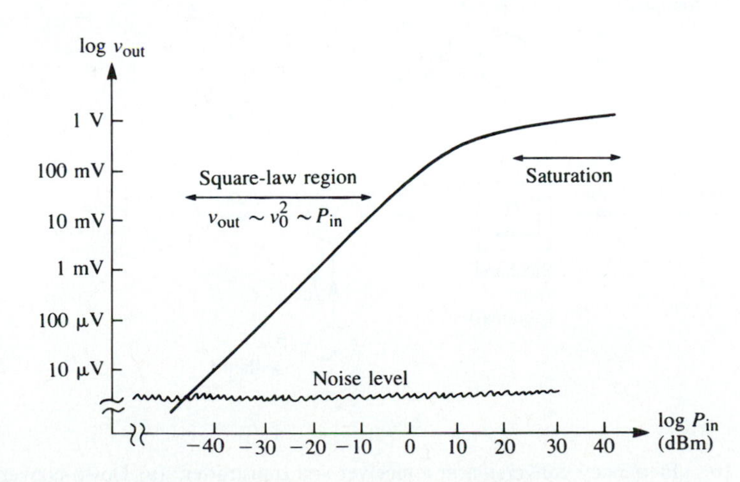

FIGURE 11.15 Square-law region for a typical diode detector.

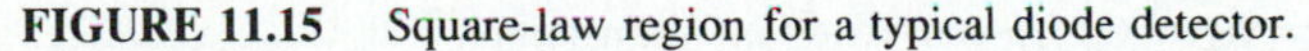

an intermediate frequency (IF), $f_{IF} = f_{RF} - f_{LO}$, and a much higher frequency, $f_{RF} + f_{LO}$, which is filtered out. See Figure 11.16a. The IF signal usually has a frequency between 10 and 100 MHz, and can be amplified with a low-noise amplifier. This is called a heterodyne receiver, and is useful because it has much better sensitivity and noise characteristics (using an IF amplifier minimizes $1/f$ noise) than the direct detection scheme discussed in the previous section. A heterodyne system also has the advantage of being able to tune over a band by simply changing the LO frequency, without the need for a high-gain, wideband RF amplifier.

As shown in Figure 11.16b, a mixer can also be used in a transmitter to offset the frequency of an RF signal by an amount equal to f_{IF}. This is a convenient technique, as it allows the use of identical local oscillators in the transmitter and receiver; a single oscillator may serve this purpose in a radar or transceiver system.

There are several types of mixer circuits, but the simplest is the *single-ended mixer*; single-ended mixers often are used as part of more sophisticated mixers. A typical single-ended mixer circuit is shown in Figure 11.17, where an RF signal,

$$v_{RF}(t) = v_r \cos \omega_r t, \tag{11.36}$$

is combined with an LO signal,

$$v_{LO}(t) = v_0 \cos \omega_0 t, \tag{11.37}$$

and fed into a diode. The combiner may be a simple T-junction combiner, or a directional coupler. An RF matching circuit may precede the diode, and the diode may be biased

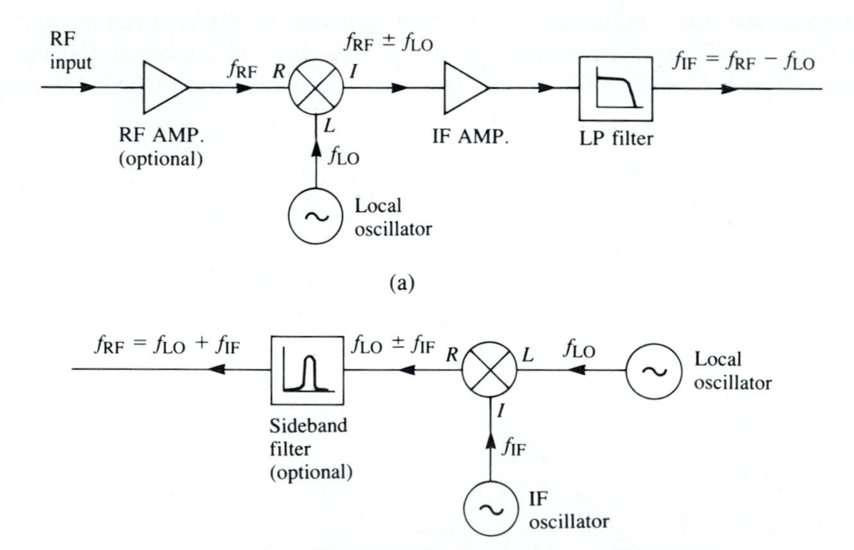

FIGURE 11.16 Frequency conversion in a receiver and transmitter. (a) Down-conversion in a heterodyne receiver. (b) Up-conversion in a transmitter.

through chokes that allow DC to pass while blocking RF. From (11.29), the diode current will consist of a constant DC bias term, and RF and LO signals of frequencies ω_r and ω_0, due to the term which is linear in v. The v^2 term will give rise to the following output current:

$$\begin{aligned} i &= \frac{G'_d}{2}(v_r \cos \omega_r t + v_0 \cos \omega_0 t)^2 \\ &= \frac{G'_d}{2}(v_r^2 \cos^2 \omega_r t + 2v_r v_0 \cos \omega_r t \cos \omega_0 t + v_0^2 \cos^2 \omega_0 t) \\ &= \frac{G'_d}{4}[v_r^2 + v_0^2 + v_r^2 \cos 2\omega_r t + v_0^2 \cos 2\omega_0 t \\ &\quad + 2v_r v_0 \sin(\omega_r - \omega_0)t + 2v_r v_0 \sin(\omega_r + \omega_0)t]. \end{aligned} \tag{11.38}$$

The DC terms can be ignored, and the $2\omega_r$ and $2\omega_0$ terms will be filtered out. The most important terms are those of frequency $\omega_r \pm \omega_0$.

For a receiver or down-converter, the $\omega_r - \omega_0$ term will become the IF signal. Note that, for a given local oscillator frequency, there will be two RF frequencies that will mix down to the same IF frequency. If the RF frequency is $\omega_r = \omega_0 + \omega_i$, then the output frequencies of the mixer will be $\omega_r \pm \omega_0 = 2\omega_0 + \omega_i$, and ω_i; if the RF frequency is $\omega_r = \omega_0 - \omega_i$, the mixer output frequencies will be $\omega_r \pm \omega_0 = 2\omega_0 - \omega_i$, and $-\omega_i$. This latter output is called the *image response* of the mixer, and is indistinguishable from the direct response. It can be eliminated by RF filtering at the input of the mixer, but this is difficult because the desired RF frequency $(\omega_0 + \omega_i)$ is relatively close to the spurious image frequency at $(\omega_0 - \omega_i)$, since generally $\omega_i << \omega_0$. Another way to eliminate the image response is by using an image rejection mixer.

In an up-converter, or modulation, application the two inputs will usually be a local oscillator and an IF oscillator, as in Figure 11.16b. The IF signal would be modulated with the desired information signal. Then the output will be $\omega_0 \pm \omega_i$, where ω_i is the IF frequency. The frequency $\omega_0 + \omega_i$ is called the *upper sideband* (USB), while $\omega_0 - \omega_i$ is called the *lower sideband* (LSB). Double sideband (DSB) modulation retains both sidebands, while single sideband (SSB) modulation removes one of the sidebands by filtering or by using an image rejection mixer (also called a single sideband modulator).

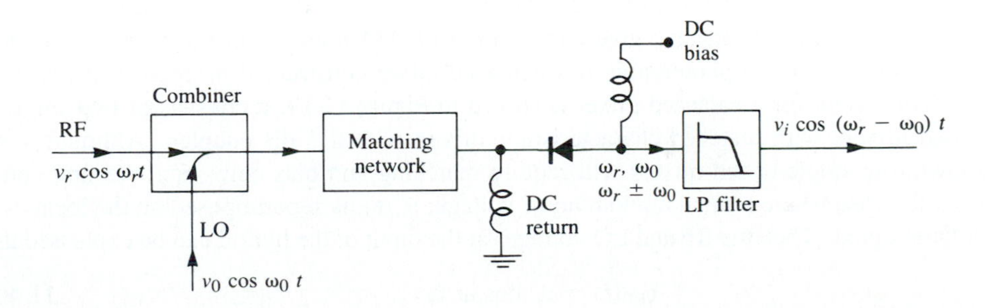

FIGURE 11.17 Single-ended mixer circuit.

Mixer design involves impedance matching the three ports, which is complicated by the fact that several frequencies and their harmonics are involved. Undesired harmonic power can be dissipated in resistive terminations, or blocked with reactive terminations. Resistive loads increase the loss of the mixer, and reactive loads are usually very frequency sensitive. An important figure of merit for a mixer is the *conversion loss*, defined as

$$L_C = 10 \log \frac{\text{available RF input power}}{\text{IF output power}} \text{ dB.} \tag{11.39}$$

Practical mixers usually have a conversion loss between 4 and 7 dB. One factor that strongly affects the conversion loss of a mixer is the local oscillator signal (or *pump*) power level; minimum conversion loss usually occurs for LO powers between 0 and 10 dBm. This power level is large enough to violate the small signal approximation of (11.29), so results using such a model may not be very accurate. Precise design requires numerical solution of the nonlinear equation that describes the diode characteristics [2].

Because a mixer is often the first or second component in a receiver system, its noise characteristics can be of critical importance. When specifying the noise figure of a mixer (or a receiver that uses a mixer), a distinction must be made as to whether the input is a single sideband signal or a double sideband signal. This is because the mixer will produce an IF output for two RF frequencies ($\omega_0 \pm \omega_i$), and therefore collect noise power at both frequencies. When used with a DSB input, the mixer will have desired signals at both RF frequencies, while a SSB input provides the desired signal only at one of these frequencies. Thus the DSB noise figure will be 3 dB lower than the SSB noise figure.

Besides conversion loss, there are several other characteristics that describe mixer performance. Impedance matching at the RF and LO inputs is important for good signal sensitivity and noise figure. In many applications it is desirable to have good isolation between the RF and LO ports so that, for example, LO power will not be radiated out the receive antenna. Other factors include the cancellation of AM noise from the LO, and supression of higher-order harmonics. The single-ended mixer performs reasonably well in terms of all these characteristics, but the mixer designs discussed below can be used to obtain substantially better performance for some specific characteristics.

Balanced Mixer

A *balanced mixer* combines two or more identical single-ended mixers with a 3 dB hybrid junction (90° or 180°) to give either better input SWR or better RF/LO isolation. The balanced mixer can also give cancellation of AM noise from the local oscillator. Figure 11.18 shows a photograph of a balanced mixer constructed in microstrip form.

The circuit for a balanced mixer is shown in Figure 11.19; it consists of two single-ended mixers with matched characteristics, driven with a 3 dB coupler. Although not shown, the single-ended mixers will require matching and bias networks. We first consider the case where a small random noise voltage, $v_n(t)$, is superimposed on the local oscillator signal. Then the RF and LO voltages at the input of the hybrid can be expressed as

$$v_{\text{RF}}(t) = v_r \cos \omega_r t, \tag{11.40}$$

$$v_{\text{LO}}(t) = [v_0 + v_n(t)] \cos \omega_0 t, \tag{11.41}$$

FIGURE 11.18 A balanced mixer constructed in microstrip form. The mixer uses a Lange coupler for the RF and LO inputs, followed by two diode chips. At the right of the circuit can be seen the bypass capacitor chips, DC bias lines, and IF output.
Courtesy of M/A-COM Inc.

where $v_r << v_0$, and $v_n(t) << v_0$. If we have a 90° hybrid, the voltages across the two diodes are

$$v_1(t) = v_r \cos(\omega_r t - 90°) + (v_0 + v_n)\cos(\omega_0 t - 180°)$$
$$= v_r \sin\omega_r t - (v_0 + v_n)\cos\omega_0 t, \qquad 11.42a$$
$$v_2(t) = v_r \cos(\omega_r t - 180°) + (v_0 + v_n)\cos(\omega_0 t - 90°)$$
$$= -v_r \cos\omega_r t + (v_0 + v_n)\sin\omega_0 t. \qquad 11.42b$$

The quadratic term of the diode V-I characteristic will give rise to the desired mixer products, so we will only consider this term and assume identical diodes so that diode currents can be represented as

$$i_1 = kv_1^2, \qquad 11.43a$$
$$i_2 = -kv_2^2, \qquad 11.43b$$

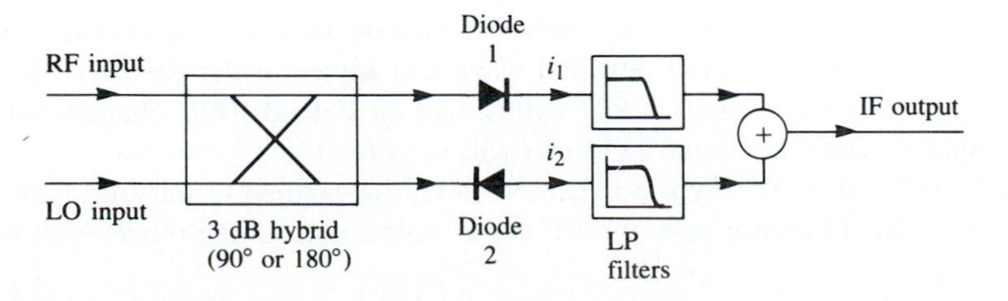

FIGURE 11.19 Balanced mixer circuit.

where the negative sign in (11.43b) accounts for the reversed polarity of the diodes. Using (11.42) in (11.43) gives the diode currents as

$$i_1 = k[v_r^2 \sin^2 \omega_r t + (v_0 + v_n)^2 \cos^2 \omega_0 t - 2v_r(v_0 + v_n) \sin \omega_r t \cos \omega_0 t],$$

$$i_2 = -k[v_r^2 \cos^2 \omega_r t + (v_0 + v_n)^2 \sin^2 \omega_0 t - 2v_r(v_0 + v_n) \cos \omega_r t \sin \omega_0 t].$$

After low-pass filtering, the remaining terms will be DC, noise, and IF frequency terms:

$$i_1 = k\left[\frac{1}{2}v_r^2 + \frac{1}{2}(v_0 + v_n)^2 - v_r(v_0 + v_n) \sin \omega_i t\right], \tag{11.44a}$$

$$i_2 = -k\left[\frac{1}{2}v_r^2 + \frac{1}{2}(v_0 + v_n)^2 + v_r(v_0 + v_n) \sin \omega_i t\right], \tag{11.44b}$$

where $\omega_i = \omega_r - \omega_0$ is the IF frequency. Combining these currents gives the IF output as

$$i_{\text{IF}} = i_1 + i_2 = -2kv_r(v_0 + v_n) \sin \omega_i t \simeq -2kv_r v_0 \sin \omega_i t, \tag{11.45}$$

since $v_n << v_0$. This result shows that the first-order terms in the noise voltage are cancelled by the mixer, while the desired IF signals combine in phase. Practical mixers can give from 15 to 30 dB of AM noise rejection.

Now consider reflection of the input RF and LO signals from the diodes. If we have a balanced mixer with a 90° hybrid, the input RF signal will give rise to the following reflected waves (phasors) from the diodes:

$$V_{\Gamma 1} = \Gamma V_1 = \frac{\Gamma V_r}{\sqrt{2}}, \tag{11.46a}$$

$$V_{\Gamma 2} = \Gamma V_2 = -j\frac{\Gamma V_r}{\sqrt{2}}, \tag{11.46b}$$

where Γ is the reflection coefficient of each diode, and V_r is the phasor RF input voltage. These two reflections will then arrive and combine back at the RF and LO input ports with the following amplitudes:

$$V_\Gamma^{\text{RF}} = \frac{V_{\Gamma 1}}{\sqrt{2}} - j\frac{V_{\Gamma 2}}{\sqrt{2}} = \frac{1}{2}\Gamma V_r - \frac{1}{2}\Gamma V_r = 0, \tag{11.47a}$$

$$V_\Gamma^{\text{LO}} = \frac{V_{\Gamma 2}}{\sqrt{2}} - j\frac{V_{\Gamma 1}}{\sqrt{2}} = -\frac{1}{2}j\Gamma V_r - \frac{1}{2}j\Gamma V_r = -j\Gamma V_r. \tag{11.47b}$$

Thus the RF input is matched, but the reflected wave appears at the LO port. Similarly, when the LO port is driven, the reflected wave will appear at the RF port. So the RF and LO inputs of a mixer using a 90° hybrid will have good SWR characteristics, but the isolation between the RF and LO ports will be poor.

Alternatively, if a 180° hybrid is used with the RF applied to the sum port and the LO applied to the difference port, the RF waves reflected from the diodes will be

$$V_{\Gamma 1} = V_{\Gamma 2} = \frac{\Gamma V_r}{\sqrt{2}}. \tag{11.48}$$

Then the reflections back at the sum and difference ports will be

$$V_\Gamma^\Sigma = \frac{V_{\Gamma 1}}{\sqrt{2}} + \frac{V_{\Gamma 2}}{\sqrt{2}} = \Gamma V_r \tag{11.49a}$$

$$V_\Gamma^\Delta = \frac{V_{\Gamma 1}}{\sqrt{2}} - \frac{V_{\Gamma 2}}{\sqrt{2}} = 0. \tag{11.49b}$$

The LO waves reflected from the diodes will be

$$V_{\Gamma 1} = -V_{\Gamma 2} = \frac{\Gamma V_r}{\sqrt{2}}, \tag{11.50}$$

and the reflections back at the sum and difference ports will be

$$V_\Gamma^\Sigma = \frac{V_{\Gamma 1}}{\sqrt{2}} + \frac{V_{\Gamma 2}}{\sqrt{2}} = 0, \tag{11.51a}$$

$$V_\Gamma^\Delta = \frac{V_{\Gamma 1}}{\sqrt{2}} - \frac{V_{\Gamma 2}}{\sqrt{2}} = \Gamma V_r. \tag{11.51b}$$

In both cases, the mismatch appears at the corresponding input port, while the RF and LO ports are isolated.

Other Types of Mixers

Like the balanced mixers described above, there are several other mixer circuits that can be used to enhance or reduce various modulation products and harmonics [2]. Some of these are briefly described below.

Antiparallel diode mixer. A circuit that is often used for subharmonically pumped mixers for millimeter wave applications uses a back-to-back pair of diodes, as shown in Figure 11.20. In operation, the local oscillator frequency is one-half of the usual LO frequency ($\omega_r - \omega_i$), and the diode nonlinearity generates a second harmonic of the LO frequency to mix with ω_r and produce the desired output frequency. Actually, most mixers can be used in this manner, but the antiparallel diode pair creates a symmetrical V-I characteristic that suppresses the fundamental mixing product of the RF and LO

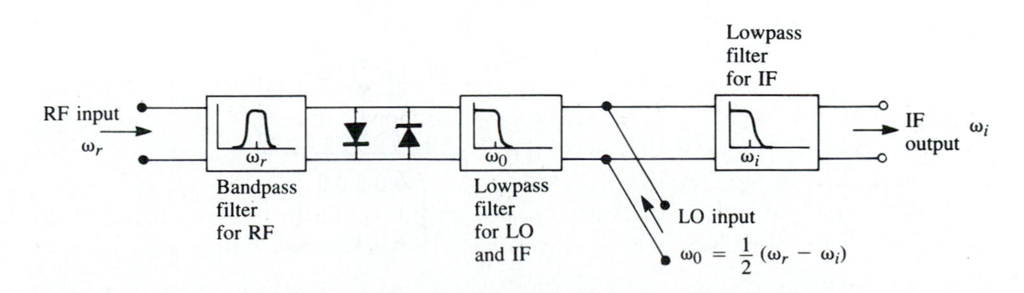

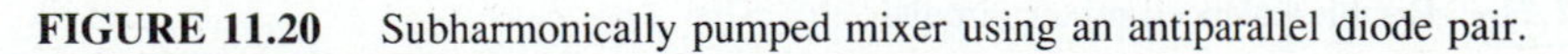
FIGURE 11.20 Subharmonically pumped mixer using an antiparallel diode pair.

signals and leads to a better conversion loss. This configuration also suppresses AM noise from the local oscillator.

Double-balanced mixer. The single-ended mixer has an output consisting of all harmonic combinations of the RF and LO signals. The balanced mixer using a 180° hybrid suppresses all even harmonics of the LO. The double-balanced mixer, shown in Figure 11.21, can suppress even harmonics of both the LO and RF signals. This leads to a very low conversion loss. It uses two 180° hybrids, so it has good RF/LO isolation, but poor input SWR. It uses four diodes in a ring configuration, although a "star" arrangement can also be used.

Image rejection mixer. We have already noted that two distinct RF signals, $\omega_r = \omega_0 \pm \omega_i$, can produce the same IF frequency, ω_i, when mixed with a local oscillator of frequency ω_0. These two signals can be thought of as the upper and lower sidebands of ω_0 modulated by ω_i, and are usually referred to as the real (desired) and image (undesired) mixer responses. The real response can be arbitrarily selected as either the USB or the LSB. The image rejection mixer of Figure 11.22 can be used to isolate these two responses into separate LSB ($\omega_r = \omega_0 - \omega_i$) and USB ($\omega_r = \omega_0 + \omega_i$) signals. When used as an up-converter or modulator, this mixer can produce a single-sideband output signal.

The operation of the image rejection mixer can be explained as follows. Let the input RF signal consist of both upper and lower sidebands:

$$v_r = v_U \cos(\omega_0 + \omega_i)t + v_L \cos(\omega_0 - \omega_i)t. \quad 11.52$$

Then the input to the two mixers is

$$v_r^A = \frac{v_U}{\sqrt{2}} \cos(\omega_0 + \omega_i)t + \frac{v_L}{\sqrt{2}} \cos(\omega_0 - \omega_i)t, \quad 11.53a$$

$$v_r^B = \frac{v_U}{\sqrt{2}} \cos[(\omega_0 + \omega_i)t - 90°] + \frac{v_L}{\sqrt{2}} \cos[(\omega_0 - \omega_i)t - 90°]. \quad 11.53b$$

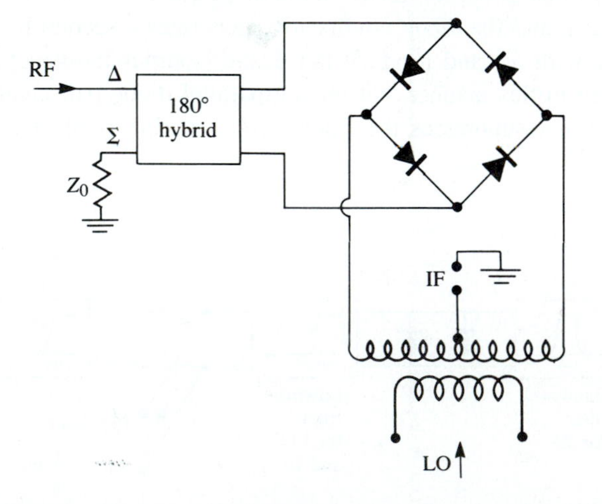

FIGURE 11.21 Double-balanced mixer circuit.

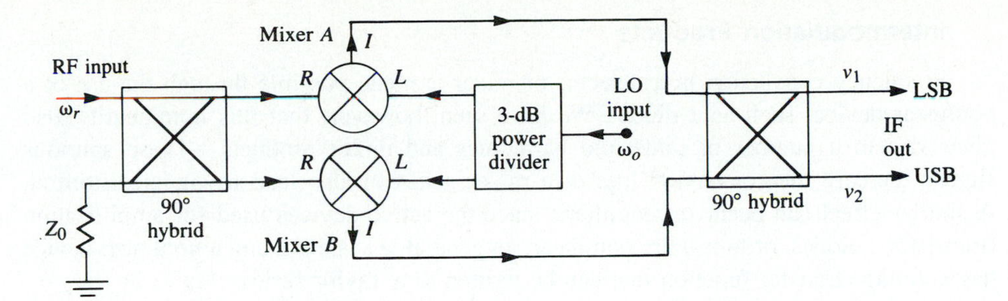

FIGURE 11.22 Image rejection mixer circuit.

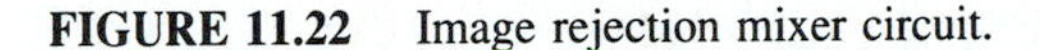

After mixing with an LO signal of $\cos \omega_0 t$, the IF outputs of the mixers are

$$v_i^A = kv_U \sin \omega_i t - kv_L \sin \omega_i t, \qquad 11.54a$$

$$v_i^B = kv_U \sin(\omega_i t - 90°) - kv_L \sin(\omega_i t + 90°). \qquad 11.54b$$

Combining these two signals in the 90° hybrid at the IF output gives the top output signal as

$$v_1 = \frac{k}{\sqrt{2}}[v_U \sin \omega_i t - v_L \sin \omega_i t + v_U \sin(\omega_i t - 180°) - v_L \sin \omega_i t]$$

$$= -\sqrt{2}kv_L \sin \omega_i t, \qquad 11.55a$$

which is the LSB component. The bottom IF output is

$$v_2 = \frac{k}{\sqrt{2}}[v_U \sin(\omega_i t - 90°) - v_L \sin(\omega_i t - 90°)$$

$$+ v_U \sin(\omega_i t - 90°) - v_L \sin(\omega_i t + 90°)]$$

$$= -\sqrt{2}kv_U \cos \omega_i t, \qquad 11.55b$$

which is the USB component. Image rejection or isolation ratios of 20 dB or more are typical. Table 11.2 summarizes some of the basic characteristics of the various mixers we have discussed.

TABLE 11.2 Basic Characteristics of Some Mixers

Mixer Type	Number of Diodes	RF SWR	RF/LO Isolation	L_C	Third-Order Intercept
Single ended	1	Poor	Fair	Good	13 dBm
Balanced (90°)	2	Good	Poor	Good	13 dBm
Balanced (180°)	2	Fair	Excellent	Good	13 dBm
Double balanced	4	Poor	Excellent	Excellent	18 dBm
Image rejection	8	Good	Good	Good	15 dBm

Intermodulation Products

Frequency conversion in a detector or mixer is made possible through the use of a nonlinear device, such as a diode. We have seen, however, that this nonlinearity also gives rise to a number of undesired harmonics and mixer products. These spurious signals increase the conversion loss of a mixer, and can also lead to signal distortion. A similar effect can occur in amplifiers, since the active devices used for amplification (transistors, diodes, or tubes) are nonlinear. In general, a system using a nonlinear device has a voltage transfer function that can be written as a Taylor series:

$$v_{out} = a_0 + a_1 v_{in} + a_2 v_{in}^2 + a_3 v_{in}^3 + \cdots . \qquad 11.56$$

For a detector or mixer, the a_0 term corresponds to the DC bias voltage, while the desired detected or mixed output is part of the v_{in}^2 term. For an amplifier, the linear v_{in} term provides the desired response. The operation of a subharmonically pumped mixer depends on the v_{in}^3 term. Thus, depending on the application, one of these terms provides the desired output, while the remaining terms produce undesired spurious signals.

If the input to the system consists of a single frequency (or *tone*), say $v_{in} = \cos \omega_1 t$, then the output voltage given by (11.56) will consist of all harmonics, $m\omega_1$, of the input signal. These harmonics are classified by their *order*, which is equal to m. Thus, for an amplifier, the first-order harmonic (fundamental) is the desired response, and the presence of higher-order harmonics is called harmonic distortion. If an amplifier had a bandwidth of an octave or more, the second-order distortion product of a low-frequency signal could be in the passband of the amplifier. In a mixer application, single-tone distortion products are generally eliminated by filtering.

More serious problems arise when the input to the system consists of two relatively closely spaced frequencies (*two–tone*), say $v_{in} = \cos \omega_1 t + \cos \omega_2 t$. Then the output spectrum will consist of all harmonics of the form $m\omega_1 + n\omega_2$, where m and n may be positive or negative integers; the order of a given product is then defined as $|m| + |n|$. The v_{in}^2 term of (11.56) will produce harmonics at the frequencies $2\omega_1$, $2\omega_2$, $\omega_1 - \omega_2$, and $\omega_1 + \omega_2$, which are all second order products. These frequencies are generally far away from the fundamentals ω_1 and ω_2, and so can easily be filtered. Such filtering may be impossible for a broadband amplifier, or receiver system, however. The $\omega_1 - \omega_2$ product is usually the desired result for a mixer. The v_{in}^3 term will lead to third-order products such as $3\omega_1$, $3\omega_2$, $2\omega_1 + \omega_2$, and $2\omega_2 + \omega_1$, which can be filtered, and to the products $2\omega_1 - \omega_2$ and $2\omega_2 - \omega_1$, which generally cannot be filtered, even in a narrow-band system. Such products that arise from mixing two input signals are called intermodulation distortion; the third-order two-tone intermodulation products $2\omega_1 - \omega_2$ and $2\omega_2 - \omega_1$ are especially important because they may set the dynamic range or bandwidth of the system. Higher-order terms in (11.56) may also contribute such harmonics, but usually the dominant contributions come from the lowest-order terms.

A measure of the second- or third-order intermodulation distortion is given by the intercept points, which are points on the graph of output power versus input power for the nonlinear component or system under consideration. Such an intercept diagram is shown in Figure 11.23. A plot of output signal ($\omega_1 \pm \omega_2$ for mixer, or ω_1 and ω_2 for an amplifier) power versus input power has a slope of unity for small signal levels. As the input power increases, saturation sets in, causing clipping of the output waveform and

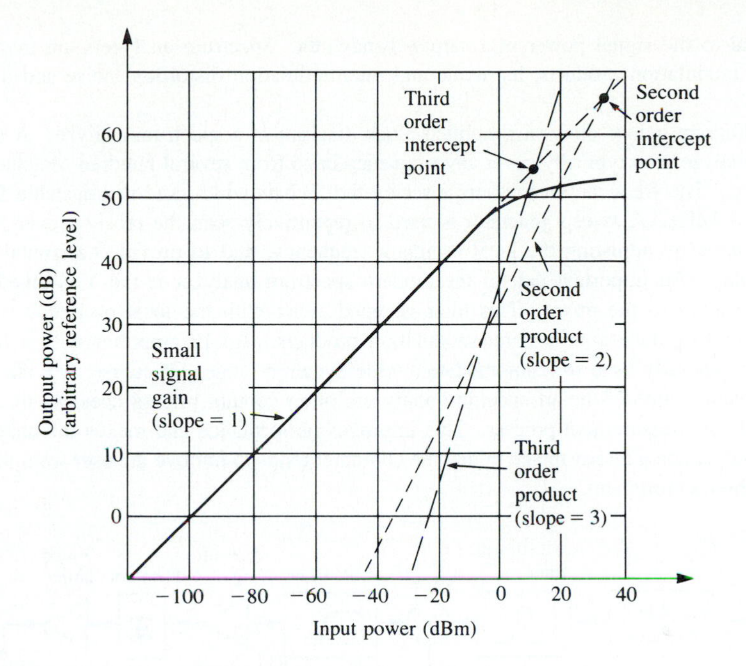

FIGURE 11.23 Intercept diagram for a nonlinear component.

signal distortion. This distortion manifests itself by diverting part of the input power to various harmonics. Equation (11.56) shows that the power in a second order product varies as v_{in}^2, so the curve of output power for this product will have a slope of two. If the linear part of the small signal gain curve is extended, it will intercept the second order product power curve at the second order intercept point. This point can be specified by either the input or the output power at the intersection, and is a measure of the amount of second order intermodulation distortion. The component would actually be operated well below this point. A similar intercept point is defined for third-order intermodulation distortion. Because this product is due to the v_{in}^3 term, its curve has a slope of three. Besides the fact that this type of distortion is more difficult to filter than other distortions, its intercept point usually occurs at a lower power level than the second order intercept point. If two components having individual intercept output powers of I_1 and I_2, and gains G_1 and G_2, are connected in cascade, the total intercept output power, I_T, can be shown to be given as

$$I_T = \left(\frac{1}{G_2 I_1} + \frac{1}{I_2} \right)^{-1}. \tag{11.57}$$

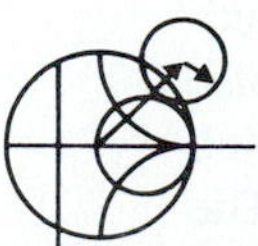

POINT OF INTEREST: The Spectrum Analyzer

A spectrum analyzer gives a frequency-domain representation of an input signal, displaying the average power density versus frequency. Thus, its function is dual to that of the oscilloscope, which displays a time-domain representation of an input signal. A spectrum analyzer is basically a sensitive receiver that tunes over a specified frequency band and gives a video output that is proportional to the signal power in a narrow bandwidth. Spectrum analyzers are invaluable for measuring modulation products, harmonic and intermodulation distortion, noise and interference effects.

The diagram below shows a simplified block diagram of a spectrum analyzer. A microwave spectrum analyzer can typically cover any frequency band from several hundred Megahertz to tens of Gigahertz. The frequency resolution is set by the IF bandwidth, and is adjustable from about 100 Hz to 1 MHz. A sweep generator is used to repetitively scan the receiver over the desired frequency band by adjusting the local oscillator frequency, and to provide horizontal deflection of the display. An important part of the modern spectrum analyzer is the YIG-tuned bandpass filter at the input to the mixer. This filter is tuned along with the local oscillator, and acts as a preselector to reduce spurious intermodulation products. An IF amplifier with a logarithmic response is generally used to accommodate a wide dynamic range. Of course, like many modern test instruments, state-of-the-art spectrum analyzers often contain microprocessors to control the system and the measurement process. This improves performance and makes the analyzer more versatile, but can be a disadvantage in that the computer tends to remove the user from the physical reality of the measurement.

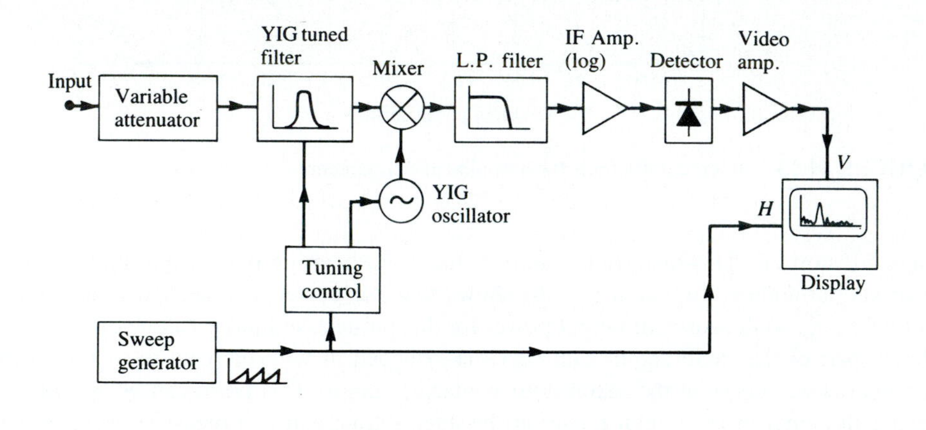

11.3 TRANSISTOR AMPLIFIER DESIGN

Amplification is one of the most basic and prevalent microwave circuit functions. In the past, microwave amplifiers generally used tubes, or diodes in a negative resistance reflection-type circuit, but today most microwave amplifiers use gallium arsenide field effect transistors (GaAs FETs). Bipolar transistors are limited to low microwave frequency amplifiers, but bipolar device performance is improving rapidly. Transistor amplifiers are rugged, low-cost, and very reliable, and can be used up to 60 GHz for both low-noise and medium-power applications. In this section we will give an introduction to the design of narrowband, low-power transistor amplifiers, including low-noise design; there are several good sources in the literature [4], [5], [6], [7], and [8], which can be consulted for design techniques for broadband and high-power amplifiers. The design method presented here is based on the S parameters of the amplifier circuit and the transistor. It can be used with bipolar transistors as well as FETs. A photograph of a laboratory prototype GaAs FET amplifier is shown in Figure 11.24.

FIGURE 11.24 A two-stage FET amplifier prototype, operating at 1.4 GHz. Photograph courtesy of Millitech Corporation, S. Deerfield, Mass.

Transducer Gain and Stability

A single-stage microwave transistor amplifier can be modelled by the circuit of Figure 11.25, where a matching network is used on both sides of the transistor. As discussed in Section 5.6, there are several definitions of power gain that can be applied to this type of circuit, but the most useful for the present purpose is the transducer power gain of (5.89),

$$\begin{aligned} G_T &= \frac{1-|\Gamma_S|^2}{|1-\Gamma_{\text{in}}\Gamma_S|^2}|S_{21}|^2\frac{1-|\Gamma_L|^2}{|1-S_{22}\Gamma_L|^2} \\ &= \frac{1-|\Gamma_S|^2}{|1-S_{11}\Gamma_S|^2}|S_{21}|^2\frac{1-|\Gamma_L|^2}{|1-\Gamma_{\text{out}}\Gamma_L|^2}, \end{aligned} \qquad 11.58$$

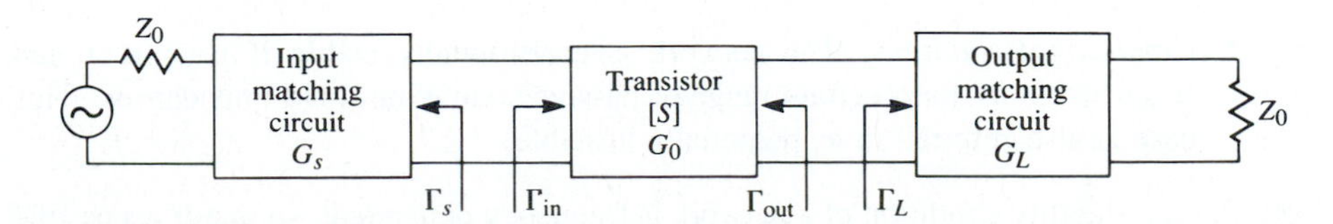

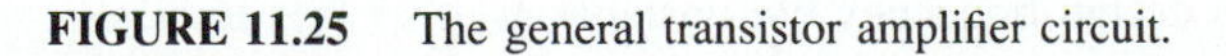

FIGURE 11.25 The general transistor amplifier circuit.

where the reflection coefficients on the input and output sides of the transistor are defined as

$$\Gamma_{\text{in}} = S_{11} + \frac{S_{12}S_{21}\Gamma_L}{1 - S_{22}\Gamma_L}, \tag{11.59a}$$

$$\Gamma_{\text{out}} = S_{22} + \frac{S_{12}S_{21}\Gamma_S}{1 - S_{11}\Gamma_S}. \tag{11.59b}$$

The transducer gain is the ratio of the power delivered to the load to the power available from the source, and so depends on the input match as well as the output match. Thus, we can define separate effective gain factors for the input (source) matching network, the transistor, and the output (load) matching network as follows:

$$G_S = \frac{1 - |\Gamma_S|^2}{|1 - \Gamma_{\text{in}}\Gamma_S|^2}, \tag{11.60a}$$

$$G_0 = |S_{21}|^2, \tag{11.60b}$$

$$G_L = \frac{1 - |\Gamma_L|^2}{|1 - S_{22}\Gamma_L|^2}. \tag{11.60c}$$

Then the overall transducer gain is $G_T = G_S G_0 G_S$. The effective gains from G_S and G_L are due to the impedance matching of the transistor.

If the transistor is *unilateral*, so that $S_{12} = 0$ or is small enough to be ignored, (11.59) reduces to $\Gamma_{\text{in}} = S_{11}$, $\Gamma_{\text{out}} = S_{22}$, and the unilateral transducer gain reduces to $G_{TU} = G_S G_0 G_L$, where

$$G_S = \frac{1 - |\Gamma_S|^2}{|1 - S_{11}\Gamma_S|^2}, \tag{11.61a}$$

$$G_0 = |S_{21}|^2, \tag{11.61b}$$

$$G_L = \frac{1 - |\Gamma_L|^2}{|1 - S_{22}\Gamma_L|^2}. \tag{11.61c}$$

Next we must consider the stability of the amplifier. In the circuit of Figure 11.25, oscillation is possible if either the input or output port impedance has a negative real part; this would then imply that $|\Gamma_{\text{in}}| > 1$ or $|\Gamma_{\text{out}}| > 1$. Because Γ_{in} and Γ_{out} depend on the source and load matching networks, the stability of the amplifier depends on Γ_S and Γ_L as presented by the matching networks. Thus, we define two types of stability:

1. *Unconditional stability*: The network is unconditionally stable if $|\Gamma_{\text{in}}| < 1$ and $|\Gamma_{\text{out}}| < 1$ for all passive source and load impedances (i.e., $|\Gamma_S| < 1$ and $|\Gamma_L| < 1$).
2. *Conditional stability*: The network is conditionally stable if $|\Gamma_{\text{in}}| < 1$ and $|\Gamma_{\text{out}}| < 1$ only for a certain range of passive source and load impedances. This case is also referred to as potentially unstable.

Note that the stability condition of a network is frequency dependent, so that it is possible for an amplifier to be stable at its design frequency but unstable at other frequencies.

Applying the above requirements for unconditional stability to (11.59) gives the following conditions that must be satisfied by Γ_S and Γ_L if the amplifier is to be unconditionally stable:

$$|\Gamma_{in}| = \left| S_{11} + \frac{S_{12}S_{21}\Gamma_L}{1 - S_{22}\Gamma_L} \right| < 1, \quad 11.62a$$

$$|\Gamma_{out}| = \left| S_{22} + \frac{S_{12}S_{21}\Gamma_S}{1 - S_{11}\Gamma_S} \right| < 1. \quad 11.62b$$

If the device is unilateral ($S_{12} = 0$), these conditions reduce to the simple results that $|S_{11}| < 1$ and $|S_{22}| < 1$ are sufficient for unconditional stability. Otherwise, the inequalities of (11.62) define a range of values for Γ_S and Γ_L where the amplifier will be stable. Finding this range for Γ_S and Γ_L can be facilitated by using the Smith chart, and plotting the input and output *stability circles*. The stability circles are defined as the loci in the Γ_L (or Γ_S) plane for which $|\Gamma_{in}| = 1$ (or $|\Gamma_{out}| = 1$). The stability circles then define the boundaries between stable and potentially unstable regions of Γ_S and Γ_L. Γ_S and Γ_L must lie on the Smith chart ($|\Gamma_S| < 1$, $|\Gamma_L| < 1$ for passive matching networks).

We can derive the equation for the output stability circle as follows (the procedure for the input stability circle is the same—just interchange S_{11} and S_{22}). First use (11.62a) to express the condition that $|\Gamma_{in}| = 1$ as

$$\left| S_{11} + \frac{S_{12}S_{21}\Gamma_L}{1 - S_{22}\Gamma_L} \right| = 1, \quad 11.63$$

or

$$|S_{11}(1 - S_{22}\Gamma_L) + S_{12}S_{21}\Gamma_L| = |1 - S_{22}\Gamma_L|.$$

Now define Δ (the determinant of the scattering matrix) as

$$\Delta = S_{11}S_{22} - S_{12}S_{21}. \quad 11.64$$

Then we can write the above result as

$$|S_{11} - \Delta\Gamma_L| = |1 - S_{22}\Gamma_L|. \quad 11.65$$

Now square both sides and simplify to obtain,

$$|S_{11}|^2 + |\Delta|^2|\Gamma_L|^2 - (\Delta\Gamma_L S_{11}^* + \Delta^*\Gamma_L^* S_{11}) = 1 + |S_{22}|^2|\Gamma_L|^2 - (S_{22}\Gamma_L^* + S_{22}\Gamma_L)$$

$$(|S_{22}|^2 - |\Delta|^2)\Gamma_L\Gamma_L^* - (S_{22} - \Delta S_{11}^*)\Gamma_L - (S_{22}^* - \Delta^* S_{11})\Gamma_L^* = |S_{11}|^2 - 1$$

$$\Gamma_L\Gamma_L^* - \frac{(S_{22} - \Delta S_{11}^*)\Gamma_L + (S_{22}^* - \Delta^* S_{11})\Gamma_L^*}{|S_{22}|^2 - |\Delta|^2} = \frac{|S_{11}|^2 - 1}{|S_{22}|^2 - |\Delta|^2}. \quad 11.66$$

Now complete the square by adding $|S_{22} - \Delta S_{11}^*|^2/(|S_{22}|^2 - |\Delta|^2)^2$ to both sides:

$$\left| \Gamma_L - \frac{(S_{22} - \Delta S_{11}^*)^*}{|S_{22}|^2 - |\Delta|^2} \right|^2 = \frac{|S_{11}|^2 - 1}{|S_{22}|^2 - |\Delta|^2} + \frac{|S_{22} - \Delta S_{11}^*|^2}{(|S_{22}|^2 - |\Delta|^2)^2},$$

or

$$\left| \Gamma_L - \frac{(S_{22} - \Delta S_{11}^*)^*}{|S_{22}|^2 - |\Delta|^2} \right| = \left| \frac{S_{12}S_{21}}{|S_{22}|^2 - |\Delta|^2} \right|. \quad 11.67$$

In the complex Γ plane, an equation of the form $|\Gamma - C| = R$ represents a circle with center at C (a complex number) and a radius R (a real number). Thus, (11.67) defines the output stability circle with a center C_L and radius R_L, where

$$C_L = \frac{(S_{22} - \Delta S_{11}^*)^*}{|S_{22}|^2 - |\Delta|^2} \quad \text{(center)}, \tag{11.68a}$$

$$R_L = \left| \frac{S_{12}S_{21}}{|S_{22}|^2 - |\Delta|^2} \right| \quad \text{(radius)}. \tag{11.68b}$$

Similar results can be obtained for the input stability circle by interchanging S_{11} and S_{22}:

$$C_S = \frac{(S_{11} - \Delta S_{22}^*)^*}{|S_{11}|^2 - |\Delta|^2} \quad \text{(center)}, \tag{11.69a}$$

$$R_S = \left| \frac{S_{12}S_{21}}{|S_{11}|^2 - |\Delta|^2} \right| \quad \text{(radius)}. \tag{11.69b}$$

Given the S parameters of the device, we can plot the input and output stability circles to define where $|\Gamma_{in}| = 1$ and $|\Gamma_{out}| = 1$. On one side of the input stability circle we will have $|\Gamma_{out}| < 1$, while on the other side we will have $|\Gamma_{out}| > 1$. Similarly, we will have $|\Gamma_{in}| < 1$ on one side of the output stability circle, and $|\Gamma_{in}| > 1$ on the other side. So we now need to determine which areas on the Smith chart represent the stable region, for which $|\Gamma_{in}| < 1$ and $|\Gamma_{out}| < 1$.

Consider the output stability circles plotted in the Γ_L lane for $|S_{11}| < 1$ and $|S_{11}| > 1$, as shown in Figure 11.26. If we set $Z_L = Z_0$, then $\Gamma_L = 0$ and (11.62a) shows that $|\Gamma_{in}| = |S_{11}|$. Now if $|S_{11}| < 1$, then $|\Gamma_{in}| < 1$, so $\Gamma_L = 0$ must be in a stable region. This means that the center of the Smith chart ($\Gamma_L = 0$) is in the stable region, so all of the Smith chart ($|\Gamma_L| < 1$) that is exterior to the stability circle defines the stable range

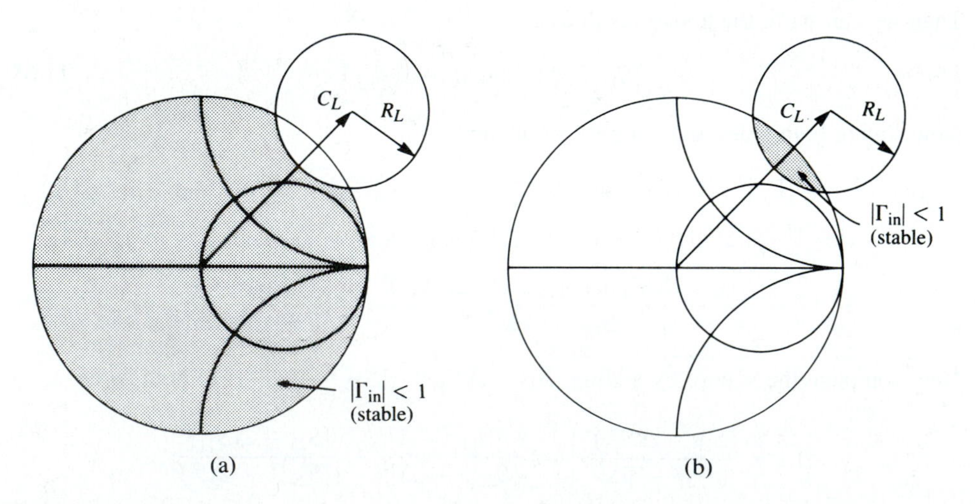

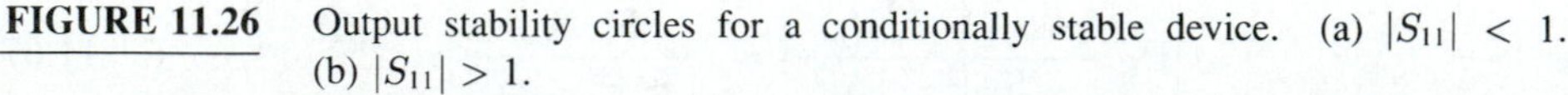
FIGURE 11.26 Output stability circles for a conditionally stable device. (a) $|S_{11}| < 1$. (b) $|S_{11}| > 1$.

for Γ_L. This region is shaded in Figure 11.26a. Alternatively, if we set $Z_L = Z_0$ but have $|S_{11}| > 1$, then $|\Gamma_{\text{in}}| > 1$ for $\Gamma_L = 0$ and the center of the Smith chart must be in an unstable region. In this case the stable region is the inside region of the stability circle that intersects the Smith chart, as illustrated in Figure 11.26b. Similar results apply to the input stability circle.

If the device is unconditionally stable, the stability circles must be completely outside (or totally enclose) the Smith chart. We can state this result mathematically as

$$\big||C_L| - R_L\big| > 1, \qquad \text{for } |S_{11}| < 1, \tag{11.70a}$$

$$\big||C_S| - R_S\big| > 1, \qquad \text{for } |S_{22}| < 1. \tag{11.70b}$$

If $|S_{11}| > 1$ or $|S_{22}| > 1$, the amplifier cannot be unconditionally stable because we can always have a source or load impedance of Z_0 leading to $\Gamma_S = 0$ or $\Gamma_L = 0$, thus causing $|\Gamma_{\text{in}}| > 1$ or $|\Gamma_{\text{out}}| > 1$.

Alternatively, it can be shown that the amplifier will be unconditionally stable if the following necessary and sufficient conditions are met:

$$K = \frac{1 - |S_{11}|^2 - |S_{22}|^2 + |\Delta|^2}{2|S_{12}S_{21}|} > 1, \tag{11.71}$$

and

$$|\Delta| < 1. \tag{11.72}$$

We can prove this result by starting with the inequalities of (11.62). If the amplifier is to be unconditionally stable, the first condition that must be met is, from (11.62a),

$$|\Gamma_{\text{in}}| = \left| S_{11} + \frac{S_{12}S_{21}\Gamma_L}{1 - S_{22}\Gamma_L} \right| < 1,$$

for all $|\Gamma_L| < 1$. This inequality can be rewritten as

$$\left| \frac{1}{S_{22}} \left(S_{22}S_{11} + \frac{S_{12}S_{21}S_{22}\Gamma_L}{1 - S_{22}\Gamma_L} \right) \right| < 1,$$

or

$$\left| \frac{1}{S_{22}} \left(\Delta + \frac{S_{12}S_{21}}{1 - S_{22}\Gamma_L} \right) \right| < 1. \tag{11.73}$$

Now let $\Gamma_L = |\Gamma_L|e^{j\theta}$; allowable values of Γ_L for a passive load and matching network must lie within the unit circle defined by $|\Gamma_L| = 1$. The factor $1/(1 - S_{22}\Gamma_L)$ maps this circle into a new circle with a center at $1/(1 - |S_{22}|^2)$ and a radius of $|S_{22}|/(1 - |S_{22}|^2)$. This can be seen by setting $|\Gamma_L| = 1$ and writing this factor as

$$\frac{1}{1 - S_{22}e^{j\theta}} = \frac{1}{1 - |S_{22}|e^{j\phi}},$$

and noting that the maximum and minimum excursions of this factor occur for $\phi = 0$ and π, respectively, and are given as $1/(1-|S_{22}|)$ and $1/(1+|S_{22}|)$, respectively. The center of the circle is then the average of these values, $1/2\left[1/(1 - |S_{22}|) + 1/(1 + |S_{22}|)\right] = 1/(1 - |S_{22}|^2)$ while the radius is given by half the difference: $1/2\left[1/(1 - |S_{22}|) - 1/(1 + |S_{22}|)\right] = |S_{22}|/(1 - |S_{22}|^2)$. Thus we have that $1/(1 - S_{22}e^{j\theta}) = 1/(1 - |S_{22}|^2) + (|S_{22}|e^{j\psi})/\,(1 - |S_{22}|^2)$.

Using this result allows us to rewrite (11.73) as

$$\frac{1}{|S_{22}|}\left|\left(\Delta + \frac{S_{12}S_{21}}{1-|S_{22}|^2}\right) + \frac{S_{12}S_{21}|S_{22}|e^{j\psi}}{1-|S_{22}|^2}\right| < 1, \qquad 11.74$$

which must be true for all values of the angle, ψ. The left-hand side is maximum for that value of ψ that makes the phase angles of the two terms identical, so (11.74) can be simplified to

$$\frac{1}{|S_{22}|}\left|\Delta + \frac{S_{12}S_{21}}{1-|S_{22}|^2}\right| + \frac{|S_{12}S_{21}|}{1-|S_{22}|^2} < 1,$$

or

$$0 \le \frac{1}{|S_{22}|}\left|\Delta + \frac{S_{12}S_{21}}{1-|S_{22}|^2}\right| < 1 - \frac{|S_{12}S_{21}|}{1-|S_{22}|^2}. \qquad 11.75$$

Squaring both sides of this inequality and simplifying gives

$$|S_{12}S_{21}|^2 + |\Delta|^2(1-|S_{22}|^2) + \Delta S_{12}^*S_{21}^* + \Delta^* S_{12}S_{21} < |S_{22}|^2(1 - 2|S_{12}S_{21}| - |S_{22}|^2).$$

Using the result that $\Delta S_{12}^*S_{21}^* + \Delta^* S_{12}S_{21} = |S_{11}|^2|S_{22}|^2 - |S_{12}S_{21}|^2 - |\Delta|^2$ then gives

$$2|S_{12}S_{21}| < 1 - |S_{11}|^2 - |S_{22}|^2 + |\Delta|^2, \qquad 11.76$$

which is identical to the condition stated in (11.71). Since K remains unchanged after an interchange of S_{11} and S_{22}, we can conclude that (11.71) applies to $|\Gamma_{\text{out}}| < 1$, as well as $|\Gamma_{\text{in}}| < 1$.

Next, the right-hand side of (11.75) indicates that

$$0 < 1 - |S_{22}|^2 - |S_{12}S_{21}|, \qquad 11.77a$$

and the corresponding result obtained from (11.62b) is

$$0 < 1 - |S_{11}|^2 - |S_{12}S_{21}|. \qquad 11.77b$$

Adding these two inequalities gives

$$2|S_{12}S_{21}| < 2 - |S_{11}|^2 - |S_{22}|^2.$$

From the triangle inequality we know that

$$|\Delta| = |S_{11}S_{22} - S_{12}S_{21}| \le |S_{11}S_{22}| + |S_{12}S_{21}|,$$

so the above result can be reduced to

$$2(|\Delta| - |S_{11}S_{22}|) < 2 - |S_{11}|^2 - |S_{22}|^2,$$

$$|\Delta| < 1 - \frac{1}{2}(|S_{11}|^2 - 2|S_{11}S_{22}| + |S_{22}|^2),$$

$$|\Delta| < 1 - \frac{1}{2}(|S_{11}| - |S_{22}|)^2 < 1, \qquad 11.78$$

which is identical to (11.72). Thus, we have shown that a two-port network will be unconditionally stable if and only if $K > 1$ and $|\Delta| < 1$.

If the device is only conditionally stable, operating points for Γ_S and Γ_L must be chosen in the stable region, and it is good practice to check the stability at several frequencies near the design frequency. If it is possible to accept a design with less than maximum gain, the transistor can usually be made to be unconditionally stable by using resistive loading [4].

EXAMPLE 11.5

The S parameters for the HP HFET-102 GaAs FET at 2 GHz with a bias voltage $V_{gs} = 0$ are given as follows ($Z_0 = 50\,\Omega$):

$$S_{11} = 0.894\angle{-60.6^\circ},$$
$$S_{21} = 3.122\angle{123.6^\circ},$$
$$S_{12} = 0.020\angle{62.4^\circ},$$
$$S_{22} = 0.781\angle{-27.6^\circ}.$$

Determine the stability of this transistor by calculating K and $|\Delta|$, and plot the stability circles.

Solution

From (11.71) and (11.72) we compute K and $|\Delta|$ as

$$\Delta = S_{11}S_{22} - S_{12}S_{21} = 0.696\angle{-83^\circ},$$

$$K = \frac{1 + |\Delta|^2 - |S_{11}|^2 - |S_{22}|^2}{2|S_{12}S_{21}|} = 0.607.$$

We have $|\Delta| = 0.696 < 1$, but $K < 1$, so the device is potentially unstable. The centers and radii of the stability circles are given by (11.68) and (11.69):

$$C_L = \frac{(S_{22} - \Delta S_{11}^*)^*}{|S_{22}|^2 - |\Delta|^2} = 1.361\angle{47^\circ},$$

$$R_L = \frac{|S_{12}S_{21}|}{|S_{22}|^2 - |\Delta|^2} = 0.50,$$

$$C_S = \frac{(S_{11} - \Delta S_{22}^*)^*}{|S_{11}|^2 - |\Delta|^2} = 1.132\angle{68^\circ},$$

$$R_S = \frac{|S_{12}S_{21}|}{|S_{11}|^2 - |\Delta|^2} = 0.199.$$

This data can be used to plot the input and output stability circles, as shown in Figure 11.27. Since $|S_{11}| < 1$ and $|S_{22}| < 1$, the central part of the Smith chart represents the stable operating region for Γ_S and Γ_L. The unstable regions are darkened. ○

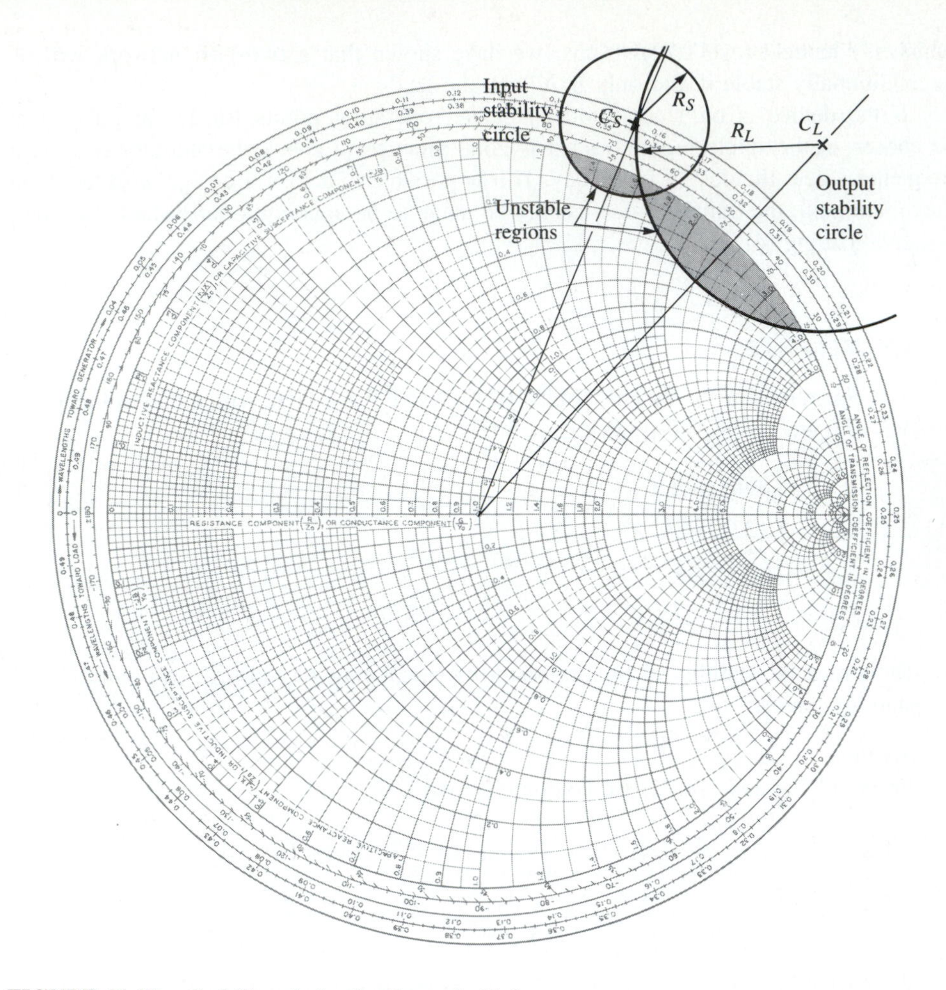

FIGURE 11.27 Stability circles for Example 11.5.

Design for Maximum Gain (Conjugate Matching)

After the stability of the transistor has been determined, and the stable regions for Γ_S and Γ_L have been located on the Smith chart, the input and output matching sections can be designed. Since G_0 of (11.61b) is fixed for a given transistor, the overall gain of the amplifier will be controlled by the gains, G_S and G_L, of the matching sections. Maximum gain will be realized when these sections provide a conjugate match between the amplifier source or load impedance and the transistor. Because most transistors appear as a significant impedance mismatch (large $|S_{11}|$ and $|S_{22}|$), the resulting frequency response will be narrowband. In the next section we will discuss how to design for less than maximum gain, with a corresponding improvement in bandwidth.

With reference to Figure 11.25 and our discussion in Section 3.6 on conjugate impedance matching, we know that maximum power transfer from the input matching network to the transistor will occur when

$$\Gamma_{\text{in}} = \Gamma_S^*, \tag{11.79a}$$

and the maximum power transfer from the transistor to the output matching network will occur when

$$\Gamma_{\text{out}} = \Gamma_L^*. \tag{11.79b}$$

Then, assuming lossless matching sections, these conditions will maximize the overall transducer gain. From (11.58), this maximum gain will be given by

$$G_{T_{\max}} = \frac{1}{1 - |\Gamma_S|^2}|S_{21}|^2 \frac{1 - |\Gamma_L|^2}{|1 - S_{22}\Gamma_L|^2}. \tag{11.80}$$

In the general case with a bilateral transistor, Γ_{in} is affected by Γ_{out}, and vice versa, so that the input and output sections must be matched simultaneously. Using (11.79) in (11.59) gives the necessary equations:

$$\Gamma_S^* = S_{11} + \frac{S_{12}S_{21}\Gamma_L}{1 - S_{22}\Gamma_L}, \tag{11.81a}$$

$$\Gamma_L^* = S_{22} + \frac{S_{12}S_{21}\Gamma_S}{1 - S_{11}\Gamma_S}. \tag{11.81b}$$

We can solve for Γ_S by first rewriting these equations as follows:

$$\Gamma_S = S_{11}^* + \frac{S_{12}^* S_{21}^*}{1/\Gamma_L^* - S_{22}^*},$$

$$\Gamma_L^* = \frac{S_{22} - \Delta\Gamma_S}{1 - S_{11}\Gamma_S},$$

where $\Delta = S_{11}S_{22} - S_{12}S_{21}$. Substituting this expression for Γ_L^* into the expression for Γ_S and expanding gives

$$\Gamma_S(1 - |S_{22}|^2) + \Gamma_S^2(\Delta S_{22}^* - S_{11}) = \Gamma_S(\Delta S_{11}^* S_{22}^* - |S_{11}|^2 - \Delta S_{12}^* S_{21}^*) + S_{11}^*(1 - |S_{22}|^2) + S_{12}^* S_{21}^* S_{22}.$$

Using the result that $\Delta(S_{11}^* S_{22}^* - S_{12}^* S_{21}^*) = |\Delta|^2$ allows this to be rewritten as a quadratic equation for Γ_S:

$$(S_{11} - \Delta S_{22}^*)\Gamma_S^2 + (|\Delta|^2 - |S_{11}|^2 + |S_{22}|^2 - 1)\Gamma_S + (S_{11}^* - \Delta^* S_{22}) = 0. \tag{11.82}$$

The solution is,

$$\Gamma_S = \frac{B_1 \pm \sqrt{B_1^2 - 4|C_1|^2}}{2C_1}. \tag{11.83a}$$

Similarly, the solution for Γ_L can be written as

$$\Gamma_L = \frac{B_2 \pm \sqrt{B_2^2 - 4|C_2|^2}}{2C_2}. \tag{11.83b}$$

The variables B_1, C_1, B_2, C_2 are defined as

$$B_1 = 1 + |S_{11}|^2 - |S_{22}|^2 - |\Delta|^2, \tag{11.84a}$$

$$B_2 = 1 + |S_{22}|^2 - |S_{11}|^2 - |\Delta|^2, \tag{11.84b}$$

$$C_1 = S_{11} - \Delta S_{22}^*, \tag{11.84c}$$

$$C_2 = S_{22} - \Delta S_{11}^*. \tag{11.84d}$$

The results are much simpler for the unilateral case. When $S_{12} = 0$, (11.81) shows that $\Gamma_S = S_{11}^*$ and $\Gamma_L = S_{22}^*$, and then maximum transducer gain of (11.80) reduces to

$$G_{TU_{\max}} = \frac{1}{1 - |S_{11}|^2} |S_{21}|^2 \frac{1}{1 - |S_{22}|^2}. \tag{11.85}$$

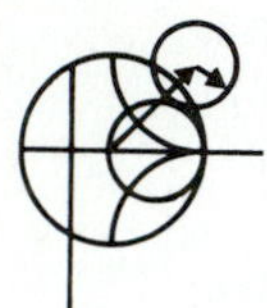

EXAMPLE 11.6

Design an amplifier for maximum gain at 4.0 GHz using single-stub matching sections. Calculate and plot the input return loss and the gain from 3 to 5 GHz. The GaAs FET has the following S parameters ($Z_0 = 50\,\Omega$):

f (GHz)	S_{11}	S_{21}	S_{12}	S_{22}
3.0	$0.80\angle -89^\circ$	$2.86\angle 99^\circ$	$0.03\angle 56^\circ$	$0.76\angle -41^\circ$
4.0	$0.72\angle -116^\circ$	$2.60\angle 76^\circ$	$0.03\angle 57^\circ$	$0.73\angle -54^\circ$
5.0	$0.66\angle -142^\circ$	$2.39\angle 54^\circ$	$0.03\angle 62^\circ$	$0.72\angle -68^\circ$

Solution

We first check the stability of the transistor by calculating Δ and K at 4.0 GHz:

$$\Delta = S_{11}S_{22} - S_{12}S_{21} = 0.488\angle -162^\circ,$$

$$K = \frac{1 - |S_{11}|^2 - |S_{22}|^2 + |\Delta|^2}{2|S_{12}S_{21}|} = 1.195.$$

Since $|\Delta| < 1$ and $K > 1$, the transistor is unconditionally stable at 4.0 GHz. There is no need to plot the stability circles.

For maximum gain, we should design the matching sections for a conjugate match to the transistor. Thus, $\Gamma_S = \Gamma_{\text{in}}^*$ and $\Gamma_L = \Gamma_{\text{out}}^*$, and Γ_S, Γ_L can be determined from (11.83):

$$\Gamma_S = \frac{B_1 \pm \sqrt{B_1^2 - 4|C_1|^2}}{2C_1} = 0.872\angle 123^\circ$$

$$\Gamma_L = \frac{B_2 \pm \sqrt{B_2^2 - 4|C_2|^2}}{2C_2} = 0.876\angle 61^\circ.$$

Then the effective gain factors of (11.60) can be calculated as

$$G_S = \frac{1}{1 - |\Gamma_S|^2} = 4.17 = 6.20 \text{ dB},$$

$$G_0 = |S_{21}|^2 = 6.76 = 8.30 \text{ dB},$$

$$G_L = \frac{1 - |\Gamma_L|^2}{|1 - S_{22}\Gamma_L|^2} = 1.67 = 2.22 \text{ dB}.$$

So the overall transducer gain will be

$$G_{T_{\max}} = 6.20 + 8.30 + 2.22 = 16.7 \text{ dB}.$$

The matching networks can easily be determined using the Smith chart. For the input matching section, we first plot Γ_S, as shown in Figure 11.28a. The impedance, Z_S, represented by this reflection coefficient is the impedance seen looking into the matching section toward the source impedance, Z_0. Thus, the matching section must transform Z_0 to the impedance Z_S. There are several ways of doing this, but we will use an open-circuited shunt stub followed by a length of line. Thus we convert to the normalized admittance y_s, and work backward (toward the load on the Smith chart) to find that a line of length 0.120λ will bring us to the $1 + jb$ circle. Then we see that the required stub admittance is $+j3.5$, for an open-circuited stub length of 0.206λ. A similar procedure gives a line length of 0.206λ and a stub length of 0.206λ for the output matching circuit.

The final amplifier circuit is shown in Figure 11.28b. This circuit only shows the RF components; the amplifier will also require some bias circuitry. The return loss and gain were calculated using a CAD package, interpolating the necessary S parameters from the table given above. The results are plotted in Figure 11.28c, and show the expected gain of 16.7 dB at 4.0 GHz, with a very good return loss. The bandwidth where the gain drops by 1 dB is about 2.5%. ○

Constant Gain Circles and Design for Specified Gain (Unilateral Device)

In many cases it is preferable to design for less than the maximum obtainable gain, to improve bandwidth or to obtain a specific value of amplifier gain. This can be done by designing the input and output matching sections to have less than maximum gains; in other words, mismatches are purposely introduced to reduce the overall gain. The design procedure is facilitated by plotting *constant gain circles* on the Smith chart, to represent loci of Γ_S and Γ_L that give fixed values of gain (G_S and G_L). To simplify our discussion, we will only treat the case of a unilateral device; the more general case of a bilateral device must sometimes be considered in practice and is discussed in detail in references [4], [5], and [6].

In many practical cases $|S_{12}|$ is small enough to be ignored, and the device can then be assumed to be unilateral. This greatly simplifies the design procedure. The error in the transducer gain caused by approximating $|S_{12}|$ as zero is given by the ratio G_T/G_{TU}.

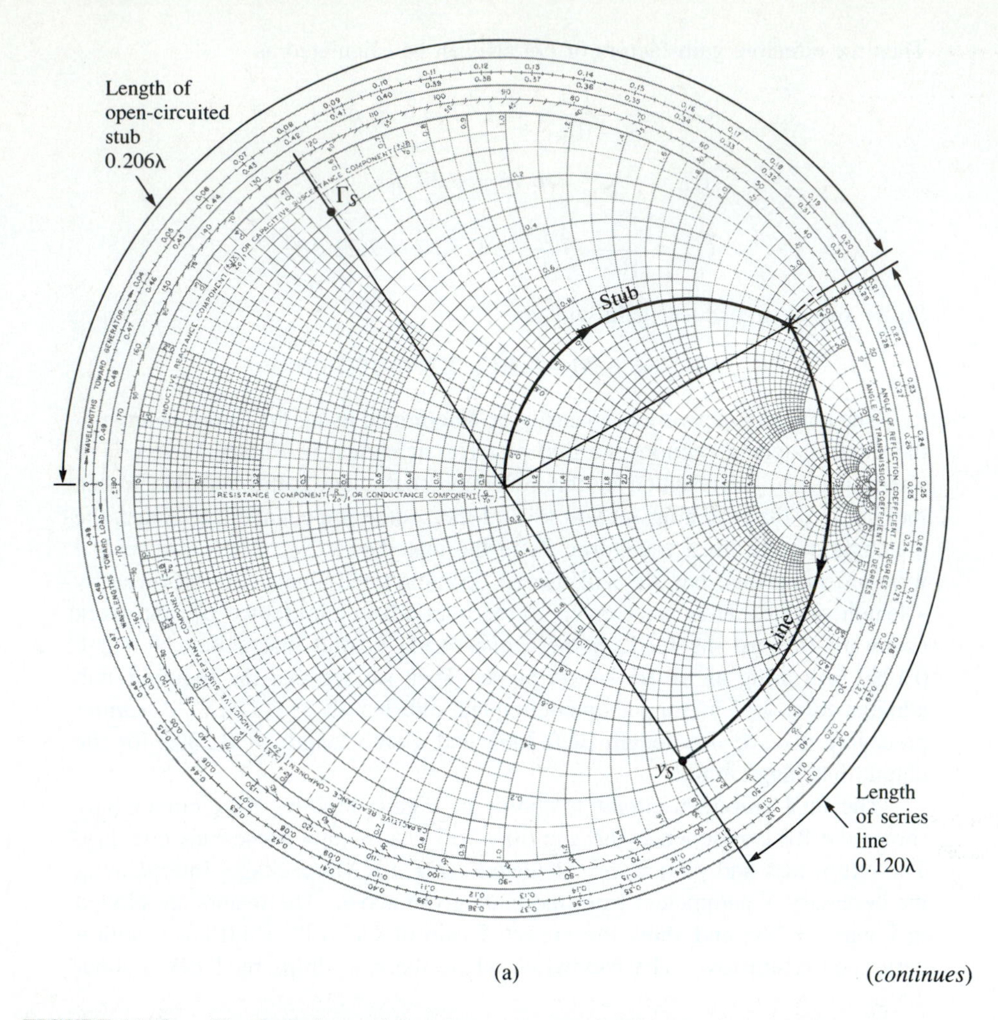

(a) *(continues)*

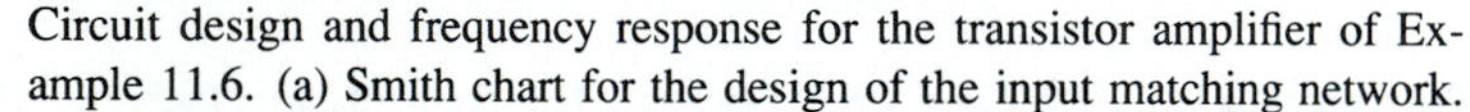

FIGURE 11.28 Circuit design and frequency response for the transistor amplifier of Example 11.6. (a) Smith chart for the design of the input matching network.

It can be shown that this ratio is bounded by

$$\frac{1}{(1+U)^2} < \frac{G_T}{G_{TU}} < \frac{1}{(1-U)^2}, \tag{11.86}$$

where U is defined as the *unilateral figure of merit*,

$$U = \frac{|S_{12}||S_{21}||S_{11}||S_{22}|}{(1-|S_{11}|^2)(1-|S_{22}|^2)}. \tag{11.87}$$

Usually an error of a few tenths of a dB or less justifies the unilateral assumption.

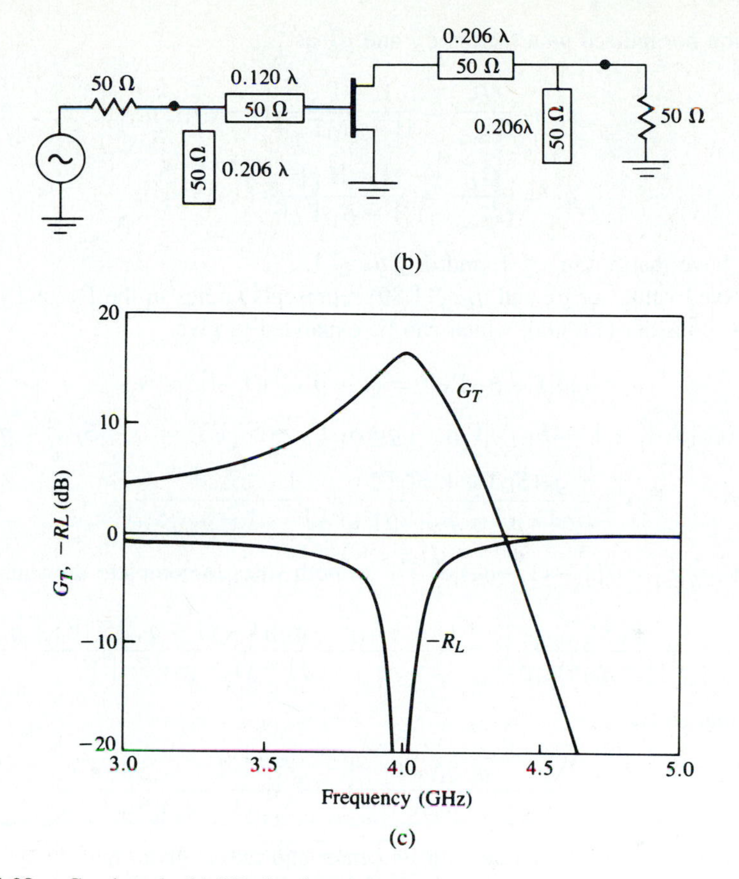

FIGURE 11.28 Continued. (b) RF circuit. (c) Frequency response.

The expression for G_S and G_L for the unilateral case are given by (11.61a) and (11.61c):

$$G_S = \frac{1 - |\Gamma_S|^2}{|1 - S_{11}\Gamma_S|^2},$$

$$G_L = \frac{1 - |\Gamma_L|^2}{|1 - S_{22}\Gamma_L|^2}.$$

These gains are maximized when $\Gamma_S = S_{11}^*$ and $\Gamma_L = S_{22}^*$, resulting in the maximum values given by

$$G_{S_{\max}} = \frac{1}{1 - |S_{11}|^2}, \qquad 11.88a$$

$$G_{L_{\max}} = \frac{1}{1 - |S_{22}|^2}. \qquad 11.88b$$

Now define normalized gain factors g_S and g_L as

$$g_S = \frac{G_S}{G_{S_{max}}} = \frac{1 - |\Gamma_S|^2}{|1 - S_{11}\Gamma_S|^2}(1 - |S_{11}|^2), \tag{11.89a}$$

$$g_L = \frac{G_L}{G_{L_{max}}} = \frac{1 - |\Gamma_L|^2}{|1 - S_{22}\Gamma_L|^2}(1 - |S_{22}|^2). \tag{11.89b}$$

Then we have that $0 \leq g_S \leq 1$, and $0 \leq g_L \leq 1$.

For fixed values of g_S and g_L, (11.89) represents circles in the Γ_S or Γ_L plane. To show this, consider (11.89a), which can be expanded to give

$$g_S|1 - S_{11}\Gamma_S|^2 = (1 - |\Gamma_S|^2)(1 - |S_{11}|^2),$$

$$(g_S|S_{11}|^2 + 1 - |S_{11}|^2)|\Gamma_S|^2 - g_S(S_{11}\Gamma_S + S_{11}^*\Gamma_S^*) = 1 - |S_{11}|^2 - g_S,$$

$$\Gamma_S\Gamma_S^* - \frac{g_S(S_{11}\Gamma_S + S_{11}^*\Gamma_S^*)}{1 - (1 - g_S)|S_{11}|^2} = \frac{1 - |S_{11}|^2 - g_S}{1 - (1 - g_S)|S_{11}|^2}. \tag{11.90}$$

Now add $(g_S^2|S_{11}|^2)/[1 - (1 - g_S)|S_{11}|^2]^2$ to both sides to complete the square:

$$\left|\Gamma_S - \frac{g_S S_{11}^*}{1 - (1 - g_S)|S_{11}|^2}\right|^2 = \frac{(1 - |S_{11}|^2 - g_S)[1 - (1 - g_S)|S_{11}|^2] + g_S^2|S_{11}|^2}{[1 - (1 - g_S)|S_{11}|^2]^2}.$$

Simplifying gives,

$$\left|\Gamma_S - \frac{g_S S_{11}^*}{1 - (1 - g_S)|S_{11}|^2}\right| = \frac{\sqrt{1 - g_S}(1 - |S_{11}|^2)}{1 - (1 - g_S)|S_{11}|^2}, \tag{11.91}$$

which is the equation of a circle with its center and radius given by

$$C_S = \frac{g_S S_{11}^*}{1 - (1 - g_S)|S_{11}|^2}, \tag{11.92a}$$

$$R_S = \frac{\sqrt{1 - g_S}(1 - |S_{11}|^2)}{1 - (1 - g_S)|S_{11}|^2}. \tag{11.92b}$$

The results for the constant gain circles of the output section can be shown to be,

$$C_L = \frac{g_L S_{22}^*}{1 - (1 - g_L)|S_{22}|^2}, \tag{11.93a}$$

$$R_L = \frac{\sqrt{1 - g_L}(1 - |S_{22}|^2)}{1 - (1 - g_L)|S_{22}|^2}. \tag{11.93b}$$

The centers of each family of circles lie along straight lines give by the angle of S_{11}^* or S_{22}^*. Note that when g_S (or g_L) = 1 (maximum gain), the radius R_S (or R_L) = 0, and the center reduces to S_{11}^* (or S_{22}^*), as expected. Also, it can be shown that the 0 dB gain circles ($G_S = 1$ or $G_L = 1$) will always pass through the center of the Smith chart. These results can be used to plot a family of circles of constant gain for the input and output sections. Then Γ_S and Γ_L can be chosen along these circles to provide the desired gains. The choices for Γ_S and Γ_L are not unique, but it makes sense to choose points close to the center of the Smith chart to minimize the mismatch and thus maximize the bandwidth. Alternatively, as we will see in the next section, the input network mismatch can be chosen to provide a low-noise design.

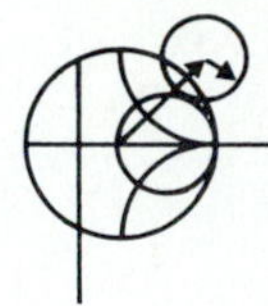

EXAMPLE 11.7

Design an amplifier to have a gain of 11 dB at 4.0 GHz. Plot constant gain circles for $G_S = 2$ dB and 3 dB, and $G_L = 0$ dB and 1 dB. Calculate and plot the input return loss and overall amplifier gain from 3 to 5 GHz. The FET has the following S parameters ($Z_0 = 50\,\Omega$):

f (GHz)	S_{11}	S_{21}	S_{12}	S_{22}
3	$0.80\angle{-90^\circ}$	$2.8\angle{100^\circ}$	0	$0.66\angle{-50^\circ}$
4	$0.75\angle{-120^\circ}$	$2.5\angle{80^\circ}$	0	$0.60\angle{-70^\circ}$
5	$0.71\angle{-140^\circ}$	$2.3\angle{60^\circ}$	0	$0.58\angle{-85^\circ}$

Solution

Since $S_{12} = 0$ and $|S_{11}| < 1$ and $|S_{22}| < 1$, the transistor is unilateral and unconditionally stable. From (11.88) we calculate the maximum matching section gains as

$$G_{S_{\max}} = \frac{1}{1 - |S_{11}|^2} = 2.29 = 3.6 \text{ dB},$$

$$G_{L_{\max}} = \frac{1}{1 - |S_{22}|^2} = 1.56 = 1.9 \text{ dB}.$$

The gain of the mismatched transistor is

$$G_o = |S_{21}|^2 = 6.25 = 8.0 \text{ dB},$$

so the maximum unilateral transducer gain is

$$G_{TU_{\max}} = 3.6 + 1.9 + 8.0 = 13.5 \text{ dB}.$$

Thus we have 2.5 dB more gain than is required by the specifications.

We use (11.89), (11.92), and (11.93) to calculate the following data for the constant gain circles:

$G_S = 3$ dB	$g_S = 0.875$	$C_S = 0.706\angle{120^\circ}$	$R_S = 0.166$
$G_S = 2$ dB	$g_S = 0.691$	$C_S = 0.627\angle{120^\circ}$	$R_S = 0.294$
$G_L = 1$ dB	$g_L = 0.806$	$C_L = 0.520\angle{70^\circ}$	$R_L = 0.303$
$G_L = 0$ dB	$g_L = 0.640$	$C_L = 0.440\angle{70^\circ}$	$R_S = 0.440$

The constant gain circles are shown in Figure 11.29a. We choose $G_S = 2$ dB and $G_L = 1$ dB, for an overall amplifier gain of 11 dB. Then we select Γ_S and Γ_L along these circles as shown, to minimize the distance from the center

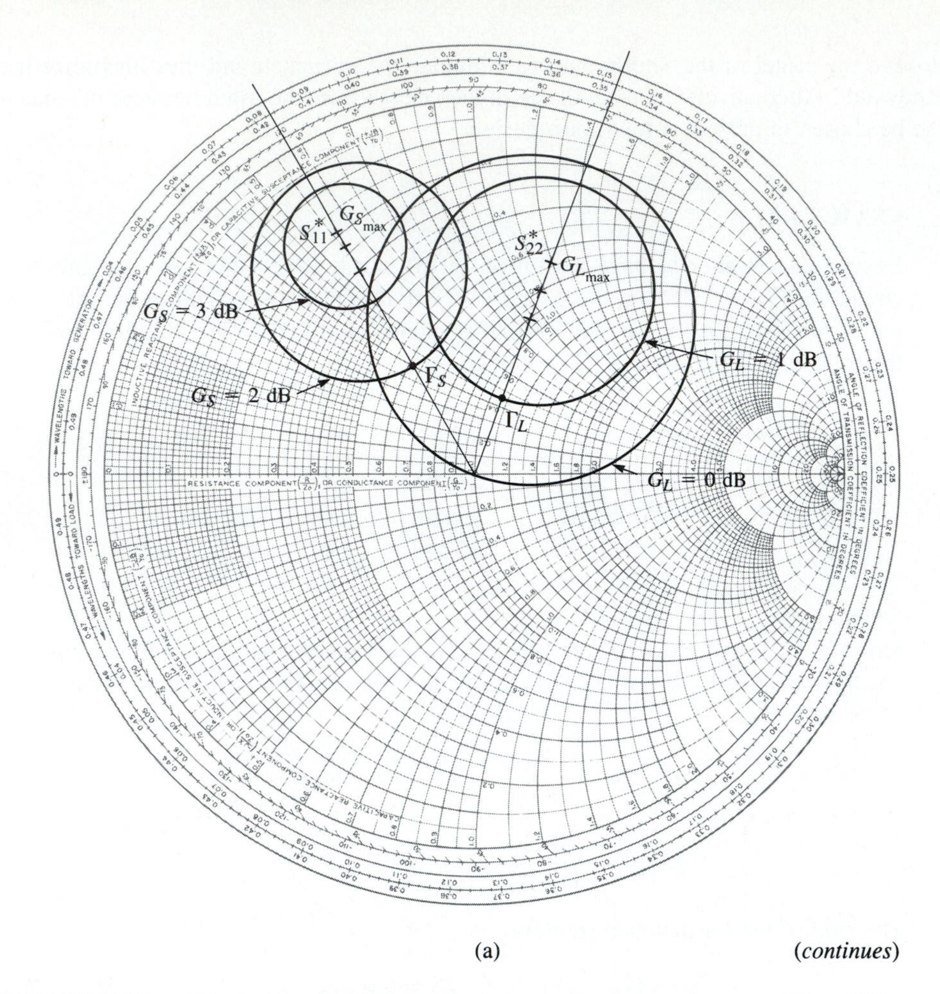

(a) *(continues)*

FIGURE 11.29 Circuit design and frequency response for the transistor amplifier of Example 11.7. (a) Constant gain circles.

of the chart (this places Γ_S and Γ_L along the radial lines at 120° and 70°, respectively). Thus, $\Gamma_S = 0.33\angle 120°$ and $\Gamma_L = 0.22\angle 70°$, and the matching networks can be designed using shunt stubs as in Example 11.5.

The final amplifier circuit is shown in Figure 11.29b. The response was calculated using CAD software, with interpolation of the given S parameter data. The results are shown in Figure 11.29c, where it is seen the desired gain of 11 dB is achieved at 4.0 GHz. The bandwidth over which the gain varies by ±1 dB or less is about 25%, which is considerably better than the bandwidth of the maximum gain design Example 11.5. The return loss, however, is not very good, being only about 5 dB at the design frequency. This is due to the deliberate mismatch introduced into the matching sections to achieve the specified gain. ○

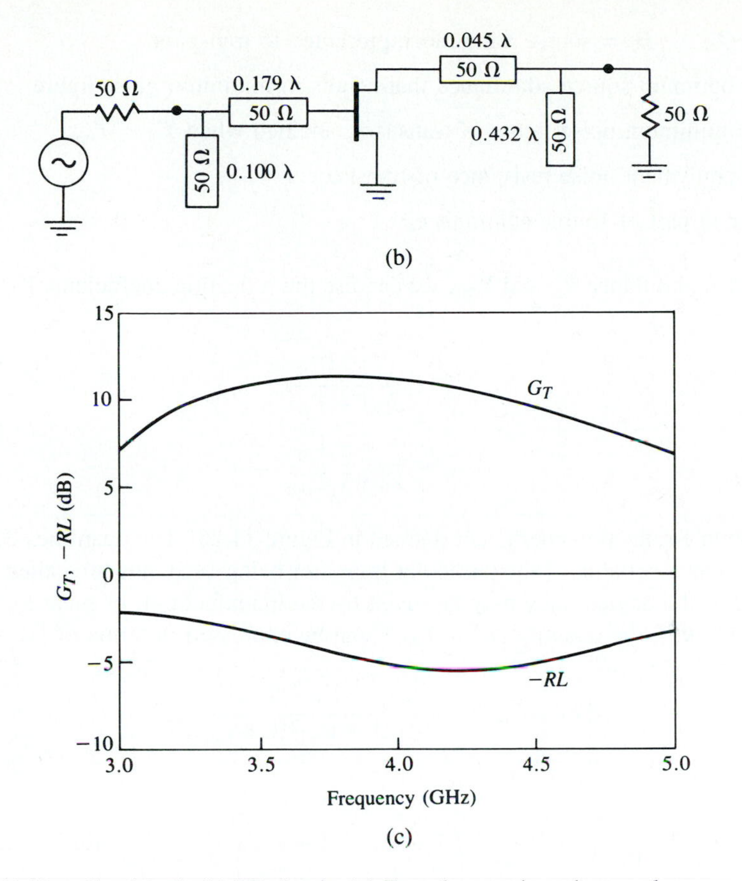

FIGURE 11.29 Continued. (b) RF circuit. (c) Transducer gain and return loss.

Constant Noise Figure Circles and Design for Low Noise

Besides stability and gain, another important design consideration for a microwave amplifier is its noise figure. In receiver applications especially, it is often required to have a preamplifier with as low a noise figure as possible since, as we saw in Section 11.1, the first stage of a receiver front end has the dominant effect on the noise performance of the overall system. Generally it is not possible to obtain both minimum noise figure and maximum gain for an amplifier, so some sort of compromise must be made. This can be done by using constant gain circles and *circles of constant noise figure* to select a usable trade-off between noise figure and gain. Here we will derive the equations for constant noise figure circles, and show how they are used in transistor amplifier design.

As derived in references [6] and [7], the noise figure of a two-port amplifier can be expressed as

$$F = F_{\text{min}} + \frac{R_N}{G_S}|Y_S - Y_{\text{opt}}|^2, \tag{11.94}$$

where the following definitions apply:

$Y_S = G_S + jB_S$ = source admittance presented to transistor.

Y_{opt} = optimum source admittance that results in minimum noise figure.

F_{min} = minimum noise figure of transistor, attained when $Y_S = Y_{opt}$.

R_N = equivalent noise resistance of transistor.

G_S = real part of source admittance.

Instead of the admittance Y_S and Y_{opt}, we can use the reflection coefficients Γ_S and Γ_{opt}, where

$$Y_S = \frac{1}{Z_0}\frac{1-\Gamma_S}{1+\Gamma_S}, \tag{11.95a}$$

$$Y_{opt} = \frac{1}{Z_0}\frac{1-\Gamma_{opt}}{1+\Gamma_{opt}}. \tag{11.95b}$$

Γ_S is the source reflection coefficient defined in Figure 11.25. The quantities F_{min}, Γ_{opt}, and R_N are characteristics of the particular transistor being used, and are called the *noise parameters* of the device; they may be given by the manufacturer, or measured.

Using (11.95), the quantity $|Y_S - Y_{opt}|^2$ can be expressed in terms of Γ_S and Γ_{opt}:

$$|Y_S - Y_{opt}|^2 = \frac{4}{Z_0^2}\frac{|\Gamma_S - \Gamma_{opt}|^2}{|1+\Gamma_S|^2|1+\Gamma_{opt}|^2}. \tag{11.96}$$

Also,

$$G_S = \text{Re}\{Y_S\} = \frac{1}{2Z_0}\left(\frac{1-\Gamma_S}{1+\Gamma_S} + \frac{1-\Gamma_S^*}{1+\Gamma_S^*}\right) = \frac{1}{Z_0}\frac{1-|\Gamma_S|^2}{|1+\Gamma_S|^2}. \tag{11.97}$$

Using these results in (11.94) gives the noise figure as

$$F = F_{min} + \frac{4R_N}{Z_0}\frac{|\Gamma_S - \Gamma_{opt}|^2}{(1-|\Gamma_S|^2)|1+\Gamma_{opt}|^2}. \tag{11.98}$$

For a fixed noise figure, F, we can show that this result defines a circle in the Γ_S plane. First define the *noise figure parameter*, N, as

$$N = \frac{|\Gamma_S - \Gamma_{opt}|^2}{1-|\Gamma_S|^2} = \frac{F - F_{min}}{4R_N/Z_0}|1+\Gamma_{opt}|^2, \tag{11.99}$$

which is a constant, for a given noise figure and set of noise parameters. Then rewrite (11.99) as

$$(\Gamma_S - \Gamma_{opt})(\Gamma_S^* - \Gamma_{opt}^*) = N(1-|\Gamma_S|^2),$$

$$\Gamma_S\Gamma_S^* - (\Gamma_S\Gamma_{opt}^* + \Gamma_S^*\Gamma_{opt}) + \Gamma_{opt}\Gamma_{opt}^* = N - N|\Gamma_S|^2,$$

$$\Gamma_S\Gamma_S^* - \frac{(\Gamma_S\Gamma_{opt}^* + \Gamma_S^*\Gamma_{opt})}{N+1} = \frac{N-|\Gamma_{opt}|^2}{N+1}.$$

Now add $|\Gamma_{opt}|^2/(N+1)^2$ to both sides to complete the square to obtain

$$\left|\Gamma_S - \frac{\Gamma_{opt}}{N+1}\right| = \frac{\sqrt{N(N+1-|\Gamma_{opt}|^2)}}{(N+1)}. \qquad 11.100$$

This result defines circles of constant noise figure with centers at

$$C_F = \frac{\Gamma_{opt}}{N+1}, \qquad 11.101a$$

and radii of

$$R_F = \frac{\sqrt{N(N+1-|\Gamma_{opt}|^2)}}{N+1}. \qquad 11.101b$$

EXAMPLE 11.8

A GaAs FET is biased for minimum noise figure, and has the following S parameters and noise parameters at 4 GHz ($Z_0 = 50\,\Omega$): $S_{11} = 0.6\angle{-60^\circ}$, $S_{21} = 1.9\angle{81^\circ}$, $S_{12} = 0.05\angle{26^\circ}$, $S_{22} = 0.5\angle{-60^\circ}$; $F_{min} = 1.6$ dB, $\Gamma_{opt} = 0.62\angle{100^\circ}$, $R_N = 20\,\Omega$. For design purposes, assume the device is unilateral, and calculate the maximum error in G_T resulting from this assumption. Then design an amplifier having a 2.0 dB noise figure with the maximum gain that is compatible with this noise figure.

Solution

We first compute the unilateral figure of merit from (11.87):

$$U = \frac{|S_{12}S_{21}S_{11}S_{22}|}{(1-|S_{11}|^2)(1-|S_{22}|^2)} = 0.059.$$

Then from (11.86) the ratio G_T/G_{TU} is bounded as

$$\frac{1}{(1+U)^2} < \frac{G_T}{G_{TU}} < \frac{1}{(1-U)^2},$$

or

$$0.891 < \frac{G_T}{G_{TU}} < 1.130.$$

In dB,

$$-0.50 < G_T - G_{TU} < 0.53 \text{ dB},$$

where G_T and G_{TU} are now in dB. Thus, we should expect less than about ± 0.5 dB error in gain.

Next, we use (11.99) and (11.101) to compute the center and radius of the 2 dB noise figure circle:

$$N = \frac{F - F_{\min}}{4R_N/Z_0}|1 + \Gamma_{\text{opt}}|^2 = \frac{1.58 - 1.445}{4(20/50)}|1 + 0.62\angle 100^\circ|^2$$

$$= 0.0986,$$

$$C_F = \frac{\Gamma_{\text{opt}}}{N+1} = 0.56\angle 100^\circ$$

$$R_F = \frac{\sqrt{N(N + 1 - |\Gamma_{\text{opt}}|^2)}}{N+1} = 0.24.$$

This noise figure circle is plotted in Figure 11.30a. Minimum noise figure ($F_{\min} = 1.6$ dB) occurs for $\Gamma_S = \Gamma_{\text{opt}} = 0.62\angle 100^\circ$.

Next we calculate data for several input section constant gain circles. From (11.92),

G_S(dB)	g_S	C_S	R_S
1.0	0.805	$0.52\angle 60^\circ$	0.300
1.5	0.904	$0.56\angle 60^\circ$	0.205
1.7	0.946	$0.58\angle 60^\circ$	0.150

These circles are also plotted in Figure 11.30a. We see that the $G_S = 1.7$ dB gain circle just intersects the $F = 2$ dB noise figure circle, and that any higher gain will result in a worse noise figure. From the Smith chart the optimum solution is then $\Gamma_S = 0.53\angle 75^\circ$, yielding $G_S = 1.7$ dB and $F = 2.0$ dB.

For the output section we choose $\Gamma_L = S_{22}^* = 0.5\angle 60^\circ$ for a maximum G_L of

$$G_L = \frac{1}{1 - |S_{22}|^2} = 1.33 = 1.25 \text{ dB}.$$

The transistor gain is

$$G_0 = |S_{21}|^2 = 3.61 = 5.58 \text{ dB},$$

so the overall transducer gain will be

$$G_{TU} = G_S + G_0 + G_L = 8.53 \text{ dB}.$$

A complete AC circuit for the amplifier, using open-circuited shunt stubs in the matching sections, is shown in Figure 11.30b. A computer analysis of the circuit gave a gain of 8.36 dB. ○

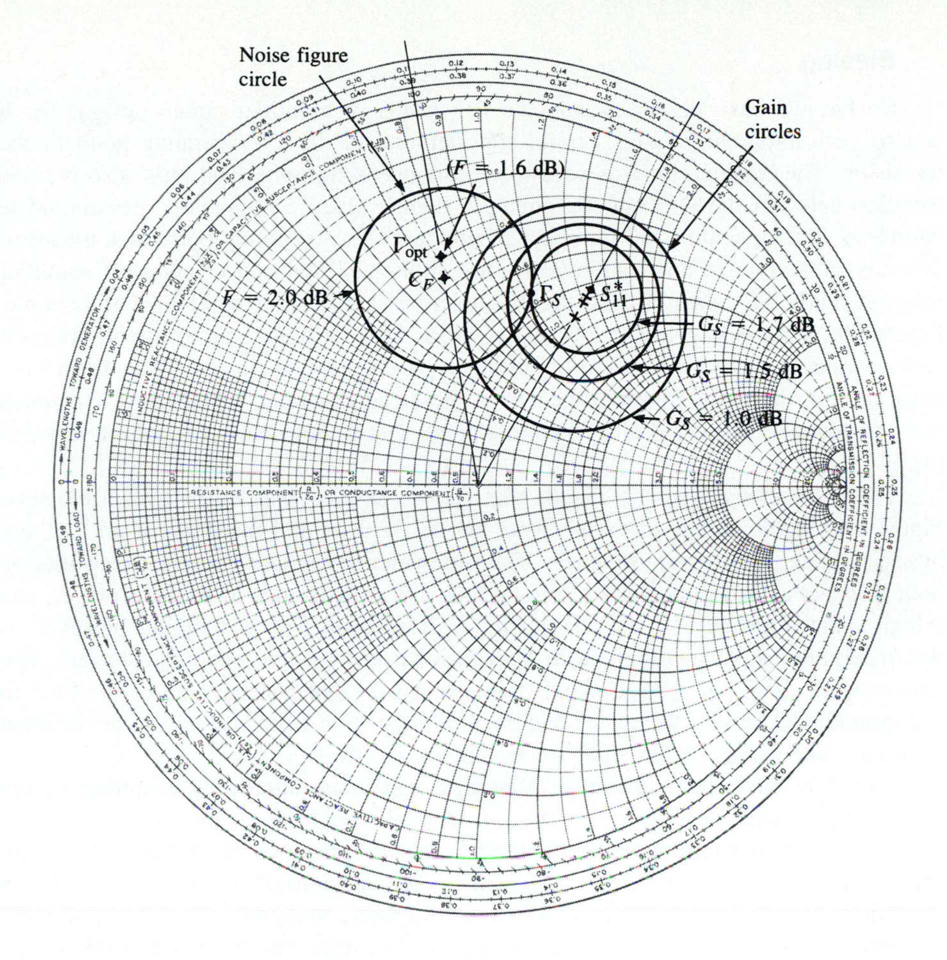

(a)

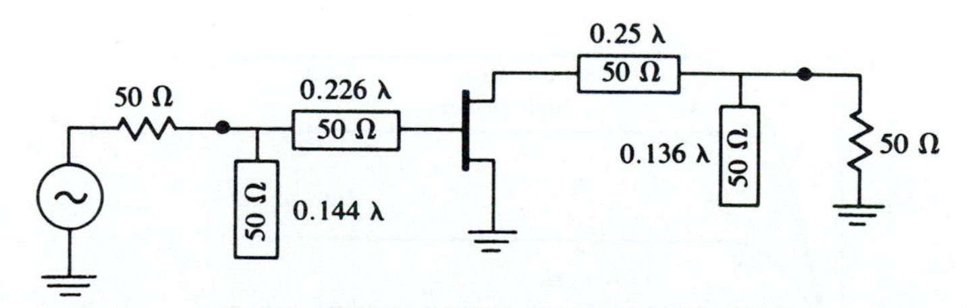

(b)

FIGURE 11.30 Circuit design for the transistor amplifier of Example 11.8. (a) Constant gain and noise figure circles. (b) RF circuit.

Biasing

So far, we have discussed only the RF details of amplifier circuit design, but it is also very important to give careful consideration to the DC operating point of the transistor. The biasing circuit is used to set this operating point, and must also provide isolation between the DC and AC circuits. It may also be required to accommodate variations in temperature and transistor parameters. The operating point of a transistor depends on the type of application (low-noise, high-gain, high-power), class of amplifier (class A, class AB, class B), and the type of transistor (bipolar or FET). For example, Figure 11.31 shows a typical family of I_{DS} versus V_{DS} (drain-to-source current, drain-to-source voltage), with the location of several typical operating points. Thus, for a low-noise design, the drain current is generally chosen to be about 15% of I_{DSS} (saturated drain-to-source current). Similar operating points can be selected for bipolar transistors [4], [6].

There are many types of circuits that can be used to bias microwave transistors. Figure 11.32a shows one of the simplest, where the gate and the drain of an FET are biased with separate power supplies. The series-inductor, shunt capacitor networks on both lines act as low-pass filters to offer a low series impedance to DC bias power, and a high shunt impedance to the RF signal. A disadvantage of this circuit, however, is that it requires a power supply with both positive and negative output voltages. The bias network of Figure 11.32b avoids this problem by holding the gate of the FET to DC ground, and biasing the source and drain with positive voltages that can be obtained from the same supply. Many other possibilities exist [4], [6].

There are many other important aspects to solid-state microwave amplifier design that we are unable to treat here. Thus, a more thorough treatment would cover the case of a potentially unstable bilateral device, and the concepts of maximum stable gain and operating power gain circles. Then there are specialized designs for broadband amplifiers, reflection amplifiers, high-power amplifiers, multistage amplifiers, and the interesting topic of distributed amplifier design. We refer the interested reader to the literature [4], [5], [6], and [7] for further discussion of these topics.

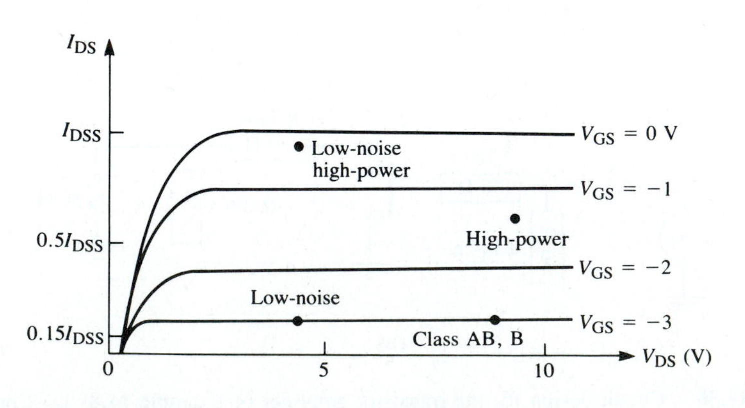

FIGURE 11.31 Typical operating points for a GaAs FET.

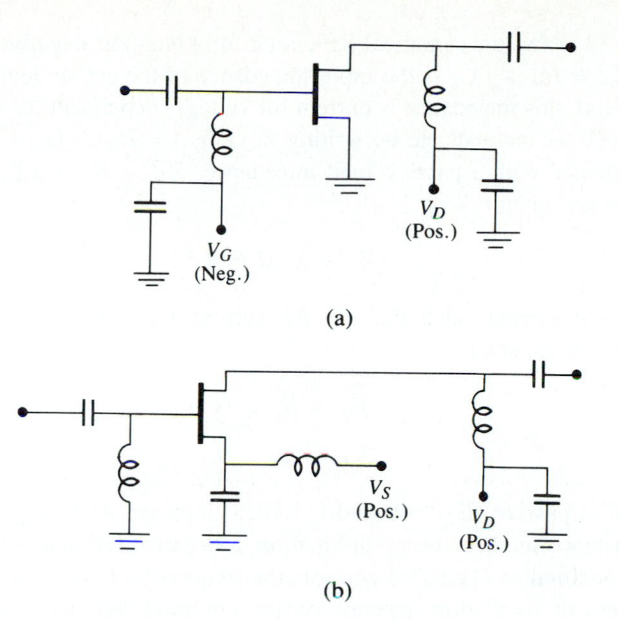

FIGURE 11.32 FET biasing circuits. (a) Using a bipolar supply. (b) Using a single-polarity supply.

11.4 OSCILLATOR DESIGN

A microwave oscillator converts DC power to RF power, and so is one of the most basic and essential components in a microwave system. A solid-state oscillator uses an active device, such as a diode or transistor, in conjunction with a passive circuit to produce a sinusoidal steady-state RF signal. At startup, however, oscillation is triggered by transients or noise, after which a properly designed oscillator will reach a stable oscillation state. This process requires that the active device be nonlinear. In addition, since the device is producing RF power, it must have a negative resistance. Because of this active and nonlinear element, the complete analysis of oscillator operation is very difficult.

We will first discuss the operation and design of one-port negative resistance oscillators. Such circuits represent oscillators that use IMPATT or Gunn diodes. Then we treat transistor oscillators, where either an FET or bipolar device is operated with a passive termination that produces a negative resistance at its input port. That is, the transistor is operated in an unstable region, in contrast to an amplifier application where a stable operating point is required.

One-Port Negative Resistance Oscillators

Here we present some of the basic principles of operation for one-port negative oscillators; much of this material will also apply to two-port (transistor) oscillators.

Figure 11.33 shows the canonical RF circuit for a one-port negative-resistance oscillator, where $Z_{\text{in}} = R_{\text{in}} + jX_{\text{in}}$ is the input impedance of the active device (e.g., a biased diode). In general, this impedance is current (or voltage) dependent, as well as frequency dependent, which we can indicate by writing $Z_{\text{in}}(I, j\omega) = R_{\text{in}}(I, j\omega) + jX_{\text{in}}(I, j\omega)$. The device is terminated with a passive load impedance, $Z_L = R_L + jX_L$. Applying Kirchoff's voltage law gives

$$(Z_L + Z_{\text{in}})I = 0. \tag{11.102}$$

If oscillation is occurring, such that the RF current I is nonzero, then the following conditions must be satisfied:

$$R_L + R_{\text{in}} = 0, \tag{11.103a}$$

$$X_L + X_{\text{in}} = 0. \tag{11.103b}$$

Since the load is passive, $R_L > 0$ and (11.103a) indicates that $R_{\text{in}} < 0$. Thus, while a positive resistance implies energy dissipation, a negative resistance implies an energy source. The condition of (11.103b) controls the frequency of oscillation.

The process of oscillation depends on the nonlinear behavior of Z_{in}, as follows. Initially, it is necessary for the overall circuit to be unstable at a certain frequency, that is, $R_{\text{in}}(I, j\omega) + R_L < 0$. Then any transient excitation or noise will cause an oscillation to build up at the frequency, ω. As I increases, $R_{\text{in}}(I, j\omega)$ must become less negative until the current I_0 is reached such that $R_{\text{in}}(I_0, j\omega_0) + R_L = 0$, and $X_{\text{in}}(I_0, j\omega_0) + X_L(j\omega_0) = 0$. Then the oscillator is running in a stable state. The final frequency, ω_0, generally differs from the startup frequency because X_{in} is current dependent, so that $X_{\text{in}}(I, j\omega) \neq X_{\text{in}}(I_0, j\omega_0)$.

Thus we see that the conditions of (11.103) are not enough to guarantee a stable state of oscillation. In particular, stability requires that any perturbation in current or frequency will be damped out, allowing the oscillator to return to its original state. This condition can be quantified by considering the effect of a small change, δI, in the current and a small change, δs, in the complex frequency $s = \alpha + j\omega$. If we let $Z_T(I, s) = Z_{\text{in}}(I, s) + Z_L(s)$, then we can write a Taylor series for $Z_T(I, s)$ about the

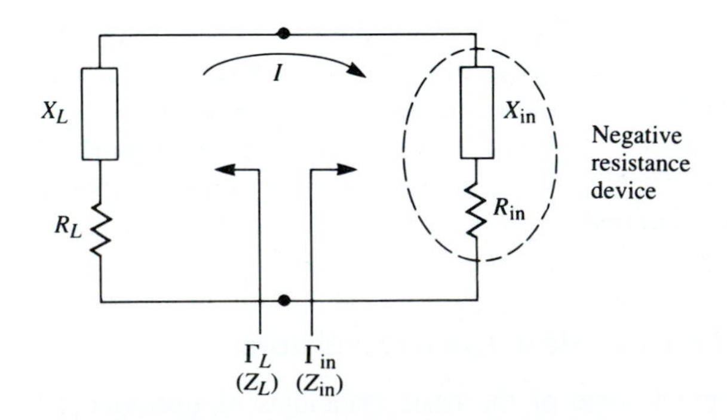

FIGURE 11.33 Circuit for a one-port negative-resistance oscillator.

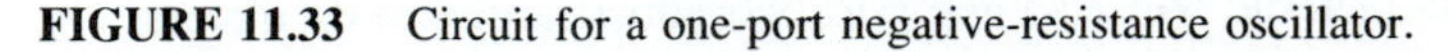

operating point I_0, ω_0 as

$$Z_T(I,s) = Z_T(I_0,s_0) + \left.\frac{\partial Z_T}{\partial s}\right|_{s_0,I_0} \delta s + \left.\frac{\partial Z_T}{\partial I}\right|_{s_0,I_0} \delta I = 0, \tag{11.104}$$

since $Z_T(I,s)$ must still equal zero if oscillators are occurring. In (11.104), $s_0 = j\omega_0$ is the complex frequency at the original operating point. Now use the fact that $Z_T(I_0, s_0) = 0$, and that $\frac{\partial Z_T}{\partial s} = -j\frac{\partial Z_T}{\partial \omega}$, to solve (11.104) for $\delta s = \delta\alpha + j\delta\omega$:

$$\delta s = \left.\frac{-\partial Z_T/\partial I}{\partial Z_T/\partial s}\right|_{s_0,I_0} \delta I = \frac{-j(\partial Z_T/\partial I)(\partial Z_T^*/\partial \omega)}{|\partial Z_T/\partial \omega|^2}\delta I. \tag{11.105}$$

Now if the transient caused by δI and $\delta\omega$ is to decay, we must have $\delta\alpha < 0$ when $\delta I > 0$. Equation (11.105) then implies that

$$I_m\left\{\frac{\partial Z_T}{\partial I}\frac{\partial Z_T^*}{\partial \omega}\right\} < 0,$$

or

$$\frac{\partial R_T}{\partial I}\frac{\partial X_T}{\partial \omega} - \frac{\partial X_T}{\partial I}\frac{\partial R_T}{\partial \omega} > 0. \tag{11.106}$$

For a passive load, $\partial R_L/\partial I = \partial X_L/\partial I = \partial R_L/\partial \omega = 0$, so (11.106) reduces to

$$\frac{\partial R_{\text{in}}}{\partial I}\frac{\partial}{\partial \omega}(X_L + X_{\text{in}}) - \frac{\partial X_{\text{in}}}{\partial I}\frac{\partial R_{\text{in}}}{\partial \omega} > 0. \tag{11.107}$$

As discussed above, we usually have that $\partial R_{\text{in}}/\partial I > 0$ [8]. So (11.107) can be satisfied if $\partial(X_L + X_{\text{in}})/\partial\omega >> 0$, which implies that a high-Q circuit will result in maximum oscillator stability. Cavity and dielectric resonators are often used for this purpose.

Effective oscillator design requires the consideration of several other issues, such as the selection of an operating point for stable operation and maximum power output, frequency-pulling, large-signal effects, and noise characteristics. But we must leave these topics to more advanced texts.

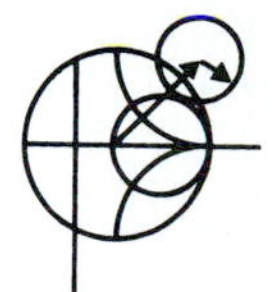

EXAMPLE 11.9

A one-port oscillator uses a negative-resistance diode having $\Gamma_{\text{in}} = 1.25\angle 40^\circ$ ($Z_0 = 50\ \Omega$) at its desired operating point, for $f = 6$ GHz. Design a load matching network for a 50 Ω load impedance.

Solution

From either the Smith chart (see Problem 11.23), or by direct calculation, we find the input impedance as

$$Z_{\text{in}} = -44 + j123\ \Omega.$$

Then, by (11.103), the load impedance must be

$$Z_L = 44 - j123\ \Omega.$$

A shunt stub and series section of line can be used to convert 50 Ω to Z_L, as shown in the circuit of Figure 11.34. ○

Transistor Oscillators

In a transistor oscillator, a negative-resistance one-port is effectively created by terminating a potentially unstable transistor with an impedance designed to drive the device in an unstable region. The circuit model is shown in Figure 11.35; the actual power output port can be on either side of the transistor. In the case of an amplifier, we preferred a device with a high degree of stability—ideally, an unconditionally stable device. For an oscillator, we require a device with a high degree of instability. Typically, common source or common gate FET configurations are used (common emitter or common base for bipolar devices), often with positive feedback to enhance the instability of the device.

After the transistor configuration is selected, the output stability circle can be drawn in the Γ_T plane, and Γ_T selected to produce a large value of negative resistance at the input to the transistor. Then the load impedance Z_L can be chosen to match Z_{in}. Because such a design uses the small-signal S parameters, and because R_{in} will become less negative as the oscillator power builds up, it is necessary to choose R_L so that $R_L + R_{\text{in}} < 0$. Otherwise, oscillation will cease when the increasing power increases R_{in} to the point where $R_L + R_{\text{in}} > 0$. In practice, a value of,

$$R_L = \frac{-R_{\text{in}}}{3}, \qquad 11.108a$$

is typically used. The reactive part of Z_L is chosen to resonate the circuit,

$$X_L = -X_{\text{in}}. \qquad 11.108b$$

When oscillation occurs between the load network and the transistor, oscillation will simultaneously occur at the output port, which we can show as follows. For steady-state oscillation at the input port, we must have $\Gamma_L \Gamma_{\text{in}} = 1$. Then from (11.59a) (after replacing Γ_L with Γ_T), we have

$$\frac{1}{\Gamma_L} = \Gamma_{\text{in}} = S_{11} + \frac{S_{12}S_{21}\Gamma_T}{1 - S_{22}\Gamma_T} = \frac{S_{11} - \Delta\Gamma_T}{1 - S_{22}\Gamma_T}, \qquad 11.109$$

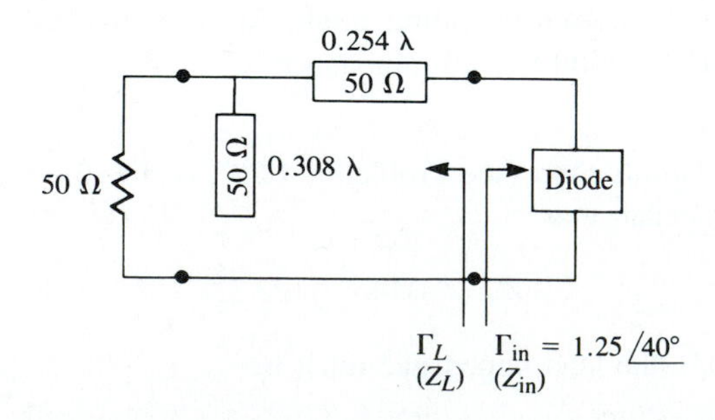

FIGURE 11.34 Load matching circuit for the one-port oscillator of Example 11.9.

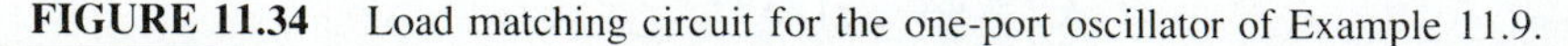

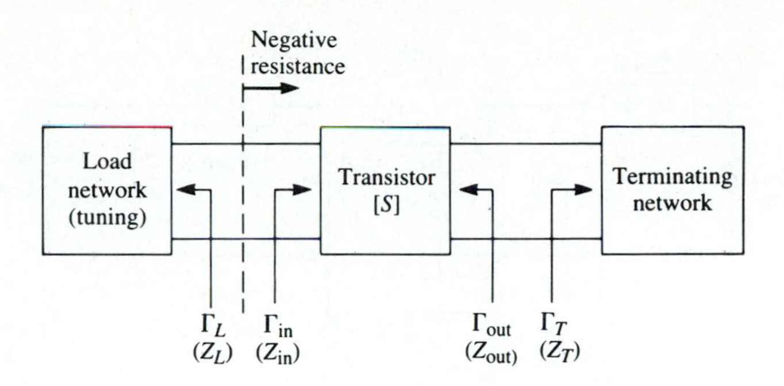

FIGURE 11.35 Circuit for a two-port transistor oscillator.

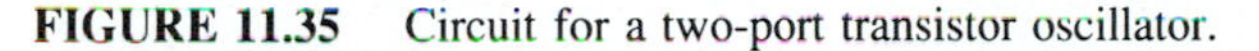

where $\Delta = S_{11}S_{22} - S_{12}S_{21}$. Solving for Γ_T gives,

$$\Gamma_T = \frac{1 - S_{11}\Gamma_L}{S_{22} - \Delta\Gamma_L}. \tag{11.110}$$

Then from (11.59b) we have that

$$\Gamma_{out} = S_{22} + \frac{S_{12}S_{21}\Gamma_L}{1 - S_{11}\Gamma_L} = \frac{S_{22} - \Delta\Gamma_L}{1 - S_{11}\Gamma_L}, \tag{11.111}$$

which shows that $\Gamma_T\Gamma_{out} = 1$. Thus, the condition for oscillation of the terminating network is satisfied. Note that the appropriate S parameters to use in the above development are generally the large-signal parameters of the transistor.

EXAMPLE 11.10

Design a transistor oscillator at 4 GHz using a GaAs FET in a common gate configuration, with a 5 nH inductor in series with the gate to increase the stability. Choose a terminating network to match to a 50 Ω load, and an appropriate tuning network. The S parameters of the transistor in a common source configuration are (Z_0 = 50 Ω): $S_{11} = 0.72\angle{-116^\circ}$, $S_{21} = 2.60\angle{76^\circ}$, $S_{12} = 0.03\angle{57^\circ}$, $S_{22} = 0.73\angle{-54^\circ}$.

Solution

The first step is to convert the common source S parameters to the S parameters that apply to the transistor in a common gate configuration with a series inductor. (See Figure 11.36a.) This is most easily done using a microwave CAD package. The new S parameters are

$$S'_{11} = 2.18\angle{-35^\circ},$$
$$S'_{21} = 2.75\angle{96^\circ},$$
$$S'_{12} = 1.26\angle{18^\circ},$$
$$S'_{22} = 0.52\angle{155^\circ}.$$

Note that $|S'_{11}|$ is significantly greater than $|S_{11}|$, which suggests that the configuration of Figure 11.36a is more unstable than the common source configuration. Calculating the output stability circle (Γ_T plane) parameters from

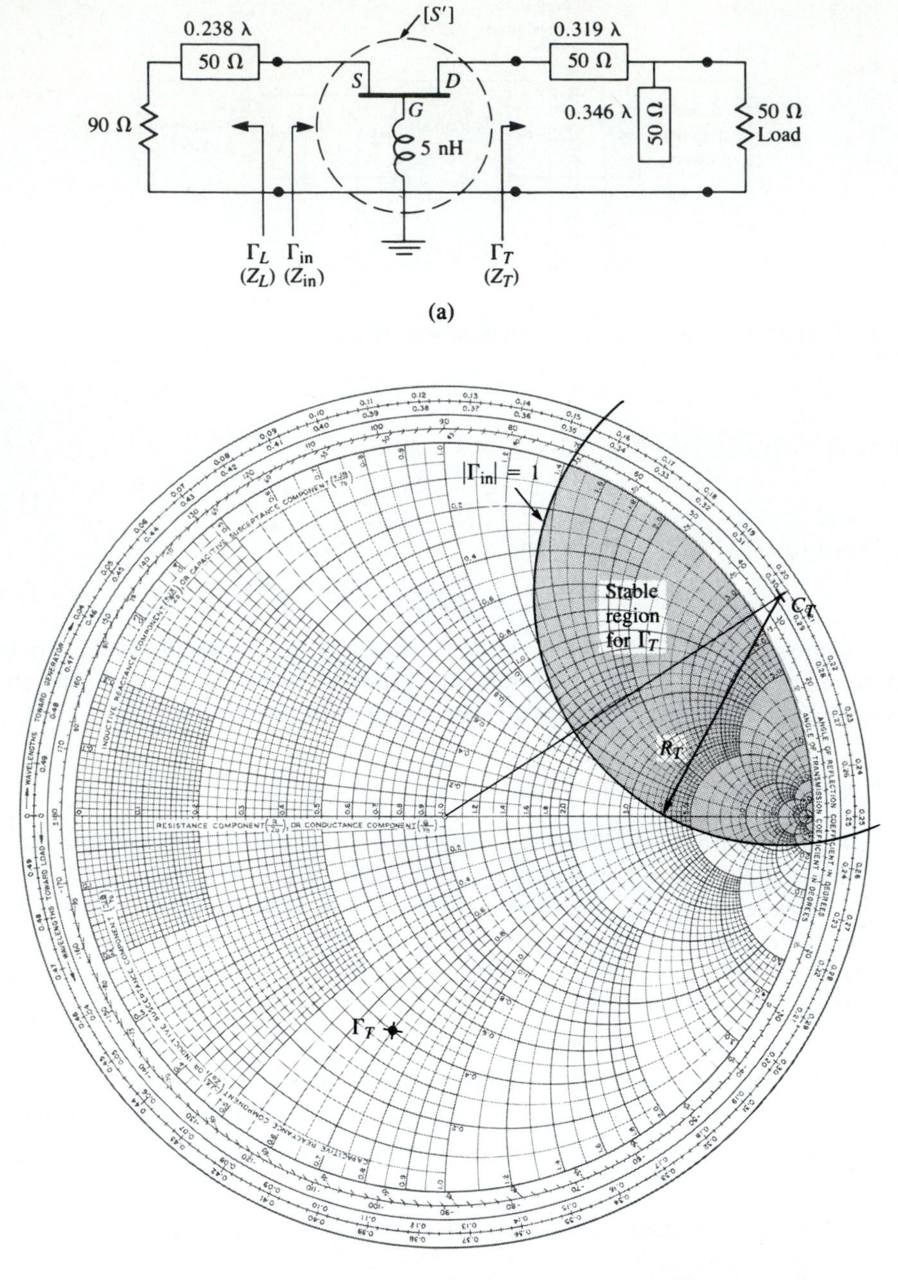

FIGURE 11.36 Circuit design for the transistor oscillator of Example 11.10. (a) Oscillator circuit. (b) Smith chart for determining Γ_T.

(11.68) gives

$$C_T = \frac{(S'_{22} - \Delta' S'^{*}_{11})^*}{|S'_{22}|^2 - |\Delta'|^2} = 1.08\angle 33^\circ,$$

$$R_T = \left| \frac{S'_{12} S'_{21}}{|S'_{22}|^2 - |\Delta'|^2} \right| = 0.665.$$

Since $|S'_{11}| = 2.18 > 1$, the stable region is inside this circle, as shown in the Smith chart in Figure 11.36b.

There is a great amount of freedom in our choice for Γ_T, but one objective is to make $|\Gamma_{\text{in}}|$ large. Thus we try several value of Γ_T located on the opposite side of the chart from the stability circle, and select $\Gamma_T = 0.59\angle -104^\circ$. Then we can design a single-stub matching network to convert a 50 Ω load to $Z_T = 20 - j35$ Ω, as shown in Figure 11.36a.

For the given value of Γ_T, we calculate Γ_{in} as,

$$\Gamma_{\text{in}} = S'_{11} + \frac{S'_{12} S'_{21} \Gamma_T}{1 - S'_{22} \Gamma_T} = 3.94\angle -2.4^\circ,$$

or $Z_{\text{in}} = -84 - j1.9$ Ω. Then, from (11.108), we find Z_L as

$$Z_L = \frac{-R_{\text{in}}}{3} - jX_{\text{in}} = 28 + j1.9 \ \Omega.$$

Using $R_{\text{in}}/3$ should ensure enough instability for the startup of oscillation. The easiest way to implement the impedance Z_L is to use a 90 Ω load with a short length of line, as shown in the figure. It is likely that the steady-state oscillation frequency will differ from 4 GHz because of the nonlinearity of the transistor parameters. ○

11.5 PIN DIODE CONTROL CIRCUITS

Switches are used extensively in microwave systems, for directing signal or power flow between other components. Switches can also be used to construct other types of control circuits, such as phase shifters and attenuators. Mechanical switches can be made in waveguide or coaxial form, and can handle high powers, but are bulky and slow. PIN diodes, however, can be used to construct an electronic switching element easily integrated with planar circuitry and capable of high-speed operation. (Switching speeds of ten nanoseconds or less are typical.) FETs can also be used as switching elements.

The PIN diode has V-I characteristics that make it a good RF switching element. When reverse biased, a small series junction capacitance leads to a relatively high diode impedance, while a forward bias current removes the junction capacitance and leaves the diode in a low impedance state [3]. Equivalent circuits for these two states are shown in Figure 11.37. Typical values for the parameters are: $C_j = 1$ pF, or less; $L_i = 0.5$ nH, or less; $R_r = 5$ Ω, or less; $R_f = 1$ Ω, or less. The equivalent circuits do not include parasitic effects due to packaging, which may be important. The forward bias current is typically 10–30 mA, and the reverse bias voltage is typically 40–60 V. The bias signal

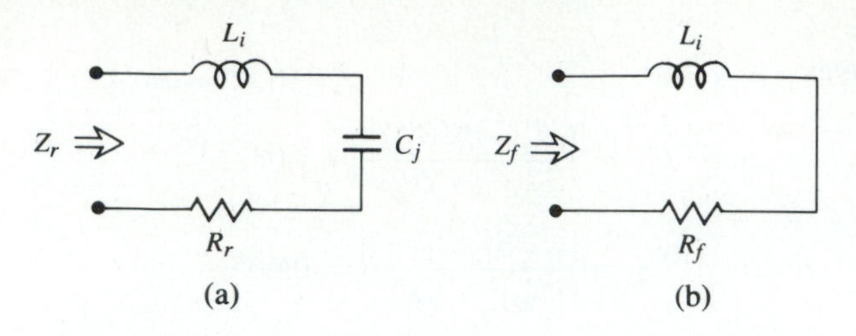

FIGURE 11.37 Equivalent circuits for the ON and OFF states of a PIN diode. (a) Reverse bias (OFF) state. (b) Forward bias (ON) state.

must be applied to the diode with RF chokes and DC blocks to isolate it from the RF signal.

Single-Pole Switches

A PIN diode can be used in either a series or a shunt configuration to form a single-pole, single-throw RF switch. These circuits are shown in Figure 11.38, with bias networks. In the series configuration of Figure 11.38a, the switch is on when the diode is forward biased, while in the shunt configuration the switch is on when the diode is reversed biased. In both cases, input power is reflected when the switch is in the OFF state. The DC blocks should have a very low impedance at the RF operating frequency, while the RF choke inductors should have a very high RF impedance. In some designs, high impedance quarter-wavelength lines can be used in place of the chokes, to provide RF blocking.

Ideally, a switch would have zero insertion loss in the ON state, and infinite attenuation in the OFF state. Realistic switching elements, of course, result in some insertion loss for the ON state, and finite attenuation for the OFF state. Knowing the diode parameters for the equivalent circuits of Figure 11.37 allows the insertion loss for the ON and OFF states to be calculated for the series and shunt switches. With reference to the

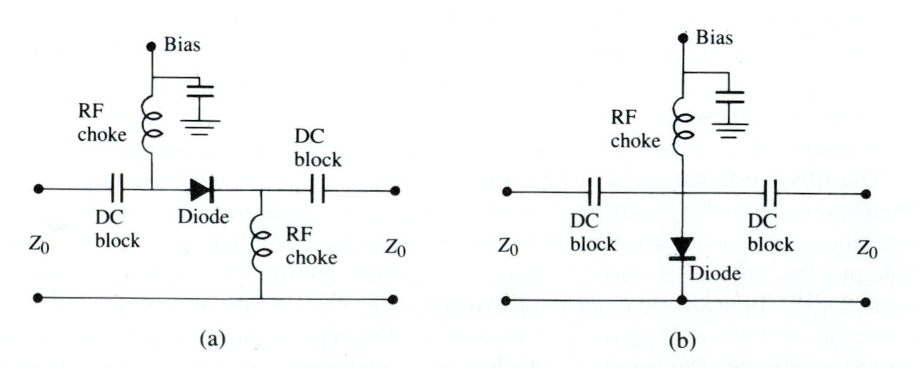

FIGURE 11.38 Single-pole PIN diode switches. (a) Series configuration. (b) Shunt configuration.

simplified switch circuits of Figure 11.39, we can define the insertion loss in terms of the actual load voltage, V_L, and V_0, which is the load voltage which would appear if the switch (Z_d) were absent:

$$IL = -20 \log \left| \frac{V_L}{V_0} \right|. \tag{11.112}$$

Simple circuit analysis applied to the two cases of Figure 11.39 gives the following results:

$$IL = -20 \log \left| \frac{2Z_0}{2Z_0 + Z_d} \right| \quad \text{(series switch)}, \tag{11.113a}$$

$$IL = -20 \log \left| \frac{2Z_d}{2Z_d + Z_0} \right| \quad \text{(shunt switch)}. \tag{11.113b}$$

In both cases, Z_d is the diode impedance for either the reverse or forward bias state. Thus,

$$Z_d = \begin{cases} Z_r = R_r + j(\omega L_i - 1/\omega C_j) & \text{for reverse bias} \\ Z_f = R_f + j\omega L_i & \text{for forward bias.} \end{cases} \tag{11.114}$$

The ON state or OFF state insertion loss of a switch can usually be improved by adding an external reactance in series or in parallel with the diode, to compensate for the reactance of the diode. This technique usually reduces the bandwidth, however.

Several single-throw switches can be combined to form a variety of multiple-pole and/or multiple-throw configurations [8]. Figure 11.40 shows series and shunt circuits for a single-pole, double-throw switch; such a switch requires at least two switching elements. In operation, one diode is biased in the low impedance state, with the other diode biased in the high impedance state. The input signal is switched from one output

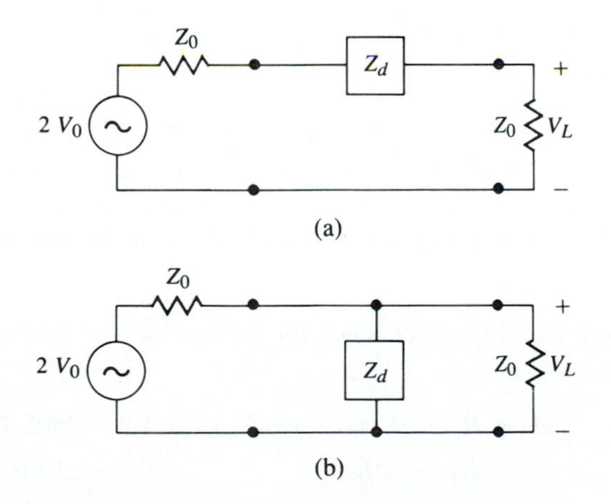

FIGURE 11.39 Simplified equivalent circuits for the series and shunt single-pole PIN diode switches. (a) Series switch. (b) Shunt switch.

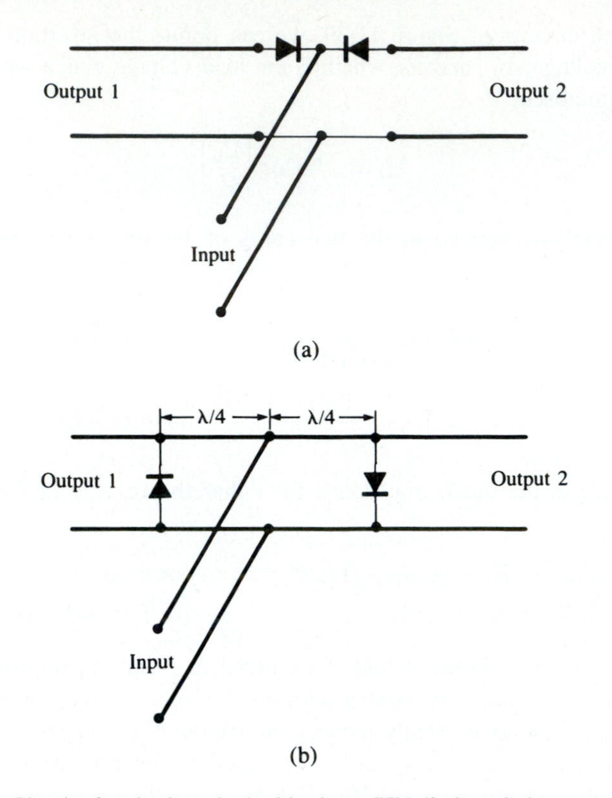

FIGURE 11.40 Circuits for single-pole double-throw PIN diode switches. (a) Series. (b) Shunt.

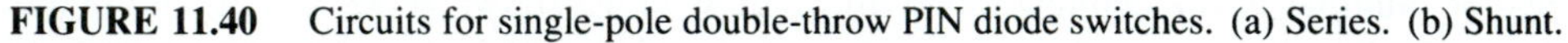

to the other by reversing the diode states. The quarter-wave lines of the shunt circuit limit the bandwidth of this configuration.

EXAMPLE 11.11

A single-pole switch is to be constructed using a PIN diode with the following parameters: $C_j = 0.1$ pF, $R_r = 1\ \Omega$, $R_f = 5.0\ \Omega$, $L_i = 0.4$ nH. If the operating frequency is 5 GHz, and $Z_0 = 50\ \Omega$, what circuit (series or shunt) should be used to obtain the greatest ratio of off-to-on attenuation.

Solution

We first compute the diode impedance for the reverse and forward bias states, using (11.114):

$$Z_d = \begin{cases} Z_r = R_r + j(\omega L_i - 1/\omega C_j) & = 1.0 - j305.7\ \Omega \\ Z_f = R_f + j\omega L_i & = 0.5 + j12.6\ \Omega. \end{cases}$$

Then using (11.113) gives the insertion losses for the ON and OFF states of the series and shunt switches as follows:

For the series circuit,

$$IL_{\text{on}} = -20 \log \left| \frac{2Z_0}{2Z_0 + Z_f} \right| = 0.11 \text{ dB},$$

$$IL_{\text{off}} = -20 \log \left| \frac{2Z_0}{2Z_0 + Z_r} \right| = 10.16 \text{ dB}.$$

For the shunt circuit,

$$IL_{\text{on}} = -20 \log \left| \frac{2Z_r}{2Z_r + Z_0} \right| = 0.03 \text{ dB},$$

$$IL_{\text{off}} = -20 \log \left| \frac{2Z_f}{2Z_f + Z_0} \right| = 7.07 \text{ dB}.$$

So the series configuration has the greatest difference in attenuation between the ON and OFF states, but the shunt circuit has the lowest ON insertion loss. ○

PIN Diode Phase Shifters

Several types of microwave phase shifters can be constructed with PIN diode switching elements. Compared with ferrite phase shifters, diode phase shifters have the advantages of small size, integrability with planar circuitry, and high-speed. The power requirements for diode phase shifters, however, are generally greater than those for a latching ferrite phase shifter, because diodes require continuous bias current while the latching ferrite device only requires a pulsed current to change its state. There are basically three types of PIN diode phase shifters: *switched line, loaded line*, and *reflection type*.

The switched-line phase shifter is the most straightforward type, using two single-pole double-throw switches to route the signal flow between one of two transmission lines of different length. See Figure 11.41. The differential phase shift between the two paths is given by

$$\Delta\phi = \beta(\ell_2 - \ell_1), \tag{11.115}$$

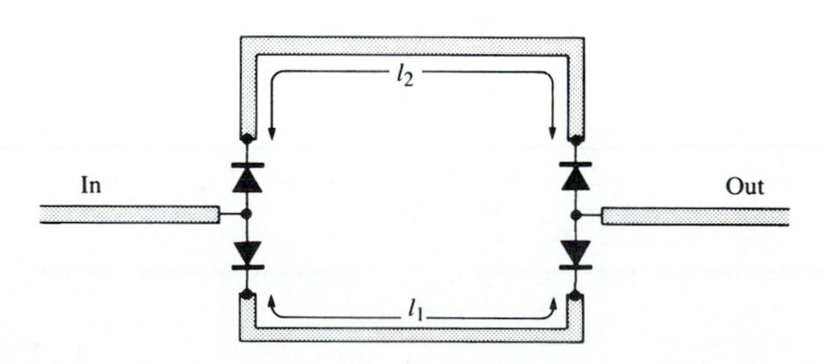

FIGURE 11.41 A switched-line phase shifter.

where β is the propagation constant of the line. If the transmission lines are TEM (or quasi-TEM, like microstrip), this phase shift is a linear function of frequency, which implies a true time delay between the input and output ports. This is a useful feature in broadband systems, because distortion is minimized. This type of phase shifter is also inherently reciprocal, and can be used for both receive and transmit functions. The insertion loss of the switched line phase shifter is equal to the loss of the SPDT switches plus line losses.

Like many other types of phase shifters, the switched-line phase shifter is usually designed for binary phase shifts of $\Delta\phi = 180°$, $90°$, $45°$, etc. One potential problem with this type of phase shifter is that resonances can occur in the OFF line, if its length is near a multiple of $\lambda/2$. The resonant frequency will be slightly shifted due to the series junction capacitances of the reversed biased diodes, so the lengths ℓ_1 and ℓ_2 should be chosen with this effect taken into account.

A design that is useful for small amounts of phase shift (generally 45°, or less) is the loaded-line phase shifter. The basic principle of this type of phase shifter can be illustrated with the circuit of Figure 11.42a, which shows a transmission line loaded with a shunt susceptance, jB. The reflection and transmission coefficients can be written as

$$\Gamma = \frac{1-(1+jb)}{1+(1+jb)} = \frac{-jb}{2+jb}, \tag{11.116a}$$

$$T = 1 + \Gamma = \frac{2}{2+jb}, \tag{11.116b}$$

where $b = BZ_0$ is the normalized susceptance. Thus the phase shift in the transmitted wave introduced by the load is

$$\Delta\phi = \tan^{-1}\frac{b}{2}, \tag{11.117}$$

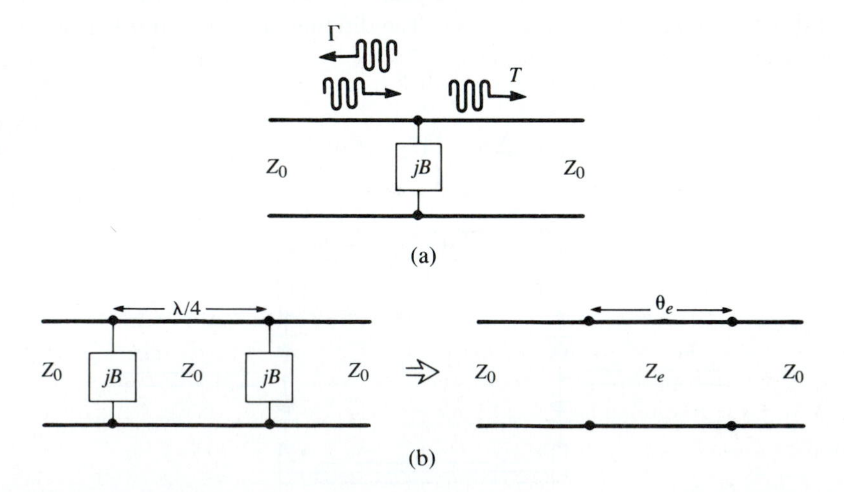

FIGURE 11.42 Loaded-line phase shifters. (a) Basic circuit. (b) Practical loaded-line phase shifter and its equivalent circuit.

which can be made positive or negative, depending on the sign of b. A disadvantage is the insertion loss that is inherently present, due to the reflection from the shunt load. And increasing b to obtain a larger $\Delta\phi$ entails a greater insertion loss, as seen from (11.116a).

The reflections from the shunt susceptance can be reduced by using the circuit of Figure 11.42b, where two shunt loads are separated by a $\lambda/4$ length of line. Then the partial reflection from the second load will be 180° out of phase with the partial reflection from the first load, leading to a cancellation. We can analyze this circuit by calculating its *ABCD* matrix and comparing it to the *ABCD* matrix of an equivalent line having a length θ_e and characteristic impedance Z_e. Thus, for the loaded line,

$$\begin{bmatrix} A & B \\ C & D \end{bmatrix} = \begin{bmatrix} 1 & 0 \\ jB & 1 \end{bmatrix} \begin{bmatrix} 0 & jZ_0 \\ j/Z_0 & 0 \end{bmatrix} \begin{bmatrix} 1 & 0 \\ jB & 1 \end{bmatrix}$$

$$= \begin{bmatrix} -BZ_0 & jZ_0 \\ j(1/Z_0 - B^2 Z_0) & -BZ_0 \end{bmatrix}, \qquad 11.118a$$

while the equivalent transmission line has an *ABCD* matrix given by

$$\begin{bmatrix} A & B \\ C & D \end{bmatrix} = \begin{bmatrix} \cos\theta_e & jZ_e \sin\theta_e \\ j\sin\theta_e / Z_e & \cos\theta_e \end{bmatrix}. \qquad 11.118b$$

So we have that

$$\cos\theta_e = -BZ_0 = -b, \qquad 11.119a$$

$$Z_e = Z_0 \cos\theta_e = \frac{Z_0}{\sqrt{1-b^2}}. \qquad 11.119b$$

For small b, θ_e will be close to $\pi/2$, and these results reduce to

$$\theta_e \simeq \frac{\pi}{2} + b, \qquad 11.120a$$

$$Z_e \simeq Z_0 \left(1 + \frac{b}{2}\right). \qquad 11.120b$$

The susceptance, B, can be implemented with a lumped inductor or capacitor, or with a stub, and switched between two states with a SPST diode switch.

The third type of PIN diode phase shifter is the reflection phase shifter, which uses a SPST switch to control the path length of a reflected signal. Usually a quadrature hybrid is used to provide a two-port circuit, although other types of hybrids, or even a circulator, could be used for this purpose.

Figure 11.43 shows a reflection-type phase shifter using a quadrature hybrid. In operation, an input signal divides equally among the two right-hand ports of the hybrid. The diodes are both biased in the same state (forward or reverse biased), so the waves reflected from the two terminations will add in phase at the indicated output port. Turning the diodes on or off changes the total path length for both reflected waves by $\Delta\phi$, producing a phase shift of $\Delta\phi$ at the output. Ideally, the diodes would look like short circuits in their on state, and open-circuits in their off state, so that the reflection coefficients at the right side of the hybrid can be written as $\Gamma = e^{j\phi}$ for the diodes in their ON

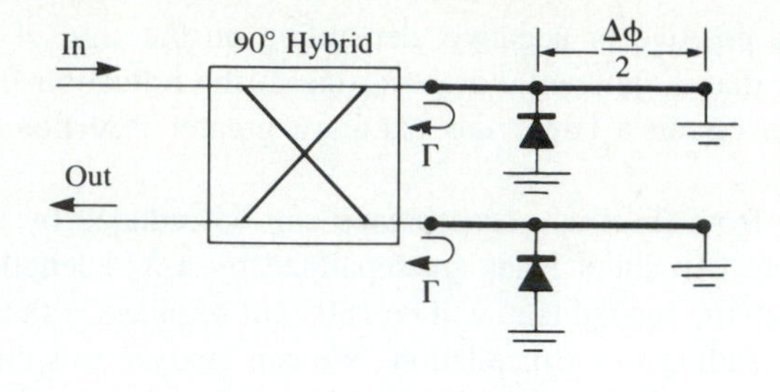

FIGURE 11.43 A reflection phase shifter using a quadrature hybrid.

state, and $\Gamma = e^{j(\phi+\Delta\phi)}$ for the diodes in their OFF state. There are infinite numbers of choices of line lengths that give the desired $\Delta\phi$ (that is, the value of ϕ is a degree of freedom), but it can be shown that bandwidth is optimized if the reflection coefficients for the two states are phase conjugates. Thus, if $\Delta\phi = 180°$, the best bandwidth will be obtained if $\Gamma = \pm j$, or $\phi = 0,\ 2\pi$, etc.

A good input match for the reflection-type phase shifter requires that the diodes be well-matched. The insertion loss is limited by the loss of the hybrid, as well as the forward and reverse resistances of the diodes. Impedance transformation sections can be used to improve performance in this regard.

11.6 MICROWAVE INTEGRATED CIRCUITS

The trend of any maturing electrical technology is toward smaller size, lighter weight, lower cost, and increased complexity. Microwave technology has been moving in this direction for the last 10–20 years, with the development of microwave integrated circuits. This technology serves to replace bulky and expensive waveguide and coaxial components with small and inexpensive planar components, and is analogous to the digital integrated circuitry that has led to the rapid increase in sophistication of computer systems. Microwave integrated circuitry (MIC) can incorporate transmission lines, discrete resistors, capacitors, and inductors, and active devices such as diodes and transistors. MIC technology has advanced to the point where complete microwave subsystems, such as receiver front ends and radar transmit/receive modules, can be integrated on a chip that is only a few square millimeters in size.

There are two distinct types of microwave integrated circuits. *Hybrid* MICs have one layer of metallization for conductors and transmission lines, with discrete components (resistors, capacitors, transistors, diodes, etc.) bonded to the substrate. In a thin-film hybrid MIC, some of the simpler components are deposited on the substrate. Hybrid MICs were first developed in the 1960s, and still provide a very flexible and cost-effective means for circuit implementation. *Monolithic* microwave integrated circuits (MMICs) are a more recent development, where the active and passive circuit elements are grown on the substrate. The substrate is a semiconductor material, and several layers of metal, dielectric, and resistive films are used. Below we will briefly describe these two types of MICs, in terms of the materials and fabrication processes that are required and the relative merits of each type of circuitry.

Hybrid Microwave Integrated Circuits

Material selection is an important consideration for any type of MIC; characteristics such as electrical conductivity, dielectric constant, loss tangent, thermal transfer, mechanical strength, and manufacturing compatability must be evaluated. Generally the substrate material is of primary importance. For hybrid MICs, alumina, quartz, and teflon fiber are commonly used for substrates. Alumina is a rigid ceramic-like material with a dielectric constant of about 9–10. A high dielectric constant is often desirable for lower frequency circuits because it results in a smaller circuit size. At higher frequencies, however, the substrate thickness must be decreased to prevent radiation loss and other spurious effects; then the transmission lines (typically microstrip, slotline, or coplanar waveguide) can become too narrow to be practical. Quartz has a lower dielectric constant (~ 4) which, with its rigidity, makes it useful for higher frequency (> 20 GHz) circuits. Teflon and similar types of soft plastic substrates have dielectric constants ranging from 2 to 10, and can provide a large substrate area at a low cost, as long as rigidity and good thermal transfer are not required. Transmission line conductors for hybrid MICs are typically copper or gold.

After the circuit has been designed, the next step is to lay out the components and their interconnections, and make a mask for the metallization layer. This step is usually performed with a CAD system for integrated circuit layout, such as MICAD or CALMA®. The mask itself may be made on rubylith (a soft mylar film), usually at a magnified scale ($2\times$, $5\times$, $10\times$, etc.) for a high accuracy. Then an actual-size mask is made on a thin sheet of glass or quartz. The metalized substrate is coated with photoresist, covered with the mask, and exposed to a light source. The substrate can then be etched to remove the unwanted areas of metal. Plated-through, or via, holes can be made by evaporating a layer of metal inside a hole that has been drilled in the substrate. Finally, the discrete components are soldered or wire-bonded to the conductors. This is generally the most labor-intensive part of hybrid MIC fabrication, and therefore the most expensive part of the process.

Then the MIC can be tested. Often provision is made for variations in component values and other circuit tolerances by providing tuning or trim stubs that can be manually trimmed for each circuit. This increases circuit yield, but also increases cost since trimming involves labor at a highly skilled level. The layout of a typical hybrid MIC circuit is shown in Figure 11.44. A photograph of a hybrid MIC FET amplifier is shown in Figure 11.45.

Monolithic Microwave Integrated Circuits

Progress in GaAs material processing and device development since the late 1970s has led to the feasibility of the monolithic microwave integrated circuit, where all passive and active components required for a given circuit can be grown or implanted in the

®Registered trademark of General Electric Company.

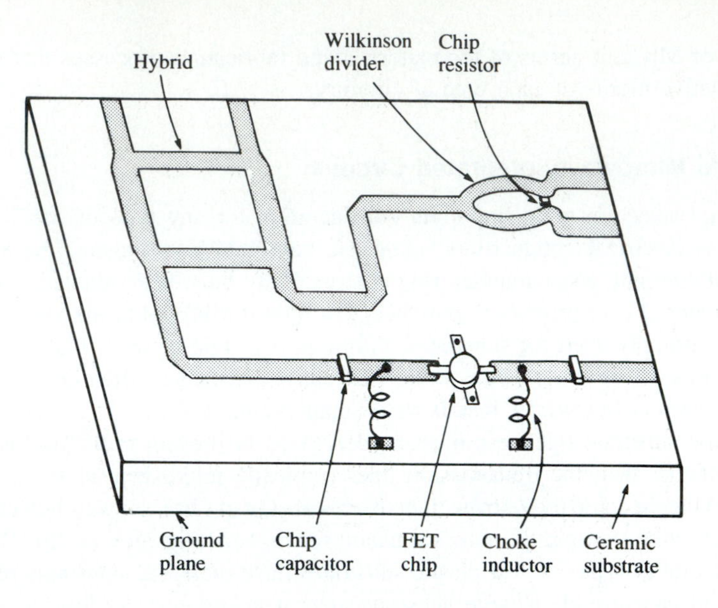

FIGURE 11.44 Layout of a hybrid microwave integrated circuit.

FIGURE 11.45 A hybrid MIC FET amplifier. This circuit uses two identical two-stage amplifiers in parallel with Lange couplers at the input and output.
Courtesy of M/A-COM Inc.

substrate. Potentially, the MMIC can be made at low cost because the manual labor required for fabricating hybrid MICs is eliminated. In addition, a single wafer can contain a large number of circuits, all of which can be processed and fabricated simultaneously.

The substrate of an MMIC must be a semiconductor material to accommodate the fabrication of active devices; the type of devices and the frequency range dictate the type of substrate material. Thus, silicon bipolar transistors can be used up to several Gigahertz, silicon-on-sapphire (SOS) MESFETs can be used up to several Gigahertz, and submicron gate-length GaAs FETs have been used up to 60 GHz. The GaAs FET is a very versatile circuit element, finding applications in low-noise amplifiers, high-gain amplifiers, broadband amplifiers, mixers, oscillators, phase shifters, and switches. Thus, GaAs is probably the most common substrate for MMICs, but silicon, silicon-on-sapphire, and indium phosphide (InP) are sometimes used.

Transmission lines and other conductors are usually made with gold metallization. To improve adhesion of the gold to the substrate, a thin layer of chromium or titanium is generally deposited first. These metals are relatively lossy, so the gold layer must be made at least several skin depths thick to reduce attenuation. Capacitors and overlaying lines require insulating dielectric films, such as SiO, SiO_2, Si_2N_4, and Ta_2O_5. These materials have a high dielectric constant and low loss, and are compatible with integrated circuit processing. Resistors require the deposition of lossy films; NiCr, Ta, Ti, and doped GaAs are commonly used.

Designing a MMIC requires extensive use of CAD software, for circuit design and optimization as well as mask generation. Careful consideration must be given to the circuit design to allow for component variations and tolerances, and the fact that circuit trimming after fabrication will be difficult, or impossible (and defeats the goal of low-cost production). Thus, effects such as transmission line discontinuities, bias networks, spurious coupling, and package resonances must be taken into account.

After the circuit design has been finalized, the masks can be generated. One or more masks are generally required for each processing step. Processing begins by forming an active layer in the semiconductor substrate for the necessary active devices; this can be done by ion implantation or by epitaxial techniques. Then active areas are isolated by etching or additional implantation, leaving mesas for the active devices. Next, ohmic contacts are made to the active device areas by alloying a gold or gold/germanium layer onto the substrate. FET gates are then formed with a titanium/platinum/gold compound deposited between the source and drain areas. At this time the active device processing has been essentially completed, and intermediate tests can be made to evaluate the wafer. If it meets specifications, the next step is to deposit the first layer of metallization for contacts, transmission lines, inductors, and other conducting areas. Then resistors are formed by depositing resistive films, and the dielectric films required for capacitors and overlays are deposited. A second layer of metallization completes the formation of capacitors and any remaining interconnections. The final processing steps involve the bottom, or backside, of the substrate. First it is lapped to the required thickness, then via holes are formed by etching and plating. Via holes provide ground connections to the circuitry on the top side of the substrate, and provide a heat dissipation path from the active devices to the ground plane. After the processing has been completed, the individual circuits can be cut from the wafer, and tested. Figure 11.46 shows the structure of a typical MMIC, and Figure 11.47 shows a photograph of a two-stage MMIC FET amplifier.

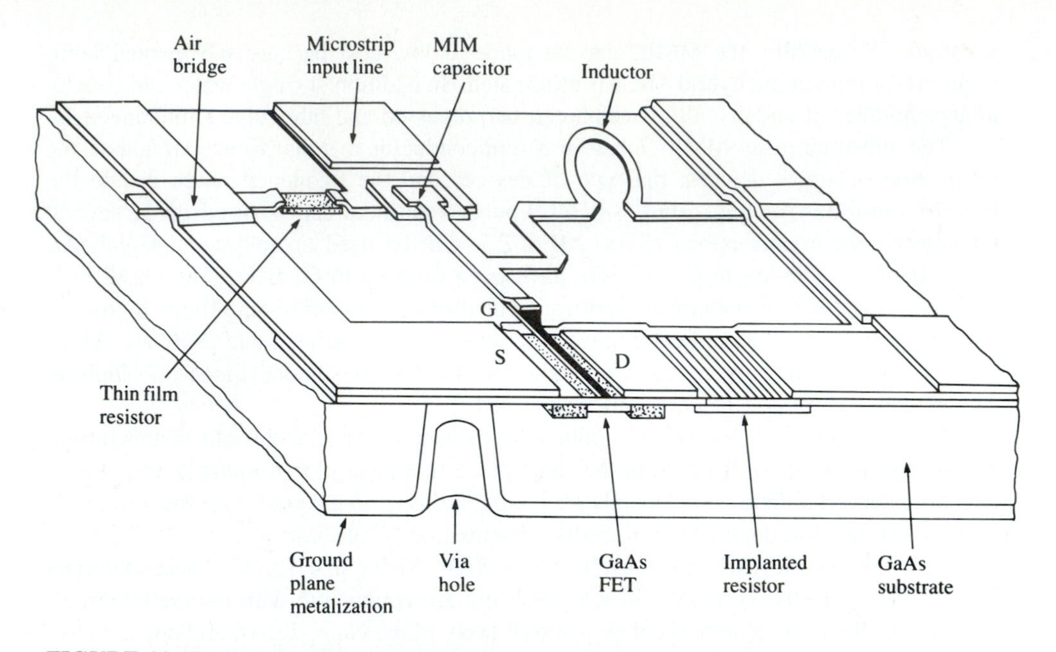

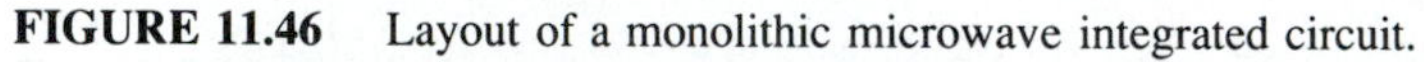

FIGURE 11.46 Layout of a monolithic microwave integrated circuit.

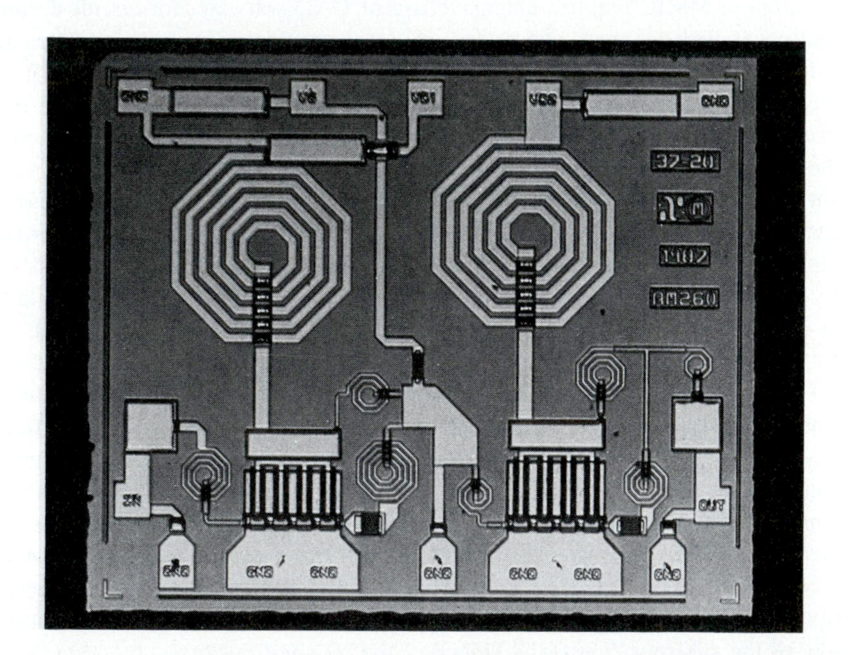

FIGURE 11.47 Photograph of a two-stage 2–8 GHz GaAs FET amplifier. Note the use of several spiral inductors for RF bypass.

Courtesy of Adams-Russell Semiconductor Center, M/A-COM Inc.

Monolithic microwave integrated circuits are not without some disadvantages, when compared with hybrid MICs or other type of circuitry. First, MMICs tend to waste large areas of relatively expensive semiconductor substrate for components such as transmission lines and hybrids. Also, the processing steps and required tolerances for an MMIC are very critical, resulting in low yields. These factors tend to make MMICs expensive, especially when made in small quantities (less than several hundred). MMICs generally require a more thorough design procedure to include effects such as component tolerances and discontinuities, and debugging, tuning, or trimming after fabrication is difficult. Because their small size limits heat dissipation, MMICs cannot be used for circuits requiring more than moderate power levels. And high-Q resonators and filters are difficult to implement in MMIC form because of the inherent resistive losses in MMIC materials.

Besides the obvious features of small size and weight, MMICs have some unique advantages over other types of circuits. Since it is very easy to fabricate additional FETs in an MMIC design, circuit flexibility and performance can often be enhanced with little additional cost. Also, monolithically integrated devices have much less parasitic reactance than discrete packaged devices, so MMIC circuits can often be made with broader bandwidth than hybrid circuits. And MMICs generally give very reproducible results, especially for circuits from the same wafer.

11.7 SUMMARY OF MICROWAVE SOURCES

A source of microwave power is obviously essential for any microwave system. Communication and radar systems generally use a relatively high-power source for the transmitter, and one or more low-power sources for local oscillator and down conversion functions in the receiver. Radar transmitters are often operated in a pulsed mode, and peak powers that are much greater than the continuous power rating of the source can then be attained. Electronic warfare systems use sources in much the same way as a radar system, with the additional requirement for tunability over a wide bandwidth. Radiometer and radio astronomy receiver systems require low-power sources for local oscillators (although it can be argued that the primary source of microwave power for such systems is the radiation emitted from the hot body under observation). Test and measurement systems usually require a low-power microwave source, often tunable over a wide bandwidth. And the microwave oven, that most common of all microwave systems, requires a single-frequency high-power source.

At present, these requirements are met with a variety of solid state and microwave tube sources. Generally the division is between solid state sources for low power and low frequencies, and tubes for high power and/or high frequencies. Figure 11.48 illustrates the power versus frequency performance for these two types of sources. Solid state sources have the advantages of small size, ruggedness, low cost, and compatability with microwave integrated circuits, and so are usually preferred whenever they can meet the necessary power and frequency requirements. But very high power applications are dominated by microwave tubes, and even though the power and frequency performance of solid state sources is steadily improving, it appears that the need for microwave tubes will not be eliminated any time soon.

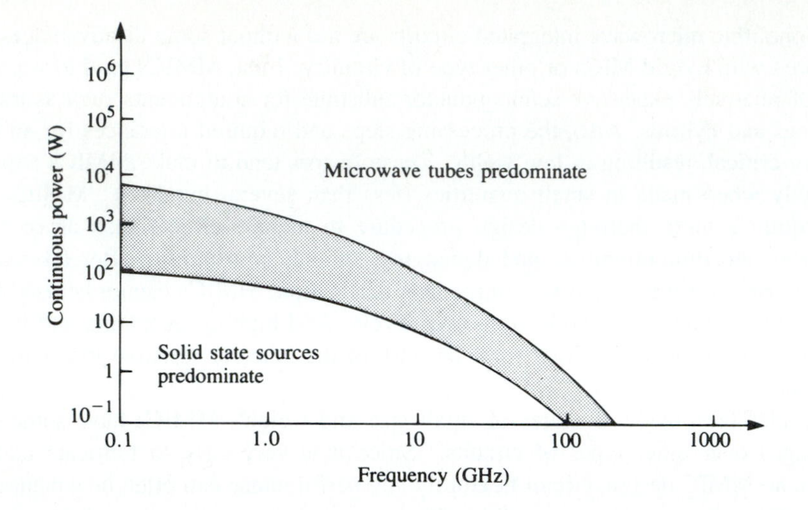

FIGURE 11.48 Power versus frequency performance of solid state sources and microwave tubes.

Presenting the thorough discussion of the operation and design of microwave solid state and tube sources that these components deserve is well beyond the scope of this book. Instead, we will briefly describe and summarize the performance of several of the most common types of solid state and microwave tube sources.

Solid State Sources

Solid state microwave sources can be categorized as two-terminal devices (diodes), or three-terminal devices (transistor oscillators). The most common diode sources are the Gunn diode and the IMPATT diode, both of which directly convert a DC bias to RF power in the frequency range of about 2 to 100 GHz. The Gunn diode is a transferred-electron device that uses a bulk semiconductor (usually GaAs or InP), as opposed to a pn junction [3]. This effect leads to a negative-resistance characteristic that can be employed with an external resonator to produce a stable oscillator. DC to RF efficiencies are generally less than 10%. Figure 11.49 shows the power (continuous and pulsed) versus frequency performance for a variety of commercially available Gunn sources. Gunn diodes can also be used as negative-resistance reflection-type amplifiers. Figure 11.50 shows a photograph of two commercially available Gunn diode sources.

The IMPATT (IMPact ionization Avalanche Transit Time) diode uses a reverse-biased pn junction to generate microwave power [3]. The material is usually silicon or gallium arsenide, and the diode is operated with a relatively high voltage (70–100 V) to achieve a reverse-biased avalanche breakdown current. When coupled with a high-Q resonator and biased at an appropriate operating point, a negative-resistance effect can be achieved at the RF operating frequency, and oscillation will occur. Oscillator circuit design for IMPATT diodes essentially follows the procedure outlined in Section 11.4 for negative-resistance oscillators. IMPATT sources are generally more noisy than sources using Gunn diodes, but are capable of higher powers and higher DC to RF conversion

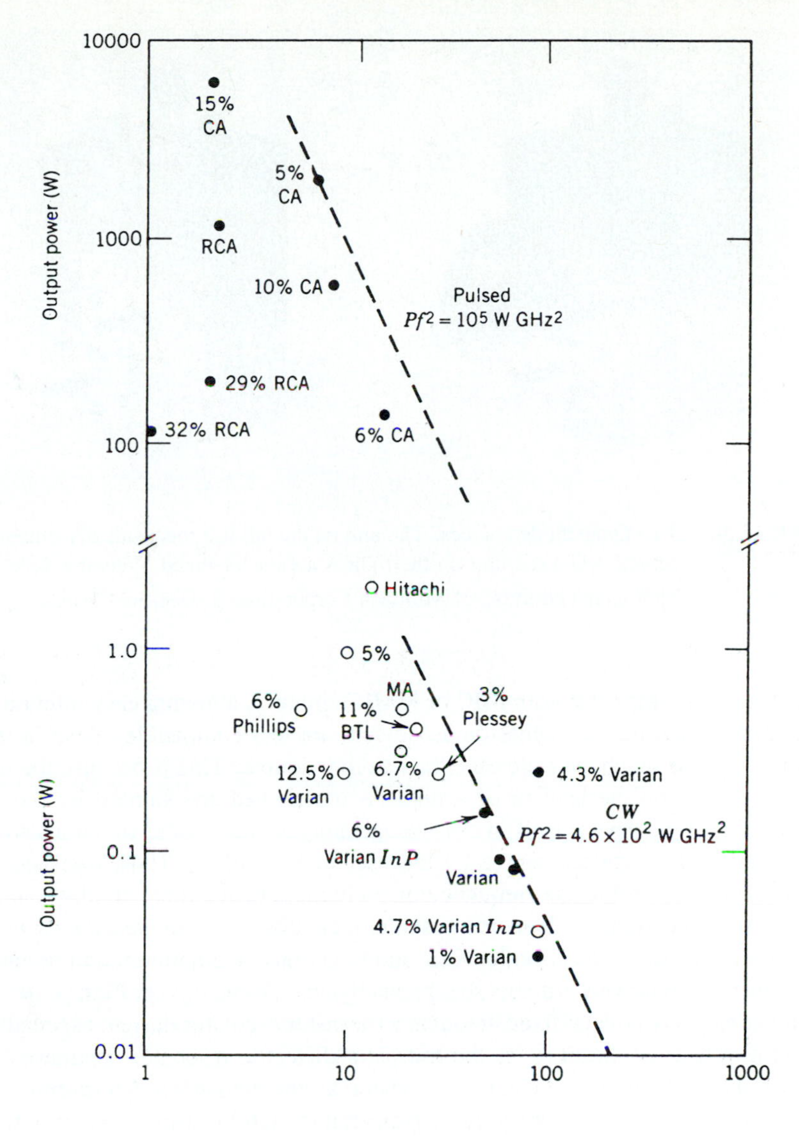

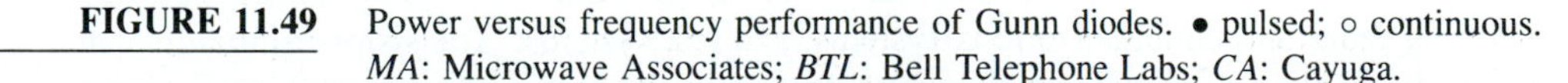

FIGURE 11.49 Power versus frequency performance of Gunn diodes. • pulsed; ○ continuous. *MA*: Microwave Associates; *BTL*: Bell Telephone Labs; *CA*: Cayuga.

efficiencies. IMPATT's also have better temperature stability than Gunn diodes. Figure 11.51 shows the power versus frequency performance for typical commercial IMPATT sources. IMPATT diodes can also be used as negative-resistance amplifiers.

We have already discussed transistor oscillator design in Section 11.4. Such sources generally have lower frequency and power capabilities when compared to Gunn or IMPATT sources, but do offer several advantages over diodes. First, oscillators using GaAs

FIGURE 11.50 Two Gunn diode sources. The unit on the left is a mechanically tunable E-band source, while the unit on the right is a varactor-tuned V-band source. Photograph courtesy of Millitech Corporation, S. Deerfield, Mass.

FETs are readily compatible with MIC or MMIC circuitry, allowing easy integration with FET amplifiers and mixers, while diode devices are less compatible. Also, a transistor oscillator circuit is much more flexible than a diode source. This is because the negative-resistance oscillation mechanism of a diode is determined and limited by the physical characteristics of the device itself, while the operating characteristics of a transistor source can be adjusted to a greater degree by the oscillator circuitry. Thus, transistor oscillators allow more control of the frequency of oscillation, temperature stability, and output noise than do diode sources. Transistor oscillator circuits also lend themselves well to frequency tuning, phase or injection locking, and to various modulation requirements. Transistor sources are relatively efficient, but presently not capable of very high power outputs.

An increasingly popular fixed-frequency transistor oscillator design, especially useful for local oscillator applications, is the transistor dielectric resonator oscillator (TDRO). This uses a small, high-Q dielectric resonator as the frequency-determining load for an FET oscillator circuit. A very good temperature stability can be achieved, and the resonator is compatible with MIC design, leading to a compact and low-cost unit.

Tunable sources are necessary in many types of electronic warfare systems, frequency-hopping radar and communications systems, and test systems. Transistor oscillators can be made tunable by using an adjustable element in the resonant load, such as a varactor diode or a magnetically-biased YIG sphere [8]. When reverse-biased, the junction capacitance of a varactor diode can be controlled with the DC bias voltage. Thus, a voltage-controlled oscillator (VCO) can be made by using a reverse-biased varactor diode in the tank circuit of a transistor oscillator. In a YIG-tuned oscillator (YTO), a single-crystal YIG sphere is used to control the inductance of a coil in the tank circuit of the oscillator. Since YIG is a ferrimagnetic material, its effective permeability can be controlled with an external DC magnetic bias field, thus controlling the oscillator frequency. YIG oscillators can be made to tune over a decade or more of bandwidth, while

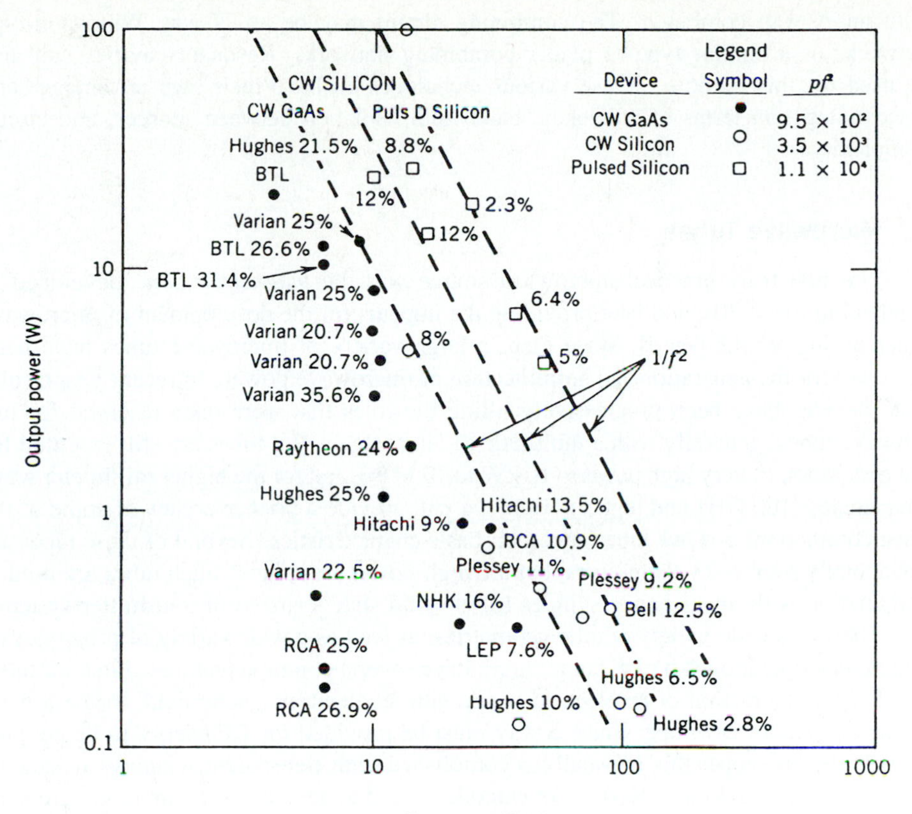

FIGURE 11.51 Power versus frequency performance of IMPATT diodes.

varactor-tuned oscillators are limited to a tuning range of about an octave. YIG-tuned oscillators, however, cannot be tuned as fast as varactor oscillators.

In many applications the RF power requirement exceeds the power capacity of a single solid state source. But because of the many advantages offered by solid state sources, a lot of effort has been directed towards increasing output power through the use of various power combining techniques. Thus, the outputs of two or more sources are combined in phase, effectively multiplying the output power of a single source by the number of individual sources being used. In principle, an unlimited amount of RF power can be generated in this manner; in practice, however, factors such as high-order modes and combiner losses limit the multiplication to about 10–20 dB.

Power combining can be done by combining powers at the device level or at the circuit level. In some applications, power can be combined spatially by using an array of antennas, where each element is fed with a separate source. At the device level, several diode or transistor junctions are essentially connected in parallel over an electrically small region, and used as a single device. This technique is thus limited to a relatively few device junctions. At the circuit level, the power output from N devices can be combined with an N-way combiner. The combining circuit may be an N-way Wilkinson-type network, or a similar type of planar combining network. Resonant cavities can also be used for this purpose. These various techniques all have their own advantages and disadvantages in terms of efficiency, bandwidth, isolation between sources, and circuit complexity.

Microwave Tubes

The first truly practical microwave source was the *magnetron tube*, developed in England in the 1930s, and later providing the impetus for the development of microwave radar during World War II. Since then, a large variety of microwave tubes have been designed for the generation and amplification of microwave power. In recent years, solid state devices have been progressively filling the roles that were once reserved for microwave tubes, generally with a multitude of advantages. But tubes are still essential for the generation of very high powers (10 kW to 10 MW), and for the higher millimeter wave frequencies (100 GHz and higher). Here we will provide a brief overview of some of the most common microwave tubes, and their basic characteristics. Several of these tubes are not actually sources by themselves, but are high-power amplifiers. Such tubes are used in conjunction with lower power sources (often solid state sources) in transmitter systems.

There is a wide variety of tube geometries, as well as a wide variety of principles on which tube operation is based, but all tubes have several common features. First, all tubes involve the interaction of an electron beam with an electromagnetic field, inside a glass or metal vacuum envelope. Thus, a way must be provided for RF energy to be coupled outside the envelope; this is usually accomplished with transparent windows or coaxial coupling probes or loops. Next, a hot cathode is used to generate a stream of electrons by thermionic emission. Cathodes are usually fabricated from a barium oxide-coated metal surface, or an impregnated tungsten surface. The electron stream is then focused into a narrow beam by a focusing anode with a high voltage bias. Alternatively, a solenoidal electromagnet can be used to focus the electron beam. For pulsed operation, a beam modulating electrode is used between the cathode and anode. A positive bias voltage will attract electrons from the cathode, and turn the beam on, while a negative bias will turn the beam off. After the electron beam leaves the region of the tube where the desired interaction with the RF field takes place, a collector element is used to provide a complete current path back to the cathode power supply. The assembly of the cathode, focusing anode, and modulating electrode is called the *electron gun*. Because of the requirement for a high vacuum, and the need to dissipate large amounts of heat, microwave tubes are generally very large and bulky. In addition, tubes often require large, heavy biasing magnets, and high voltage power supplies. Factors to consider when choosing a particular type of tube include power output, frequency, bandwidth, tuning range, and noise.

Microwave tubes can be grouped into two categories, depending on the type of electron beam-field interaction. In *linear-beam*, or "O," type tubes the electron beam traverses the length of the tube, and is parallel to the electric field. In the *crossed-field*, or "m," type tube the focusing field is perpendicular to the accelerating electric field. Microwave tubes can also be classified as either oscillators or amplifiers.

The *klystron* is a linear-beam tube that is widely used as both an amplifier and an oscillator. In a klystron amplifier, the electron beam passes through two or more resonant cavities. The first cavity accepts an RF input and modulates the electron beam by bunching it into high-density and low-density regions. The bunched beam then travels to the next cavity, which accentuates the bunching effect. At the final cavity the RF power is extracted, at a highly amplified level. Two cavities can produce up to about 20 dB of gain, while using four cavities (about the practical limit) can give 80–90 dB gain. Klystrons are capable of peak powers in the megawatt range, with RF output/DC input power conversion efficiencies of 30–50%.

The reflex klystron is a single-cavity klystron tube which operates as an oscillator by using a reflector electrode after the cavity to provide positive feedback via the electron beam. It can be tuned by mechanically adjusting the cavity size. The major disadvantage of klystrons is their narrow bandwidth, which is a result of the high-Q cavities required for electron bunching. Klystrons have very low AM and FM noise levels.

The narrow bandwidth of the klystron amplifier is overcome in the *traveling wave tube* (TWT). The TWT is a linear-beam amplifier that uses an electron gun and a focusing magnet to accelerate a beam of electrons through an interaction region. Usually the interaction region consists of a slow-wave helix structure, with an RF input at the electron gun end, and an RF output at the collector end. The helical structure slows down the propagating RF wave so that it travels at the same velocity as the wave and beam travel along the interaction region, and amplification is effected. Then the amplified signal is coupled from the end of the helix. The TWT has the highest bandwidth of any amplifier tube, ranging from 30 to 120%; this makes it very useful for electronic warfare systems, which require high power over broad bandwidths. It has a power rating of several hundred watts (typically), but this can be increased to several kilowatts by using an interaction region consisting of a set of coupled cavities; the bandwidth will be reduced, however. The efficiency of the TWT is relatively small, typically ranging from 20 to 40%.

A variation of the TWT is the *backward wave oscillator* (BWO). The difference between a TWT and the BWO is that in the BWO, the RF wave travels along the helix from the collector toward the electron gun. Thus the signal for amplification is provided by the bunched electron beam itself, and oscillation occurs. A very useful feature of the BWO is that its output frequency can be tuned by varying the DC voltage between the cathode and the helix; tuning ranges of an octave or more can be achieved. The power output of the BWO, however, is relatively low (typically less than 1 W), so these tubes are generally being replaced with solid state sources.

Another type of linear-beam oscillator tube is the *extended interaction oscillator* (EIO). The EIO is very similar to a klystron, and uses an interaction region consisting of several cavities coupled together, with positive feedback to support oscillation. It has a narrow tuning bandwidth, and a moderate efficiency, but it can supply high powers at frequencies up to several hundred GHz. Only the gyratron can deliver more power.

Crossed-field tubes include the *magnetron*, the *crossed-field amplifier*, and the *gyratron*. As previously mentioned, the magnetron was the first high-power microwave source. It consists of a cylindrical cathode surrounded by a cylindrical anode with several cavity resonators along the inside of its periphery. A magnetic bias field is applied parallel to the cathode-anode axis. In operation, a cloud of electrons is formed which rotates around the cathode in the interaction region. As with linear-beam devices, electron bunching occurs, and energy is transferred from the electron beam to the RF wave. RF power can be coupled out of the tube with a probe, loop, or aperture window.

Magnetrons are capable of very high power outputs—on the order of several kilowatts. And the magnetron has an efficiency of 80% or more. A significant disadvantage, however, is that they are very noisy, and cannot maintain frequency or phase coherency when operated in a pulsed mode. These factors are important for high-performance pulsed radars, where processing techniques operate on a sequence of returned pulses. (Modern radars of this type today generally use a stable low-noise solid state source, followed by a TWT for power amplification.) The application of magnetrons is now primarily for microwave cooking.

The crossed-field amplifier (CFA) has a geometry similar to a TWT, but employs a crossed-field interaction that is similar to that of the magnetron. The RF input is applied to a slow-wave structure in the interaction region of the CFA, but the electron beam is deflected by a negatively biased electrode, called the sole, to force the beam perpendicular to the slow-wave structure. In addition, a magnetic bias field is applied perpendicular to this electric field, and perpendicular to the electron beam direction. The magnetic field exerts a force on the electron beam that counteracts the field from the sole. In the absence of an RF input, the electric and magnetic fields are adjusted so that their effects on the electron beam cancel, leaving the beam to travel parallel to the slow-wave structure. Applying an RF field causes velocity modulation of the beam, and bunching occurs. The beam is also periodically deflected toward the slow-wave circuit, producing an amplified signal. Crossed-field amplifiers have very good efficiencies—up to 80%, but the gain is limited to 10–15 dB. Also, the CFA has a noisier output than either a klystron amplifier or TWT. Its bandwidth can be up to 40%.

Another crossed-field tube is the gyratron, which can be used as an amplifier or an oscillator. This tube consists of an electron gun with input and output cavities along the axis of the electron beam, similar to a klystron amplifier. But the gyratron also has a solenoidal bias magnet that provides an axial magnetic field. This field forces the electrons to travel in tight spirals down the length of the tube. The electron velocity is high enough so that relativistic effects are important. Bunching occurs, and energy from the transverse component of the electron velocity is coupled to the RF field.

A significant feature of the gyratron is that the frequency of operation is determined by the bias field strength and the electron velocity, as opposed to the dimensions of the tube itself. This makes the gyratron especially useful for millimeter wave frequencies; it offers the highest output power (10–100 kW) of any tube in this frequency range. It also has a high efficiency for tubes in the millimeter wave range. The gyratron is a relatively new type of tube, but it is rapidly replacing tubes such as reflex klystrons and EIOs as sources of millimeter wave power.

Figures 11.52 and 11.53 summarize the power versus frequency performance of microwave tube oscillators and amplifiers.

REFERENCES

[1] F. T. Ulaby, R. K. Moore, and A. K. Fung, *Microwave Remote Sensing: Active and Passive, Volume I, Microwave Remote Sensing, Fundamentals and Radiometry*. Addison-Wesley, Reading, Mass, 1981.

[2] S. A. Maas, *Microwave Mixers*, Artech House, Dedham, Mass, 1986.

[3] S. Y. Yngvesson, *Microwave Semiconductor Devices*, Kluwer Academic Publishers, 1991.

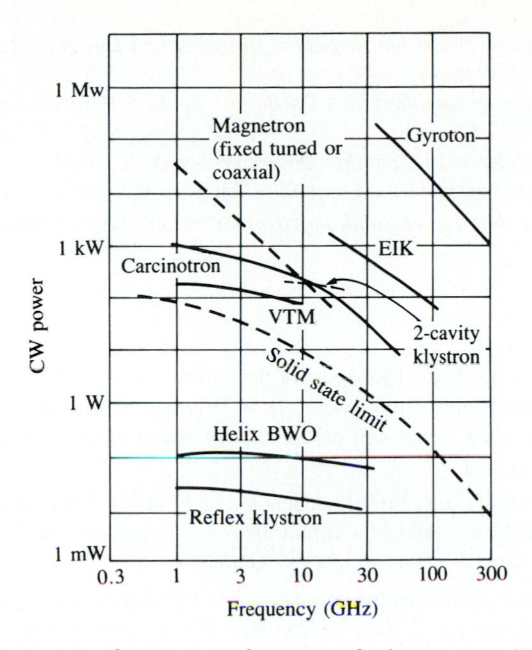

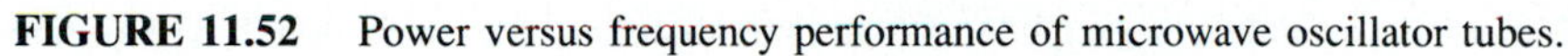

FIGURE 11.52 Power versus frequency performance of microwave oscillator tubes.

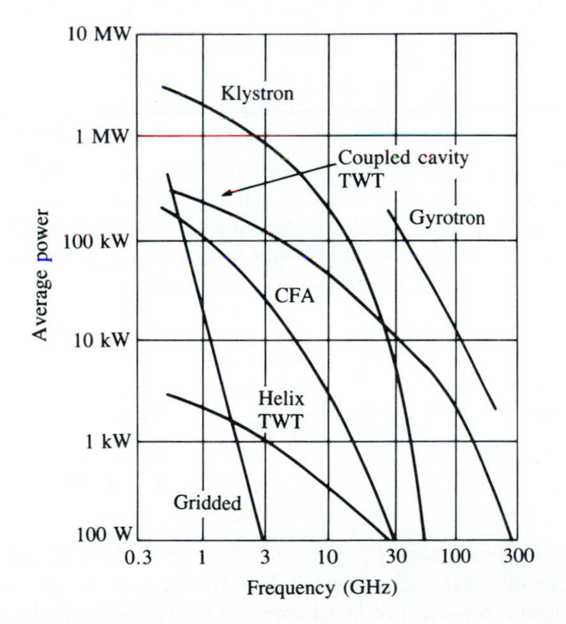

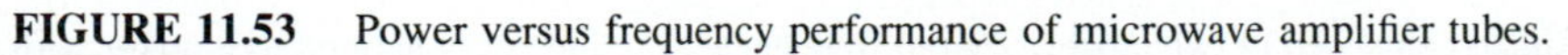

FIGURE 11.53 Power versus frequency performance of microwave amplifier tubes.

[4] G. Gonzalez, *Microwave Transistor Amplifiers, Analysis and Design*, Prentice-Hall, Englewood Cliffs, N. J., 1984.
[5] G. D. Vendelin, *Design of Amplifiers and Oscillators by the S-Parameter Method*, Wiley, N. Y., 1982.
[6] T. T. Ha, *Solid-State Microwave Amplifier Design*, Wiley, N. Y., 1981.
[7] C. Gentile, *Microwave Amplifiers and Oscillators*, McGraw-Hill, N. Y., 1987.
[8] I. Bahl and P. Bhartia, *Microwave Solid-State Circuit Design*, Wiley Interscience, N. Y. 1988.

PROBLEMS

11.1 The Y-factor method is to be used to measure the equivalent noise temperature of a component. A hot load of $T_1 = 300$ K and a cold load of $T_2 = 77$ K will be used. If the noise temperature of the amplifier is $T_e = 250$ K, what will be the ratio of power meter readings at the output of the component for the two loads?

11.2 Assume that measurement errors limit the accuracy of Y in a Y-factor measurement to $\pm 5\%$. If $T_1 = 290$ K and $T_2 = 77$ K, calculate and plot the resulting percent error in T_e, for $T_e = 50$ K to 250 K. What is the value of T_e that results in minimum error?

11.3 It is necessary to connect an antenna to a low-noise receiver with a transmission line. The frequency is 10 GHz, and the distance is 2 meters. The choices are: copper X-band waveguide, RG-8/U coaxial cable, or copper circular waveguide with an inner diameter of 2.0 cm. Which type of line should be used for the best noise figure? Disregard impedance mismatch.

11.4 A certain transmission line has a noise figure $F = 1$ dB at a temperature $T_0 = 290$ K. Calculate and plot the noise figure of this line as its physical temperature ranges from $T = 0$ K to 1000 K.

11.5 An amplifier with a bandwidth of 1 GHz has a gain of 15 dB and a noise temperature of 250 K. If the 1 dB compression point occurs for an input power level of -10 dBm, what is the dynamic range of the amplifier?

11.6 An amplifier with a gain of 12 dB, a bandwidth of 150 MHz, and a noise figure of 4.0 dB feeds a receiver with a noise temperature of 900 K. Find the noise figure of the overall system.

11.7 Consider the microwave system shown below, where the bandwidth is 1 GHz centered at 20 GHz, and the physical temperature of the system is $T = 300$ K. What is the equivalent noise temperature of the source? What is the noise figure of the amplifier, in dB? What is the noise figure of the cascaded transmission line and amplifier, in dB? When the noisy source is connected to the system, what is the total noise power output of the amplifier, in dBm?

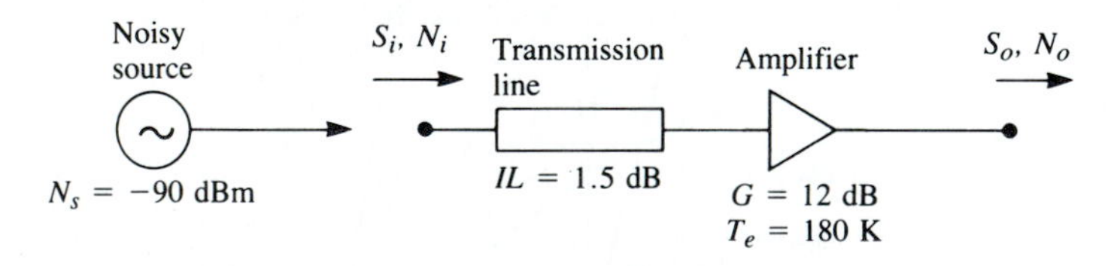

11.8 A diode has the following parameters: $C_j = 0.1$ pF, $R_s = 15\ \Omega$, $I_s = 0.1\ \mu$A, and $L_p = C_p = 0$. Compute the open-circuit voltage sensitivity at 10 GHz for $I_0 = 0$, 20, and 50 μA. Assume $\alpha = 1/25$ mV, and neglect the effect of bias current on the junction capacitance.

11.9 An input signal composed of two closely spaced frequencies, ω_1 and ω_2, is applied to a mixer with an LO frequency of ω_0. (See the input spectrum shown below.) Calculate and sketch the resulting output spectrum due to the v^2 term of the mixer response.

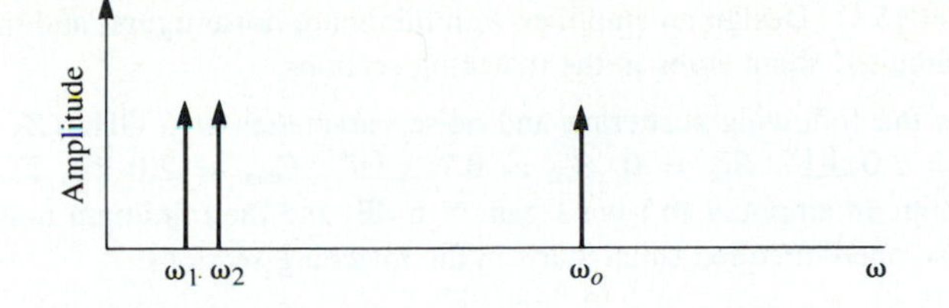

11.10 A *phase detector* produces an output signal proportional to the phase difference between two RF input signals. Let these input signals be expressed as

$$v_1 = v_0 \cos \omega t,$$

$$v_2 = v_0 \cos(\omega t + \theta).$$

If these two signals are applied to a single-balanced mixer using a 90° hybrid, show that the IF output signal, after low-pass filtering, is given by

$$i = kv_0^2 \sin \theta,$$

where k is a constant. If the mixer uses a 180° hybrid, show that the corresponding output signal is given by

$$i = kv_0^2 \cos \theta.$$

11.11 A two-tone input with a 6 dB difference in the two signal levels is applied to a nonlinear component. What is the relative power ratio of the resulting two third-order intermodulation products $2\omega_1 - \omega_2$ and $2\omega_2 - \omega_1$, if ω_1 and ω_2 are close together?

11.12 Derive (11.57) for third-order intercept points, assuming the individual distortion products add in phase at the output of the cascade.

11.13 A microwave transistor has the following S parameters: $S_{11} = 0.34\angle{-170°}$, $S_{21} = 4.3\angle{80°}$, $S_{12} = 0.06\angle{70°}$, and $S_{22} = 0.45\angle{-25°}$. Determine the stability, and plot the stability circles if the device is potentially unstable.

11.14 Repeat Problem 11.13 for the following transistor S parameters: $S_{11} = 0.8\angle{-90°}$, $S_{21} = 5.1\angle{80°}$, $S_{12} = 0.3\angle{70°}$, and $S_{22} = 0.62\angle{-40°}$.

11.15 Design an amplifier for maximum gain at 5.0 GHz with a GaAs FET that has the following S parameters ($Z_0 = 50\ \Omega$) : $S_{11} = 0.65\angle{-140°}$, $S_{21} = 2.4\angle{50°}$, $S_{12} = 0.04\angle{60°}$, $S_{22} = 0.70\angle{-65°}$. Design matching sections using open-circuited shunt stubs.

11.16 Design an amplifier with maximum G_{TU} using a transistor with the following S parameters ($Z_0 = 50\,\Omega$) at 6.0 GHz: $S_{11} = 0.61\angle{-170°}$, $S_{21} = 2.24\angle{32°}$, $S_{12} = 0$, $S_{22} = 0.72\angle{-83°}$. Design L-section matching sections using lumped elements.

11.17 Design an amplifier to have a gain of 10 dB at 6.0 GHz, using a transistor with the following S parameters ($Z_0 = 50\ \Omega$): $S_{11} = 0.61\angle{-170°}$, $S_{21} = 2.24\angle{32°}$, $S_{12} = 0$, $S_{22} = 0.72\angle{-83°}$. Plot (and use) constant gain circles for $G_S = 1$ dB and $G_L = 2$ dB. Use matching sections with open-circuited shunt stubs.

11.18 Compute the unilateral figure of merit for the transistor of Problem 11.13. What is the maximum error in the transducer gain if an amplifier is designed assuming the device is unilateral?

11.19 Show that the 0 dB gain circle for G_S ($G_S = 1$), defined by (11.92), will pass through the center of the Smith chart.

11.20 A GaAs FET has the following scattering and noise parameters at 8 GHz ($Z_0 = 50\ \Omega$): $S_{11} = 0.7\angle{-110°}$, $S_{12} = 0.02\angle{60°}$, $S_{21} = 3.5\angle{60°}$, $S_{22} = 0.8\angle{-70°}$, $F_{\min} = 2.5$ dB, $\Gamma_{\text{opt}} =$

$0.70\angle 120°$, $R_N = 15\ \Omega$. Design an amplifier with minimum noise figure, and maximum possible gain. Use open-circuited shunt stubs in the matching sections.

11.21 A GaAs FET has the following scattering and noise parameters at 6 GHz ($Z_0 = 50\,\Omega$): $S_{11} = 0.6\angle -60°$, $S_{21} = 2.0\angle 81°$, $S_{12} = 0$, $S_{22} = 0.7\angle -60°$, $F_{\min} = 2.0$ dB, $\Gamma_{\text{opt}} = 0.62\angle 100°$, $R_N = 20\ \Omega$. Design an amplifier to have a gain of 6 dB, and the minimum noise figure possible with this gain. Use open-circuited shunt stubs in the matching sections.

11.22 Repeat Problem 11.21, but design the amplifier for a noise figure of 2.5 dB, and the maximum possible gain that can be achieved with this noise figure.

11.23 Prove that the standard Smith chart can be used for negative resistances by plotting $1/\Gamma^*$ (instead of Γ). Then the resistance circle values are read as negative, while the reactance circles are unchanged.

11.24 Design a transistor oscillator at 6 GHz using an FET in a common source configuration driving a 50 Ω load on the drain side. The S parameters are ($Z_0 = 50\ \Omega$): $S_{11} = 0.9\angle -150°$, $S_{21} = 2.6\angle 50°$, $S_{12} = 0.2\angle -15°$, $S_{22} = 0.5\angle -105°$. Calculate and plot the output stability circle, and choose Γ_T for $|\Gamma_{\text{in}}| >> 1$. Design load and terminating networks.

11.25 A single-pole, single-throw switch uses a PIN diode in a shunt configuration. The frequency is 4 GHz, $Z_0 = 50\ \Omega$, and the diode parameters are, $C_j = 0.5$ pF, $R_r = 0.5\ \Omega$, $R_f = 0.3\ \Omega$, $L_i = 0.3$ nH. Find the electrical length of an open-circuited shunt stub, placed across the diode, to minimize the insertion loss for the ON state of the switch. Calculate the resulting insertion losses for the ON and OFF states.

11.26 A single-pole, single-throw switch is constructed using two identical PIN diodes in the arrangement shown below. In the ON state, the series diode is forward biased and the shunt diode is reversed biased; and vice versa for the OFF state. If $f = 6$ GHz, $Z_0 = 50\ \Omega$, $C_j = 0.1$ pF, $R_r = 0.5\ \Omega$, $R_f = 0.3\ \Omega$, and $L_i = 0.4$ nH, determine the insertion losses for the ON and OFF states.

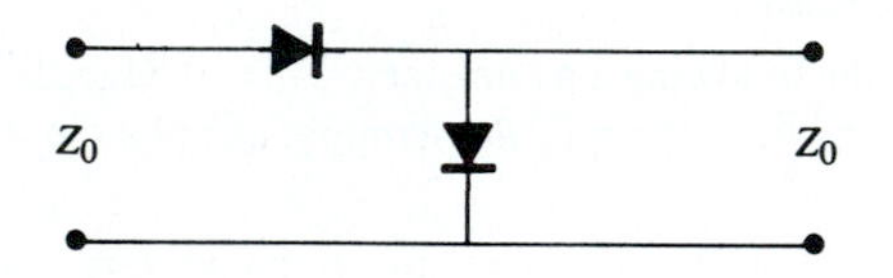

11.27 Consider the loaded-line phase shifter shown below. If $Z_0 = 50\ \Omega$, find the necessary stub lengths for a differential phase shift of 45°, and calculate the resulting insertion loss for both states of the phase shifter. Assume all lines are lossless, and that the diodes can be approximated as ideal shorts or opens.

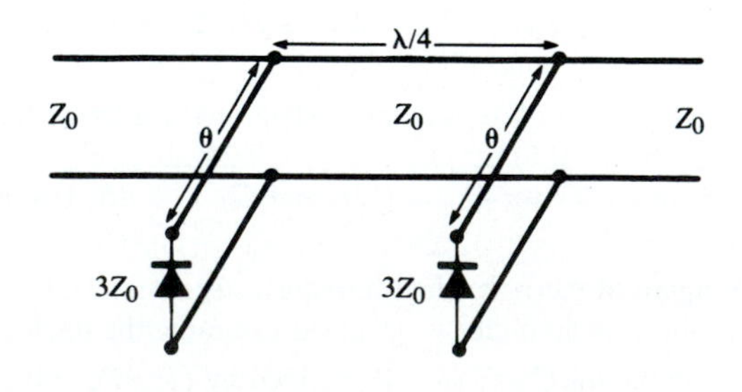

CHAPTER 12

Introduction to Microwave Systems

A microwave system consists of passive and active microwave components arranged to perform a useful function. Probably the two most important examples are microwave radar systems and microwave communication systems, but there are many others. In this chapter we will describe the basic operation of several types of microwave systems to give a general overview of the application of microwave technology, and to show how many of the subjects of earlier chapters fit into the overall scheme of complete microwave systems.

An important component in any radar or communication system is the antenna, so we will discuss some of the basic properties of antennas in Section 12.2. Then communication, radar, and radiometry systems are treated as important applications of microwave technology. Propagation effects, biological effects, and other miscellaneous applications are also briefly discussed.

All of the above topics are of sufficient depth that many books have been written for each. Thus our purpose here is not to give a complete and thorough treatment of these subjects, but instead to introduce these topics as a way of placing the other material in this book in the larger systems context. The interested reader is referred to the references at the end of the chapter for more complete treatments.

12.1 SYSTEM ASPECTS OF ANTENNAS

In this section we will discuss some of the basic characteristics of antennas that will be needed for our study of radar and communications systems. In essence, an antenna is a component that converts a wave propagating on a transmission line to a plane wave propagating in free-space (transmission), or vice versa (reception). There is a wide variety of antenna types and geometries, but all antennas can be described with a set of parameters and terms, as discussed below.

Definition of Important Antenna Parameters

Radiation pattern. The power radiated (or received) by an antenna is a function of angular position and radial distance from the antenna. At electrically large distances (many wavelengths), r, the power density drops off as $1/r^2$ in any direction. The variation of power density with angular position is determined by the type and design of the antenna, and can be graphically represented as a radiation pattern plot. Such a plot may take the form of a three-dimensional graph of power versus elevation and azimuth angles, but more commonly the radiation pattern is represented by principal plane pattern plots, such as the E-plane or H-plane where one angle is held fixed while the other is varied. The transmit and receive patterns of an antenna are identical if the antenna contains no nonreciprocal materials or components.

Far-field. This is the region away from an antenna where the radiated wave essentially takes the form of a plane wave. A commonly used criteria is $2D^2/\lambda$, where D is the maximum linear dimension of the antenna, and λ is the operating wavelength. Radiation patterns are generally assumed to be in the far-field of the antenna.

Directivity. Many antennas are used to transmit or receive power in a fixed direction, and so it may be desired to maximize the radiation pattern, or antenna response, in this direction. A quantitative measure of this response is the directive gain of the antenna, for a given direction. The maximum value of directive gain is the directivity of the antenna. The directivity depends solely on the shape of the radiation pattern.

Efficiency. Like other microwave components, an antenna may dissipate power due to conductor loss or dielectric loss, and so an antenna efficiency can be defined as the ratio of total power radiated by the antenna to the input power of the antenna.

Gain. Antenna gain is the product of efficiency and directivity, and accounts for the fact that loss reduces the power density radiated in a given direction.

Impedance. An antenna presents a driving-point impedance to the source or load to which it is connected, and so impedance mismatch with a feed line can occur. This mismatch degrades antenna performance, and is dependent on the external circuitry which is connected to the antenna.

Bandwidth. The usable frequency bandwidth of an antenna may be limited by impedance mismatch or pattern deterioration. Matching networks can sometimes be used to increase the impedance bandwidth of an antenna.

Polarization. The polarization of an antenna refers to the polarization of the electric field vector of the radiated wave. Typical polarizations include linear (vertical or horizontal), and circular (RHCP or LHCP). Some antennas are designed to operate with two polarizations.

Size/complexity. A basic characteristic of an antenna is that, for efficient operation, its size must be $\lambda/2$ or greater, and that higher gain can only be achieved with larger size. In addition, features such as pattern shaping, polarization control, or large bandwidth are generally obtained only with an increase in the complexity of the antenna.

Basic Types of Antennas

Because of the diversity implied in the above characteristics, and the fact that the operation of an antenna is very dependent on its geometry, a wide variety of antenna types and geometries have been developed. We summarize some of these below, and in Figure 12.1

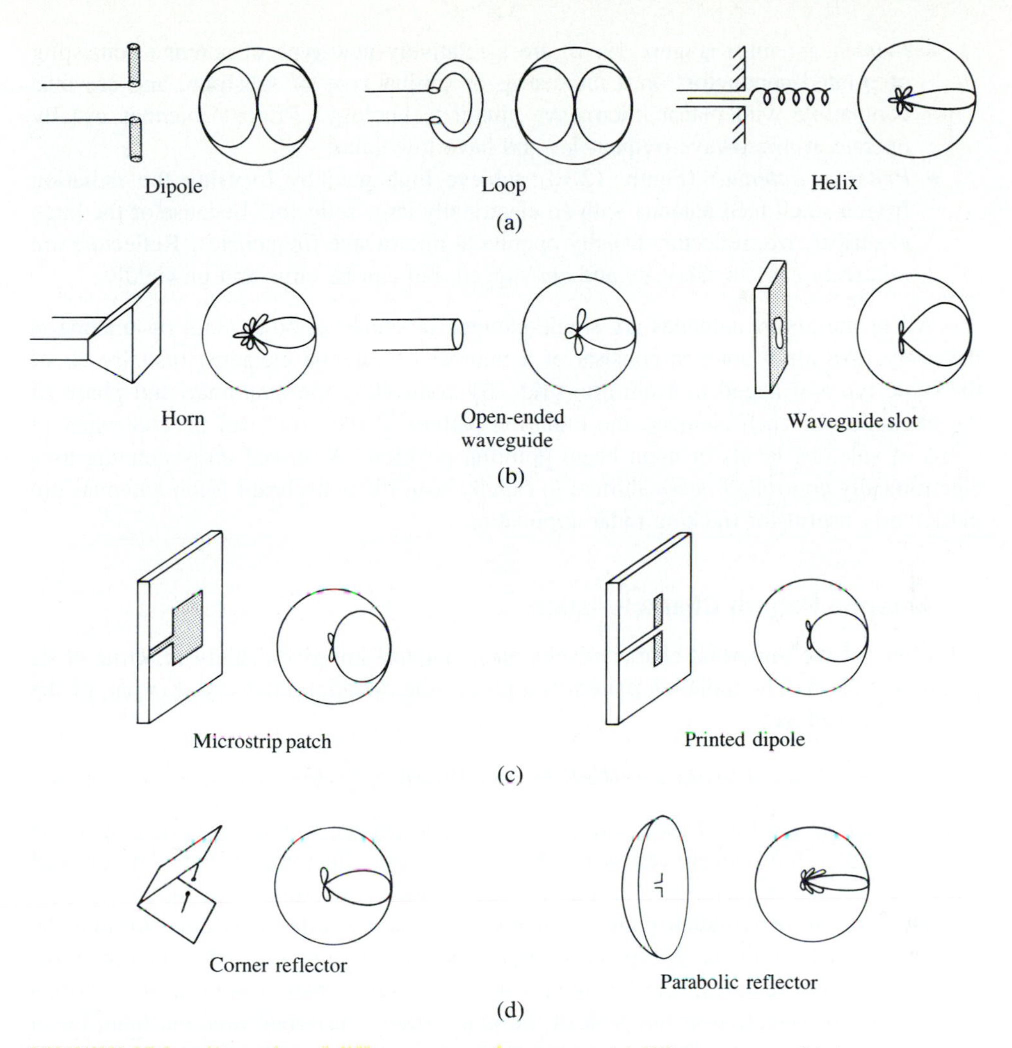

FIGURE 12.1 Examples of different types of antennas. (a) Wire antennas. (b) Aperture antennas. (c) Printed antennas. (d) Reflector antennas.

- *Wire antennas* (Figure 12.1a) are probably the simplest types of antennas, and most of the earliest antennas (e.g., as used by Hertz and Marconi) were of this type. Wire dipoles (and monopoles mounted on a ground plane) are most commonly used at lower frequencies (HF to UHF), and have relatively low gains. Wire antennas are easy to fabricate, easy to feed, and are lightweight.
- *Aperture antennas* (Figure 12.1b) are often just flared sections of waveguide (a horn), or even open-ended waveguides. Slots in a waveguide or a ground plane are additional examples. Aperture antennas are most commonly used at microwave frequencies and have moderate gains.

- *Printed antennas* (Figure 12.1c) are a relatively new type of antenna consisting of printed conductors on a microstrip or similar type of substrate, and are thus compatible with planar microwave circuit technology. Printed antennas usually operate at microwave frequencies and have low gains.
- *Reflector antennas* (Figure 12.1d) achieve high gain by focusing the radiation from a small feed antenna with an electrically large reflector. Because of the large electrical size, reflectors usually operate at microwave frequencies. Reflectors are relatively easy to fabricate and are rugged, but can be large and unwieldly.

All of the above antennas are single-element antennas; another class of antenna is the array. An array antenna consists of a number of antenna elements (usually all of the same type) arranged in a uniform grid. By controlling the amplitude and phase of the excitation of each element, the radiation pattern of the array can be controlled in terms of sidelobe levels or main beam pointing position. A phased array antenna uses electronically controlled phase shifters to rapidly scan the main beam; such antennas are particularly useful for tracking radar applications.

Antenna Pattern Characteristics

Many of the important characteristics of an antenna are given solely in terms of its radiation pattern. The radiation pattern is a plot of the radiation intensity, $F(\theta, \phi)$, of the antenna, defined as

$$F(\theta, \phi) = r^2|\bar{E}(\theta, \phi) \times \bar{H}^*(\theta, \phi)| = r^2 S(\theta, \phi), \tag{12.1}$$

where $\bar{E}$ and $\bar{H}$ are the electric and magnetic fields radiated to the far-field zone of the antenna, and S is the magnitude of the Poynting vector. Since the $\bar{E}$ and $\bar{H}$ fields will decay as $1/r$, the r^2 factor in (12.1) effectively removes this range dependence.

Principal plane radiation patterns are made by taking orthogonal cuts through the main beam of the antenna. A typical example, shown in Figure 12.2, identifies the most important pattern features. Most antennas have a radiation pattern with a well-formed main beam; the direction of the peak of the main beam is referred to as the main beam direction, or scan angle. The sharpness of the main beam is quantified by the 3 dB beamwidth; this beamwidth may be different in the two principal planes. Next to the main beam are the sidelobes; the height of the largest sidelobe relative to the main beam is called the peak sidelobe level (or just sidelobe level), and is usually given in dB. The pattern plot is usually in dB, normalized to the peak of the main beam. A rectangular plot is shown in Figure 12.2, but polar plots are also commonly used.

An antenna having a radiation pattern with relatively narrow beamwidths in both principal planes is sometimes called a *pencil-beam antenna*. Similarly, an antenna having a narrow beamwidth in one plane and a much broader beamwidth in the orthogonal plane is referred to as a *fan-beam antenna*.

Like the beamwidth, the directivity is a measure of the sharpness of the main beam, or the extent to which the pattern is focussed in a given direction. It is defined as,

$$D = \frac{4\pi F_{\max}}{P_{\text{rad}}} = \frac{4\pi F_{\max}}{\int_{\phi=0}^{2\pi} \int_{\theta=0}^{\pi} F(\theta, \phi) \sin\theta \, d\theta \, d\phi}, \tag{12.2}$$

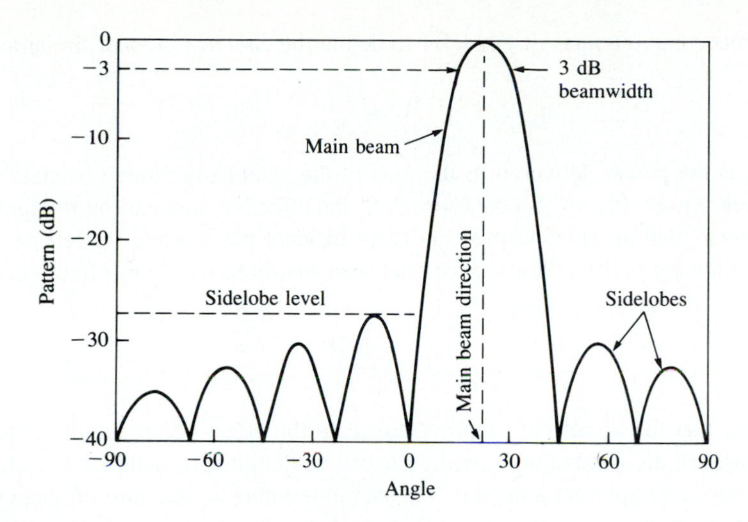

FIGURE 12.2 A typical antenna radiation pattern.

where F_{max} is the maximum value of radiation intensity, F, and P_{rad} is the total power radiated by the antenna. An isotropic antenna is a hypothetical antenna having $F(\theta, \phi) =$ constant; from (12.2) the directivity of such an antenna would be unity, or 0 dB. No real antenna has an isotropic pattern, however, so directivity will always be greater than unity. Narrower beamwidth implies a larger directivity.

Since beamwidth and directivity are measures of the same thing, we might expect a relation between them. In fact, the relationship depends on the shape of the main beam, as well as the rest of the radiation pattern, but an approximate result that is useful for many cases of practical interest is given by

$$D \simeq \frac{32,400}{\theta_1 \theta_2}, \qquad 12.3$$

where θ_1, θ_2 are the beamwidths, in degrees, of the main beam in the two principal planes.

If a transmitting antenna radiates a total power of P_{rad} with an isotropic pattern (equal power density in all directions), then the power density (Poynting vector magnitude) at a distance r from the antenna is given by P_{rad} divided by the surface area of the sphere through which all power passes: $S = P_{rad}/4\pi r^2$ W/m^2. If the antenna has a directivity D, then the power density in the direction of the main beam (the maximum power density) will be

$$S = \frac{P_{rad} D}{4\pi r^2} \text{ W/m}^2, \qquad 12.4$$

since, by the definition of (12.2), the directivity can be considered as the ratio of the peak radiation intensity to the average radiation intensity. The result in (12.4) can also be derived by combining (12.1) and (12.2).

For receiving antennas, it is useful to define the *effective area* of the antenna as

$$A_e = \frac{P_\ell}{S}, \tag{12.5}$$

where P_ℓ is the power delivered to the load of the antenna (assumed lossless), and S is the incident power density. Since $P_\ell = A_e S$, the effective area can be thought of as the "capture area" that an antenna presents to an incident plane wave. It can be shown [1] that the following exact relation exists between the directivity and effective area of an antenna:

$$D = \frac{4\pi A_e}{\lambda^2}. \tag{12.6}$$

This shows that the directivity is proportional to the effective area of an antenna. The effective area of an antenna may be greater or less than its physical cross-sectional area, but for many large aperture antennas it is useful to define an aperture efficiency, η_a, that relates the effective area to the physical area, A:

$$A_e = \eta_a A. \tag{12.7}$$

Then, using (12.6), the directivity of such an antenna can be written in terms of its physical area as,

$$D = \frac{4\pi \eta_a A}{\lambda^2}. \tag{12.8}$$

Typical aperture efficiencies for large antennas range from 0.60 to 1.0, depending on the type of antenna, excitation, and other factors.

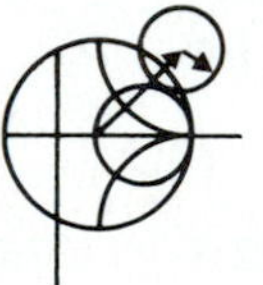

EXAMPLE 12.1

A short wire dipole of length ℓ, where $\ell << \lambda$, is oriented along the z-axis and has far-zone fields given by the following expressions:

$$E_\theta = \frac{jk_0\eta_0 I_0 \ell}{4\pi r} \sin\theta e^{-jk_0 r},$$

$$H_\phi = \frac{jk_0 I_0 \ell}{4\pi r} \sin\theta e^{-jk_0 r},$$

where I_0 is the current on the dipole (constant along the length of the dipole). Use these results to compute the radiation intensity, F, the directivity, and the effective area of the dipole.

Solution

First observe that $E_\theta = \eta_0 H_\phi$, which is characteristic of plane wave propagation. From (12.1) the radiation intensity is given as

$$F(\theta, \phi) = r^2 |\bar{E} \times \bar{H}^*| = \eta_0 \left(\frac{k_0 I_0 \ell}{4\pi}\right)^2 \sin^2\theta.$$

The dipole thus has a beam maximum at $\theta = 90°$, and the pattern is independent of ϕ. From (12.2) we compute the directivity:

$$P_{\text{rad}} = C \int_{\phi=0}^{2\pi} \int_{\theta=0}^{\pi} \sin^3 \theta \, d\theta d\phi = \frac{8\pi C}{3},$$

where C is a constant that will be cancelled in the next step. Since $F_{\max} = C$, we have

$$D = \frac{4\pi F_{\max}}{P_{\text{rad}}} = 1.5.$$

Using (12.6) gives the effective area of the dipole as

$$A_e = \frac{\lambda^2 D}{4\pi} = \frac{3\lambda^2}{8\pi}.$$

Table 12.1 lists the directivities for some common types of antennas.

Antenna Efficiency, Gain, and Temperature

The *radiation efficiency* of an antenna is defined as

$$\eta = \frac{P_{\text{rad}}}{P_{\text{in}}}, \quad 0 \leq \eta \leq 1, \qquad 12.9$$

where P_{rad} is the total power radiated by the antenna and P_{in} is the input power to the antenna. Thus, the radiation efficiency is a measure of how much power is lost in the antenna, to dielectric and conductor losses. These losses reduce the radiated power in any given direction, so in terms of the input power the power density in the direction of the main beam is

$$S = \frac{\eta P_{\text{in}} D}{4\pi r^2} \text{ W/m}^2, \qquad 12.10$$

as derived from (12.4) and (12.9). Most antennas have relatively high radiation efficiencies, typically ranging from $\eta = 0.6$ to 0.95.

TABLE 12.1 Directivities of Common Antennas

Antenna	Directivity
Short dipole	1.5
$\lambda/2$ dipole	1.6
$\lambda/4$ monopole	3.2
Rectangular aperture ($a \times b$) (uniform excitation)	$4\pi ab/\lambda^2$
Rectangular aperture ($a \times b$) (TE_{10} excitation)	$32ab/\pi\lambda^2$

Another factor which can reduce the effective radiated power is reflection loss due to impedance mismatch between the feed line and the antenna; the radiated power is reduced by a factor of $(1 - |\Gamma|^2)$, where Γ is the reflection coefficient between the line and the antenna. This loss factor is identical in effect to the radiation efficiency, but is not included in the radiation efficiency (or in the gain of the antenna) because it is dependent on the external circuitry connected to the antenna, while the radiation efficiency is a characteristic of the antenna itself.

While the directivity depends only on the shape of the radiation pattern, and not the efficiency, the gain includes the effect of loss. Thus

$$G = \eta D, \tag{12.11}$$

which shows that antenna gain never exceeds the directivity.

If the entire pattern of a lossless antenna ($\eta = 1$) sees a background with a temperature T_b, the equivalent noise temperature of the antenna will be T_b. But if the antenna is so lossy that $\eta \simeq 0$, the (matched) antenna will look like a resistor at temperature T_p, where T_p is the physical temperature of the antenna, and so will have an equivalent noise temperature T_p. For other values of efficiency, the antenna will have an equivalent noise temperature given as the weighted average of these two results:

$$T_e = \eta T_b + (1 - \eta) T_p. \tag{12.12}$$

Thus the antenna of a receiver system picks up noise power from the background at which it is aimed, as well as the thermal noise resulting from the internal losses of the antenna. Noise power can be received from the main beam of the antenna, as well as all the sidelobes, albeit at a reduced level. The temperature T_b is the equivalent temperature of the background, and is referred to as the brightness temperature. If this temperature varies over the pattern of the antenna, an integration is required [1], [2] to determine the net temperature seen by the antenna. For example, the overhead sky has a brightness temperature of about 5 K, while near the horizon the brightness temperature can increase to 100–150 K.

12.2 MICROWAVE COMMUNICATION SYSTEMS

Microwave communication links are an important practical application of microwave technology, and are used to carry voice, data, or television signals over distances ranging from intercity links to deep-space spacecraft. In this section we will give a brief introduction to microwave communications systems, concentrating on microwave radio systems.

Types of Communication Systems

Microwave communication systems can be grouped into two types: guided-wave systems, where the signal is transmitted over a low-loss cable or waveguide, and radio links, where the signal propagates through space. In both cases the information-carrying

TABLE 12.2 Typical Communication Channel Bandwidths

Type	Bandwidth
Voice	4 kHz
Television	6 MHz
Digital voice (PCM)	64 kHz
Digital data	50 kHz–1.5 MHz

signal will have a much lower bandwidth than the microwave carrier frequency. Some typical channel bandwidths are given in Table 12.2.

Using modulation and multiplexing techniques, a microwave link can carry a large number of individual channels; this is one of the principal advantages of a microwave communication link. For example, a nominal 4 GHz microwave carrier with a relatively modest 10% bandwidth can carry 100,000 voice channels, or 66 television channels.

Guided-wave communication channels may use coaxial line or waveguide, but the high attenuation of coax generally limits its application to frequencies below 1 GHz. At microwave and millimeter wave frequencies, less loss is obtained with circular waveguides. Ever since the 1930s, when waveguide was first being developed, the TE_{01} circular waveguide mode has created interest for communications systems because its attenuation decreases with frequency (see Section 4.4). The problem is that the TE_{01} mode is not the dominant mode of circular waveguide, so power can be coupled to undesired propagating modes; this spurious power effectively increases the attenuation of the guide. Consequently, very few microwave links using waveguide have been developed beyond the prototype stage, although recently there has been interest in millimeter wave waveguide links [3]. Instead, fiber optic cables (which are essentially waveguides at optical frequencies), are being heavily used for long-distance telephone traffic, as they offer low-loss, very high bandwidths, and a high degree of ruggedness.

Because of loss, the power level on any transmission line or waveguide (including optical fiber) decreases exponentially with distance, as $e^{-2\alpha z}$, where α is the attenuation constant of the line, and z is the distance from the source. As we saw in the previous section, however, the power radiated from an antenna falls off as $1/R^2$, where R is the distance from the antenna. Thus, in the absence of other effects (e.g., propagation loss, polarization mismatch, obstructions), a radio link will intrinsically have less path loss than a guided-wave link, for large distances. A radio link is also advantageous in that right-of-ways on land are not required, unlike guided-wave systems. Of course, radio links are mandatory for satellite or spacecraft applications.

Microwave radio propagation is essentially *line-of-sight*, meaning that microwave signals travel in straight lines, and do not follow the contour of the earth or reflect off the ionosphere, as do lower frequency signals. So for long-distance links on the ground, repeater stations are required at frequent intervals, to receive and retransmit the signals (usually in both directions). A communications satellite is a repeater also, connecting widely separated earth stations. Figure 12.3 shows a photograph of a NAVSTAR navigational satellite.

FIGURE 12.3 Photograph of a NAVSTAR global positioning satellite, showing the solar cell panels and the twelve-element array of helix antennas. This is an L-band system that provides highly precise navigational information to users on the ground, in the air, and on the sea.

Courtesy of Satellite and Space Electronics Division, Rockwell International, Seal Beach, Calif.

The Friis Power Transmission Formula

A basic microwave radio link is shown in Figure 12.4, where P_t is the transmitter power, G_t and G_r are the transmit and receive antenna gains, and R is the distance between the antennas. It is desired to find P_r, the power received at the receiver.

If we assume that the main beams of both antennas are aligned with each other, then the radiated power density at the receive antenna can be found from (12.10) and (12.11)

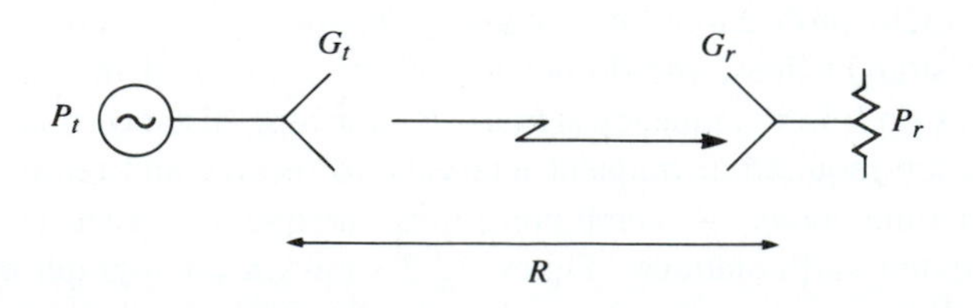

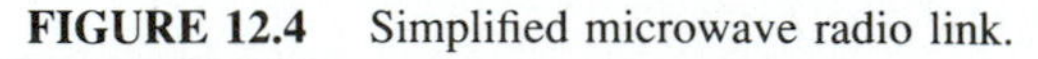

FIGURE 12.4 Simplified microwave radio link.

as

$$S = \frac{P_t G_t}{4\pi R^2} \text{ W/m}^2. \qquad 12.13$$

Then, from (12.5), the power collected by the receive antenna is

$$P_r = S A_e,$$

where A_e is the effective area of the receive antenna. We can express A_e in terms of the gain of the receive antenna using (12.6) and (12.11):

$$A_e = \frac{\lambda^2 G_r}{4\pi}, \qquad 12.14$$

which then includes the effect of losses in the receive antenna. Combining these results gives the received power as

$$P_r = P_t \frac{G_t G_r \lambda^2}{(4\pi R)^2}, \qquad 12.15$$

which is known as the *Friis power transmission equation* [1]. This result does not include the effects of antenna impedance mismatch, polarization mismatch, or losses due to propagation effects.

Note that the Friis equation shows that the received power is proportional to the gain of either antenna, and decays as $1/R^2$. Figure 12.5 illustrates the difference in path loss between free-space propagation and guided wave propagation by plotting the attenuation versus path length for three types of transmission lines and free-space. The

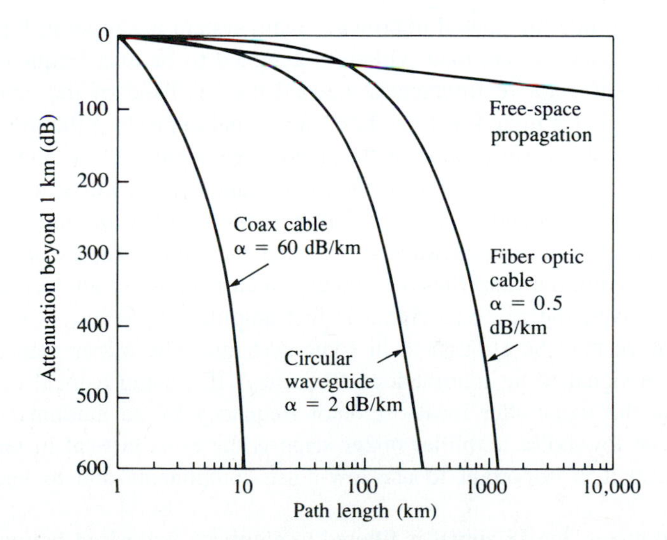

FIGURE 12.5 Attenuation for various transmission schemes.

three transmission lines include a coaxial cable, a circular waveguide (TE_{01} mode), and an optical fiber, each with an attenuation constant that is typical for such a line. The curves are normalized to 0 dB at $R = 1$ km, to eliminate the dependences on frequency and antenna gain. Observe that the path loss for the transmission lines becomes very severe for large distances, but that free-space path loss behaves quite differently. Of course, signals on transmission lines would not be susceptible to propagation effects, weather conditions, and external interference.

EXAMPLE 12.2

An earth station with a transmitter power of 120 W, a frequency of 6 GHz, and an antenna gain of 42 dB transmits to a satellite repeater. The receiver antenna on the satellite has a gain of 31 dB, and the satellite is in a synchronous orbit 35,900 km above the earth. What is the received power, in dBm?

Solution
The numerical antenna gains are

$$G_t = 10^{42/10} = 15849,$$

$$G_r = 10^{31/10} = 1259.$$

The wavelength is $\lambda = 0.05$m. Then from (12.15) the received power is

$$P_r = P_t \frac{G_t G_r \lambda^2}{(4\pi R)^2} = \frac{120(15849)(1259)(0.05)^2}{(4\pi)^2(35.9 \times 10^6)^2} = 2.94 \times 10^{-11}\text{W}$$

$$= -75.3 \text{ dBm}.$$

○

Microwave Transmitters and Receivers

A basic amplitude-modulated microwave radio system is shown in Figure 12.6. The input baseband signal (voice, data, video) is assumed to be at a frequency f_m; a low-pass filter serves to remove frequencies beyond the passband of the channel. Next, a local oscillator signal is mixed with the baseband signal to produce the modulated carrier. The mixer essentially performs a product function (see Section 11.2), and yields a double sideband signal. That is, the output of the mixer contains a lower sideband, $f_{\text{LO}} - f_m$, as well as an upper sideband, $f_{\text{LO}} + f_m$. This process is called *up-conversion.* The local oscillator signal, f_{LO}, is a microwave signal, and so is much higher in frequency than f_m. The power amplifier amplifies the signal, which is then radiated by the antenna.

At the receiver, the received signal is first amplified by a low-noise amplifier, although this stage may be eliminated in some systems. The mixer then *down-converts* the microwave signal to an intermediate frequency (IF), using a local oscillator which is offset from the transmitter local oscillator frequency by an amount equal to the IF frequency. The low-noise amplifier-mixer stage is the most critical in terms of overall noise figure, so it is important to use low-noise components and to keep losses to a minimum.

After the mixer, the IF signal is filtered to eliminate undesired harmonics, and amplified by the IF amplifier. The IF amplifier/filter has high gain with a narrow bandwidth

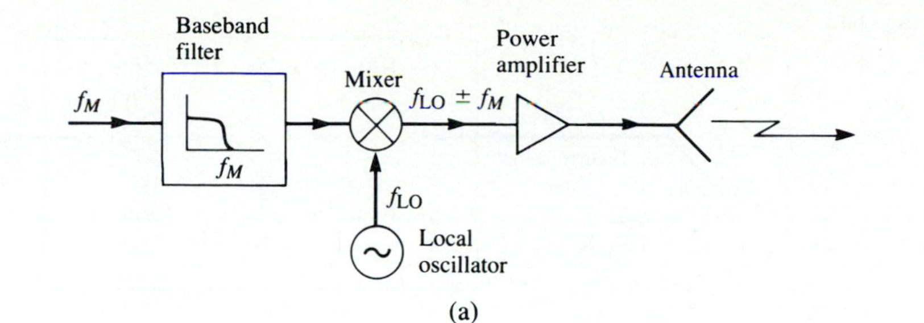

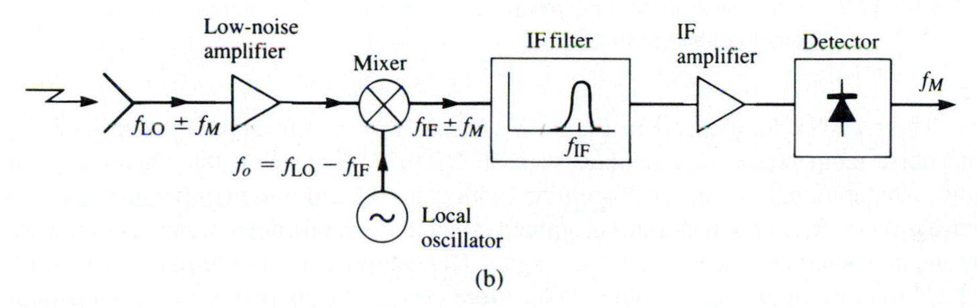

FIGURE 12.6 Block diagrams of an AM microwave transmitter and receiver. (a) Microwave radio transmitter. (b) Microwave radio receiver.

(about $2f_m$), and leads to less noise power than if a high-gain RF amplifier were used alone. Using an IF amplifier also minimizes the effect of $1/f$ noise. In addition, the receiver can easily be tuned by changing the LO frequency. This system is known as a superheterodyne receiver. The output of the IF amplifier goes to the detector, from which the baseband signal, f_m, is recovered.

There are many variations on the above design, with the most common being the use of different modulation schemes. Single sideband (SSB) modulation generates a signal with only one sideband (either $f_{LO} - f_m$ or $f_{LO} + f_m$), which uses one-half the bandwidth of the double-sideband case. A single sideband signal can be generated using a single-sideband mixer, or by filtering out one of the sidebands from a double sideband signal. Better S/N ratios can be obtained with frequency modulation (FM), where the frequency of the RF carrier is varied according to the modulation voltage variation.

Noise Characterization of a Microwave Receiver

Let us now use the results of Section 11.1 to analyze the noise characteristics of a complete antenna-transmission line-receiver front end, as shown in Figure 12.7. In this system the total noise power at the output of the receiver, N_o, will be due to contributions from the antenna pattern, the loss in the antenna, the loss in the transmission line, and from the receiver components. This noise power will determine the minimum detectable signal level for the receiver and, for a given transmitter power, the maximum range of the communication link.

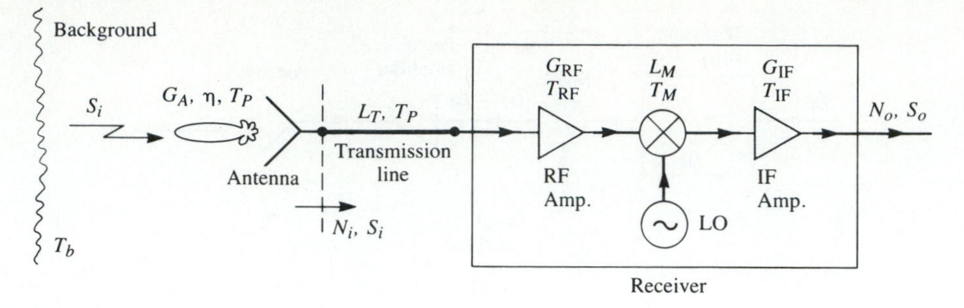

FIGURE 12.7 Noise analysis of a microwave receiver front end, including antenna and transmission line contributions.

The receiver components in Figure 12.7 consist of an RF amplifier with gain G_{RF} and noise temperature T_{RF}, a mixer with an RF-to-IF conversion loss factor L_M and noise temperature T_M, and an IF amplifier with gain G_{IF} and noise temperature T_{IF}. The noise effects of later stages can be ignored, since the overall noise figure is dominated by the characteristics of the first few stages. The component noise temperatures can be related to noise figures as $T = (F - 1)T_0$. From (11.22) the equivalent noise temperature of the receiver can be found as

$$T_{REC} = T_{RF} + \frac{T_M}{G_{RF}} + \frac{T_{IF} L_M}{G_{RF}} . \qquad 12.16$$

The transmission line connecting the antenna to the receiver has a loss L_T, and is at a physical temperature T_p. So from (11.15) its equivalent noise temperature is,

$$T_{TL} = (L_T - 1)T_p. \qquad 12.17$$

Again using (11.22), the noise temperature of the transmission line (TL) and receiver cascade is

$$\begin{aligned} T_{TL+REC} &= T_{TL} + L_T T_{REC} \\ &= (L_T - 1)T_p + L_T T_{REC}. \end{aligned} \qquad 12.18$$

This noise temperature is defined at the antenna terminals (the input to the transmission line).

As discussed in Section 12.1, the entire antenna pattern can collect noise power. If the antenna has a reasonably high gain with relatively low sidelobes, we can assume that all noise power comes via the main beam, so that the noise temperature of the antenna is given by (12.12):

$$T_A = \eta T_b + (1 - \eta)T_p, \qquad 12.19$$

where η is the efficiency of the antenna, T_p is its physical temperature, and T_b is the equivalent brightness temperature of the background seen by the main beam. (One must be careful with this approximation, as it is quite possible for the noise power collected by the sidelobes to exceed the noise power collected by the main beam, if the sidelobes are aimed at a hot background.) The noise power at the antenna terminals, which is also the noise power delivered to the transmission line, is

$$N_i = kBT_A = kB[\eta T_b + (1-\eta)T_p], \tag{12.20}$$

where B is the system bandwidth. If S_i is the received power at the antenna terminals, then the input signal-to-noise ratio at the antenna terminals is S_i/N_i.

The output signal power is

$$S_o = S_i L_T G_{RF} L_M G_{IF} = S_i G_{SYS}, \tag{12.21}$$

where G_{SYS} has been defined as a system power gain. The output noise power is

$$\begin{aligned} N_o &= [N_i + kB(T_{TL} + T_{REC})]G_{SYS} \\ &= kB(T_A + T_{TL} + T_{REC})G_{SYS} \\ &= kB[\eta T_b + (1-\eta)T_p + (L_T - 1)T_p + L_T T_{REC}]G_{SYS} \\ &= kBT_{SYS}G_{SYS}, \end{aligned} \tag{12.22}$$

where T_{SYS} has been defined as the overall system noise temperature. The output signal-to-noise ratio is,

$$\frac{S_o}{N_o} = \frac{S_i}{kBT_{SYS}} = \frac{S_i}{kB[\eta T_b + (1-\eta)T_p + (L_T - 1)T_p + L_T T_{REC}]}. \tag{12.23}$$

It may be possible to improve this signal-to-noise ratio by various signal processing techniques. Note that it may appear to be convenient to use an overall system noise figure to calculate the degradation in signal-to-noise ratio from input to output for the above system, but one must be very careful with such an approach because noise figure is defined for $N_i = kT_0B$, which is not the case here. It is often less confusing to work directly with noise temperatures and powers, as we did above.

EXAMPLE 12.3

A microwave receiver like that of Figure 12.7 has the following parameters:

$f = 4.0$ GHz, $G_{RF} = 20$ dB,
$B = 1$ MHz, $F_{RF} = 3.0$ dB,
$G_A = 26$ dB, $L_M = 6.0$ dB,
$\eta = 0.90$, $F_M = 7.0$ dB,
$T_p = 300$ K, $G_{IF} = 30$ dB,
$T_b = 200$ K, $F_{IF} = 30$ dB.
$L_T = 1.5$ dB,

If the received power at the antenna terminals is $S_i = -80$ dBm, calculate the input and output signal-to-noise ratios.

Solution
We first convert the above dB quantities to numerical values, and noise figures to noise temperatures:

$$G_{\text{RF}} = 10^{20/10} = 100,$$
$$G_{\text{IF}} = 10^{30/10} = 1000,$$
$$L_T = 10^{1.5/10} = 1.41,$$
$$L_M = 10^{6/10} = 4.0,$$
$$T_M = (F_M - 1) = (10^{7/10} - 1)(290) = 1163 \text{ K},$$
$$T_{\text{RF}} = (F_{\text{RF}} - 1)T_0 = (10^{3/10} - 1)(290) = 289 \text{ K},$$
$$T_{\text{IF}} = (F_{\text{IF}} - 1)(10^{1.1/10} - 1)(290) = 84 \text{ K}.$$

Then from (12.16), (12.17), and (12.19) the noise temperatures of the receiver, transmission line, and antenna are

$$T_{\text{REC}} = T_{\text{RF}} + \frac{T_M}{G_{\text{RF}}} + \frac{T_{\text{IF}}L_M}{G_{\text{RF}}} = 289 + \frac{1163}{100} + \frac{84(4.0)}{100} = 304 \text{ K},$$
$$T_{\text{TL}} = (L_T - 1)T_p = (1.41 - 1)300 = 123 \text{ K},$$
$$T_A = \eta T_b + (1 - \eta)T_p = 0.9(200) + (1 - 0.9)(300) = 210 \text{ K}.$$

Then the input noise power, from (12.20), is

$$N_i = kBT_A = 1.38 \times 10^{-23}(10^6)(210) = 2.9 \times 10^{-15} \text{ W} = -115 \text{ dBm}.$$

So the input signal-to-noise ratio is,

$$\frac{S_i}{N_i} = -80 + 115 = 34 \text{ dB}.$$

From (12.22) the total system noise temperature is

$$T_{\text{SYS}} = T_A + T_{\text{TL}} + L_T T_{\text{REC}} = 210 + 123 + (1.41)(304) = 762 \text{ K}.$$

This result clearly shows the noise contributions of the various components. The output signal-to-noise ratio is found from (12.23) as

$$\frac{S_o}{N_o} = \frac{S_i}{kBT_{\text{SYS}}},$$

$$kBT_{\text{SYS}} = 1.38 \times 10^{-23}(10^6)(762) = 1.05 \times 10^{-14} \text{ W} = -110 \text{ dBm},$$

so

$$\frac{S_o}{N_o} = -80 + 110 = 30\text{dB}.$$

○

Frequency-Multiplexed Systems

To take advantage of the wide absolute bandwidth of a microwave channel, a large number of narrow bandwidth channels can be multiplexed together. This is essentially a process of up-converting groups of voice, data, or video channels to consecutive frequency bands in the RF spectrum of the transmitter, and then down-converting at the receiver. The procedure is known as frequency-division multiplexing (FDM), and is in contrast to time-division multiplexing (TDM), where small "time-slices" of a large number of input signals are transmitted sequentially.

A simplified FDM radio is shown in Figure 12.8. At the transmitter, several narrowband baseband channels are first multiplexed together to form a group. For example, a standard grouping for voice channels is to multiplex twelve 4 kHz (nominal) bandwidth channels into a 60–108 kHz frequency band. This can be done using single-sideband mixers and a set of precision local oscillators. Several of these groups can then be combined into one RF channel, and several RF channels may be accommodated on the same RF carrier. Two RF channels are shown in Figure 12.8. To avoid wasted bandwidth, all mixing should be single-sideband.

The resulting multiplexed signal is then amplified and transmitted. At the receiver, RF channel dropping filters are used to separate the RF channels. These filters are usually three- or four-port components, where the desired channel is filtered out through one port, and the remaining spectrum is passed through to the next channel dropping filter. Each RF channel is then down-converted, IF filtered and amplified, and demultiplexed into the original baseband channels. Practical FDM systems have capacities of 10^5 or more telephone channels. A frequency-hopping system is similar to a FDM system, but uses a psuedo-random sequence of transmission frequencies to provide immunity from interference and jamming.

12.3 RADAR SYSTEMS

Radar, or Radio Detection And Ranging, is probably the most prevalent application of microwave technology. In its basic operation, a transmitter sends out a signal which is partly reflected by a distant target, and then detected by a sensitive receiver. If a narrow beam antenna is used, the target's direction can be accurately given by the position of the antenna. The distance to the target is determined by the time required for the signal to travel to the target and back, and the radial velocity of the target is related to the Doppler shift of the return signal. Below are listed some of the typical applications of radar systems.

Civilian Applications

- Airport surveillance
- Marine navigation
- Weather radar
- Altimetry
- Aircraft landing
- Burglar alarms
- Speed measurement (police radar)

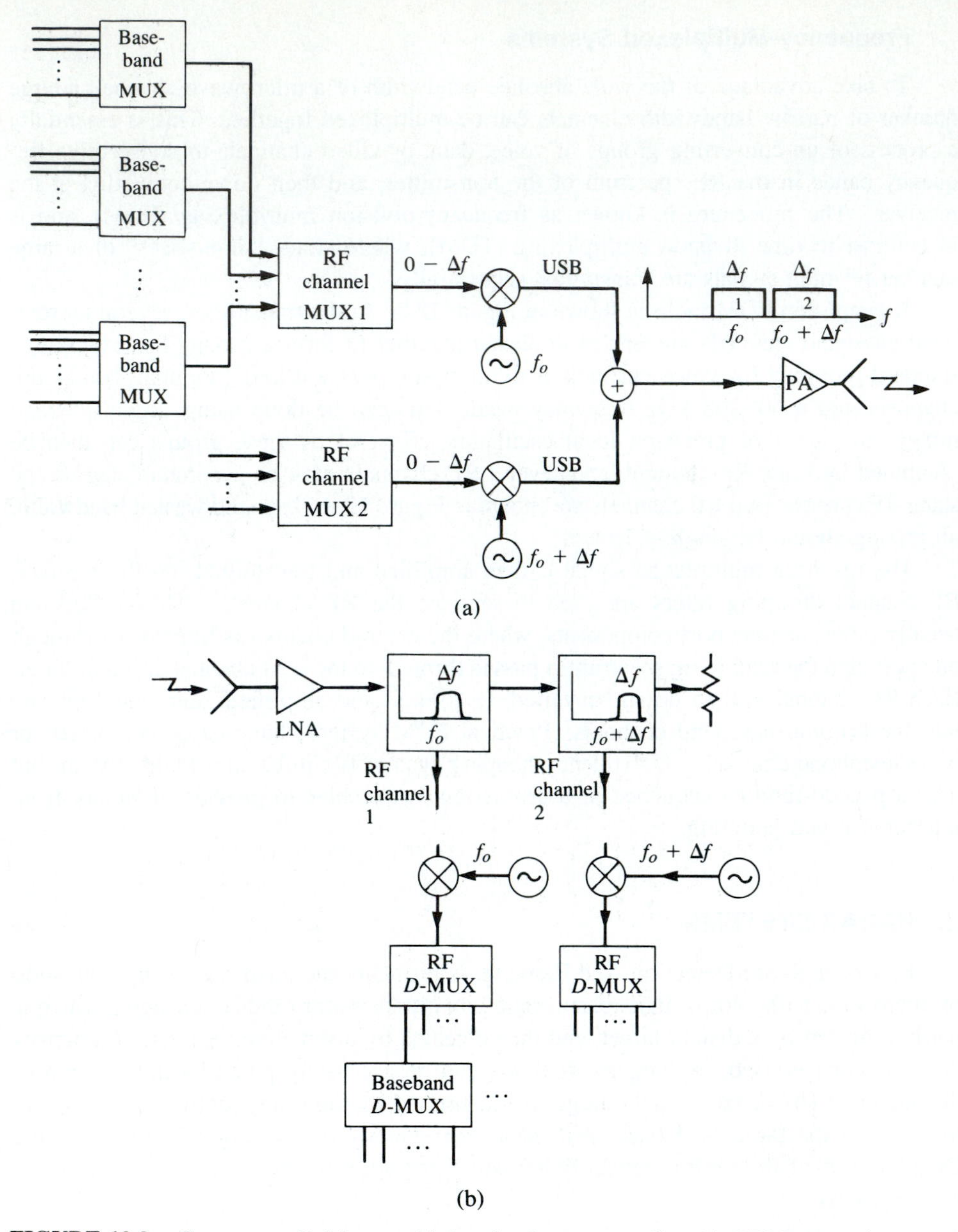

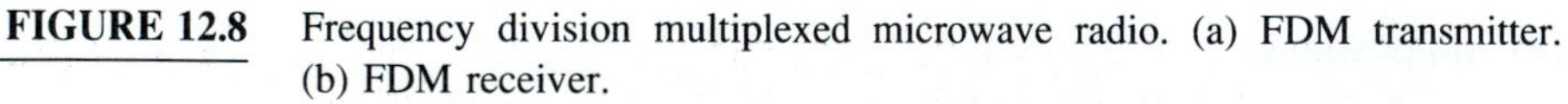
FIGURE 12.8 Frequency division multiplexed microwave radio. (a) FDM transmitter. (b) FDM receiver.

- Mapping

Military Applications

- Air and marine navigation
- Detection and tracking of aircraft, missiles, spacecraft

- Missile guidance
- Fire control for missiles and artillery
- Weapon fuses
- Reconnaissance

Scientific Applications

- Astronomy
- Mapping and imaging
- Precision distance measurement
- Remote sensing of natural resources

Figure 12.9 shows a photograph of the phased array radar for the PATRIOT missile system. We will now derive the radar equation, which governs the basic operation of most radars, and then describe some of the more common types of radar systems.

The Radar Equation

Two basic radar systems are illustrated in Figure 12.10; in the *monostatic system* the same antenna is used for both transmit and receive, while the *bistatic system* uses two separate antennas for these functions. Most radars are of the monostatic type, but

FIGURE 12.9 Photograph of the PATRIOT phased array radar. This is a C-band multifunction radar that provides tactical air defense, including target search and tracking, and missile fire control. The phased array antenna uses 5000 ferrite phase shifters to electronically scan the antenna beam.

Photo provided by Raytheon Company.

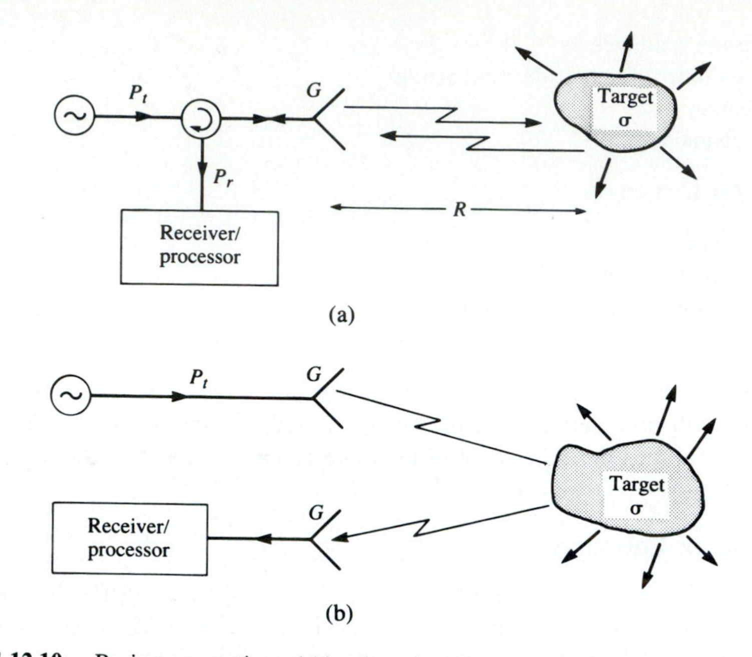

FIGURE 12.10 Basic monostatic and bistatic radar systems. (a) Monostatic radar system. (b) Bistatic radar system.

in some applications (such as missile fire control) the target is illuminated by a separate transmit antenna. Separate antennas are also sometimes used to achieve the necessary isolation between transmitter and receiver.

Here we will consider the monostatic case, but the bistatic case is very similar. If the transmitter radiates a power P_t through an antenna of gain G, the power density incident on the target is, from (12.4) and (12.11),

$$S_t = \frac{P_t G}{4\pi R^2}, \tag{12.24}$$

where R is the distance to the target. It is assumed that the target is in the main beam direction of the antenna. The target will scatter the incident power in various directions; the ratio of the scattered power in a given direction to the incident power density is defined as the *radar cross-section*, σ, of the target. Mathematically,

$$\sigma = \frac{P_s}{S_t}, \tag{12.25}$$

where P_s is the total power scattered by the target. The radar cross-section thus has the dimensions of area, and is a property of the target itself. It depends on the incident and reflection angles, as well as the polarization of the incident wave.

Since the target acts as a finite-sized source, the power density of the reradiated field must decay as $1/4\pi R^2$ away from the target. Thus the power density of the scattered field back at the receive antenna must be

$$S_r = \frac{P_t G \sigma}{(4\pi R^2)^2}. \tag{12.26}$$

Then using (12.14) for the effective area of the antenna gives the received power as

$$P_r = \frac{P_t G^2 \lambda^2 \sigma}{(4\pi)^3 R^4}. \tag{12.27}$$

This is the *radar equation*. Note that the received power varies as $1/R^4$, which implies that a sensitive low-noise receiver is needed to detect targets at long ranges.

Because of noise received by the antenna and generated in the receiver, there will be some minimum detectable power that can be discriminated by the receiver. If this power is P_{min}, then (12.27) can be rewritten to give the maximum range as

$$R_{\text{max}} = \left[\frac{P_t G^2 \sigma \lambda^2}{(4\pi)^3 P_{\text{min}}}\right]^{1/4}. \tag{12.28}$$

Signal processing can effectively reduce the minimum detectable signal, and so increase the usable range. One very common processing technique used with pulse radars is pulse integration, where a sequence of N received pulses are integrated over time. The effect is to reduce the noise level, which has a zero mean, relative to the returned pulse level, resulting in an improvement factor of approximately N [4].

Of course, the above results seldom describe the performance of an actual radar system. Factors such as propagation effects, the statistical nature of the detection process, and external interference often serve to reduce the usable range of a radar system.

EXAMPLE 12.4

A pulsed radar operating at 10 GHz has an antenna with a gain of 28 dB, and a transmitter power of 2 kW (pulse power). If it is desired to detect a target with a cross-section of 12 m^2, and the minimum detectable signal is $P_{\text{min}} = -90$ dBm, what is the maximum range of the radar?

Solution

The required numerical values are

$$G = 10^{28/10} = 631,$$

$$P_{\text{min}} = 10^{-90/10} \text{ mW} = 10^{-12} \text{ W},$$

$$\lambda = 0.03 \text{ m}.$$

Then the radar range equation of (12.28) gives the maximum range as

$$R_{\text{max}} = \left[\frac{(2\times 10^3)(631)^2(12)(.03)^2}{(4\pi)^3(10^{-12})}\right]^{1/4}$$

$$= 8114 \text{ m}.$$

○

Pulse Radar

A pulse radar determines target range by measuring the round-trip time of a pulsed microwave signal. Figure 12.11 shows a typical pulse radar system block diagram. The transmitter portion consists of a single-sideband mixer used to frequency offset a microwave oscillator of frequency f_0 by an amount equal to the IF frequency. After power amplification, pulses of this signal are transmitted by the antenna. The transmit/receive switch is controlled by the pulse generator to give a transmit pulse width τ, with a pulse repetition frequency (PRF) of $f_r = 1/T_r$. The transmit pulse thus consists of a short burst of a microwave signal at the frequency $f_0 + f_{\text{IF}}$. Typical pulse durations range from 100 ms to 50 ns; shorter pulses give better range resolution, but longer pulses result in a better signal-to-noise ratio after receiver processing. Typical pulse repetition frequencies range from 100 Hz to 100 kHz; higher PRFs give more returned pulses per unit time, which improves performance, but lower PRFs avoid range ambiguities that can occur when $R > cT_r/2$.

In the receive mode, the returned signal is amplified and mixed with the local oscillator of frequency f_0 to produce the desired IF signal. The local oscillator is used for both up-conversion in the transmitter as well as down-conversion in the receiver; this simplifies the system and avoids the problem of frequency drift, which would be a

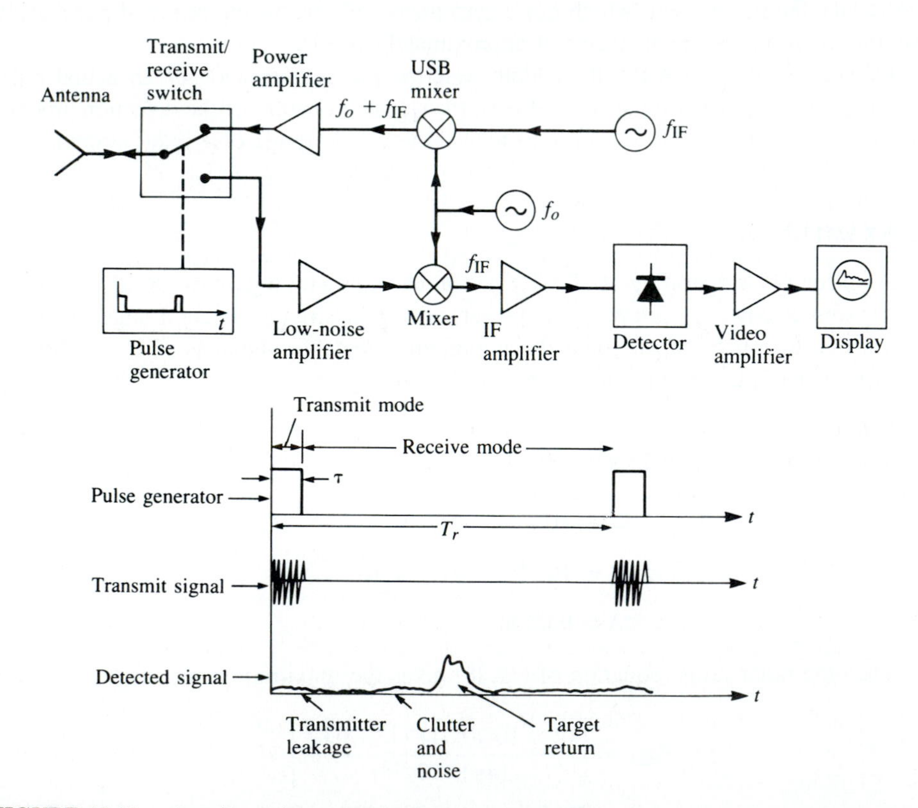

FIGURE 12.11 A pulse radar system and timing diagram.

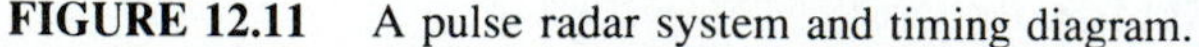

consideration if separate oscillators were used. The IF signal is amplified, detected, and fed to a video amplifier/display. Search radars often use a continuously rotating antenna for 360° azimuthal coverage; in this case the display shows a polar plot of target range versus angle. Many modern radars use a computer for the processing of the detected signal and display of target information.

The transmit/receive (T/R) switch in the pulse radar actually performs two functions: forming the transmit pulse train, and switching the antenna between the transmitter and receiver. This latter function is also known as *duplexing*. In principle, the duplexing function could be achieved with a circulator, but an important requirement is that a high degree of isolation (about 80–100 dB) be provided between the transmitter and receiver, to avoid transmitter leakage into the receiver which would drown the target return (or possibly damage the receiver). As circulators typically achieve only 20–30 dB of isolation, some type of switch, with high isolation, is required. If necessary, further isolation can be obtained by using additional switches along the path of the transmitter circuit.

Doppler Radar

If the target has a velocity component along the line-of-sight of the radar, the returned signal will be shifted in frequency relative to the transmitted frequency, due to the doppler effect. If the transmitted frequency is f_o, and the radial target velocity is v, then the shift in frequency, or the doppler frequency, will be

$$f_d = \frac{2vf_o}{c}, \tag{12.29}$$

where c is the velocity of light. The received frequency is then $f_o \pm f_d$, where the plus sign corresponds to an approaching target and the minus sign corresponds to a receding target.

Figure 12.12 shows a basic doppler radar system. Observe that it is much simpler than a pulse radar, since a continuous wave signal is used, and the transmit oscillator can also be used as a local oscillator for the receive mixer, because the received signal is frequency offset by the doppler frequency. The filter following the mixer should have a passband corresponding to the expected minimum and maximum target velocities. It is important that the filter have high attenuation at zero frequency, to eliminate the effect of clutter return and transmitter leakage at the frequency f_0, as these signals would down-convert to zero frequency. Then a high degree of isolation is not necessary between transmitter and receiver, and a circulator can be used. This type of filter response also helps to reduce the effect of $1/f$ noise.

The above radar cannot distinguish between approaching and receding targets, as the sign of f_d is lost in the detection process. Such information can be recovered, however, by using a mixer that produces separately the upper and lower sideband products.

Since the return of a pulse radar from a moving target will contain a doppler shift, it is possible to determine both the range and velocity (and position, if a narrow beam antenna is used) of a target with a single radar. Such a radar is known as a *pulse-doppler radar*, and offers several advantages over pulse or doppler radars. One problem with a pulse radar is that it is impossible to distinguish between a true target and clutter returns from the ground, trees, buildings, etc. Such clutter returns may be picked up from the antenna

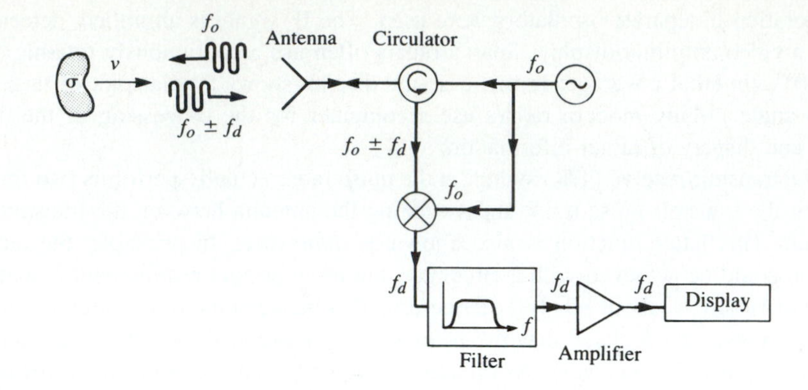

FIGURE 12.12 Doppler radar system.

sidelobes. But if the target is moving (e.g., as in an airport surveillance radar application), the doppler shift can be used to separate its return from clutter, which is stationary.

Radar Cross-Section

A radar target is characterized by its radar cross section, as defined in (12.25), which gives the ratio of scattered power to incident power density. The cross-section of a target depends on the frequency and polarization of the incident wave, and on the incident and reflected angles relative to the target. Thus we can define a monostatic cross-section (incident and reflected angles identical), and a bistatic cross-section (incident and reflected angles different).

For simple shapes the radar cross-section can be calculated as an electromagnetic boundary value problem; more complex targets require numerical techniques, or measurement to find the cross-section. The radar cross-section of a conducting sphere can be calculated exactly; the monostatic result is shown in Figure 12.13, normalized to πa^2, the physical cross-sectional area of the sphere. Note that the cross-section increases very quickly with size for electrically small spheres ($a << \lambda$). This region is called the *Rayleigh region*, and it can be shown that σ varies as $(a/\lambda)^4$ in this region. (This strong dependence on frequency explains why the sky is blue, as the blue component of sunlight scatters more strongly from atmospheric particles than do the lower frequency red components.)

For electrically large spheres, where $a >> \lambda$, the radar cross-section of the sphere is equal to its physical cross-section, πa^2. This is the *optical region*, where geometrical optics are valid. Many other shapes, such as flat plates at normal incidence, also have cross-sections that approach the physical area for electrically large sizes.

Between the Rayleigh region and the optical region is the *resonance region*, where the electrical size of the sphere is on the order of a wavelength. Here the cross-section is oscillating with frequency, due to phase addition and cancellation of various scattered field components. Of particular note is the fact that the cross-section may reach quite high values in this region.

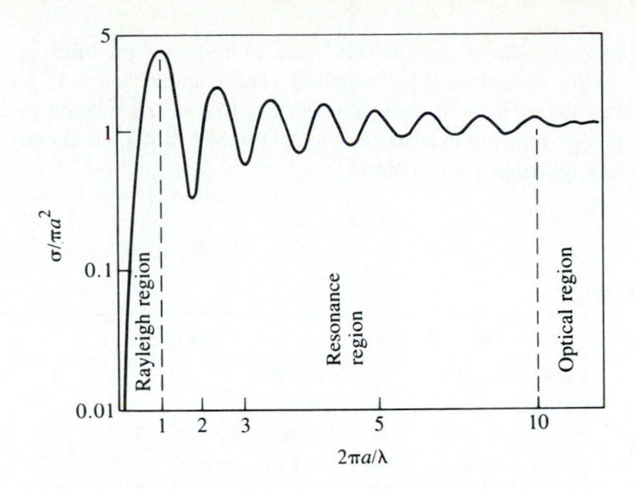

FIGURE 12.13 Monostatic radar cross-section of a conducting sphere.

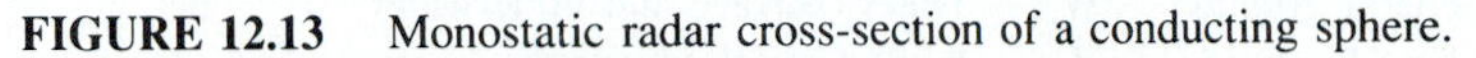

Complex targets such as aircraft or ships generally have cross-sections that vary rapidly with frequency and aspect angle. In military applications it is often desirable to minimize the radar cross-section of vehicles, to reduce detectability. This can be accomplished by using radar absorbing materials (lossy dielectrics) in the construction of the vehicle. Table 12.3 lists the approximate radar cross-sections of a variety of different targets.

POINT OF INTEREST: Letter Designations for Military Electronics Systems

United States military electronic systems are usually identified with a letter/number system called the Joint Electronics Type Designated System. The code begins with the letters AN (orginally this stood for "Army-Navy," but the designations now include Air Force as well), followed by three letters that, respectively, identify the platform, system type, and its function. Then the model

TABLE 12.3 Typical Radar Cross Sections

Target	σ (m^2)
Bird	0.01
Missile	0.5
Person	1.
Small plane	1–2
Bicyle	2
Small boat	2
Fighter plane	3–8
Bomber	30–40
Large airliner	100
Truck	200

number appears, possibly followed by "A," "B," etc. to designate modifications of the original model. The tables below define the letter identifiers. For example, the F-15 jet fighter uses the AN/APG-63 multipurpose radar for navigation, target acquisition, and weapon guidance. From the tables we see that this designation indicates that the platform is a piloted aircraft, the system is a radar, and the primary function is fire control.

System Platform

Code	Category	Code	Category
A	Piloted aircraft	P	Portable
B	Submarine	S	Water
D	Pilotless carrier	U	General Utility
F	Fixed ground	V	Vehicle, ground
G	General ground	W	Water (surface and underwater)
K	Amphibious	Z	Piloted-pilotless aircraft combination
M	Mobile		

System Type

Code	Category	Code	Category
A	Light, heat, radiation	P	Radar
C	Carrier	Q	Sonar
D	Radiac	R	Radio
G	Telegraph, teletype	S	Special
I	Intercom, public address	T	Telephone
J	Mechanical wire	V	Visual
K	Telemetering	W	Armament
L	Countermeasure	Y	Data processing
M	Meteorological		
N	Sound in air		

System Function

Code	Category	Code	Category
B	Bombing	N	Navigation
C	Communications	Q	Special
D	Direction finding	R	Receiving
E	Ejection, release	S	Detecting, ranging
G	Fire control	T	Transmitting
H	Recording	W	Automatic or remote flight control
K	Computing	X	Identification, recognition
M	Maintenance, test		

12.4 RADIOMETRY

A radar system obtains information about a target by transmitting a signal and receiving the echo from the target, and thus can be described as an active remote sensing system. Radiometry, however, is a passive technique which develops information about a target solely from the microwave portion of the blackbody radiation (noise) that it either emits directly or reflects from surrounding bodies. A *radiometer* is a sensitive receiver specially designed to measure this noise power.

Theory and Applications of Radiometry

As discussed in Section 11.1, a body in thermodynamic equilibrium at a temperature T radiates energy according to Planck's radiation law. In the microwave region this result reduces to $P = kTB$, where k is Boltzmann's constant, B is the system bandwidth, and P is the radiated power. This result strictly applies only to a *blackbody*, which is defined as an idealized material which absorbs all incident energy, and reflects none; a blackbody also radiates energy at the same rate as it absorbs energy, thus maintaining thermal equilibrium. A nonideal body will partially reflect incident energy, and so does not radiate as much power as would a blackbody at the same temperature. A measure of the power radiated by a body relative to that radiated by an ideal blackbody at the same temperature is the emissivity, e, defined as

$$e = \frac{P}{kTB}, \qquad 12.30$$

where P is the power radiated by the nonideal body, and kTB is the power that would be emitted by a perfect blackbody. Thus, $0 \leq e \leq 1$, and $e = 1$ for a perfect blackbody.

As we saw in Section 11.1, noise power can also be quantified in terms of equivalent temperature. Thus for radiometric purposes we can define a brightness temperature, T_B, as

$$T_B = eT, \qquad 12.31$$

where T is the physical temperature of the body. This shows that, radiometrically, a body always looks cooler than its actual temperature, since $0 \leq e \leq 1$.

Now consider Figure 12.14, which shows the antenna of a radiometer receiving noise powers from various sources. The antenna is pointed at a region of the earth which has an apparent brightness temperature T_B. The atmosphere emits radiation in all directions; the component radiated directly toward the antenna is T_{AD}, while the power reflected from the earth to the antenna is T_{AR}. There may also be noise powers that enter the sidelobes of the antennas, from the sun or other sources. Thus, we can see that the total brightness temperature seen by the radiometer is a function of the scene under observation, as well as the observation angle, frequency, polarization, attenuation of the atmosphere, and the antenna pattern. The objective of radiometry is to infer information about the scene from the measured brightness temperature, and an analysis of the radiometric mechanisms that relate brightness temperature to physical conditions of the scene. For example, the power reflected from a uniform layer of snow over soil can be treated as plane wave reflection from a multilayer dielectric region, leading to the

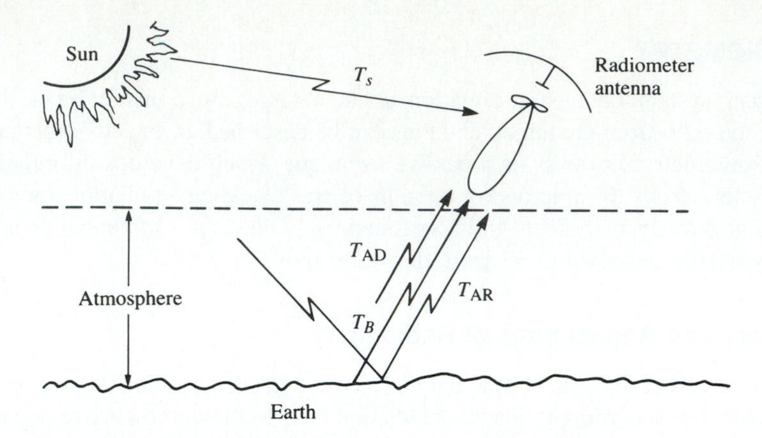

FIGURE 12.14 Noise power sources in a typical radiometer application.

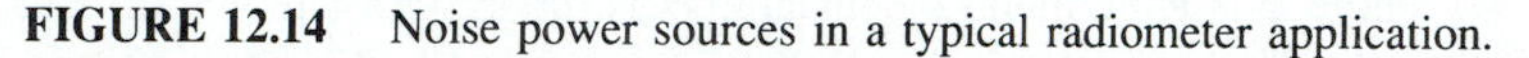

development of an algorithm that gives the thickness of the snow in terms of measured brightness temperature at various frequencies.

Microwave radiometry is a relatively new area of technology, and one which is strongly interdisciplinary, drawing on results from fields such as electrical engineering, oceanography, geophysics, and atmospheric and space sciences, to name a few. Below, some of the more typical applications of microwave radiometry are listed.

Environmental Applications

- Measurement of soil moisture
- Flood mapping
- Snow cover/Ice cover mapping
- Ocean surface windspeed
- Atmospheric temperature profile
- Atmospheric humidity profile

Military Applications

- Target detection
- Target recognition
- Surveillance
- Mapping

Astronomy Applications

- Planetary mapping
- Solar emission mapping
- Mapping of galactic objects
- Measurement of cosmological background radiation

Figure 12.15 shows a photograph of a radiometer used to measure the water vapor profile of the atmosphere.

FIGURE 12.15 Photograph of a microwave radiometer used to measure the humidity or water vapor profile of the atmosphere. This system operates from 20 to 24 GHz to sense the water resonance at 22.2 GHz, and uses an autocorrelation technique to cancel noise.
Courtesy of the Microwave Remote Sensing Laboratory, University of Massachusetts.

Total Power Radiometer

The aspect of radiometry that is of most interest to the microwave engineer is the design of the radiometer itself. The basic problem is to build a receiver that can distinguish between the desired radiometric noise and the inherent noise of the receiver, even though the radiometric power is usually less than the receiver noise power. Although it is not a very practical instrument, we will first consider the total power radiometer, because it represents a simple and direct approach to the problem and serves to illustrate the difficulties involved in radiometer design.

The block diagram of a typical total power radiometer is shown in Figure 12.16. The front end of the receiver is a standard superheterodyne circuit consisting of an RF amplifier, a mixer/local oscillator, and an IF stage. The IF filter determines the system bandwidth, B. The detector is generally a square-law device, so that its output voltage is proportional to the input power. The integrator is essentially a low-pass filter with a cutoff frequency of $1/\tau$, and serves to smooth out short-term variations in the noise power. For simplicity, we assume that the antenna is lossless, although in practice antenna loss will affect the apparent temperature of the antenna, as given in (12.12).

If the antenna is pointed at a background scene with a brightness temperature T_B, the antenna power will be $P_A = kT_BB$; this is the desired signal. The receiver contributes noise which can be characterized as a power $P_R = kT_RB$ at the receiver input, where T_R is the overall noise temperature of the receiver. Thus the output voltage of the radiometer is

$$V_o = G(T_B + T_R)kB, \tag{12.32}$$

where G is the overall gain constant of the radiometer. Conceptually, the system is calibrated by replacing the antenna input with two calibrated noise sources, from which the system constants GkB and GT_RkB can be determined. (This is similar to the Y-factor method for measuring noise temperature.) Then the desired brightness temperature, T_B, can be measured with the system.

Two types of errors occur with this radiometer. First is an error, ΔT_N, in the measured brightness temperature due to noise fluctuations. Since noise is a random process, the measured noise power may vary from one integration period to the next. The integrator (or low-pass filter) acts to smooth out ripples in V_o with frequency components greater than $1/\tau$. It can be shown that the remaining error is [2]

$$\Delta T_N = \frac{T_B + T_R}{\sqrt{B\tau}}. \tag{12.33}$$

This result shows that if a longer measurement time, τ, can be tolerated, the error due to noise fluctuation can be reduced to a negligible value.

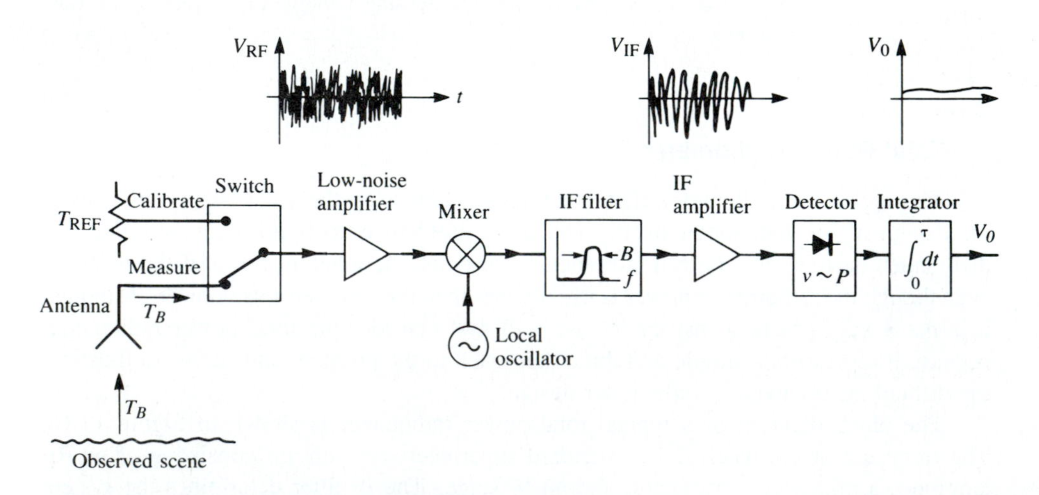

FIGURE 12.16 Total power radiometer block diagram.

A more serious error is due to random variations in the system gain, G. Such variations generally occur in the RF amplifier, mixer, or IF amplifier, over a period of one second or longer. So if the system is calibrated with a certain value of G, which changes by the time a measurement is made, an error will occur, as given in reference [2] as

$$\Delta T_G = (T_B + T_R)\frac{\Delta G}{G}, \tag{12.34}$$

where ΔG is the rms change in the system gain, G.

It will be useful to consider some typical numbers at this time. For example, a 10 GHz total power radiometer may have a bandwidth of 100 MHz, a receiver temperature of $T_R = 500$ K, an integrator time constant of $\tau = 0.01$ s, and a system gain variation $\Delta G/G = 0.01$. If the antenna temperature is $T_B = 300$ K, (12.33) gives the error due to noise fluctuations as, $\Delta T_N = 0.8$ K, while (12.34) gives the error due to gain variations as $\Delta T_G = 8$ K. These results, which are based on reasonably realistic data, show that gain variation is the most detrimental factor affecting the accuracy of the total power radiometer.

The Dicke Radiometer

We have seen that the dominant factor affecting the accuracy of the total power radiometer is the variation of gain of the overall system. Since such gain variations have a relatively long time constant ($>$ 1 second), it is conceptually possible to eliminate this error by repeatedly calibrating the radiometer at rapid rate. This is the principle behind the operation of the Dicke null-balancing radiometer.

A system diagram is shown in Figure 12.17. The superheterodyne receiver is identical to the total power radiometer, but the input is periodically switched between the antenna and a variable power noise source; this switch is called the Dicke switch. The output of the square-law detector drives a synchronous demodulator, which consists of a switch and a difference circuit. The demodulator switch operates in synchronism with the Dicke switch, so that the output of the subtractor is proportional to the difference between the noise powers from the antenna, T_B, and the reference noise source, T_{REF}. The output of the subtractor is then used as an error signal to a feedback control circuit, which controls the power level of the reference noise source so that V_o approaches zero. In this balanced state, $T_B = T_{\text{REF}}$, and T_B can be determined from the control voltage, V_c. The square-wave sampling frequency, f_s, is chosen to be much faster than the drift time of the system gain, so that this effect is virtually eliminated. Typical sampling frequencies range from 10 to 1000 Hz.

A typical radiometer would measure the brightness temperature T_B over a range of about 50–300 K; this then implies that the reference noise source would have to cover this same range, which is difficult to do in practice. Thus, there are several variations on the above design, differing essentially in the way that the reference noise power is controlled or added to the system. One possible method is to use a constant T_{REF} which is somewhat hotter than the maximum T_B to be measured. The amount of reference noise power delivered to the system is then controlled by varying the pulse width of the sampling waveform. Another approach is to use a constant reference noise power, and

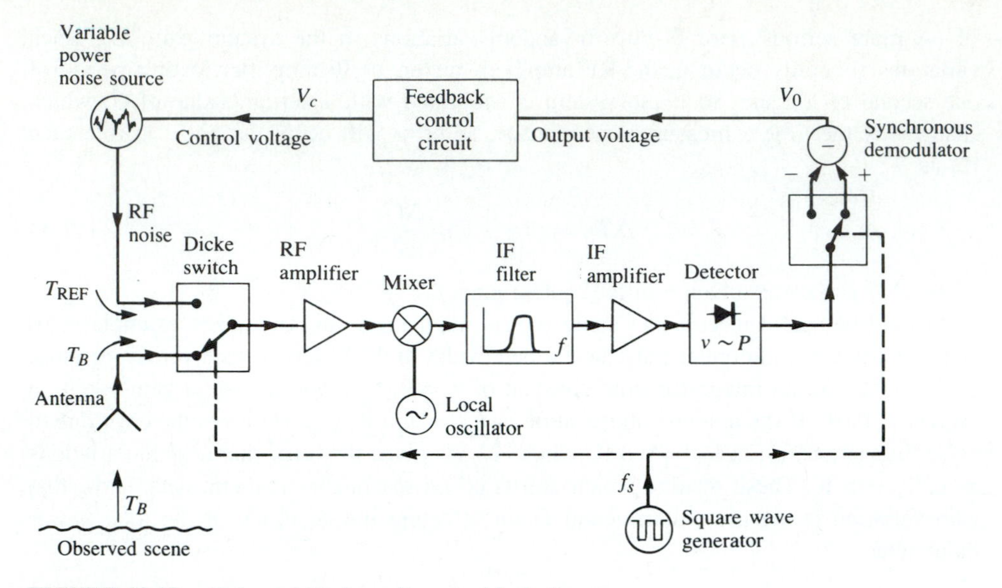

FIGURE 12.17 Balanced Dicke radiometer block diagram.

vary the gain of the IF stage during the reference sample time to achieve a null output. Other possibilities, including alternatives to the Dicke radiometer, are discussed in the literature [2].

12.5 MICROWAVE PROPAGATION

In free-space electromagnetic waves propagate in straight lines without attenuation or other adverse effects. Free-space, however, is an idealization that is only approximated when microwave energy propagates through the atmosphere or in the presence of the earth. In practice the performance of a communication, radar, or radiometry system may be seriously affected by propagation effects such as reflection, refraction, attenuation, or diffraction. Below we discuss some specific propagation phenomenon that can influence the operation of microwave systems. It is important to realize that propagation effects generally cannot be quantified in any exact or rigorous sense, but can only be described in terms of their statistics.

Atmospheric Effects

The relative permittivity of the atmosphere is close to unity, but is actually a function of air pressure, temperature, and humidity. An empirical result which is useful at microwave frequencies is given by [5]

$$\epsilon_r = \left[1 + 10^{-6}\left(\frac{79P}{T} - \frac{11V}{T} + \frac{3.8 \times 10^5 V}{T^2}\right)\right]^2, \qquad 12.35$$

where P is the barometric pressure in millibars, T is the temperature in kelvin, and V is the water vapor pressure in millibars. This result shows that permittivity generally decreases (approaches unity) as altitude increases, since pressure and humidity decrease with height faster than does temperature. This change in permittivity with altitude causes radio waves to bend toward the earth, as depicted in Figure 12.18. Such refraction of radio waves can sometimes be useful, since it may extend the range of radar and communication systems beyond the limit imposed by the presence of the earth's horizon.

If an antenna is at a height, h, above the earth, simple geometry gives the line-of-sight distance to the horizon as

$$d = \sqrt{2Rh}, \qquad 12.36$$

where R is the radius of the earth. From Figure 12.18 we see that the effect of refraction on range can be accounted for by using an effective earth radius kR, where $k > 1$. A value commonly used [4] is $k = 4/3$, but this is only an average value which changes with weather conditions. In a radar system, refraction effects can lead to errors when determining the elevation of a target close to the horizon.

Weather conditions can sometimes produce a temperature inversion, where the temperature increases with altitude. Equation (12.35) then shows that the atmospheric permittivity will decrease much faster than normal, with increasing altitude. This condition can sometimes lead to *ducting* (also called trapping, or anomalous propagation), where a radio wave can propagate long distances parallel to the earth's surface, via the duct created by the layer of air along the temperature inversion. The situation is very similar to propagation in a dielectric waveguide. Such ducts can range in height from 50–500 feet, and may be near the earth's surface, or higher in altitude.

Another atmospheric effect is attenuation, caused primarily by the absorption of microwave energy by water vapor and molecular oxygen. Maximum absorption occurs when the frequency coincides with one of the molecular resonances of water or oxygen, thus atmospheric attenuation has distinct peaks at these frequencies. Figure 12.19 shows the atmospheric attenuation vs. frequency. At frequencies below 10 GHz the atmosphere has very little effect on the strength of a signal. At 22.2 and 183.3 GHz, resonance peaks

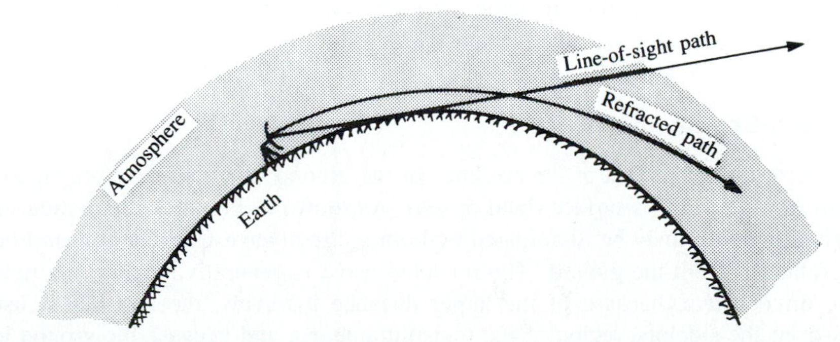

FIGURE 12.18 Refraction of radio waves by the atmosphere.

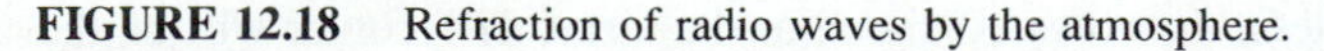

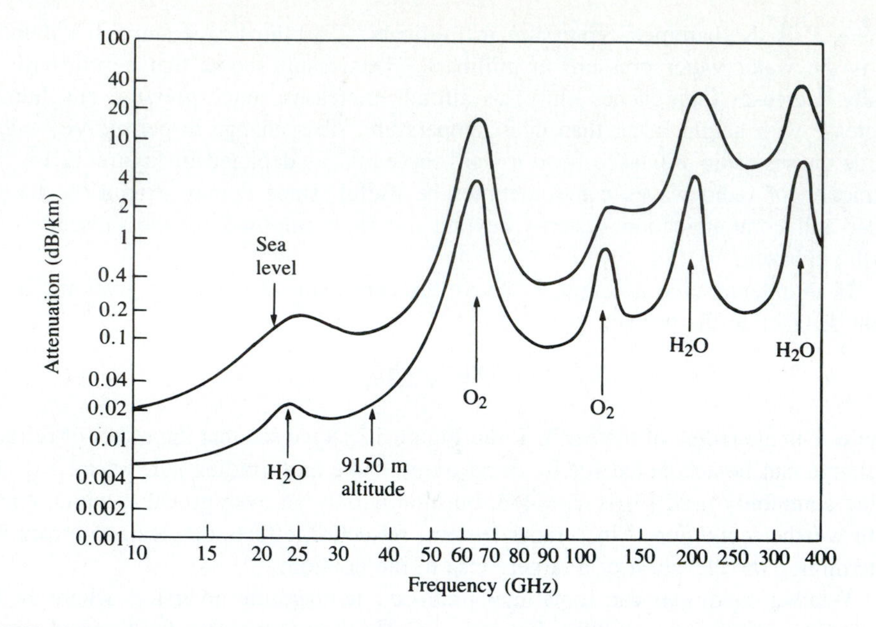

FIGURE 12.19 Average atmospheric attenuation versus frequency (horizontal polarization).

occur due to water vapor resonances, while resonances of molecular oxygen cause peaks at 60 and 120 GHz. Thus there are "windows" in the millimeter wave band near 35, 94, and 135 GHz where radar and communication systems can operate with minimum loss. Precipitation such as rain, snow, or fog will increase the attenuation, especially at higher frequencies. The effect of atmospheric attenuation can be included in system design when using the Friis transmission equation or the radar equation.

In some instances the system frequency may be chosen at a point of maximum atmospheric attenuation. Remote sensing of the atmosphere (temperature, water vapor, rain rate) is often done with radiometers operating near 20 or 55 GHz, to maximize the sensing of atmospheric conditions (see Figure 12.15). Another interesting example is spacecraft-to-spacecraft communication at 60 GHz. This millimeter wave frequency has the advantages of a large bandwidth and small antennas with high gains and, since the atmosphere is very lossy at this frequency, the possibilities of interference, jamming, and eavesdropping from earth are greatly reduced.

Ground Effects

The most obvious effect of the presence of the ground on microwave propagation is reflection from the earth's surface (land or sea). As shown in Figure 12.20, a radar target (or receiver antenna) may be illuminated by both a direct wave from the transmitter and a wave reflected from the ground. The reflected wave is generally smaller in amplitude than the direct wave, because of the larger distance it travels, the fact that it usually radiates from the sidelobe region of the transmit antenna, and because the ground is not a perfect reflector. Nevertheless, the received signal at the target or receiver will be the vector sum of the two wave components and, depending on the relative phases of the two waves, may be greater or less than the direct wave alone. Because the distances involved are usually very large in terms of the electrical wavelength, even a small variation in the permittivity of the atmosphere can cause *fading* (long term fluctuations) or *scintillation* (short term fluctuations) in the signal strength. These effects can also be caused by reflections from inhomogeneities in the atmosphere.

In communication systems such fading can sometimes be reduced by making use of the fact that the fading of two communication channels having different frequencies, polarizations, or physical locations is essentially independent. Thus a communication link can reduce fading by combining the outputs of two (or more) such channels; this is called a diversity system.

Another ground effect is *diffraction*, whereby a radio wave scatters energy in the vicinity of the line-of-sight boundary at the horizon, thus giving a range slightly beyond the horizon. This effect is usually very small at microwave frequencies. Of course, when obstacles such as hills, mountains, or buildings are in the path of propagation, diffraction effects can be stronger.

In a radar system, unwanted reflections often occur from terrain, vegetation, trees, buildings, and the surface of the sea. Such clutter echoes generally degrade or mask the return of a true target, or show up as a false target, in the context of a surveillance or tracking radar. In mapping or remote sensing applications such clutter returns may actually constitute the desired signal.

Plasma Effects

A *plasma* is a gas consisting of ionized particles. The ionosphere consists of spherical layers of atmosphere with particles which have been ionized by solar radiation, and thus forms a plasma region. A very dense plasma is formed on a spacecraft as it reenters the atmosphere, due to the high temperatures produced by friction. Plasmas are also produced by lightning, meteor showers, and nuclear explosions.

A plasma is characterized by the number of ions per unit volume; depending on this density and the frequency, a wave might be reflected, absorbed, or transmitted by the plasma medium. An effective permittivity can be defined for a uniform plasma region

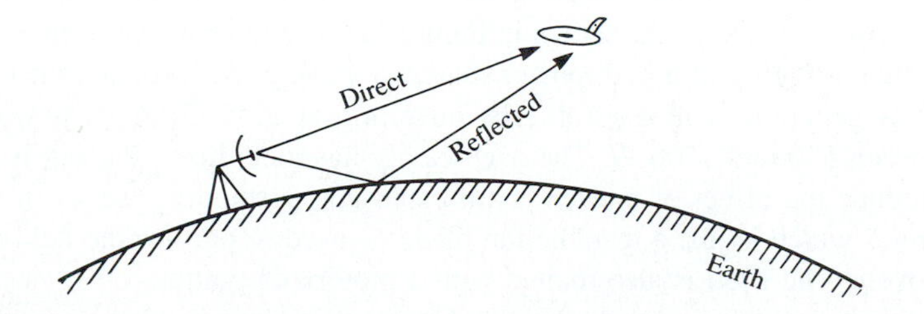

FIGURE 12.20 Direct and reflected waves over the earth's surface.

as

$$\epsilon_e = \epsilon_o \left(1 - \frac{\omega_p^2}{\omega^2}\right), \tag{12.37}$$

where

$$\omega_p = \sqrt{\frac{Nq^2}{m\epsilon_0}} \tag{12.38}$$

is the plasma frequency. In (12.38), q is the charge of the electron, m is the mass of the electron, and N is the number of ionized particles per unit volume. By studying the solution of Maxwell's equations for plane wave propagation in such a medium, it can be shown that wave propagation through a plasma is only possible for $\omega > \omega_p$. Lower frequency waves will be totally reflected. If a magnetic field is present, the plasma becomes anisotropic, and the analysis is more complicated. The earth's magnetic field may be strong enough to produce such an anisotropy in some cases.

The ionosphere consists of several different layers with varying ion densities; in order of increasing ion density, these layers are referred to as D, E, F_1, and F_2. The characteristics of these layers depends on seasonal weather and solar cycles, but the average plasma frequency is about 8 MHz. Thus, signals at frequencies less than 8 MHz (e.g., short-wave radio) can reflect off the ionosphere to travel distances well beyond the horizon. Higher frequency signals, however, will pass through the ionosphere.

A similar effect occurs with a spacecraft entering the atmosphere. The high velocity of the spacecraft causes a very dense plasma to form around the vehicle. The electron density is high enough so that, from (12.38), the plasma frequency is very high, thus inhibiting communication with the spacecraft until its velocity has decreased. Besides this blackout effect, the plasma layer may also cause a large impedance mismatch between the antenna and its feed line.

12.6 OTHER APPLICATIONS AND TOPICS

Microwave Heating

To the average consumer, the term "microwave" connotes a microwave oven, which is used in many households for heating food; industrial and medical applications also exist for microwave heating. As shown in Figure 12.21, a microwave oven is a relatively simple system consisting of a high-power source, a waveguide feed, and the oven cavity. The source is generally a magnetron tube operating at 2.45 GHz; its power output is usually between 500 and 1500 W. The oven cavity has metallic walls, and is electrically large. To reduce the effect of uneven heating caused by standing waves in the oven, a "mode stirrer," which is just a metallic fan blade, is used to perturb the field distribution inside the oven. The food is also rotated with a motorized platter.

In a conventional oven a gas or charcoal fire, or an electric heating element, generates heat outside of the material to be heated. The outside of the material then gets heated

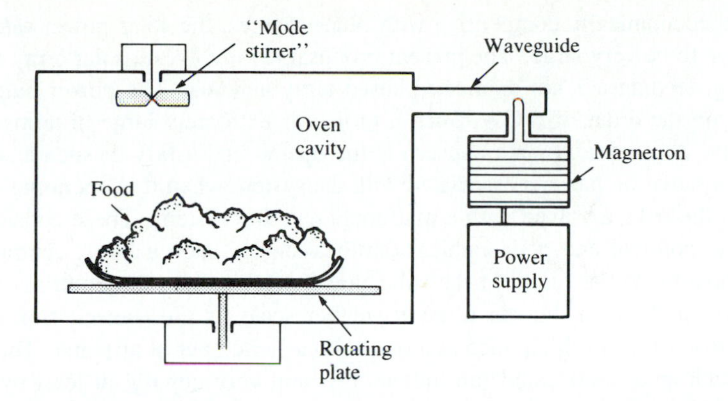

FIGURE 12.21 A microwave oven.

by convection, and the inside of the material by conduction. In microwave heating, by contrast, the inside of the material gets heated first. The process through which this occurs involves the resonance of water molecules and conduction losses in materials with large loss tangents [5], [6]. An interesting fact is that the loss tangents of many foods decrease with increasing temperature, so that microwave heating is to some extent self-regulating. The result is that microwave cooking generally gives faster and more uniform heating of food, as compared with conventional cooking. The efficiency of a microwave oven, when defined as the ratio of power converted to heat (in the food) to the power supplied to the oven, is generally less than 50%; this is usually greater than the cooking efficiency of a conventional oven, however.

The most critical issue in the design of a microwave oven is safety. Since a very high power source is used, leakage levels must be very small to avoid exposing the user to harmful radiation. Thus the magnetron, feed waveguide, and oven cavity must all be carefully shielded. The door of the oven requires particular attention; besides close mechanical tolerances, the joint around the door usually employs RF absorbing material and a $\lambda/4$ choke flange to reduce leakage to an acceptable level.

Energy Transfer

Electrical power transmission lines are a very efficient and convenient way to transfer energy from one point to another, as they have relatively low loss and initial costs, and can be easily routed. There are applications, however, where it is inconvenient or impossible to use such power lines. In such cases it is conceivable that electrical power can be transmitted without wires by a well-focused microwave beam [7].

One example is the solar satellite power station, where it has been proposed that electricity be generated in space by a large orbiting array of solar cells, and transmitted to a receiving station on earth by a microwave beam. We would thus be provided with a virtually inexhaustible source of electricity. Placing the solar arrays in space has the advantage of power delivery uninterrupted by darkness, clouds, or precipitation, which are problems encountered with earth-based solar arrays.

To be economically competitive with other sources, the solar power satellite station would have to be very large. The present proposal [5] involves a solar array about 5×10 km in size, feeding a 1 km diameter phased array antenna. The power output on earth would be on the order of 5 GW. Such a project is extremely large in terms of cost and complexity. Also of legitimate concern is the operational safety of such a scheme, both in terms of radiation hazards associated with the system when it is operating as designed, as well as the risks involved with a malfunction of the system. These considerations, as well as the political and philosophical ramifications of such a large, centralized power system, have made the future of the solar power satellite station doubtful.

Similar in concept, but on a much smaller scale, is the transmission of electrical power from earth to a vehicle such as a small drone helicopter or airplane. The advantages are that such an aircraft could run indefinitely, and very quietly, at least over a limited area. Battlefield surveillance and weather prediction would be some possible applications. The concept has been demonstrated with several projects involving small pilotless aircraft.

A very high power pulsed microwave source and a high-gain antenna can be used to deliver an intense burst of energy to a target, and thus used as a weapon. The pulse may be intense enough to do physical damage to the target, or it may act to overload and destroy sensitive electronic systems.

Electronic Warfare

Military radar and communication systems may be limited or prevented from performing their intended function by deliberate means, such as interference, jamming, and other countermeasures. We will give a brief outline of this subject, which generally goes under the title of *electronic warfare*. Most of the discussion applies to radar systems, although some techniques, such as jamming, apply to communication systems as well. Electronic warfare can be divided into three major headings.

1. *Electronic support measures* (ESM). These involve the use of "threat warning" receivers to detect the presence of a search or tracking radar, or the presence of a jamming signal. Such receivers are used on aircraft, ships, and ground vehicles, and usually give characteristics of the threat radar such as operating frequency, pulse waveform, and direction.
2. *Electronic countermeasures* (ECM). These include the use of both active and passive techniques to either confuse or deceive a radar or communication system.
3. *Electronic counter-countermeasures* (ECCM). Any radar or communication system can be rendered ineffective by a forceful and determined ECM effect. The purpose of ECCM is to make such a goal too costly to be achieved.

We will discuss electronic countermeasures in more detail. Passive countermeasures are methods to reduce the probability that a target will be detected by a radar, or to present false targets to the radar. Passive countermeasures do not involve the emission of any signals or noise. Examples include chaff, which are clouds of thin foil strips ejected by an aircraft to present a false target to a radar; decoys, which are dummy vehicles that appear to a radar as an aircraft or missile; radar cross-section (RCS) reduction, by proper shaping of the vehicle, or by coating it with lossy materials.

Active countermeasures involve the radiation of a signal to confuse or deceive the radar or communication system, and generally take the form of jammers. A noise jammer radiates a relatively large power in the operating band of the radar, to mask the true target return, or to overload the receiver front end. An effective ECCM strategy here is to design the receiver with overload protection, and to provide frequency agility, whereby the radar changes its operating frequency over a large range in a pseudo-random manner. A barrage jammer radiates noise signals over a wide band of frequencies, to encompass the tuning range of a frequency agile system. The jamming power over a given radar bandwidth is then less than that of the noise jammer operating at one frequency. If the jamming signal is being transmitted from a location different than the target, its effect can be reduced by using an adaptive array antenna on the radar to form a pattern with a null in the direction of the jammer. A repeater jammer receives the radar pulse, delays it (or otherwise modifies it), and retransmits it to give the radar false range or velocity information.

It is relatively straightforward to use the radar equation and Friis power transmission formula to quantitatively describe some of the most common jamming scenarios. In the *self-screening jammer* (SSJ) case, the interogating radar is jammed by a jammer at the target itself, as illustrated in Figure 12.22a. The objective of the jammer is to radiate enough power at the radar to overwhelm the target return. The jammer has the advantage of only a one-way path loss, but it generally must radiate over a broader bandwidth and larger angular sector than the radar, to be sure of blanking its signal. The important parameter is the jammer-to-signal (J/S) power ratio; effective jamming requires a J/S ratio of 0 to 20 dB, depending on the radar characteristics. From the radar equation of (12.27), the received power at the radar due to reflection from the target is

$$P_r = P_t \frac{G_r^2 \lambda^2 \sigma}{(4\pi)^3 R^4}, \qquad 12.39$$

where P_t is the radar transmitter power, G_r is the radar antenna gain (monostatic case assumed), and σ is the radar cross-section of the target. Using the Friis power transmission formula of (12.15) gives the jammer noise power at the radar as

$$P_n = P_j \frac{G_j G_r \lambda^2}{(4\pi R)^2} \left(\frac{B_r}{B_j} \right), \qquad 12.40$$

where P_j is the radiated power of the jammer, and G_j is the gain of the jammer antenna. The factor B_r/B_j is the ratio of the radar bandwidth to the bandwidth of the jammer, and accounts for the fact that the jammer must radiate power over a frequency band large enough to account for an uncertainty in the radar frequency. Using these two results gives the J/S ratio as,

$$\frac{J}{S} = \frac{P_n}{P_r} = \frac{P_j}{P_t} \frac{4\pi R^2 G_j}{\sigma G_r} \left(\frac{B_r}{B_j} \right) \quad \text{(SSJ)}. \qquad 12.41$$

In a *stand-off jammer* (SOJ) scenario, the jammer is at a different location than the target. The target lies in the main beam of the radar antenna, but the jammer radiates power into the sidelobes, as illustrated in Figure 12.22b. The radar return from the target is still

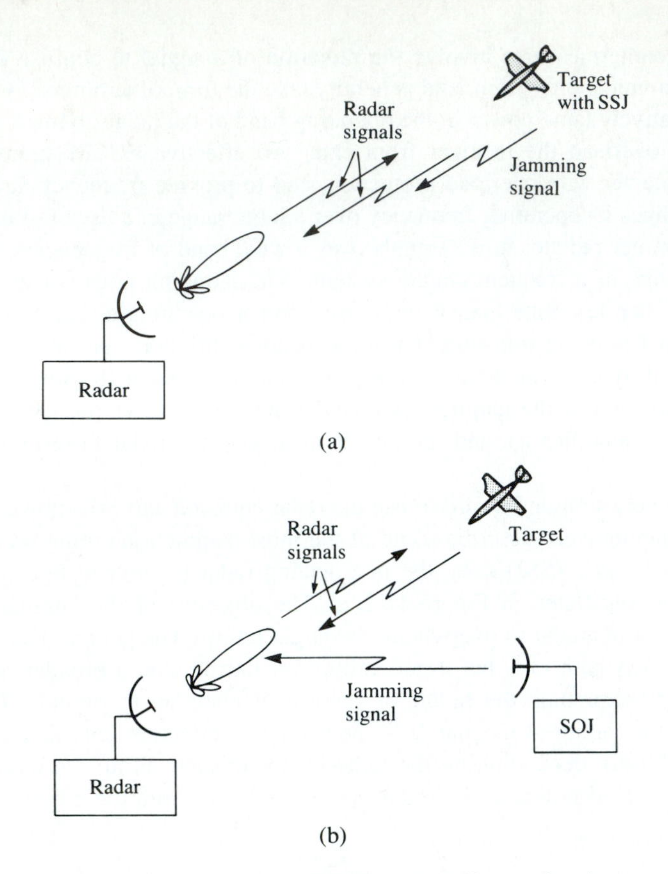

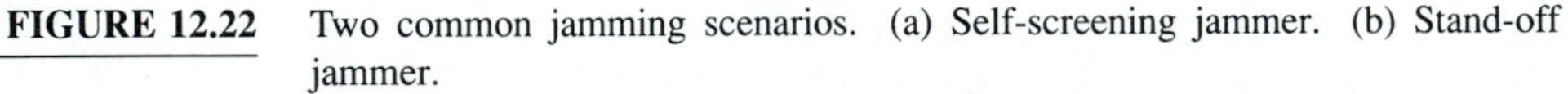
FIGURE 12.22 Two common jamming scenarios. (a) Self-screening jammer. (b) Stand-off jammer.

given by (12.39), but the noise power received from the jammer is now given by

$$P_n = P_j \frac{G_j G_r^{\text{SL}} \lambda^2}{(4\pi R_j)^2} \left(\frac{B_r}{B_j}\right), \qquad 12.42$$

where G_r^{SL} is the gain in the sidelobe (SL) region of the radar antenna, and R_j is the distance from the jammer to the radar. Then the J/S ratio for this case is

$$\frac{J}{S} = \frac{P_n}{P_r} = \frac{P_j}{P_t} \frac{4\pi R^4 G_j G_r^{\text{SL}}}{R_j^2 G_r^2 \sigma} \left(\frac{B_r}{B_j}\right) \quad \text{(SOJ)}. \qquad 12.43$$

EXAMPLE 12.5

A 10 GHz tracking radar has a transmitter power of 100 kW, an antenna gain of 35 dB, and an IF bandwidth of 1 MHz. It is attempting to track an aircraft with a cross-section of 4 m^2 at a range of 3 km. A self-screening jammer aboard the aircraft has a transmitter power of 1 kW over a 20 MHz bandwidth, with an antenna gain of 10 dB. Find the resulting J/S ratio, and compare with J/S for the same jammer used in a stand-off configuration at a distance of 10 km from the radar. Assume a 20 dB sidelobe level for the radar antenna.

Solution

The radar antenna gain is

$$G_r = 10^{35/10} = 3162,$$

and the jammer antenna gain is

$$G_j = 10^{10/10} = 10.$$

The wavelength is $\lambda = 0.03$ m. Then from (12.41), the J/S ratio for the SSJ case is

$$\frac{J}{S} = \frac{P_j}{P_t}\frac{4\pi R^2 G_j}{\sigma G_r}\left(\frac{B_r}{B_j}\right) = \frac{(1000)(4\pi)(3000)^2(10)(1)}{(10^5)(4)(3162)(20)} = 45 = 16.5 \text{ dB} \quad \text{(SSJ)}.$$

The radar antenna gain in its sidelobe region is

$$G_r^{\text{SL}} = 10^{(35-20)/10} = 31.6,$$

so the J/S ratio for the SOJ case can be found from (12.43) as

$$\frac{J}{S} = \frac{P_j}{P_t}\frac{4\pi R^4 G_j G_r^{\text{SL}}}{R_j^2 G_r^2 \sigma}\left(\frac{B_r}{B_j}\right) = \frac{(1000)(3000)^4(10)(31.6)(1)}{(10^5)(10,000)^2(3162)^2(20)}$$
$$= 0.013 = -19 \text{ dB} \qquad \text{(SOJ)}.$$

So we see that the self-screening jammer should be reasonably effective, but the stand-off jammer will not be very useful. ○

Biological Effects and Safety

The proven dangers of exposure to microwave radiation are due to thermal effects. The body absorbs RF and microwave energy and converts it to heat; as in the case of a microwave oven, this heating occurs within the body, and may not be felt at low levels. Such heating is most dangerous in the brain, the eye, the genitals, and the stomach organs. Excessive radiation can lead to cataracts, sterility, or cancer. Thus it is important to determine a safe radiation level standard, so that users of microwave equipment will not be exposed to harmful power levels.

The quantity of interest is power density (Poynting vector magnitude), in watts per unit area. On a clear day the sun radiates a power density of about 100 mW/cm^2, but the effect of this radiation is much less severe than a corresponding level of microwave radiation because the sun heats the outside of the body, and much of the heat is absorbed by the air, while microwave power heats from inside the body.

The United States and much of Western Europe have adopted a maximum radiation level standard of 5 mW/cm^2. The Soviet Union recommends a maximum level of 1 mW/cm^2, and even lower levels for prolonged exposure to microwave radiation. A separate standard applies to microwave ovens in the United States: law requires that all ovens be tested to ensure that the power level at 5 cm from any point on the oven does not exceed 1 mW/cm^2.

Most experts feel that the above limits represent safe levels, with a reasonable margin. Some researchers, however, feel that health hazards may occur due to nonthermal effects of long-term exposure to even low levels of microwave radiation.

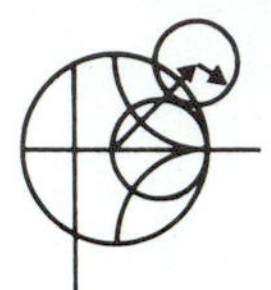

EXAMPLE 12.6

A 6 GHz common-carrier microwave communications link uses a tower-mounted antenna with a gain of 40 dB, and a transmitter power of 5 W. To evaluate the radiation hazard of this system, calculate the power density at a distance of 20 m from the antenna. Do this for a position in the main beam of the antenna, and for a position in the sidelobe region of the antenna. Assume a worst-case sidelobe level of -10 dB.

Solution

The numerical gain of the antenna is

$$G = 10^{40/10} = 10^4$$

Then from (12.10) and (12.11), the power density in the main beam of the antenna at a distance of $R = 20$ m is

$$S = \frac{P_{\text{in}}G}{4\pi R^2} = \frac{5 \times 10^4}{4\pi(20)^2} = 10.\ \text{W/m}^2 = 1.0\ \text{mW/cm}^2.$$

The worst-case power density in the sidelobe region would be 10 dB below this, or 0.10 mW/cm^2.

Thus we see that the power density in the main beam at 20 m is below the United States standard, and right at the maximum level recommended in the Soviet Union. The power density in the sidelobe region is well below both limits. These power densities will diminish rapidly with increasing distance, due to the $1/r^2$ dependence. ○

REFERENCES

[1] C. A. Balanis, *Antenna Theory: Analysis and Design*, Harper and Row, N. Y., 1982.

[2] F. T. Ulaby, R. K. Moore, and A. K. Fung, *Microwave Remote Sensing: Active and Passive, Volume I, Microwave Remote Sensing, Fundamentals and Radiometry*. Addison-Wesley, Reading, Mass., 1981.

[3] K. Miyauchi, "Millimeter-Wave Communications," in *Infrared and Millimeter Waves*, vol. 9, K. J. Button, Ed., Academic Press, N. Y., 1983.

[4] M. I. Skolnik, *Introduction to Radar Systems*, McGraw-Hill, N. Y., 1962.

[5] F. E. Gardiol, *Introduction to Microwaves*, Artech House, Dedham, Mass., 1984.

[6] E. C. Okress, *Microwave Power Engineering*, Academic Press, N. Y., 1968.

[7] W. C. Brown, "The History of Power Transmission by Radio Waves," *IEEE Trans. Microwave Theory and Techniques*, vol. MTT-32, pp. 1230–1242 September 1984.

PROBLEMS

12.1 At a distance of 5 km from an antenna, the radiated power density in the main beam is measured to be 1.27×10^{-6} W/m^2. If the input power to the antenna is 100 W, what is the gain of the antenna?

12.2 A short monopole antenna on a ground plane has a far-zone electric field given by

$$E_\theta = \frac{C}{r} \sin\theta e^{-jk_0 r}, \qquad \text{for } 0 \leq \theta \leq 90°,$$

and is zero elsewhere (i.e., below the ground plane). Determine the directivity of this antenna.

12.3 A power density of 2×10^{-6} W/m^2 is incident in the main beam direction of an antenna having a gain of 12 dB. If the frequency is 8 GHz, what is the power delivered to a matched load at the antenna terminals?

12.4 When a certain antenna, having a physical temperature of 295 K, is aimed at the overhead sky an equivalent noise temperature of 34.5 K is measured. If we assume that the antenna has enough gain so that it sees an essentially uniform brightness temperature of 5 K, what is the efficiency of the antenna?

12.5 A microwave radio link at 4.9 GHz uses transmit and receive antennas with gains of 38 dB. If the distance between transmitter and receiver is 27 km, and it is desired to have a minimum received power level of −60 dBm, what is the required transmitter power?

12.6 A microwave relay transmitter has a power of 10 W. If the antenna has a gain of 30 dB, and we assume a worst-case sidelobe level of −15 dB (below the main beam), what are the power densities in the main beam and sidelobe regions of the antenna, at a distance of 30 m?

12.7 Derive the radar equation for the bistatic case, where the transmit and receive antennas have gains of G_t and G_r, and are at distances R_t and R_r from the target, respectively.

12.8 A pulse radar has a pulse repetition frequency $f_r = 1/T_r$. Determine the maximum unambiguous range of the radar. (Range ambiguity occurs when the round-trip time of a return pulse is greater than the pulse repetition time, so it becomes unclear as to whether a given return pulse belongs to the last transmitted pulse, or some earlier transmitted pulse.)

12.9 A doppler radar operating at 12 GHz is intended to detect target velocities ranging from 1 m/sec to 20 m/sec. What is the required passband of the doppler filter?

12.10 A pulse radar operates at 2 GHz and has a per-pulse power of 1 kW. If it is to be used to detect a target with $\sigma = 20$ m^2 at a range of 10 km, what should be the minimum isolation between the transmitter and receiver, so that the leakage signal from the transmitter is at least 10 dB below the received signal? Assume an antenna gain of 30 dB.

12.11 An antenna, having a gain G, is shorted at its terminals. What is the minimum monostatic radar cross-section in the direction of the main beam?

12.12 A self-screening jammer has an output power of 100 W, an antenna gain of 15 dB, and a bandwidth of 10 MHz. It is used against a radar having a transmitter power of 10 kW, with an antenna gain of 30 dB and an IF bandwidth of 0.5 MHz. If the operating frequency is 12 GHz and the target RCS is 5 m^2, compute the minimum range for which J/S = 1.

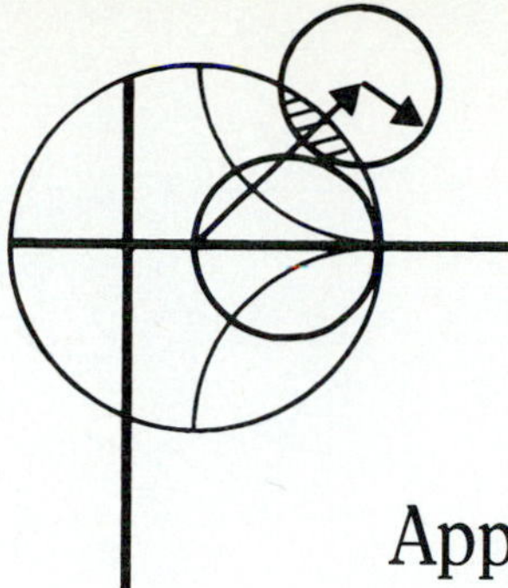

Appendices

Appendix A PREFIXES

Multiplying Factor	Prefix	Symbol
10^{12}	tera	T
10^{9}	giga	G
10^{6}	mega	M
10^{3}	kilo	k
10^{2}	hecto	h
10^{1}	deka	da
10^{-1}	deci	d
10^{-2}	centi	c
10^{-3}	milli	m
10^{-6}	micro	μ
10^{-9}	nano	n
10^{-12}	pico	p
10^{-15}	femto	f

Appendix B VECTOR ANALYSIS

Coordinate Transformations

Rectangular to cylindrical:

	$\hat{x}$	$\hat{y}$	$\hat{z}$
$\hat{\rho}$	$\cos\phi$	$\sin\phi$	0
$\hat{\phi}$	$-\sin\phi$	$\cos\phi$	0
$\hat{z}$	0	0	1

Rectangular to spherical:

	$\hat{x}$	$\hat{y}$	$\hat{z}$
$\hat{r}$	$\sin\theta\cos\phi$	$\sin\theta\sin\phi$	$\cos\theta$
$\hat{\theta}$	$\cos\theta\cos\phi$	$\cos\theta\sin\phi$	$-\sin\theta$
$\hat{\phi}$	$-\sin\phi$	$\cos\phi$	0

Cylindrical to spherical:

	$\hat{\rho}$	$\hat{\phi}$	$\hat{z}$
$\hat{r}$	$\sin\theta$	0	$\cos\theta$
$\hat{\theta}$	$\cos\theta$	0	$-\sin\theta$
$\hat{\phi}$	0	1	0

These tables can be used to transform unit vectors as well as vector components; e.g.,

$$\hat{\rho} = \hat{x}\cos\phi + \hat{y}\sin\phi$$

$$A_\rho = A_x\cos\phi + A_y\sin\phi$$

Vector Differential Operators

Rectangular coordinates:

$$\nabla f = \hat{x}\frac{\partial f}{\partial x} + \hat{y}\frac{\partial f}{\partial y} + \hat{z}\frac{\partial f}{\partial z}$$

$$\nabla \cdot \bar{A} = \frac{\partial A_x}{\partial x} + \frac{\partial A_y}{\partial y} + \frac{\partial A_z}{\partial z}$$

$$\nabla \times \bar{A} = \hat{x}\left(\frac{\partial A_z}{\partial y} - \frac{\partial A_y}{\partial z}\right) + \hat{y}\left(\frac{\partial A_x}{\partial z} - \frac{\partial A_z}{\partial x}\right) + \hat{z}\left(\frac{\partial A_y}{\partial x} - \frac{\partial A_x}{\partial y}\right)$$

$$\nabla^2 f = \frac{\partial^2 f}{\partial x^2} + \frac{\partial^2 f}{\partial y^2} + \frac{\partial^2 f}{\partial z^2}$$

$$\nabla^2 \bar{A} = \hat{x}\nabla^2 A_x + \hat{y}\nabla^2 A_y + \hat{z}\nabla^2 A_z$$

Cylindrical coordinates:

$$\nabla f = \hat{\rho}\frac{\partial f}{\partial \rho} + \hat{\phi}\frac{1}{\rho}\frac{\partial f}{\partial \phi} + \hat{z}\frac{\partial f}{\partial z}$$

$$\nabla \cdot \bar{A} = \frac{1}{\rho}\frac{\partial}{\partial \rho}(\rho A_\rho) + \frac{1}{\rho}\frac{\partial A_\phi}{\partial \phi} + \frac{\partial A_z}{\partial z}$$

$$\nabla \times \bar{A} = \hat{\rho}\left(\frac{1}{\rho}\frac{\partial A_z}{\partial \phi} - \frac{\partial A_\phi}{\partial z}\right) + \hat{\phi}\left(\frac{\partial A_\rho}{\partial z} - \frac{\partial A_z}{\partial \rho}\right) + \hat{z}\frac{1}{\rho}\left[\frac{\partial(\rho A_\phi)}{\partial \rho} - \frac{\partial A_\rho}{\partial \phi}\right]$$

$$\nabla^2 f = \frac{1}{\rho}\frac{\partial}{\partial \rho}\left(\rho\frac{\partial f}{\partial \rho}\right) + \frac{1}{\rho^2}\frac{\partial^2 f}{\partial \phi^2} + \frac{\partial^2 f}{\partial z^2}$$

$$\nabla^2 \bar{A} = \nabla(\nabla \cdot \bar{A}) - \nabla \times \nabla \times \bar{A}$$

Spherical coordinates:

$$\nabla f = \hat{r}\frac{\partial f}{\partial r} + \hat{\theta}\frac{1}{r}\frac{\partial f}{\partial \theta} + \frac{\hat{\phi}}{r\sin\theta}\frac{\partial f}{\partial \phi}$$

$$\nabla \cdot \bar{A} = \frac{1}{r^2}\frac{\partial}{\partial r}(r^2 A_r) + \frac{1}{r\sin\theta}\frac{\partial}{\partial \theta}(\sin\theta A_\theta) + \frac{1}{r\sin\theta}\frac{\partial A_\phi}{\partial \phi}$$

$$\nabla \times \bar{A} = \frac{\hat{r}}{r\sin\theta}\left[\frac{\partial}{\partial \theta}(A_\phi \sin\theta) - \frac{\partial A_\theta}{\partial \phi}\right] + \frac{\hat{\theta}}{r}\left[\frac{1}{\sin\theta}\frac{\partial A_r}{\partial \phi} - \frac{\partial}{\partial r}(rA_\phi)\right] + \frac{\hat{\phi}}{r}\left[\frac{\partial}{\partial r}(rA_\theta) - \frac{\partial A_r}{\partial \theta}\right]$$

$$\nabla^2 f = \frac{1}{r^2}\frac{\partial}{\partial r}\left(r^2\frac{\partial f}{\partial r}\right) + \frac{1}{r^2\sin\theta}\frac{\partial}{\partial \theta}\left(\sin\theta\frac{\partial f}{\partial \theta}\right) + \frac{1}{r^2\sin^2\theta}\frac{\partial^2 f}{\partial \phi^2}$$

$$\nabla^2 \bar{A} = \nabla\nabla \cdot \bar{A} - \nabla \times \nabla \times \bar{A}$$

Vector identities:

$$\bar{A} \cdot \bar{B} = |A||B| \cos\theta, \qquad \text{where } \theta \text{ is the angle between } \bar{A} \text{ and } \bar{B} \tag{B.1}$$

$$|\bar{A} \times \bar{B}| = |A||B| \sin\theta, \qquad \text{where } \theta \text{ is the angle between } \bar{A} \text{ and } \bar{B}. \tag{B.2}$$

$$\bar{A} \cdot \bar{B} \times \bar{C} = \bar{A} \times \bar{B} \cdot \bar{C} = \bar{C} \times \bar{A} \cdot \bar{B} \tag{B.3}$$

$$\bar{A} \times \bar{B} = -\bar{B} \times \bar{A} \tag{B.4}$$

$$\bar{A} \times (\bar{B} \times \bar{C}) = (\bar{A} \cdot \bar{C})\bar{B} - (\bar{A} \cdot \bar{B})\bar{C} \tag{B.5}$$

$$\nabla(fg) = g\nabla f + f\nabla g \tag{B.6}$$

$$\nabla \cdot (f\bar{A}) = \bar{A} \cdot \nabla f + f\nabla \cdot \bar{A} \tag{B.7}$$

$$\nabla \cdot (\bar{A} \times \bar{B}) = (\nabla \times \bar{A}) \cdot \bar{B} - (\nabla \times \bar{B}) \cdot \bar{A} \tag{B.8}$$

$$\nabla \times (f\bar{A}) = (\nabla f) \times \bar{A} + f\nabla \times \bar{A} \tag{B.9}$$

$$\nabla \times (\bar{A} \times \bar{B}) = \bar{A}\nabla \cdot \bar{B} - \bar{B}\nabla \cdot \bar{A} + (\bar{B} \cdot \nabla)\bar{A} - (\bar{A} \cdot \nabla)\bar{B} \tag{B.10}$$

$$\nabla \cdot (\bar{A} \cdot \bar{B}) = (\bar{A} \cdot \nabla)\bar{B} + (\bar{B} \cdot \nabla)\bar{A} + A \times (\nabla \times \bar{B}) + \bar{B} \times (\nabla \times \bar{A}) \tag{B.11}$$

$$\nabla \cdot \nabla \times \bar{A} = 0 \tag{B.12}$$

$$\nabla \times (\nabla f) = 0 \tag{B.13}$$

$$\nabla \times \nabla \times \bar{A} = \nabla\nabla \cdot \bar{A} - \nabla^2 \bar{A} \tag{B.14}$$

Note: the term $\nabla^2 \bar{A}$ only has meaning for rectangular components of $\bar{A}$.

$$\int_V \nabla \cdot \bar{A}\, dv = \oint_S \bar{A} \cdot d\bar{s} \qquad \text{(divergence theorem)} \tag{B.15}$$

$$\int_S (\nabla \times \bar{A}) \cdot d\bar{s} = \oint_C \bar{A} \cdot d\bar{\ell} \qquad \text{(Stokes' theorem)} \tag{B.16}$$

Appendix C BESSEL FUNCTIONS

Bessel functions are solutions to the differential equation,

$$\frac{1}{\rho}\frac{d}{d\rho}\left(\rho\frac{df}{d\rho}\right) + \left(k^2 - \frac{n^2}{\rho^2}\right) f = 0 \tag{C.1}$$

where k^2 is real and n is an integer. The two independent solutions to this equation are called ordinary Bessel functions of the first and second kind, written as $J_n(k\rho)$ and $Y_n(k\rho)$, and so the general solution to (C.1) is

$$f(\rho) = AJ_n(k\rho) + BY_n(k\rho) \tag{C.2}$$

where A and B are arbitrary constants to be determined from boundary conditions.

These functions can be written in series form as

$$J_n(x) = \sum_{m=0}^{\infty} \frac{(-1)^m (x/2)^{n+2m}}{m!(n+m)!} \tag{C.3}$$

$$Y_n(x) = \frac{2}{\pi}\left(\gamma + \ln\frac{x}{2}\right) J_n(x) - \frac{1}{\pi}\sum_{m=0}^{n-1}\frac{(n-m-1)!}{m!}\left(\frac{2}{x}\right)^{n-2m} - \frac{1}{\pi}\sum_{m=0}^{\infty}\frac{(-1)^m (x/2)^{n+2m}}{m!(n+m)!}$$
$$\times\left(1 + \frac{1}{2} + \frac{1}{3} + \cdots + \frac{1}{m} + 1 + \frac{1}{2} + \cdots + \frac{1}{n+m}\right) \quad \text{C.4}$$

where $\gamma = 0.5772\ldots$ is Euler's constant, and $x = k\rho$. Note that Y_n becomes infinite at $x = 0$, due to the ln term. From these series expressions, small argument formulas can be obtained as

$$J_n(x) \sim \frac{1}{n!}\left(\frac{x}{2}\right)^n \quad \text{C.5}$$

$$Y_0(x) \sim \frac{2}{\pi}\ln x \quad \text{C.6}$$

$$Y_n(x) \sim \frac{-1}{\pi}(n-1)!\left(\frac{x}{2}\right)^n, \qquad n > 0 \quad \text{C.7}$$

Large argument formulas can be derived as

$$J_n(x) \sim \sqrt{\frac{2}{\pi x}}\cos\left(x - \frac{\pi}{4} - \frac{n\pi}{2}\right) \quad \text{C.8}$$

$$Y_n(x) \sim \sqrt{\frac{2}{\pi x}}\sin\left(x - \frac{\pi}{4} - \frac{n\pi}{2}\right) \quad \text{C.9}$$

Figure C.1 shows graphs of a few of the lowest order Bessel functions of each type.

Recurrence formulas relate Bessel functions of different orders:

$$Z_{n+1}(x) = \frac{2n}{x} Z_n(x) - Z_{n-1}(x) \quad \text{C.10}$$

$$Z_n'(x) = \frac{-n}{x} Z_n(x) + Z_{n-1}(x) \quad \text{C.11}$$

$$Z_n'(x) = \frac{n}{x} Z_n(x) - Z_{n+1}(x) \quad \text{C.12}$$

$$Z_n'(x) = \frac{1}{2}[Z_{n-1}(x) - Z_{n+1}(x)] \quad \text{C.13}$$

where $Z_n = J_n$ or Y_n. The following integral relations involving Bessel functions are useful:

$$\int_0^x Z_m^2(kx) x\,dx = \frac{x^2}{2}\left[Z_n'^2(kx) + \left(1 - \frac{n^2}{k^2x^2}\right) Z_n^2(kx)\right] \quad \text{C.14}$$

$$\int_0^x Z_n(kx) Z_n(\ell x) x\,dx = \frac{x}{k^2 - \ell^2}[k Z_n(\ell x) Z_{n+1}(kx) - \ell Z_n(kx) Z_{n+1}(\ell x)] \quad \text{C.15}$$

$$\int_0^{p_{nm}}\left[J_n'^2(x) + \frac{n^2}{x^2} J_n^2(x)\right] x\,dx = \frac{p_{nm}^2}{2} J_n'^2(p_{nm}) \quad \text{C.16}$$

$$\int_0^{p'_{nm}}\left[J_n'^2(x) + \frac{n^2}{x^2} J_n^2(x)\right] x\,dx = \frac{(p'_{nm})^2}{2}\left(1 - \frac{n^2}{(p'_{nm})^2}\right) J_n^2(p'_{nm}) \quad \text{C.17}$$

where $J_n(p_{nm}) = 0$, and $J_n'(p'_{nm}) = 0$. The zeros of $J_n(x)$ and $J_n'(x)$ are on pages 712 and 713.

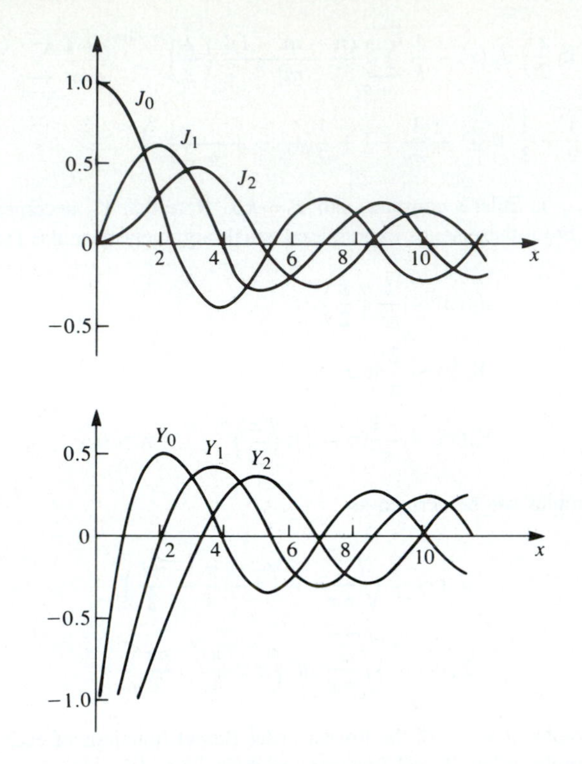

FIGURE C.1 Bessel functions of the first and second kind.

Zeros of Bessel functions of first kind: $J_n(x) = 0$ for $0 < x < 25$.

n	1	2	3	4	5	6	7	8
0	2.4048	5.5200	8.6537	11.7951	14.9309	18.0710	21.2116	24.3524
1	3.8317	7.0155	10.1743	13.3236	16.4706	19.6158	22.7600	
2	5.1356	8.4172	11.6198	14.7959	17.9598	21.1170	24.2711	
3	6.3801	9.7610	13.0152	16.2234	19.4094	22.5827		
4	7.5883	11.0647	14.3725	17.6160	20.8269	24.1990		
5	8.7714	12.3386	15.7001	18.9801	22.2178			
6	9.9361	13.5892	17.0038	20.3208	23.5861			
7	11.0863	14.8212	18.2876	21.6416	24.9349			
8	12.2250	16.0378	19.5545	22.9452				
9	13.3543	17.2412	20.8070	24.2339				
10	14.4755	18.4335	22.0470					
11	15.5898	19.6160	23.2759					
12	16.6983	20.7899	24.4949					
13	17.8014	21.9562						
14	18.9000	23.1158						
15	19.9944	24.2692						
16	21.0851							
17	22.1725							
18	23.2568							
19	24.3383							

Extrema of Bessel functions of first kind: $dJ_n(x)/dx = 0$ for $0 < x < 25$.

n	1	2	3	4	5	6	7	8
0	3.8317	7.0156	10.1735	13.3237	16.4706	19.6159	22.7601	25.9037
1	1.8412	5.3314	8.5363	11.7060	14.8636	18.0155	21.1644	24.3113
2	3.0542	6.7061	9.9695	13.1704	16.3475	19.5129	22.6721	
3	4.2012	8.0152	11.3459	14.5859	17.7888	20.9724	24.1469	
4	5.3175	9.2824	12.6819	15.9641	19.1960	22.4010		
5	6.4156	10.5199	13.9872	17.3128	20.5755	23.8033		
6	7.5013	11.7349	15.2682	18.6374	21.9318			
7	8.5778	12.9324	16.5294	19.9419	23.2681			
8	9.6474	14.1156	17.7740	21.2291	24.5872			
9	10.7114	15.2868	19.0045	22.5014				
10	11.7709	16.4479	20.2230	23.7608				
11	12.8265	17.6003	21.4309					
12	13.8788	18.7451	22.6293					
13	14.9284	19.8832	23.8194					
14	15.9754	21.0154						
15	17.0203	22.1423						
16	18.0683	23.2644						
17	19.1045	24.3819						
18	20.1441							
19	21.1823							
20	22.2192							
21	23.2548							
22	24.2894							

Appendix D OTHER MATHEMATICAL RESULTS

Useful Integrals

$$\int_0^a \cos^2 \frac{n\pi x}{a} dx = \int_0^a \sin^2 \frac{n\pi x}{a} dx = \frac{a}{2}, \qquad \text{for } n \geq 1 \qquad \text{D.1}$$

$$\int_0^a \cos \frac{m\pi x}{a} \cos \frac{n\pi x}{a} dx = \int_0^a \sin \frac{m\pi x}{a} \sin \frac{n\pi x}{a} dx = 0, \qquad \text{for } m \neq n \qquad \text{D.2}$$

$$\int_0^a \cos \frac{m\pi x}{a} \sin \frac{n\pi x}{a} dx = 0 \qquad \text{D.3}$$

Taylor Series

$$f(x) = f(x_0) + (x - x_0) \left.\frac{df}{dx}\right|_{x=x_0} + \frac{(x - x_0)^2}{2!} \left.\frac{d^2 f}{dx^2}\right|_{x=x_0} + \cdots \qquad \text{D.4}$$

$$e^x = 1 + x + \frac{x^2}{2!} + \frac{x^3}{3!} + \cdots \qquad \text{D.5}$$

$$\frac{1}{1-x} = 1 + x + x^2 + x^3 + \cdots, \qquad \text{for } |x| < 1 \qquad \text{D.6}$$

$$\sqrt{1+x} = 1 + \frac{x}{2} - \frac{x^2}{8} + \cdots, \qquad \text{for } |x| < 1 \qquad \text{D.7}$$

$$\ln x = 2\left(\frac{x-1}{x+1}\right) + \frac{2}{3}\left(\frac{x-1}{x+1}\right)^3 + \cdots, \qquad \text{for } x > 0 \qquad \text{D.8}$$

$$\sin x = x - \frac{x^3}{3!} + \frac{x^5}{5!} + \cdots \qquad \text{D.9}$$

$$\cos x = 1 - \frac{x^2}{2!} + \frac{x^4}{4!} + \cdots \qquad \text{D.10}$$

Appendix E PHYSICAL CONSTANTS

- Permittivity of free-space = $\epsilon_0 = 8.854\times10^{-12}$ F/m
- Permeability of free-space = $\mu_0 = 4\pi\times10^{-7}$ H/m
- Impedance of free-space = $\eta_0 = 376.7\ \Omega$
- Velocity of light in free-space = c = 2.998×10^8 m/s
- Charge of electron = $q = 1.602\times10^{-19}$ C
- Mass of electron = $m = 9.107\times10^{-31}$ kg
- Boltzmann's constant = $k = 1.380\times10^{-23}$ J/°K
- Planck's constant = $\hbar = 1.054 \times 10^{-34}$ J-s
- Gyromagnetic ratio = $\gamma = 1.759 \times 10^{11}$ C/Kg (for $g = 2$)

Appendix F CONDUCTIVITIES FOR SOME MATERIALS

Material	Conductivity S/m (20°C)	Material	Conductivity S/m (20°C)
Aluminum	3.816×10^7	Nichrome	1.0×10^6
Brass	2.564×10^7	Nickel	1.449×10^7
Bronze	1.00×10^7	Platinum	9.52×10^6
Chromium	3.846×10^7	Sea water	3-5
Copper	5.813×10^7	Silicon	4.4×10^{-4}
Distilled water	2×10^{-4}	Silver	6.173×10^7
Germanium	2.2×10^6	Steel (silicon)	2×10^6
Gold	4.098×10^7	Steel (stainless)	1.1×10^6
Graphite	7.0×10^4	Solder	7.0×10^6
Iron	1.03×10^7	Tungsten	1.825×10^7
Mercury	1.04×10^6	Zinc	1.67×10^7
Lead	4.56×10^6		

Appendix G DIELECTRIC CONSTANTS AND LOSS TANGENTS FOR SOME MATERIALS

Material	Frequency	ϵ_r	$\tan\delta$ (25°C)
Alumina (99.5%)	10 GHz	9.5–10.	0.0003
Barium tetratitanate	6 GHz	37±5%	0.0005
Beeswax	10 GHz	2.35	0.005
Beryllia	10 GHz	6.4	0.0003
Ceramic (A-35)	3 GHz	5.60	0.0041
Fused quartz	10 GHz	3.78	0.0001
Gallium arsenide	10 GHz	13.	0.006
Glass (pyrex)	3 GHz	4.82	0.0054
Glazed ceramic	10 GHz	7.2	0.008
Lucite	10 GHz	2.56	0.005
Nylon (610)	3 GHz	2.84	0.012
Parafin	10 GHz	2.24	0.0002
Plexiglass	3 GHz	2.60	0.0057
Polyethylene	10 GHz	2.25	0.0004
Polystyrene	10 GHz	2.54	0.00033
Porcelain (dry process)	100 MHz	5.04	0.0078
Rexolite (1422)	3 GHz	2.54	0.00048
Silicon	10 GHz	11.9	0.004
Styrofoam (103.7)	3 GHz	ecomm	0.0001
Teflon	10 GHz	2.08	0.0004
Titania (D-100)	6 GHz	96±5%	0.001
Vaseline	10 GHz	2.16	0.001
Water (distilled)	3 GHz	76.7	0.157

Appendix H PROPERTIES OF SOME MICROWAVE FERRITE MATERIALS

Material	Trans-Tech Number	$4\pi Ms$ G	ΔH Oe	ϵ_r	$\tan\delta$	T_c °C	$4\pi Mr$ G
Magnesium ferrite	TT1-105	1750	225	12.2	0.00025	225	1220
Magnesium ferrite	TT1-390	2150	540	12.7	0.00025	320	1288
Magnesium ferrite	TT1-3000	3000	190	12.9	0.0005	240	2000
Nickel ferrite	TT2-101	3000	350	12.8	0.0025	585	1853
Nickel ferrite	TT2-113	500	150	9.0	0.0008	120	140
Nickel ferrite	TT2-125	2100	460	12.6	0.001	560	1426
Lithium ferrite	TT73-1700	1700	<400	16.1	0.0025	460	1139
Lithium ferrite	TT73-2200	2200	<450	15.8	0.0025	520	1474
Yttrium garnet	G-113	1780	45	15.0	0.0002	280	1277
Aluminum garnet	G-610	680	40	14.5	0.0002	185	515

Appendix I STANDARD RECTANGULAR WAVEGUIDE DATA

Band*	Recommended Frequency Range (GHz)	TE_{10} Cutoff Frequency (GHz)	EIA Designation WR-XX	Inside Dimensions Inches (cm)	Outside Dimensions Inches (cm)
L	1.12–1.70	0.908	WR-650	6.500×3.250 (16.51×8.255)	6.660×3.410 (16.916×8.661)
R	1.70–2.60	1.372	WR-430	4.300×2.150 (10.922×5.461)	4.460×2.310 (11.328×5.867)
S	2.60–3.95	2.078	WR-284	2.840×1.340 (7.214×3.404)	3.000×1.500 (7.620×3.810)
H (G)	3.95–5.85	3.152	WR-187	1.872×0.872 (4.755×2.215)	2.000×1.000 (5.080×2.540)
C (J)	5.85–8.20	4.301	WR-137	1.372×0.622 (3.485×1.580)	1.500×0.750 (3.810×1.905)
W (H)	7.05–10.0	5.259	WR-112	1.122×0.497 (2.850×1.262)	1.250×0.625 (3.175×1.587)
X	8.20–12.4	6.557	WR-90	0.900×0.400 (2.286×1.016)	1.000×0.500 (2.540×1.270)
Ku (P)	12.4–18.0	9.486	WR-62	0.622×0.311 (1.580×0.790)	0.702×0.391 (1.783×0.993)
K	18.0–26.5	14.047	WR-42	0.420×0.170 (1.07 ×0.43)	0.500×0.250 (1.27 ×0.635)
Ka (R)	26.5–40.0	21.081	WR-28	0.280×0.140 (0.711×0.356)	0.360×0.220 (0.914×0.559)
Q	33.0–50.5	26.342	WR-22	0.224×0.112 (0.57 ×0.28)	0.304×0.192 (0.772×0.488)
U	40.0–60.0	31.357	WR-19	0.188×0.094 (0.48 ×0.24)	0.268×0.174 (0.681×0.442)
V	50.0–75.0	39.863	WR-15	0.148×0.074 (0.38 ×0.19)	0.228×0.154 (0.579×0.391)
E	60.0–90.0	48.350	WR-12	0.122×0.061 (0.31 ×0.015)	0.202×0.141 (0.513×0.356)
W	75.0–110.0	59.010	WR-10	0.100×0.050 (0.254×0.127)	0.180×0.130 (0.458×0.330)
F	90.0–140.0	73.840	WR-8	0.080×0.040 (0.203×0.102)	0.160×0.120 (0.406×0.305)
D	110.0–170.0	90.854	WR-6	0.065×0.0325 (0.170×0.083)	0.145×0.1125 (0.368×0.2858)
G	140.0–220.0	115.750	WR-5	0.051×0.0255 (0.130×0.0648)	0.131×0.1055 (0.333×.2680)

* Letters in parentheses denote alternative designations.

Appendix J STANDARD COAXIAL CABLE DATA

Cable Type	Impedance (Ω)	Dielectric Material†	Overall Diameter (In.)	Dielectric Diameter (In.)	Maximum Operating Voltage
RG–8A/U	52	P	0.405	0.285	5000
RG–9B/U	50	P	0.425	0.285	5000
RG–55/U	54	P	0.216	0.116	1900
RG–58/U	50	P	0.195	0.116	1900
RG–59/U	75	P	0.242	0.146	2300
RG–141/U	50	T	0.190	0.116	1900
RG–142/U	50	T	0.206	0.116	1900
RG–174/U	50	—	0.100	0.060	1500
RG–178/U	50	T	0.075	0.036	750
RG–179/U	75	T	0.090	0.057	750
RG–180/U	95	T	0.137	0.103	750
RG–187/U	75	T	0.110	0.060	1200
RG–188/U	50	—	0.110	0.060	—
RG–195/U	95	T	0.155	0.102	1500
RG–213/U	50	P	0.405	0.285	5000
RG–214/U	50	P	0.425	0.285	5000
RG–223/U	50	P	0.216	0.116	1900
RG–401	50	T	0.250	0.208	—
RG–402	50	T	0.141	0.118	—
RG–405	50	T	0.087	0.066	—

† P: Polyethlene, T: Teflon

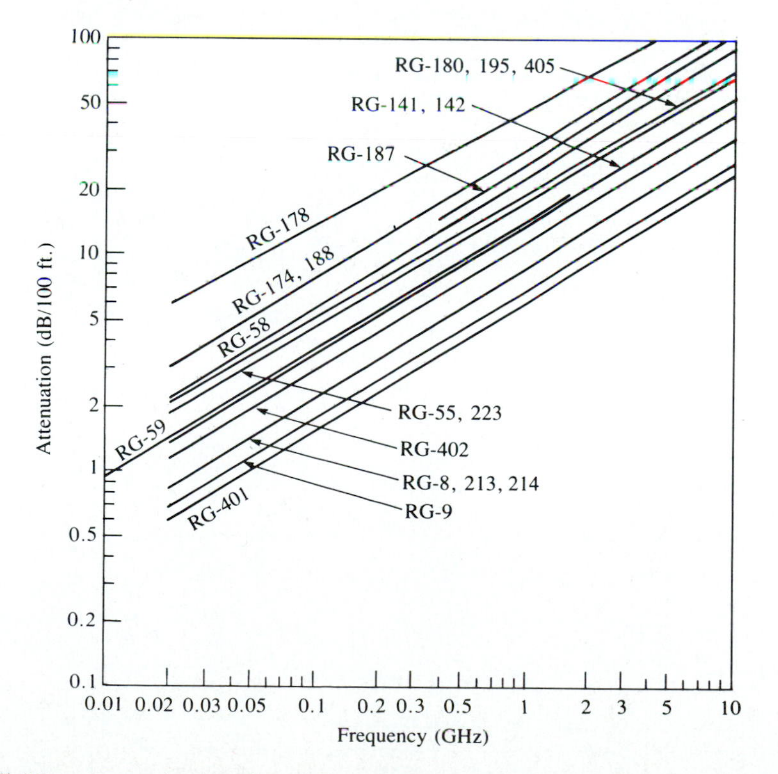

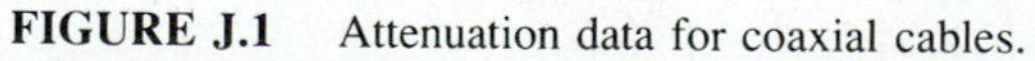
FIGURE J.1 Attenuation data for coaxial cables.

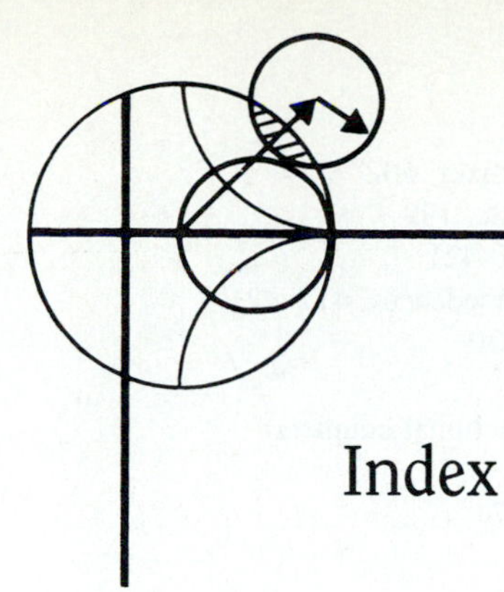

Index